Fundamentals of Engineering Drawing

Fourth Edition

Cecil Jensen

Former Technical Director
R. S. McLaughlin Collegiate and
   Vocational Institute
Oshawa, Ontario, Canada

Jay D. Helsel

Professor and Chairman
Department of Industry
   and Technology
California University of Pennsylvania
California, Pennsylvania

GLENCOE

McGraw-Hill

New York, New York   Columbus, Ohio   Mission Hills, California   Peoria, Illinois

Photo Credit: Cover photo courtesy of Wayne Eastep

Jensen, Cecil Howard
  [Engineering drawing and design. 5th ed. Pt. 1-3]
  Fundamentals of engineering drawing and design / Cecil Jensen,
    Jay D. Helsel.
        p.   cm.
  Consists of the first three parts of Engineering drawing and
design, 5th ed.
  Includes index.
  ISBN 0-02-801800-1
  1. Mechanical drawing.   2. Engineering design.   I. Helsel, Jay D.
II. Title.
T353.J47   1996b
604.2—dc20   94–11372
                                                   CIP

Second printing (February 1996) includes references to ASME Y14.5M–1994.

Printed in the United States of America.

Send all inquiries to:
Glencoe/McGraw-Hill
936 Eastwind Drive
Westerville, OH 43081

ISBN 0-02-801800-1

    3 4 5 6 7 8 9   004   03 02 01 00 99

CONTENTS

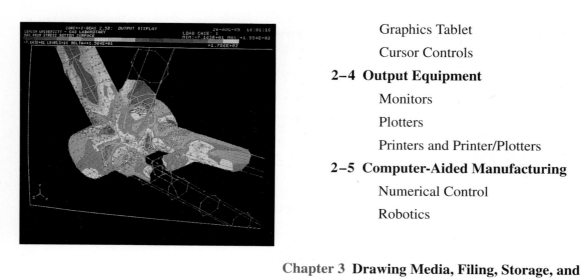

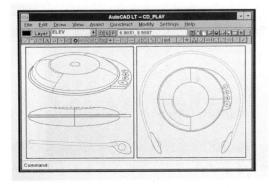

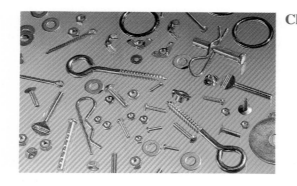

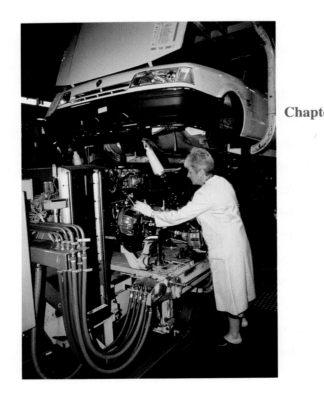

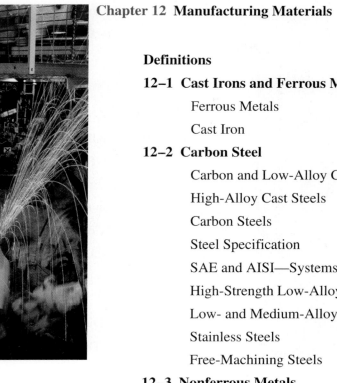

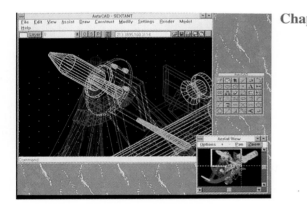

ABOUT THE AUTHORS

CECIL JENSEN is the author or coauthor of many successful technical books, including *Engineering Drawing and Design, Fundamentals of Engineering Drawing, Fundamentals of Engineering Graphics* (formerly called *Drafting Fundamentals*), *Interpreting Engineering Drawings, Geometric Dimensioning and Tolerancing for Engineering and Manufacturing Technology, Architectural Drawing and Design for Residential Construction, Home Planning and Design,* and *Interior Design.* Some of these books are printed in three languages and are used in many countries.

He has 27 years of teaching experience in mechanical and architectural drafting, and was a technical director for a large vocational school in Canada. He has also been responsible for the supervision of the teaching of technical courses for General Motors apprentices in Oshawa, Canada. Before entering the teaching profession, Mr. Jensen gained several years of design experience in the industry.

Mr. Jensen is a member of the Canadian Standards Committee (CSA) on Technical Drawings (which includes both mechanical and architectural drawing), and is chairman of the Committee on Dimensioning and Tolerancing. Mr. Jensen is Canada's representative on the American (ANSI) Standards for Dimensioning and Tolerancing, and has represented Canada at two world (ISO) conferences in Oslo (Norway) and Paris on the standardization of technical drawings.

He took an early retirement from the teaching profession in order to devote his full attention to writing.

JAY D. HELSEL is a professor of industry and technology at California University of Pennsylvania. He completed his undergraduate work at California State College and was awarded a master's degree from Pennsylvania State University. He has done advanced graduate work at West Virginia and at the University of Pittsburgh, where he completed a doctoral degree in educational communications and technology. In addition, Dr. Helsel holds a certificate in airbrush techniques and technical illustration from the Pittsburgh Art Institute.

He has worked in industry and has taught drafting, metalworking, woodworking, and a variety of laboratory and professional courses at both the secondary and college levels. During the past 25 years, he has also worked as a free-lance artist and illustrator. His work appears in many technical publications.

Dr. Helsel is coauthor of *Engineering Drawing and Design, Fundamentals of Engineering Drawing, Programmed Blueprint Reading, Mechanical Drawing,* and *Computer-Aided Engineering Drawing.*

ACKNOWLEDGMENTS

The authors are indebted to the members of ANSI Y14.5M Dimensioning and Tolerancing for the countless hours they have contributed to making this such a successful standard.

We also wish to thank Donald Voisinet, Professor of Engineering Technology and Coordinator of Design and Drafting at the Niagara County Community College, for his assistance in developing Chapter 2 Computer-Aided Drafting (CAD). The authors are indebted to Dennis Short, Associate Professor in the Technical Graphics Department of the School of Technology at Purdue University, for his development of the instructional material to be included in the *Instructor's Management System.*

The authors thank the many people who helped in the preparation of this edition including John Beck, the executive editor, Freida O'Neil-Robinson, the editor, and Jennifer King, the senior production editor.

The authors and staff of Glencoe/McGraw-Hill wish to express their appreciation to the following individuals for their professional review of the text and sample design.

Kenneth Arnold
Tulsa Equipment Manufacturing Co.

Tom Brennan
ITT Technical Institute

Judith Dalton
ITT Technical Institute

Thomas Eddins
Clayton State College

Melvin Freeman
Houston Community College

Joseph Greenfield
Suny College of Technology

Mel Hartley
Bessmer State Technical

Michael Holler
Paragon Films, Inc.

Stanley Hopkins
New England Institute of Technology

Tommy Justice
John Patterson State Technical College

Hamid Khan
Ball State University

Harold Lott
Calhoun Community College

Walter Reed
Oregon Polytechnic Institute

Deb Rosenweig
York Technical Institute

M. Peter Saxon, III
Porter and Chester Institute

Dan Steinke
ITT Technical Institute

Mostafa Tossi
Penn State Worthington Scranton

George Voll
ITT Technical Institute

David Webb
Salt Lake Community College

PREFACE

The fourth edition of *Fundamentals of Engineering Drawing* builds upon the success of the previous edition. Not all students will choose a drafting career. However, an understanding of the language of graphics is necessary for anyone who plans to work in any of the fields of technology.

Drafting, like all technical areas, is constantly changing. The computer has revolutionized the way in which drawing and parts are made. Thus, in this new edition, the authors have made every effort to translate the most current technical information available into the most useful form from the standpoint of both instructor and student. The latest developments and current practices in all areas of graphic communication, CAD, functional drafting, material representation, shop processes, and metrication have been incorporated into this edition. The approach used synthesizes, simplifies, and converts complex drafting standards and procedures into understandable instructional units.

Before beginning to work on this edition, a questionnaire was mailed to a number of users and non-users of the text requesting their input on text material and format. In response to the reviewers' suggestions and recommendations, we have incorporated the following changes in the fourth edition. The suggestions and changes are:

1. Using a two-column format, rather than a three-column format, which is easier for students to read.
2. Updating the photographs of drafting and CAD equipment. The first three chapters of this edition provide up-to-date color photographs of drafting equipment.
3. A greater selection of drawing and design projects. For this edition, we have added numerous projects throughout the text.
4. Deleting CAD icons and applications throughout the existing text. In this edition, Chapter 2 covers CAD equipment including processing and storage equipment, input equipment, and output equipment. Other chapters feature basic CAD commands pertaining to a particular topic, following the information on manual drafting.
5. Continuing to use the unit approach to teach the subject matter. Reviewers find this approach to be a real bonus. It allows them to readily put together a customized program that suits the needs of their students and local industry by choosing the appropriate units. This edition continues to divide chapters into mini-teaching units.
6. Continuing to cover current ANSI and ISO drawing practices better than every other text. It is a basic requirement of any engineering drawing text to keep the instructors aware of the latest drawing standards and practices. In this edition, we again included the latest drawing standards on:

- Methods of representation (Unit 6–1)
- Symbols representation for installed rivets used in aerospace equipment (Unit 11–5)
- Simplified representation on drawing (Unit 14–1), a new standard being prepared by ISO and ANSI

These simplified representations have been used for years by American industries as a cost-saving feature. They are covered in this edition of *Fundamentals of Engineering Drawing* as they have been since the second edition.

In addition to the major changes incorporated from the reviewers' recommendations, this edition also provides other new features including key terms and a new two-color design. Each chapter begins with the important terms and their definitions. The new two-color design is used to highlight the text's features and enhance its appearance.

Fundamentals of Engineering Drawing, 4th edition is supported by ancillary products:

- *Instructor's Management System.* This comprehensive solutions manual details solutions to many of the end-of-chapter drawing problems. It provides instructors with course objectives, instructional tips, teaching transparencies, and chapter tests.
- *Problems Workbook for Fundamentals of Engineering Drawing.* Many of the reviewers suggested that we include a problems workbook to enable their students to have more practice. A correlated problems workbook is now available.

Comments and suggestions concerning this and future editions of the text are most welcomed.

Cecil Jensen and Jay Helsel

BASIC DRAWING AND DESIGN

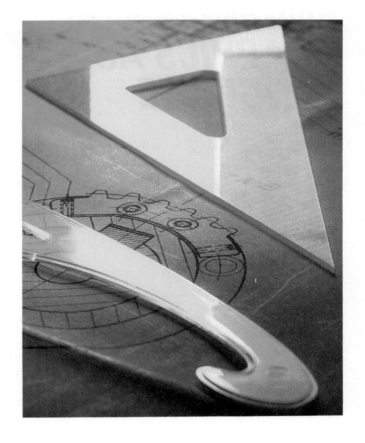

CHAPTER 1

ENGINEERING GRAPHICS AS A LANGUAGE

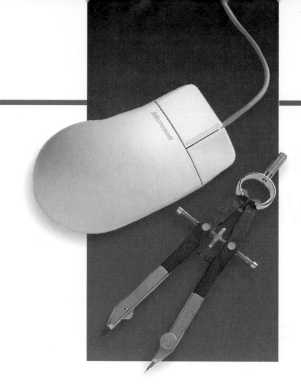

Definitions

Artistic drawing   The expression of real or imagined ideas of a cultural nature.

CADD   An acronym for computer-aided drafting and design.

Computer-aided drafting (CAD)   Drafting and design done with the use of a computer.

Drafting   A language using pictures to communicate thoughts and ideas.

End-product drawings   Drawings which consist of detail or part drawings and assembly or subassembly drawings, but do not include supplementary drawings.

Engineering drawing   The main method of communication between all people concerned with the design and manufacture of parts.

Graphic representation   The expression of ideas through lines or marks impressed on a surface.

Instrument or **manual drawings**   Drawings made with the use of instruments.

Layouts   Drawings made to scale of the object to be built.

Sketches   Drawings made with a pencil or pen but without the assistance of instruments or computers.

Technical drawing   The expression of technical ideas or ideas of a practical nature.

1-1 THE LANGUAGE OF INDUSTRY

Since earliest times people have used drawings to communicate and record ideas so that they would not be forgotten. Figure 1-1-1 shows builders in an early civilization following technical drawings to construct a building.

Graphic representation means dealing with the expression of ideas by lines or marks impressed on a surface. A drawing is a graphic representation of a real thing. Drafting, therefore, is a graphic language, because it uses pictures to communicate thoughts and ideas. Because these pictures are understood by people of different nations, drafting is referred to as a "universal language."

Drawing has developed along two distinct lines, with each form having a different purpose. On the one hand, artistic drawing is concerned mainly with the expression of real or imagined ideas of a cultural nature. Technical drawing, on the

FIG. 1-1-1   Early use of drawing in constructing a building.

TYPICAL BRANCHES OF ENGINEERING GRAPHICS	ACTIVITIES	PRODUCTS	SPECIALIZED AREAS
MECHANICAL 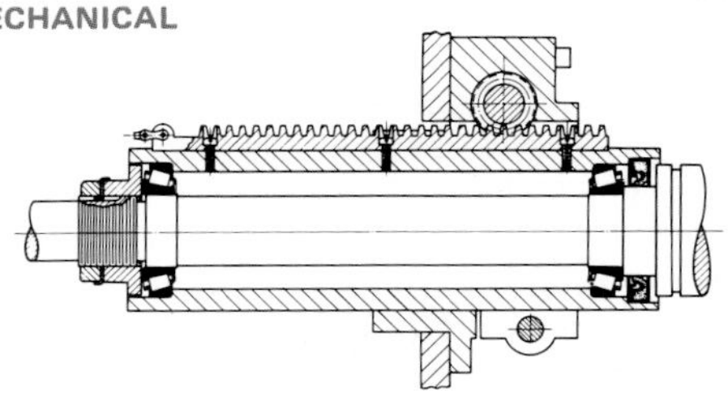	DESIGNING TESTING MANUFACTURING MAINTENANCE CONSTRUCTION	MATERIALS MACHINES DEVICES	POWER GENERATION TRANSPORTATION MANUFACTURING POWER SERVICES ATOMIC ENERGY MARINE VESSELS
ARCHITECTURAL 	PLANNING DESIGNING SUPERVISING	BUILDINGS ENVIRONMENT LANDSCAPE	COMMERCIAL BUILDINGS RESIDENTIAL BUILDINGS INSTITUTIONAL BUILDINGS ENVIRONMENTAL SPACE FORMS
ELECTRICAL 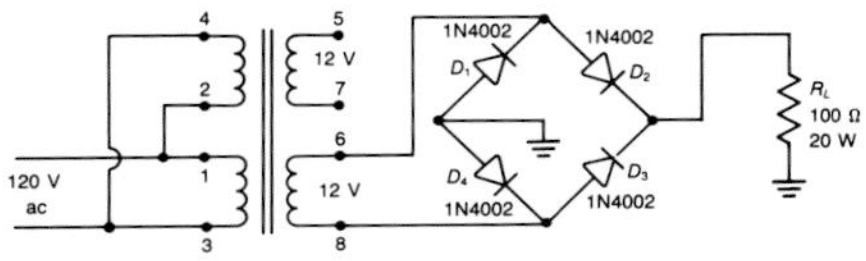	DESIGNING DEVELOPING SUPERVISING PROGRAMMING	COMPUTERS ELECTRONICS POWER ELECTRICAL	POWER GENERATION POWER APPLICATION TRANSPORTATION ILLUMINATION INDUSTRIAL ELECTRONICS COMMUNICATIONS INSTRUMENTATION MILITARY ELECTRONICS
AEROSPACE 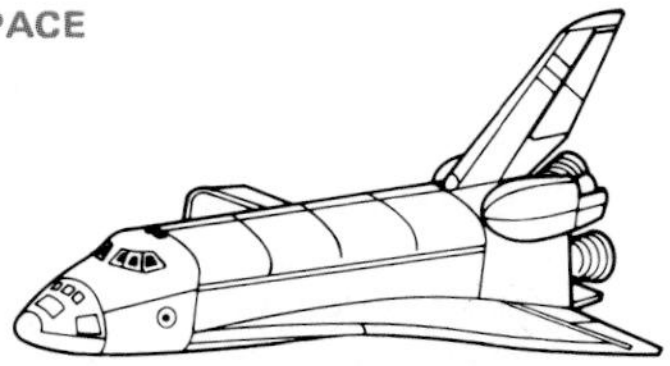	PLANNING DESIGNING TESTING	MISSILES PLANES SATELLITES ROCKETS	AERODYNAMICS STRUCTURAL DESIGN INSTRUMENTATION PROPULSION SYSTEMS MATERIALS RELIABILITY TESTING PRODUCTION METHODS
PIPING 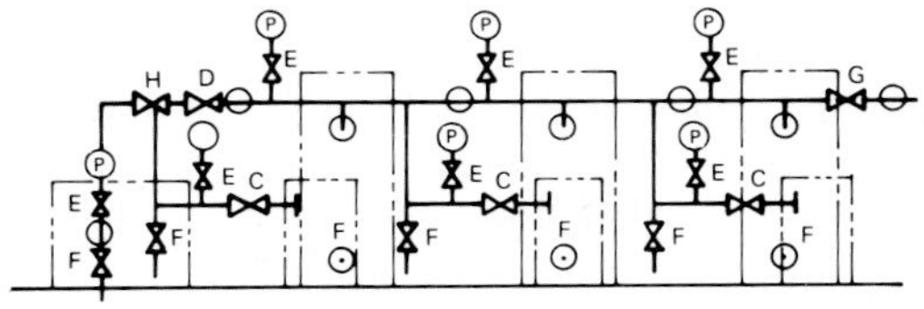	DESIGNING TESTING MANUFACTURING MAINTENANCE CONSTRUCTION	BUILDINGS HYDRAULICS PNEUMATICS PIPE LINES	LIQUID TRANSPORTATION MANUFACTURING POWER SERVICES HYDRAULICS PNEUMATICS
STRUCTURAL 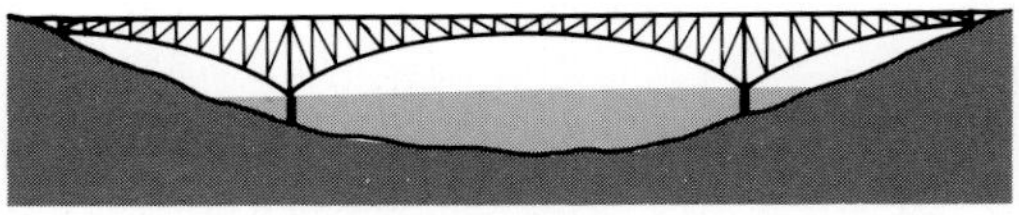	PLANNING DESIGNING MANUFACTURING CONSTRUCTION	MATERIALS BUILDINGS MACHINES VEHICLES BRIDGES	STRUCTURAL DESIGNS BUILDINGS PLANES SHIPS AUTOMOBILES BRIDGES
TECHNICAL ILLUSTRATING 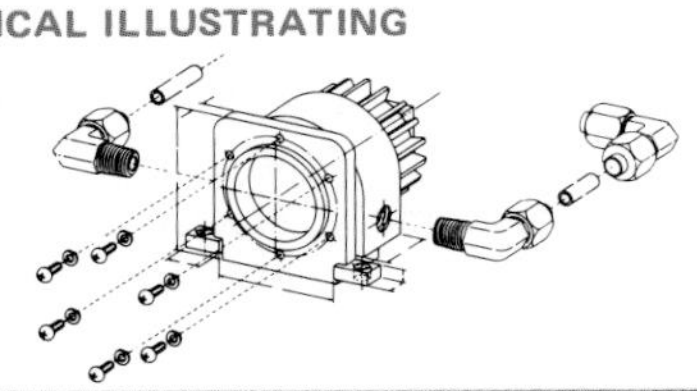	PROMOTION DESIGNING ILLUSTRATING	CATALOGS MAGAZINES DISPLAYS	NEW PRODUCTS ASSEMBLY INSTRUCTIONS PRESENTATIONS COMMUNITY PROJECTS RENEWAL PROGRAMS

FIG. 1-1-2   Various fields of drafting.

other hand, is concerned with the expression of technical ideas or ideas of a practical nature, and it is the method used in all branches of technical industry.

Even highly developed word languages are inadequate for describing the size, shape, and relationship of physical objects. For every manufactured object there are drawings that describe its physical shape completely and accurately, communicating engineering concepts to manufacturing. For this reason, drafting is called the "language of industry."

Drafters translate the ideas, rough sketches, specifications, and calculations of engineers, architects, and designers into working plans that are used in making a product (Fig. 1-1-2, pg. 3). Drafters may calculate the strength, reliability, and cost of materials. In their drawings and specifications, they describe exactly what materials workers are to use on a particular job. To prepare their drawings, drafters use either computer-aided drafting and design (CADD) systems or manual drafting instruments, such as compasses, dividers, protractors, templates, and triangles, as well as drafting machines that combine the functions of several devices. They also may use engineering handbooks, tables, and calculators to assist in solving technical problems.

Drafters are often classified according to their type of work or their level of responsibility. Senior drafters (designers) use the preliminary information provided by engineers and architects to prepare design "layouts" (drawings made to scale of the object to be built). Detailers (junior drafters) make drawings of each part shown on the layout, giving dimensions, material, and any other information necessary to make the detailed drawing clear and complete. Checkers carefully examine drawings for errors in computing or recording sizes and specifications.

Drafters may also specialize in a particular area, such as mechanical, electrical, electronic, aeronautic, structural design, piping, or architectural drafting.

Drawing Standards

Throughout the long history of drafting, many drawing conventions, terms, abbreviations, and practices have come into common use. It is essential that different drafters use the same practices if drafting is to serve as a reliable means of communicating technical theories and ideas.

In the interest of worldwide communication, the International Organization of Standardization (ISO) was established in 1946. One of its committees, ISO TCIO, was formed to deal with the subject of technical drawings. Its goal was to come up with a universally accepted set of drawing standards. Today most countries have adopted, either in full or with minor changes, the standards set up by this committee, making drafting a truly universal language.

The American National Standards Institute (ANSI) is the governing body that establishes the standards for the United States through its ANSI Y14.5 committee, made up of selected personnel from industry, technical organizations, and education. Members from the ANSI Y14.5 also serve on the ISO TCIO subcommittee.

The standards used throughout this text reflect the current thinking of the ANSI committee. These standards apply primarily to end-product drawings. *End-product drawings* usually consist of detail or part drawings and assembly or subassembly

drawings, and are not intended to fully cover other supplementary drawings, such as checklists, item lists, schematic diagrams, electrical wiring diagrams, flowcharts, installation drawings, process drawings, architectural drawings, and pictorial drawings.

The information and illustrations presented here have been revised to reflect current industrial practices in the preparation and handling of engineering documents. The increased use of reduced-size copies of engineering drawings made from microfilm and the reading of microfilm require the proper preparation of the original engineering document regardless of whether the drawing was made manually or by computer (CAD). All future drawings should be prepared for eventual photographic reduction or reproduction. The observance of the drafting practices described in this text will contribute substantially to the improved quality of photographically reproduced engineering drawings.

1-2 CAREERS IN ENGINEERING GRAPHICS

The Student

While students are learning basic drafting skills (Fig. 1-2-1), they will also be increasing their general technical knowledge, learning about some of the engineering and manufacturing processes involved in production. Not all students will choose a drafting career. However, an understanding of this graphic language is necessary for anyone who works in any of the fields of technology, and is essential for those who plan to enter the skilled trades or become a technician, technologist, or engineer.

Because a drawing is a set of instructions that the worker will follow, it must be accurate, clean, correct, and complete. When drawings are made with the use of instruments, they are called *instrument* (or *manual*) *drawings*. When they are developed with the use of a computer, they are known as *computer-aided drawings*. When made without instruments or the aid of a computer, drawings are referred to as *sketches*. The ability to sketch ideas and designs and to produce accurate drawings is a basic part of drafting skills.

In everyday life, a knowledge of engineering graphics is helpful in understanding house plans and assembly, maintenance, and operating instructions for many manufactured or hobby products.

Places of Employment

There are well over 300,000 people working in drafting positions in the United States. A significant number are women. About 9 out of 10 drafters are employed in private industry. Manufacturing industries that employ a large number of drafters are those making machinery, electrical equipment, transportation equipment, and fabricated metal products. Nonmanufacturing industries employing a large number of drafters are engineering and architectural consulting firms, construction companies, and public utilities.

4

(A)

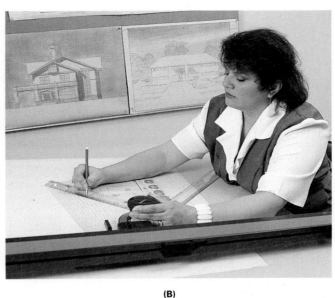

(B)

FIG. 1-2-1   College drafting room. *(Left—Glencoe file photo; right— Doug Martin.)*

Drafters also work for the government; the majority work for the armed services. Drafters employed by state and local governments work chiefly for highway and public works departments. Several thousand drafters are employed by colleges and universities and by other nonprofit organizations.

Training, Qualifications, and Advancement

There are many design careers available at different technical levels of performance. Most companies are in need of design and drafting services for growth in technical development, construction, or production. Any person interested in becoming a drafter can acquire the necessary training from a number of sources, including junior and community colleges, extension divisions of universities, vocational/technical schools, and correspondence schools. Others may qualify for drafting positions through on-the-job training programs combined with part-time schooling.

The prospective drafter's training in post-high school drafting programs should include courses in mathematics and physical sciences, as well as in CAD and CADD. Studying fabrication practices and learning some trade skills are also helpful, since many higher-level drafting jobs require knowledge of manufacturing or construction methods. This is especially true in the mechanical discipline due to the implementation of CAD/CAM (computer-aided design/computer-aided manufacturing). Many technical schools offer courses in structural design, strength of materials, physical metallurgy, CAM, and robotics.

As drafters gain skill and experience, they may advance to higher-level positions such as checkers, senior drafters, designers, supervisors, and managers (Fig. 1-2-2). Drafters who take additional courses in engineering and mathematics are often able to qualify for engineering positions.

Qualifications for success as a drafter include the ability to visualize objects in three dimensions and the development of

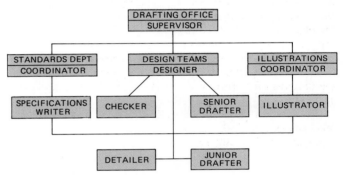

FIG. 1-2-2   Positions within the drafting office.

problem-solving design techniques. Since the drafter is the one who finalizes the details on drawings, attentiveness to detail is a valuable asset.

Employment Outlook

Employment opportunities for drafters are expected to remain stable as a result of the complex design problems of modern products and processes. The need for drafters will, however, fluctuate with local and national economics. Since drafting is a part of manufacturing, job opportunities in this field will also rise or drop in accordance with various manufacturing industries. The demand for drafters will be high in some areas and low in others as a result of high-tech expansion or a slump in sales. In addition, computerization is creating many new products, and support and design occupations, including drafters, will continue to grow. On the other hand, photoreproduction of drawings and expanding use of CAD have eliminated many routine tasks done by drafters. This development will probably reduce the need for some less skilled drafters.

REFERENCES AND SOURCE MATERIALS

1. Charles Bruning Co.
2. *Occupational Outlook Handbook.*

1-3 THE DRAFTING OFFICE

Drafting room technology has progressed at the same rapid pace as the economy of the country. Many changes have taken place in the modern drafting room compared to the typical drafting room scene before CAD, as shown in Fig. 1-3-1. Not only is there far more equipment, but it is of much higher quality. Noteworthy progress has been and continues to be made.

The drafting office is the starting point for all engineering work. Its product, the *engineering drawing,* is the main method of communication among all people concerned with the design and manufacture of parts. Therefore the drafting office must provide accommodations and equipment for the drafters, from designer and checker to detailer or tracer; for the personnel who make copies of the drawings and file the originals; and for the secretarial staff who assist in the preparation of the drawings. Typical drafting workstations are shown in Figs. 1-3-2 and 1-3-3.

Fewer engineering departments now rely on manual drafting methods. Computers are replacing drafting boards at a steady pace because of increased productivity. However, where a high volume of finished or repetitive work is not necessary, manual drafting does the job adequately and inexpensively. CAD and manual drafting can serve as full partners in the design process, enabling the designer to do jobs that are simply not possible or feasible with manual equipment alone.

Besides increasing the speed with which a job is done, a CAD system can perform many of the tedious and repetitive

(A) THE DRAFTING OFFICE AT THE TURN OF THE CENTURY. (Bettman Archives, Inc.)

(B) MANUAL DRAFTING. (Vemco Corp.)

FIG. 1-3-2   Manual drafting office. *(Doug Martin)*

(C) CAD DRAFTING. (Computervision Corp.)

FIG. 1-3-1   Evolution of the drafting office. *(A—Bettman Archive, B—Vemos Corp., C—courtesy of International Business Machines Corporation.)*

FIG. 1-3-3   CAD drafting office. *(Courtesy of International Business Machines Corporation)*

skills ordinarily required of a drafter, such as lettering and differentiating line weights. CAD thus frees the drafter to be more creative while it quickly performs the mundane tasks of drafting. It is conservatively estimated that CAD has been responsible for an improvement of at least 30 percent in production in terms of time spent on drawing.

A CAD system by itself cannot create. A drafter must create the drawing, and thus a strong design and drafting background remains essential.

It may not be practical to handle all the workload in a design or drafting office on a CAD system. Although most design and drafting work most certainly can benefit from it, some functions will continue to be performed by traditional means. Thus some companies will use CAD for only a portion of the workload. Still others use CAD almost exclusively. Whatever the percentage of CAD use, one fact is certain: It has had, and will continue to have, a dramatic effect on design and drafting careers.

Once a CAD system has been installed, the required personnel must be hired or trained. Trained personnel generally originate from one of three popular sources: educational institutions, CAD equipment manufacturer training courses, and individual company programs.

1-4 MANUAL DRAFTING

Over the years, the designer's chair and drafting table have evolved into a drafting station that provides a comfortable, integrated work area. Yet much of the equipment and supplies employed years ago are still in use today, although vastly improved.

Drafting Furniture

Special tables and desks are manufactured for use in single-station or multi-station design offices. Typical are desks with attached drafting boards (Fig. 1-4-1). The boards may be used by the occupant of the desk to which it is attached, in which case it may swing out of the way when not in use, or may be reversed for use by the person in the adjoining station.

In addition to such special workstations, a variety of individual desks, chairs, tracing tables, filing cabinets, and special storage devices for equipment are available (Fig. 1-4-2).

The simplest manually adjustable tables typically consist of a hinged surface riding on a vertical rod secured by a hand knob screw. The hand knob screw is loosened, the top is set at the desired angle, and the hand knob screw is retightened.

Drafting Equipment

See Fig. 1-4-3 (pg. 8) for a variety of drafting equipment.

FIG. 1-4-2   Drafting office furniture. *(The Mayline Company)*

Drawing Boards

The drawing sheet is attached directly to the surface of a drafting table or a portable drawing board (Fig. 1-4-4, pg. 8). Drafting boards are used in schools and at home and generally have a smaller work surface than that found on drafting tables. They are designed to stay flat and have straight guiding edges. Most professional drafting tables have a special overlay drawing surface material that "recovers" from minor pinholes and dents.

FIG. 1-4-1   Drafting workstations.

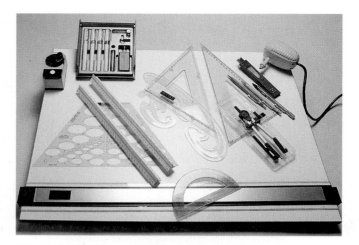

FIG. 1-4-3   Drafting equipment. *(STUDIOHIO)*

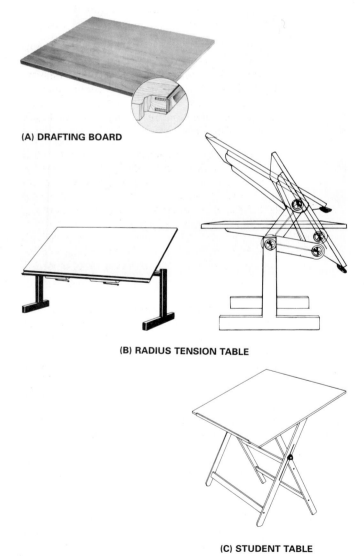

(A) DRAFTING BOARD

(B) RADIUS TENSION TABLE

(C) STUDENT TABLE

FIG. 1-4-4   Drafting tables and boards. *(Norman Wade Co. Ltd.)*

Drafting Machines

In the well-equipped engineering department, where the designer is expected to do accurate drafting, the T square has largely been replaced by the drafting machine. This device, which combines the functions of T square, triangles, scale, and protractor, is estimated to save up to 50 percent of the user's time. All positioning is done with one hand, while the other hand is free to draw.

Drafting machines may be attached to any drafting board or table. Two types are currently available (Fig. 1-4-5). In the track type, a vertical beam carrying the drafting instruments rides along a horizontal beam fastened to the top of the table. In

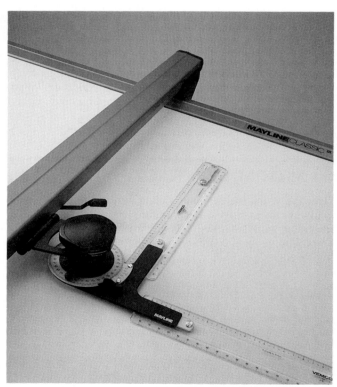

(A) TRACK TYPE

(B) ARM-TYPE

FIG. 1-4-5   Drafting machines. *(A—The Mayline Company, B—Doug Martin)*

the arm (or elbow) type, two arms pivot from the top of the machine and are relative to each other.

The track-type machine has several advantages over the arm-type. It is better suited for large drawings and is normally more stable and accurate. The track type also allows the drafting table to be positioned at a steeper angle and permits locking in the vertical and horizontal positions.

Some track-type drafting machines provide a digital display of angles, the *X-Y* coordinates, and a memory function.

Parallel Slide

The parallel slide, also called the parallel bar, is used in drawing horizontal lines and for supporting triangles when vertical and sloping lines are being drawn (Fig. 1-4-6). It is fastened on each end to cords, which pass over pulleys. This arrangement permits movement up and down the board while maintaining the parallel slide in a horizontal position.

T Squares

Although the T square has been replaced in drafting offices and schools by drafting machines, parallel slides, and computers, it is still used by some students for drafting at home. For this reason it is included in this text.

The T square (Fig. 1-4-7) performs the same function as the parallel slide. T squares are made of various materials, the

FIG. 1-4-6   Drafting table with parallel slide. *(Doug Martin)*

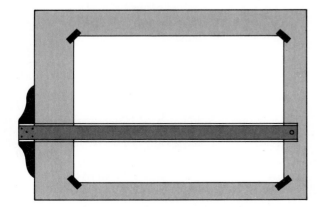

FIG. 1-4-7   T square.

more popular being plastic-edged wood blades with heads made from wood or plastic.

The head of the T square is placed on the left side of a drawing board for use by right-handed people and on the right side of the drawing board for use by left-handed people.

Triangles

Triangles are used together with the parallel slide or T square when you are drawing vertical and sloping lines (Fig. 1-4-8, pg. 10). The triangles most commonly used are the 30/60° and the 45° triangles. Singly or in combination, these triangles can be used to form angles in multiples of 15°. For other angles, the protractor (Fig. 1-4-9, pg. 10) is used. All angles can be drawn with the adjustable triangle (Fig. 1-4-10, pg. 11); this instrument replaces the two common triangles and the protractor.

Scales

Scale may refer to the measuring instrument or the size to which a drawing is to be made.

Measuring Instrument   Shown in Fig. 1-4-11 (pg. 11) are the common shapes of scales used by drafters to make measurements on their drawings. Scales are used only for measuring and are not to be used as a straightedge for drawing lines. It is important that drafters draw accurately to scale. The scale to which the drawing is made must be given in the title block or strip which is part of the drawing.

Sizes to Which Drawings Are Made   When an object is drawn at its actual size, the drawing is called *full scale* or *scale 1:1*. Many objects, however, such as buildings, ships, or airplanes, are too large to be drawn full scale, so they must be drawn to a reduced scale. An example would be the drawing of a house to a scale of ¼ in. = 1 ft or 1:48.

Frequently, objects such as small wristwatch parts are drawn larger than their actual size so that their shape can be seen clearly. Such a drawing has been drawn to an enlarged scale. The minute hand of a wristwatch, for example, could be drawn to a scale of 5:1.

Many mechanical parts are drawn to half scale, 1:2, and quarter scale, 1:4, or nearest metric scale, 1:5. Notice that the scale is expressed as an equation. With reference to the 1:5 scale, the left side of the equation represents a unit of the size drawn; the right side represents the equivalent five units of measurement of the actual object.

Scales are made with a variety of combined scales marked on their surfaces. This combination of scales spares the drafter the necessity of calculating the sizes to be drawn when working to a scale other than full size.

Metric Scales   The linear unit of measurement for mechanical drawings is the millimeter. Scale multipliers and divisors of 2 and 5 are recommended (Fig. 1-4-12, pg. 11).

The numbers shown indicate the difference in size between the drawing and the actual part. For example, the ratio 10:1 shown on the drawing means that the drawing is 10 times the actual size of the part, whereas a ratio of 1:5 on the drawing means the object is 5 times as large as it is shown on the drawing.

FIG. 1-4-8   Triangles.

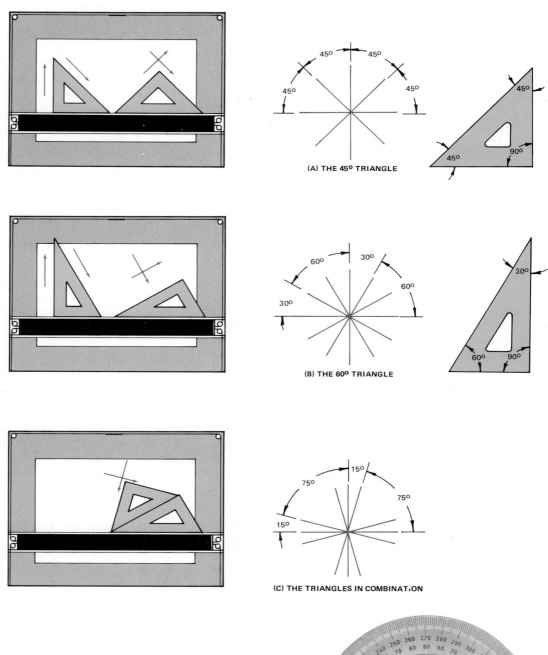

(A) THE 45° TRIANGLE

(B) THE 60° TRIANGLE

(C) THE TRIANGLES IN COMBINATION

FIG. 1-4-9   Protractors.
*(Staedtler
Mars Ltd)*

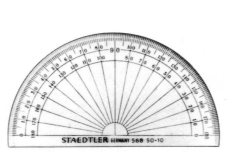

(A) HALF-CIRCLE PROTRACTOR

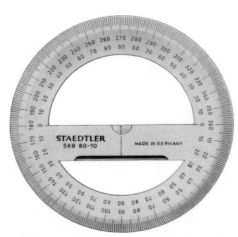

(B) FULL-CIRCLE PROTRACTOR

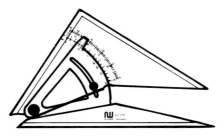

FIG. 1-4-10   Adjustable triangle. *(Norman Wade Co. Ltd.)*

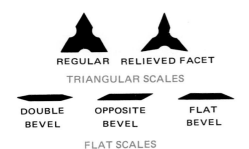

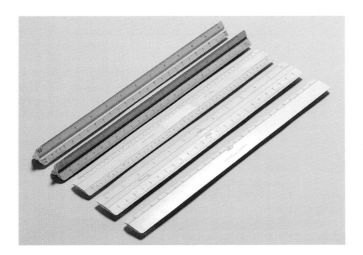

REGULAR   RELIEVED FACET

TRIANGULAR SCALES

DOUBLE   OPPOSITE   FLAT
BEVEL   BEVEL   BEVEL

FLAT SCALES

FIG. 1-4-11   Drafting scales. *(STUDIOHIO)*

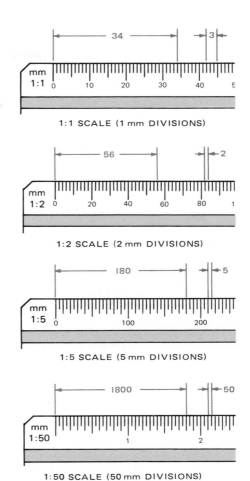

1:1 SCALE (1 mm DIVISIONS)

1:2 SCALE (2 mm DIVISIONS)

1:5 SCALE (5 mm DIVISIONS)

1:50 SCALE (50 mm DIVISIONS)

ENLARGED	SIZE AS	REDUCED
1000 : 1	1 : 1	1 : 2
500 : 1		1 : 5
200 : 1		1 : 10
100 : 1		1 : 20
50 : 1		1 : 50
20 : 1		1 : 100
10 : 1		1 : 200
5 : 1		1 : 500
2 : 1		1 : 1000

FIG. 1-4-12   Metric scales.

The units of measurement for architectural drawings are the meter and millimeter. The same scale multipliers and divisors used for mechanical drawings are used for architectural drawings.

Inch (U.S. Customary) Scales

Inch Scales   There are three types of scales that show various values that are equal to 1 inch (in.) (Fig. 1-4-13, pg. 12). They are the decimal inch scale, the fractional inch scale, and the scale that has divisions of 10, 20, 30, 40, 50, 60, and 80

parts to the inch. The last scale is known as the civil engineer's scale. It is used for making maps and charts. The divisions, or parts of an inch, can be used to represent feet, yards, rods, or miles. This scale is also useful in mechanical drawing when the drafter is dealing with decimal dimensions.

On fractional inch scales, multipliers or divisors of 2, 4, 8, and 16 are used, offering such scales as full size, half size, and quarter size.

Foot Scales   These scales are used mostly in architectural work (Fig. 1-4-14, pg. 12). They differ from the inch scales in that each major division represents a foot, not an inch, and end units are subdivided into inches or parts of an inch. The more common scales are ⅛ in. = 1 ft, ¼ in. = 1 ft, 1 in. = 1 ft, and 3 in. = 1 ft. The most commonly used inch and foot scales are shown in Fig. 1-4-15.

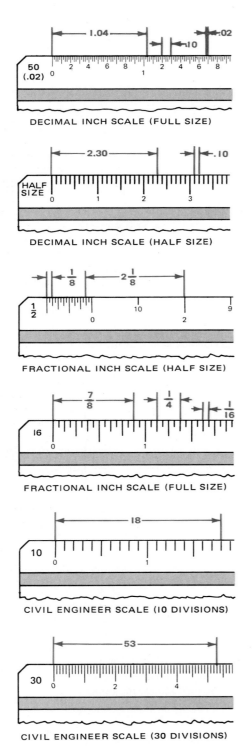

FIG. 1-4-13   Inch scales.

Compasses

The compass is used for drawing circles and arcs. Several basic types and sizes are available (Fig. 1-4-16).

- *Friction head compass,* standard in most drafting sets.
- *Bow compass,* which operates on the jackscrew or ratchet principle by turning a large knurled nut.
- *Drop bow compass,* used mostly for drawing small circles. The center rod contains the needle point and remains stationary while the pencil or pen leg revolves around it (Fig. 1-4-17).

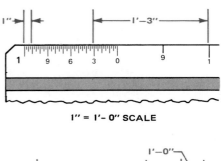

1″ = 1′– 0″ SCALE

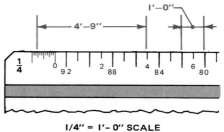

1/4″ = 1′– 0″ SCALE

FIG. 1-4-14   Foot and inch scales.

DECIMALLY DIMENSIONED DRAWINGS	FRACTIONALLY DIMENSIONED DRAWINGS	DIMENSIONED IN FEET AND INCHES	
		EQUIVALENT	
		SCALE	RATIO
10:1	8:1	6 IN. = 1 FT	1:2
5:1	4:1	3 IN. = 1 FT	1:4
2:1	2:1	1½ IN. = 1 FT	1:8
1:1	1:1	1 IN. = 1 FT	1:12
1:2	1:2	¾ IN. = 1 FT	1:16
1:5	1:4	½ IN. = 1 FT	1:24
1:10	1:8	⅜ IN. = 1 FT	1:32
1:20	1:16	¼ IN. = 1 FT	1:48
ETC.	ETC.	3/16 IN. = 1 FT	1:64
		⅛ IN. = 1 FT	1:96
		1/16 IN. = 1 FT	1:192

FIG. 1-4-15   Recommended drawing scales.

- *Beam compass,* a bar with an adjustable needle and pencil-and-pen attachment for drawing large arcs or circles.
- *Adjustable arc,* a device used to accurately draw any radius from 7 to 20 in. (200 to 5000 mm).

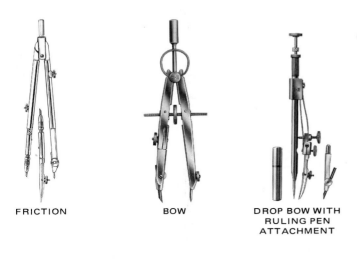

FRICTION   BOW   DROP BOW WITH RULING PEN ATTACHMENT

BEAM WITH RULING PEN ATTACHMENT

FIG. 1-4-16   Compasses. *(Keuffel & Esser Co.)*

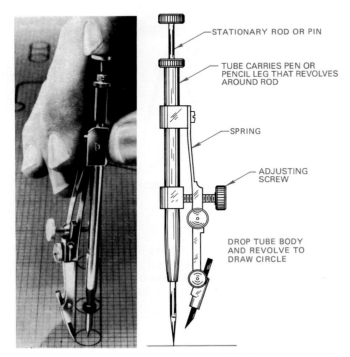

STATIONARY ROD OR PIN

TUBE CARRIES PEN OR PENCIL LEG THAT REVOLVES AROUND ROD

SPRING

ADJUSTING SCREW

DROP TUBE BODY AND REVOLVE TO DRAW CIRCLE

FIG. 1-4-17   The drop-bow compass is used for drawing very small circles.

The bow compass is adjusted by turning a screw whose knurled head is located either in the center or to one side. The bow compass can be used and adjusted with one hand as shown in Fig. 1-4-18 (pg. 14). The proper technique is:

1. Adjust the compass to the correct radius.
2. Hold the compass between the thumb and finger.
3. With greater pressure on the leg with the needle located on the intersection of the center lines, rotate the compass in a clockwise direction. The compass should be slightly tipped in the direction of motion.

Dividers

Lines are divided and distances transferred (moved from one place to another) with dividers. The basic types of dividers are shown in Fig. 1-4-19 (pg. 14).

Dividers have a steel pin insert in each leg and come in a variety of sizes and designs, similar to the compasses. A compass can be used as a divider by replacing its lead point with a steel pin.

One type of divider, known as a proportional divider, is used to enlarge or reduce line segments or an object without the need of mathematical calculations or the use of a scale. An adjustable center point permits you to set the desired proportion. Setting the divider points on one side of the dividers to the original line length or distance automatically determines the reduced or enlarged size on the opposite divider points.

Pencils and Leads

Pencils   As with all other equipment, advances in pencil design have made drawing lines and lettering easier. The new automatic pencils are designed to hold leads of one width, thus eliminating the need to sharpen the lead. These pencils (Fig. 1-4-20, pg. 15) are available in several different lead sizes (color-coded for easy identification) and hardnesses. Leads are designed for use on paper or drafting film, or both. Thus a drafter will have several automatic pencils, each having a selected line width, lead hardness, and make, for performing particular line or lettering tasks on film or paper.

Another type of drafting pencil, often referred to as a mechanical pencil or lead holder, advances a uniformly sized lead that periodically requires sharpening. The leads for the mechanical pencils are usually sharpened in an electric lead pointer, which produces a tapered point. A sandpaper block is used to sharpen compass leads.

Leads   Because of the drawing media used and the type of reproduction required, pencil manufacturers have marketed three types of lead for the preparation of engineering drawings.

Graphite Lead   This is the conventional type of lead, which has been used for years. It is available in a variety of grades or hardnesses—9H, 8H, 7H, 6H (hard); 5H and 4H (medium hard); 3H and 2H (medium); H and F (medium soft); and HB, B, 2B, 3B, 4B, 5B, and 6B (very soft), the very soft grades not being recommended for drafting. The selection of the proper grade of lead is important. A hard lead might penetrate the drawing paper while a soft lead will smear.

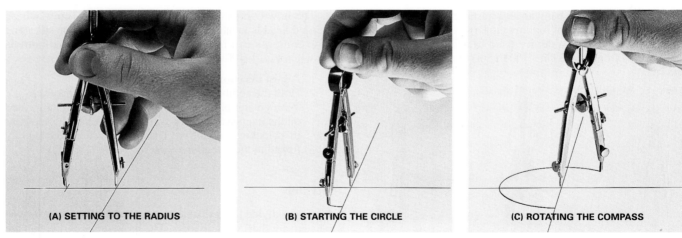

FIG. 1-4-18 Adjusting the radius and drawing a circle with the bow compass.

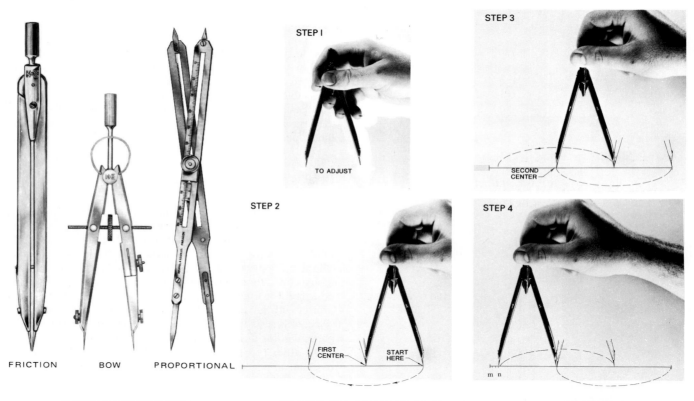

FIG. 1-4-19 Dividers. *(Keuffel & Esser Co.)*

The next two types of drafting leads were developed as a result of the introduction of film as a drawing medium. A limited number of grades are available in these leads, and they do not correspond to the grades used for graphite lead.

Plastic Lead This type of lead is designed for use only on film. It has good microform reproduction characteristics.

Plastic-Graphite Lead As the name implies, this lead is made of plastic and graphite. There are two basic types: fired and extruded. They are designed for use only on film, erase well, do not readily smear, and produce a good opaque line that is suitable for microform reproduction. The main drawback with this type of lead is that it does not hold a point well.

Erasers and Cleaners

Erasers A variety of erasers have been designed to do special jobs—remove surface dirt, minimize surface damage on film or vellum, and remove ink or pencil lines (Fig. 1-4-21).

Erasing Machines   Very little pressure is required when using the electrically powered erasing machine because the high-speed rotation of the shaft actually does the clean-up job rapidly and flawlessly. These machines make erasures with pinpoint accuracy (Fig. 1-4-22).

Cleaners   An easy way to clean tracings is to sprinkle them lightly with gum eraser particles while working. Then triangles, scales, etc., stay spotless and clean the surface automatically as they are moved back and forth. The particles contain no grit or abrasive, and will actually improve the lead- or ink-taking quality of the drafting surface.

Erasing Shields   These thin pieces of metal or plastic (Fig. 1-4-23) have a variety of openings to permit the erasure of fine detail lines or lettering without disturbing nearby work that is to be left on the drawing. With this device, erasures can be made quickly and accurately.

Brushes

A light brush (Fig. 1-4-24) is used to keep the drawing area clean. By using a brush to remove eraser particles and any accumulated dirt, the drafter avoids smudging the drawing.

Templates

To save time, drafters use templates (Fig. 1-4-25, pg. 16) for drawing circles and arcs. Templates are available with standard hole sizes ranging from small to 6.00 in. (150 mm) in diameter. Templates are also used for drawing standard square, hexagonal, triangular, and elliptical shapes and standard electrical and architectural symbols.

(A) AUTOMATIC PENCILS

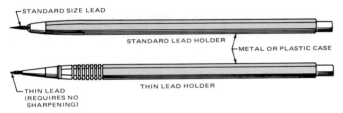

STANDARD SIZE LEAD

STANDARD LEAD HOLDER

METAL OR PLASTIC CASE

THIN LEAD (REQUIRES NO SHARPENING)

THIN LEAD HOLDER

(B) MECHANICAL PENCILS

FIG. 1-4-20   Drafting pencils. *(Koh-I-Noor)*

FIG. 1-4-22   Erasing machine. *(Koh-I-Noor)*

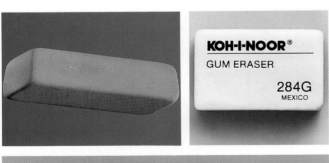

FIG. 1-4-21   Erasers. *(A—STUDIOHIO, B & C—Koh-I-Noor)*

FIG. 1-4-23   Erasing shields. *(STUDIOHIO)*

FIG. 1-4-24   Drafter's brush. *(J.S. Staedtler, Inc.)*

FIG. 1-4-25   Templates. *(Teledyne Post)*

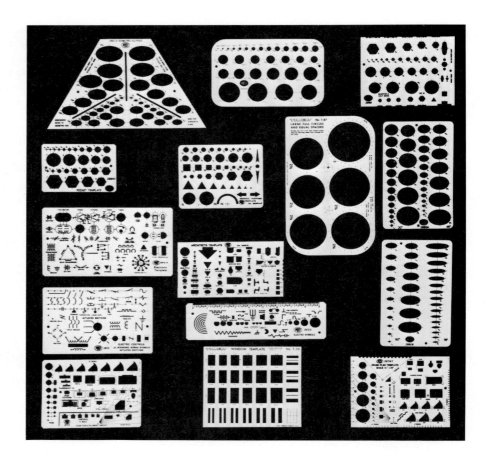

Irregular Curves

For drawing curved lines in which, unlike circular arcs, the radius of curvature is not constant, a tool known as an irregular or French curve (Fig. 1-4-26) is used. The patterns for these curves are based on various combinations of ellipses, spirals, and other mathematical curves. The curves are available in a variety of shapes and sizes. Normally, the drafter plots a series of points of intersection along the desired path and then uses the French curve to join these points so that a smooth-flowing curve results.

Curved Rules and Splines

Curved rules and splines (Fig. 1-4-27) solve the problem of ruling a smooth curve through a given set of points. They lie flat on the board and are as easy to use as a triangle; yet they can be bent to fit any contour to a 3 in. (75 mm) minimum radius and will hold the position without support. A clear, plastic ruling edge stands away from the board just far enough to prevent ink lines from smearing.

Inking Equipment

Although most production drawings are drawn with pencil, in the last few years the number of ink drawings has been on the increase. The use of this type of drawing for technical illustrations and the demand for good, clear drawings for microform

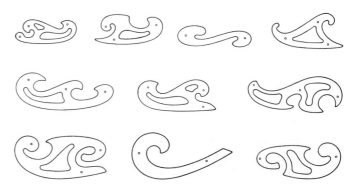

FIG. 1-4-26   Irregular curves.

reproduction have brought about the introduction of new and improved inking methods and techniques.

The three types of inking pens used by drafters today are the ruling pen and the refillable and disposable technical pens. The ruling pen has been around for years and is still used by many drafters. Technical pens are newer and are more commonly used.

Refillable Technical Pens   Technical pens have made the inking of technical drawings relatively easy. These pens have points of different sizes for drawing different line widths

FIG. 1-4-27   Curved rule and spline. *(Doug Martin)*

(Fig. 1-4-28). Technical pen points produce uniform line widths because their design provides a steady flow of ink that does not usually clog.

Some technical pens have a refillable cartridge for storing ink. Others have a cartridge that is used once and then replaced.

Technical pen points are made of different materials for use on different surfaces. There are three main kinds of points: hard-chrome stainless steel (for paper or vellum), tungsten-carbide (for the longest wear on film, vellum, and paper), and jewel point for drafting on any media. Technical pen points have a shoulder to prevent smudging. The shoulder is barrel-shaped for use with curved or straight guiding instruments.

The refillable technical pen has some definite advantages over the ruling pen:

1. It is easy to produce lines of uniform width since no line adjustments are required.
2. The barrel-shaped shoulder prevents smudging because point and shoulder are offset.
3. The cartridge makes it easy to load ink. Also, this pen needs refilling less often than a ruling pen.
4. The ink laid down by a technical pen can dry more uniformly because it flows more evenly.
5. The technical pen does not often need to be cleaned because it has a weighted cleaning wire inside the capillary tube of the point.
6. A syringe can be used to flush out the pen point with a cleaning solution.

The technical pen can be stored in a humidified container that prevents the ink from drying out. In some technical pens, a seal forms in the cap when the cap is tightened with a firm twist. If you are careful in filling and cleaning pens, your inking will be better.

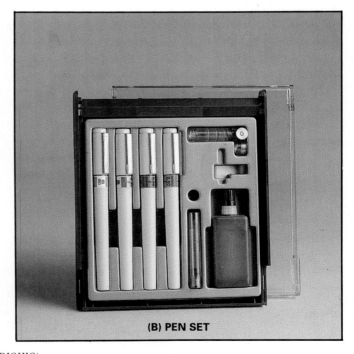

00	0	1	2	
0.25	0.35	0.40	0.50	mm
.010	.014	.017	.020	in.

(A) LINE WIDTHS

(B) PEN SET

FIG. 1-4-28   Examples of technical pen nibs and technical pens. *(STUDIOHIO)*

Disposable Technical Pens   Disposable technical pens are similar to refillable pens except that the pen is discarded once the ink within the pen is used. They have the advantage over the other pens because their fast-drying ink prevents smearing. They are available in a variety of line widths and the drawing of lines, circles, and arcs is done in the same way as refillable technical pens.

Ruling Pens   The ruling pen is steadily being replaced by the technical pens. Where technical pens may be used for line work, drawing arrowheads, and lettering, the ruling pen, as the name implies, is used only for ruling lines.

The ruling pen (Fig. 1-4-29) has blades that can be adjusted to draw lines of different widths. It can be used to draw both straight and curved lines. Ruling pens are sometimes called all-purpose pens. They can be filled from a cartridge tube, a squeeze bottle, or a dropper cap.

Calculators

Calculators, such as the one shown in Fig. 1-4-30, are used by drafters to make fast mathematical calculations using division, multiplication, and extractions of square roots, and to solve problems involving areas, volumes, masses, strengths of materials, pressures, etc.

Basic Equipment

The following is a list of items (Fig. 1-4-31) commonly used in drafting.

   Drawing board
   T square, parallel-ruling straightedge (parallel slide), or
      drafting machine

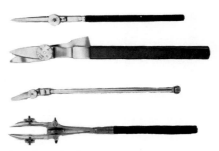

FIG. 1-4-29   A variety of ruling pens.

   Drawing sheets (paper or film)
   Drafting tape
   Drafting pencils
   Erasers
   Erasing shield
   Triangles, 45° and 30/60° (not required with drafting
      machine)
   Scales
   Templates
   Irregular curve
   Inking pen
   Brush
   Protractor
   Cleaning powder
   Calculator

Your drafting instructor can tell you exactly what equipment will be needed for your course.

ASSIGNMENTS

See Assignments 1 through 4 for Unit 1-4 on pages 20–21.

FIG. 1-4-30   Calculator.
(Doug Martin)

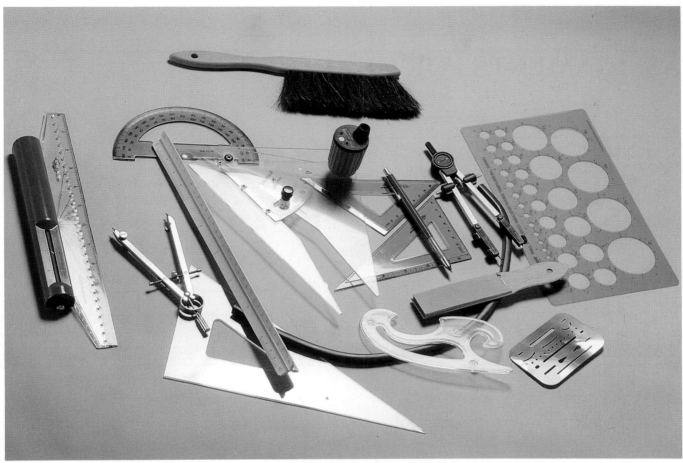

(A) INSTRUMENTS

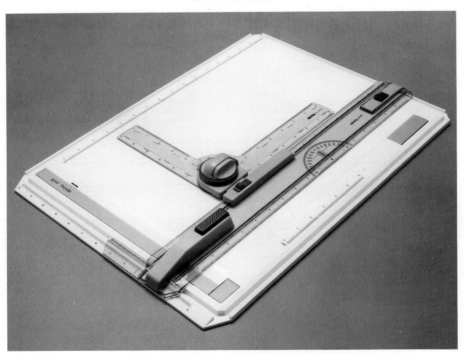

(B) PORTABLE DRAWING BOARD AND DRAFTING MACHINE

FIG. 1-4-31   Basic drafting equipment. *(A—STUDIOHIO, B—Norman Wade Co. Ltd.)*

ASSIGNMENTS FOR CHAPTER 1

ASSIGNMENTS FOR UNIT 1-4, MANUAL DRAFTING

1. Using the scales shown in Fig. 1-4-A determine lengths A through K.
2. Metric measurements assignment. With reference to Fig. 1-4-B and using the scale:

1:1	measure distances A through E
1:2	measure distances F through K
1:5	measure distances L through P
1:10	measure distances Q through U
1:50	measure distances V through Z

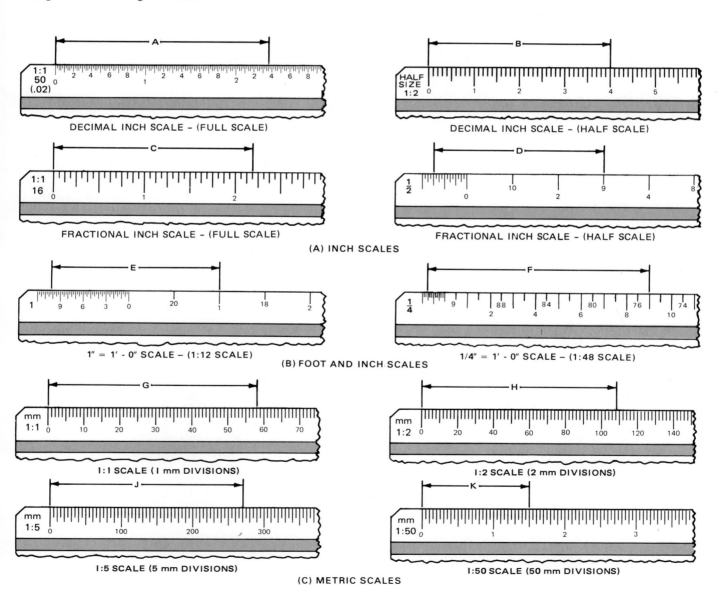

DECIMAL INCH SCALE – (FULL SCALE)

DECIMAL INCH SCALE – (HALF SCALE)

FRACTIONAL INCH SCALE – (FULL SCALE)

FRACTIONAL INCH SCALE – (HALF SCALE)

(A) INCH SCALES

1″ = 1′ - 0″ SCALE – (1:12 SCALE)

1/4″ = 1′ - 0″ SCALE – (1:48 SCALE)

(B) FOOT AND INCH SCALES

1:1 SCALE (1 mm DIVISIONS)

1:2 SCALE (2 mm DIVISIONS)

1:5 SCALE (5 mm DIVISIONS)

1:50 SCALE (50 mm DIVISIONS)

(C) METRIC SCALES

FIG. 1-4-A   Reading drafting scales.

3. Inch measurement assignment. With reference to Fig. 1-4-B and using the scale:
    1:1 decimal inch scale; measure distances A through F
    1:1 fractional inch scale; measure distances G through M
    1:2 decimal inch scale; measure distances N through T
    1:2 fractional inch scale; measure distances U through Z

4. Foot and inch measurement assignment. With reference to Fig. 1-4-B and using the scale:
    1" = 1' – 0", measure distances A through F
    3" = 1' – 0", measure distances G through M
    ¼" = 1' – 0", measure distances N through T
    ⅜" = 1' – 0", measure distances U through Z

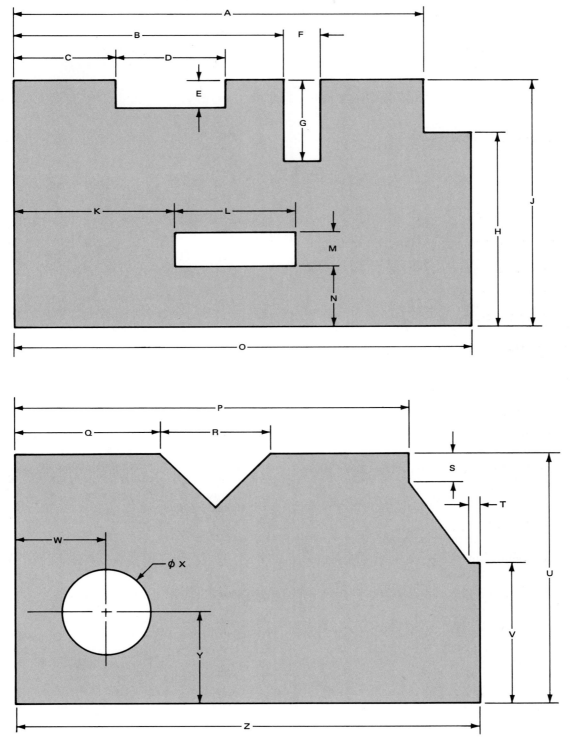

FIG. 1-4-B   Scale measurement assignment.

CHAPTER 2

COMPUTER-AIDED DRAFTING (CAD)

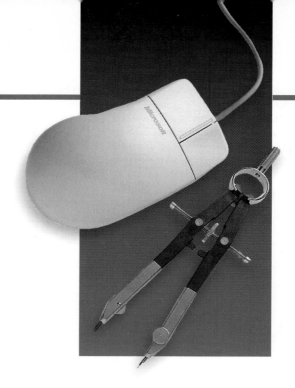

Definitions

Bit   A binary digit; the smallest unit of information recognized by a computer.

Byte   A unit of computer memory consisting of 8 bits.

Central processing unit (CPU)   Part of the computer processing equipment.

CIM   The acronym for computer-integrated manufacturing.

Computer   The combination of a CPU and a terminal.

File server   The central unit holding both the programs and files for a network of computers or terminals.

Hard copy   A paper copy of a finished drawing created on a computer.

Networked system   A group of computers that can communicate with each other.

Personal computer   A microcomputer unit usually consisting of a CPU, a keyboard, and a monitor.

Pixel   A "dot" of color that appears on a computer screen.

Program   A group of written instructions for a computer, logically arranged to perform a task.

Random-access memory (RAM)   Temporary storage locations for data entered into the computer.

Read-only memory (ROM)   The permanent memory in a computer.

Terminal   The combination of a keyboard and a graphics display monitor.

Write protect   A safety feature which prevents the loss of data from storage disks.

X-Y linear coordinate system   A system for specifying locations in a drawing based on horizontal and vertical axes.

2-1 INTRODUCTION TO COMPUTER-AIDED DRAFTING (CAD)

The last 20 years have brought great changes to the drafting room. Its physical appearance, furnishings, even its drafters and engineers have moved quickly from their battered domain of old into the information age. This era is often referred to as the "technical revolution."

These changes were brought about largely by the integrated-circuit (IC) chip. It has, in fact, revolutionized the way we work and play. This era has seen dramatic changes in worldwide communications at all levels—personal, professional, industrial—and in every facet of modern-day life. The microchip is on our wrists (quartz digital watches). It is used to help solve math problems (hand-held calculators). It entertains (video games) and helps to run businesses (computers). The technical changes it has launched have affected many careers, and retraining to upgrade job skills has become commonplace. Drafting and design have been at the forefront of the changes. CAD and CADD are familiar acronyms that have swept through the profession.

Components of a CAD System

A CAD system is made up of various combinations of devices. This holds true for small-, medium-, and large-system applications. The specific package selected largely depends on the needs of the user. Various types of drawings, such as check prints or finished drawings, referred to as *hard copy*, may be preferred by certain companies. Other companies that are fully automated may not require any drawing whatsoever. This means that one company will choose a piece of equipment that prepares a drawing one way. Another company will select equipment that uses another method for preparation of

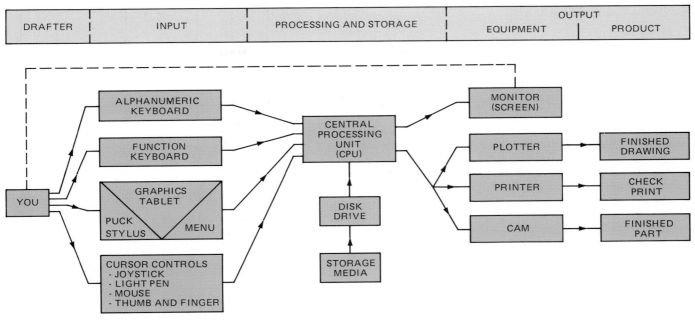

DRAFTER	INPUT	PROCESSING AND STORAGE	OUTPUT	
			EQUIPMENT	PRODUCT

FIG. 2-1-1   Operational flowchart of a CAD system.

drawings. Still a third company will not utilize any equipment to produce hard copies.

Generally, each system element can be categorized as one of the following types:

- Processing and storage
- Input
- Output

This unit analyzes each major component within these categories. Figure 2-1-1 shows an overall diagram of a complete micro-CAD system, including the various input and output devices. The purpose and function of each piece, as well as its relationship to the complete system, will be described. The items shown are typical components found in any system. It would be unlikely, however, to find all of these items in one particular system. The central processing unit (CPU) is considered part of the processing equipment. An alphanumeric (letters and numbers) keyboard is used to manually input data and is normally found with the graphics display monitor as one unit. In combination, a keyboard and graphics display monitor are commonly referred to as a *terminal*. A CPU is normally with the terminal; this combination is commonly referred to as a *computer*. Thus, a computer is composed of a CPU, alphanumeric keyboard, and graphics display monitor.

The typical system arrangement is *interactive*. This means that a person must cause the interaction between the CPU and the graphics display shown on the monitor screen. An alphanumeric keyboard or other input device will aid this process. After the design and/or drawing on the CAD unit has been completed, the information may be transferred to various output devices.

Figure 2-1-2 illustrates one type of system arrangement. To the right is a workstation that includes a monitor, a keyboard and a mouse for inputting data, and a CPU with disk drives. In the background to the left is a plotter.

FIG. 2-1-2   CAD system equipment. *(Courtesy of International Business Machines Corporation)*

2-2 PROCESSING AND STORAGE EQUIPMENT

The CAD program, drawings, and symbols are stored on disk. The CPU collects input information and places the lines and letters in such a way as to produce the required drawings and data.

Central Processing Unit

Bits and Bytes

The CPU is the computing portion of the system. A large number of integrated-circuit (IC) chips are combined into a microprocessor. This is where the number crunching occurs, i.e., the performance of fundamental computations. The number of computations, or the capacity of the unit, is designated by the number of bytes. *Byte* is the base term used by the system manufacturers, describing a character of memory containing 8 bits. A *bit* is a binary digit, the smallest unit of information recognized by a computer.

The size of the CPU normally determines the type of CAD system. *Micro* is the term used for small systems; *networked systems* and *mainframe*, for large systems.

Micro System

The micro unit has a typical user (or dynamic) memory capacity up to 32 megabytes (MB) or greater. This means there is enough space for about 32 million characters. Due to binary numbering limitations, 1K is not exactly 1K in the computer. It is the closest (1024) multiple. To get a feel for memory size, keep in mind that 640K roughly equals 400 typewritten pages. The term "larger and smaller" has been used to describe micro units. This means that larger and larger capacity is put into smaller and smaller units. The chart shown in Fig. 2-2-1 illustrates the rapid advancement in microcomputing technology. The associated cost per bit (or byte) has fallen at a rate inversely proportional to this increased capacity. As a result, the microcomputer revolution was launched.

A single or dual monitor and alphanumeric keyboard are normally part of a micro unit, as shown in Fig. 2-2-2. These units are known as *personal computers* or *PCs*. PC systems are economically priced, are readily available, and are used for a multitude of drafting applications. With their greatly increased capacity, they can accomplish a large amount of work.

Complex, expensive systems need not be tied up with drafting requirements. The data gathered on the PC (or system) can later be "translated" to a larger system. CAD has thus become accessible to a growing number of designers and drafters.

Networked System

AutoCAD may be utilized on multiple systems that are tied together (networked). A *file server* is the central unit holding both the program and drawing files. The network provides the full power of the system to each terminal. All of the software and drawing information may be shared resulting in reduced cost and standardization.

Mainframe

A mainframe system has a huge CPU (or host computer). CAD is only one of a multitude of functions that it can execute. Work such as the creation of payroll spreadsheets may be part of its operation. It offers more capability than the PC and networked systems. Mainframes were available long before CAD

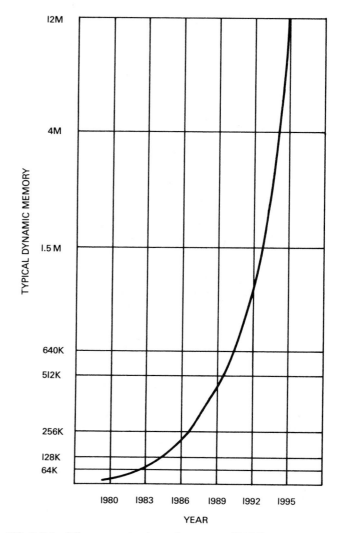

FIG. 2-2-1   Microcomputer dynamic memory (RAM).

became popular; consequently, in the early days of CAD, the mainframe was the most commonly used type of system for computer graphics. Mainframe terminals are normally found in a remote location of the workplace and are not combined into a single unit, as is the PC. For example, several terminals, each of which may be in a remote location, are still connected to the same processing unit, as shown in Fig. 2-2-3.

Software

Program Language

Software is the set of instructions that cause hardware (machines) to function. A *program* is synonymous with software and can refer to the original source code or to the executable (machine language) version. The instructions tell the computer what to do and when to do it. They are used to input information into the system. CAD programs are written in a variety of languages. Common ones are LISP (List Processing

(A)   **(B)**

FIG. 2-2-2   A variety of microcomputers. *(Courtesy of International Business Machines Corporation)*

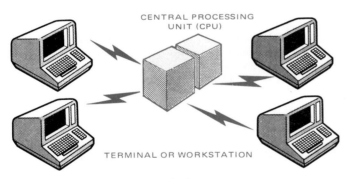

FIG. 2-2-3   Mainframe or networked system.

Language for Symbolic Computing), BASIC (*Be*ginners' *A*ll-purpose *S*ymbolic *I*nstruction *C*ode), and C. BASIC is popular since it uses Englishlike and mathlike "easy-to-use" language. The notations are made up of statements rather than sentences.

No matter which language is used, drafters and designers need not necessarily develop a knowledge of it. A drafter or designer will normally serve as the user of programs rather than the developer. Some proficiency, however, may be desired, and an introductory programming course or LISP workshop may be taken for this purpose.

Software has been extensively developed over the years. A computer programmer will spend thousands of work hours to develop a program. The computer explosion has dropped hundreds of CAD software and hardware products onto the markets and a multitude of complete programs are available. The recommendation with software is, "Buy—don't build." If the desired software to do the job can be purchased, why invest thousands of work hours to develop a new program?

Storage Medium

CAD systems will magnetically store programs and drawings on disk. A disk economically stores drawings and programs and allows their speedy retrieval.

Two types of disks are available: *hard* (fixed) and *floppy* (removable). Disks are also single or dual. This means that data may be stored on one side (single) or on both sides (dual) of the disk.

Two standard diameters for floppy disks are 5.25 and 3.50 in. (Fig. 2-2-4, pg. 26). Note the protective covering over both disks. Disks must be handled gently. Even a small scratch can ruin the contents of the disk. The cover will protect the disk from dust, dirt, and accidental scratching. It cannot, however, prevent mishandling. A disk, for example, should not be exposed to heat or magnetic fields. The contents will immediately become damaged.

Notice the small tabs shown in Fig. 2-2-4D. They are used to protect disk contents. It is possible to accidentally write over the data on the disk with new data. If this occurs, the original data is destroyed. The small tab covers the slot in order to prevent the loss of data. This safety feature is known as *write protect*.

Areas of the disk are divided into tracks and sectors. *Tracks* are circles that surround the disk. *Sectors* are arc-shaped sections of tracks. Tracks and sectors divide the disk so that information can be stored in findable places (Fig. 2-2-5, pg. 26).

The number of sides and density of a disk are given as:

- DSDD—Double sided, double density.
- DSHD—Double sided, high density.

25

(A) 5.25 IN. COVERED FLOPPY DISK   **(B) 5.25 IN. UNCOVERED FLOPPY DISK**

(C) 3.50 IN. FLOPPY DISK

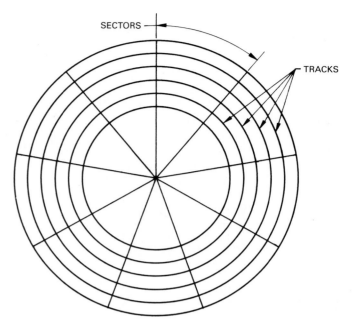

FIG. 2-2-5   Tracks and sectors.

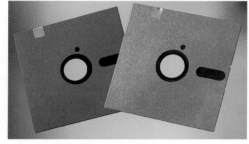

(D) WRITE-PROTECT TABS FOR 5.25 IN. FLOPPY DISKS

FIG. 2-2-4   Floppy disks. *(A & C—courtesy of International Business Machines Corporation; B, C, & D—STUDIOHIO)*

Choose floppy disks for which your computer's disk drive was designed. A lower quality diskette or higher quality diskette may not function properly. High-density disks are the most common.

The compact disc (CD) distributes a larger amount of data than a floppy disk (Fig. 2-2-6). This 12 cm optical disc can store numerical data in excess of 600 Megabytes on a single disc. This is equivalent to approximately 500 times the storage capacity of a standard 5¼" floppy diskette. As a mass storage device for the computer, CD-ROM (Read-Only Memory) uses an optical system for reading data. This read-only system allows only

FIG. 2-2-6   Compact discs.
        (Mak-I Photo Design)

prerecorded disc to be read out. The user cannot directly store or modify the data. The CD-ROM has become an integral part of the new multimedia systems that combine the best features of both television and computer. The CD-ROM features include animation, video, sound, and user interaction. CD-ROM is finding its greatest use wherever the access and storage of large files are required. Applications in this area include CAD/CAM, imaging, and graphics. It is this growing list of applications that makes the CD-ROM a major success.

Storage Units

Disk Drives

Additional devices are required to allow input to, or output from, the processing equipment. Among these are disk drives and memories. Systems that use the floppy disk require equipment to drive the software. A disk drive (Fig. 2-2-7) receives the floppy disk directly and may be used as permanent storage. The information on the disk may be a program, symbols, or drawings. When loading a disk into a drive, be certain to insert it correctly. The label is to be up, the slotted portion is placed in first, and the disk carefully slid into the drive.

FIG. 2-2-7 Floppy disk drive. *(Mak-I Photo Design)*

Memory Systems

Computers have memory systems to store programs and data. The memory may be either permanent or temporary. Permanent memory is referred to as *read-only memory* (ROM). ROM may also be referred to as *firmware*. Firmware is software inside of hardware. Temporary memory is referred to as *dynamic*, *user*, or *random-access memory* (RAM). RAM provides temporary storage locations for entries made by any input device, allowing a drawing to be developed. A complete set of statements (a program), up to system capacity, is stored in RAM. Computers are generally designated by system capacity. A 640K unit, for example, has 640,000 bytes of dynamic memory. The contents in dynamic memory are destroyed when power is lost or when a power surge occurs. Thus power surge protectors should be used on all systems. In the long run, this will prevent many hours of work from being lost.

The contents stored on a disk are not lost as a result of a power interruption. Remember, though, not to remove a floppy disk while placing information onto it (writing) or taking information from it (reading). A glitch (discontinuity) may result. The drive busy light on the disk drive will be lit when information is being read from, or written to, the disk.

Computer systems allow you to see what data are held on the floppy disk. Data related to an individual program or drawing make up a *file*. All the files on the disk are listed in a *directory*. To see what files are on the disk, insert the disk in the drive and type in the directory command.

2-3 INPUT DEVICES

Skills in using and operating input devices are replacing the skills of manual drafting. This equipment instructs the computer to draw lines and circles of various widths and sizes and to apply such graphics as symbols, dimensions, and notes to a drawing.

Alphanumeric Keyboard

An alphanumeric keyboard allows you to communicate directly with the CPU. It may be used to manually input data for:

- Nongraphic work.
- Text and notes additions.
- Exact coordinate input.
- Command selection.

Every CAD system provides a means to select the part of a program that allows the use of a particular command. One way to effect this is to key in the appropriate letter(s) after the COMMAND prompt. For example, LINE is the command for line creation. To select that part of the program, key in LINE or L and press the ENTER key.

The keyboard is an extended version of a standard typewriter keyboard (Fig. 2-3-1). The alphanumeric keys are the same. Usually, though, there are additional keys that allow some specialized command options to be quickly accomplished. "Alpha" refers to the keys that input letters of the alphabet. "Numeric" refers to the other keys, each of which inputs a number. You may type in an alphanumeric instruction, and complete it by pressing the ENTER key. Knowledge of the keyboard is indispensable for anyone involved with CAD. The speed of inputting data with the keyboard is a function of the ability to use it.

Function Keypad

The function keypad is an input device that is used to retrieve a program or part of a program. (The term *program function keypad* is probably more descriptive, but for simplicity, the term *function keypad* will be used to describe this input device.) A function keypad contains several buttons or keys. A part of a program is electronically connected to one of the buttons or keys, and this is operated during the execution of a particular function.

Graphics Tablet

The graphics tablet is a flat surface area electronically sensitized beneath the surface. The tablet is available in a wide range of

FIG. 2-3-1 Alphanumeric keyboard. *(STUDIOHIO)*

sizes. It may vary from a small surface [11 × 11 in. or 279 × 279 millimeters (mm)] to one that exceeds an E-size drawing (36 in. × 48 in. or 910 × 1220 mm). Beneath the surface lies a grid pattern of many horizontal and vertical sensors. When proper contact is made on the surface, electrical impulses are transmitted to the computer. The information that is transmitted provides the programmed instructions to the CPU.

Digitizing

The graphics tablet is another type of input device. In terms of graphics creation, it is more important than the keyboard. Also, it has many purposes. One use is quick and accurate graphic conversion. A rough sketch can be converted to a finished drawing by the transferring of point and line locations to the screen. Information is based on the *X-Y* (horizontal-vertical) linear coordinate system. These are entered quickly and efficiently into the computer, using this so-called "electronic drawing board" (the graphics tablet). The result of the input data is graphically displayed on a monitor screen. This method is commonly known as *digitizing*. Consequently, a graphics tablet is also referred to as a *digitizer*. Other functions, such as

symbol input, may be performed with the aid of a tablet. Figure 2-3-2 shows a graphics tablet menu.

Tablet Menu

CAD systems are menu-driven. This means that a selection is made from a preprogrammed menu (tablet or screen) by the user to call up a particular part of a program. Graphics tablets are provided with menus having a variety of options. A menu may be placed on the tablet surface. A graphics tablet menu is shown in Fig. 2-3-3. Each of the small boxes, or cells, illustrates the list of available choices. Many of the terms are abbreviated to fit into the box. To make a selection, place and activate the stylus or puck over the desired item, such as a polygon, as shown in Fig. 2-3-3. Notice both a name and an icon within each cell.

Any selection on the tablet will retrieve a portion of the program corresponding to the desired title or symbol. For example, if you wish to draw lines, select the menu labeled LINE with the line *icon* (graphical representation), as shown in Fig. 2-3-2. This will call up the part of the program that allows the creation of lines.

FIG. 2-3-2 Graphics tablet menu. *(Autodesk Inc.)*

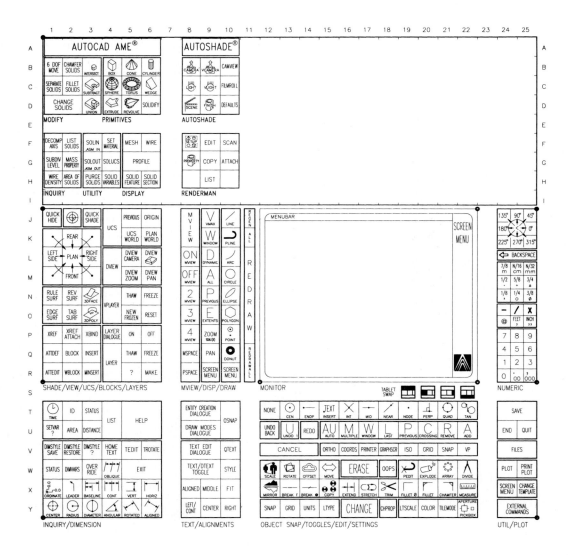

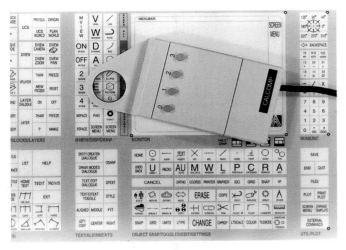

FIG. 2-3-3   Graphics tablet menu and puck.

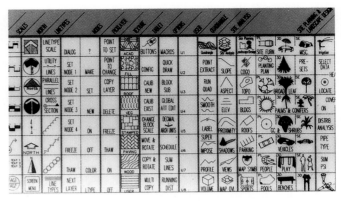

FIG. 2-3-4   Partial landscape design menu. *(LandCADD)*

A menu limits the design drafter to certain types of commands. Both the software and the menu, however, can be changed on many of these systems. One way is to insert a symbol disk or a menu disk, or enter LOAD LISP commands. Completely different applications are thus programmed into the unit. A different mask, or overlay, corresponding to a particular program, such as the partial landscape menu shown in Fig. 2-3-4, can be used. The option to switch menus is particularly useful when one wishes to create various schematic diagrams or other specialized graphics.

Cursor Controls

Several pieces of equipment are used in conjunction with the graphics tablet. These may include a stylus or pen, a push-button cursor or puck, and a power module or console.

Puck and Stylus

A *stylus* (electric pen) or a puck is used to select the desired horizontal-vertical coordinate locations on the graphics tablet. These locations are transferred to the computer. Figure 2-3-3 illustrates a tablet with a push-button puck. Several styles of pucks and styluses are available. Figure 2-3-5 illustrates these cursor controls.

FIG. 2-3-5   Puck and styluses.
(Doug Martin)

Each puck has fine black cross hairs that are used for positioning. The buttons on a puck may be used for various purposes, depending on the manner in which the software has been created. Pressing the appropriate button causes horizontal (X) and vertical (Y) data to be accurately sent to the CPU by an electrical signal. The result is displayed on the monitor screen as a bright mark, or *cursor*. This selection process is repeated as often as necessary to complete the drawing.

A stylus essentially operates the same way as a single-button puck. The position is selected by the tip of the pen. As you move the pen tip across the tablet surface, the position will change correspondingly on the monitor. It will appear as a cursor on the monitor. After the desired position has been located, it is digitized by activating the stylus, usually by pressing down on it.

Other Cursor Controls

The Mouse   Figure 2-3-6A shows another example of a puck or *mouse*, which is the most popular type of cursor control. The mouse rolls across a rectangularly shaped pad which relates to the shape of the monitor. A mouse generally has one or more buttons that are used to "click" the cursor into place on the screen.

Light Pen   A light pen is used as a direct-entry input pointing device. It is also considered a digitizer since it can change displayed points and select menu options on the screen. The pen is electronic and contains a photocell sensory element to detect the presence of light. Hence the term *light pen*. Attached to one end is a cable through which the signal is transmitted. The other end of the pen may be positioned by hand to a desired screen location. After positioning, touch the screen with the tip of the pen. Depressing it causes the pen to become activated. Light spots are sensed. A signal is sent to the system, indicating the position. By this method, any element of the graphics display may be identified to the computer. One disadvantage of the light pen is that after several hours of prolonged use a drafter may tire from holding it.

Other Devices   Several other types of input devices may be used to position the cursor. They include a joystick (Fig. 2-3-6B), touch pad, and Tracball. While these devices are normally not used, it is possible for one to be found on a particular system.

2-4 OUTPUT EQUIPMENT

The drawings that are being produced are observed on the monitor screen. At any time, a print of what is seen on the screen can readily be obtained from a printer or plotter. When the drawing or design is completed, a finished drawing with near perfect lines and lettering is made on a plotter. This is equivalent to the drawing made manually (only better). Prints or microfilm are then produced from this drawing.

If a company has a developed CAD/CAM operation, a drawing may not be required. Instead, a set of instructions in coded form is directly transmitted to the fabricating equipment. The instructions include all location and size information necessary to produce the product.

Monitors

As a drawing is being made, its image can be produced on the screen of a monitor (or graphics display station). The *monitor* displays data in both alphanumeric (written) and graphical (pictorial) form. The user can view a picture of the design as the design is being entered into the system. This display can be "called up" in a variety of ways, depending on the type of monitor used.

Cathode-Ray Tube

The most popular monitor is the cathode-ray tube (CRT). The display method is similar to a television screen. A sample screen display is shown in Fig. 2-4-1. There are several types of CRTs.

FIG. 2-3-6   Other cursor controls.
(*A—Glencoe file photo; B—STUDIOHIO.*)

FIG. 2-4-1   Monitor display. *(Autocad/Macintosh)*

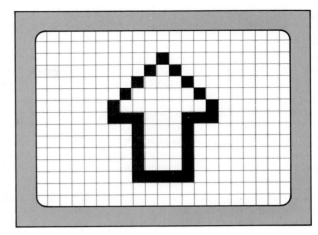

FIG. 2-4-2   Raster display on a CRT.

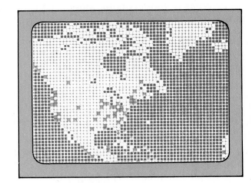

FIG. 2-4-3   Pixel representation.

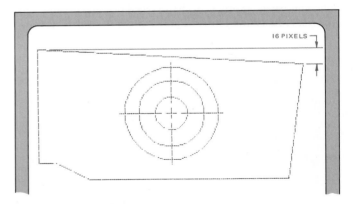

FIG. 2-4-4   Low-resolution display.

Vector   The vector-writing CRT is drawn on by an *X-Y* direction coordinate system. This is similar in principle to the popular Etch-a-Sketch toy. The computer first locates points and then connects the points.

Raster   Raster screens have become the dominant type of CRT displays. The raster uses a grid network to display the image. Each grid is either a dark or a light image that falls within a square area that appears on the screen as a dot (dark) or undot (light) in many color choices. Each dot is known as a *pixel* (picture element). An analogy for how this works is a placard pattern used by fans in the stands of a football game. The cards are used to show a graphical message, and each fan within the pattern holds up either a dark or a light card. At a sufficient distance, the combinations of dark and light spots produce a recognizable image. This phenomenon is shown by the arrow in Fig. 2-4-2.

Another example of pixel representation is shown by the map in Fig. 2-4-3. The horizontal and vertical scan lines are shown so that you can clearly see each pixel. Since the pixels are so large, this would be considered a low-resolution image. Inclined lines will always appear jagged, a phenomenon known as "jaggies" or "stairstepping" (Fig. 2-4-4).

The *resolution* (clearness) of an image depends on the closeness of lines forming the grid pattern. The smaller the pixels, the more resolution the image has. The greater the number of dots per unit area, the greater the resolution. The greater number of dots improves picture quality. For example, an enlarged photograph has a loss in resolution, or clearness, because of the reduced number of dots per unit area.

Significant advancement in raster technology has been achieved in recent years. New techniques enhance resolution. Jaggies appearing on the lines and polygon boundaries have been significantly reduced. Consequently, low resolution is virtually eliminated. The improved resolution will more likely be seen in the medium-resolution to "hi-res" (high-resolution) range, as illustrated in the art throughout the text and in Fig. 2-4-1.

Color   The types of CRTs mentioned are color-enhanced. This display screen is similar to the color television screen. Each of three electron guns emits one of the primary television colors: red, green, or blue. From a combination of these, any pattern of colors may be utilized.

Other Monitors

There are other less-popular types of monitors. One type is *plasma*. Plasma technique uses a flat, thin panel and the glow from an inert gas to display the image. Neon is the gas commonly used. The gas is ionized to emit visible light in a dot matrix pattern. The plasma technique is potentially popular since it is better for your eyes than the CRT. The CRT has low-level radiation emission and a flicker associated with it. This is eliminated with the plasma panel display, which is virtually flicker-free. The primary disadvantage of plasma display is that it produces poor resolution and the size of the panel is small.

Other types of monitors include electroilluminescent, liquid crystal, and projection techniques. Each has its associated problems, and their use is minimal.

Plotters

There are three types of plotters commonly used by industry to produce a finished drawing:

1. Pen plotter
2. Electrostatic plotter
3. Laser plotter

Pen Plotters

A pen plotter is an electromechanical graphics output device. Pen plotters (shown in Fig. 2-4-5) are the most popular type used. They move a pen in one direction across the paper medium. The second direction to produce the lines is accomplished by the paper movement. The plotter is used to produce a finished drawing. This can be any combination of lines and alphanumerics. If a CAD system is thought of as an automated drafting machine, the plotter is the part replacing the activity of "laying lead." It produces the finished original drawing that was previously developed and displayed on the monitor.

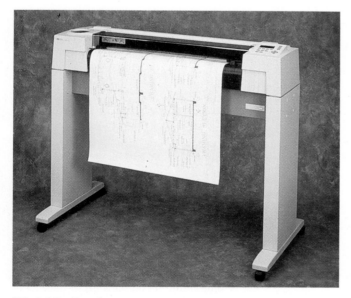

FIG. 2-4-5   Pen plotter. *(Courtesy of Hewlett-Packard Co.)*

FIG. 2-4-6   Pen holder for multiple pen plotter. *(Doug Martin)*

Regardless of screen resolution, a plotter will produce quality lines. There are more addressable dots per unit area. Thus line quality is excellent and virtually jaggie-free.

Various types of ink pens, such as wet ink, felt tip, or liquid ball, can be used. They may be a single color or multicolor. More importantly, plotter pens offer a variable line width (weight) option, usually 0.3 and 0.7 mm widths, with modern models now offering a pencil option. Different pens are inserted or removed rather quickly. The desired pen (different width or color) is inserted and the holder returned to the plotter (Fig. 2-4-6).

The surface area of a pen plot drawing may be as small as an A size (8.5 × 11 in.) and larger than an E size (34 × 44 in.) sheet or drawing format. The size of the drawing to be placed on the paper can also vary. You are able to vary the scale of the lines and characters by manually setting the plotting surface area and/or scale.

Pens will draw on various types of media. Most popular are opaque drawing paper, vellum, and polyester film. Being able to match the medium to its purpose is a distinct advantage of the pen plotter. Another advantage is that the drawing produced is of high quality and is uniform and precise. On the other hand, an average D size (22 × 34 in.) pen plot will tie up the plotter for long periods of time. Thus the plotter should not be used as a print machine. Once the drawing has been finalized, prints may be produced by whiteprint photocopy, or microfilm equipment.

The pen plotter is slow compared with other output devices. It will take from several seconds (simple drawing) to several minutes (complex drawing) to produce a drawing—a disadvantage for a user requiring large-scale production. Yet due to cost factors pen plotters will remain the most common output device for low- to medium-volume applications. This still represents a major reduction of time from manual methods of producing these drawings. In fact, a rather complicated D size drawing, which may have taken a week to create on the screen, can be plotted in less than 20 minutes.

Pen plotter technology includes the drum, microgrip, and flatbed types.

Drum   A drum plotter consists of a long cylinder and a pen carriage. The surface area is curved rather than flat and is in the shape of a cylinder. Hence the term *drum.* The drum rotates to provide one axis of movement. The carriage moves the pen(s) to provide the other axis of movement.

Microgrip   The medium is gripped at the edges with a microgrip plotter. The paper is moved back and forth. High performance is attained at a low cost.

Flatbed   Pen movement occurs in both axes with the flatbed plotter. The pen carriage is controlled in both the *X* and the *Y* axes. Motors and cable are used for control. Short digital steps, normally less than .010 in. long, produce the line. The vellum, polyester film, or other medium is held on the bed surface by electrostatic attraction. Flatbed plotters are no longer popular.

Electrostatic Plotters

Electrostatic plotters will replace pen plotters in applications requiring high production. See Fig. 2-4-7. Plots are made using an ion deposition process toner at a speed exceeding one inch of paper length per second. This means that a plot can be prepared as much as 20 times quicker. A D size drawing is plotted in less than a minute with crisp, black lines. The disadvantage of the electrostatic plotter is that it is expensive. As the cost of equipment decreases, however, a wider use is anticipated.

Laser Plotters

A laser plotter uses a moving laser beam to alter a point-by-point electrical charge on the surface of a rotating drum. The drum is exposed to dry ink, which adheres to the charged areas of the drum. This is then transferred to a hot roller where the ink is fused to the medium. Laser plotters, like electrostatic plotters, produce drawings at high speed.

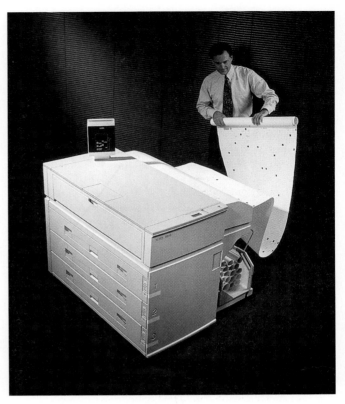

FIG. 2-4-8   Laser printer/plotter. *(Courtesy of Xerox Corporation)*

Printers and Printer/Plotters

A *print* is a preliminary drawing that is produced by a printer. Older printer models produce an image of what is seen on the monitor, complete with all jagged lines and circles. All lines are shown as one thickness. The print is used to check preliminary stages of the design. Recent technical advances, however, have vastly improved the resolution and led to the development of printer/plotters (Fig. 2-4-8).

The newer printers duplicate the screen display quickly and conveniently. The primary advantage is speed. They produce output much more quickly and less expensively than pen plotting. Complex graphic screen displays may be copied by the PRINT command. Whatever is on the screen will be copied, including any combination of graphic and nongraphic (text) displays. This procedure is called a *screen dump.* The entire process requires seconds. The copy, except for the dot matrix type, approaches the level of quality produced by plotters.

Laser Printer/Plotters

Laser technology is developed to the point where it will meet plotter quality. The primary disadvantage is that it cannot produce the large drawings required by industry.

Dot Matrix Printer

The newer printers have a much higher resolution than the old style. They do not, however, approach the quality of a plotter. Their advantage is low cost.

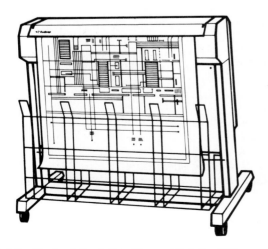

FIG. 2-4-7   Electrostatic plotter. *(Norman Wade Co. Ltd.)*

Ink-jet Printer

The ink-jet process deposits variously colored ink droplets on the medium. The result is a multiple-color copy. They are now capable of producing large drawings at a fairly reasonable cost.

Photoprinter

A typical photoprinter involves the use of fiberoptics technology to reproduce the image on dry silver paper. The unit produces a small A size copy that is satisfactory for quick preview during intermediate work steps.

2-5 COMPUTER-AIDED MANUFACTURING

Computer-aided manufacturing (CAM) uses the result of a computer-aided design. Combining CAD and CAM has had the effect of radically increasing productivity and accuracy. CAD/CAM means that an engineering drawing is no longer produced, since there is a direct, hard-wired connection between design and production. The output of the CAD system is a drawing stored in a geometric data base. This drawing is transmitted directly into the CAM equipment.

Numerical Control

One method of transmitting the design information is known as *numerical control* (NC) or *computer numerical control* (CNC). CNC can store the designs that are used with a variety of production-related processes. The result is a finished manufactured part. This means that errors will no longer occur, since a machinist will not have to interpret an engineering drawing.

True, CAD/CAM is the ultimate goal of industry. Many have CAD; many have CAM; few, however, have CAD/CAM. CAD/CAM and *computer-integrated manufacturing* (CIM) have been implemented at a slow rate. There are, however, software packages available, such as NC CODE and Smart CAM, that operate quite well in conjunction with AutoCAD.

Robotics

The other part of CAM is known as robotics. Robot machinery differs from CNC machinery in that movement is now the prime duty. Automatic manipulators are used to perform a variety of material-handling functions. The robot manipulators are arms and hands (Fig. 2-5-1). They will grasp, operate, assemble, and handle with great consistency and dependability. They are able to perform tasks that are considered too difficult, dangerous, or monotonous for human workers. This is especially true in environments that are intolerable to human beings. The range of tasks includes:

- Working with metals at extremely high temperatures (e.g., spot welding).
- Working in rooms filled with toxic fumes (e.g., spray painting).
- Exerting great forces (e.g., lifting and moving heavy, awkward products).
- Working with delicate parts (e.g., adjusting electronic systems).

FIG. 2-5-1   Robot display on a monitor screen.

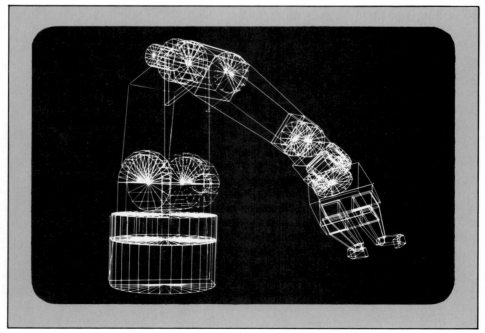

CHAPTER 3

DRAWING MEDIA, FILING, STORAGE, AND REPRODUCTION

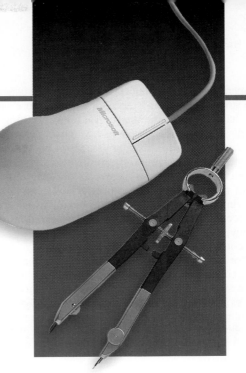

Definitions

Boot To start up the computer.

CD-ROM (Compact Disc Read-Only Memory) A high-capacity (approximately 600 MB) mass storage device that uses an optical rather than a magnetic system for reading data. CD-ROMs are frequently used to distribute large amounts of data.

Diazo (whiteprint) process A drawing reproduction method involving photosensitive paper and chemicals.

Drawing media The material on which the original drawing is made.

Drawing paper Primarily opaque paper used for drawing originals.

Magneto-optic drive An optical mass storage device similar to the CD-ROM that is erasable and rewritable. These drives allow for the recording of large amounts of data on compact, reusable, high-capacity, removable media.

Margin marks Marks made in the margins of a drawing to convey information such as folding marks or graphical scale.

Photoreproduction A drawing reproduction method using an engineering plain-paper copier.

Title block Margin notes containing the drawing number, the name of the firm or organization, the title or description, and the scale; located in the lower right-hand corner.

Tracing paper Primarily translucent paper used for drawing originals or tracing original drawings.

Zoning A method of dividing large drawings into segments for easy reference.

3-1 DRAWING MEDIA AND FORMAT

Drawing Media

The term *drawing media* in this text refers only to the material on which the original drawing is made. Most engineering drawings that are to be reproduced are prepared on a transparent material such as vellum (a drafting paper having translucent properties) or polyester film. These drawing media differ sufficiently between each other and within themselves to provide a wide choice of qualities and characteristics for selection of the perfect material for specific drawing requirements.

Paper

Drafting papers come with a wide range of qualities—strength, erasability, permanence, translucency, etc. The distinguishing feature between drawing and tracing paper is translucency.

Opaque papers are used primarily as drawing papers. Because of a change in drafting practices, the need for drawing papers today is limited to the two extremes in the scale of quality—the very highest-grade, permanent drawing papers with the best possible erasing quality for maps or master drawings, which are later photographed, and the inexpensive school type of papers for the educational market. Since master drawings often must be revised and corrected, the major consideration in high-grade drawing papers is erasability.

Vellum is an inexpensive drawing paper that has good transparency qualities. However, it is not recommended if the drawing will be subject to excessive handling or rough usage. It is suitable for ink or pencil and permits good-quality reproduction by any of the reproduction processes. Generally, the thinner papers are more translucent, but the heavier papers possess more strength, withstand repeated erasures more satisfactorily, and have greater durability in handling and filing.

Formerly, tracing papers were used almost exclusively for the ink tracing of pencil drawings made on opaque papers. Translucency was the prime requirement. Today the usual drafting practice is to develop the master drawing directly on translucent paper from which reproductions can be made, thereby saving the time, expense, and checking involved in the tracing process. This practice means that, in addition to good translucency, modern high-grade tracing paper must be able to withstand considerable handling. The paper should retain these qualities for a long time to avoid the eventual necessity of redrawing, either manually or photographically.

If electrostatic reproduction or microfilming is used to produce prints or microfilm of the CAD- or manually-produced drawing, either opaque or translucent paper can be used as the drawing medium.

Polyester Film

The advantages of polyester film as a drawing material are many. Raw polyester has natural dimensional stability, strong resistance to tearing, high transparency, age and heat resistance, nonsolubility, and waterproofness. The outstanding virtue of film over any other drafting medium is that film is almost indestructible. Its amazing permanence safeguards the important investment in engineering drawings and records. It is permanently translucent, waterproof, unaffected by aging, superb for pencil drafting, ink work, and typewriting, and has unequaled erasability. The working, or drawing, side of the film has a matte surface for accepting pencil or ink and permitting erasures.

Polyester materials, however, present some problems. The material must be sufficiently dense to avoid reflection from the copyboard in microforming, but translucent enough for backlighting or contact printing.

Should a diazo (whiteprint) machine be used to make prints of the CAD or manually produced drawing, translucent film or paper is used as the drawing medium. However, reproducible prints (diazo intermediates) can be made from the original drawing and finished prints then made from the intermediates.

Preprinted Grid Sheets

This type of drawing paper makes the job of making manually prepared drawings easier and quicker. Cross-sectional lines printed directly on the paper or used as a liner beneath the drawing paper provide an accurate guide for all drawing work. These cross-sectional lines are available in several grid sizes. The squared and pictorial styles (isometric, perspective, and oblique) are the more common preprinted grid papers used by drafters. These grids, when applied directly on the paper, as shown in Fig. 3-1-1, or on film with special nonreproducible ink, will not appear on the prints when the drawing is reproduced by diazo or photographic methods.

For freehand work, whether rough sketches in the field or in the drafting room, whether these are preliminary or to be used as the finished drawings, gridline papers are an invaluable aid. The cross-sectional patterns serve as ready-made guides for base lines, dimensioning, and angles (Fig. 3-1-2).

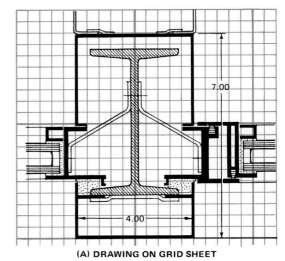

(A) DRAWING ON GRID SHEET

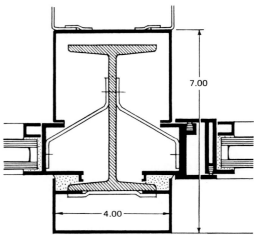

(B) PRINT OF DRAWING SHOWN IN (A)

FIG. 3-1-1   Drawing directly on preprinted grid sheet.

FIG. 3-1-2   Preprinted grid sheet being used as an underlay.
(Doug Martin)

Standard Drawing Sizes

Inches   Drawing sizes in the inch system are based on dimensions of commercial letterheads, 8.5 × 11 in., and standard rolls

36

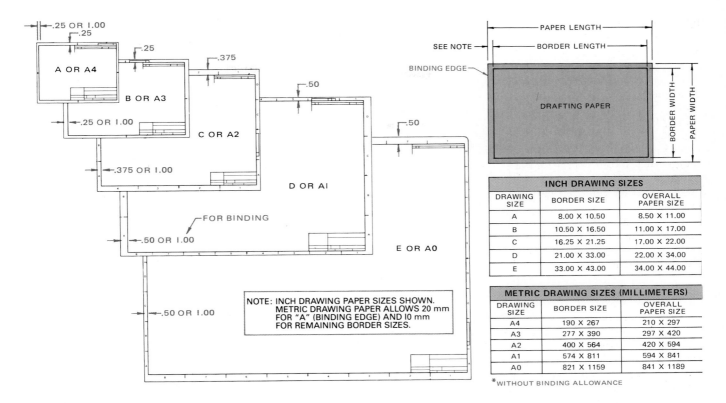

INCH DRAWING SIZES		
DRAWING SIZE	BORDER SIZE	OVERALL PAPER SIZE
A	8.00 X 10.50	8.50 X 11.00
B	10.50 X 16.50	11.00 X 17.00
C	16.25 X 21.25	17.00 X 22.00
D	21.00 X 33.00	22.00 X 34.00
E	33.00 X 43.00	34.00 X 44.00

METRIC DRAWING SIZES (MILLIMETERS)		
DRAWING SIZE	BORDER SIZE	OVERALL PAPER SIZE
A4	190 X 267	210 X 297
A3	277 X 390	297 X 420
A2	400 X 564	420 X 594
A1	574 X 811	594 X 841
A0	821 X 1159	841 X 1189

*WITHOUT BINDING ALLOWANCE

NOTE: INCH DRAWING PAPER SIZES SHOWN. METRIC DRAWING PAPER ALLOWS 20 mm FOR "A" (BINDING EDGE) AND 10 mm FOR REMAINING BORDER SIZES.

FIG. 3-1-3   Standard drawing sizes.

of paper or film, 36 and 42 in. wide. They can be cut from these standard rolls with a minimum of waste (Fig. 3-1-3).

Metric   Metric drawing sizes are based on the A0 size, having an area of 1 square meter (m^2) and a length-to-width ratio of $1:\sqrt{2}$. Each smaller size has an area half of the preceding size, and the length-to-width ratio remains constant (Fig. 3-1-4).

CAD   After *booting* the CAD system (preparing it to run an application program), the drawing size limits must be set prior to starting the drawing. These limits will be determined by the space that the object to be drawn will require. For example, a full scale single-view drawing of a small object will require only a small drawing size. A larger drawing size will be required to prepare a full-scale multiview drawing of a larger object. There are several standard drawing sizes from which to choose.

An alternate to this procedure would be to draw to full scale and then scale down the finished plot and insert it into an appropriate paper size.

A third alternative would be to draw the paper and border at whatever scale will fit the drawing and adjust the scale when plotting to the corrected plotted size.

Drawing Format

A general format for drawings is shown in Fig. 3-1-5 (pg. 38), which illustrates a drawing trimmed to size. It is recommended that preprinted drawing forms be made to the trimmed size and have rounded corners, as shown, to minimize dog-ears and tears.

Zoning System

Drawings larger than B size may be *zoned* for easy reference by dividing the space between the trimmed size and the inside border into zones measuring 4.25×5.50 in. These zones are numbered horizontally and lettered vertically, with uppercase letters, from the lower RH (right-hand) corner, as in Fig. 3-1-5, so that any area of the drawing can be identified by a letter and

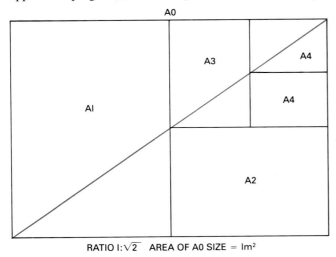

RATIO $1:\sqrt{2}$ AREA OF A0 SIZE = $1m^2$

FIG. 3-1-4   Metric drawing sizes.

FIG. 3-1-5   Drawing format.

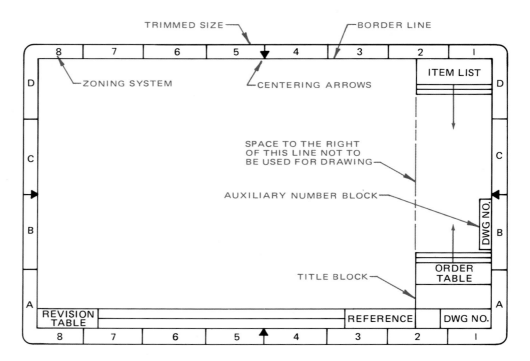

a number, such as B3, similar to reading a road map. Just as with maps, zoning is useful to locate fine detail on complex drawings.

Marginal Marking

In addition to zone identification, the margin may also carry fold marks to enable folding and graphical scale to facilitate reproduction to a specific size. In the process of microforming, it is necessary to center the drawing within rather close limits in order to meet standards. To facilitate this operation, it has become common practice to put a centering arrow or mark on at least three sides of the drawing. Most practices include the arrows on each of the four sides. If three sides are used, the arrows should be on the two sides and on the bottom. This helps the camera operator align the drawing properly since the copyboard usually contains cross hairs through the center of the board at right angles. With any three arrows aligned on the cross hairs, centering is automatic. The arrows should be on the center of the border that outlines the information area of the drawing, not at the edge of the sheet on which the drawing is made.

Title Block

Title blocks vary greatly and are usually preprinted. Drafters are rarely required to make their own.

The title block is located in the lower right-hand corner. The arrangement and size of the title block are optional, but the following information should be included:

1. Drawing number
2. Name of firm or organization
3. Title or description
4. Scale

Provision may also be made within the title block for the date of issue, signatures, approvals, sheet number, drawing size; job, order, or contract number; references to this or other documents; and standard notes, such as tolerances or finishes. An example of a typical title block is shown in Fig. 3-1-6. In classrooms, a title strip (Fig. 3-1-7) is often used on A and B size drawings.

Item (Material) List

The whole space above the title block, with the exception of the auxiliary number block, should be reserved for tabulating materials, change of order, and revision. Drawing in this space should be avoided. On preprinted forms, the right-hand inner border may be graduated to facilitate ruling for an item list. Fig. 3-1-8 shows a combined item list, order table, and title block.

Change or Revision Table

All drawings should carry a change or revision table, either down the right-hand side or across the bottom of the drawing.

NORDALE MACHINE COMPANY		
PITTSBURGH, PENNSYLVANIA		
COVER PLATE		
MATERIAL- MS		NO. REQD-4
SCALE- 1 : 2	DN BY D Scott	A - 7628
DATE- 3/6/94	CH BY B Jensen	

FIG. 3-1-6   Title block.

DRAFTING TECHNOLOGY CALIFORNIA UNIVERSITY OF PENNSYLVANIA CALIFORNIA, PENNSYLVANIA	NAME:		DWG NAME:	DWG NO.
	COURSE:			
	DATE:	APPD:	SCALE:	

FIG. 3-1-7   Title strip.

AMT	DET	STOCK SIZE		MAT.

NORDALE MACHINE COMPANY
PITTSBURGH, PENNSYLVANIA

MODEL _____

PART NAME_____
_____

PART NO. _____

OPERATION_____

FOR USE ON_____

METAL _____ DIE CLEARANCE _____

TOLERANCE: ±0.5mm UNLESS OTHERWISE SPECIFIED   METRIC

LAYOUT_____	DRAWN_____
CHECKED_____	APPROVED_____
SCALE_____	DATE_____
FROM B.P._____	DATED_____

SHEETS	SHEET	**NO.**

(A) TYPICAL SET-UP

QTY	ITEM	MATL	DESCRIPTION	PT NO.
1	BASE	G1	PATTERN #A3154	1
1	CAP	G1	PATTERN # B7156	2
1	SUPPORT	AISI-1212	.38 × 2.00 × 4.38	3
1	BRACE	AISI-1212	.25 × 1.00 × 2.00	4
1	COVER	AISI-1035	.1345 (# 10 GA USS) × 6.00 × 7.50	5
1	SHAFT	AISI-1212	Ø1.00 × 6.50	6
2	BEARINGS	SKF	RADIAL BALL # 6200Z	7
2	RETAINING CLIP	TRUARC	N5000-725	8
1	KEY	STL	WOODRUFF # 608	9
1	SET SCREW	CUP POINT	HEX SOCKET .25UNC × 1.50	10
4	BOLT-HEX HD-REG	SEMI-FIN	.38UNC × 1.50LG	11
4	NUT-REG HEX	STL	.38UNC	12
4	LOCK WASHER-SPRING	STL	.38-MED	13

NOTE: Parts 7 to 13 are purchased items

(B) MATERIAL LIST APPLICATION

FIG. 3-1-8   Combined order table, item list, and title block.

In addition to the description of drawing changes, provision may be made for recording a revision symbol zone location, issue number, date, and approval of the change. Typical revision tables are shown in Fig. 3-1-9 (pg. 40).

Auxiliary Number Blocks

An auxiliary number block, approximately 2 × .25 in. (50 × 10 mm), is placed above the title block so that after prints are folded, the number will appear close to the top RH corner of the print, as in Fig. 3-1-5. This is done to facilitate identification when the folded prints are filed on edge.

Auxiliary number blocks are usually placed within the inside border, but they may be placed in the margin outside the border line if space permits.

REFERENCES AND SOURCE MATERIAL

1. Keuffel and Esser Co.
2. *Machine Design* and National Microfilm
3. Eastman Kodak Co.

3-2 FILING AND STORAGE

One of the most common and difficult problems facing an engineering department is how to set up and maintain an efficient engineering filing area. Normal office filing methods are not considered satisfactory for engineering drawings. To properly serve its function, an engineering filing area must meet

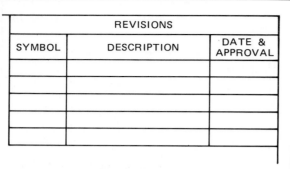

(A) VERTICAL REVISION TABLE

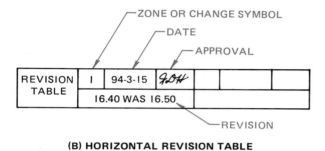

(B) HORIZONTAL REVISION TABLE

FIG. 3-1-9   Revision tables.

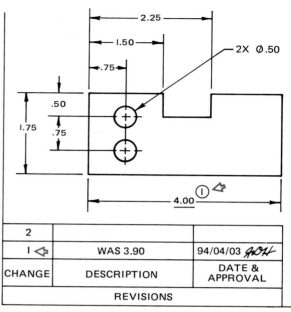

(C) APPLICATION

two important criteria: accessibility of information and protection of valuable documentation.

For this kind of system to be effective, drawings must be readily accessible. The degree of accessibility depends on whether drawings are considered active, semiactive, or inactive.

Filing Systems

Original Drawings

Unless a company has developed a full microforming system, the drafter's original drawings, produced either manually or by CAD, are kept and filed for future use or reference, and prints are made as required. To avoid crease lines, the originals, unlike prints, must *not* be folded. They are filed in either a flat or rolled position (Fig. 3-2-1).

In determining the type of equipment to use for engineering files, remember that different types of drawings require different kinds of files. Also, in planning a filing system, keep in mind that filing requirements are always increasing; unlike normal office files that can be purged each year, the more drawings produced, the more need to be stored. Therefore, any filing systems must have the flexibility of being easily expanded—usually in a minimum of space.

Microfilm Filing Systems

Although microfilming (Fig. 3-2-2) has been an established practice in many engineering offices for some time, the advent of CAD and new high-speed reproduction methods has made it less significant. It seems logical that reducing drawings to tiny images on film would make them more difficult to locate.

However, this is not the case, for while they are reduced in size, they are made more uniform. This results in improved file arrangements.

Forms of Film   One way to classify microfilm is according to the physical form in which it is used.

Roll Film   This is the form of the film after it has been removed from the camera and developed. Microfilm comes in four different widths—16, 35, 70, and 105 mm—and is stored in magazines.

Aperture Cards   Perhaps the simplest of the flat microforms is finished roll film cut into separate frames, each mounted on a card having a rectangular hole. Aperture cards are available in many sizes.

Jackets   Jackets are made of thin, clear plastic and have channels into which short strips of microfilms are inserted. They come in a variety of film-channel combinations for 16 and/or 35 mm microfilm. Like aperture cards, jackets can be viewed easily.

Microfiche   A microfiche is a sheet of clear film containing a number of microimages arranged in rows (Fig. 3-2-3). A common size is 100×150 mm, frequently arranged to contain 98 images. Microfiches are especially well suited for quantity distribution of standard information, such as parts and service lists.

CAD Storage

Original drawings in CAD are digital information and stored on magnetic media, such as tape, floppy disks, and hard disks, or

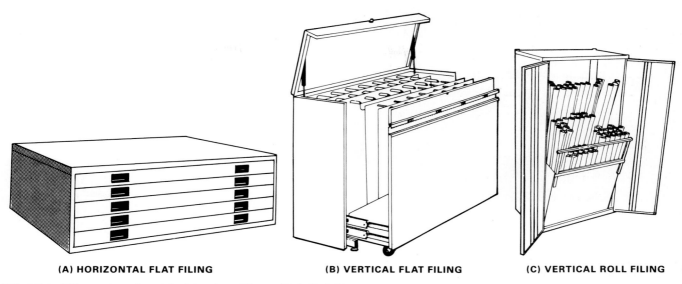

(A) HORIZONTAL FLAT FILING   **(B) VERTICAL FLAT FILING**   **(C) VERTICAL ROLL FILING**

FIG. 3-2-1   Filing systems for original drawings. *(Norman Wade Co. Ltd.)*

FIG. 3-2-2   Microfilming.
(Doug Martin)

FIG. 3-2-3   Microfiche.
(STUDIOHIO)

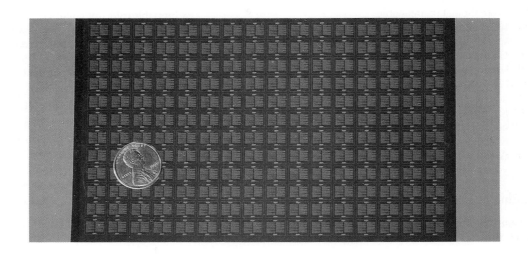

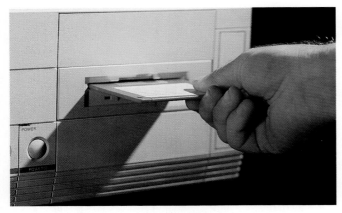

FIG. 3-2-4   Floppy disk with protective cover. *(STUDIOHIO)*

on optical media such as laser disks, CD-ROMs, or magneto-optical diskettes. Since magnetic media can be easily damaged, special procedures are needed to protect original drawings. One finished drawing plot or printer plot (often referred to as "hard copy") made on film, vellum, or paper may be stored as a permanent record in the same way that manually drawn originals are preserved. Optical disks are not easily damaged and make excellent permanent records.

Handling Diskettes

The diskettes used in personal computers and workstations for removable storage should be properly handled, labeled, and stored. The 3.5 in., double sided, high density diskette (Fig. 3-2-4) is the most common type of diskette and stores approximately 1.44 MB of data. All diskettes should be clearly labeled as to their content and ownership. Diskettes not in use should be properly stored in a diskette storage unit such as a file box or binder page. Magnetic media are not permanent, and multiple copies of diskettes with important data should be maintained.

The following are some rules for handling and storing diskettes or any other similar computer media:

1. Always properly label and store the diskette.
2. Never use a pencil to write on the diskette label. The loose graphite can damage the disk drive's read/write head.
3. Do not touch the disk surface with your fingers or anything else. This can ruin the surface of the diskette and make the data unreadable.
4. Keep diskettes away from magnetic fields such as motors, speakers, and some desk lamps.
5. Do not expose diskettes to temperature extremes. The safe temperature range for diskettes is between 50° and 140°F (10–60°C). Diskettes left in the direct sun will be destroyed by the high temperature.
6. Keep diskettes away from liquids and loose debris. Do not use a diskette that has been wet or dirty as the diskette may damage the drive.
7. Always make a backup diskette of important data (like your assignments).

Folding of Prints

To facilitate handling, mailing, and filing, prints should be folded to letter size, 8.5 × 11 in. (210 × 297 mm), in such a

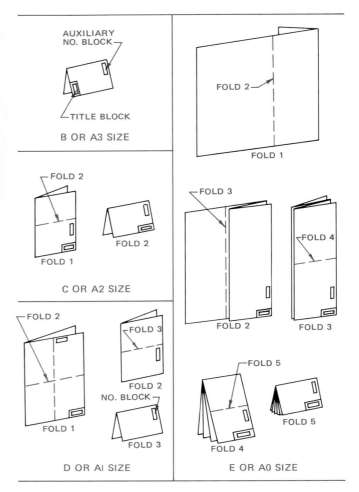

FIG. 3-2-5   Folding of prints.

way that the title block and auxiliary number always appear on the front face and the last fold is always at the top. In filing, this prevents other drawings from being pushed into the folds of filed prints. Recommended methods of folding standard-size prints are illustrated in Fig. 3-2-5.

On preprinted forms, it is recommended that fold marks be included in the margin of size B and larger drawings and be identified by number, for example, "fold 1," "fold 2." In zoned prints, the fold lines will coincide with zone boundaries, but they should nevertheless be identified.

To avoid loss of clarity by frequent folding, important details should not be placed close to fold areas. As a time-saver, some copiers are equipped to automatically fold prints.

REFERENCES AND SOURCE MATERIAL

1. Eastman Kodak Co. and "Setting Up and Maintaining an Effective Drafting Filing System," *Reprographics.*

3-3 DRAWING REPRODUCTION

A revolution in reproduction technologies and methods began several decades ago. It brought with it new equipment and supplies that have made quick copying commonplace. The introduction of high-quality, moderate-cost CAD printers and plotters in the 1990s has also influenced the choice of reproduction equipment. The new technologies make it possible to apply improved systems approaches and new information-handling techniques to all types of files, ranging from small documents to large engineering drawings (Fig. 3-3-1).

The pressures on business and government for greater efficiency, space savings, cost reductions, lower investment costs, and equally important factors provided a fertile field for the new reproduction technologies. There is no reason to believe that such pressures will diminish. In fact, as the years go by, it is certain that more and more improvements will occur, newer and better reproduction and information-handling equipment and methods will be discovered, and the advantages they offer will find ever-widening application.

The following reproduction methods apply whether the original drawing was prepared manually or by CAD (plotter).

Reproduction Equipment

Studies of reproduction facilities, existing or proposed, should first consider the nature of the demand for this service, then the processes which best satisfy the demand, and finally the particular machines which employ the processes. Factors to consider at these stages of study include:

- *Input originals*—sizes, paper mass, color, artwork
- *Quality of output copies*—depending on expected use and degree of legibility required
- *Size of copies*—same size, enlarged, reduced
- *Color*—copy paper and ink or pen
- *Registration*—in multiple-color work
- *Volume*—numbers of orders and copies per order
- *Speed*—machine productivity, convenient start-stop and load-unload
- *Cost*—direct labor, direct material, overhead, service
- *Future requirements*

Copiers

The methods used to produce copies are the diazo process, photoreproduction, copying with printer/plotters, and microfilming.

Diazo Process (Whiteprint)   In this process (Fig. 3-3-2) paper or film coated with a photosensitive diazonium salt is exposed to light passing through the original drawing made on a translucent paper or film. The exposed coated sheet is then developed by an ammonia vapor or agent. Where the light passes through the clear areas of the original drawing, it decomposes the diazonium salt, leaving a clear area on the copy (print). Where markings on the original block the light, the ammonia and the unexposed coating produce an opaque dye image of the original markings. A positive original makes a positive copy, and a negative original makes a negative copy. The three diazo processes currently used differ mainly in the way the developing agent is introduced to the diazo coat. These are ammonia vapor developing, moist developing, and pressure developing (PD). The most significant characteristic of the diazo process is that it is the most economical method of making prints. The main disadvantages of this process are that

FIG. 3-3-1   Drawing reproduction. *(Glencoe file photo)*

FIG. 3-3-2   Diazo (whiteprint) machine. *(Norman Wade Co. Ltd.)*

only full-size prints can be made and the original drawing being copied must be made on a translucent material.

Photoreproduction   Photoreproduction using an engineering plain-paper copier (Fig. 3-3-3, pg. 44) has become popular because there is no need to use transparent/translucent originals. It prints on bond paper, vellum, and drafting film, in sizes from 8.5 × 11 in. up to 36 in. wide by any manageable length. It combines high-speed productivity with many efficient, often automatic, features to rapidly deliver high-quality, large-size copies. The many advantages include:

1. No need for translucent originals. Copies are made equally well from opaque or translucent drawings.
2. No ammonia or developing agent necessary. Consequently, this method is environmentally safe.
3. High-volume copying. A standard office copier will produce clear multiple copies at the rate of several D or E size drawings per minute.

FIG. 3-3-3   Photoreproduction (plain-paper) copier. *(OCE-USA, Inc.)*

4. Large-size copies. In addition to same-size copies, there are reduction (less than 50 percent) and enlargement (up to 200 percent) options.
5. Time-saving functions. Some plain-paper copies will reduce, automatically fold, and collate the prints.
6. Ideal for cut-and-paste drafting.

The disadvantage of an engineering plain paper copier is the cost of service.

Digital Plain Paper Printer/Plotters   One of the newest and most versatile methods for reproducing engineering drawings is to plot and copy directly from CAD (Fig. 3-3-4). Using a Digital Laser system with CAD plotting allows you to bypass the plotter bottleneck. High-resolution, 600 dpi (dots per inch) engineering drawings are plotted directly from the CAD system with multiple copies produced in just seconds. A laser system, for example, is able to create a complex D size CAD plot in less than 15 seconds and identical copies at the rate of 12 prints per minute. This technology is an important step toward a fully integrated engineering/reprographics department. The disadvantage of this method is the cost of equipment. Plotters or printer/plotters should be used only when the number of reproduction copies is low and color is important.

Microfilm Equipment   Microfilm equipment includes readers and viewers and reader-printers.

Readers and Viewers   Microfilm readers magnify film images large enough to be read and project the images onto a translucent or opaque screen. Some readers accommodate only one microfilm (roll, jacket, microfiche, or aperture card). Others can be used with two or more. Scanning-type readers, having a variable-type magnification, are used when frames containing a large drawing are viewed. Generally, only parts of the drawing can be viewed at one time.

Reader-Printers   Two kinds of equipment are used to make enlarged prints from microfilm: reader-printers and enlarger-printers. The *reader-printer* (Fig. 3-3-5) is a reader that incorporates a means of making hard copy from the projected image. The *enlarger-printer* is designed only for copying and does not include the means for reading.

Scanners   Many engineering offices have converted their manual drawings into CAD drawings. Completely redrawing each one is a tedious and time-consuming process. Even using a large graphics tablet and retracing the drawing by digitizing takes time and is expensive. The most effective means to accomplish this conversion is to use a scanner. Scanning technology allows you to feed in a manual drawing and convert the vectors (line lengths, circles, etc.) into the computer data base. The conversion is not perfect, so you will have to call up each drawing and edit it as needed.

REFERENCES AND SOURCE MATERIALS

1. Oce Bruning

FIG. 3-3-4   Laser printer/plotter. *(Courtesy of Xerox Corporation)*

FIG. 3-3-5   Microfilm reader/printer. *(Doug Martin)*

CHAPTER 4

BASIC DRAFTING SKILLS

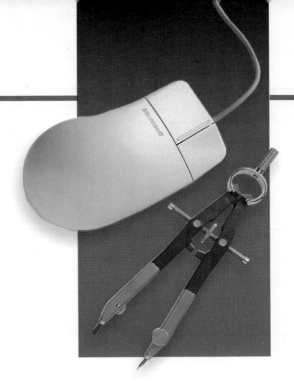

Definitions

Arcs Parts of a circle.

Center lines Lines consisting of alternating long and short dashes used to represent the axis of symmetrical parts and features, bolt circles, and paths of motion.

Concentric circles Circles of different diameters that have the same center.

Construction lines Lightly drawn lines used during the preliminary stages of a drawing.

Coordinates Numerical information used to describe a specific point in an *X-Y* linear coordinate system.

Ellipses Ovals.

Gothic-style lettering The preferred style of lettering used in drafting.

Guidelines Lines used to ensure uniform lettering.

Irregular curves Nonconcentric, nonstraight lines drawn smoothly through a series of points.

Line The most important single entity on a technical drawing.

Multiview or **orthographic projection** A type of drawing in which an object is usually shown in more than one view.

Overlay A piece of tracing paper used to redraw a portion of a sketch or drawing.

Pictorial drawing A drawing which shows the width, height, and depth of an object in one view.

Radius The distance from the center of a circle to its edge.

Refined sketch A neatly drawn, finished-looking freehand sketch.

Tangent arcs Parts of two circles that touch.

Unrefined or **rough sketch** A quickly drawn, freehand sketch.

Visible lines Clearly drawn lines used to represent visible edges or contours of objects.

4-1 STRAIGHT LINE WORK, LETTERING, AND ERASING

Manual Drafting

Line Work

A line is the fundamental, and perhaps the most important, single entity on a technical drawing. Lines are used to help illustrate and describe the shape of objects that will later become real parts. The various lines used in drawing form the "alphabet" of the drafting language. Like letters of the alphabet, they are different in appearance (see Figs. 4-1-1 and 4-1-2, pp. 46–48). The distinctive features of all lines that form a permanent part of the drawing are the differences in their width and construction. Lines must be clearly visible and stand out in sharp contrast to one another. This line contrast is necessary if the drawing is to be clear and easily understood.

The drafter first draws very light construction lines, setting out the main shape of the object in various views. Since these first lines are very light, they can be erased easily should changes or corrections be necessary. When the drafter is satisfied that the layout is accurate, the construction lines are then changed to their proper type, according to the alphabet of lines. Guidelines, used to ensure uniform lettering, are also drawn very lightly.

TYPE OF LINE	APPLICATION	DESCRIPTION
HIDDEN LINE  — — — — THIN — — — —	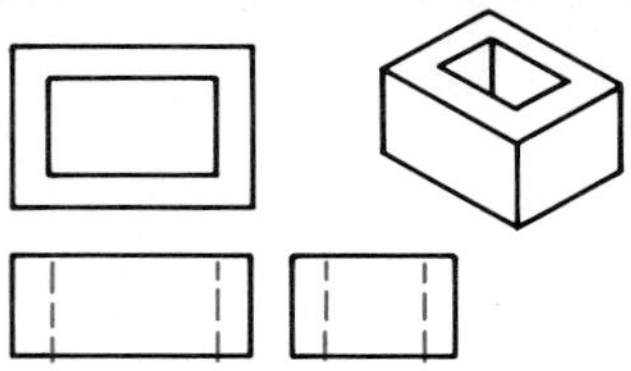	THE HIDDEN OBJECT LINE IS USED TO SHOW SURFACES, EDGES, OR CORNERS OF AN OBJECT THAT ARE HIDDEN FROM VIEW.
CENTER LINE  THIN ALTERNATE LINE AND SHORT DASHES	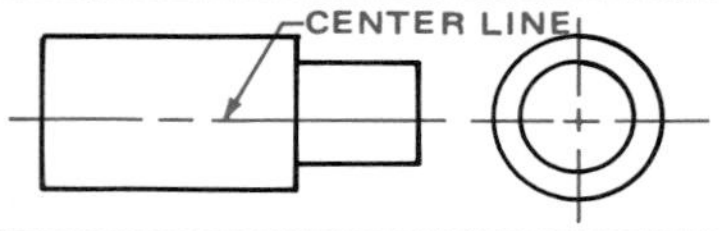 CENTER LINE	CENTER LINES ARE USED TO SHOW THE CENTER OF HOLES AND SYMMETRICAL FEATURES.
SYMMETRY LINE CENTER LINE THICK SHORT LINES 	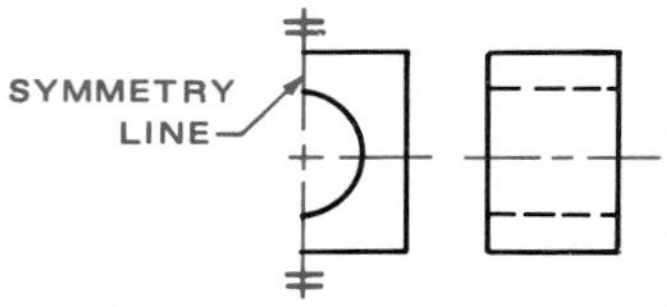 SYMMETRY LINE	SYMMETRY LINES ARE USED WHEN PARTIAL VIEWS OF SYMMETRICAL PARTS ARE DRAWN. IT IS A CENTER LINE WITH TWO THICK SHORT PARALLEL LINES DRAWN AT RIGHT ANGLES TO IT AT BOTH ENDS.
EXTENSION AND DIMENSION LINES  THIN DIMENSION LINE EXTENSION LINE	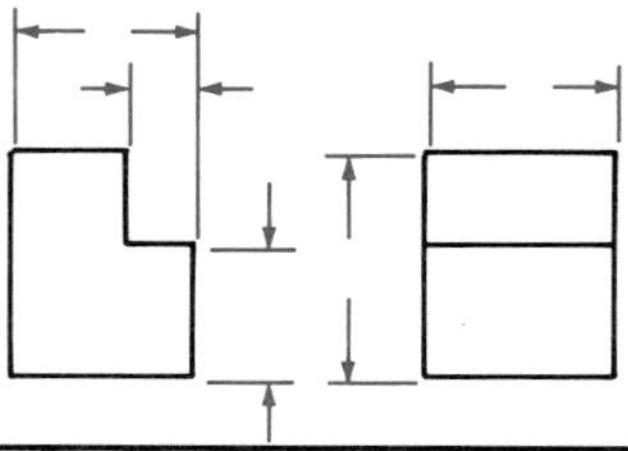	EXTENSION AND DIMENSION LINES ARE USED WHEN DIMENSIONING AN OBJECT.
LEADERS  ARROW DOT THIN	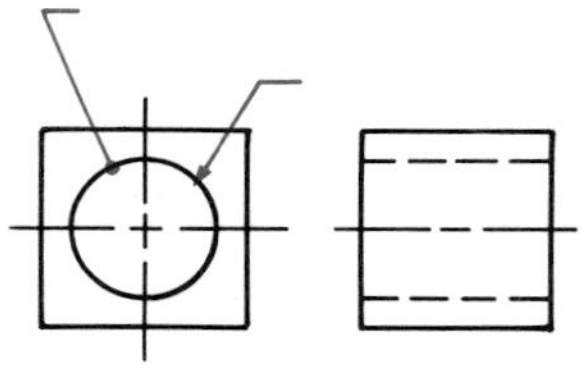	LEADERS ARE USED TO INDICATE THE PART OF THE DRAWING TO WHICH A NOTE REFERS. ARROWHEADS TOUCH THE OBJECT LINES WHILE THE DOT RESTS ON A SURFACE.
BREAK LINES 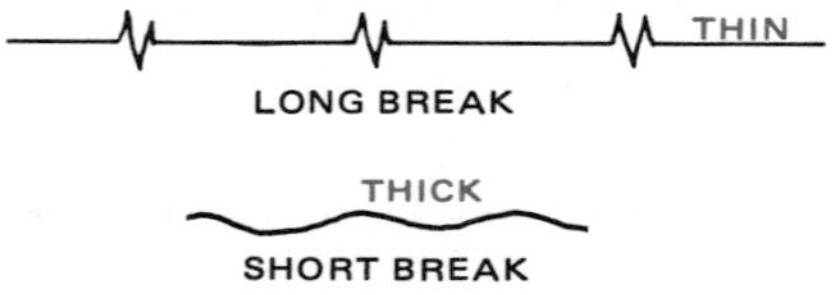 THIN LONG BREAK THICK SHORT BREAK	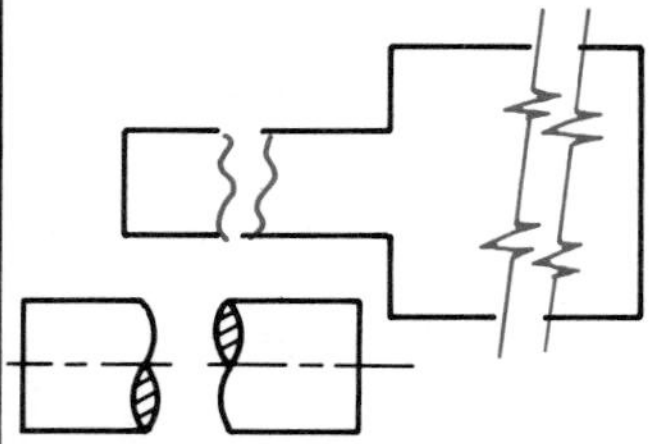	BREAK LINES ARE USED WHEN IT IS DESIRABLE TO SHORTEN THE VIEW OF A LONG PART.
CUTTING-PLANE LINE  THICK OR	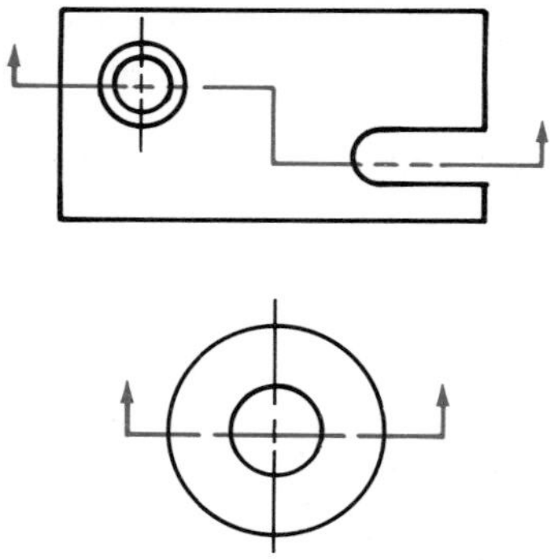	THE CUTTING-PLANE LINE IS USED TO DESIGNATE WHERE AN IMAGINARY CUTTING TOOK PLACE.

FIG. 4-1-1   Types of lines.

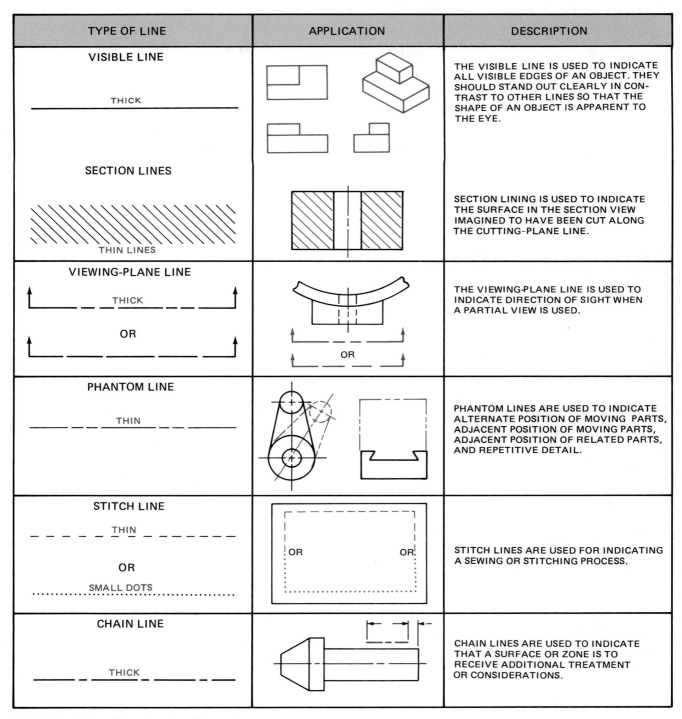

TYPE OF LINE	APPLICATION	DESCRIPTION
VISIBLE LINE THICK		THE VISIBLE LINE IS USED TO INDICATE ALL VISIBLE EDGES OF AN OBJECT. THEY SHOULD STAND OUT CLEARLY IN CONTRAST TO OTHER LINES SO THAT THE SHAPE OF AN OBJECT IS APPARENT TO THE EYE.
SECTION LINES THIN LINES		SECTION LINING IS USED TO INDICATE THE SURFACE IN THE SECTION VIEW IMAGINED TO HAVE BEEN CUT ALONG THE CUTTING-PLANE LINE.
VIEWING-PLANE LINE THICK OR		THE VIEWING-PLANE LINE IS USED TO INDICATE DIRECTION OF SIGHT WHEN A PARTIAL VIEW IS USED.
PHANTOM LINE THIN		PHANTOM LINES ARE USED TO INDICATE ALTERNATE POSITION OF MOVING PARTS, ADJACENT POSITION OF MOVING PARTS, ADJACENT POSITION OF RELATED PARTS, AND REPETITIVE DETAIL.
STITCH LINE THIN OR SMALL DOTS		STITCH LINES ARE USED FOR INDICATING A SEWING OR STITCHING PROCESS.
CHAIN LINE THICK		CHAIN LINES ARE USED TO INDICATE THAT A SURFACE OR ZONE IS TO RECEIVE ADDITIONAL TREATMENT OR CONSIDERATIONS.

FIG. 4-1-1   Types of lines. (continued)

Line Widths   Two widths of lines, thick and thin, as shown in Fig. 4-1-3 (pg. 49), are recommended for use on drawings. Thick lines are .030 to .038 in. (0.5 to 0.8 mm) wide, thin lines between .015 and .022 in. (0.3 to 0.5 mm) wide. The actual width of each line is governed by the size and style of the drawing and the smallest size to which it is to be reduced. All lines of the same type should be uniform throughout the drawing.

Spacing between parallel lines should be such that there is no "fill-in" when the copy is reproduced by available photographic methods. Spacing of no less than .12 in. (3 mm) normally meets reproduction requirements.

All lines should be sharp, clean-cut, opaque, uniform, and properly spaced for legible reproduction by all commonly used methods, including microforming, in accordance with industry

47

FIG. 4-1-2   Application of
lines. *(ANSI
Y14.2M, 1982)*

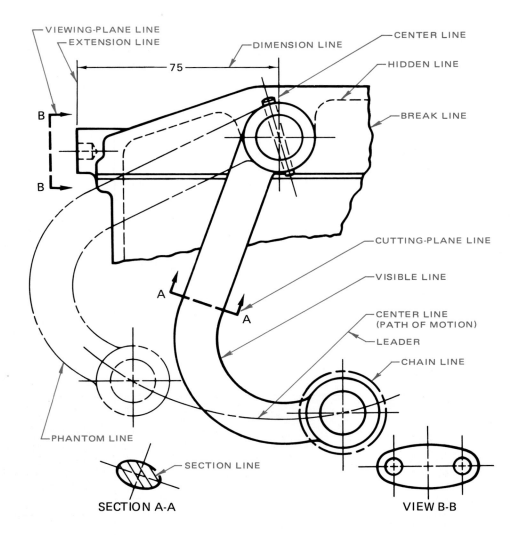

and government requirements. There should be a distinct contrast between the two widths of lines.

Visible Lines   The "visible lines" should be used for representing visible edges or contours of objects. Visible lines should be drawn so that the views they outline clearly stand out on the drawing with a definite contrast between these lines and secondary lines.

The applications of the other types of lines are explained in detail throughout this text.

Drawing Straight Lines   When using a T square to draw horizontal lines (Fig. 4-1-4), hold the head of the T square against the edge of the drawing board and slide the T square either up or down to the desired position. Firmly press down on the blade of the T square to prevent it from moving, then proceed to draw the line. When drawing vertical lines, a triangle, which rests on the top side of the T square, is moved to the desired position and both the blade of the T square and the triangle are held firmly to the drawing board with the hand not holding the pencil.

When a parallel slide is used, as in Fig. 4-1-5, it will always be in a horizontal position as the wire and rollers in the slide

move both ends of the slide simultaneously and at the same speed.

A general rule to follow when drawing straight lines is to lean the pencil in the direction of the line you are about to draw. A right-handed person would lean the pencil to the right and draw horizontal lines from left to right. The left-handed person would reverse this procedure. When drawing vertical lines, lean the pencil away from yourself, toward the top of the drafting board, and draw lines from bottom to top. Lines sloping from the bottom to the top right are drawn from bottom to top; lines sloping from the bottom to the top left are drawn from top to bottom. This procedure for drawing sloping lines would be reversed for a left-handed person.

When using a conical-shaped lead, rotate the pencil slowly between your thumb and your forefinger when drawing lines (Fig. 4-1-6). This keeps the lines uniform in width and the pencil sharp. Do not rotate a pencil having a bevel or wedge-shaped lead.

Many drafters today use the 0.5 mm mechanical pencil. Holding the pencil perpendicular to the paper, the drafter can produce uniform lines easily. The pencil is not rotated for this procedure. Pencils and leads are available from 0.2 to 0.9 mm in diameter for creating different line widths.

THICK

WIDTH .032 IN. (0.7 mm)

THIN

WIDTH .016 IN. (0.35 mm)

FIG. 4-1-3   Line widths.

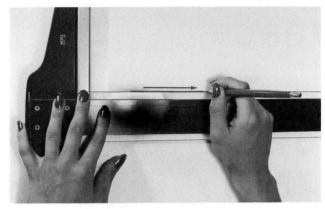

(A) DRAWING A HORIZONTAL LINE

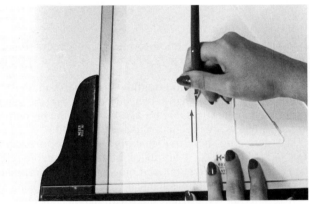

(B) DRAWING A VERTICAL LINE

FIG. 4-1-4   Drawing horizontal and vertical lines.

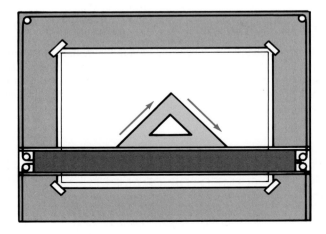

FIG. 4-1-5   Drawing horizontal and sloping lines with the aid of a parallel slide.

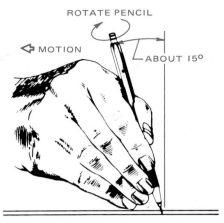

FIG. 4-1-6   Rotate the pencil when drawing lines.

Lettering

Single-Stroke Gothic Lettering   The most important requirements for lettering are legibility, reproducibility, and ease of execution. These are particularly important because of the increased use of microforming, which requires optimum clarity and adequate size of all details and lettering. It is recommended that all drawings be made to conform to these requirements and that particular attention be paid to avoid the following common faults:

1. Unnecessarily fine detail
2. Poor spacing of details
3. Carelessly drawn figures and letters
4. Inconsistent delineation
5. Incomplete erasures that leave ghost images
6. Use of differing densities, such as pencil, ink, and typescript on the same drawing

These requirements are met in the recommended single-stroke Gothic characters shown in Fig 4-1-7 (pg. 50) or adaptations thereof, which improve reproduction legibility. One such adaptation by the National Microfilm Association is the vertical Gothic style Microfont alphabet (Fig. 4-1-8, pg. 50) intended for general use.

Either inclined or vertical lettering is permissible, but only one style of lettering should be used throughout a drawing. The preferred slope for the inclined characters is 2 in 5, or approximately 68° with the horizontal.

Uppercase letters should be used for all lettering on drawings unless lowercase letters are required to conform with other established standards, equipment nomenclature, or marking.

Lettering for titles, subtitles, drawing numbers, and other uses may be made freehand, by typewriter, or with the aid of mechanical lettering devices, such as templates and lettering machines. Regardless of the method used, all characters are to conform, in general, with the recommended Gothic style and must be legible in full- or reduced-size copy by any accepted method of reproduction.

Letters in words should be spaced so that the background areas between the letters are approximately equal, and words are to be clearly separated by a space equal to the height of the lettering (Fig. 4-1-9, pg. 50). The vertical space between lines

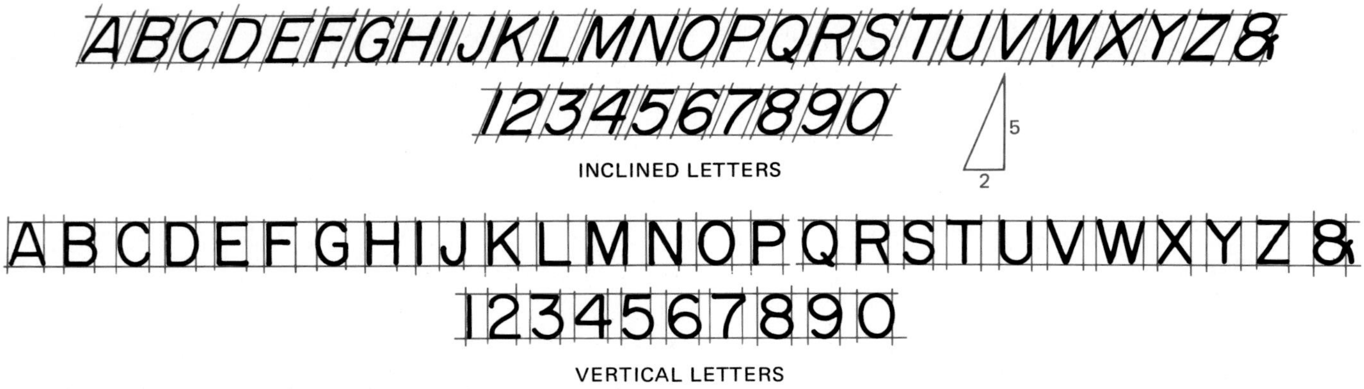

INCLINED LETTERS

VERTICAL LETTERS

FIG. 4-1-7   Approved Gothic lettering for engineering drawings.

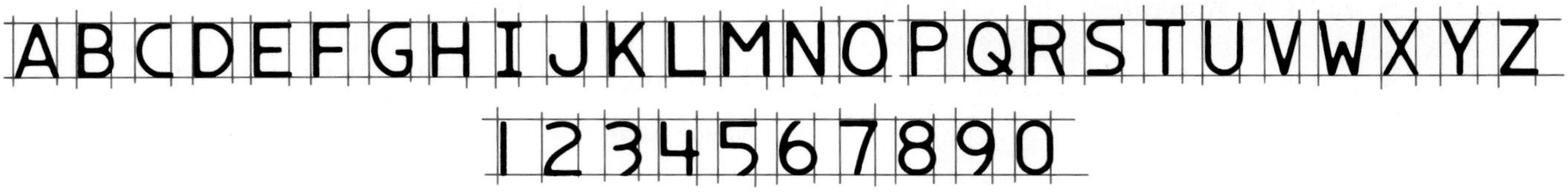

FIG. 4-1-8   Microfont letters. *(National Microfilm Assoc.)*

of lettering should be no more than the height of the lettering and no less than half the height of the lettering.

The recommended minimum freehand and mechanical letter heights for various applications are given in Fig 4-1-10. So that lettering will be uniform and of proper height, light guidelines, properly spaced, are drawn first and then the lettering is drawn between these lines.

Notes should be placed horizontally on drawings and separated vertically by spaces at least equal to double the height of the character size used, to maintain the identity of each note.

Decimal points must be uniform, dense, and large enough to be clearly visible on approved types of reduced copy. Decimal points should be placed in line with the bottom of the associated digits and be given adequate space.

Lettering should not be underlined except when special emphasis is required. The underlining should not be less than .06 in. (1.5 mm) below the lettering.

When drawings are being made for microforming, the size of the lettering is an important consideration. A drawing may be reduced to half size when microformed at 30X reduction and blown back at 15X magnification. (Most microform engineering readers and blowback equipment have a magnification

of 15X. If a drawing is microformed at 30X reduction, the enlarged blown-back image is 50 percent; at 24X, it is 62 percent of its original size.)

Standards generally do not allow characters smaller than .12 in. (3 mm) for drawings to be reduced 30X, and the trend is toward larger characters. Figure 4-1-11 shows the proportionate size of letters after reduction and enlargement.

The lettering heights, spacing, and proportions in Figs. 4-1-10 and 4-1-11 normally provide acceptable reproduction or camera reduction and blowback. However, manually, mechanically, optimechanically, or electromechanically applied lettering (typewriter, etc.) with heights, spacing, and proportions less than those recommended are acceptable when the reproducibility requirements of the accepted industry or military reproduction specifications are met.

Erasing Techniques

Revision or change practice is inherent in the method of making engineering drawings. It is much more economical to introduce changes or additions on an original drawing than to redraw the entire drawing. Consequently, erasing has become a science all

FIG. 4-1-9   Spacing of lettering. *(National Microfilm Assoc.)*

PREFERRED
(OPEN-TYPE LETTERING)

UNDESIRABLE
(CRAMPED LETTERING)

USE	INCH		METRIC mm		DRAWING SIZE
	FREEHAND	MECHANICAL	FREEHAND	MECHANICAL	
DRAWING NUMBER IN	0.250	0.240	7	7	UP TO AND INCLUDING 17 x 22 INCHES
TITLE BLOCK	0.312	0.290			LARGER THAN 17 x 22 INCHES
DRAWING TITLE	0.250	0.240	7	7	ALL
SECTION AND TABULATION LETTERS	0.250	0.240	7	7	
ZONE LETTERS AND NUMERALS IN BORDER	0.188	0.175	5	5	
DIMENSION, TOLERANCE, LIMITS, NOTES, SUBTITLES FOR SPECIAL VIEWS, TABLES, REVISIONS, AND ZONE LETTERS FOR THE BODY OF THE DRAWING	0.125	0.120	3.5	3.5	UP TO AND INCLUDING 17 x 22 INCHES
	0.156	0.140	5	5	LARGER THAN 17 x 22 INCHES

THIS IS AN EXAMPLE OF .125 IN. LETTERING

THIS IS AN EXAMPLE OF .188 IN. LETTERING

THIS IS AN EXAMPLE OF .250 IN. LETTERING

FIG. 4-1-10   Recommended lettering heights. *(ANSI Y14.2M, 1982)*

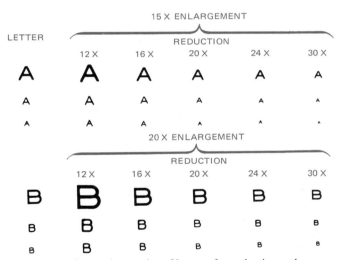

FIG. 4-1-11   Proportionate size of letters after reduction and enlargement. *(National Microfilm Assoc.)*

 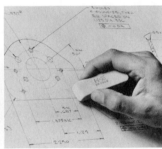

(A) USE OF ERASING SHIELD   **(B) CLEANING UP AN AREA**

FIG. 4-1-12   Erasing. *(Keuffel & Esser Co.)*

its own. Proper erasing is extremely important since some drawings are revised a great number of times. Consequently, good techniques and materials must be used that permit repeated erasures on the same area. Some recommendations follow.

1. Avoid damaging the surface of the drawing medium by selecting the proper eraser.
2. Lines not thoroughly erased produce ghostlike images on prints, resulting in reduced legibility.
3. A hard, smooth surface such as a triangle, placed under the lines being removed makes erasing easier.
4. Using an erasing shield protects the adjacent lines and lettering and also eliminates wrinkling (Fig. 4-1-12).

5. Also erase on the back side of the paper. Lines frequently pick up dirt or graphite on the underside, and if not erased, will still produce lines on the print.
6. Be sure to completely remove erasure debris from the drawing surface.
7. When extensive changes are required, it may be more economical to cut and paste or make an intermediate drawing.
8. When erasing, use no more pressure than necessary.
9. If the drawing quality of the paper or other medium has been damaged by erasing, it may be improved by sprinkling an inking powder on the surface and rubbing it with a cloth.

In addition, it is necessary to match the density of the surrounding background when erasures are made. Often, the erased area is much cleaner than the rest of the drawing. If the change is made on this clean area, the contrast between line and background is different and that area presents a problem in reproduction. It is usual practice to "smudge" the erased area so that it looks about the same shade as the surrounding area.

Removing Lines on Film　Lines on photoreproduction film fall into two classifications: photographic lines and pencil-and-ink lines. All these lines can be removed easily so that the erased area can be used for further drafting. Here are some tips for removing lines.

Erasers　There are three basic types of erasers: rubber, plastic, and liquid. Rubber and plastic erasers may tend to cause a shine on the drafting surface. When plastic erasers are used, the shiny appearance may actually be transparentizing, because of the plasticizers used in their manufacture, rather than wearing on the drafting surface. The transparentizing effect is not detrimental since it does not reduce the ability of the surface to take a pencil line. Good drafting lines can easily be drawn over areas from which lines have been erased many times. A good general rule to follow is to use a soft, nonabrasive eraser and only enough pressure to remove the line. Liquid erasers do not put a shine on the drafting surface and can be used to make many erasures in the same place.

Erasing machines require special techniques. Many different types of erasing inserts are available. To avoid burning holes in paper or melting the matte surface on film, you must use a light touch and keep the eraser moving.

When the drafting surface is affected by excessive erasures, it can be repaired by rolling a regular typewriter eraser across the smooth area or by rubbing a small amount of drafting powder into the area with a finger.

Pencil Lines　Pencil lines can be removed from all film with a soft, non-abrasive rubber or plastic eraser or with liquid eraser. To keep the drafting surface from becoming too shiny, avoid excessive pressure.

Fastening Paper to the Board

The most common method of holding the drawing paper to the drafting board is with drafting tape.

When fastening the paper to the board, line up the bottom or top edge of the paper with the top horizontal edge of the T square, parallel slide, or horizontal scale of the drafting machine (Fig. 4-1-13). When refastening a partially completed drawing, use lines on the drawing rather than the edge of the paper for alignment.

 # CAD

Skills in drawing lines and lettering are not required with CAD. Rather, skill in operating the CAD equipment becomes essential.

Before learning the fundamentals of drafting, you should become familiar with the basics of line creation.

Coordinate Input

One of the most common ways to create lines using CAD is by coordinate input (keying in distances). There are three methods of coordinate input:

1. Absolute coordinate
2. Relative coordinate
3. Polar coordinate (line length and angle)

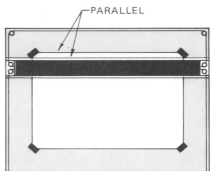

FIG. 4-1-13　Positioning the paper on the board.

Absolute Coordinates　Coordinate input is based on the rectangular (horizontal and vertical) measurement system. All absolute distances are described in terms of their distance from the drawing origin. Relative and polar coordinates may be described with respect to a particular point (position) on the drawing. This position can be at any location on the drawing and is described by two-dimensional coordinates, horizontal (X) and vertical (Y). The X axis is horizontal and is considered the first and basic reference axis. The Y axis is vertical and is 90° to the X axis. Any distance to the right of the drawing origin is considered a positive X value and any distance to the left of the drawing origin a negative X value. Distances above the drawing origin are positive Y values and distances below the drawing origin are negative Y values.

For example, four points lie in a plane, as shown in Fig. 4-1-14. The plane is divided into four quadrants. Point A lies in quadrant 1 and is located at position (6, 5), with the X coordinate first, followed by the Y coordinate. Point B lies in quadrant 2 and is located at position (−4, 3). Point C lies in quadrant 3 and is located at position (−5, −4). Point D lies in quadrant 4 and is located at position (3, −2).

Using the coordinate system to locate a point on a drawing would be greatly simplified if all parts of the drawing were located in the first quadrant because all the values would be positive and the plus and minus signs would not be required.

For this reason, on CAD systems the drawing origin is normally located at the lower left of the monitor. Thus the origin is a base reference point from which all positions on the drawing are measured. Its dimensions are $X = 0$ and $Y = 0$, referred to as 0,0. Figure 4-1-15 shows two points located on a monitor screen. Point 1 (shown as a cross) has an X coordinate of 2.50 and a Y coordinate of 6.00 with reference to the origin (0,0). Point 2 has an X coordinate of 8.35 and a Y coordinate of 2.20 with reference to the origin.

Relative Coordinates A relative coordinate is located with respect to the current access location (last cursor position selected) rather than the origin (0,0). In other words, it is located with respect to another point on your drawing. Often, both absolute and relative coordinate input are used during drawing preparation. A line may be created by combining these two methods.

An example of this type of dimensioning is shown in Fig. 4-1-16. Point 1 was located in the identical way in which point 1 in Fig. 4-1-15 was located, that is, using absolute coordinates. In Fig. 4-1-16, point 2 is located relative to point 1. Thus the relative coordinates of point 2 would be 5.85,−3.80 (5.85 to the right and 3.80 below the last position). Note the minus sign in front of the 3.80 dimension. Point 2 in both figures is in the same position relative to the origin.

Polar Coordinates A polar coordinate is similar to a relative coordinate since it is positioned with respect to the current access location. A line, however, will be specified according to its actual length and a direction rather than as X, Y coordinate distance. The direction is measured angularly in a counterclockwise direction from a horizontal line. Zero degrees is located horizontally to the right (at 3 o'clock) as shown in Fig. 4-1-17.

For example, a line 6.50 in. long to be drawn from point E to point F, as shown in Fig. 4-1-17A, would have polar coordinates of 6.50 and 45°. The line drawn from point G to point H (Fig. 4-1-17B) would have polar coordinates of 8.00 and 210°.

Line Styles, Text, and Erasing

All CAD systems have the options to create or apply line styles and text. The LINETYPE command allows you to select the desired line style. The TEXT command allows you to add

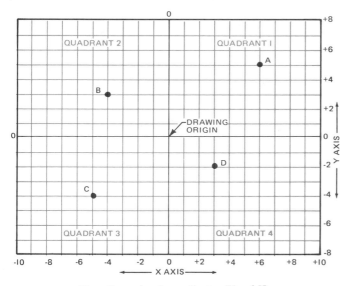

FIG. 4-1-14 Two-dimensional coordinates (X and Y).

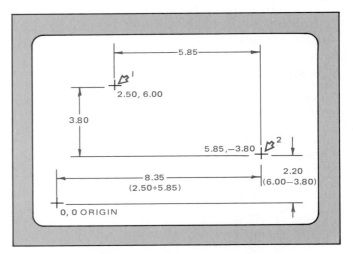

FIG. 4-1-16 Locating point 2 from point 1 using the relative coordinate option.

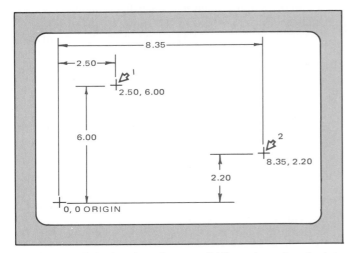

FIG. 4-1-15 Point locations shown on CAD monitor using absolute coordinates.

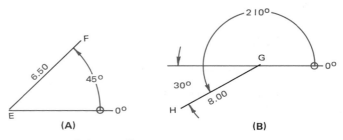

FIG. 4-1-17 Polar coordinates.

letters, numbers, words, notes, symbols, and messages to the drawing. Additionally, it can be used for size description (lengths, diameters, etc.).

As noted earlier, the need for revision or change is inherent in preparing technical drawings. Consequently, a variety of editing commands for removing unwanted lines and lettering from the drawing are available on all CAD systems.

REFERENCES AND SOURCE MATERIAL

1. National Microfilm Association
2. Keuffel and Esser Co.
3. Eastman Kodak Co.

ASSIGNMENTS

See Assignments 1 through 14 for Unit 4-1 on pages 65–70.

4-2 CIRCLES AND ARCS

Center Lines

Center lines consist of alternating long and short dashes (Fig. 4-2-1). They are used to represent the axis of symmetrical parts and features, bolt circles, and paths of motion. The long dashes of the center lines may vary in length, depending upon the size of the drawing. Center lines should start and end with long dashes and should not intersect at the spaces between dashes. Center lines should extend uniformly and distinctly a short distance beyond the object or feature of the drawing unless a longer extension is required for dimensioning or for some other purpose. They should not terminate at other lines of the drawing, nor should they extend through the space between views.

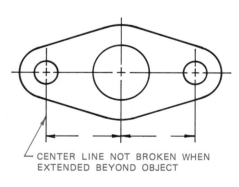

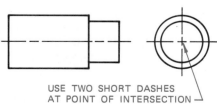

CENTER LINE NOT BROKEN WHEN EXTENDED BEYOND OBJECT

USE TWO SHORT DASHES AT POINT OF INTERSECTION

FIG. 4-2-1   Center line technique.

Very short center lines may be unbroken if no confusion results with other lines.

Center lines are used to locate the center of circles and arcs. They are first drawn as light construction lines, then finished as alternate long and short dashes, with the short dashes intersecting at the center of the circle.

 CAD

Center lines may be drawn by following the line commands explained in your CAD manual. The procedure by which you construct center lines in CAD varies greatly among different software packages.

Drawing Circles and Arcs

Circles and arcs are drawn with the aid of a compass or template. When circles and arcs are drawn with a compass, it is recommended that the compass lead be softer and blacker than the pencil lead being used on the same drawing. For example, if you are drawing with a 2H or 3H pencil, use an H compass lead. This will produce a drawing having similar line work since it is necessary to compensate for the weaker impression left on the drawing medium by the compass lead as compared with the stronger direct pressure of the pencil point. For drawing circles and arcs, see Figs. 4-2-2 and 4-2-3.

It is essential that the compass lead be reasonably sharp at all times in order to ensure proper line width. The compass lead should be sharpened to a bevel point, with the top rounded off, as shown in Fig. 4-2-4. The lead is slightly shorter than the needle point.

For drawing large circles and arcs, a beam compass, as shown in Fig. 4-2-5A (pg. 56), is used. For drawing very large arcs, the adjustable arc may be used (Fig. 4-2-5B). It accurately draws any radius from 7 to 200 m.

Drafters find it much easier and faster to use circle templates. There are sets that contain all common sizes and shapes of holes that most drafters are ever called upon to draw. When using a circle template, choose the correct diameter, line up the marks on the template with the center lines, and trace a dark thick line.

The drawing of arcs should be done before the tangent lines are made heavy. Draw light construction lines to establish the compass point location and check to make certain that the compass lead meets properly with both tangent lines before drawing the arc.

 CAD

Circles   There are several common methods used to construct circles. The common methods used to draw circles include: (1) center and radius; (2) center and diameter; (3) three-point circle; and (4) two-point circle.

Arcs   Among the 10 most common methods used to draw arcs and fillets are: (1) three points; (2) start, center, and end; (3) start, end, and radius; (4) start, center, and angle; and (5) fillet.

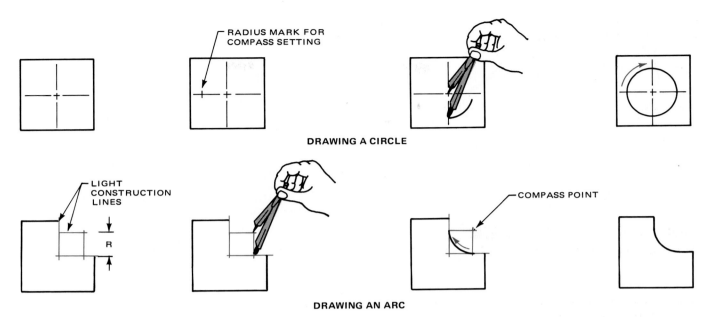

RADIUS MARK FOR COMPASS SETTING

DRAWING A CIRCLE

LIGHT CONSTRUCTION LINES

R

COMPASS POINT

DRAWING AN ARC

FIG. 4-2-2   Drawing circles and arcs.

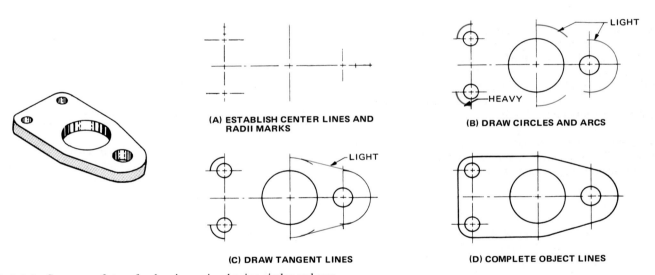

(A) ESTABLISH CENTER LINES AND RADII MARKS

(B) DRAW CIRCLES AND ARCS

LIGHT

HEAVY

LIGHT

(C) DRAW TANGENT LINES

(D) COMPLETE OBJECT LINES

FIG. 4-2-3   Sequence of steps for drawing a view having circles and arcs.

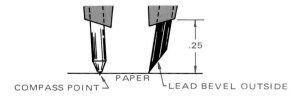

.25

COMPASS POINT   PAPER   LEAD BEVEL OUTSIDE

FIG. 4-2-4   Sharpening and setting the compass lead.

ASSIGNMENTS

See Assignments 15 through 22 for Unit 4-2 on pages 70–73.

4-3 DRAWING IRREGULAR CURVES

Curved lines may be drawn with the aid of irregular curves, flexible curves, and elliptical templates. Using an irregular curve, establish the points through which the curved line passes (Fig. 4-3-1, pg. 56), and draw a light freehand line through these points. Next, fit the irregular curve or other instrument by trial against a part of the curved line and draw a portion of line. Move the curve to match the next portion, and so forth. Each new position should fit enough of the part just drawn (overlap) to ensure continuing a smooth line. It is very important to notice whether the radius of the curved line is increasing or decreasing and to place the irregular curve in the same way.

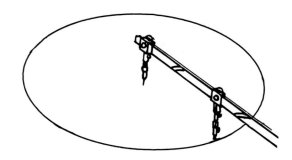

(A) USING A BEAM COMPASS

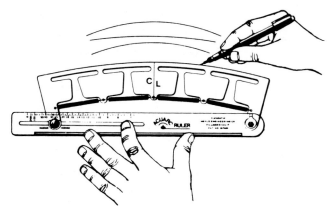

(B) USING AN ADJUSTABLE ARC

FIG. 4-2-5   Beam compass and adjustable arc.

If the curved line is symmetrical about the axis, the position of the axis may be marked on the irregular curve with a pencil for one side and then reversed to match and draw the other side.

 CAD

An irregular curve is a nonconcentric, nonstraight line drawn smoothly through a series of points. In CAD systems, it is commonly referred to as a *spline*.

ASSIGNMENTS

See Assignments 23 through 25 for Unit 4-3 on page 73.

4-4 SKETCHING

Freehand sketching is the simplest form of drawing. It is one of the quickest ways to express ideas. The drafter may use sketches to help simplify and explain thoughts and concepts to other people in discussions of mechanical parts and mechanisms. Sketching, therefore, is an important method of communication.

Freehand sketching is also a necessary part of drafting because the drafter in industry frequently sketches ideas and designs prior to making drawings with instruments. Practice in sketching helps the student develop a good sense of proportion and accuracy of observation.

Freehand drawings generally need some freehand lettering to explain features of a new idea or how a new product works. A drawing well planned may be worth a thousand words, as an old saying goes, but a few choice words well organized can explain some details.

The notes lettered in the rough sketch in Fig. 4-4-1 describe some functional features that are important to operation. Simple freehand lettering will complement an idea that is captured in a sketch, especially if the lettering is neat and carefully placed on a drawing.

Types of Sketches

Any image drawn on paper freehand (without a straightedge or other instrument) may be called a *sketch*. The sketching may be *rough*, or *unrefined*. That is, the sketch may be drawn quickly with jagged lines. The rough sketch can quickly express thoughts. Instruments are not used in the preparation of rough sketches. Figure 4-4-2 shows rough sketches that were used to develop preliminary (early) designs of a two-position automobile mirror.

The other way of sketching is *refined*. That is, the sketch is neat and looks finished. A refined sketch is carefully drawn. Templates and straightedges may be used to complement the freehand lines. It shows good proportion and excellent line values. It may be more persuasive than an unrefined sketch. Many refined sketches are based on a rough sketch that has captured the general idea.

Overlays

A very good way to *refine* (improve) a sketch is to use an overlay. Sketches are often drawn on paper that can be seen through. This paper is called translucent paper or tracing paper.

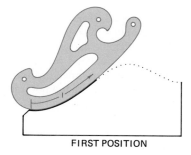

FIRST POSITION

SECOND POSITION

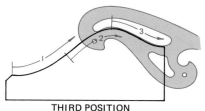

THIRD POSITION

FIG. 4-3-1   Drawing a curved line.

FIG. 4-4-1   A rough sketch with notes about important features.

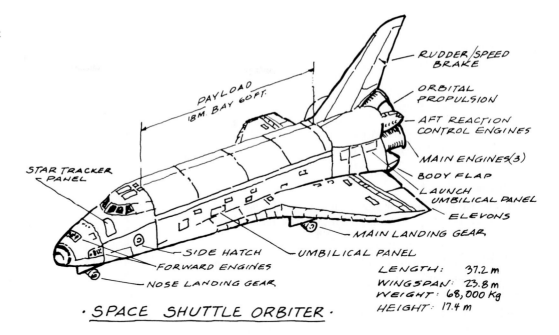

SPACE SHUTTLE ORBITER

FIG. 4-4-2   Study of a two-position mirror for a racing car.

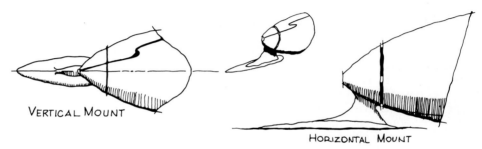

VERTICAL MOUNT

HORIZONTAL MOUNT

The best parts of a sketch may be quickly traced by putting a new piece of this paper, called the *overlay*, on top (Fig. 4-4-3). Thus, refining ideas means sketching over and over again on tracing paper until the design is right.

The overlay is used in two important ways that may seem very much alike. The first use is for reshaping an idea. This might include refining the proportions of the parts of an object or changing its shape entirely. Second, an overlay can be used to refine, or improve, the drawing itself without really changing the design. These two operations can be, and usually are, done at the same time.

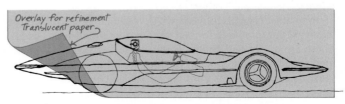

FIG. 4-4-3   The overlay can speed up the drawing process.

Nature of a Sketch

A sketch is an important form of graphic (drawn) communication. There are many levels of sketches for different uses (Fig. 4-4-4). If you know the language of mechanical drawing well, you will know how to sketch your ideas quickly, clearly, and accurately.

Temporary Sketch

Many technical sketches have short lives. They are done merely to solve an immediate problem, then thrown away.

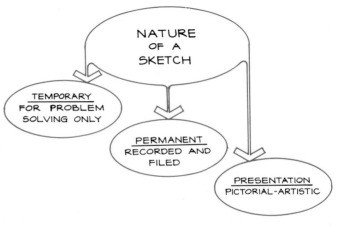

FIG. 4-4-4   Sketches are classified in three ways.

Other technical sketches are kept longer. It may take weeks or even months to study some sketches and make mechanical drawings from them. However, these sketches too may be thrown away some day.

Permanent Sketch

Sometimes the engineering department or the management of a company will include a sketch in a notice to other employees. Such a sketch is an important record and should be kept. Therefore, some sketches are filed as part of a company's permanent records.

Presentation Sketch

The sketch that is refined for presentation is usually *pictorial* (picturelike). It is used to convince a client or manager to accept and approve the ideas presented. The pictorial sketch has a three-dimensional view that can be understood easily by nontechnical people. Such sketches are usually drawn so that they appeal to the eye.

Views Needed For a Sketch

In Chap. 6, multiview projection will be discussed in full. However, you need to know some basic things about views and how they are placed in order to make sketches.

There are two types of drawings that you can sketch easily. The first is a pictorial drawing. In this type of drawing, the width, height, and depth of an object are shown in one view (Fig. 4-4-5). This type of sketching is covered in Chap. 15. The other type is called *multiview projection* or *orthographic projection*. This type of projection is fully explained in Chap. 6. In this type of drawing an object is usually shown in more than one view. You do this by drawing sides of the object and relating them to each other, as shown in Fig. 4-4-6.

If an object can be fully shown in only one view, this means that its shape can be described well in two dimensions (sizes). These are height and width. Things shown in one-view drawings generally have a depth or thickness that is uniform (the same throughout). This third dimension can be given as a note rather than being drawn. A typical one-view drawing is shown in Fig. 4-4-7A. The thickness of the stamping is shown by a note on the sketch. Many parts that are cylindrical in shape can also be shown in single views if the diameters are noted (Fig. 4-4-7B).

Materials for Sketching

Paper

You can use plain paper for sketching. If you need to refine the sketch, use tracing paper. Control proportions while sketching by using cross-section paper, also called graph paper or squared paper (Fig. 4-4-8). Graph paper should be used whenever possible because it cuts down on drawing time. The graph paper most often used has heavily ruled 1 in. squares. The 1 in. squares are then subdivided into lightly ruled $\frac{1}{10}$, $\frac{1}{8}$, $\frac{1}{4}$, or $\frac{1}{2}$ in. squares. This paper is called 10 to the inch, 8 to the inch, and so on. Graph paper is also ruled in millimeters. There are many

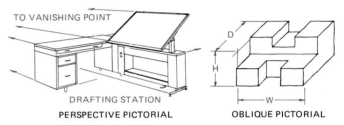

FIG. 4-4-5   Pictorial sketches.

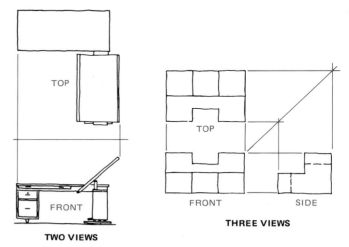

FIG. 4-4-6   Multiview sketches.

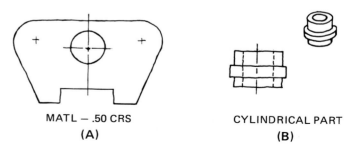

FIG. 4-4-7   One-view sketches.

specially ruled graph papers for particular kinds of drawing, such as isometric and perspective. These kinds of drawings are explained in Chap. 15.

Pencils and Erasers

Most drafters like to use soft lead pencils (grades F, H, or HB), properly sharpened. They also use an eraser that is good for soft leads, such as a plastic eraser or a kneaded-rubber eraser.

Freehand Lines

Lines drawn freehand have a natural look. They show freedom of movement because of their slight changes in direction. Hold the pencil far enough from the point that you can move your fingers easily and yet put enough pressure on the point to make dense, black lines when you need to. Draw light construction lines with very little pressure on the point. They should be light enough that they need not be erased.

Straight Lines

You can sketch lines in the following ways:

1. Draw them continuously.
2. Draw short dashes where the line should start and end. Then place the pencil point on the starting dash, draw toward it.
3. Draw a series of strokes that touch each other or are separated by very small spaces.
4. Draw a series of overlapping strokes.

Before you try to draw objects, practice sketching straight lines to improve your line technique. Draw horizontal lines from left to right if you're right-handed, and right to left if you're left-handed. Draw vertical lines from the top down.

Inclined Lines

Sketch sloped, or inclined, lines from left to right. It might be easiest to turn the paper and draw an inclined line the same way as a horizontal one. When trying to sketch a specific angle, first draw a vertical line and a horizontal line to form a right angle (90°). Divide the right angle in half to form two 45° angles. The sketched line should pass through the corners of the squares on the grid (Fig. 4-4-9A). Use an isometric grid

when angles of 30° and 60° are required (Fig. 4-4-9B). By starting with these simple angles, you can estimate (guess) other angles more exactly. Note the direction for drawing the inclined lines.

Proportions for Sketching

Sketches are not usually made to scale (exact measure). Nonetheless, it is important to keep sketches in proportion (similar to exact measure). In preparing the layout, look at the largest overall dimension, usually width, and estimate the size. Next, determine the proportion of the height to the width. Then, as the front view takes shape with the width and height, compare the smaller details with the larger ones and fill them in too (Fig. 4-4-10, pg. 60).

It is important that the design drafter, in sketching an object, have a good sense of how distances relate to each other. This ability will allow the drafter to show the width, height, and depth of an object in the right proportions.

For example, suppose that the design drafter plans a contemporary stereo cabinet to be 4 units wide. The height of the cabinet is 2 units. The depth of the cabinet is 1 unit. This is a proportion of 2:1. If you were designing this cabinet in customary units, you might choose a 15 in. width, height, and depth unit. The width overall would be 60 in. The height would be 30 in., and the depth 15 in. The proportions are developed in 2:1 units.

As previously mentioned, using a grid sheet will cut down on drafting time and result in a more accurate drawing. In the

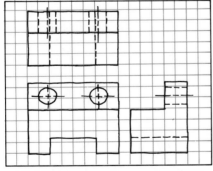

(A) COORDINATE SKETCHING PAPER

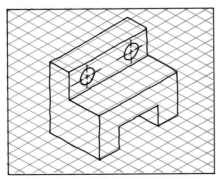

(B) ISOMETRIC SKETCHING PAPER

FIG. 4-4-8   Graph paper.

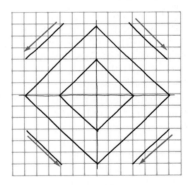

(A) SQUARE GRID

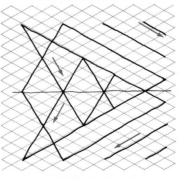

(B) ISOMETRIC GRID

FIG. 4-4-9   Using graph paper to help in sketching sloped lines.

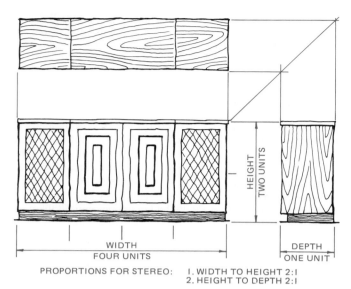

FIG. 4-4-10   Sketching a contemporary stereo with proportional units.

example given in Fig. 4-4-10, using a 4-to-the-inch, or .25 grid, and making each space equal to 3 in., a unit would be equal to five spaces on the grid sheet.

Circles and Arcs

There are two ways to sketch circles. For the first way (Fig. 4-4-11A), draw very light horizontal and vertical lines. Next, estimate the length of the *radius* (the distance from the center of the circle to its edge; plural, *radii*) and mark it off. Then draw a square in which you can sketch the circle. For the second way (Fig. 4-4-11B), first draw very light center lines. Then draw *bisecting* (halving) lines through the center at convenient angles. Next, mark off estimated radii of the same length on all lines. The bottom of the curve is generally easier to form, so draw it first. Then turn the paper so that the rest of the circle to be drawn is on the bottom. Complete the circle. You can sketch *arcs* (parts of a circle); *tangent arcs* (parts of two circles that touch); and *concentric circles* (circles of different diameters that have the same center) in the same way you sketch circles (Fig. 4-4-12). Use light, straight, construction lines to block in the area of the figure.

For large circles and arcs and for *ellipses* (ovals), use a scrap of paper with the radius marked off along one edge (Fig. 4-4-13). Put one end of the marked-off radius on the center of the circle. Draw the arc by placing a pencil at the other end of the radius and turning the scrap paper. Note that two radii are needed for an ellipse. To draw the ellipse, sketch both center lines. Keep both radius points on the center lines as you draw with the pencil at the other end.

Getting Started

When you are starting to sketch a view (or views), first lightly sketch the overall size as a rectangular or square shape, carefully estimating its proportions. Then add lines for the details of the shape and thicken all lines forming the view (Fig. 4-4-14).

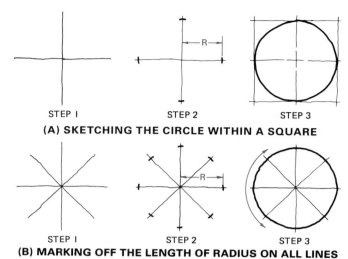

(A) SKETCHING THE CIRCLE WITHIN A SQUARE

(B) MARKING OFF THE LENGTH OF RADIUS ON ALL LINES

FIG. 4-4-11   Sketching circles.

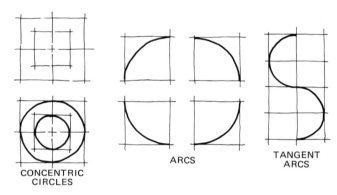

CONCENTRIC CIRCLES   ARCS   TANGENT ARCS

FIG. 4-4-12   Sketching arcs and concentric circles.

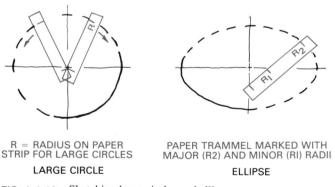

R = RADIUS ON PAPER STRIP FOR LARGE CIRCLES
LARGE CIRCLE

PAPER TRAMMEL MARKED WITH MAJOR (R2) AND MINOR (RI) RADII
ELLIPSE

FIG. 4-4-13   Sketching large circles and ellipses.

Figure 4-4-15 shows the two methods of sketching circles. Figure 4-4-16 illustrates, both pictorially and orthographically, the use of graph paper for sketching a machine part.

🖥 CAD

Sketching is still a necessary part of drafting because the drafter in industry frequently sketches ideas and designs prior

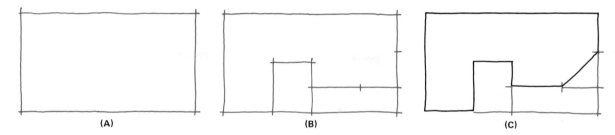

FIG. 4-4-14   Sketching a view having straight lines.

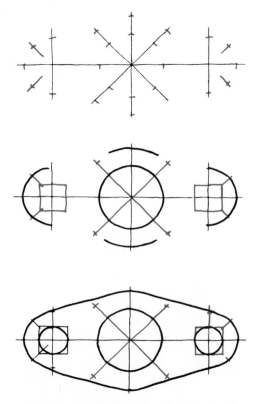

FIG. 4-4-15   Sketching a figure having circles and arcs.

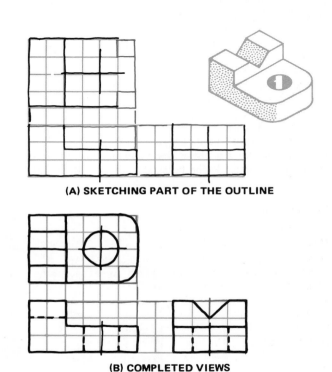

(A) SKETCHING PART OF THE OUTLINE

(B) COMPLETED VIEWS

FIG. 4-4-16   Usual procedure for sketching three views.

to making a CAD drawing. Sketching may be accomplished manually or on a CAD monitor using the SKETCH command.

ASSIGNMENTS

See Assignments 26 through 31 for Unit 4-4 on pages 74–75.

4-5 INKING

Most technical drawings are made with pencil on a good-quality tracing paper (vellum) or on drafting film. However, ink drawings have seen an increase in the past few decades. Ink is often used to make high-quality tracings that can be copied (whiteprinted, photographed, or microfilmed). The need for

highly efficient ink plotters for CAD brought about improvements in ink and inking equipment for manual drafting. For example, the technical pen with refillable cartridges, cartridge ink supplies, erasable ink, accessories, and attachments has made inking more efficient and improved the quality of the finished drawing.

Drawing Ink

Ink used for technical drawings is also called *india ink*. It must have some special characteristics. For example, it must flow freely and yet dry quickly. It must adhere (stick) well and not chip, crack, peel, or smear after drying. It must also be completely opaque in order to produce good uniform line tone and yet be erasable on all drafting media.

Both waterproof and nonwaterproof ink can be used on high-quality paper, cloth, polyester film, and illustration board.

The waterproof ink is best suited for mechanical drawings. Nonwaterproof ink is best for fine-line pictorial drawings. Both types of ink produce opaque lines that reproduce well when they are photographed or copied.

Using Ink Equipment

Figure 4-5-1 shows the correct position for drawing straight lines with a technical pen. Note the direction of the stroke and the angle of the pen. Hold the technical pen in a nearly vertical position (perpendicularly to the medium) to get the most uniform line.

For making circles and arcs, templates and compasses can be used with technical pens, as shown in Fig. 4-5-2. When inking with a ruling-pen type of compass, allow the compass to lean about 15° in the direction of the rotating motion.

Templates having built-in risers for inking are available (Fig. 4-5-3). Regular templates may also be used by building up one side of the template with small pieces of masking tape. These act as spacers between the inking surface of the drawing medium and the bottom side of the template.

Irregular (French) curves can be used to guide the pen when curves that are not circular arcs are being inked.

The ruling pen has blades that you can adjust to draw lines of different widths. It can be used to draw both straight and curved lines. Ruling pens are sometimes called all-purpose pens. They can be filled from a cartridge tube, a squeeze bottle, or a dropper cap.

The simple rules listed below will help you produce high-quality ink drawings using the ruling pen. Study them carefully.

1. Do not hold the pen over the drawing when you are filling it.
2. Do not fill the pen too full; about .10 to .20 in. (3 to 6 mm) is usually enough.
3. Never dip the pen into the ink bottle or allow ink to get on the outside of the blades.
4. Keep the blades clean by wiping them often with a soft towel, tissue, or pen cleaner.
5. While inking a line, keep both nibs (points) of the blades in contact with the drawing surface.
6. Keep the blades slanted in the direction of the pen stroke (Fig. 4-5-4).
7. Do not press the nibs hard against the straightedge. Pressing too hard will make line widths uneven.
8. Do not press the nibs too hard against the drawing surface. The blades may get dull or damage the drawing surface.

Many compasses have ruling-pen attachments that are used for inking arcs and circles. To use a compass for inking, remove

FIG. 4-5-1   The position of a technical pen is important when drawing lines. *(Teledyne Post)*

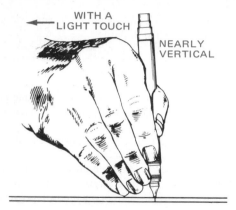

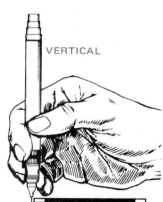

(A) USING TEMPLATES

(B) USING A LENGTHENING BAR FOR LARGE CIRCLES

(C) USING A COMPASS

FIG. 4-5-2   Drawing circles and arcs with a technical pen.

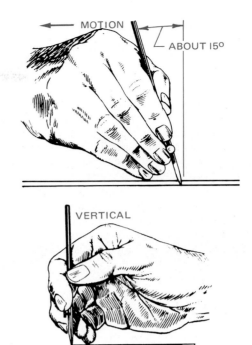

FIG. 4-5-4   Inking straight lines with a ruling pen.

FIG. 4-5-3   A template with built-in risers for inking. *(STUDIOHIO)*

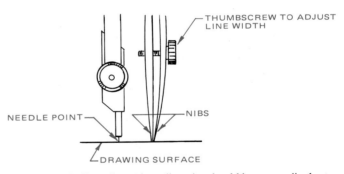

FIG. 4-5-5   Ruling nibs and needle point should be perpendicular to drawing surface.

the pen is ready to use again. To use an ultrasonic cleaner, just dip the pen point into a well containing cleaning fluid. From 60,000 to 80,000 cycles per second of sound energy act to loosen the dried ink on the point. The normal high range of human hearing is 15,000 cycles per second.

To clean a technical pen thoroughly, you may have to take it apart. Remove the cap and ink cartridge and soak the point in a special cleaning fluid. After removing the dried ink from the point, flush the parts in a stream of cold water. Dry all the parts before using the pen again. If the parts are not dry, the ink will be diluted (mixed with water) and lines will not be dark enough. If care is taken in handling and storing inking equipment, it will work well and need very little repair.

the pencil leg from the compass and replace it with the ruling-pen leg. Adjust the needle point carefully, as shown in Fig. 4-5-5, so that the legs are perpendicular to the drawing surface. Always draw the circle in one stroke. The compass should be slanted slightly when a ruling pen is being used. The lengthening bar, or beam compass, can be used for large circles.

Pen Cleaning

The technical pen and ruling pen sometimes need cleaning. The ultrasonic cleaner is very useful because it works so rapidly. The cleaning usually takes from 20 to 30 seconds, and

Lettering, Arrowheads, and Symbols

Lettering and drawing of arrowheads and symbols are done in the same manner as with a pencil except that either a dry or wet technical pen is used.

Erasing Techniques

The ink used on polyester drafting film is waterproof. However, you can easily remove ink from the film by rubbing it with a moistened plastic eraser. Do not use any pressure in rubbing. The polyester film does not absorb ink, and therefore all the ink dries on top of its highly finished surface. Remove ink from other surfaces, such as tracing vellum or illustration board, with regular ink erasers or chemically treated ink erasers that absorb ink. But be very careful. Press lightly with strokes in the direction of the line to remove ink caked on the surface. Too much pressure damages the surface and makes it hard to revise the drawing. Different erasing techniques work better on different surfaces. Try different techniques to find the best one for each surface.

Order of Inking

Smooth joints and tangents, sharp corners, and neat fillets give a drawing a professional look and make it easy to read. Good inking requires careful practice and a definite order of working procedures.

The usual order of inking or tracing is the same order as for pencil drawings (Fig. 4-5-6). First, ink the arcs, centered over the pencil lines, as in (A). Ink the horizontal lines next, as in (B). Complete the drawing with the vertical lines, as in (C). Then add the dimension lines, arrowheads, finish marks, and so on, and fill in the dimensions, as in (D).

Inking continues to be a prime professional procedure for preparing engineering documents. Inked drawings will produce excellent reproductions and, of course, enhance the corporate image for presentation drawings. The drafting technician with manual inking skills will be able to complement those drawings prepared with CAD systems using multipen plotters.

ASSIGNMENTS

See Assignments 32 and 33 for Unit 4-5 on page 75.

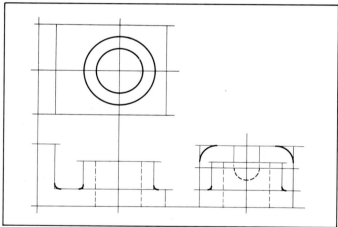

(A) DRAW CIRCLES AND ARCS

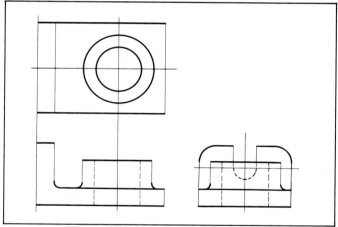

(B) DRAW HORIZONTAL OBJECT LINES

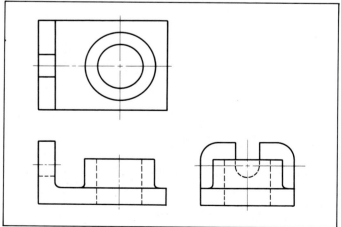

(C) DRAW VERTICAL OBJECT LINES

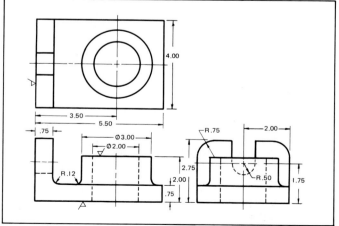

(D) ADD CENTER LINES, DIMENSION LINES, LETTERING, AND SYMBOLS

FIG. 4-5-6   Sequence of steps in developing a finished ink drawing.

ASSIGNMENTS FOR CHAPTER 4

NOTE ABOUT DRAWING ASSIGNMENTS

The assignments throughout this text are designed to teach the student certain aspects of technical drawings. It does not matter whether the drawing is made manually or by CAD. However, in order to simplify set-up instructions, the X, Y, and Z axes are shown in lieu of an arrow to indicate the viewing direction of the front view on all pictorial drawing assignments.

NOTE ABOUT DUAL DIMENSIONING

The dual dimensions shown in this book, especially in the assignment sections, are neither hard nor soft conversions. Instead, the sizes are those that would be most commonly used in the particular dimensioning units and so are only approximately equal. Dual dimensioning this way avoids awkward sizes and allows instructor and student to be confident when using either set of dimensions. Where dual dimensioning is shown, the dimensions given first or placed above are inch units of measurement. The other values shown are in millimeters.

ASSIGNMENTS FOR UNIT 4-1, STRAIGHT LINE WORK, LETTERING, AND ERASING

1. Lettering assignment. Set up a B (A3) size sheet similar to that shown in Fig. 4-1-A. Using uppercase Gothic lettering shown in Fig. 4-1-7, complete each line. Each letter and number is to be drawn several times to the three recommended lettering heights shown. Very light guidelines must be drawn first.

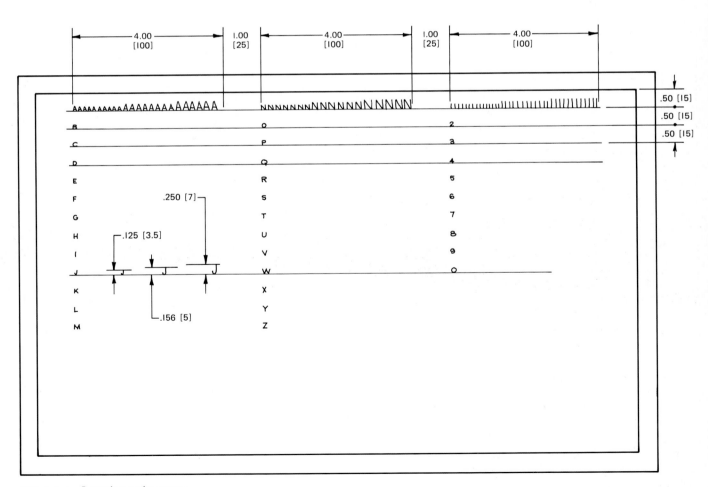

FIG. 4-1-A   Lettering assignment.

2. On a B (A3) size sheet draw one of the templates shown in Figs. 4-1-B or 4-1-C. Scale 1:1. Do not dimension.

3. On a B (A3) size sheet draw the shearing blank shown in Fig. 4-1-D. Scale 1:1. Do not dimension.

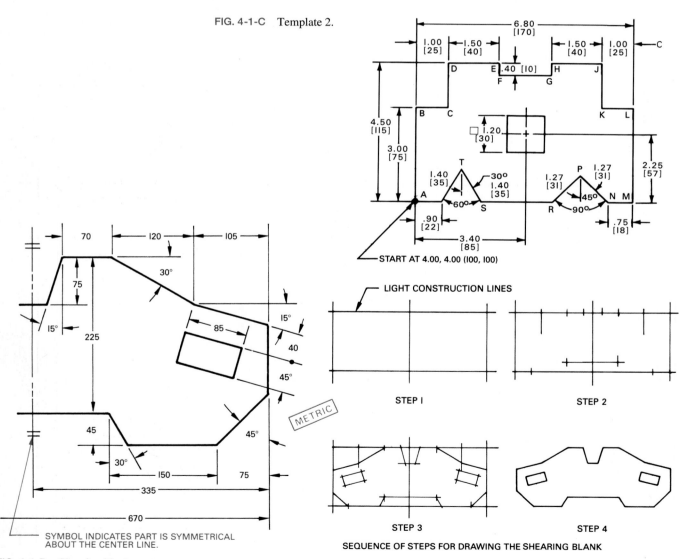

FIG. 4-1-B   Template 1.

FIG. 4-1-C   Template 2.

FIG. 4-1-D   Shearing blank.

SYMBOL INDICATES PART IS SYMMETRICAL ABOUT THE CENTER LINE.

SEQUENCE OF STEPS FOR DRAWING THE SHEARING BLANK

4. On a B (A3) size sheet draw the three parts shown in Fig. 4-1-E. Scale 1:1. Do not dimension.

5. On a B (A3) size sheet draw the three parts shown in Fig. 4-1-F. Use light construction lines as parts of each line will not be required. Scale 1:1. Do not dimension.

6. On a B (A3) size sheet draw any two of the three parts shown in Fig. 4-1-G. Scale 1:1. Do not dimension.

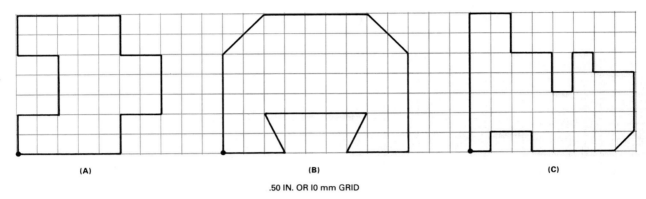

.50 IN. OR IO mm GRID

FIG. 4-1-E   Line drawing assignment.

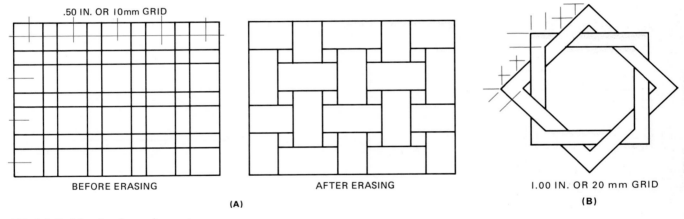

FIG. 4-1-F   Line drawing assignment.

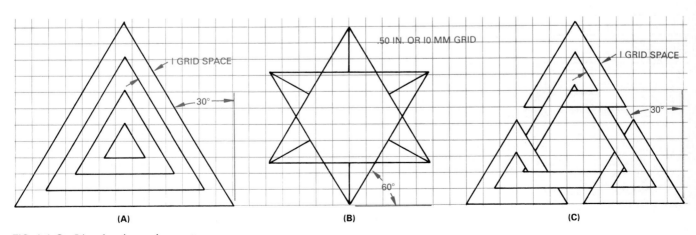

FIG. 4-1-G   Line drawing assignment.

7. On a B (A3) size sheet draw the three patterns shown in Fig. 4-1-H. Scale 1:1. Do not dimension.
8. On a B (A3) size sheet draw the designs shown in Fig. 4-1-J. Scale 1:1. Do not dimension.
9. On a B (A3) size sheet draw the designs shown in Fig. 4-1-K. Scale 1:1. Do not dimension.

10. Using absolute coordinates draw Figs. 4-1-L and 4-1-M on a B (A3) size format. The bottom left corner of the drawing (point 1) is the starting point. Scale 1:1.
11. Using relative coordinates, draw Fig. 4-1-N on a B (A3) size format. The bottom left corner of the drawing (point 1) is the starting point. Scale 1:1.

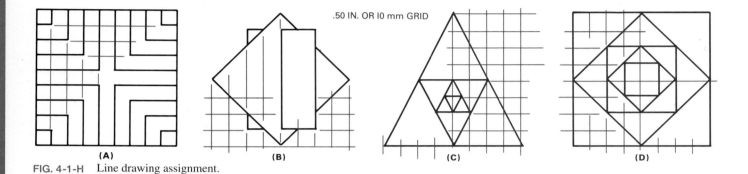

FIG. 4-1-H   Line drawing assignment.

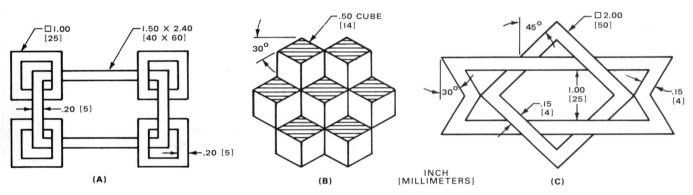

FIG. 4-1-J   Inlay designs.

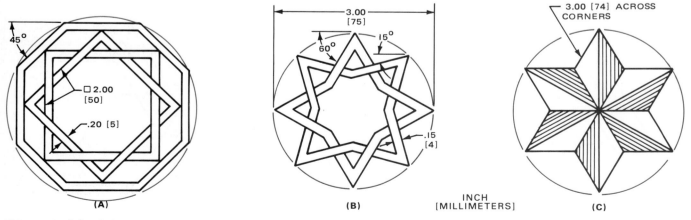

FIG. 4-1-K   Inlay designs.

ABSOLUTE COORDINATES (INCHES)		
Point	X Axis	Y Axis
1	.25	.25
2	7.00	.25
3	8.50	1.00
4	8.50	2.25
5	7.00	3.00
6	8.50	3.75
7	8.50	5.25
8	7.00	6.25
9	5.50	6.25
10	5.50	4.50
11	4.75	4.50
12	4.75	6.25
13	3.50	6.25
14	3.50	5.50
15	1.50	5.50
16	1.50	6.25
17	.75	6.25
18	.25	5.50
19	.25	.25
New Start		
20	.75	1.50
21	2.25	3.50
22	.75	3.50
23	.75	1.50
New Start		
24	3.00	2.25
25	5.75	2.25
26	5.75	3.75
27	3.00	3.75
28	3.00	2.25
New Start		
29	2.75	.75
30	6.00	.75
31	5.25	1.50
32	3.50	1.50
33	2.75	.75

FIG. 4-1-L   Absolute coordinates.

ABSOLUTE COORDINATES (MILLIMETERS)		
Point	X Axis	Y Axis
1	10	10
2	50	10
3	50	20
4	120	20
5	120	10
6	150	10
7	180	30
8	220	30
9	220	100
10	160	100
11	160	130
12	140	130
13	140	160
14	110	160
15	120	140
16	90	140
17	70	100
18	40	120
19	60	160
20	40	160
21	10	140
22	10	80
23	20	40
24	10	40
25	10	10
New Start		
26	40	50
27	160	50
28	120	90
29	50	70
30	40	50

FIG. 4-1-M Absolute coordinates.

RELATIVE COORDINATES (INCHES)		
Point	X Axis	Y Axis
1	0	0
2	4.50	0
3	0	.75
4	−.75	0
5	0	.75
6	−.75	0
7	0	.75
8	−3.00	0
9	0	−2.25
New Start—Solid		
10	0	.75
11	3.75	0
New Start		
12	−.75	.75
13	−3.00	0
New Start—Solid		
14	0	1.50
15	4.50	0
16	0	2.25
17	−4.50	0
18	0	−2.25
New Start—Solid		
19	0	.75
20	3.75	0
21	0	1.50
New Start—Solid		
22	−.75	0
23	0	−.75
24	−3.00	0
New Start—Solid		
25	5.25	−4.50
26	2.25	0
27	0	2.25
28	−.75	0
29	0	−.75
30	−.75	0
31	0	−.75
32	−.75	0
33	0	−.75
New Start—Solid		
34	.75	.75
35	1.50	0
New Start—Solid		
36	0	.75
37	−.75	0

FIG. 4-1-N Relative coordinates.

RELATIVE COORDINATES (MILLIMETERS)		
Point	X Axis	Y Axis
1	0	0
2	30	0
3	0	10
4	10	0
5	0	−10
6	30	0
7	0	50
8	−10	0
9	0	−15
10	−50	0
11	0	15
12	−10	0
13	0	−50
New Start—Solid		
14	5	10
15	15	0
16	0	20
17	−15	0
18	0	−20
New Start—Solid		
19	45	0
20	15	0
21	0	20
22	−15	0
23	0	−20

FIG. 4-1-P Relative coordinates.

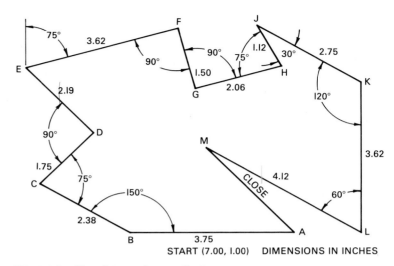

FIG. 4-1-R Template—polar coordinates.

12. Using relative coordinates, draw Fig. 4-1-P on a B (A3) size format. The bottom left corner of the drawing (point 1) is the starting point. Scale 1:1.
13. On a B (A3) size format draw the template shown in Fig. 4-1-R. Start at point A. Scale 1:1.
14. On a chart list show the polar coordinates for the templates shown in Figs. 4-1-R, 4-1-B, and 4-1-C. Move in a clockwise direction starting at point A.

ASSIGNMENTS FOR UNIT 4-2, CIRCLES AND ARCS

15. On a B (A3) size format, draw the dial indicator shown in Fig. 4-2-A. Scale 2:1. Do not dimension but add the word DEGREES and the degree numbers shown.
16. On a B (A3) size format, draw the dart board shown in Fig. 4-2-B. Scale 1:2. Use diagonal line shading and add the numbers. Do not dimension.

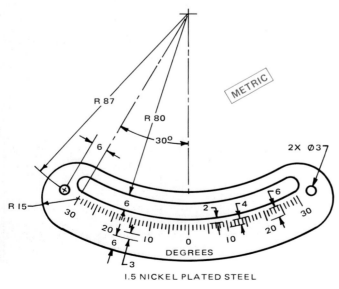

FIG. 4-2-A Dial indicator.

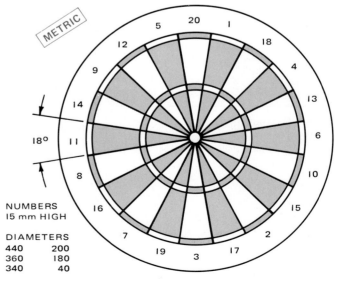

FIG. 4-2-B Dart board.

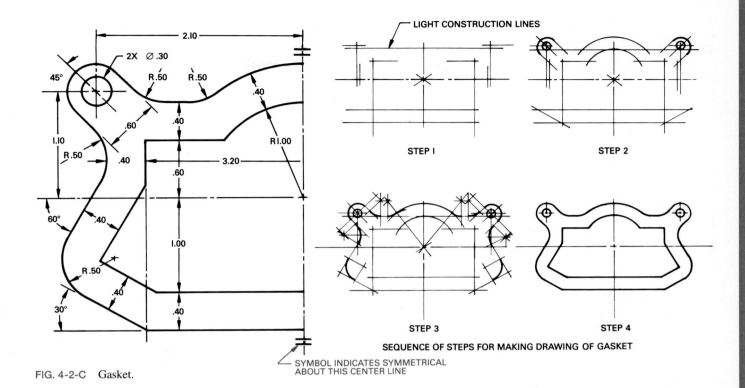

LIGHT CONSTRUCTION LINES

STEP I

STEP 2

STEP 3

STEP 4

SEQUENCE OF STEPS FOR MAKING DRAWING OF GASKET

SYMBOL INDICATES SYMMETRICAL
ABOUT THIS CENTER LINE

FIG. 4-2-C   Gasket.

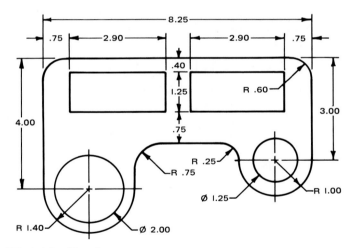

FIG. 4-2-D   Template.

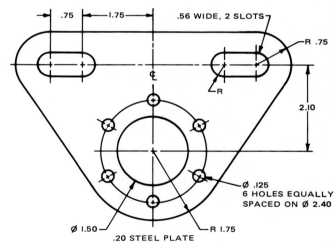

FIG. 4-2-F   Shaft support.

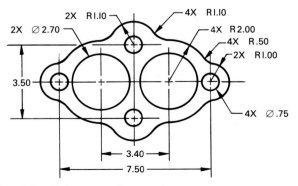

FIG. 4-2-E   Carburetor gasket.

17. On an A (A4) size format, draw the gasket shown in Fig. 4-2-C. Scale 1:1. Do not dimension.

18. On a B (A3) size format, draw the template shown in Fig. 4-2-D. Scale 1:1. Do not dimension.

19. On a B (A3) size format draw one of the parts shown in Figs. 4-2-E and 4-2-F. Scale 1:1. Do not dimension.

20. On an A (A4) size format draw one of the parts shown in Figs. 4-2-G to 4-2-J. Scale 1:1. Do not dimension.
21. On an A (A4) size format draw one of the parts shown in Figs. 4-2-K to 4-2-M. Scale 1:1. Do not dimension.
22. On an A (A4) size format draw the reel shown in Fig. 4-2-N. Scale 1:1. Do not dimension.

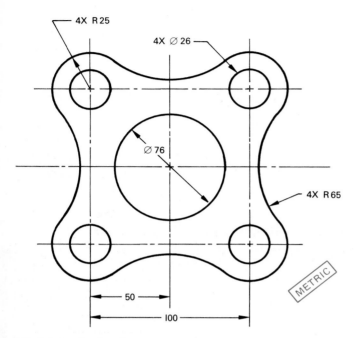

FIG. 4-2-G Anchor plate.

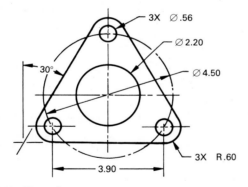

FIG. 4-2-H Base plate.

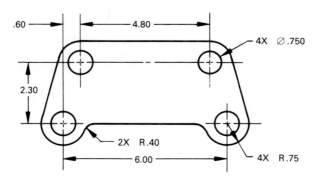

FIG. 4-2-J Cover plate.

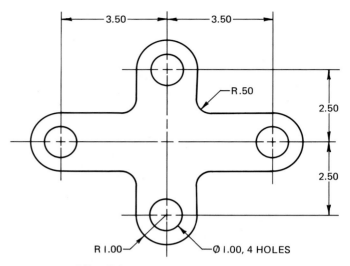

FIG. 4-2-K Offset link.

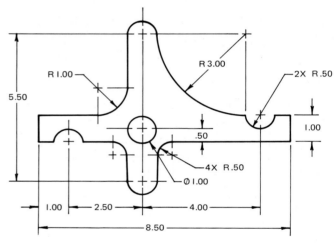

FIG. 4-2-L Pawl.

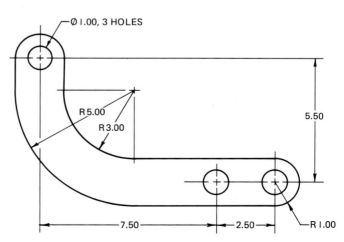

FIG. 4-2-M Rod support.

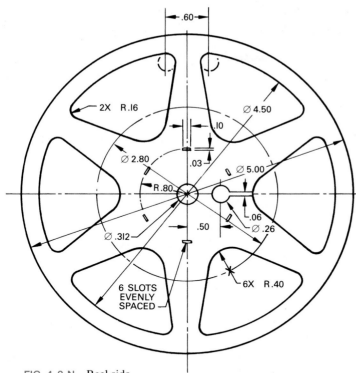

FIG. 4-2-N   Reel side.

ASSIGNMENTS FOR UNIT 4-3, IRREGULAR CURVES

23. On a B (A3) size format, lay out the pattern for the table leg shown in Fig. 4-3-A to the scale 1:2.
24. Using grid paper or creating a grid on the monitor, draw the furniture patterns shown in Fig. 4-3-B. Use .50 in. or 10 mm grid.
25. Using grid paper or creating a grid on the monitor, draw the line graph shown in Fig. 4-3-C. Use .25 in. or 5 mm grid.

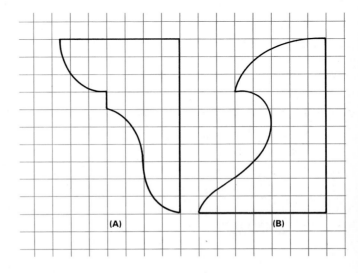

(A)   (B)

FIG. 4-3-B   Furniture patterns.

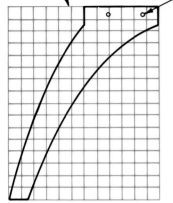

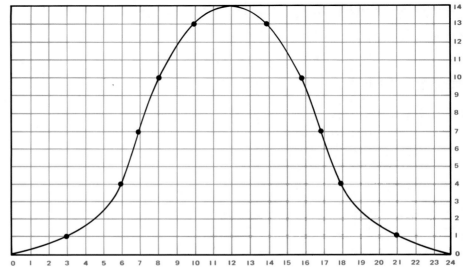

FIG. 4-3-A   Table leg.

FIG. 4-3-C   Line graph.

ASSIGNMENTS FOR UNIT 4-4, SKETCHING

26. Using grid paper, sketch the template shown in Fig. 4-2-D.

27. Using grid paper, sketch the shaft support shown in Fig. 4-2-F.
28. Using grid paper, sketch the patterns shown in Fig. 4-4-A.
29. Using grid paper, sketch the patterns shown in Fig. 4-4-B.

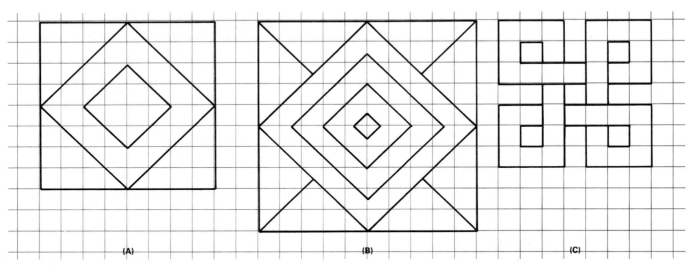

(A)   (B)   (C)

FIG. 4-4-A   Sketching assignment.

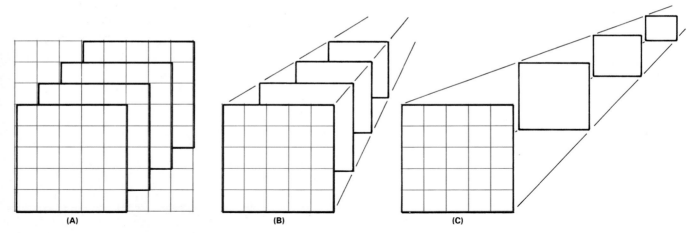

(A)   (B)   (C)

FIG. 4-4-B   Sketching assignment.

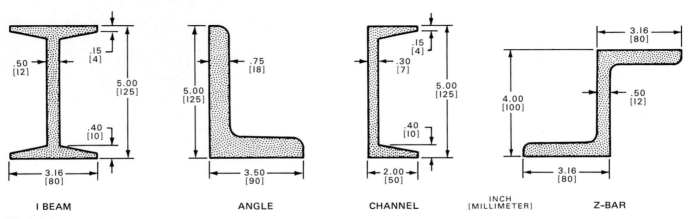

I BEAM   ANGLE   CHANNEL   INCH [MILLIMETER]   Z-BAR

FIG. 4-4-C   Structural steel shapes.

FIG. 4-4-D Sketching lines, circles, and arcs.

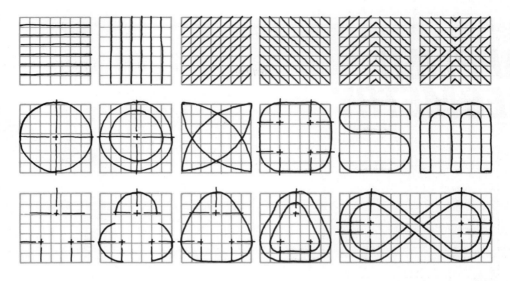

30. Using grid paper, sketch the structural steel shapes shown in Fig. 4-4-C. The shapes do not have to be drawn to scale, but should be drawn in proportion.
31. Using grid paper, sketch the patterns shown in Fig. 4-4-D.

ASSIGNMENTS FOR UNIT 4-5, INKING

32. As assigned by your instructor, make an ink drawing of any of the assignments for Units 4-1 through 4-4.
33. Make an ink drawing of one of the parts shown in Figs. 4-5-A to 4-5-C. Scale 1:1. Do not dimension.

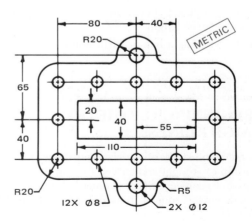

.50 IN. OR 10 mm GRID

FIG. 4-5-B Inlay design.

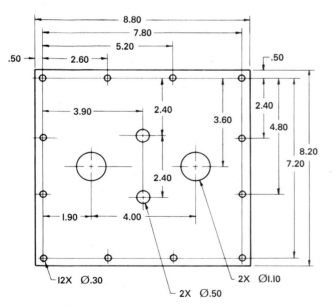

FIG. 4-5-A Cover plate.

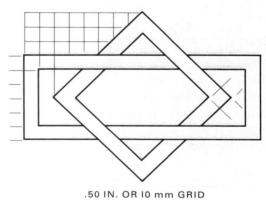

FIG. 4-5-C Gasket.

CHAPTER 5

APPLIED GEOMETRY

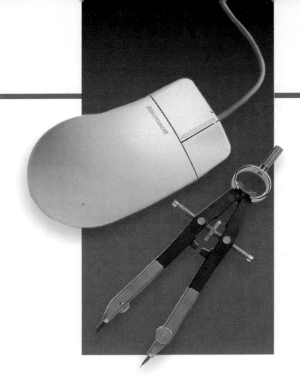

Definitions

Angle The shape formed by two lines that extend from the same point.

Bisect To cut or divide into two.

Circumscribe To place a figure around another, touching it at points but not cutting it.

Geometry The study of the size, shape, and relationship of objects.

Helix The curve generated by a point that revolves uniformly around and up or down the surface of a cylinder.

Inscribe To place a figure within another so that all angular points of it lie on the boundary (circumference).

Parabola A plane curve generated by a point that moves along a path equidistant from a fixed line and fixed point.

Parallel Continuously equidistant lines or surfaces.

Perpendicular At right angles (90°) to a line or surface.

Polygon A figure bounded by five or more straight lines, possibly of uneven length.

Tangent Meeting a line or surface at a point but not intersecting it.

5-1 BEGINNING GEOMETRY: STRAIGHT LINES

Geometry is the study of the size and shape of objects. The relationship of straight and curved lines in drawing shapes is also a part of geometry. Some geometric figures used in drafting include circles, squares, triangles, hexagons, and octagons (Fig. 5-1-1).

Geometric constructions are made of individual lines and points drawn in proper relationship to one another. Accuracy is extremely critical.

Geometric constructions are very important to drafters, surveyors, engineers, architects, scientists, mathematicians, and designers. Geometric constructions have important uses, both in making drawings and in solving problems with graphs and diagrams. Sometimes it is necessary to use geometric constructions, particularly if, when doing manual drafting, the drafter does not have the advantages afforded by a drafting machine, an adjustable triangle, or templates for drawing hexagonal and elliptical shapes. Therefore, nearly everyone in all technical fields needs to know the constructions explained in this chapter.

All of the lines and shapes shown in this chapter can be drawn using CAD commands. This chapter covers manual drafting using the instruments and equipment described in Chap. 4. The following exercises provide practice in geometric constructions.

To Draw a Line or Lines Parallel to and at a Given Distance from an Oblique Line

1. Given line *AB* (Fig. 5-1-2), erect a perpendicular *CD* to *AB*.
2. Space the given distance from the line *AB* by scale measurement or by an arc along line *CD*.
3. Position a triangle, using a second triangle or a T square as base, so that one side of the triangle is parallel with the given line.
4. Slide this triangle along the base to the point at the desired distance from the given line, and draw the required line.

To Draw a Straight Line Tangent to Two Circles

Place a T square or straightedge so that the top edge just touches the edges of the circles, and draw the tangent line (Fig. 5-1-3). Perpendiculars to this line from the centers of the circles give the tangent points T_1 and T_2.

To Bisect a Straight Line

1. Given line *AB* (Fig. 5-1-4, pg. 78), set the compass to a radius greater than ½ *AB*.

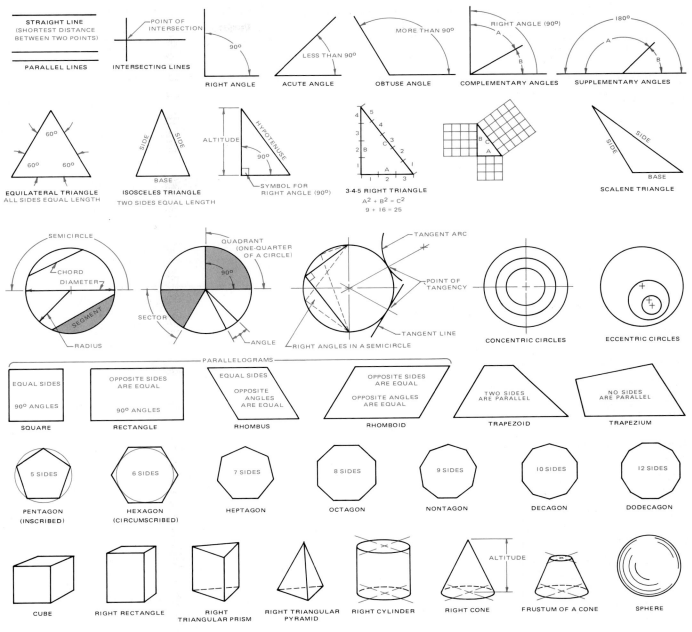

FIG. 5-1-1   Dictionary of drafting geometry.

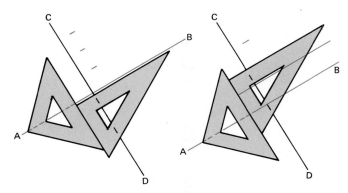

FIG. 5-1-2   Drawing parallel lines with the use of triangles.

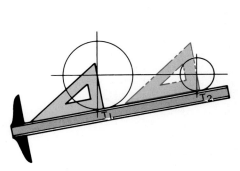

FIG. 5-1-3   Drawing a straight line tangent to two circles.

2. Using centers at *A* and *B*, draw intersecting arcs above and below line *AB*. A line *CD* drawn through the intersections will divide *AB* into two equal parts and will be perpendicular to line *AB*.

To Bisect an Arc

1. Given arc *AB* (Fig. 5-1-5), set the compass to a radius greater than ½ *AB*.
2. Using points *A* and *B* as centers, draw intersecting arcs above and below arc *AB*. A line drawn through the intersections *C* and *D* will divide the arc *AB* into two equal parts.

To Bisect an Angle

1. Given angle *ABC*, with center *B* and a suitable radius (Fig. 5-1-6), draw an arc to cut *BC* at *D* and *BA* at *E*.
2. With centers *D* and *E* and equal radii, draw arcs to intersect at *F*.
3. Join *B* and *F* and extend to *G*. Line *BG* is the required bisector.

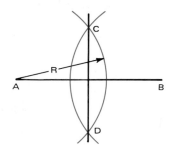

FIG. 5-1-4   Bisecting a line.

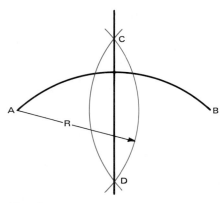

FIG. 5-1-5   Bisecting an arc.

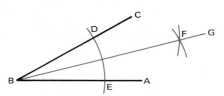

FIG. 5-1-6   Bisecting an angle.

To Divide a Line into a Given Number of Equal Parts

1. Given line *AB* and the number of equal divisions desired (12, for example), draw a perpendicular from *A*.
2. Place the scale so that the desired number of equal divisions is conveniently included between *B* and the perpendicular. Then mark these divisions, using short vertical marks from the scale divisions, as in Fig. 5-1-7.
3. Draw perpendiculars to line *AB* through the points marked, dividing the line *AB* as required.

ASSIGNMENTS

See Assignments 1 through 3 for Unit 5-1 on pages 84–85.

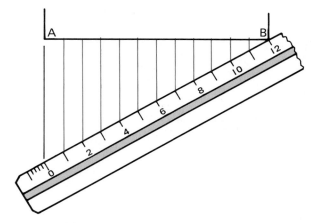

FIG. 5-1-7   Dividing a straight line into equal parts.

5-2 ARCS AND CIRCLES

To Draw an Arc Tangent to Two Lines at Right Angles to Each Other

Given radius R of the arc (Fig. 5-2-1):

1. Draw an arc having radius *R* with center at *B*, cutting the lines *AB* and *BC* at *D* and *E*, respectively.
2. With *D* and *E* as centers and with the same radius *R*, draw arcs intersecting at *O*.
3. With center *O*, draw the required arc. The tangent points are *D* and *E*.

To Draw an Arc Tangent to the Sides of an Acute Angle

Given radius *R* of the arc (Fig. 5-2-2):

1. Draw lines inside the angle, parallel to the given lines, at distance *R* away from the given lines. The center of the arc will be at *C*.

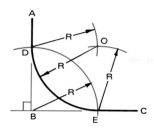

FIG. 5-2-1   Arc tangent to two lines at right angles to each other.

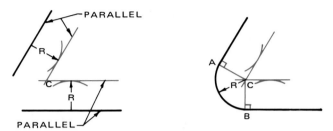

FIG. 5-2-2   Drawing an arc tangent to the sides of an acute angle.

2. Set the compass to radius *R*, and with center *C* draw the arc tangent to the given sides. The tangent points *A* and *B* are found by drawing perpendiculars through point *C* to the given lines.

To Draw an Arc Tangent to the Sides of an Obtuse Angle

Given radius *R* of the arc (Fig. 5-2-3):

1. Draw lines inside the angle, parallel to the given lines at distance *R* away from the given lines. The center of the arc will be at *C*.
2. Set the compass to radius *R*, and with center *C* draw the arc tangent to the given sides. The tangent points *A* and *B* are found by drawing perpendiculars through point *C* to the given lines.

To Draw a Circle on a Regular Polygon

1. Given the size of the polygon (Fig. 5-2-4), bisect any two sides, for example, *BC* and *DE*. The center of the polygon is where bisectors *FO* and *GO* intersect at point *O*.
2. The inner circle radius is *OH*, and the outer circle radius is *OA*.

To Draw a Reverse, or Ogee, Curve Connecting Two Parallel Lines

1. Given two parallel lines *AB* and *CD* and distances *X* and *Y* (Fig. 5-2-5), join points *B* and *C* with a line.
2. Erect a perpendicular to *AB* and *CD* from points *B* and *C*, respectively.
3. Select point *E* on line *BC* where the curves are to meet.
4. Bisect *BE* and *EC*.
5. Points *F* and *G* where the perpendiculars and bisectors meet are the centers for the arcs forming the ogee curve.

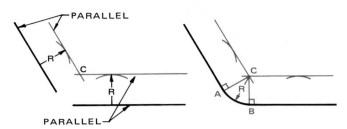

FIG. 5-2-3   Drawing an arc tangent to the sides of an obtuse angle.

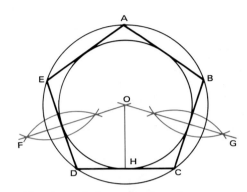

FIG. 5-2-4   Drawing a circle on a regular polygon.

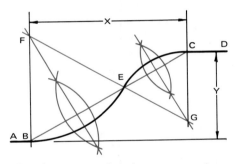

FIG. 5-2-5   Drawing a reverse (ogee) curve connecting two parallel lines.

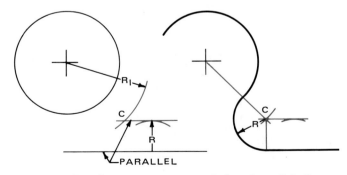

FIG. 5-2-6   Drawing an arc tangent to a circle and a straight line.

To Draw an Arc Tangent to a Given Circle and Straight Line

1. Given *R*, the radius of the arc (Fig. 5-2-6), draw a line parallel to the given straight line between the circle and the line at distance *R* away from the given line.

2. With the center of the circle as center and radius R_1 (radius of the circle plus R), draw an arc to cut the parallel straight line at C.
3. With center C and radius R, draw the required arc tangent to the circle and the straight line.

To Draw an Arc Tangent to Two Circles

1. Given the radius of arc R (Fig. 5-2-7A), with the center of circle A as center and radius R_2 (radius of circle A plus R), draw an arc in the area between the circles.
2. With the center of circle B as center and radius R_3 (radius of circle B plus R), draw an arc to cut the other arc at C.
3. With center C and radius R, draw the required arc tangent to the given circles.

As an alternative:

1. Given radius of arc R (Fig. 5-2-7B), with the center of circle A as center and radius $R - R_2$, draw an arc in the area between the circles.
2. With the center of circle B as center and radius $R - R_3$, draw an arc to cut the other arc at C.
3. With center C and radius R, draw the required arc tangent to the given circles.

To Draw an Arc or Circle Through Three Points Not in a Straight Line

1. Given points A, B, and C (Fig. 5-2-8), join points A, B, and C as shown.
2. Bisect lines AB and BC and extend bisecting lines to intersect at O. Point O is the center of the required circle or arc.
3. With center O and radius OA draw an arc.

ASSIGNMENTS

See Assignments 4 through 8 for Unit 5-2 on pages 86–87.

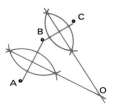
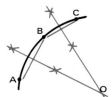

FIG. 5-2-8   Drawing an arc or circle through three points not in a straight line.

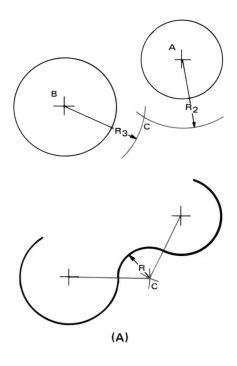

(A)

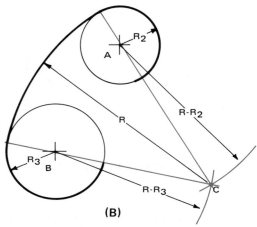

(B)

FIG. 5-2-7   Drawing an arc tangent to two circles.

5-3 POLYGONS

A *polygon* is a plane figure bounded by five or more straight lines not necessarily of equal length. A *regular polygon* is a plane figure bounded by five or more straight lines of equal length and containing angles of equal size.

To Draw a Hexagon, Given the Distance Across the Flats

1. Establish horizontal and vertical center lines for the hexagon (Fig. 5-3-1).
2. Using the intersection of these lines as center, with radius one-half the distance across the flats, draw a light construction circle.
3. Using the 60° triangle, draw six straight lines, equally spaced, passing through the center of the circle.
4. Draw tangents to these lines at their intersection with the circle.

To Draw a Hexagon, Given the Distance Across the Corners

1. Establish horizontal and vertical center lines, and draw a light construction circle with radius one-half the distance across the corners (Fig. 5-3-2).
2. With a 60° triangle, establish points on the circumference 60° apart.
3. Draw straight lines connecting these points.

To Draw an Octagon, Given the Distance Across the Flats

1. Establish horizontal and vertical center lines and draw a light construction circle with radius one-half the distance across the flats (Fig. 5-3-3).
2. Draw horizontal and vertical lines tangent to the circle.
3. Using the 45° triangle, draw lines tangent to the circle at a 45° angle from the horizontal.

To Draw an Octagon, Given the Distance Across the Corners

1. Establish horizontal and vertical center lines and draw a light construction circle with radius one-half the distance across the corners (Fig. 5-3-4).
2. With the 45° triangle, establish points on the circumference between the horizontal and vertical center lines.

3. Draw straight lines connecting these points to the points where the center lines cross the circumference.

To Draw a Regular Polygon, Given the Length of the Sides

As an example, let a polygon have seven sides.
1. Given the length of side *AB* (Fig. 5-3-5), with radius *AB* and *A* as center, draw a semicircle and divide it into seven equal parts using a protractor.

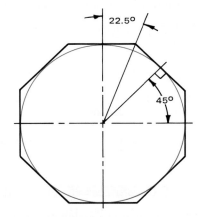

FIG. 5-3-3   Constructing an octagon, given the distance across flats.

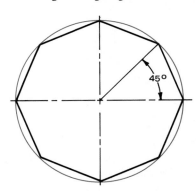

FIG. 5-3-4   Constructing an octagon, given the distance across corners.

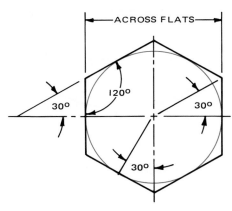

FIG. 5-3-1   Constructing a hexagon, given the distance across flats.

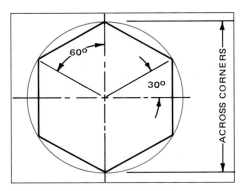

FIG. 5-3-2   Constructing a hexagon, given the distance across corners.

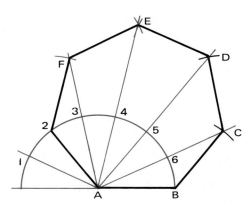

FIG. 5-3-5   Constructing a regular polygon, given the length of one side.

2. Through the second division from the left, draw radial line A2.
3. Through points 3, 4, 5, and 6 extend radial lines as shown.
4. With *AB* as radius and *B* as center, cut line A6 at *C*. With the same radius and *C* as center, cut line A5 at *D*. Repeat at *E* and *F*.
5. Connect these points with straight lines.

These steps can be followed in drawing a regular polygon with any number of sides.

To Inscribe a Regular Pentagon in a Given Circle

1. Given circle with center *O* (Fig. 5-3-6), draw the circle with diameter *AB*.
2. Bisect line *OB* at *D*.
3. With center *D* and radius *DC*, draw arc *CE* to cut the diameter at *E*.
4. With *C* as center and radius *CE*, draw arc *CF* to cut the circumference at *F*. Distance *CF* is one side of the pentagon.
5. With radius *CF* as a chord, mark off the remaining points on the circle. Connect the points with straight lines.

ASSIGNMENTS

See Assignments 9 through 11 for Unit 5-3 on pages 88–89.

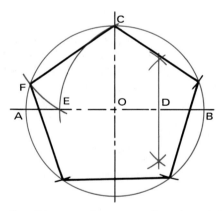

FIG. 5-3-6 Inscribing a regular pentagon in a given circle.

5-4 ELLIPSE

The *ellipse* is the plane curve generated by a point moving so that the sum of the distances from any point on a curve to two fixed points, called *foci*, is a constant.

Often a drafter is called upon to draw oblique and inclined holes and surfaces that take the approximate form of an ellipse. Several methods, true and approximate, are used for its construction. The terms *major diameter* and *minor diameter* will be used in place of *major axis* and *minor axis* so the reader won't become confused with the mathematical *X* and *Y* axes.

To Draw an Ellipse—Two-Circle Method

1. Given the major and minor diameters (Fig. 5-4-1), construct two concentric circles with diameters equal to *AB* and *CD*.
2. Divide the circles into a convenient number of equal parts. Figure 5-4-1 shows 12.
3. Where the radial lines intersect the outer circle, as at 1, draw lines parallel to line *CD* inside the outer circle.
4. Where the same radial line intersects the inner circle, as at 2, draw a line parallel to axis *AB* away from the inner circle. The intersection of these lines, as at 3, gives points on the ellipse.
5. Draw a smooth curve through these points.

To Draw an Ellipse—Four-Center Method

1. Given the major diameter *CD* and the minor diameter *AB* (Fig. 5-4-2), join points *A* and *C* with a line.
2. Draw an arc with point *O* as the center and radius *OC* and extend line *OA* to locate point *E*.
3. Draw an arc with point *A* as the center and radius *AE* to locate point *F*.

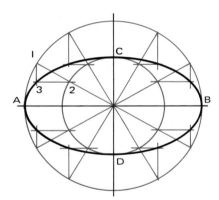

FIG. 5-4-1 Drawing an ellipse—two-circle method.

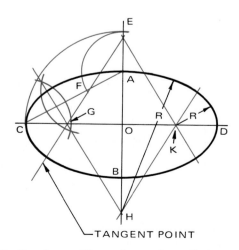

FIG. 5-4-2 Drawing an ellipse—four-center method.

4. Draw the perpendicular bisector of line *CF* to locate points *G* and *H*.
5. Draw arcs with *G* and *K* as centers and radii *HA* and *EB* to complete the ellipse.

To Draw an Ellipse—Parallelogram Method

1. Given the major diameter *CD* and minor diameter *AB* (Fig. 5-4-3), construct a parallelogram.
2. Divide *CO* into a number of equal parts. Divide *CE* into the same number of equal parts. Number the points from *C*.
3. Draw a line from *B* to point 1 on line *CE*. Draw a line from *A* through point 1 on *CO*, intersecting the previous line. The point of intersection will be one point on the ellipse.
4. Proceed in the same manner to find other points on the ellipse.
5. Draw a smooth curve through these points.

ASSIGNMENTS

See Assignments 12 and 13 for Unit 5-4 on page 90.

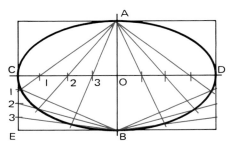

FIG. 5-4-3   Drawing an ellipse—parallelogram method.

FIG. 5-5-1   Drawing a cylindrical helix.

5-5 HELIX AND PARABOLA

Helix

The *helix* is the curve generated by a point that revolves uniformly around and up or down the surface of a cylinder. The *lead* is the vertical distance that the point rises or drops in one complete revolution.

To Draw a Helix

1. Given the diameter of the cylinder and the lead (Fig. 5-5-1), draw the top and front views.
2. Divide the circumference (top view) into a convenient number of parts (use 12) and label them.
3. Project lines down to the front view.
4. Divide the lead into the same number of equal parts and label them as shown in Fig. 5-5-1.
5. The points of intersection of lines with corresponding numbers lie on the helix. *Note*: Since points 8 to 12 lie on the back portion of the cylinder, the helix curve starting at point 7 and passing through points 8, 9, 10, 11, 12 to point 1 will appear as a hidden line.
6. If the development of the cylinder is drawn, the helix will appear as a straight line on the development.

Parabola

The *parabola* is a plane curve generated by a point that moves along a path equidistant from a fixed line (*directrix*) and a fixed point (*focus*). Again, these methods produce an approximation of the true conic section.

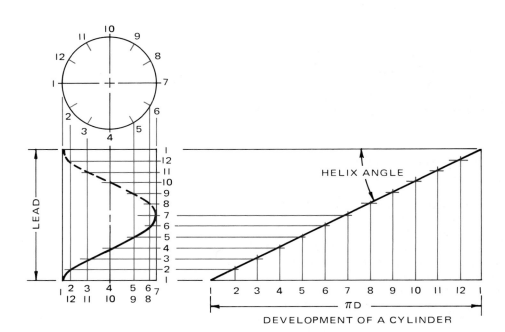

DEVELOPMENT OF A CYLINDER

To Construct a Parabola—Parallelogram Method

1. Given the sizes of the enclosing rectangle, distances *AB* and *AC* (Fig. 5-5-2A), construct a parallelogram.
2. Divide *AC* into a number of equal parts. Number the points as shown. Divide distance A–O into the same number of equal parts.
3. Draw a line from *O* to point 1 on line *AC*. Draw a line parallel to the axis through point 1 on line *AO*, intersecting the previous line *O*1. The point of intersection will be one point on the parabola.
4. Proceed in the same manner to find other points on the parabola.
5. Connect the points using an irregular curve.

To Construct a Parabola—Offset Method

1. Given the sizes of the enclosing rectangle, distances *AB* and *AC* (Fig. 5-5-2B), construct a parallelogram.
2. Divide *OA* into four equal parts.
3. The offsets vary in length as the square of their distances from *O*. Since *OA* is divided into four equal parts, distance *AC* will be divided into 4^2, or 16, equal divisions. Thus since *O*1 is one-fourth the length of *OA*, the length of line 1-1_1 will be $(\frac{1}{4})^2$, or $\frac{1}{16}$, the length of *AC*.
4. Since distance *O*2 is one-half the length of *OA*, the length of line 2-2_1 will be $(\frac{1}{2})^2$, or $\frac{1}{4}$, the length of *AC*.
5. Since distance *O*3 is three-fourths the length of *OA*, the length of line 3-3_1 will be $(\frac{3}{4})^2$, or $\frac{9}{16}$, the length of *AC*.
6. Complete the parabola by joining the points with an irregular curve.

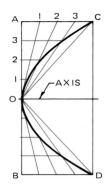

PARALLELOGRAM METHOD

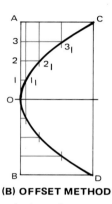

(B) OFFSET METHOD

FIG. 5-5-2   Common methods used to construct a parabola.

ASSIGNMENTS

See Assignments 14 through 16 for Unit 5-5 on pages 90–91.

ASSIGNMENTS FOR CHAPTER 5

Note: In the preparation of the following drawings, it is advisable to begin by using light construction lines. This will permit overruns and miscalculations without damage to the worksheet. After the drawing has been roughed out, make final lines of the appropriate type and thickness.

ASSIGNMENTS FOR UNIT 5-1, STRAIGHT LINES

1. Divide a B (A3) size sheet as in Fig. 5-1-A. In the designated areas draw the geometric constructions. Scale 1:1.

2. On an A (A4) size sheet, draw the geometric constructions in Fig. 5-1-B. For part C use light construction circles to develop the squares. Scale 1:1. Do not dimension.

3. On an A (A4) size sheet, draw the geometric constructions in Fig. 5-1-C. Use light construction circles to develop the figures. Scale 1:1. Do not dimension.

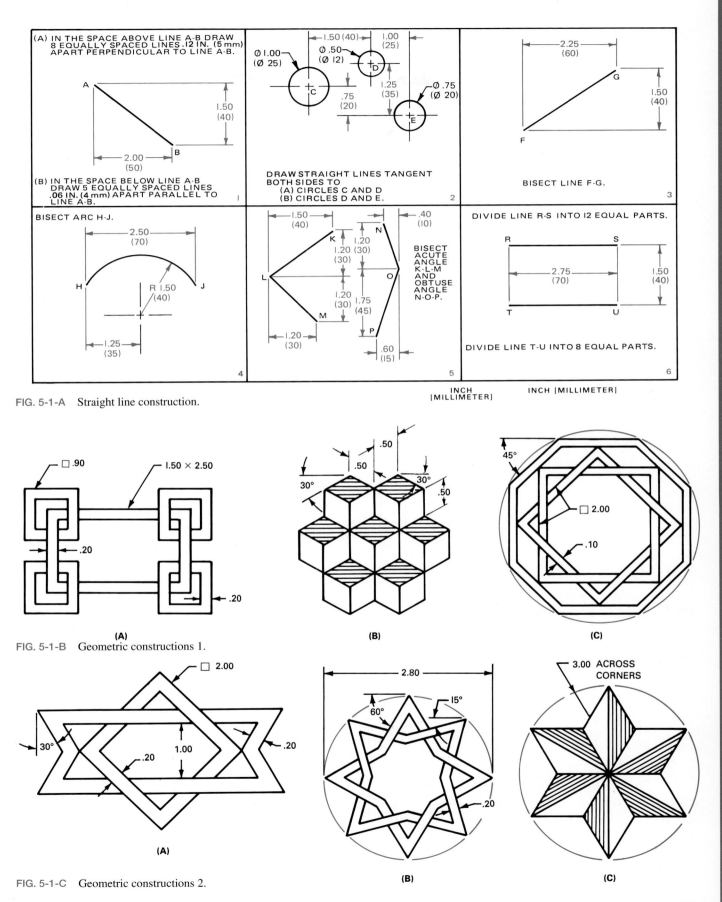

(A) IN THE SPACE ABOVE LINE A-B DRAW 8 EQUALLY SPACED LINES .12 IN. (5 mm) APART PERPENDICULAR TO LINE A-B.

1.50 (40)
2.00 (50)

(B) IN THE SPACE BELOW LINE A-B DRAW 5 EQUALLY SPACED LINES .06 IN. (4 mm) APART PARALLEL TO LINE A-B.

Ø 1.00 (Ø 25)
Ø .50 (Ø 12)
Ø .75 (Ø 20)
1.50 (40)
1.00 (25)
1.25 (35)
.75 (20)

DRAW STRAIGHT LINES TANGENT BOTH SIDES TO
(A) CIRCLES C AND D
(B) CIRCLES D AND E.

2.25 (60)
1.50 (40)

BISECT LINE F-G.

BISECT ARC H-J.

2.50 (70)
R 1.50 (40)
1.25 (35)

BISECT ACUTE ANGLE K-L-M AND OBTUSE ANGLE N-O-P.

1.50 (40)
1.20 (30)
1.20 (30)
1.20 (30)
1.20 (30)
.40 (10)
1.75 (45)
.60 (15)

DIVIDE LINE R-S INTO 12 EQUAL PARTS.

2.75 (70)
1.50 (40)

DIVIDE LINE T-U INTO 8 EQUAL PARTS.

INCH [MILLIMETER]   INCH [MILLIMETER]

FIG. 5-1-A   Straight line construction.

(A) □ .90   1.50 × 2.50   .20   .20

(B) .50   .50   30°   30°   .50

(C) 45°   □ 2.00   .10

FIG. 5-1-B   Geometric constructions 1.

(A) □ 2.00   30°   1.00   .20   .20

(B) 2.80   60°   15°   .20

(C) 3.00 ACROSS CORNERS

FIG. 5-1-C   Geometric constructions 2.

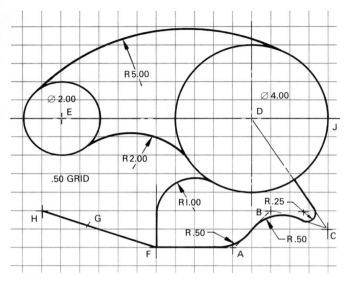

FIG. 5-2-A Geometric constructions 3.

FIG. 5-2-B Adjustable fork.

FIG. 5-2-C Rocker arm.

ASSIGNMENTS FOR UNIT 5-2, ARCS AND CIRCLES

4. On an A (A4) size sheet, complete the drawing shown in Fig. 5-2-A, given the following additional information. Points *A* to *J* are located on the .50 in. grid.
   - Draw straight lines between the points shown, then draw the arcs.
   - Divide the upper right-hand quadrant of the Ø4.00 into six 15° segments.
   - Divide the upper left-hand quadrant of the Ø4.00 into nine equal segments.
   - Bisect the arc *JDC*.
   - Line *GF* is twice as long as line *HG*. Draw a reverse ogee curve passing through points *H*, *G*, and *F*. Scale 1:1. Do not dimension.
5. Make a drawing of one of the parts shown in Fig. 5-2-B or 5-2-C. Scale 1:1. Do not dimension.

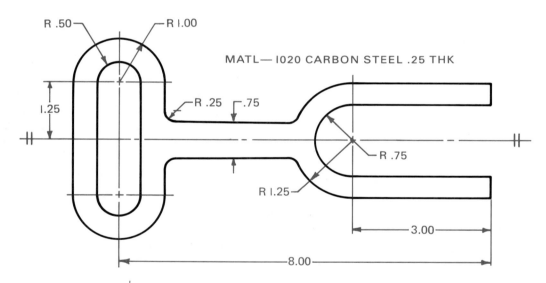

MATL—1020 CARBON STEEL .25 THK

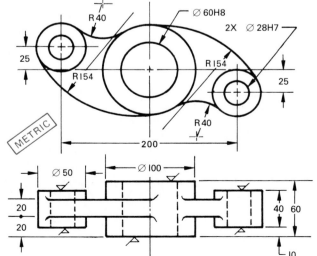

6. Divide a B (A3) size sheet as shown in Fig. 5-2-D. In the designated areas draw the geometric constructions. Scale 1:1.
7. Make a drawing of the belt drive shown in Fig. 5-2-E. The idler pulley is located midway between the B and driver shafts. The diameters of the pulley are shown in inches on the drawing. Given the RPM of the driver pulley, calculate the RPM of the other pulleys A to D. Scale: 1 in. = 1 ft.
8. Make a drawing of PT 2 shown in Fig. 5-2-F. Scale 1:1.

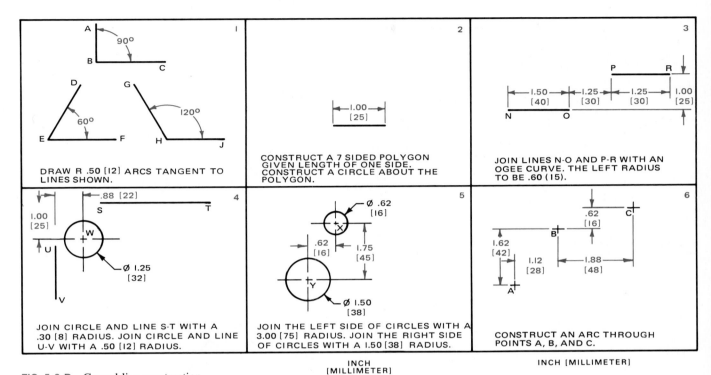

FIG. 5-2-D   Curved-line construction.

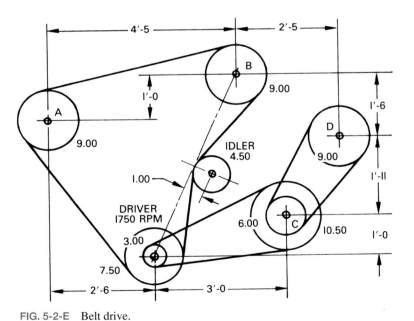

FIG. 5-2-E   Belt drive.

FIG. 5-2-F   Wire rope hook.

PT	A	B	C	D	E
1	4.90	3.20	.90	1.06	.84
2	5.40	3.50	1.00	1.10	.90
3	6.25	4.10	1.10	1.24	1.10
4	6.90	4.54	1.24	1.40	1.30

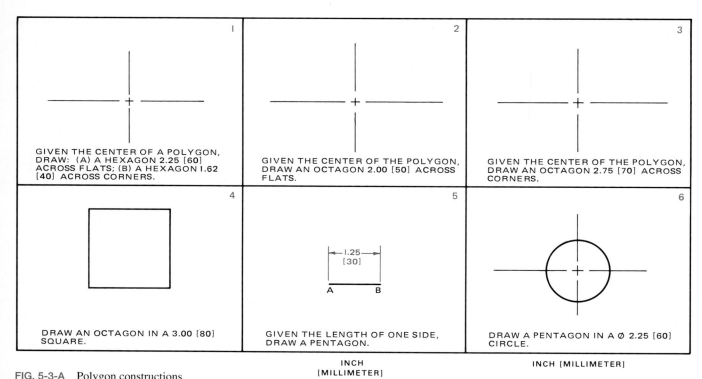

1 GIVEN THE CENTER OF A POLYGON, DRAW: (A) A HEXAGON 2.25 [60] ACROSS FLATS; (B) A HEXAGON 1.62 [40] ACROSS CORNERS.	**2** GIVEN THE CENTER OF THE POLYGON, DRAW AN OCTAGON 2.00 [50] ACROSS FLATS.	**3** GIVEN THE CENTER OF THE POLYGON, DRAW AN OCTAGON 2.75 [70] ACROSS CORNERS.
4 DRAW AN OCTAGON IN A 3.00 [80] SQUARE.	**5** GIVEN THE LENGTH OF ONE SIDE, DRAW A PENTAGON. INCH [MILLIMETER]	**6** DRAW A PENTAGON IN A Ø 2.25 [60] CIRCLE. INCH [MILLIMETER]

FIG. 5-3-A   Polygon constructions.

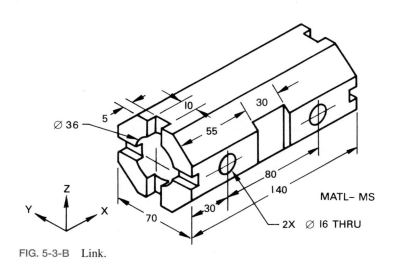

FIG. 5-3-B   Link.

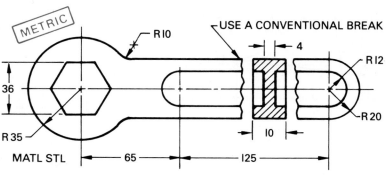

FIG. 5-3-C   Hex wrench.

ASSIGNMENTS FOR UNIT 5-3, POLYGONS

9. Divide a B (A3) size sheet as shown in Fig. 5-3-A. In the designated areas, draw the geometric constructions.
10. Make a working drawing of one of the parts shown in Figs. 5-3-B to 5-3-D. Scale 1:1.
11. Make a working drawing of one of the parts shown in Figs. 5-3-E and 5-3-F. Scale 1:1.

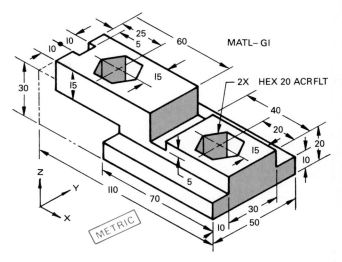

FIG. 5-3-D   Control guide.

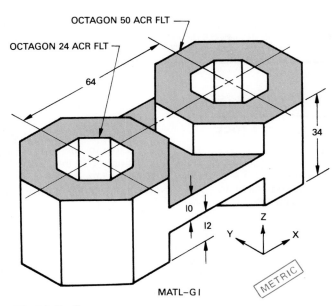

FIG. 5-3-E   Spacer.

FIG. 5-3-F   Template.

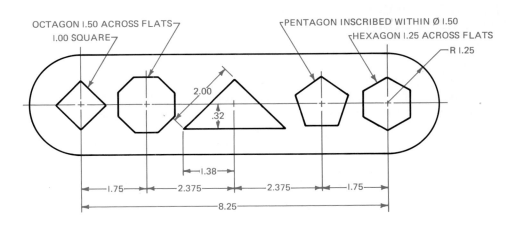

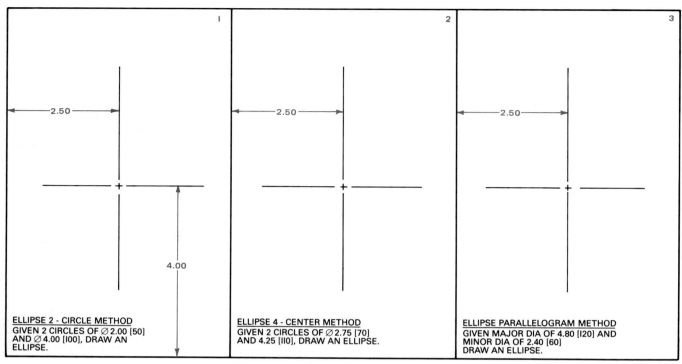

FIG. 5-4-A   Ellipse constructions.

ELLIPSE 2 - CIRCLE METHOD
GIVEN 2 CIRCLES OF ⌀ 2.00 [50]
AND ⌀ 4.00 [I00], DRAW AN
ELLIPSE.

ELLIPSE 4 - CENTER METHOD
GIVEN 2 CIRCLES OF ⌀ 2.75 [70]
AND 4.25 [IIO], DRAW AN ELLIPSE.

ELLIPSE PARALLELOGRAM METHOD
GIVEN MAJOR DIA OF 4.80 [I20] AND
MINOR DIA OF 2.40 [60]
DRAW AN ELLIPSE.

INCH [MILLIMETERS]

ASSIGNMENTS FOR UNIT 5-4, ELLIPSE

12. Divide a B (A3) size sheet as shown in Fig. 5-4-A. In the designated areas draw the geometric constructions.
13. Draw two concentric circles of 4.00 and 6.00 in. diameter. Using these circles, construct an ellipse using these two diameters. The smaller circle represents the minor diameter and the larger circle the major diameter of the ellipse. Inscribe a regular pentagon within the 4.00 in. diameter. Within the pentagon, draw a circle with the diameter tangent to the sides of the pentagon. Scale 1:1.

ASSIGNMENTS FOR UNIT 5-5, HELIX AND PARABOLA

14. On an A (A4) size sheet make a working drawing of the fan base shown in Fig. 5-5-A. Leave on the construction lines for developing the parabolic curves. Use the parallelogram method. Scale 1:1.

15. On a B (A3) size sheet lay out the three angles shown in Fig. 5-5-B and develop a parabolic curve using the parallelogram method on each. Each line has ten equal divisions of 10 mm. Scale 1:1.
16. Divide a B (A3) size sheet, as shown in Fig. 5-5-C. In the designated areas, draw the geometric constructions. Scale 1:1.

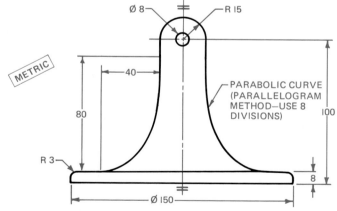

FIG. 5-5-A   Fan base.

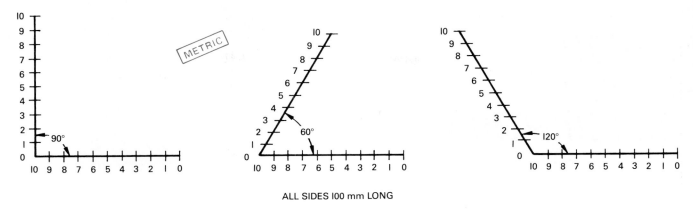

ALL SIDES 100 mm LONG

FIG. 5-5-B   Parabolic curves.

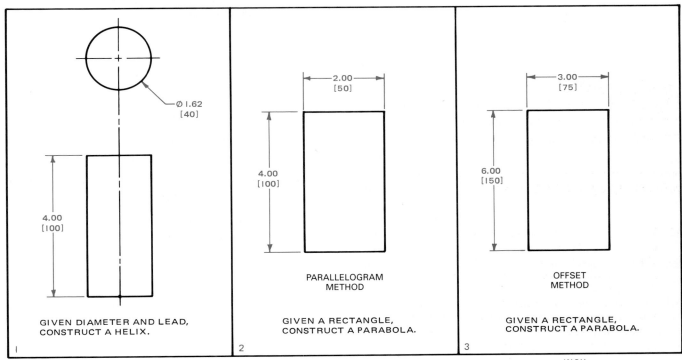

Ø 1.62 [40]	2.00 [50]	3.00 [75]
4.00 [100]	4.00 [100]	6.00 [150]
	PARALLELOGRAM METHOD	OFFSET METHOD
GIVEN DIAMETER AND LEAD, CONSTRUCT A HELIX.	GIVEN A RECTANGLE, CONSTRUCT A PARABOLA.	GIVEN A RECTANGLE, CONSTRUCT A PARABOLA.
1	2	3

INCH
[MILLIMETER]

FIG. 5-5-C   Helix and parabola constructions.

CHAPTER 6

THEORY OF SHAPE DESCRIPTION

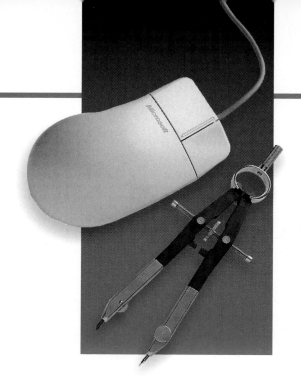

Definitions

Auxiliary view   An additional view used to depict an inclined surface that must be shown clearly and without distortion.

First-angle projection   A projection method in which the object to be represented appears between the observer and the coordinate viewing planes on which the object is orthogonally projected.

Hidden lines   Lines used to indicate the hidden edges of an object in a drawing.

Knurling   An operation that puts patterned indentations in the surface of a metal part.

Mirrored orthographic representation   The projection method in which the object to be represented is a reproduction of the image in a mirror (face up).

Miter line   A line used to quickly and accurately construct the third view of an object after two views are established.

Orthogonal projection   A projection method in which more than one view is used to define an object.

Reference arrows layout   A projection method which permits the various views to be freely positioned.

Runout   A type of curve that describes the point at which one feature blends into another feature.

Side or **end view**   An additional view used to depict cylindrically shaped surfaces containing special features.

Third-angle projection   A projection method in which the object appears behind the coordinate viewing planes on which the object is orthographically projected.

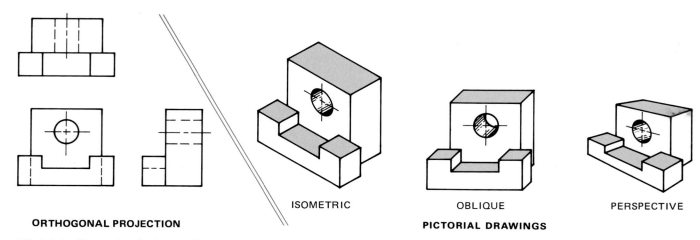

ORTHOGONAL PROJECTION

ISOMETRIC   OBLIQUE   PERSPECTIVE

PICTORIAL DRAWINGS

FIG. 6-1-1   Types of projection used in drafting.

6-1 ORTHOGRAPHIC REPRESENTATIONS

Theory of Shape Description

In the broad field of technical drawings, various projection methods are used to represent objects. Each method has its advantages and disadvantages.

The normal technical drawing is often shown in *orthogonal projection*, in which more than one view is used to draw and completely define an object (Fig. 6-1-1). The drawing of two-dimensional representations, however, requires an understanding of both the projection method and its interpretation so that the reader of the drawing, looking at two-dimensional views, will be able to visualize a three-dimensional object.

For many technical fields and their stages of development, the drafter must supply the viewer with an easily understood drawing. Pictorial drawings provide a three-dimensional view of an object as it would appear to the observer, as shown in Fig. 6-1-1. These pictorial drawings are described in detail in Chap. 15.

The steady increase in global technical intercommunication, and the interchange of drawings with different countries, as well as the evolution of methods of computer-aided design and drafting with their various types of three-dimensional representation, require that today's drafters have a knowledge of all methods of representation.

Orthographic Representations

Orthographic representation is obtained by means of parallel orthogonal projections and results in flat, two-dimensional views systematically positioned relative to each other. To show the object completely, the six views in the directions *A*, *B*, *C*, *D*, *E*, and *F* may be necessary (Fig. 6-1-2).

The most informative view of the object to be represented is normally chosen as the principal view (front view). This is view *A* according to the direction of viewing *a* and usually shows the object in the functioning, manufacturing, or mounting position. The position of the other views relative to the principal view in the drawing depends on the projection method (third-angle, first-angle, reference arrows). In practice, not all six views (*A* to *F*) are needed. When views other than the principal view are necessary, these should be selected in order to:

- Limit the number of views and sections to the minimum necessary to fully represent the object without ambiguity.
- Avoid unnecessary repetition of detail.

Methods of Representation

There are four methods of orthographic representation: third-angle projection, first-angle projection, reference arrows layout, and mirrored orthographic representation. Third-angle projection is used in the United States, Canada, and many other countries throughout the world. First-angle projection is used mainly in European and Asiatic countries.

Third-Angle Projection

The *third-angle projection* method is an orthographic representation in which the object to be represented and seen by the observer appears behind the coordinate viewing planes on which the object is orthographically projected (Fig. 6-1-3B, pg. 94). On each projection plane, the object is represented as if seen orthogonally from in front of each plane.

The positions of the various views relative to the principal (front) view are then rotated or positioned so that they lie on the same plane (drawing surface) on which the front view *A* is projected.

FIG. 6-1-2   Designation of views.

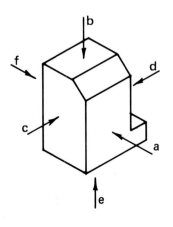

DIRECTION OF OBSERVATION		DESIGNATION OF VIEW
VIEW IN DIRECTION	VIEW FROM	
a	THE FRONT	A
b	ABOVE	B
c	THE LEFT	C
d	THE RIGHT	D
e	BELOW	E
f	THE REAR	F

Therefore in Fig. 6-1-3C with reference to the principal view *A*, the other views are arranged as follows:

- View *B*—The view from above is placed above.
- View *E*—The view from below is placed underneath.
- View *C*—The view from the left is placed on the left.
- View *D*—The view from the right is placed on the right.
- View *F*—The view from the rear may be placed on the left or on the right, as convenient.

The letters *A* to *F* are shown here only to identify the location of views when third-angle projection is used. On working drawings these letters would not be shown.

The identifying symbol for this method of representation is shown in Fig. 6-1-3D.

First-Angle Projection

The *first-angle projection* method is an orthographic representation in which the object to be represented appears between the observer and the coordinate viewing planes on which the object is orthogonally projected (Fig. 6-1-4B).

The position of the various views relative to the principal (front) view *A* are then rotated or positioned so that they lie on

the same plane (drawing surface) on which the front view *A* is projected.

Therefore, in Fig. 6-1-4C with reference to the principal view *A*, the other views are arranged as follows:

- View *B*—The view from above is placed underneath.
- View *E*—The view from below is placed above.
- View *C*—The view from the left is placed on the right.
- View *D*—The view from the right is placed on the left.
- View *F*—The view from the rear is placed on the right or on the left, as convenient.

The letters *A* to *F* are shown here only to identify the location of views when first-angle projection is used. On working drawings these letters would not be shown.

The identifying symbol of this method of representation is shown in Fig. 6-1-4D.

Reference Arrows Layout

In those cases where it is advantageous not to position the views according to the strict pattern of the third- or the first-angle projection method, *reference arrows layout* permits the various views to be freely positioned.

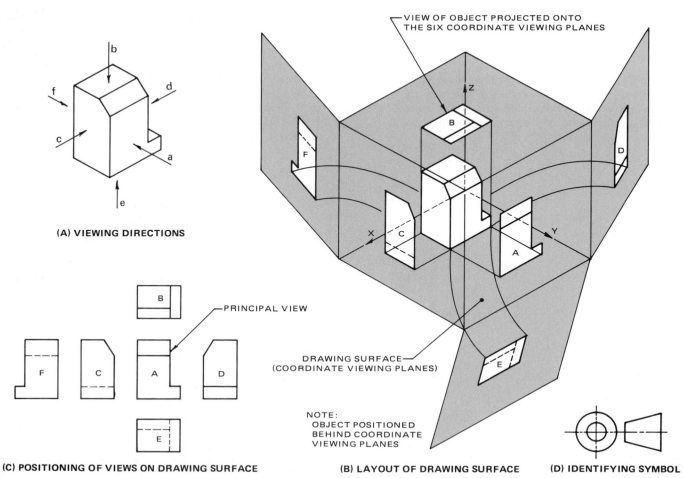

(A) VIEWING DIRECTIONS

(C) POSITIONING OF VIEWS ON DRAWING SURFACE

VIEW OF OBJECT PROJECTED ONTO THE SIX COORDINATE VIEWING PLANES

PRINCIPAL VIEW

DRAWING SURFACE (COORDINATE VIEWING PLANES)

NOTE:
OBJECT POSITIONED BEHIND COORDINATE VIEWING PLANES

(B) LAYOUT OF DRAWING SURFACE

(D) IDENTIFYING SYMBOL

FIG. 6-1-3   Third-angle projection.

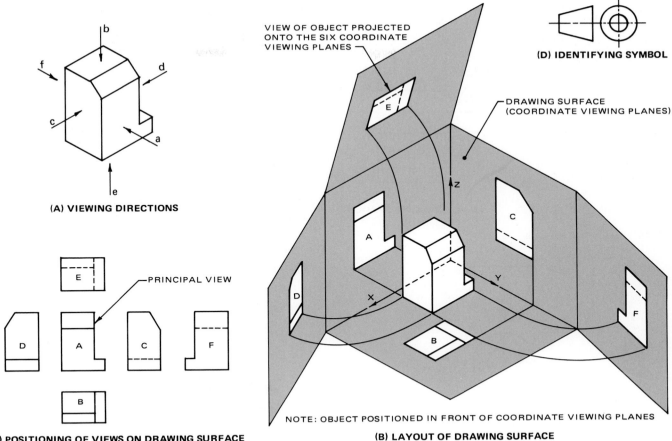

(A) VIEWING DIRECTIONS

(D) IDENTIFYING SYMBOL

VIEW OF OBJECT PROJECTED ONTO THE SIX COORDINATE VIEWING PLANES

DRAWING SURFACE (COORDINATE VIEWING PLANES)

(C) POSITIONING OF VIEWS ON DRAWING SURFACE

PRINCIPAL VIEW

NOTE: OBJECT POSITIONED IN FRONT OF COORDINATE VIEWING PLANES

(B) LAYOUT OF DRAWING SURFACE

FIG. 6-1-4   First-angle projection.

FIG. 6-1-5   Reference arrows layout.

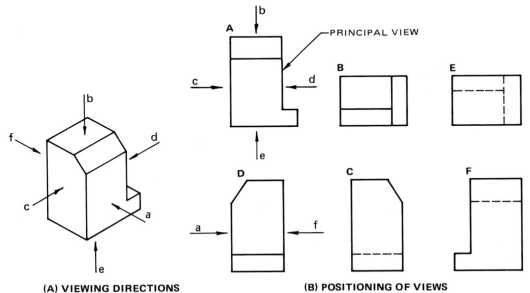

PRINCIPAL VIEW

(A) VIEWING DIRECTIONS

(B) POSITIONING OF VIEWS

With the exception of the principal view, each view is identified by a letter (Fig. 6-1-5B). A lowercase letter on the principal view, and where required on one of the side views, indicates the direction of observation of the other views, which are identified by the corresponding capital letter placed immediately above the view on the left.

The identified views may be located irrespective of the principal view. Whatever the direction of the observer, the capital letters identifying the views should always be positioned to be read from the direction from which the drawing is normally viewed. No symbol is needed on the drawing to identify this method.

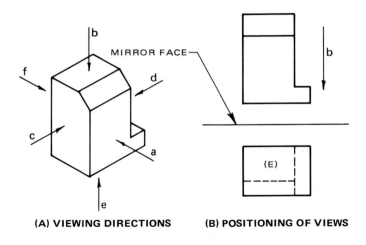

(A) VIEWING DIRECTIONS   **(B) POSITIONING OF VIEWS**

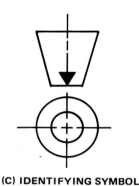

(C) IDENTIFYING SYMBOL

FIG. 6-1-6   Mirrored orthographic projection.

Mirrored Orthographic Representation

Mirrored orthographic representation is the method preferred for use in construction drawings. In this method the object to be represented is a reproduction of the image in a mirror (face up), positioned parallel to the horizontal planes of the object (Fig. 6-1-6).

The identifying symbol for this method is shown in Fig. 6-1-6C.

Identifying Symbols

The symbol used to identify the method of representation should be shown on all drawings, preferably in the lower right-hand corner of the drawing, adjacent to the title block (Fig. 6-1-7).

CAD Coordinate Input for Orthographic Representation

In Unit 4-1 you learned how to locate points and lines using coordinate input. The positions were described on the drawing

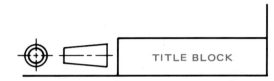

FIG. 6-1-7   Location of identifying symbol for method of representation.

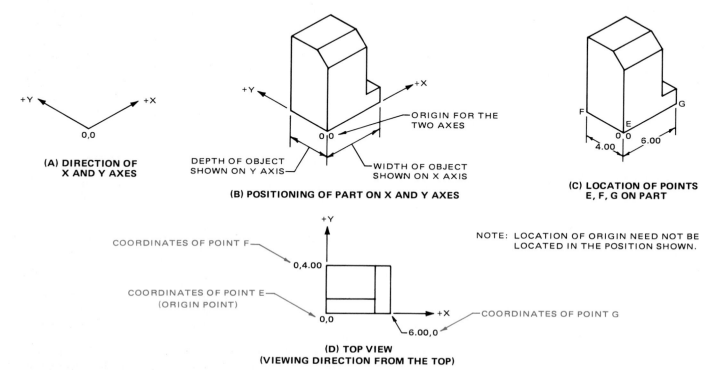

(A) DIRECTION OF X AND Y AXES

(B) POSITIONING OF PART ON X AND Y AXES

(C) LOCATION OF POINTS E, F, G ON PART

NOTE: LOCATION OF ORIGIN NEED NOT BE LOCATED IN THE POSITION SHOWN.

(D) TOP VIEW (VIEWING DIRECTION FROM THE TOP)

FIG. 6-1-8   Locating points on a part by two-axis (X and Y) coordinate input.

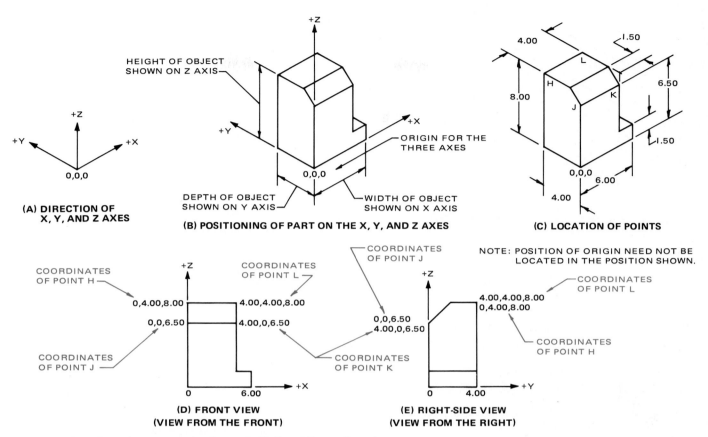

FIG. 6-1-9   Locating points on a part by three-axis (*X*, *Y*, and *Z*) coordinate input.

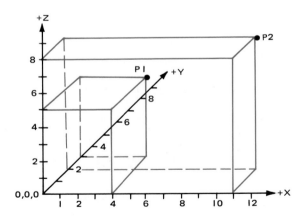

FIG. 6-1-10   Points in space.

by two-dimensional coordinates, horizontal (*X*) and vertical (*Y*). The *X* axis is horizontal and is considered the first and basic reference axis. The *Y* axis is vertical and is 90° to the *X* axis.

When third-angle orthographic representation is used to show a part, the view from above is known as the top view. The *X* and *Y* axes and coordinates are identified with this view, and width and depth features are shown (Fig. 6-1-8).

With the exceptions of the views from the top and below, all the other views require information regarding the height of features. This is provided by introducing a third axis, called the *Z* axis, to the system. The coordinates for the origin of the three axes are then identified by the numbers 0, 0, 0, the last coordinate representing distances on the *Z* axis (Fig. 6-1-9A). As previously mentioned, the point of origin may be located at any convenient location on the drawing. The coordinates for points *H*, *J*, *K*, and *L* shown in Fig. 6-1-9C would be 0, 4.00, 8.00 (point *H*); 0, 0, 6.50 (point *J*); 4.00, 0, 6.50 (point *K*); and 4.00, 4.00, 8.00 (point *L*). Note that the coordinates for a point remain the same, regardless of the view on which they are shown.

Coordinate Input to Locate Points in Space

A point in space can be described by its *X*, *Y*, and *Z* coordinates. For example, *P*1 in Fig. 6-1-10 can be described by its (*X*, *Y*, *Z*) coordinates as (4, 3, 5), and *P*2 as (11, 2, 8).

A pictorial drawing of a part can be described as lines joining a series of points in space (Fig. 6-1-11, pg. 98). The 0, 0, 0 reference indicates the absolute *X*, *Y*, *Z* coordinate origin. It has been designated to be the lower left-front corner position of the front view. The lower right-front position is labeled 12, 0, 0. This means that the coordinate location for that point is 12 units (in.) to the right and has the same elevation (height) and depth as the coordinate origin. All other positions are interpreted in the same manner.

A symbol similar to that shown in Fig. 6-1-9A is used in the pictorial drawings that follow to designate the direction of the *X*, *Y*, and *Z* axes for that particular part.

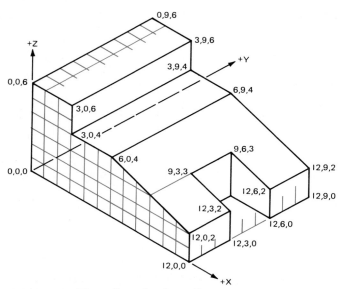

FIG. 6-1-11   Three-dimensional coordinates.

REFERENCES AND SOURCE MATERIAL

1. ISO 5456   *Technical Drawings—Projection Methods*

ASSIGNMENTS

See Assignments 1 through 4 for Unit 6-1 on pages 110–114.

6-2   ARRANGEMENT AND CONSTRUCTION OF VIEWS

Spacing the Views

For clarity and good appearance the views should be well balanced on the drawing paper, whether the drawing shows one view, two views, three views, or more. The drafter must anticipate the approximate space required. This is determined from the size of the object to be drawn, the number of views, the scale used, and the space between views. Ample space should be provided between views to permit placement of dimensions on the drawing without crowding. Space should also be allotted so that notes can be added. However, space between views should not be excessive.

Figure 6-2-1 shows how to balance the views for a three-view drawing. For a drawing with two or more views, follow these guidelines:

1. Decide on the views to be drawn and the scale to be used, e.g., 1:1 or 1:2.
2. Make a sketch of the space required for each of the views to be drawn, showing these views in their correct location. A simple rectangle for each view will be adequate (Fig. 6-2-1B).
3. Put on the overall drawing sizes for each view. (These sizes are shown as *W*, *D*, and *H*.)
4. Decide upon the space to be left between views. These spaces should be sufficient for the parallel dimension

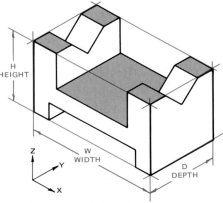

(A) DECIDING THE VIEWS TO BE DRAWN AND THE SCALE TO BE USED

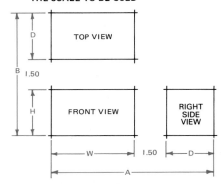

(B) CALCULATING DISTANCES A AND B

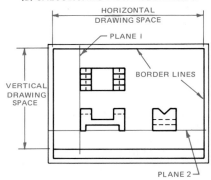

(C) ESTABLISHING LOCATION OF PLANES 1 AND 2 ON DRAWING PAPER OR CRT MONITOR

FIG. 6-2-1   Balancing the drawing on the drawing paper or monitor.

lines to be placed between views. For most drawing projects, 1.50 in. (40 mm) is sufficient.

5. Total these dimensions to get the overall horizontal distance (*A*) and overall vertical distance (*B*).
6. Select the drawing sheet that best accommodates the overall size of the drawing with suitable open space around the views.
7. Measure the "drawing space" remaining after all border lines, title strip or title block, etc., are in place (Fig. 6-2-1C).
8. Take one-half of the difference between distance *A* and the horizontal drawing space to establish plane 1.
9. Take one-half of the difference between distance *B* and the vertical drawing space to establish plane 2.

Use of a Miter Line

The use of a miter line provides a fast and accurate method of constructing the third view once two views are established (Fig. 6-2-2).

Using a Miter Line to Construct the Right Side View

1. Given the top and front views, project lines to the right of the top view.
2. Establish how far from the front view the side view is to be drawn (distance *D*).
3. Construct the miter line at 45° to the horizon.
4. Where the horizontal projection lines of the top view intersect the miter line, drop vertical projection lines.
5. Project horizontal lines to the right of the front view and complete the side view.

Using a Miter Line to Construct the Top View

1. Given the front and side views, project vertical lines up from the side view.
2. Establish how far away from the front view the top view is to be drawn (distance *D*).
3. Construct the miter line at 45° to the horizon.
4. Where the vertical projection lines of the side view intersect the miter line, project horizontal lines to the left.
5. Project vertical lines up from the front view and complete the top view.

 CAD

In a CAD environment, construction lines are usually placed on a separate working layer, and the geometry on that layer is given an identifying color. This layer can be echoed off when a plot is generated, leaving only the finished drawing. In more complicated drawings, several different construction layers may be required.

The working area of a drawing is established by a LIMITS command. The limits of the working drawing area are normally expressed as the lower left and upper right corners of the drawing and correspond to the size of the drawing form. When the drawing is plotted, the limits are used to determine the overall size of the sheet that is required.

ASSIGNMENTS

See Assignments 5 through 8 for Unit 6-2 on pages 114–115.

6-3 ALL SURFACES PARALLEL AND ALL EDGES AND LINES VISIBLE

To help you fully appreciate the shape and detail of views drawn in third-angle orthographic projection, the units for this chapter have been designed according to the types of surfaces generally found on objects. These surfaces can be divided into flat surfaces parallel to the viewing planes with and without hidden features; flat surfaces that appear inclined in one plane and parallel to the other two principal reference planes (called *inclined surfaces*); flat surfaces that are inclined in all three reference planes (called *oblique surfaces*); and surfaces that have diameters or radii. These drawings are so designed that only the top, front, and right side views are required.

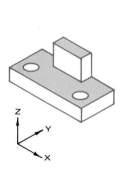
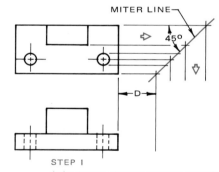
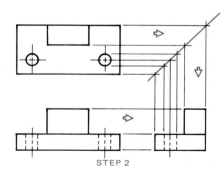

(A) ESTABLISHING WIDTH LINES ON SIDE VIEW

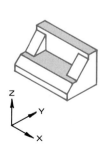
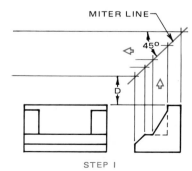

(B) ESTABLISHING WIDTH LINES ON TOP VIEW

FIG. 6-2-2   Use of a miter line.

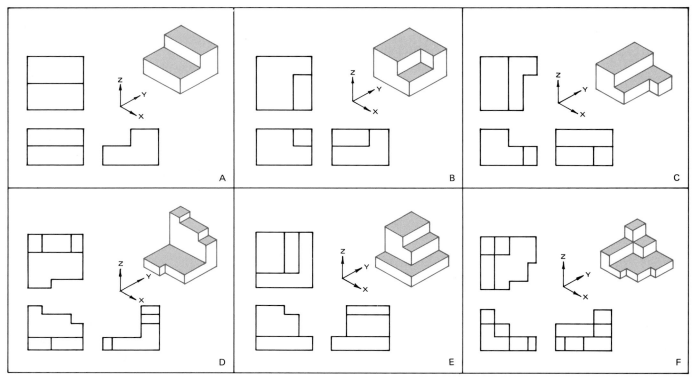

NOTE: ARROWS INDICATE DIRECTION OF SIGHT WHEN LOOKING AT THE FRONT VIEW.

FIG. 6-3-1   Illustrations of objects drawn in third-angle orthographic projection.

All Surfaces Parallel to the Viewing Planes and All Edges and Lines Visible   When a surface is parallel to the viewing planes, that surface will show as a surface on one view and a line on the other views. The lengths of these lines are the same as the lines shown on the surface view. Figure 6-3-1 shows examples.

ASSIGNMENTS

See Assignments 9 and 10 for Unit 6-3 on pages 115–116.

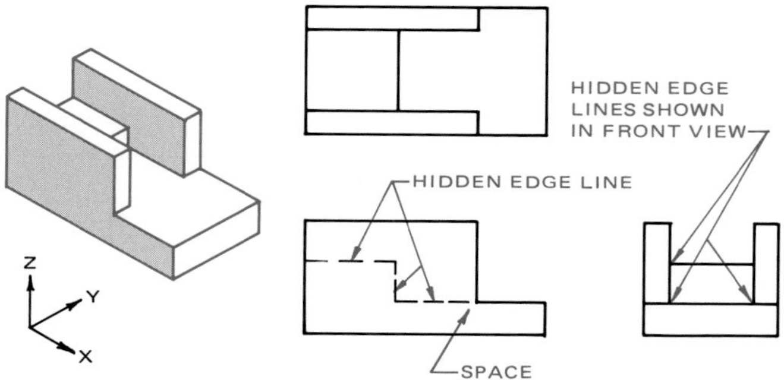

FIG. 6-4-1   Hidden lines.

HIDDEN EDGE LINES SHOWN IN FRONT VIEW

HIDDEN EDGE LINE

SPACE

6-4 HIDDEN SURFACES AND EDGES

Most objects drawn in engineering offices are more complicated than the one shown in Fig. 6-4-1. Many features (lines, holes, etc.) cannot be seen when viewed from outside the piece. These hidden edges are shown with hidden lines and are normally required on the drawing to show the true shape of the object.

Hidden lines consist of short, evenly spaced dashes. They should be omitted when not required to preserve the clarity of the drawing. The length of dashes may vary slightly in relation to the size of the drawing.

Lines depicting hidden features and phantom details should always begin and end with a dash in contact with the line at which they start and end, except when such a dash would form

a continuation of a visible detail line. Dashes should join at corners. Arcs should start with dashes at the tangent points (Fig. 6-4-2). Figure 6-4-3 shows additional examples of objects requiring hidden lines.

All CAD systems have the option to create different line styles. On large systems, these options are found on the auxiliary menu. On smaller systems, the line style selection is made directly from the tablet menu. Any style of line may be set using the LINETYPE command.

ASSIGNMENTS

See Assignments 11 through 15 for Unit 6-4 on pages 116 through 119.

FIG. 6-4-2   Application of hidden lines.

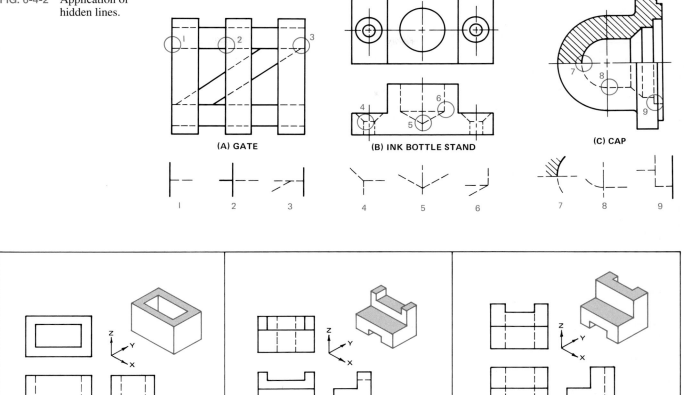

(A) GATE   (B) INK BOTTLE STAND   (C) CAP

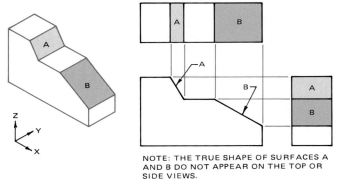

FIG. 6-4-3   Illustrations of objects having hidden features.

6-5 INCLINED SURFACES

If the surfaces of an object lie in either a horizontal or a vertical position, the surfaces appear in their true shapes in one of the three views, and these surfaces appear as a line in the other two views.

When a surface is inclined or sloped in only one direction, that surface is not seen in its true shape in the top, front, or side view. It is, however, seen in two views as a distorted surface. On the third view it appears as a line.

The true length of surfaces A and B in Fig. 6-5-1 is seen in the front view only. In the top and side views, only the width of surfaces A and B appears in its true size. The length of these surfaces is foreshortened. Figure 6-5-2 (pg. 102) shows additional examples.

Where an inclined surface has important features that must be shown clearly and without distortion, an *auxiliary*, or helper, view must be used. This type of view will be discussed in detail in Chap. 7.

ASSIGNMENTS

See Assignments 16 through 21 for Unit 6-5 on pages 120 through 124.

NOTE: THE TRUE SHAPE OF SURFACES A AND B DO NOT APPEAR ON THE TOP OR SIDE VIEWS.

FIG. 6-5-1   Sloping surfaces.

6-6 CIRCULAR FEATURES

Typical parts with circular features are illustrated in Fig. 6-6-1 (pg. 102). Note that the circular feature appears circular in one view only and that no line is used to show where a curved surface joins a flat surface. Hidden circles, like hidden flat surfaces, are represented on drawings by a hidden line.

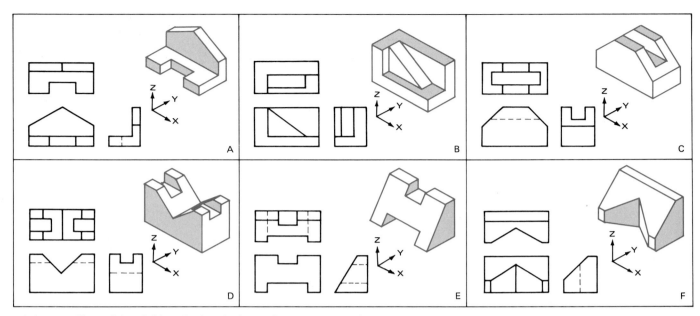

FIG. 6-5-2   Illustrations of objects having sloping surfaces.

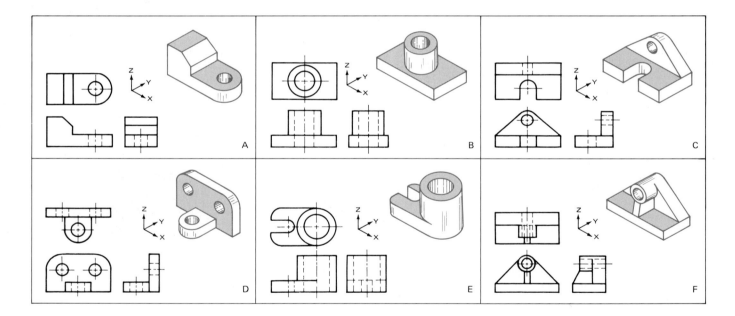

FIG. 6-6-1   Illustrations of objects having circular features.

The intersection of unfinished surfaces, such as found on cast parts, that are rounded or filleted at the point of theoretical intersection may be indicated conventionally by a line (see Unit 6-15).

Center Lines

A center line is drawn as a thin, broken line of long and short dashes, spaced alternately. Such lines may be used to locate center points, axes of cylindrical parts, and axes of symmetry, as shown in Fig. 6-6-2. Solid center lines are often used when the circular features are small. Center lines should project for a short distance beyond the outline of the part or feature to which they refer. They must be extended for use as extension lines for dimensioning purposes, but in this case the extended portion is not broken.

On views showing the circular features, the point of intersection of the two center lines is shown by the two intersecting short dashes.

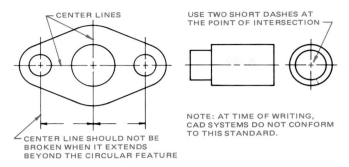

FIG. 6-6-2   Center line applications.

ASSIGNMENTS

See Assignments 22 through 26 for Unit 6-6 on pages 125 through 128.

6-7 OBLIQUE SURFACES

When a surface is sloped so that it is not perpendicular to any of the three viewing planes, it will appear as a surface in all three views but never in its true shape. This is referred to as an *oblique surface* (Fig 6-7-1). Since the oblique surface is not perpendicular to the viewing planes, it cannot be parallel to them and consequently appears foreshortened. If a true view is

required for this surface, two auxiliary views—a primary and a secondary view—need to be drawn. This is discussed in detail in Unit 7-4, "Secondary Auxiliary Views." Figure 6-7-2 shows additional examples of objects having oblique surfaces.

ASSIGNMENTS

See Assignments 27 through 29 for Unit 6-7 on pages 129 to 130.

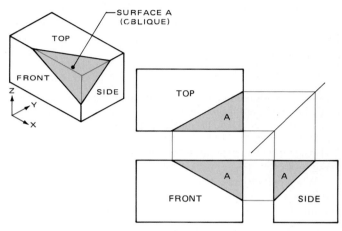

FIG. 6-7-1   Oblique surface is not its true shape in any of the three views.

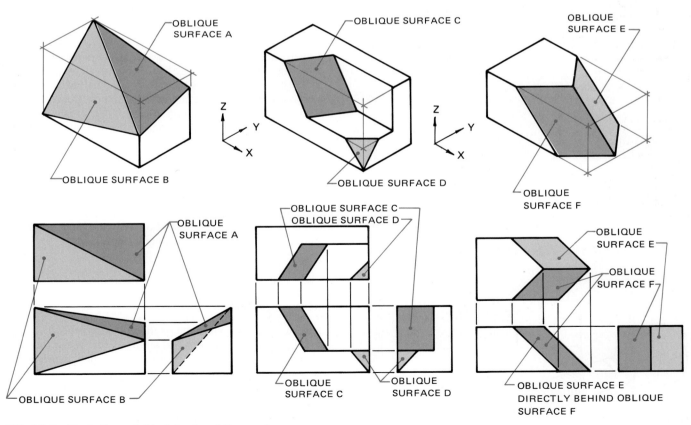

FIG. 6-7-2   Illustrations of objects having oblique surfaces.

6-8 ONE- AND TWO-VIEW DRAWINGS

View Selection

Views should be chosen that will best describe the object to be shown. Only the minimum number of views that will completely portray the size and shape of the part should be used. They should also be chosen to avoid hidden feature lines whenever possible, as shown in Fig. 6-8-1.

Except for complex objects of irregular shape, it is seldom necessary to draw more than three views. For representing simple parts, one- or two-view drawings will often be adequate.

One-View Drawings

In one-view drawings, the third dimension, such as thickness, may be expressed by a note or by descriptive words or abbreviations, such as DIA, Ø, or HEX ACRFLT. Square sections may be indicated by light, crossed, diagonal lines. This applies whether the face is parallel or inclined to the drawing plane (Fig. 6-8-2).

When cylindrically shaped surfaces include special features, such as a keyseat, a side view (often called an *end view*) is required.

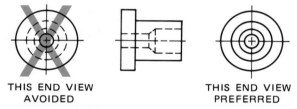

THIS END VIEW AVOIDED   THIS END VIEW PREFERRED

FIG. 6-8-1   Avoidance of hidden-line features.

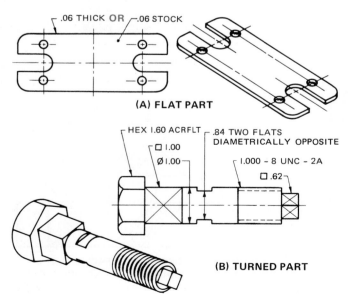

(A) FLAT PART

(B) TURNED PART

FIG. 6-8-2   One-view drawings.

Two-View Drawings

Frequently the drafter will decide that only two views are necessary to explain fully the shape of an object (Fig. 6-8-3). For this reason, some drawings consist of two adjacent views such as the top and front views only or front and right side views only. Two views are usually sufficient to explain fully the shape of cylindrical objects; if three views were used, two of them would be identical, depending on the detail structure of the part.

ASSIGNMENT

See Assignment 30 for Unit 6-8 on pages 130–131.

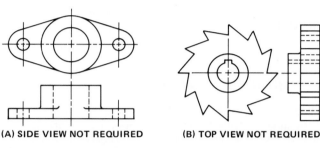

(A) SIDE VIEW NOT REQUIRED   (B) TOP VIEW NOT REQUIRED

FIG. 6-8-3   Two-view drawings.

6-9 SPECIAL VIEWS

Partial Views

Symmetrical objects may often be adequately portrayed by half views (Fig. 6-9-1A). A center line is used to show the axis of symmetry. Two short thick lines, above and below the view of the object, are drawn at right angles to, and on, the center line to indicate the line of symmetry.

Partial views, which show only a limited portion of the object with remote details omitted, should be used, when necessary, to clarify the meaning of the drawing (Fig. 6-9-1B). Such views are used to avoid the necessity of drawing many hidden features.

On drawings of objects where two side views can be used to better advantage than one, each need not be complete if together they depict the shape. Show only the hidden lines of features immediately behind the view (Fig. 6-9-1C).

Rear Views and Enlarged Views

Placement of Views

When views are placed in the relative positions shown in Fig. 6-1-3, it is rarely necessary to identify them. When they are placed in other than the regular projected position, the removed view must be clearly identified.

Whenever appropriate, the orientation of the main view on a detail drawing should be the same as on the assembly drawing. To avoid the crowding of dimensions and notes, ample space must be provided between views.

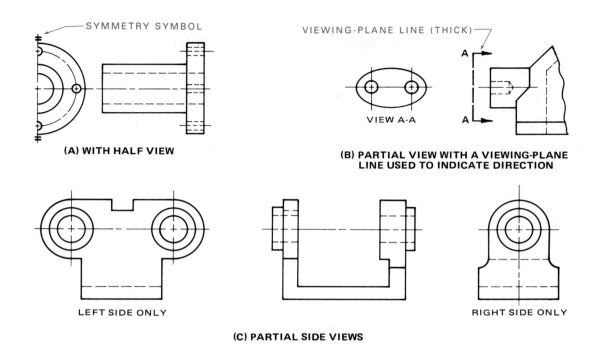

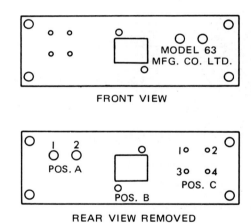

FIG. 6-9-1   Partial views.

Rear Views

Rear views are normally projected to the right or left. When this projection is not practical because of the length of the part, particularly for panels and mounting plates, the rear view must not be projected up or down. Doing so would result in the part being shown upside down. Instead, the view should be drawn as if it were projected sideways but located in some other position, and it should be clearly labeled REAR VIEW REMOVED (Fig. 6-9-2). Alternately the reference arrows layout method of representation may be used as explained in Unit 6-1.

Enlarged Views

Enlarged views are used when it is desirable to show a feature in greater detail or to eliminate the crowding of details or dimensions (Fig. 6-9-3). The enlarged view should be oriented

FIG. 6-9-2   Removed rear views.

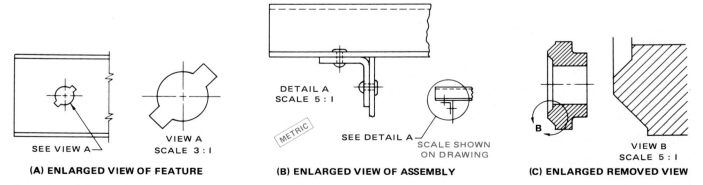

FIG. 6-9-3   Enlarged views.

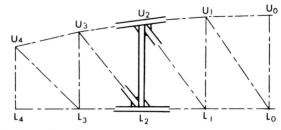

FIG. 6-9-4   Key plan.

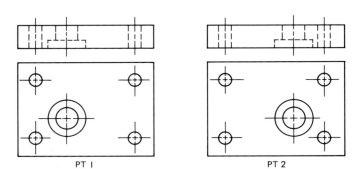

(A) TWO DRAWINGS

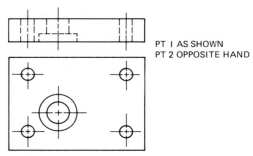

PT 1 AS SHOWN
PT 2 OPPOSITE HAND

(B) ONE DRAWING REPLACES TWO VIEWS

FIG. 6-9-5   Opposite-hand views.

in the same manner as the main view. However, if an enlarged view is rotated, state the direction and the amount of rotation of the detail. The scale of enlargement must be shown, and both views should be identified by one of the three methods shown.

Key Plans

A method particularly applicable to structural work is to include a small *key plan*, using bold lines, on each sheet of a drawing series that shows the relationship of the detail on that sheet to the whole work, as in Fig. 6-9-4.

Opposite-Hand Views

Where parts are symmetrically opposite, such as for right- and left-hand usage, one part is drawn in detail and the other is described by a note such as PART B SAME EXCEPT

OPPOSITE HAND. It is preferable to show both part numbers on the same drawing (Fig. 6-9-5).

ASSIGNMENTS

See Assignments 31 through 33 for Unit 6-9 on pages 131 to 132.

6-10 CONVENTIONAL REPRESENTATION OF COMMON FEATURES

To simplify the representation of common features, a number of conventional drafting practices are used. Many conventions are deviations from the true projection for the purpose of clarity; others are used to save drafting time. These conventions must be executed carefully, for clarity is even more important than speed.

Many drafting conventions, such as those used on thread, gear, and spring drawings, appear in various chapters throughout the text. Only the conventions not described in those chapters appear here.

Repetitive Details

Repetitive features, such as gear and spline teeth, are depicted by drawing a partial view, showing two or three of these features, with a phantom line or lines to indicate the extent of the remaining features (Figs. 6-10-1A and B). Alternatively, gears and splines may be shown with a solid thick line representing the basic outline of the part and a thin line representing the root of the teeth. This is essentially the same convention that is used for screw threads. The pitch line may be added by using the standard center line.

Knurls

Knurling is an operation that puts patterned indentations in the surface of a metal part to provide a good finger grip (Figs. 6-10-1C and D). Commonly used types of knurls are straight, diagonal, spiral, convex, raised diamond, depressed diamond, and radial. The pitch refers to the distance between corresponding indentations, and it may be a straight pitch, a circular pitch, or a diametral pitch. For cylindrical surfaces, the latter is preferred. The pitch of the teeth for coarse knurls (measured parallel to the axis of the work) is 14 teeth per inch (TPI) or about 2 mm; for medium knurls, 21 TPI or about 1.2 mm; and for fine knurls, 33 TPI or 0.8 mm. The medium-pitch knurl is the most commonly used.

As a time-saver, the knurl symbol is shown on only a part of the surface being knurled.

Holes

A series of similar holes is indicated by drawing one or two holes and showing only the center for the others (Figs. 6-10-1E and F).

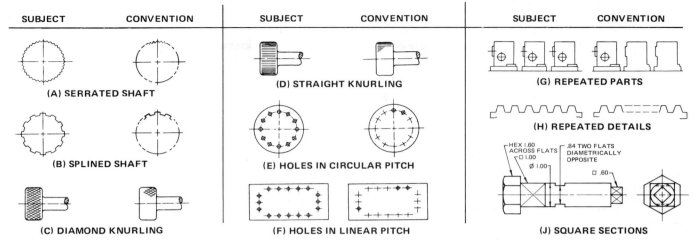

FIG. 6-10-1   Conventional representation of common features.

Repetitive Parts

Repetitive parts, or intricate features, are shown by drawing one in detail and the others in simple outline only. A covering note is added to the drawing (Figs. 6-10-1G and H).

Square Sections

Square sections on shafts and similar parts may be illustrated by thin, crossed, diagonal lines, as shown in Fig. 6-10-1J.

ASSIGNMENT

See Assignment 34 for Unit 6-10 on page 133.

6-11 CONVENTIONAL BREAKS

Long, simple parts, such as shafts, bars, tubes, and arms, need not be drawn to their entire length. Conventional breaks located at a convenient position may be used and the true length indicated by a dimension. Often a part can be drawn to a larger scale to produce a clearer drawing if a conventional break is used. Two types of break lines are generally used (Fig. 6-11-1A). Thick freehand lines are used for short breaks. The thin line with freehand zig-zag is recommended for long breaks and may be used for solid details or for assemblies containing open spaces.

Special break lines, shown in Fig. 6-11-1B, are used when it is desirable to indicate the shape of the features.

ASSIGNMENT

See Assignment 35 for Unit 6-11 on page 133.

6-12 MATERIALS OF CONSTRUCTION

Symbols used to indicate materials in sectional views are shown in Fig. 9-1-6 (pg. 229). Those shown for concrete, wood, and transparent materials are also suitable for outside

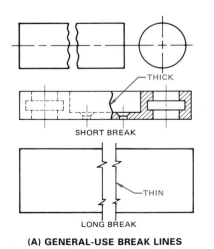

(A) GENERAL-USE BREAK LINES

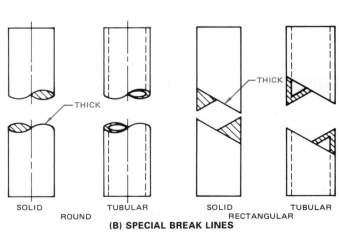

(B) SPECIAL BREAK LINES

FIG. 6-11-1   Conventional breaks.

views. Other symbols that may be used to indicate areas of different materials are shown in Fig. 6-12-1. It is not necessary to cover the entire area affected with such symbolic lining as long as the extent of the area is shown on the drawing.

Transparent Materials

These should generally be treated in the same manner as opaque materials; i.e., details behind them are shown with hidden lines if such details are necessary.

ASSIGNMENT

See Assignment 36 for Unit 6-12 on page 134.

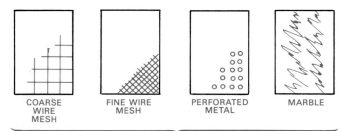

FIG. 6-12-1   Symbols to indicate materials of construction.

6-13 CYLINDRICAL INTERSECTIONS

The intersections of rectangular and circular contours, unless they are very large, are shown conventionally, as in Figs. 6-13-1 and 6-13-2. The same convention may be used to show the intersection of two cylindrical contours, or the curve of intersection may be shown as a circular arc.

ASSIGNMENT

See Assignment 37 for Unit 6-13 on page 134.

6-14 FORESHORTENED PROJECTION

When the true projection of a feature would result in confusing foreshortening, it should be rotated until it is parallel to the line of the section or projection (Fig. 6-14-1).

Holes Revolved to Show True Distance from Center

Drilled flanges in elevation or section should show the holes at their true distance from the center, rather than the true projection.

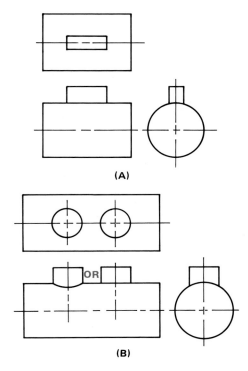

FIG. 6-13-1   Conventional representation of external intersections.

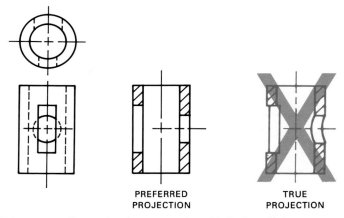

FIG. 6-13-2   Conventional representation of holes in cylinders.

ASSIGNMENT

See Assignment 38 for Unit 6-14 on page 135.

6-15 INTERSECTIONS OF UNFINISHED SURFACES

The intersections of unfinished surfaces that are rounded or filleted may be indicated conventionally by a line coinciding with the theoretical line of intersection. The need for this convention is demonstrated by the examples shown in Fig. 6-15-1, where the upper top views are shown in true projection. Note that in each example the true projection would be misleading. In the case of the large radius, such as shown in Fig. 6-15-1C, no line is drawn. Members such as ribs and arms that blend into other features terminate in curves called *runouts*. With manual drafting small runouts are usually drawn freehand. Large runouts are drawn with an irregular curve, template, or compass (Fig. 6-15-2, pg. 110).

ASSIGNMENT

See Assignment 39 for Unit 6-15 on page 135.

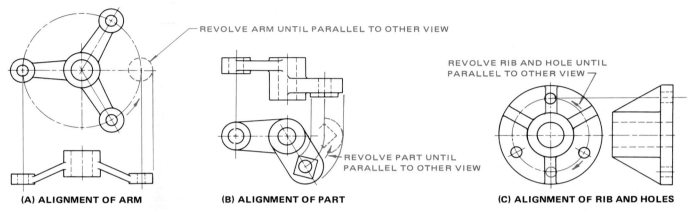

(A) ALIGNMENT OF ARM   **(B) ALIGNMENT OF PART**   **(C) ALIGNMENT OF RIB AND HOLES**

FIG. 6-14-1   Alignment of parts and holes to show true relationship.

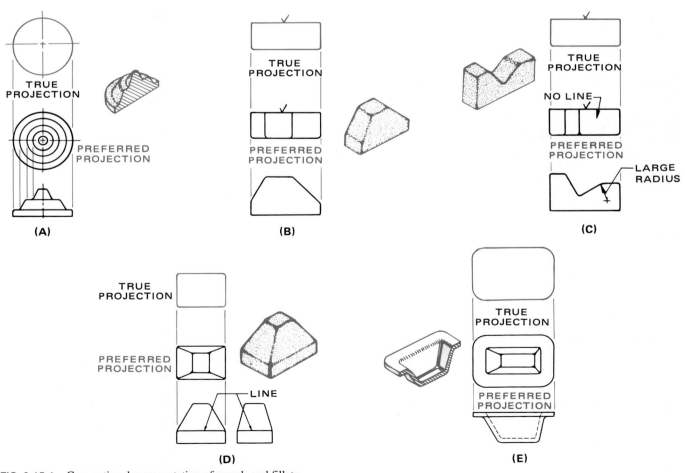

FIG. 6-15-1   Conventional representation of rounds and fillets.

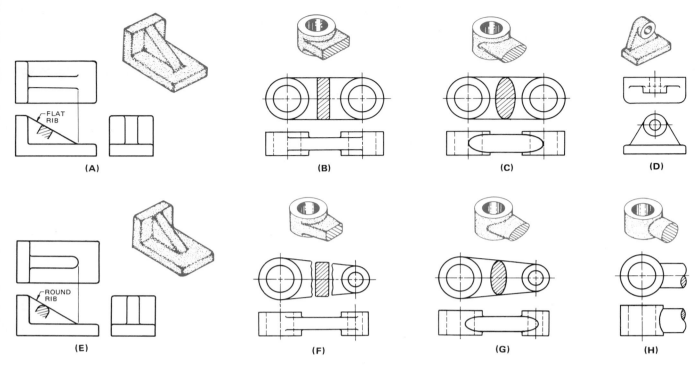

FIG. 6-15-2   Conventional representation of runouts.

ASSIGNMENTS FOR CHAPTER 6

Notes: (1) CAD may be substituted for manual drafting for any assignments in this chapter. (2) Unless otherwise specified all drawings are to be drawn in third-angle projection.

ASSIGNMENTS FOR UNIT 6-1, ORTHOGRAHIC REPRESENTATIONS

1. Draw the six views for any two parts shown in Figs. 6-1-A through 6-1-E using the following methods of representation: (A) third-angle projection; (B) first-angle projection; (C) reference arrows layout. Show only what can be seen when viewing the object. Do not attempt to show any hidden features. Viewing in the direction of axis *Y* will represent the principal view. Identify the views as shown in Figs. 6-1-3 through 6-1-5.

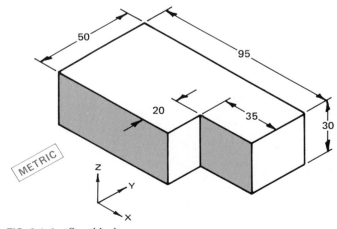

FIG. 6-1-A   Stop block.

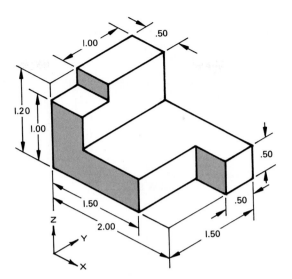

FIG. 6-1-B   Angle bracket.

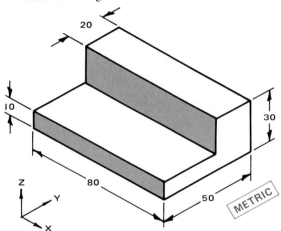

FIG. 6-1-C   Step block.

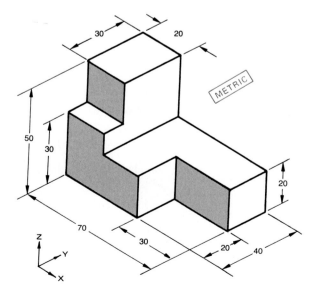

FIG. 6-1-D   Corner bracket.

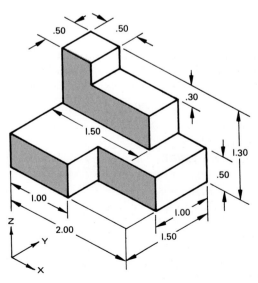

FIG. 6-1-E   Locating block.

2. Using squared graph paper of 4 or 5 squares to the inch (one square representing 1.00 in.) or 10 mm squares (one square representing 10 mm) or the grid on the CAD monitor, sketch or plot the views using the two-dimensional absolute coordinates shown in Figs. 6-1-F through 6-1-H. Scale 1:1.

ABSOLUTE COORDINATES (IN.)		
Point	X Axis	Y Axis
1	0	0
2	3.50	0
3	3.50	1.00
4	2.00	1.00
5	2.00	2.00
6	0	2.00
7	0	0
NEW START		
8	0	.50
9	3.50	.50
NEW START		
10	1.50	0
11	1.50	.50
NEW START		
12	0	2.50
13	1.50	2.50
14	1.50	3.00
15	3.50	3.00
16	3.50	4.50
17	0	4.50
18	0	2.50
NEW START		
19	0	3.50
20	3.50	3.50
NEW START		
21	2.00	3.50
22	2.00	4.50
NEW START		
23	4.00	0
24	6.00	0
25	6.00	2.00
26	5.00	2.00
27	5.00	.50
28	4.00	.50
29	4.00	0
NEW START		
30	5.00	1.00
31	6.00	1.00
NEW START		
32	4.50	0
33	4.50	.50

FIG. 6-1-F   Absolute coordinate (inches) assignment.

ABSOLUTE COORDINATES (IN.)		
Point	X Axis	Y Axis
1	0	0
2	2.50	0
3	2.50	.50
4	0	.50
NEW START		
5	2.00	.50
6	2.00	1.00
7	0	1.00
NEW START		
8	1.50	1.00
9	1.50	1.50
10	0	1.50
11	0	0
NEW START		
12	0	2.00
13	2.50	2.00
14	2.50	3.50
15	0	3.50
16	0	2.00
NEW START		
17	0	2.50
18	2.00	2.50
19	2.00	3.50
NEW START		
20	0	3.00
21	1.50	3.00
22	1.50	3.50
NEW START		
23	4.50	.50
24	3.00	.50
25	3.00	0
26	4.50	0
27	4.50	1.50
28	4.00	1.50
29	4.00	1.00
NEW START		
30	4.50	1.00
31	3.50	1.00
32	3.50	.50

FIG. 6-1-G   Absolute coordinate (inches) assignment.

ABSOLUTE COORDINATES (mm)		
Point	X Axis	Y Axis
1	0	0
2	90	0
3	90	10
4	0	10
NEW START		
5	0	0
6	0	40
7	50	40
8	50	10
NEW START		
9	70	0
10	70	10
NEW START		
11	0	50
12	70	50
13	70	70
14	90	70
15	90	90
16	0	90
17	0	50
NEW START		
18	0	80
19	50	80
20	50	90
NEW START		
21	140	10
22	100	10
23	100	0
24	140	0
25	140	40
26	130	40
27	130	10
NEW START		
28	120	0
29	120	10

FIG. 6-1-H   Absolute coordinate (metric) assignment.

3. Using squared graph paper of 4 or 5 squares to the inch (one square representing 1.00 in.) or 10 mm squares (one square representing 10 mm) or the grid on the CAD monitor, sketch or plot the views using the two-dimensional relative coordinates shown in Figs. 6-1-J through 6-1-L. Scale 1:1.

RELATIVE COORDINATES (IN.)		
Point	X Axis	Y Axis
1	0	0
2	3.00	0
3	0	.50
4	−2.00	0
5	0	1.50
6	−1.00	0
7	0	−2.00
NEW START		
8	0	1.50
9	1.00	0
NEW START		
10	1.50	0
11	0	.50
NEW START		
12	0	2.50
13	1.50	0
14	0	.50
15	1.50	0
16	0	1.50
17	−3.00	0
18	0	−2.00
NEW START		
19	1.00	2.50
20	0	2.00
NEW START		
21	0	3.50
22	1.00	0
NEW START		
23	3.50	0
24	2.00	0
25	0	2.00
26	−1.00	0
27	0	−.50
28	−1.00	0
29	0	−1.50
NEW START		
30	3.50	.50
31	2.00	0
NEW START		
32	4.00	0
33	0	.50

FIG. 6-1-J   Relative coordinate (inches) assignment.

RELATIVE COORDINATES (IN.)		
Point	X Axis	Y Axis
1	0	0
2	3.00	0
3	0	1.00
4	−1.00	0
5	0	.50
6	−2.00	0
7	0	−1.50
NEW START		
8	1.50	0
9	0	.50
10	−1.00	0
11	0	1.00
NEW START		
12	0	2.00
13	0	1.50
14	3.00	0
15	0	−.50
16	−2.50	0
17	0	−1.00
NEW START		
18	0	2.00
19	1.50	0
20	0	1.00
NEW START		
21	2.00	3.00
22	0	.50
NEW START		
23	3.50	0
24	1.50	0
25	0	1.50
26	−1.50	0
27	0	−1.50
NEW START		
28	4.50	0
29	0	1.50
NEW START		
30	3.50	.50
31	1.00	0
NEW START		
32	4.50	1.00
33	.50	0

FIG. 6-1-K   Relative coordinate (inches) assignment.

RELATIVE COORDINATES (mm)		
Point	X Axis	Y Axis
1	0	0
2	70	0
3	0	20
4	−70	0
NEW START		
5	0	0
6	0	30
7	30	0
8	0	−10
NEW START		
9	20	30
10	0	−10
NEW START		
11	40	20
12	0	−20
NEW START		
13	0	40
14	40	0
15	0	20
16	30	0
17	0	20
18	−70	0
19	0	−40
NEW START		
20	0	60
21	20	0
22	0	10
23	10	0
24	0	10
NEW START		
25	120	20
26	−40	0
27	0	−20
28	40	0
29	0	30
30	−20	0
31	0	−30
NEW START		
32	110	20
33	0	10

FIG. 6-1-L   Relative coordinate (metric) assignment.

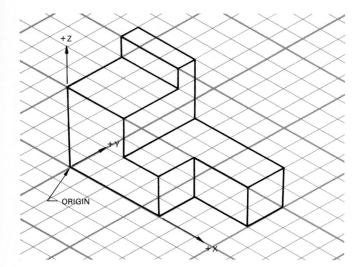

FIG. 6-1-M Stand.

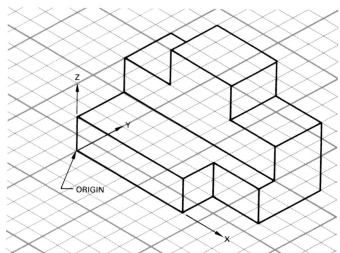

FIG. 6-1-P Slide bracket.

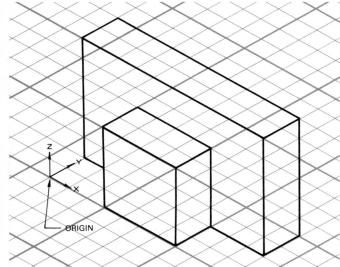

FIG. 6-1-N Spacer.

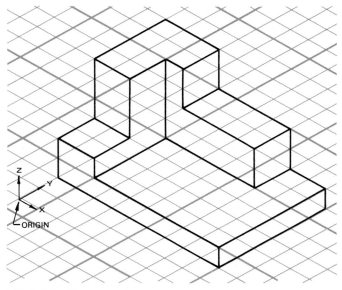

FIG. 6-1-R Corner stand.

4. Using isometric graph paper sketch (copy) any three of the parts shown in Figs. 6-1-M through 6-1-S. Each square on the grid should represent .50 in. or 10 mm. After completing the views, add the X, Y, and Z coordinates where the lines intersect one another. Identify only those intersections that can be seen. Note the location of the origin for each part.

ASSIGNMENTS FOR UNIT 6-2, ARRANGEMENT AND CONSTRUCTION OF VIEWS

5. Make a three-view sketch similar to Figs. 6-2-1B and C and establish the distance between plane 1 and the left border line and between plane 2 and the bottom border line, given the following: scale 1:1; drawing space

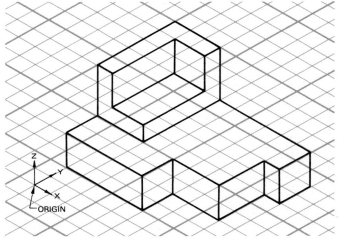

FIG. 6-1-S Guide bracket.

8.00 × 10.50 in.; part size: W = 4.10, H = 1.40, D = 2.10; space between views to be 1.00 in.

6. Make a three-view sketch similar to Figs. 6-2-1B and C and establish the distance between plane 1 and the left border line and between plane 2 and the bottom border line, given the following: scale 1:2; drawing space 8.00 × 10.50 in.: part size: W = 8.50, H = 4.90, D = 4.50; space between views to be 1.00 in.

7. Angle bracket, Fig. 6-1-B, sheet size A (A4), scale 1:1. Make a three-view drawing using a miter line to complete the right side view. Space between views to be 1.00 in.

8. Locating block, Fig. 6-1-E, sheet size A (A4), scale 1:1. Make a three-view drawing using a miter line to complete the top view. Space between views to be 1.00 in.

ASSIGNMENTS FOR UNIT 6-3, ALL SURFACES PARALLEL AND ALL EDGES AND LINES VISIBLE

9. On preprinted grid paper (.25 in. or 10 mm grid) sketch three views of each of the objects shown in Fig. 6-3-A. Each square shown on the objects represents one square on the grid paper. Allow one grid space between views and a minimum of two grid spaces between objects. Identify the type of projection used by placing the identifying symbol at the bottom of the drawing.

10. Draw three views of one of the parts shown in Figs. 6-3-B through 6-3-E (here and on pg. 116). Allow 1.00 in. or 25 mm between views. Scale full or 1:1. Do not dimension.

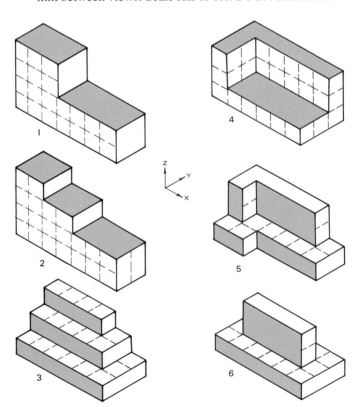

FIG. 6-3-A   Sketching assignment.

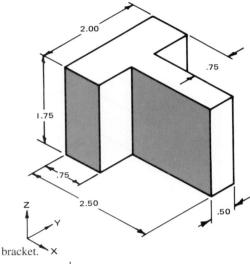

FIG. 6-3-B   T bracket.

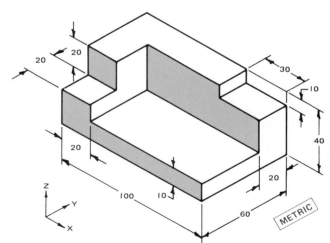

FIG. 6-3-C   Step support.

FIG. 6-3-D   Corner block.

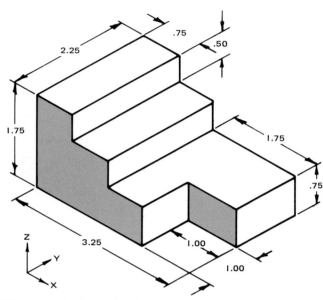

FIG. 6-3-E   Angle step bracket.

ASSIGNMENTS FOR UNIT 6-4, HIDDEN SURFACES AND EDGES

11. On preprinted grid paper (.25 in. or 10 mm grid) sketch three views of each of the objects shown in Figs. 6-4-A and 6-4-B. Each square shown on the objects represents one square on the grid paper. Allow one grid space between views and a minimum of two spaces between objects. Identify the type of projection by placing the identifying symbol at the bottom of the drawing.
12. Sketch the views needed for a multiview drawing of the parts shown in Fig. 6-4-C. Choose your own sizes and estimate proportions.
13. Match the pictorial drawings to the orthographic drawings shown in Fig. 6-4-D.

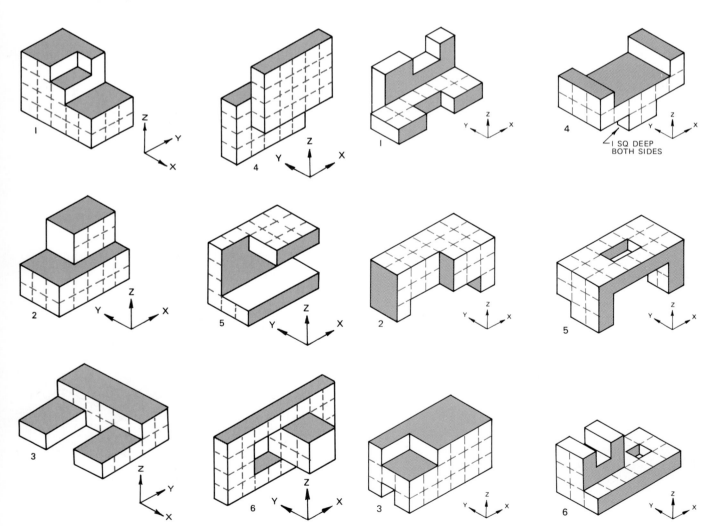

FIG. 6-4-A   Sketching assignment.

FIG. 6-4-B   Sketching assignment.

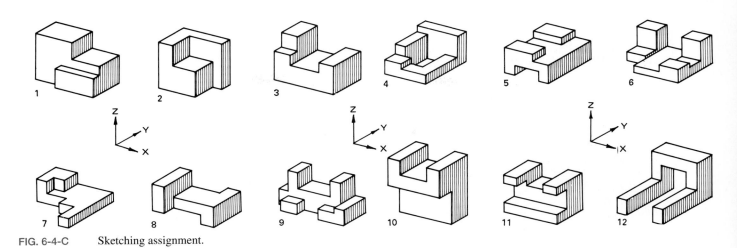

FIG. 6-4-C   Sketching assignment.

FIG. 6-4-D   Matching test.

14. Make a three-view drawing of one of the parts shown in Figs. 6-4-E through 6-4-K. Allow 1.00 in. or 25 mm between views. Do not dimension.

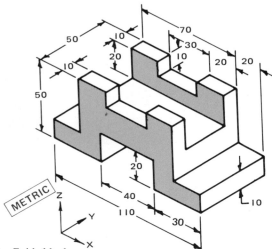

FIG. 6-4-E Guide block.

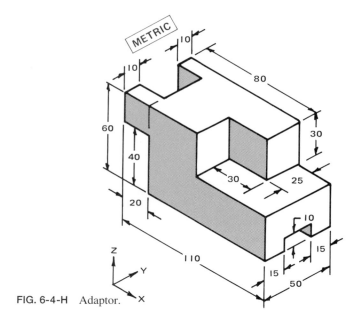

FIG. 6-4-H Adaptor.

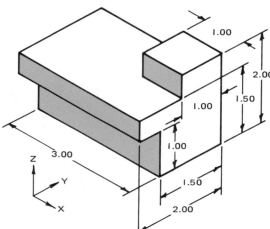

FIG. 6-4-F Bracket.

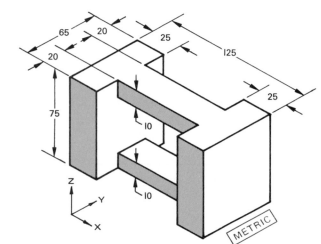

FIG. 6-4-J Bracket.

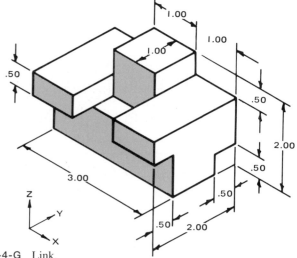

FIG. 6-4-G Link.

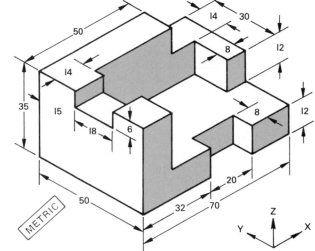

FIG. 6-4-K Adjusting guide.

15. Make a three-view drawing of one of the parts shown in Fig. 6-4-L through 6-4-S. Allow 1.00 in. or 25 mm between views. Do not dimension.

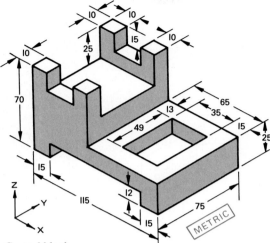

FIG. 6-4-L   Control block.

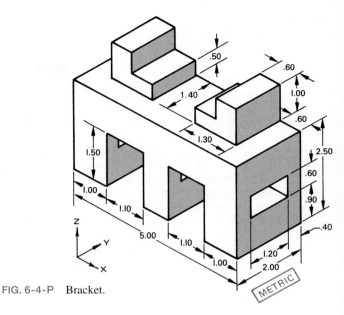

FIG. 6-4-P   Bracket.

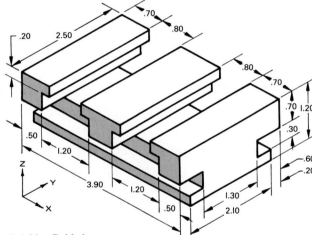

FIG. 6-4-M   Guide bar.

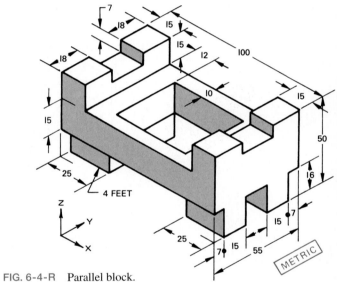

FIG. 6-4-R   Parallel block.

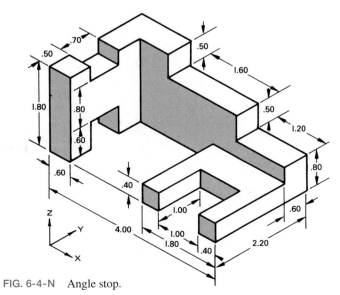

FIG. 6-4-N   Angle stop.

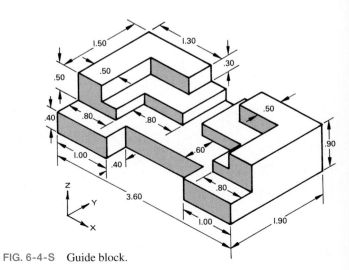

FIG. 6-4-S   Guide block.

ASSIGNMENTS FOR UNIT 6-5, INCLINED SURFACES

16. On preprinted grid paper (.25 in or 10 mm grid using 5 mm squares) sketch three views of each of the objects shown in Figs. 6-5-A or 6-5-B. Each square shown on the objects represents one square on the grid paper. Allow one grid space between views and a minimum of two grid spaces between objects. The sloped (inclined) surfaces on each of the three objects are identified by a letter. Identify the sloped surfaces on each of the three views with a corresponding letter. Also identify the type of projection used by placing the appropriate identifying symbol at the bottom of the drawing.

17. Sketch the views needed for a multiview drawing of the parts shown in Fig. 6-5-C. Choose your own sizes and estimate proportions.

18. Make three-view sketches of the parts shown in Figs. 6-5-D through 6-5-G. Follow the same instructions shown for Assignment 16.

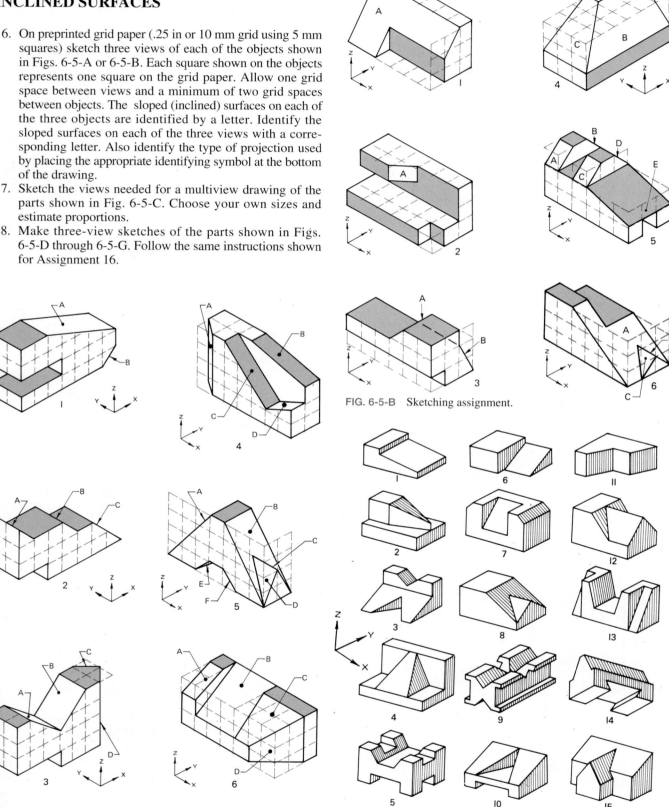

FIG. 6-5-A   Sketching assignment.

FIG. 6-5-B   Sketching assignment.

FIG. 6-5-C   Sketching assignment.

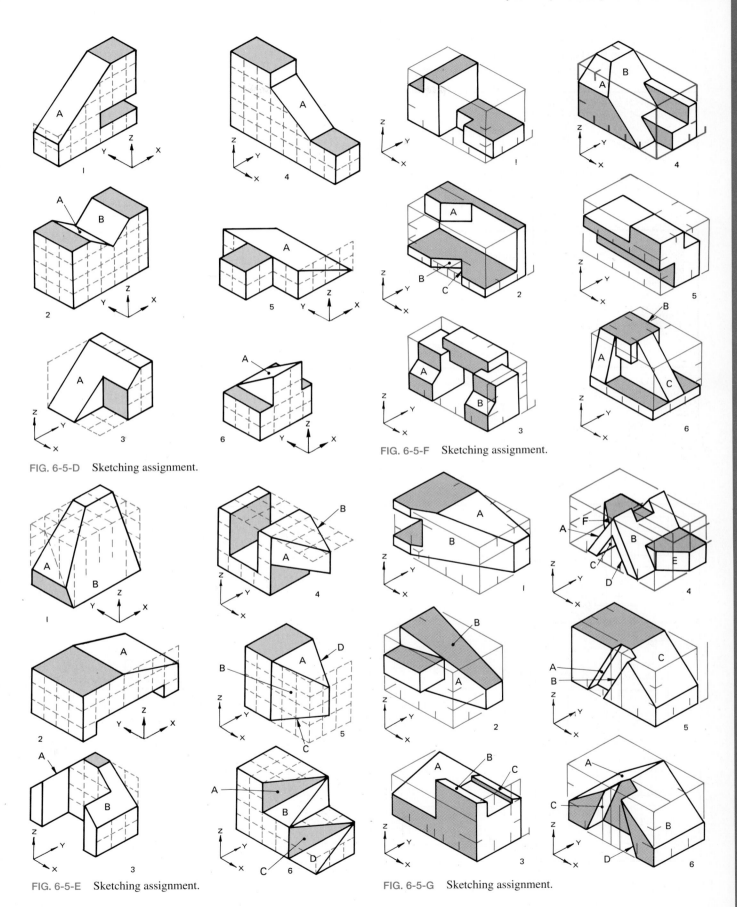

FIG. 6-5-D   Sketching assignment.

FIG. 6-5-E   Sketching assignment.

FIG. 6-5-F   Sketching assignment.

FIG. 6-5-G   Sketching assignment.

19. Match the pictorial drawings to the orthographic drawings shown in Fig. 6-5-H.

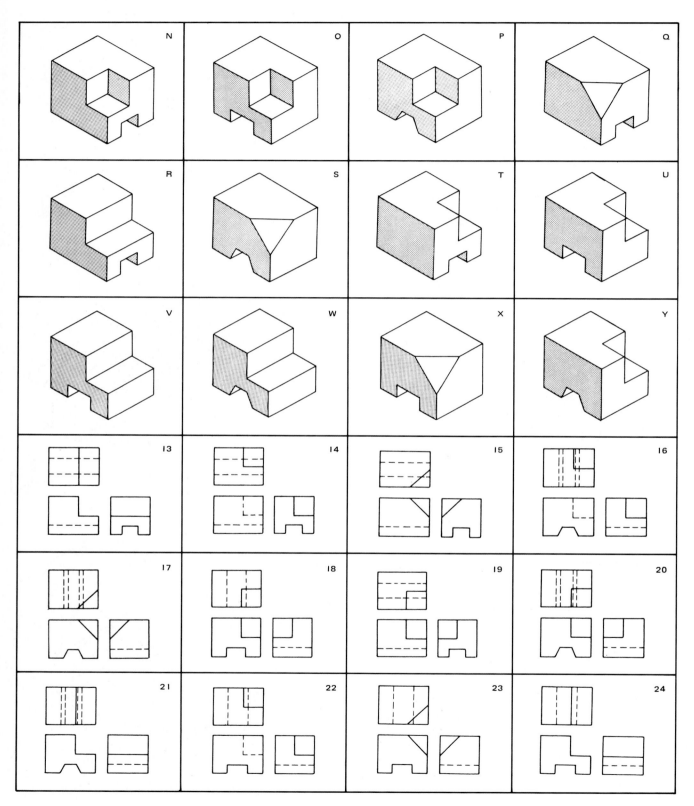

FIG. 6-5-H Matching test.

20. Make a three-view drawing of one of the parts shown in Figs. 6-5-J through 6-5-P. Allow 1.00 in. or 25 mm between views. Do not dimension.

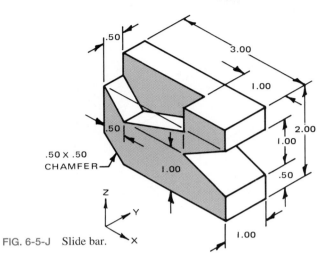

FIG. 6-5-J   Slide bar.

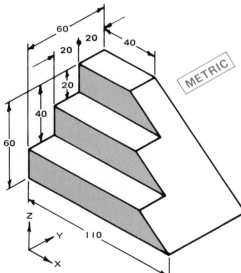

FIG. 6-5-K   Adjusting guide.

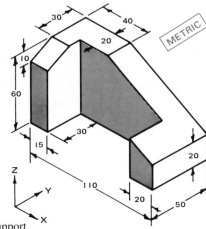

FIG. 6-5-L   Flanged support.

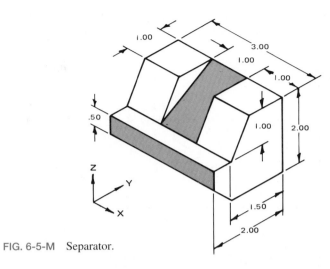

FIG. 6-5-M   Separator.

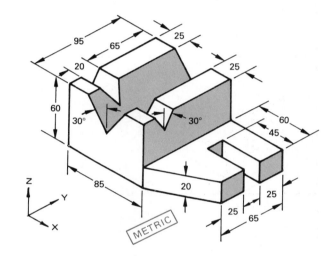

FIG. 6-5-N   Guide block.

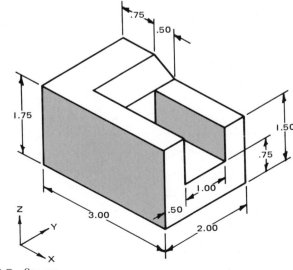

FIG. 6-5-P   Spacer.

21. Make a three-view drawing of one of the parts shown in Figs. 6-5-R through 6-5-W. Allow 1.00 in. or 25 mm between views.

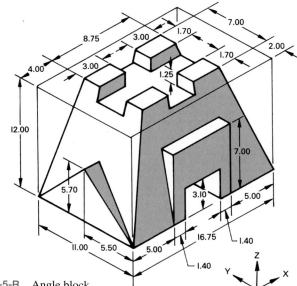

FIG. 6-5-R   Angle block.

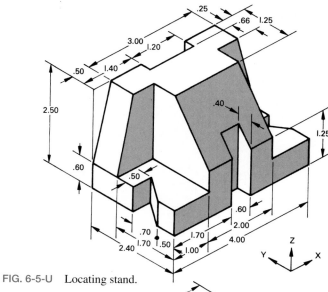

FIG. 6-5-U   Locating stand.

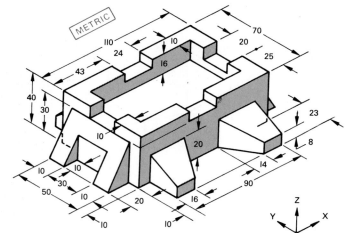

FIG. 6-5-S   Base plate.

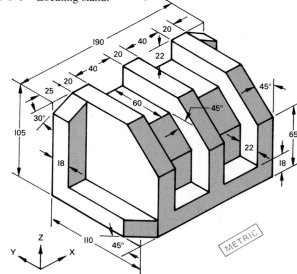

FIG. 6-5-V   Base.

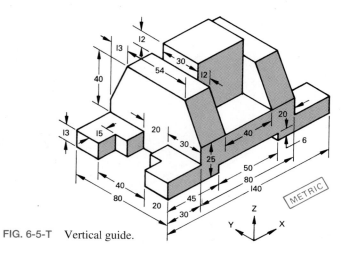

FIG. 6-5-T   Vertical guide.

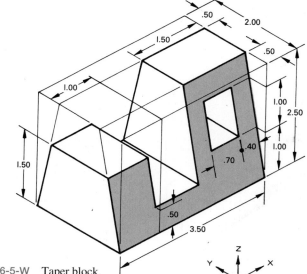

FIG. 6-5-W   Taper block.

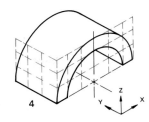

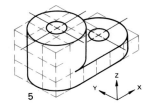

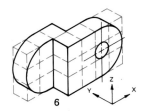

ASSIGNMENTS FOR UNIT 6-6, CIRCULAR FEATURES

22. On preprinted grid paper (.25 in. or 10 mm grid) sketch three views of each of the objects shown in Fig. 6-6-A or 6-6-B. Each square shown on the objects represents one square on the grid paper. Allow one grid space between views and a minimum of two grid spaces between objects. Identify the type of projection used by placing the appropriate identifying symbol at the bottom of the drawing.

23. Sketch the views needed for a multiview drawing of the parts shown in Fig. 6-6-C. Choose your own sizes and estimate proportions.

FIG. 6-6-A   Sketching assignment.

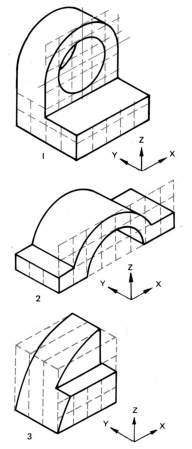

FIG. 6-6-B   Sketching assignment.

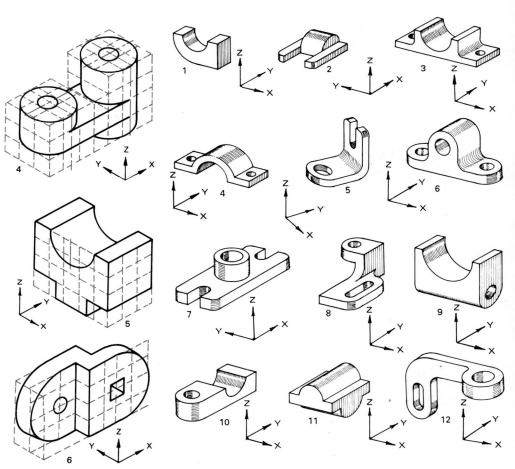

FIG. 6-6-C   Sketching assignment.

24. Sketch the views needed for a multiview drawing of the parts shown in Fig. 6-6-D or 6-6-E. Choose your own sizes and estimate proportions.

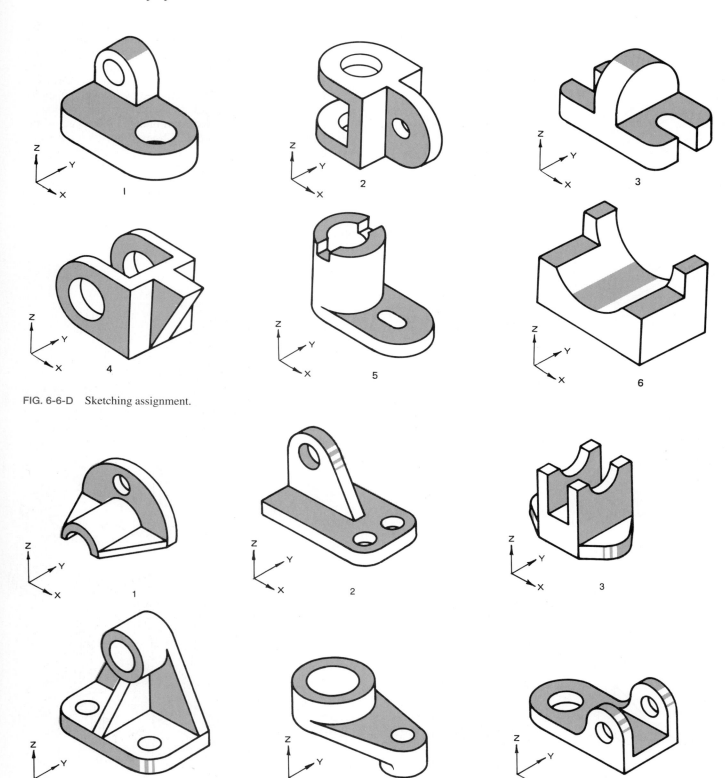

FIG. 6-6-D   Sketching assignment.

FIG. 6-6-E   Sketching assignment.

25. Make a three-view drawing of one of the parts shown in Figs. 6-6-F through 6-6-L. Allow 1.00 in. or 25 mm between views. Scale 1:1. Do not dimension.

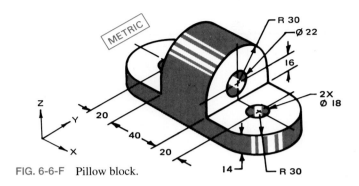

FIG. 6-6-F   Pillow block.

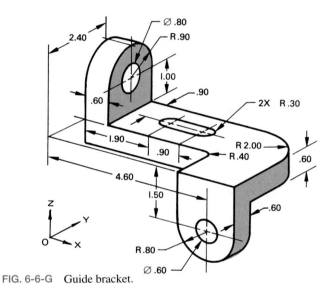

FIG. 6-6-G   Guide bracket.

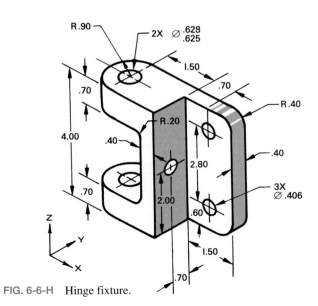

FIG. 6-6-H   Hinge fixture.

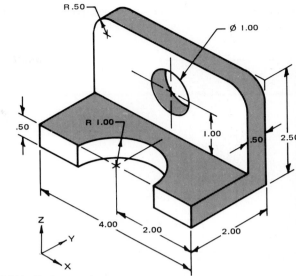

FIG. 6-6-J   Rod support.

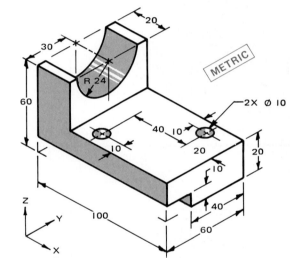

FIG. 6-6-K   Cradle support.

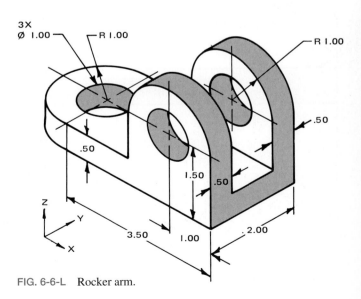

FIG. 6-6-L   Rocker arm.

26. Sketch the missing views for the parts shown in Fig. 6-6-M.

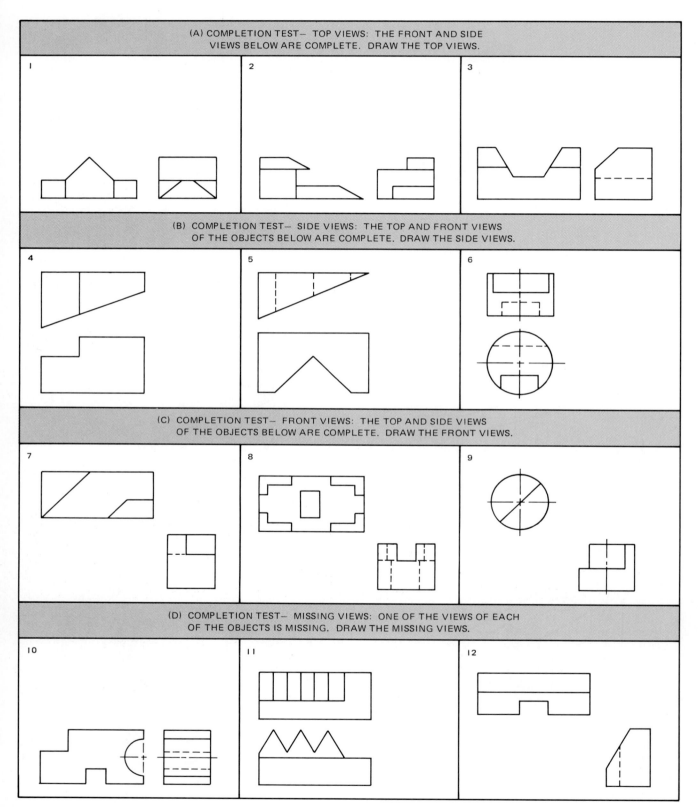

FIG. 6-6-M   Completion tests.

ASSIGNMENTS FOR UNIT 6-7, OBLIQUE SURFACES

27. On preprinted grid paper (.25 in. or 10 mm grid) sketch three views of each of the objects shown in Fig. 6-7-A or 6-7-B. Each square on the objects represents one square on the grid paper. Allow one grid space between views and a minimum of two grid spaces between objects. The oblique surfaces on the objects are identified by a letter. Identify the oblique surfaces on each of the three views with a corresponding letter. Also identify the type of projection used by placing the appropriate identifying symbol on the drawing.

28. Make a three-view drawing of one of the parts shown in Figs. 6-7-C and 6-7-D. Allow 1.20 in. (30 mm) between views. Do not dimension. The oblique surfaces on the objects are identified by a letter. Identify the oblique surfaces on each of the three views with a corresponding letter.

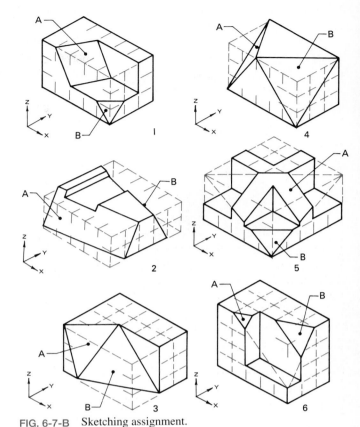

FIG. 6-7-B   Sketching assignment.

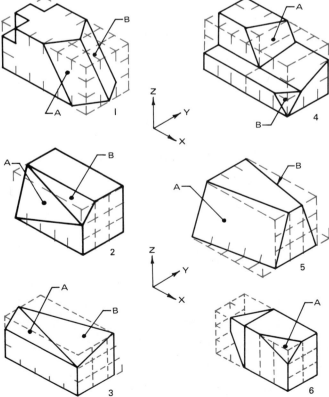

FIG. 6-7-A   Sketching assignment.

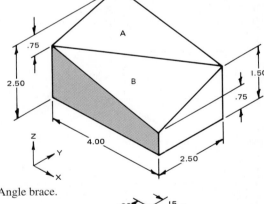

FIG. 6-7-C   Angle brace.

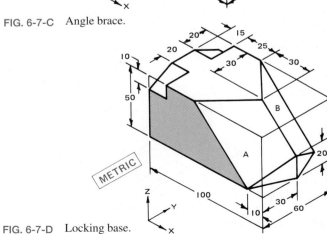

FIG. 6-7-D   Locking base.

29. Make a three-view drawing of one of the parts shown in Figs. 6-7-E through 6-7-G. Allow 1.20 in. (30 mm) between views. Do not dimension. The oblique surfaces on each part are identified by a letter. Identify the oblique surfaces on each of the three views with a corresponding letter.

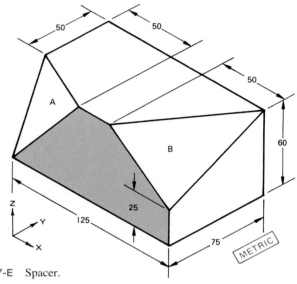

FIG. 6-7-E   Spacer.

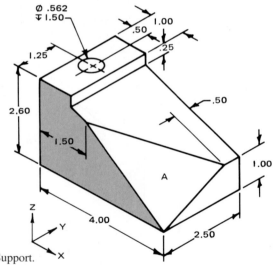

FIG. 6-7-F   Support.

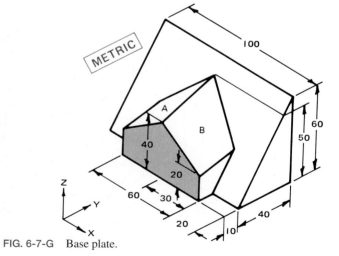

FIG. 6-7-G   Base plate.

ASSIGNMENT FOR UNIT 6-8, ONE- AND TWO-VIEW DRAWINGS

30. Select any six of the objects shown in Fig. 6-8-A and draw only the necessary views that will completely describe

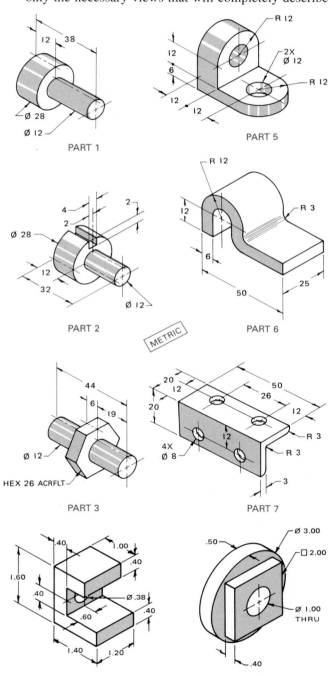

FIG. 6-8-A   Drawing assignment.

130

each part. Use symbols or abbreviations where possible. The drawings need not be to scale but should be drawn in proportion to the illustrations shown.

ASSIGNMENTS FOR UNIT 6-9, SPECIAL VIEWS

31. Select one of the objects shown in Figs. 6-9-A through 6-9-D and draw only the necessary views (full and partial) that will completely describe each part. Add dimensions and machining symbols where required. Scale 1:1.

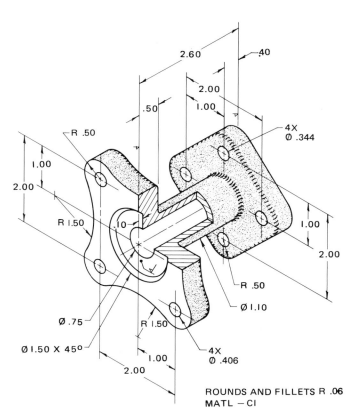

FIG. 6-9-A   Round flange.

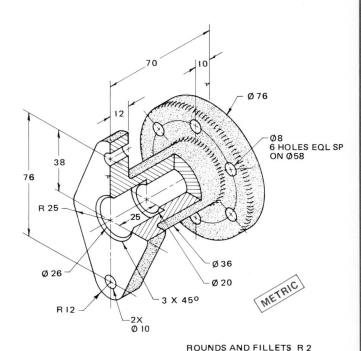

FIG. 6-9-C   Flanged coupling.

ROUNDS AND FILLETS R 2
MATL — CI

FIG. 6-9-B   Flanged adaptor.

ROUNDS AND FILLETS R .06
MATL — CI

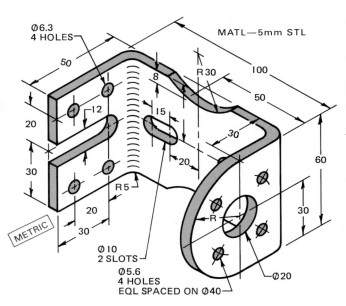

FIG. 6-9-D   Connector.

32. Select one of the panels shown in Figs. 6-9-E and 6-9-F and make a detail drawing of the part. Enlarged views are recommended. Panels such as these, where labeling is used to identify the terminals, are used extensively in the electrical and electronics industry.

33. With most truss drawings, the scale used on the overall assembly is such that intricate detail cannot be clearly shown. As a result, enlarged detail views are added. With this type of assembly, many parts are opposite-hand to their counterparts.

Bolting Data: All members to be bolted together with five .375 high-strength bolts. Spacing is 1.50 in. from end and 3.00 in. center to center.

On a B size sheet, draw the enlarged views of the gusset assemblies shown in Fig. 6-9-G. Scale 1:12.

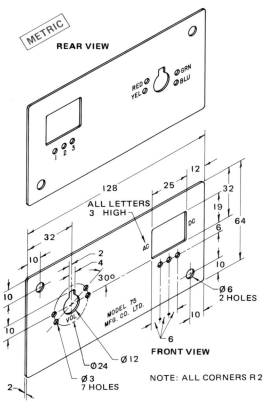

FIG. 6-9-E   Radio cover plate.

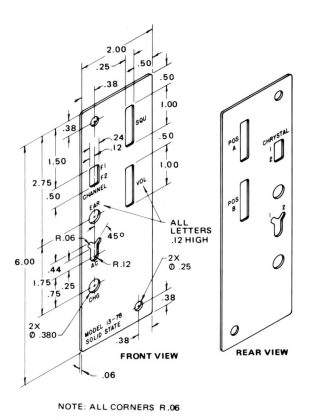

FIG. 6-9-F   Transceiver cover plate.

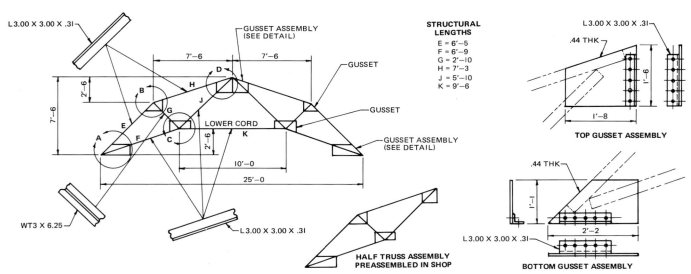

FIG. 6-9-G   Crescent truss.

ASSIGNMENT FOR UNIT 6-10, CONVENTIONAL REPRESENTATION OF COMMON FEATURES

34. Make a working drawing of one of the parts shown in Fig. 6-10-A or 6-10-B. Wherever possible, simplify the

drawing by using conventional representation of features and symbolic dimensioning (including symmetry). Scale 10:1.

ASSIGNMENT FOR UNIT 6-11, CONVENTIONAL BREAKS

35. Make a working drawing of one of the parts shown in Fig. 6-11-A or 6-11-B. Use conventional breaks to shorten the length of the part. An enlarged view is also recommended where the detail cannot be clearly shown at full scale. Apply the symmetry symbol and use symbolic dimensioning wherever possible. Scale 1:1.

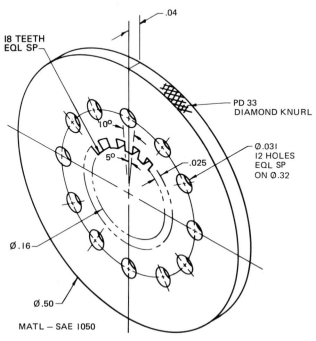

FIG. 6-10-A Adjustable locking plate.

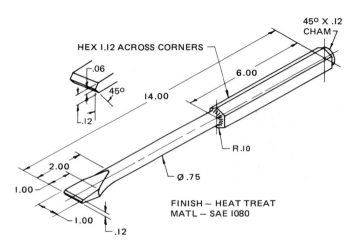

FIG. 6-11-A Hand chisel.

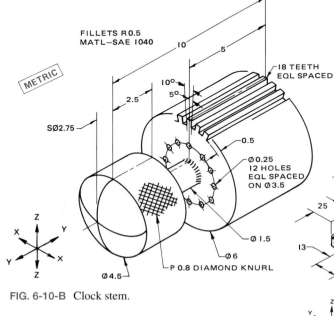

FIG. 6-10-B Clock stem.

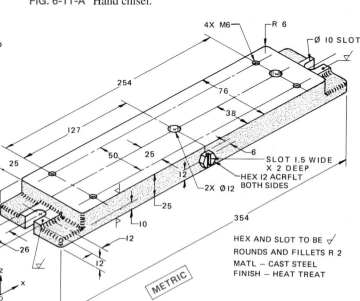

FIG. 6-11-B Fixture base.

ASSIGNMENT FOR UNIT 6-12, MATERIALS OF CONSTRUCTION

36. Make a detailed assembly drawing of one of the assemblies shown in Fig. 6-12-A or 6-12-B. Enlarged details are recommended for the steel mesh and joints. Use conventional breaks to shorten the length. Scale 1:5.

ASSIGNMENT FOR UNIT 6-13, CYLINDRICAL INTERSECTIONS

37. Make a working drawing of one of the parts shown in Fig. 6-13-A or 6-13-B. For 6-13-A a bushing is to be pressed (H7/s6) into the large hole and the stepped smaller hole is to have a running fit (H8/f7) with its respective shaft. These sizes are to be given as limit dimensions. All other finished surfaces are to have a 3.2 μm (micrometers) finish. For Fig. 6-13-B an LN3 fit is required for the two large holes. Finished surfaces are to have a 63 μin. (microinches) finish with a .06 in. material-removal allowance. Use your judgment in selecting the number of views required and deciding whether some form of sectional view would be desirable to improve the readability of the drawing. Scale 1:1.

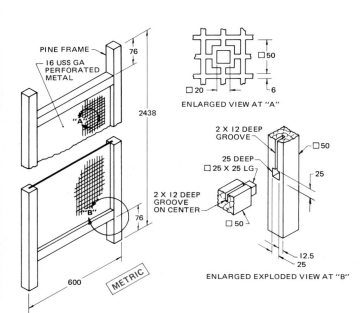

FIG. 6-12-A   Room divider.

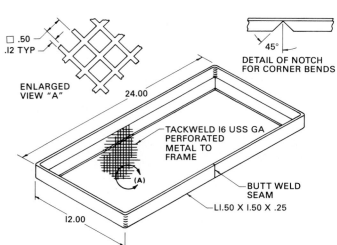

FIG. 6-12-B   Barbecue grill.

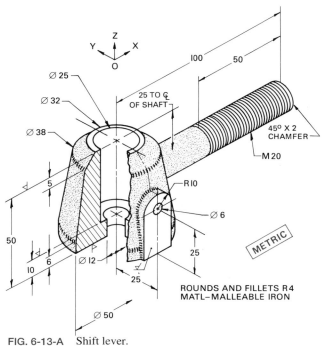

FIG. 6-13-A   Shift lever.

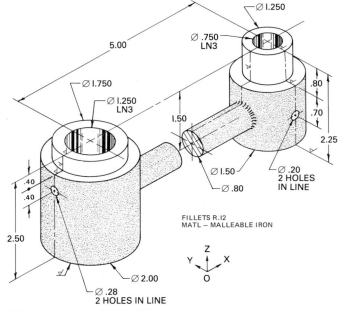

FIG. 6-13-B   Steering knuckle.

ASSIGNMENT FOR UNIT 6-14, FORESHORTENED PROJECTION

38. Make a working drawing of one of the parts shown in Fig. 6-14-A or 6-14-B. All surface finishes are to be 1.6 μm or 63 μin. Keyed holes will have H9/d9 or RC6 fits with shafts. Where required, rotate the features to show their true distances from the centers and edges. To show the true shape of the ribs or arms, a revolved section is recommended. Dimension the keyseat as per Chap. 11 and the Appendix. Scale 1:1.

ASSIGNMENT FOR UNIT 6-15, INTERSECTIONS OF UNFINISHED SURFACES

39. Make a three-view detail drawing of one of the problems shown in Fig. 6-15-A or 6-15-B. Scale 1:1. Surface finish requirements are essential for all parts. Use symbolic dimensioning wherever possible. For Fig. 6-15-A the T slot surfaces should have a maximum roughness of 0.8 μm and a maximum waviness of 0.05 mm for a 25 mm length. The back surface should have a maximum roughness of 3.2 μm with no restrictions on waviness. For Fig. 6-15-B the back surface and notch should have an equivalent control as the T slot in Fig. 6-15-A. The faces on the boss should have a maximum roughness of 125 μin. with no restrictions on waviness.

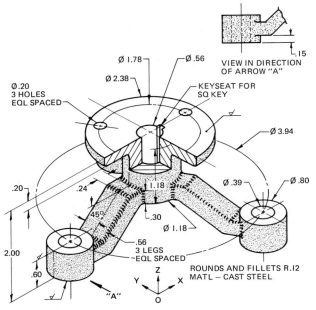

FIG. 6-14-A   Clutch.

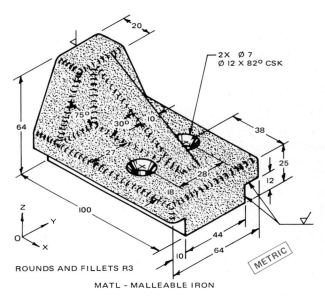

FIG. 6-15-A   Cutoff stop.

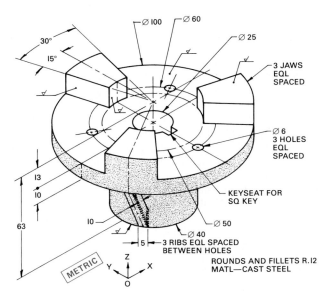

FIG. 6-14-B   Mounting bracket.

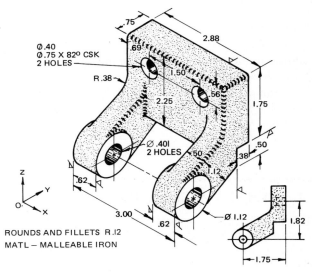

FIG. 6-15-B   Sparker bracket.

AUXILIARY VIEWS AND REVOLUTIONS

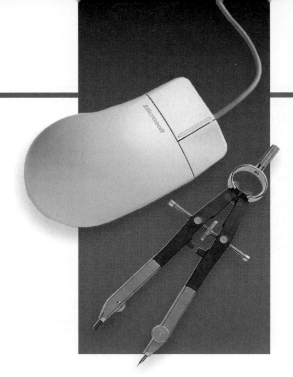

Definitions

Descriptive geometry   The use of graphic representations to solve mathematical problems.

Inclined line   In descriptive geometry, a line which appears as a true-length line in one view but is foreshortened in the other two views.

Inclined plane   In descriptive geometry, a plane which appears distorted in two views and as a line in the third view.

Multi-auxiliary view   Several additional views used to depict surfaces that must be shown clearly and without distortion.

Normal line   In descriptive geometry, a line which appears as a point in one view and as a true-length line in the other two views.

Normal plane   In descriptive geometry, a plane which appears in its true shape in one view and as a line in the other two views.

Oblique line   In descriptive geometry, a line which appears inclined in all views.

Oblique plane   In descriptive geometry, a plane which appears distorted in all views.

Revolutions   A method of representing an object through a series of views that have been "revolved" or turned on an imaginary axis.

7-1 PRIMARY AUXILIARY VIEWS

Many machine parts have surfaces that are not perpendicular, or at right angles, to the plane of projection. These are referred to as *sloping* or *inclined surfaces*. In the regular orthographic views, such surfaces appear to be distorted and their true shape is not shown. When an inclined surface has important characteristics that should be shown clearly and without distortion, an auxiliary (additional or helper) view is used so that the drawing completely and clearly explains the shape of the object. In many cases, the auxiliary view will replace one of the regular views on the drawing, as illustrated in Fig. 7-1-1.

One of the regular orthographic views will have a line representing the edge of the inclined surface. The auxiliary view is projected from this edge line, at right angles, and is drawn parallel to the edge line.

Only the true-shape features on the views need be drawn, as shown in Fig. 7-1-2. Since the auxiliary view shows only the true shape and detail of the inclined surface or features, a partial auxiliary view is all that is necessary. Likewise, the distorted features on the regular views may be omitted. Hidden lines are usually omitted unless required for clarity. This procedure is recommended for functional and production drafting where drafting costs are an important consideration. However, the drafter may be called upon to draw the complete views of the part. This type of drawing is often used for catalog and standard parts drawing.

Additional examples of auxiliary view drawings are shown in Fig. 7-1-3 (pg. 138).

Figure 7-1-4 (pg. 138) shows how to make an auxiliary view of a symmetrical object. Figure 7-1-4A shows the object in a pictorial view. In this illustration the center plane is used as the reference plane. In Fig. 7-1-4B the center plane is drawn parallel to the inclined surface shown in the front view. The edge view of this plane appears as a center line, line XY, on the top view. Number the points of intersection between the inclined surface and the vertical lines on the top view. Then transfer these numbers to the edge view of the inclined surface on the front view, as shown. Parallel to this edge view and at a convenient distance from it, draw the line $X'Y'$, as in Fig. 7-1-4C. Now, in the top view, find the distances D_1 and D_2 from the numbered points to the center line. These are the depth measurements. Transfer them onto the corresponding construction lines that you have just drawn, measuring them off on either

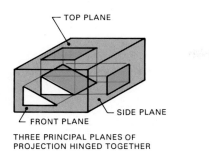

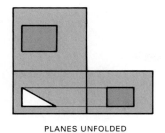

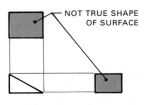

TOP PLANE

FRONT PLANE

SIDE PLANE

THREE PRINCIPAL PLANES OF
PROJECTION HINGED TOGETHER

PLANES UNFOLDED

NOT TRUE SHAPE
OF SURFACE

PLANES REMOVED SHOWING THREE
REGULAR (TOP, FRONT, SIDE) VIEWS

NOTE: IN NONE OF THESE VIEWS DOES THE SLANTED (COLORED) SURFACE APPEAR IN ITS TRUE SHAPE.

(A) WEDGED BLOCK SHOWN IN THREE REGULAR VIEWS

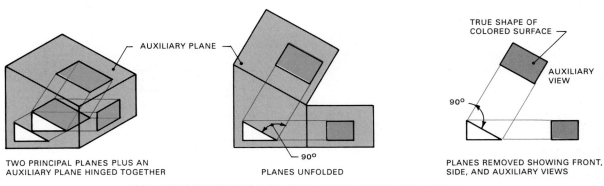

AUXILIARY PLANE

TRUE SHAPE OF
COLORED SURFACE

AUXILIARY
VIEW

90°

TWO PRINCIPAL PLANES PLUS AN
AUXILIARY PLANE HINGED TOGETHER

90°

PLANES UNFOLDED

PLANES REMOVED SHOWING FRONT,
SIDE, AND AUXILIARY VIEWS

NOTE: IN THIS EXAMPLE THE AUXILIARY PLANE REPLACED THE TOP PLANE IN
ORDER THAT THE SLANTED (COLORED) SURFACE MAY BE SHOWN IN ITS TRUE SHAPE.

(B) REPLACING THE TOP PLANE WITH AN AUXILIARY PLANE

FIG. 7-1-1   Relationship of the auxiliary plane to the three principal planes.

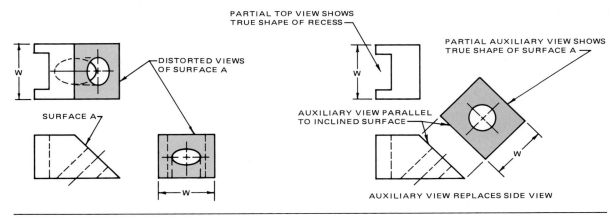

PARTIAL TOP VIEW SHOWS
TRUE SHAPE OF RECESS

DISTORTED VIEWS
OF SURFACE A

SURFACE A

W

W

PARTIAL AUXILIARY VIEW SHOWS
TRUE SHAPE OF SURFACE A

AUXILIARY VIEW PARALLEL
TO INCLINED SURFACE

W

AUXILIARY VIEW REPLACES SIDE VIEW

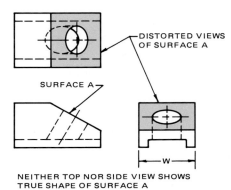

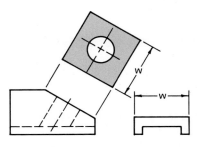

DISTORTED VIEWS
OF SURFACE A

SURFACE A

W

NEITHER TOP NOR SIDE VIEW SHOWS
TRUE SHAPE OF SURFACE A

PARTIAL VIEWS SHOWING ONLY THE
NECESSARY DETAILS ARE RECOMMENDED

W

W

AUXILIARY VIEW REPLACES TOP VIEW

FIG. 7-1-2   Auxiliary views replacing regular views.

NOTE: ONLY CONVENTIONAL BREAK ON PROJECTED
SURFACE NEED BE SHOWN ON PARTIAL VIEWS.

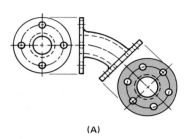

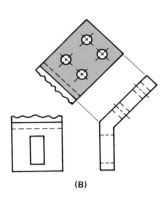

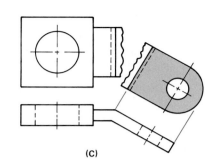

(A)   (B)   (C)

FIG. 7-1-3   Examples of auxiliary view drawings.

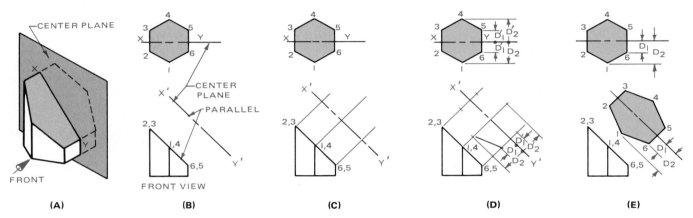

(A)   (B)   (C)   (D)   (E)

FIG. 7-1-4   Drawing an auxiliary view using the center plane reference.

side of line *X'Y'*, as shown in Fig. 7-1-4D. The result will be a set of points on the construction lines. Connect and number these points, as shown in Fig. 7-1-4E, and the front auxiliary view of the inclined surface results. The remaining portions of the object may also be projected from the center reference plane.

Dimensioning Auxiliary Views

One of the basic rules of dimensioning is to dimension the feature where it can be seen in its true shape and size. Thus the auxiliary view will show only the dimensions pertaining to those features for which the auxiliary view was drawn. The recommended dimensioning method for engineering drawings is the *unidirectional system* (Fig. 7-1-5).

ASSIGNMENT

See Assignment 1 for Unit 7-1 on pages 158–159.

7-2   CIRCULAR FEATURES IN AUXILIARY PROJECTION

As mentioned in Unit 7-1, at times it is necessary to show the complete views of an object. If circular features are involved

in auxiliary projection, the surfaces appear elliptical, not circular, in one of the views.

The method most commonly used to draw the true-shape projection of the curved surface is the plotting of a series of points on the line, the number of points being governed by the accuracy of the curved line required.

Figure 7-2-1 illustrates an auxiliary view of a truncated cylinder. The shape seen in the auxiliary view is an ellipse.

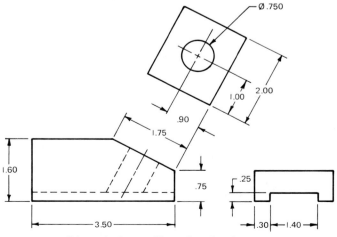

FIG. 7-1-5   Dimensioning auxiliary view drawings.

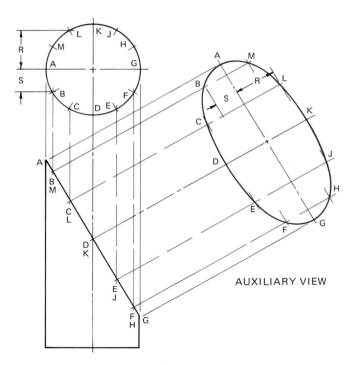

FIG. 7-2-1   Establishing true shape of truncated cylinder.

This shape is drawn by plotting lines of intersection. The perimeter of the circle in the top view is divided to give a number of equally spaced points—in this case, 12 points, *A* to *M*, spaced 30° apart (360°/12 = 30°). These points are projected down to the edge line on the front view, then at right angles to the edge line to the area where the auxiliary view will be drawn. A center line for the auxiliary view is drawn parallel

to the edge line, and width settings taken from the top view are transferred to the auxiliary view. Note width setting *R* for point *L*. Because the illustration shows a true cylinder and the point divisions in the top view are all equal, the width setting *R* taken at *L* is also the correct width setting for *C*, *E*, and *J*. Width setting *S* for *B* is also the correct width setting for *F*, *H*, and *M*. When all the width settings have been transferred to the auxiliary view, the resulting points of intersection are connected by means of an irregular curve to give the desired elliptical shape.

It is often necessary to construct the auxiliary view first in order to complete the regular views, shown in Fig. 7-2-2.

ASSIGNMENT

See Assignment 2 for Unit 7-2 on pages 159–160.

7-3 MULTI-AUXILIARY-VIEW DRAWINGS

Some objects have more than one surface not perpendicular to the plane of projection. In working drawings of these objects, an auxiliary view may be required for each surface. Naturally, this would depend upon the amount and type of detail lying on these surfaces. This type of drawing is often referred to as the *multi-auxiliary-view* drawing (Fig. 7-3-1, pg. 140).

One can readily see the advantage of using the unidirectional system of dimensioning for dimensioning an object such as the one shown in Fig. 7-3-2 (pg. 140).

ASSIGNMENTS

See Assignments 3 and 4 for Unit 7-3 on pages 161–162.

FIG. 7-2-2   Constructing the true shape of a curved surface by plotting method.

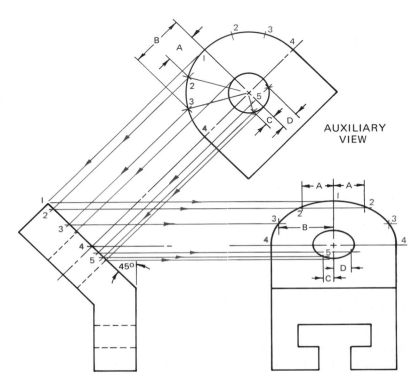

FIG. 7-3-1   Auxiliary views added to regular views to show true shape of features.

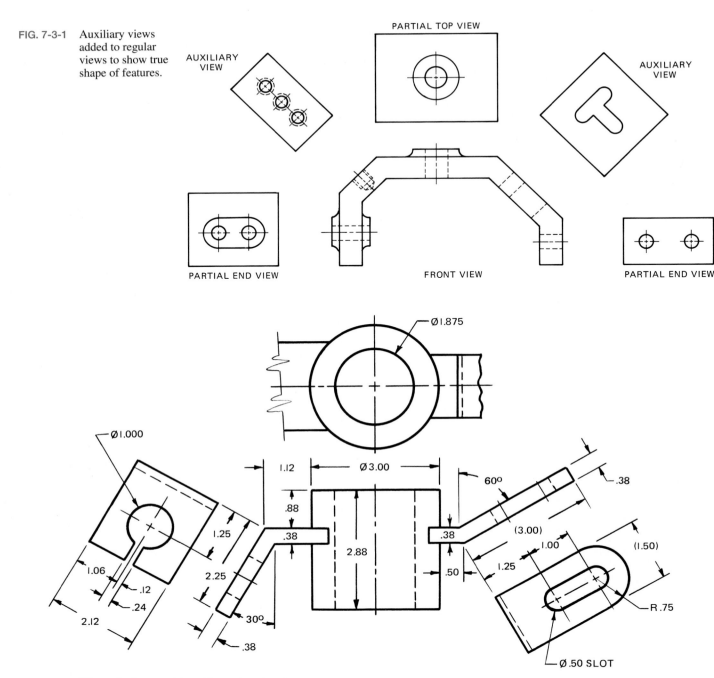

FIG. 7-3-2   Dimensioning a multi-auxiliary-view drawing.

7-4 SECONDARY AUXILIARY VIEWS

Some objects, because of their shape, require a secondary auxiliary view to show the true shape of the surface or feature. The surface or feature is usually oblique (inclined) to the principal planes of projection. In order to draw a secondary auxiliary view, such as the one shown in Fig. 7-4-1, only portions of the front and top views are drawn first (step 1). The remainder of these two views can be drawn only after the location and size of the inclined features are established by the primary and secondary views.

The primary auxiliary view is the next view to be made. It is drawn perpendicular to the inclined surface (surface *M*) in the top view (step 2).

The secondary auxiliary view is then projected perpendicular from the inclined surface (surface *N*) of the primary auxiliary view (step 3). Once the secondary auxiliary view is drawn, the size and location of the features on the secondary auxiliary view, in this case the hexagon hole, can be located on the primary auxiliary view.

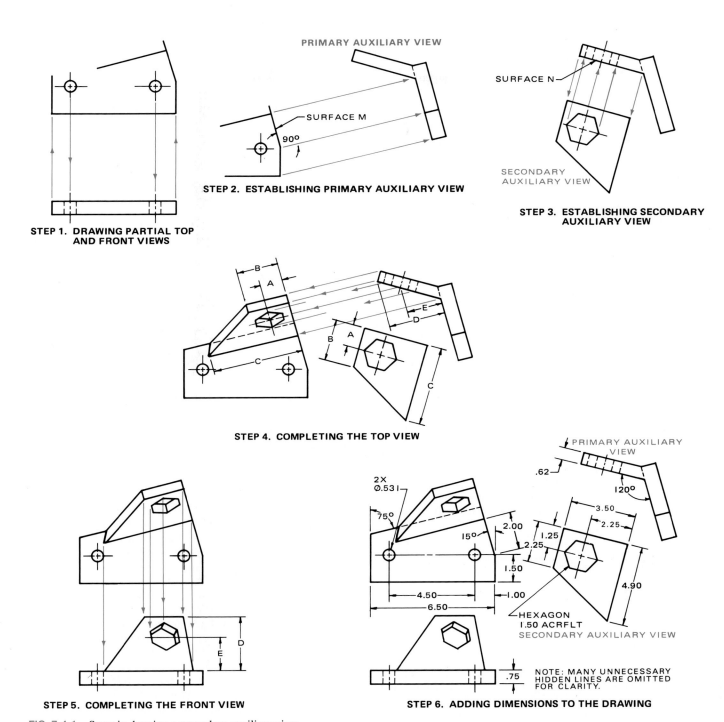

PRIMARY AUXILIARY VIEW

SURFACE M

90°

STEP 2. ESTABLISHING PRIMARY AUXILIARY VIEW

STEP 1. DRAWING PARTIAL TOP AND FRONT VIEWS

SURFACE N

SECONDARY AUXILIARY VIEW

STEP 3. ESTABLISHING SECONDARY AUXILIARY VIEW

B
A
E
D
B
A
C
C

STEP 4. COMPLETING THE TOP VIEW

D
E

STEP 5. COMPLETING THE FRONT VIEW

PRIMARY AUXILIARY VIEW
.62
120°
2X Ø.531
75°
2.00
1.25
15°
2.25
1.50
3.50
2.25
4.90
4.50
1.00
6.50
.75
HEXAGON 1.50 ACRFLT
SECONDARY AUXILIARY VIEW
NOTE: MANY UNNECESSARY HIDDEN LINES ARE OMITTED FOR CLARITY.

STEP 6. ADDING DIMENSIONS TO THE DRAWING

FIG. 7-4-1   Steps in drawing a secondary auxiliary view.

Next the top view is completed (step 4) by projecting lines and points from the primary auxiliary view and distances, such as A, B, and C, from the secondary auxiliary view.

Last, the front view is completed by projecting lines and points from the top view and distances, such as D and E, from the primary auxiliary view (step 5), then adding dimensions (step 6).

Another example of using auxiliary views in establishing the true shape and size of an oblique surface is shown in Fig. 7-4-2 (pg. 142). Surface 1-2-3-4 is inclined to the three regular planes (front, top, and side) of projection. In this example, the three regular views are drawn first. The primary auxiliary view is drawn next by projecting lines parallel to line 1-2 in the top view. Note that in this view, points 1, 2, 3, and 4 appear as a line or edge view of the surface 1-2-3-4. The secondary auxiliary view is drawn next by projecting lines perpendicular to line 1-2-3-4 in the primary auxiliary view in order to show the true shape and size of surface 1-2-3-4.

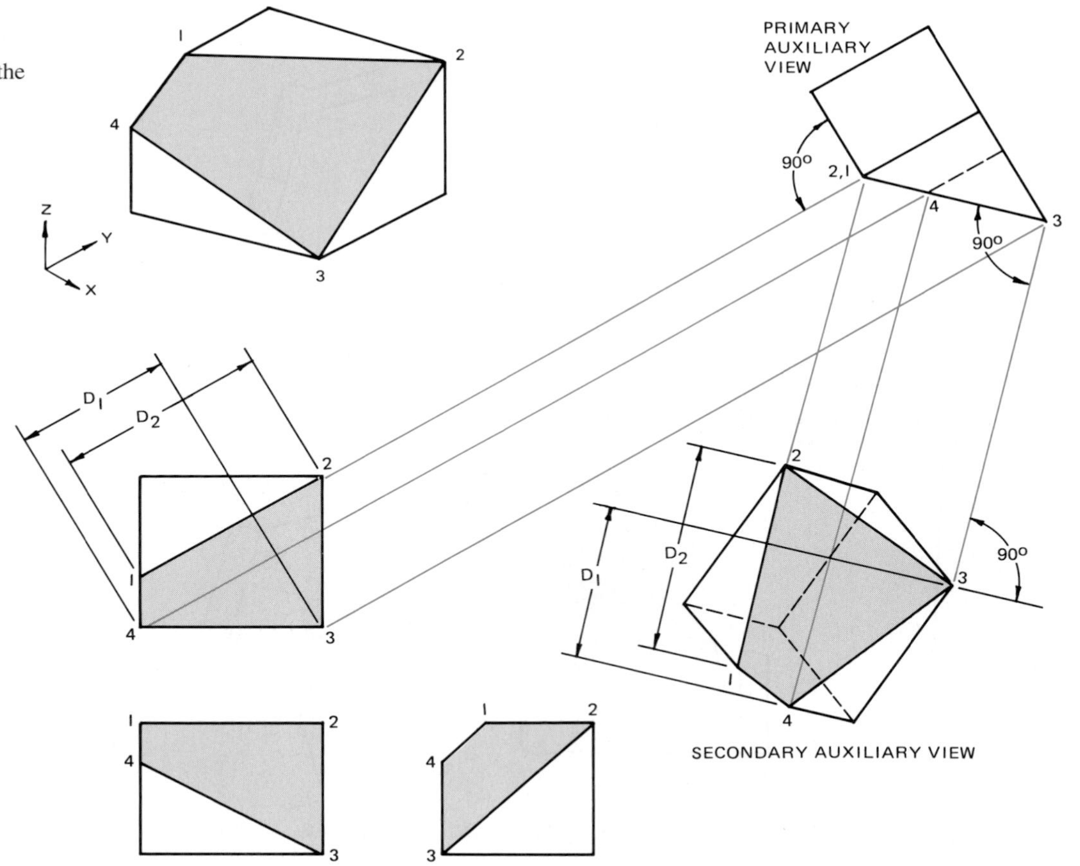

FIG. 7-4-2   Secondary auxiliary view required to find the true shape of surface 1-2-3-4.

When preparing auxiliary view drawings, be sure to allow sufficient space between views to ensure that the views and dimensions that have to be drawn later will fit in the allotted space.

ASSIGNMENT

See Assignment 5 for Unit 7-4 on page 163.

7-5 REVOLUTIONS

A major problem in technical drawing and design is the creation of projections for finding the true views of lines and planes. The following is a brief review of the principles of descriptive geometry involved in the solution of such problems. The designer, working along with an engineering team, can solve problems graphically with geometric elements. Structures that occupy space have three-dimensional forms made up of a combination of geometric elements (Fig. 7-5-1).

The graphic solutions of three-dimensional forms require an understanding of the space relations that points, lines, and planes share in forming any given shape. Problems that many times require mathematical solutions can often be solved graphically with an accuracy that will allow manufacturing and construction. *Basic descriptive geometry* is one of the designer's methods of thinking through and solving problems.

Reference Planes

In Unit 6-1 reference planes were used to show how the six basic views of an object were positioned on a flat surface. Unfolding these reference planes forms a two-dimensional surface that a drafter uses to construct views and solve problems.

To identify the different planes being used on a drawing, an identification code is needed for these planes. One such system is to identify the top or horizontal reference plane by the letter *T*; identify the front or vertical reference plane by the letter *F*; and identify the side or profile reference plane by the letter *S*. Thus point 1 on a part, line, or plane would be identified as 1*F* on the front reference plane, 1*T* on the top reference plane, and 1*S* on the side reference plane.

The folding lines shown on the box are referred to as reference lines on the drawing, as shown in Fig. 7-5-2. Other reference planes and reference lines are drawn and labeled as required.

Revolutions

As we have seen, when the true size and shape of an inclined surface do not show in a drawing, one solution is to make an auxiliary view. Another, however, is to keep using the regular reference planes while imagining that the object has been revolved (turned) as shown in Fig. 7-5-3. Remember, in auxiliary views, you set up new reference planes in order to look at objects from new directions. Understanding *revolutions*

FIG. 7-5-1 Geometric space-frame structure.

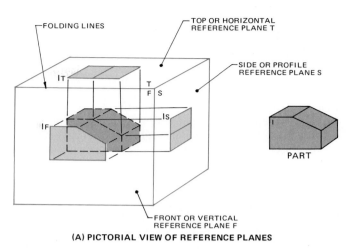

(A) PICTORIAL VIEW OF REFERENCE PLANES

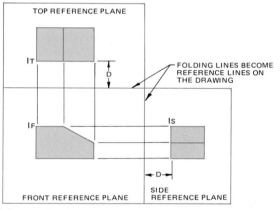

(B) UNFOLDING OF THE THREE REFERENCE PLANES

FIG. 7-5-2 Reference lines.

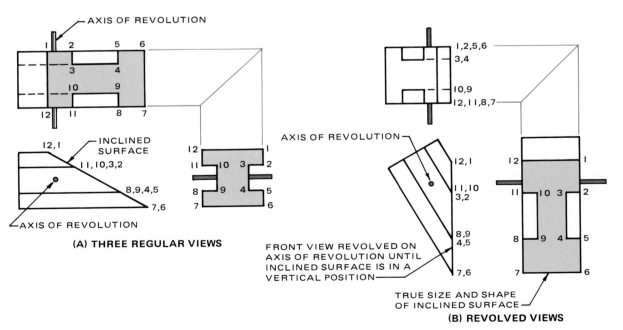

(A) THREE REGULAR VIEWS

(B) REVOLVED VIEWS

FIG. 7-5-3 Revolving front view to obtain true size and shape of inclined surface.

(ways of revolving objects) should help you better understand auxiliary views.

Axis of Revolution

An easy way to picture an object being revolved is to imagine that a shaft or an axis has been passed through it. Imagine, also, that this axis is perpendicular to one of the principal planes. In Fig. 7-5-4 the three principal planes are shown with an axis passing through each one and through the object beyond.

An object can be revolved to the right (clockwise) or to the left (counterclockwise) about an axis perpendicular to either the vertical or the horizontal plane. The object can be revolved forward (counterclockwise) or backward (clockwise) about an axis perpendicular to the profile plane. As we have seen, an axis of revolution can be perpendicular to the vertical, horizontal, or profile plane. In Fig. 7-5-5A the usual front and top views of an object are shown at the left. To the right the same views of the object are shown after the object has been revolved 45° counterclockwise about an axis perpendicular to the vertical plane. Notice that the front view is still the same in

size and shape as before except that it has a new position. The new top view has been made by projecting up from the new front view and across from the old top view. Note that the depth remains the same from one top view to the other.

In Fig. 7-5-5B, a second object is shown at the left in the usual top and front views. To the right the same views of the object are shown after it has been revolved 60° clockwise about an axis perpendicular to the horizontal plane. The new top view is the same in size and shape as before. The new front view has been made by projecting down from the new top view and across from the old front view. Note that the height remains the same from the original front view to the revolved front view.

In Fig. 7-5-5C a third object is shown at the top in the usual front and side views. Below, the same views of the object are shown after it has been revolved forward (counterclockwise) 30° about an axis perpendicular to the profile plane. The new front view has been made by projecting across from the new side view and down from the old front view in space 1. Note that the width remains the same from one front view to the other. Revolution can be clockwise, as in Fig. 7-5-5B, or it can be counterclockwise, as in A and C.

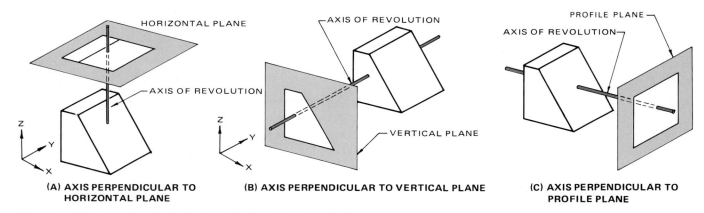

(A) AXIS PERPENDICULAR TO HORIZONTAL PLANE   **(B) AXIS PERPENDICULAR TO VERTICAL PLANE**   **(C) AXIS PERPENDICULAR TO PROFILE PLANE**

FIG. 7-5-4   The axis of revolution is perpendicular to the principal planes.

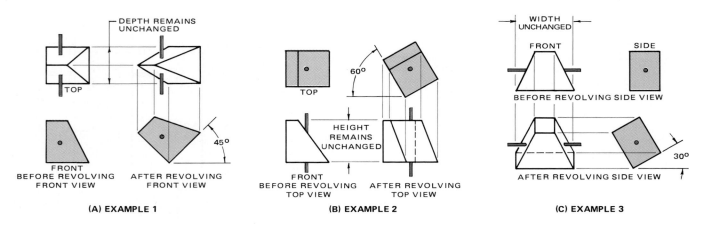

(A) EXAMPLE 1   **(B) EXAMPLE 2**   **(C) EXAMPLE 3**

FIG. 7-5-5   Single revolutions about the three axes.

The Rule of Revolution

The rule of revolution has two parts:

1. The view that is perpendicular to the axis of revolution stays the same except in position. (This is true because the axis is perpendicular to the plane on which it is projected.)
2. Distances parallel to the axis of revolution stay the same. (This is true because they are parallel to the plane or planes on which they are projected.)

Figure 7-5-6 illustrates the two parts of the rule of revolution.

True Shape of an Oblique Surface Found by Successive Revolutions

A surface shows its true shape when it is parallel to one of the principal planes. In Fig. 7-5-7A an object is shown pictorially and in orthographic projection. Note that surface 1-2-3-4 is an oblique surface because it is inclined in all three of the normal views. To find the true shape and size of this surface by revolutions, the following rotations must be made.

First Revolution Rotate the top view until line 1-2 is in a vertical position (Fig. 7-5-7B). By projection, complete the front and side views. Note that surface 1-2-3-4 now appears as line 1-3 in the front view (frontal plane).

Second Revolution Next rotate the front view until line 1-3 is in a vertical position (Fig. 7-5-7C). The true shape of surface 1-2-3-4 can now be found by revolving the part about an axis perpendicular to the frontal plane until the surface is parallel to the profile plane. The depth distances in the side and top views are identical to the depth distances in the side view shown in Fig. 7-5-7B. Note that this is the same part shown in Fig. 7-4-2, in which a secondary auxiliary view was used to establish the true size and shape of the surface.

Auxiliary Views and Revolved Views

You can show the true size of an inclined surface by either an auxiliary view (Fig. 7-5-8A, pg. 146) or a revolved view (Fig. 7-5-8B). In a revolved view, the inclined surface is turned until it is parallel to one of the principal planes. The revolved view in B is similar to the auxiliary view in A.

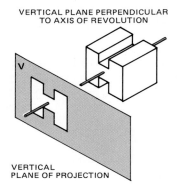

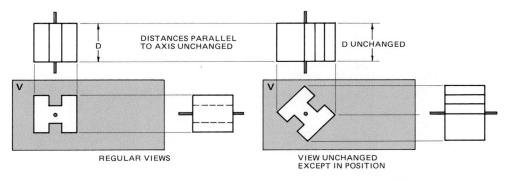

NOTE THE H-SHAPE IN THE FRONT VIEW HAS CHANGED ONLY IN POSITION.

FIG. 7-5-6   The rule of revolution.

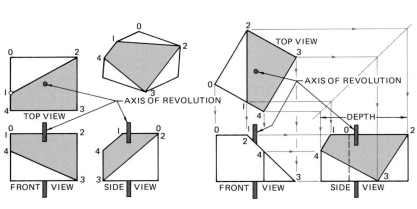

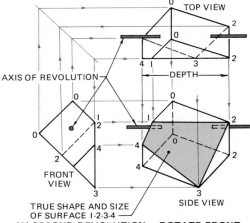

(A) THREE REGULAR VIEWS AND PICTORIAL OF PART

(B) FIRST REVOLUTION — ROTATE TOP VIEW UNTIL LINE 1-2 IS VERTICAL

(C) SECOND REVOLUTION — ROTATE FRONT VIEW UNTIL LINE REPRESENTING SURFACE 1-2-3-4 IS VERTICAL

FIG. 7-5-7   The true shape of surface 1-2-3-4 is obtained by successive revolutions.

In the auxiliary view, it is as if the observer has changed position to look at the object from a new direction. Conversely, in the revolved view, it is as if the object has changed position. Both revolutions and auxiliaries improve your ability to visualize objects. They also work equally well in solving problems.

True Length of a Line

Since an auxiliary view shows the true size and shape of an inclined surface, it can also be used to find the true length of a line. In Fig. 7-5-9A the line *OA* does not show its true length in

the top, front, or side view because it is inclined to all three of these planes of projection. In the auxiliary view in Fig. 7-5-9B, however, it does show its true length (*TL*), because the auxiliary plane is parallel to the surface *OAB*.

Figure 7-5-9C shows another way to show the true length (*TL*) of line *OA*. In this case, revolve the object about an axis perpendicular to the vertical plane until surface *OAB* is parallel to the profile plane. The side view will then show the true size of surface *OAB* and also the true length of line *OA*. A shorter method of showing the true length of *OA* is to revolve only the surface *OAB*, as shown in Fig. 7-5-9D.

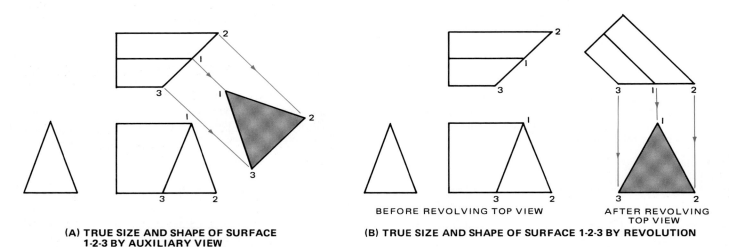

(A) TRUE SIZE AND SHAPE OF SURFACE 1-2-3 BY AUXILIARY VIEW

(B) TRUE SIZE AND SHAPE OF SURFACE 1-2-3 BY REVOLUTION

FIG. 7-5-8   True size of a surface obtained by using auxiliary and revolved views.

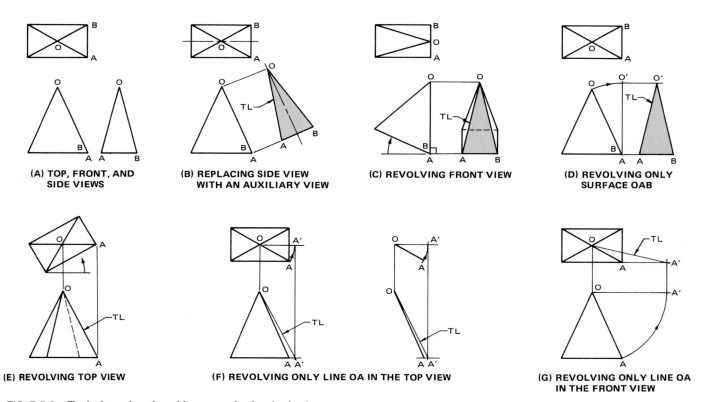

FIG. 7-5-9   Typical true-length problems examined and solved.

In Fig. 7-5-9E the object is revolved in the top view until line *OA* in that view is horizontal. The front view now shows line *OA* in its true length because this line is now parallel to the vertical plane.

In Fig. 7-5-9F still another method is shown. In this case, instead of the whole object being revolved, just line *OA* is turned in the top view until it is horizontal at *OA*. The point A_1 can then be projected to the front view. There, OA_1 will show line *OA* at its true length.

You can revolve a line in any view to make it parallel to any one of the three principal planes. Projecting the line on the plane to which it is parallel will show its true length. In Fig. 7-5-9G the line has been revolved parallel to the horizontal plane. The true length of line *OA* then shows in the top view.

Fig. 7-5-10 shows a simple part with one view revolved in each of the examples. Space 1 shows a three-view drawing of a

block in its simplest position. In space 2 (upper right) the block is shown after being revolved from the position in space 1 through 45° about an axis perpendicular to the frontal plane. The front view was drawn first, copying the front view in space 1. The top view was obtained by projecting up from the front view and across from the top view of space 1.

In space 3 (lower left) the block has been revolved from position 1 through 30° about an axis perpendicular to the horizontal plane. The top view was drawn first, copied from the top view of space 1.

In space 4 the block has been tilted from position 2 about an axis perpendicular to the side plane 30°. The side view was drawn first, copied from the side view in space 2. The widths of the front and top views were projected from the front view of space 2.

ASSIGNMENT

See Assignment 6 for Unit 7-5 on page 164.

7-6 LOCATING POINTS AND LINES IN SPACE

Points in Space

A point can be considered physically real and can be located by a small dot or a small cross. It is normally identified by two or more projections. In Fig. 7-6-1A point *A* and *B* are located on all three reference planes. Notice that the unfolding of the three planes forms a two-dimensional surface with the fold lines remaining. The fold lines are labeled as shown to indicate that *F* represents the front view, *T* represents the top view, and *S* represents the profile or right-side view. In Fig. 7-6-1B the

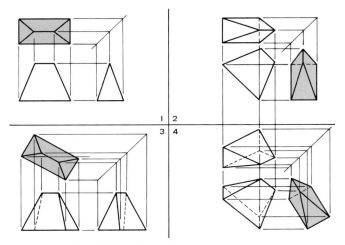

FIG. 7-5-10   Revolving a view of a part.

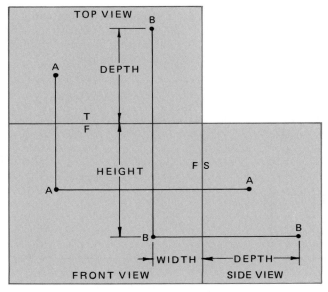

(A) POINTS A AND B IDENTIFIED ON UNFOLDED REFERENCE PLANE

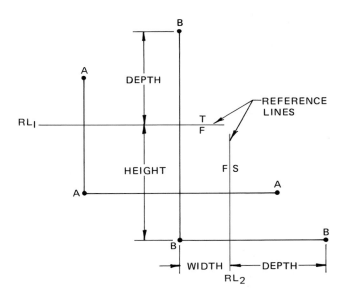

(B) POINTS A AND B IDENTIFIED BY REFERENCE LINES

FIG. 7-6-1   Points in space.

planes are replaced with reference lines RL_1 and RL_2, which are placed in the same position as the fold lines in Fig. 7-6-1A.

Lines in Space

Lines in descriptive geometry are grouped into three classes depending on how they are positioned in relation to the reference lines.

Normal Lines   A line that is perpendicular to the reference plane will project as a point on that plane. In Fig. 7-6-2A line AB is perpendicular to the front reference plane. As such, it is shown as a point (A_FB_F) in the front view and as a true-length line in the top and side views (lines A_TB_T and A_SB_S, respectively).

Inclined Lines   A line that appears inclined in one plane, as shown in Fig. 7-6-2B, and is parallel to the other two principal views which will appear foreshortened in the other two views. The inclined line shown in the front view will be the true length of line AB.

Oblique Lines   A line that appears inclined in all three views is an oblique line. It is neither parallel nor perpendicular to any of the three planes. The true length of the line is not shown in any of these views (Fig. 7-6-2C).

True Length of an Oblique Line by Auxiliary Projection

Since a normal line and an inclined line have projections parallel to a principal plane, the true length of each can be seen in that projection. Since an oblique line is not parallel to any of the three principal reference planes, an auxiliary reference line RL_3 can be placed parallel to any one of the oblique lines, as shown in Fig. 7-6-3. Transfer distances M and N shown in the regular views to the auxiliary view, locating points A_1 and B_1, respectively. Join points A_1 and B_1 with a line, obtaining the true length of line AB.

Point on a Line

The line A_FB_F in the front view of Fig. 7-6-4A contains a point C. To place point C on the line in the other two views, it is necessary to project construction lines perpendicular to the reference lines RL_1 and RL_2, as shown in Fig. 7-6-4B. The construction lines are projected to line A_TB_T in the top view and line A_SB_S in the side view, locating point C on the line in these views.

If point C is to be located on the true length of line AB, another reference line, such as RL_3, is required, and the distances N and M in the front view are then used to locate the true length of line A_1B_1 in the auxiliary view. Position C is then projected perpendicular to line A_SB_S in the side view to locate C on the true-length line.

Point-On-Point View of a Line

When the front view A_FB_F and top view A_TB_T are given, as in Fig. 7-6-5, and the point-on-point view of line AB is required, the procedure is as follows:

- Draw a true-length view of AB by the method described in "True Length of an Oblique Line by Auxiliary Projection" and shown in Fig. 7-6-3.
- Draw a reference line perpendicular to the true length of line A_1B_1, and label it RL_3.
- The next adjacent, second auxiliary view A_2B_2 will be a point-on-point view of line AB.

ASSIGNMENTS

See Assignments 7 and 8 for Unit 7-6 on page 165.

7-7 PLANES IN SPACE

Planes for practical studies are considered to be without thickness and can be extended without limit. A plane may be represented or determined by intersecting lines, two parallel lines, a line and a point, three points, or a triangle.

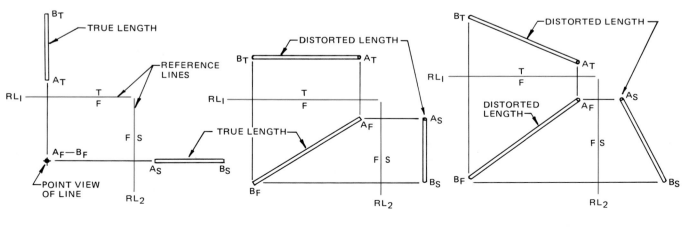

(A) A NORMAL LINE   (B) INCLINED LINE   (C) OBLIQUE LINES

FIG. 7-6-2   Lines in space.

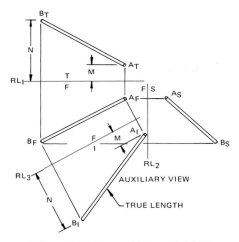

(A) REFERENCE LINE RL$_3$ PLACED PARALLEL TO FRONT VIEW

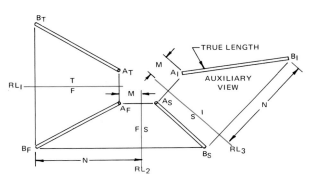

(B) REFERENCE LINE RL$_3$ PLACED PARALLEL TO SIDE VIEW

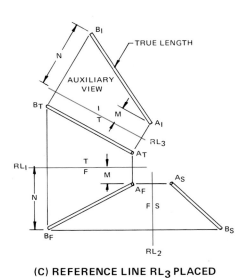

(C) REFERENCE LINE RL$_3$ PLACED PARALLEL TO TOP VIEW

FIG. 7-6-3   The length of an oblique line by auxiliary view projection.

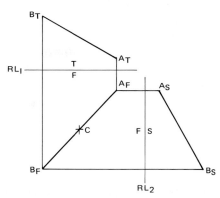

(A) PROBLEM—TO LOCATE POINT C ON LINE A-B IN OTHER VIEWS

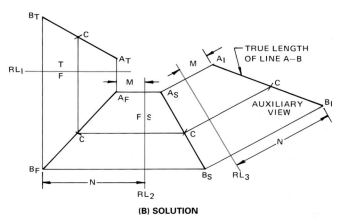

(B) SOLUTION

FIG. 7-6-4   Point on a line.

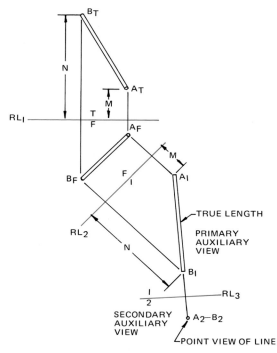

FIG. 7-6-5   Point-on-point view of a line.

The three basic planes, referred to as the *normal plane*, *inclined plane*, and *oblique plane*, are identified by their relationship to the three principal reference planes. Figure 7-7-1 illustrates the three basic planes, each plane being triangular in shape.

Normal Plane   A plane whose surface, in this case a triangular surface, appears in its true shape in the front view and as a line in the other two views.

Inclined Plane   This results when the shape of the triangular plane appears distorted in two views and as a line in the other view.

Oblique Plane   A plane whose shape appears distorted in all three views.

Locating a Line in a Plane

The top and front views in Fig. 7-7-2A show a triangular plane *ABC* and lines *RS* and *MN*, each located in one of the views.

To find their location in the other views, refer to Fig. 7-7-2B and the following procedures.

To locate line *RS* in the front view:

- Line $R_T S_T$ crosses over lines $A_T B_T$ and $A_T C_T$ at points D_T and E_T, respectively.
- Project points D_T and E_T to front view, locating points D_F and E_F.
- Extend a line through points D_F and E_F.
- The length of the line can be found by projecting points R_T and S_T to the front view, locating the end points R_F and S_F.

To locate line *MN* in the top view:

- Extend line $M_F N_F$ in the front view, locating points H_F and G_F on lines $A_F B_F$ and $A_F C_F$, respectively.
- Project points H_F and G_F to the top view, locating points H_T and G_T.
- Draw a line through points H_T and G_T.
- Project points M_F and N_F to top view, locating points on line $M_T N_T$.

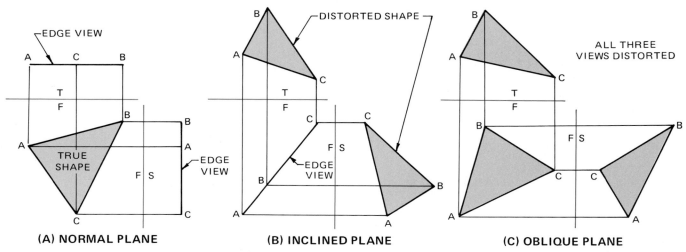

(A) NORMAL PLANE   **(B) INCLINED PLANE**   **(C) OBLIQUE PLANE**

FIG. 7-7-1   Planes in space.

FIG. 7-7-2   Locating a line on a plane.

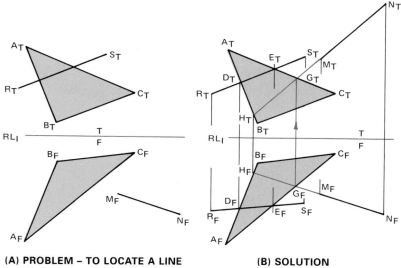

(A) PROBLEM – TO LOCATE A LINE IN THE OTHER VIEW   **(B) SOLUTION**

Locating a Point on a Plane

The top and front views shown in Fig. 7-7-3A show a triangular plane *ABC* and points *R* and *S* each located in one of the views. To find their location in the other views, refer to Fig. 7-7-3B and the following procedures.

To locate point *R* in the front view:

- Draw a line from A_T passing through point R_T to a point M_T on line B_TC_T.
- Project point M_T to front view, locating point M_F.
- Join points A_F and M_F with a line.
- Project point R_T to front view, locating point R_F.

To locate point *S* in the top view:

- Draw a line between points B_F and S_F, locating point N_F on line A_FC_F.
- Project point N_F to top view, locating point N_T.
- Draw a line through points B_T and N_T.
- Project point S_F to top view, locating point S_T.

Locating the Piercing Point of a Line and a Plane—Cutting-Plane Method

The top and front views shown in Fig. 7-7-4 show a line *UV* passing somewhere through plane *ABC*. The piercing point of the line through the plane is found as follows:

- Locate points D_T and E_T in the top view.
- Project points D_T and E_T to the front view, locating points D_F and E_F.
- The intersection of lines D_FE_F and U_FV_F is the piercing point, labeled O_F.
- Project point O_F to top view, locating point O_T.

Locating the Piercing Point of a Line and a Plane—Auxiliary View Method

The top and front views in Fig. 7-7-5 show a line *UV* passing somewhere through plane *ABC*. The piercing point of the line through the plane is found as follows:

- Draw line A_TD_T in the top view parallel to reference line RL_1.
- Project point D_T to front view, locating point D_F.

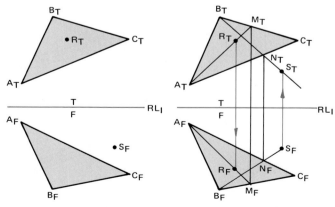

(A) PROBLEM – TO LOCATE A POINT IN THE OTHER VIEW   **(B) SOLUTION**

FIG. 7-7-3   Locating a point on a plane.

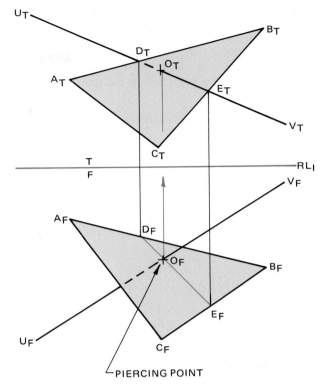

FIG. 7-7-4   Locating the piercing point of a line and a plane, cutting-plane method.

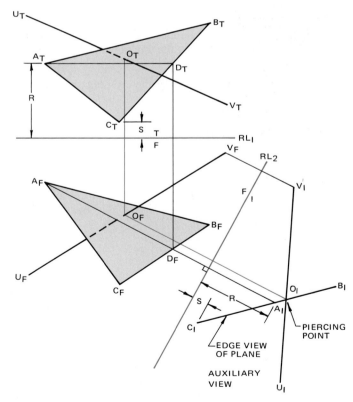

FIG. 7-7-5   Locating the piercing point of a line and a plane, auxiliary view method.

- Draw reference line RL_2 perpendicular to a line intersecting points A_F and D_F in the front view.
- Draw an auxiliary view which shows the edge view of the plane. The intersection of the plane and line locates the piercing point O_1.
- Project the piercing point to front and top views.

ASSIGNMENTS ▨▨▨▨▨▨▨▨▨▨▨▨▨▨▨

See Assignments 9 and 10 for Unit 7-7 on page 166.

7-8 ESTABLISHING VISIBILITY OF LINES IN SPACE

Visibility of Oblique Lines by Testing

In the example of the two nonintersecting pipes shown in Fig. 7-8-1A, it is not apparent which pipe is nearest the viewer at the crossing points in the two views. To establish which of the pipes lies in front of the other, the following procedure is used.

To establish the visible pipe at the crossing shown in the top view (Fig. 7-8-1B):

- Label the crossing of lines $A_T B_T$ and $C_T D_T$ as ①, ②.
- Project the crossing point to the front view, establishing points ① and ②.
- Point ① is closer to reference line RL_1, which means that line $A_T B_T$ is nearer when the top view is being observed and thus is visible.

To establish the visible pipe at the crossing shown in the front view:

- Label the crossing of lines $A_F B_F$ and $C_F D_F$ as ③, ④.
- Project the crossing point to top view, establishing points ③ and ④.

- Point ④ is closer to reference line RL_1, which means that line $C_F D_F$ is nearer when the front view is being observed and thus is visible. Figure 7-8-1C shows the correct crossings of the pipes.

Visibility of Lines and Surfaces by Testing

In cases where points or lines are approximately the same distance away from the viewer, it may be necessary to graphically check the visibility of lines and points, as in Fig. 7-8-2.

To check visibility of lines $A_T C_T$ and $B_T D_T$ in the top view:

- Label intersection of lines $A_T C_T$ and $B_T D_T$ as ①, ②.
- Project the point of intersection to front view, establishing point ① on line $A_F C_F$ and point ② on line $B_F D_F$.
- Point ① is closer to reference line RL_1, which means that line $A_T C_T$ is nearer when the top view is being observed and thus is visible.
- Point ② is farther away from reference line RL_1, which means that line $B_T D_T$ would not be seen when one is viewing from the top.

To check visibility of lines $A_F C_F$ and $B_F D_F$ in the front view:

- Label the intersection of lines $A_F C_F$ and $B_F D_F$ as ③, ④.
- Project the point of intersection to top view, establishing point ③ on line $A_T C_T$ and point ④ on line $B_T D_T$.
- Point ③ is closer to reference line RL_1, which means that line $A_F C_F$ is nearer when the front view is being observed and thus is visible.
- Point ④ is farther away from reference line RL_1, which means that line $B_F D_F$ would not be seen when one is viewing from the front. Figure 7-8-2C shows the completed top and front views of the part.

Visibility of Lines and Surfaces by Observation

In order to fully understand the shape of an object, it is necessary to know which lines and surfaces are visible in each of the views. Determining their visibility can, in most cases, be done

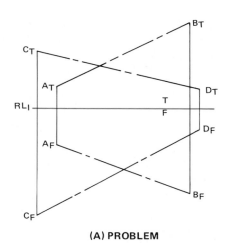

(A) PROBLEM

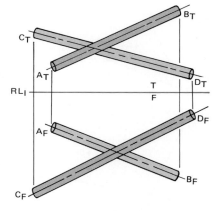

(B) ESTABLISHING LINES WHICH ARE CLOSER TO OBSERVER

(C) SOLUTION

FIG. 7-8-1   Visibility of oblique lines by testing.

by inspection. With reference to Fig. 7-8-3A, the outline of the part is obviously visible. However, the visibility of lines and surfaces within the outline must be determined. This is accomplished by determining the position of O_F in the front view. Since position O is the closest point to the reference line RL_1 in the front view, it must be the point which is closest to the observer when viewing the top view. Thus it can be seen, and the lines converging to point O_T are visible.

With reference to determining the visibility of the lines in the front view, see the top view. Plane $O_T C_T D_T$ is closest to reference line RL_1. Therefore it must be the closest surface when the observer is looking at the front view, and it must be visible. Since point B_T in the top view is farthest away from

reference line RL_1, it is the point which is farthest away when the observer is looking at the front view. Since it lies behind surface $O_F C_F D_F$, it cannot be seen.

From this example it may be stated that lines or points closest to the observer will be visible, and lines and points farthest away from the viewer but lying within the outline of the view will be hidden.

ASSIGNMENTS

See Assignments 11 and 12 for Unit 7-8 on pages 166–167.

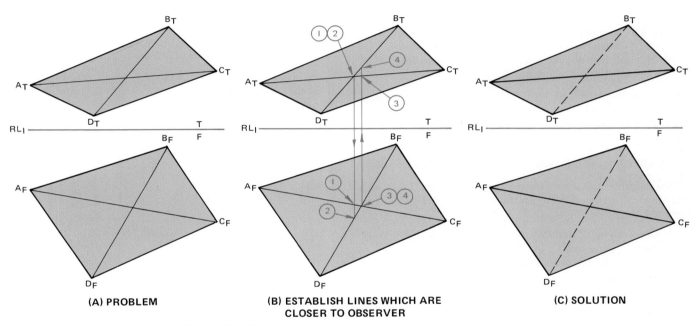

(A) PROBLEM

(B) ESTABLISH LINES WHICH ARE
CLOSER TO OBSERVER

(C) SOLUTION

FIG. 7-8-2   Establishing visibility of lines and surfaces by testing.

FIG. 7-8-3   Visibility of lines and surface by observation.

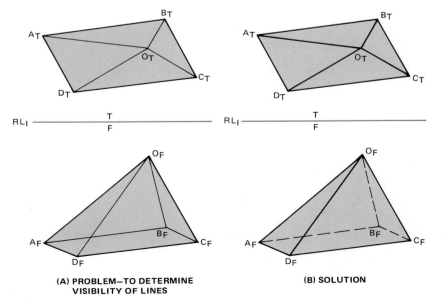

(A) PROBLEM—TO DETERMINE
VISIBILITY OF LINES

(B) SOLUTION

7-9 DISTANCES BETWEEN LINES AND POINTS

Distance from a Point to a Line

When the front and side views are given, as in Fig. 7-9-1, and the shortest distance between line AB and point P is required, the procedure is as follows:

- Draw reference line RL_2 parallel to line A_SB_S in the side view.
- Transfer distances designated as R, S, and U in the front view to the primary auxiliary view. The resulting line A_1B_1 in the auxiliary view is the true length of line AB.
- Next draw reference line RL_3 perpendicular to line A_1B_1.
- Transfer distances designated as V and W in the side view to the secondary auxiliary view, establishing points P_2 and A_2B_2, the latter being the point view of line AB.
- The shortest distance between point P and line AB is shown in the secondary auxiliary view.

Design Application   Figure 7-9-2 illustrates the application of the point-on-point view of a line to determine the clearance between a hydraulic cylinder and a clip on the wheel housing.

Shortest Distance Between Two Oblique Lines

When the front and top views are given, as in Fig. 7-9-3, and the shortest distance between the two lines AB and CD is required, the procedure is as follows:

- Draw reference line RL_2 parallel to line A_FB_F in the front view.
- Transfer the distances designated as R, S, U, and V in the top view to the primary auxiliary view. The resulting line A_1B_1 in the auxiliary view is the true length of line AB.

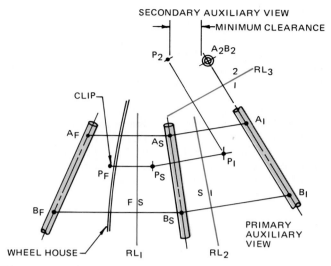

FIG. 7-9-2   Design application of distance from a point to a line.

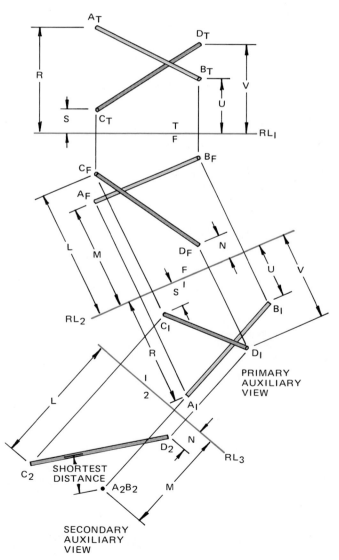

FIG. 7-9-3   Shortest distance between oblique lines.

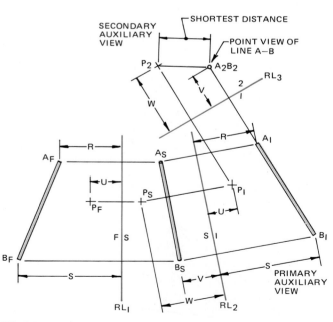

FIG. 7-9-1   Distance from a point to a line.

■ Next draw reference line RL_3 perpendicular to line A_1B_1.

■ Transfer the distances designated as L, M, and N in the front view to the secondary auxiliary view. Point A_2B_2 is a point view of line AB.

■ The shortest distance between the two lines AB and CD is shown in the secondary auxiliary view.

ASSIGNMENTS

See Assignments 13 and 14 for Unit 7-9 on page 168.

7-10 EDGE AND TRUE VIEW OF PLANES

The three primary planes of projection are horizontal, vertical (or frontal), and profile.

A plane that is not parallel to a primary plane is not visible in its true dimensions. To show a plane in true view, it must be revolved until it is parallel to a projection plane. Figure 7-10-1 shows an oblique plane ABC in the top and front views. The object is to find the true view of this plane. When the top and front views are examined carefully, no line is parallel to the reference line in either view. However, a line C_TD_T can be drawn on the plane parallel to the reference line RL_1 and projected to the front view for its true length D_FC_F. Next:

■ Draw reference line RL_2 perpendicular to line D_FC_F in the front view.

■ Transfer the distances designated R, S, and U in the top view to the primary auxiliary view. The resulting line A_1B_1 is the edge view of the plane.

■ Draw reference line RL_3 parallel to line A_1B_1.

■ Transfer the distances L, M, and N in the front view to construct the true shape of plane ABC in the secondary auxiliary view.

Design Application Figure 7-10-2 shows the application of the procedure followed for Fig. 7-10-1. Points A, B, C, and D

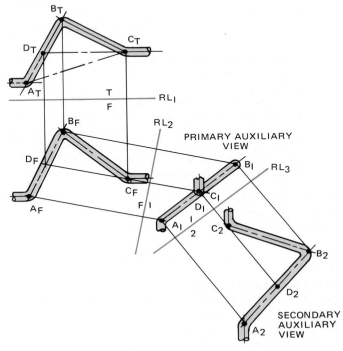

FIG. 7-10-2 Design application of true view of a plane for Fig. 7-10-1.

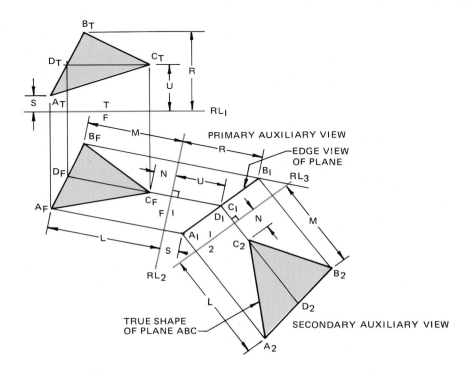

FIG. 7-10-1 True view of a plane.

correspond in both drawings, but line *AC* is omitted in Fig. 7-10-2 since it serves no practical purpose in the design.

Planes in Combination

Figure 7-10-3 demonstrates a solution where a combination of planes is involved. Note that $A_T B_T C_T$ and $A_F B_F C_F$ form one plane while $B_T C_T D_T$ and $B_F C_F D_F$ form another. Also line *BC* is common to both planes. The objective in the problem is to find the true bends at the angles *ABC* and *BCD*. The procedure is as follows:

- Construct the primary auxiliary view, which shows the true length of line *BC*.
- Construct a secondary auxiliary view, which shows *BC* as a point-on-point view. The result is the edge view of both planes *ABC* and *BCD* in the secondary auxiliary view.
- Since any view adjacent to a point-on-point view of a line must show the line in its true length, *BC* will be in true length in secondary auxiliary views 2 and 3. Therefore, projecting perpendicularly from the edge views in the

secondary auxiliary view 1 to the secondary auxiliary views 2 and 3 gives not only *BC* in true length in auxiliary views 2 and 3, but also the true angles *ABC* and *BCD*.

ASSIGNMENTS ▨▨▨▨▨▨▨▨▨▨▨▨▨▨▨▨

See Assignments 15 and 16 for Unit 7-10 on pages 168–169.

7-11 ANGLES BETWEEN LINES AND PLANES

The Angle a Line Makes with a Plane

The top and front views in Fig. 7-11-1 show a line *UV* passing somewhere through plane *ABC*. The true angle between the line and the plane will be shown in the view that shows the

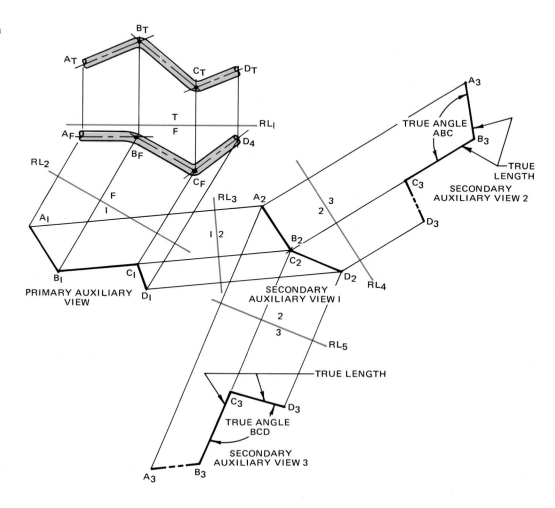

FIG. 7-10-3   Use of planes in combination.

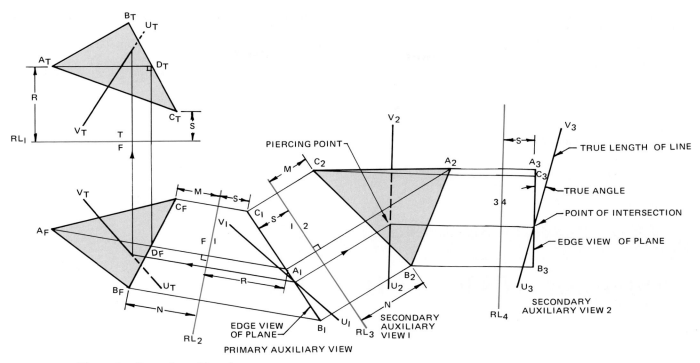

FIG. 7-11-1   The angle a line makes with a plane.

edge view of the plane and the true length of the line. This view is found as follows:

- Draw line $A_T D_T$ in the top view parallel to reference line RL_1.
- Project point D_T to front view, locating point D_F.
- Draw reference line RL_2 perpendicular to a line intersecting points A_F and D_F in the front view.
- Draw the primary auxiliary view, which shows the edge view of the plane but distorted length of line UV. The point of intersection between the line and edge view of the plane is established.
- Draw reference line RL_3 parallel to edge view of the plane shown in the auxiliary view.
- Draw the secondary auxiliary view 1, which shows the true view of the plane and location of the piercing point.
- Draw reference line RL_4 parallel to line $V_2 U_2$.
- Draw the secondary auxiliary view 2, which shows the true length of line UV and the true angle between line and edge view of plane.

Edge Lines of Two Planes

Figure 7-11-2 shows a line of intersection AB made by two planes, triangles ABC and ABD. When the top and front views are given, the point-on-point view of line AB and the true angle between the planes are found as follows:

- Draw reference line RL_2 parallel to line $A_F B_F$ in the front view.
- Transfer the distances designated R, S, and U in the top view to the primary auxiliary view. The resulting line $A_1 B_1$ in the auxiliary view is the true length of line AB.

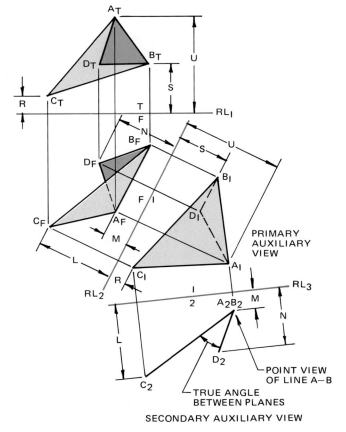

FIG. 7-11-2   Edge lines of two planes.

157

- Next draw reference line RL_3 perpendicular to line A_1B_1.
- Transfer the distances designated as L, M, and N in the front view to the secondary auxiliary view.
- Point A_2B_2 is a point-on-point view of line AB. The true angle between the two planes is seen in the secondary auxiliary view.

ASSIGNMENTS

See Assignments 17 through 19 for Unit 7-11 on pages 169 to 170.

ASSIGNMENTS FOR CHAPTER 7

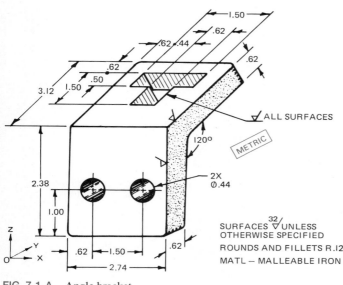

FIG. 7-1-A   Angle bracket.

SURFACES $^{32}/\nabla$ UNLESS
OTHERWISE SPECIFIED
ROUNDS AND FILLETS R.I2
MATL — MALLEABLE IRON

ASSIGNMENT FOR UNIT 7-1, PRIMARY AUXILIARY VIEWS

1. Make a working drawing of one of the parts shown in Figs. 7-1-A through 7-1-E. For Fig. 7-1-A draw the front, side, and auxiliary view. For all others draw the top, front, and auxiliary view. Partial views are to be used unless otherwise directed by your instructor. Hidden lines may be added for clarity. Scale 1:1.

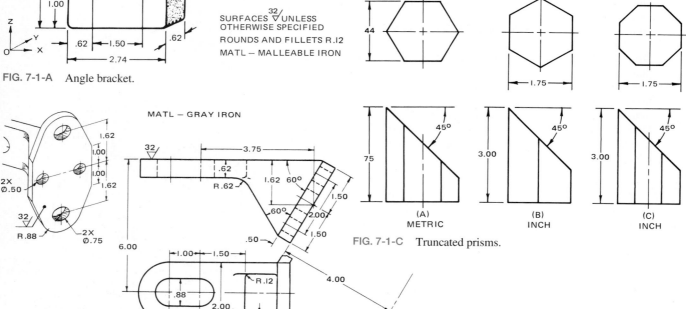

FIG. 7-1-C   Truncated prisms.

FIG. 7-1-B   Angle plate.

MATL — GRAY IRON

158

FIG. 7-1-D   Cross-slide bracket.

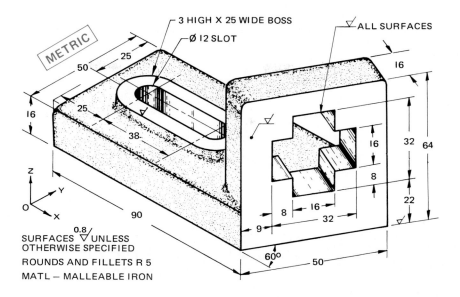

FIG. 7-1-E   Statue bases.

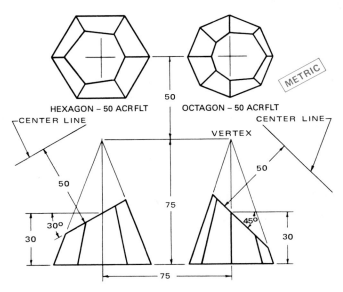

ASSIGNMENT FOR UNIT 7-2, CIRCULAR FEATURES IN AUXILIARY PROJECTION

2. Make a working drawing of one of the parts shown in Figs. 7-2-A through 7-2-D (here and on pg. 160). Refer to the drawing for set-up of views. Draw complete top and front views and a partial auxiliary view. Hidden lines may be added for clarity.

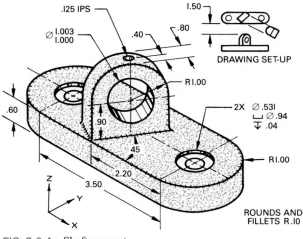

FIG. 7-2-A   Shaft support.

FIG. 7-2-B   Link.

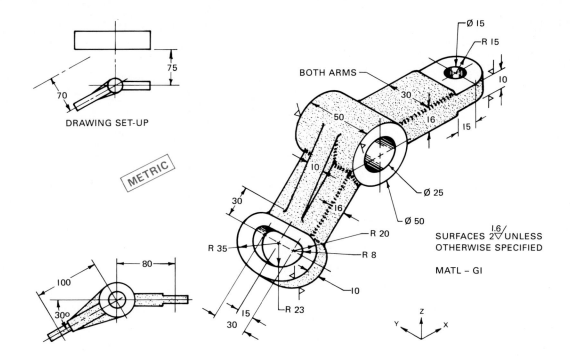

METRIC

DRAWING SET-UP

BOTH ARMS

Ø 15
R 15
30
10
50
16
15
10

SURFACES 2▽ UNLESS
OTHERWISE SPECIFIED

1.6/

MATL – GI

Ø 25
Ø 50
R 20
R 8
R 35
R 23
30
15
30
100
80
30°

Z
Y   X

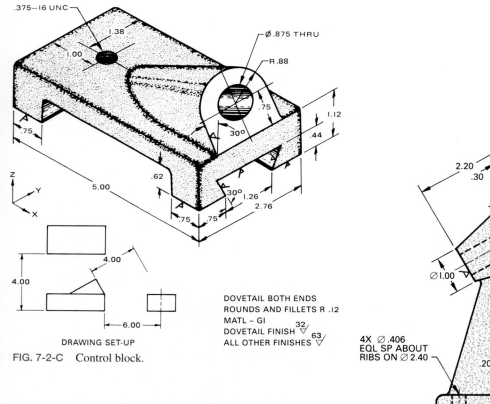

.375–16 UNC
1.38
1.00
Ø .875 THRU
R .88
.75
30°
1.12
.44
.75
.62
5.00
30° 1.26
2.76
.75   .75

Z
Y
X

4.00
4.00
6.00

DRAWING SET-UP

DOVETAIL BOTH ENDS
ROUNDS AND FILLETS R .12
MATL – GI
DOVETAIL FINISH 32▽
ALL OTHER FINISHES 63▽

FIG. 7-2-C   Control block.

10.50
3.00

DRAWING SET-UP

.70
2.20
.30
30°
.60 SQUARE HOLE,
Ø 1.10 C BORE X 1.20 DEEP
HEX
3.00 ACRFLT
Ø 2.10
1.50
Ø 1.00
4.90
4 RIBS
4X Ø .406
EQL SP ABOUT
RIBS ON Ø 2.40
.20   1.40
15°
.40
Ø 4.00

ROUNDS & FILLETS R .10

FIG. 7-2-D   Pedestal.

ASSIGNMENTS FOR UNIT 7-3, MULTI-AUXILIARY-VIEW DRAWINGS

3. Make a working drawing of one of the parts shown in Figs. 7-3-A through 7-3-C. Refer to the drawings for set-up of views. Draw complete top and front views and partial auxiliary views.

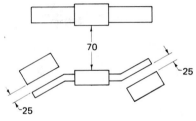

DRAWING SET-UP

MATL-.12 ALUMINUM

FIG. 7-3-A   Mounting plate.

FIG. 7-3-B   Connecting bar.

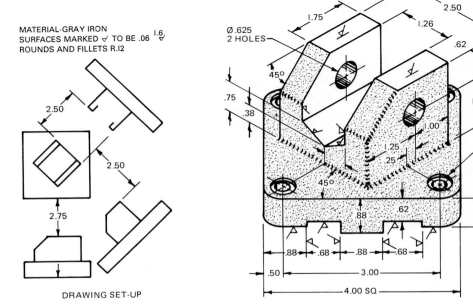

MATERIAL-GRAY IRON
SURFACES MARKED ∀ TO BE .06
ROUNDS AND FILLETS R.12

FIG. 7-3-C   Angle slide.

DRAWING SET-UP

ASSIGNMENT FOR UNIT 7-4, SECONDARY AUXILIARY VIEWS

5. Make a working drawing of one of the parts shown in Figs. 7-4-A through 7-4-C. The selection and placement of views are shown with the drawing. Only partial auxiliary views need be drawn, and hidden lines may be added to improve clarity.

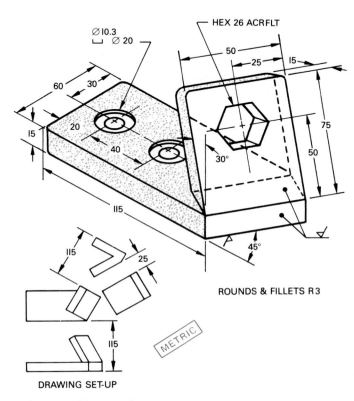

ROUNDS & FILLETS R3

METRIC

DRAWING SET-UP

FIG. 7-4-A   Hexagon slot support.

FIG. 7-4-B   Dovetail bracket.

METRIC

LOCATION OF Ø 20 HOLE ON AUXILIARY VIEW

DRAWING SET-UP

FIG. 7-4-C   Pivot arm.

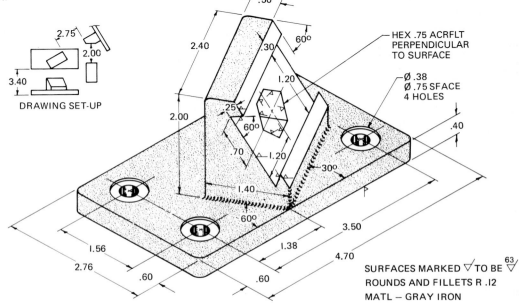

HEX .75 ACRFLT PERPENDICULAR TO SURFACE

Ø .38
Ø .75 SFACE
4 HOLES

SURFACES MARKED ▽ TO BE ⁶³▽
ROUNDS AND FILLETS R .12
MATL — GRAY IRON

DRAWING SET-UP

ASSIGNMENT FOR UNIT 7-5, REVOLUTIONS

6. Select one of the parts shown in Fig. 7-5-A and draw the views as positioned in the example given.

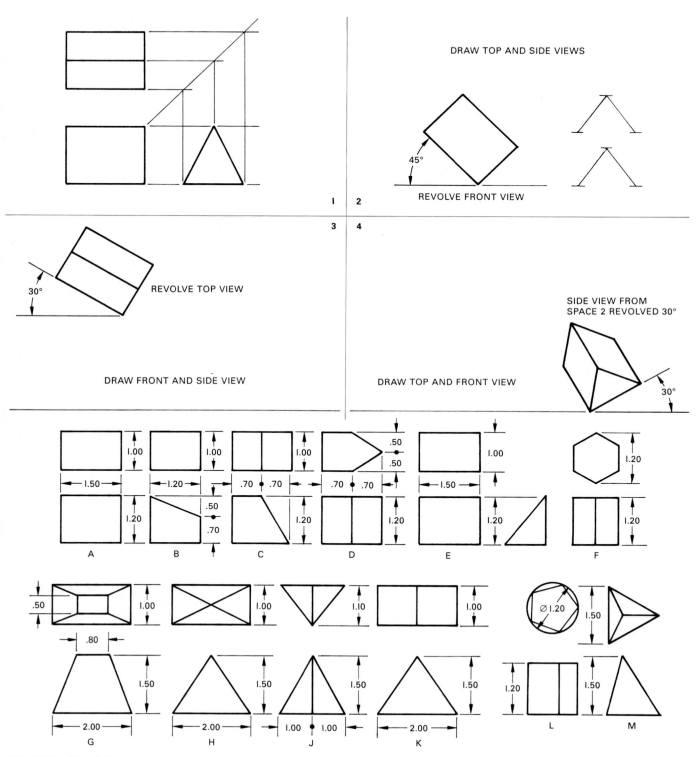

FIG. 7-5-A   Revolution assignments.

ASSIGNMENTS FOR UNIT 7-6, LOCATING POINTS AND LINES IN SPACE

7. With the use of a grid and reference lines, locate the points and/or lines in the third view for the drawings shown in Fig. 7-6-A. Scale to suit.

8. With the use of a grid and reference lines, locate the lines in the other views for the drawings shown in Fig. 7-6-B. Scale to suit.

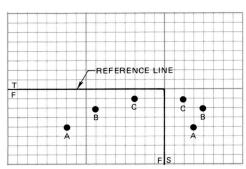

(1) LOCATE POINTS IN THE TOP VIEW

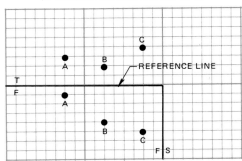

(2) LOCATE POINTS IN THE SIDE VIEW

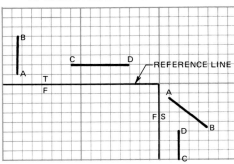

(3) LOCATE INCLINED LINES IN THE OTHER VIEW

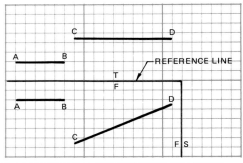

(4) LOCATE NORMAL AND INCLINED LINES IN THE OTHER VIEW

FIG. 7-6-A Assignments.

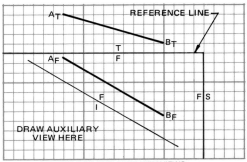

(1) DRAW SIDE AND AUXILIARY VIEWS
(INDICATE TRUE LENGTH OF LINE A–B)

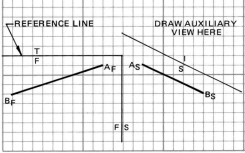

(2) DRAW TOP AND AUXILIARY VIEWS
(INDICATE TRUE LENGTH OF LINE A–B)

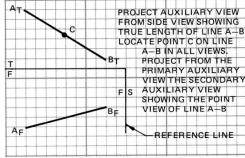

(3) DRAW SIDE AND AUXILIARY VIEWS

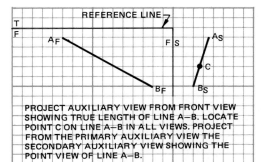

(4) DRAW TOP AND AUXILIARY VIEWS

FIG. 7-6-B Assignments.

ASSIGNMENTS FOR UNIT 7-7, PLANES IN SPACE

9. *Locating a Plane or a Line in Space.* With the use of a grid and reference lines, complete the three drawings shown in Fig. 7-7-A. Scale to suit.
10. *Locating a Point in Space and the Piercing Point of a Line and a Plane.* With the use of a grid and reference lines, complete the drawings shown in Fig. 7-7-B. Scale to suit.

ASSIGNMENTS FOR UNIT 7-8, ESTABLISHING VISIBILITY OF LINES IN SPACE

11. *Visibility of Lines and Surfaces by Observation and Testing.* With the use of a grid and reference lines, lay out the drawings shown in Fig. 7-8-A. By observation, sketch the circular pipes (drawings 1 and 2) in a manner similar to that shown in Fig. 7-8-1, showing the direction in which the pipes are sloping and which pipe is closer to the

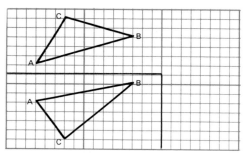

(1) DRAW THE SIDE VIEW OF TRIANGLE ABC

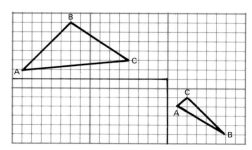

(2) DRAW THE FRONT VIEW OF TRIANGLE ABC

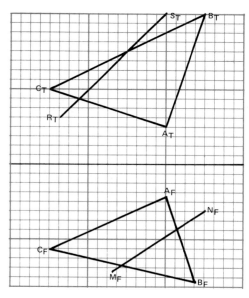

(3) LOCATING A LINE IN THE OTHER VIEW

FIG. 7-7-A   Assignments.

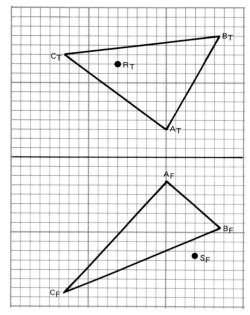

(1) LOCATE THE POINTS IN SPACE

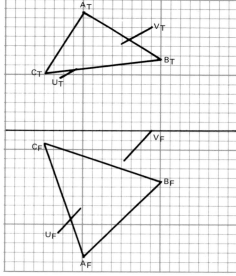

(2) LOCATE THE PIERCING POINT OF THE LINE AND PLANE AND COMPLETE LINE UV.

FIG. 7-7-B   Assignments.

observer in the two views. By testing in a manner similar to that shown in Fig. 7-8-2, establish the visibility of lines and surfaces of drawings 3 and 4. Scale to suit.

12. *Establishing Visibility of Lines and Surfaces by Testing.* With the use of a grid and reference lines, draw the four parts shown in Fig. 7-8-B. By testing in the manner shown in Fig. 7-8-2, complete the two-view drawings, showing the visible and hidden lines. Scale to suit.

FIG. 7-8-A   Assignments.

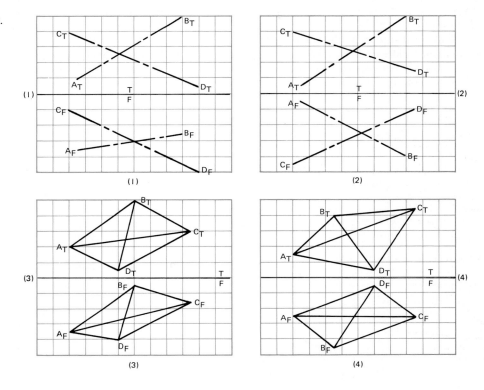

FIG. 7-8-B   Assignments.

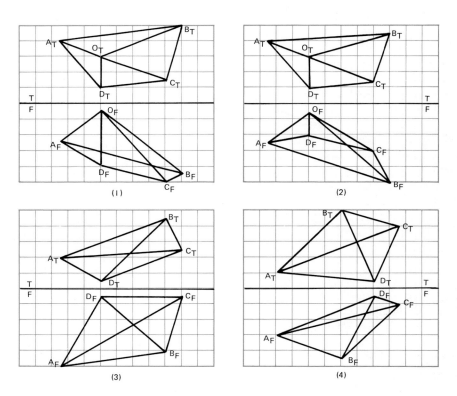

ASSIGNMENTS FOR UNIT 7-9, DISTANCES BETWEEN LINES AND POINTS

13. *Finding Distance from a Point to a Line.*With the use of a grid, reference lines and auxiliary views, find the distance from a point to a line in the two problems shown in Fig. 7-9-A. Scale to suit.
14. *Finding the Shortest Distance Between Oblique Lines.* With the use of a grid, reference lines, and auxiliary views, find the shortest distance between the oblique lines for the two problems shown in Fig. 7-9-B. Scale to suit.

ASSIGNMENTS FOR UNIT 7-10, EDGE AND TRUE VIEW OF PLANES

15. *Finding True Angles of Intersecting Planes.* With the use of grid, reference lines, and auxiliary views, find the true angles of the intersecting planes for the two problems shown in Fig. 7-10-A. Scale to suit.
16. *True View of a Plane.* With the use of a grid, reference lines, and auxiliary views, find the true view of the plane for the two problems shown in Fig. 7-10-B. Scale to suit.

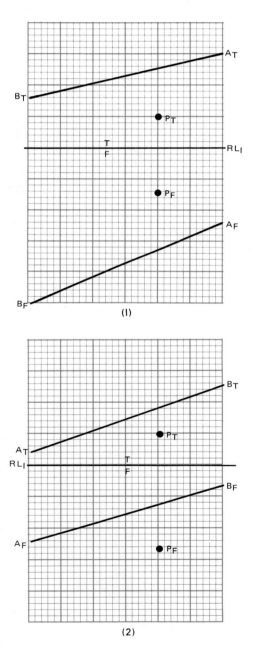

(1)

(2)

FIG. 7-9-A   Assignments.

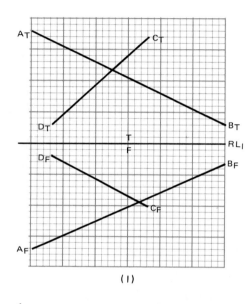

(1)

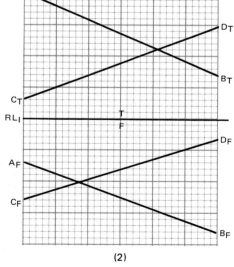

(2)

FIG. 7-9-B   Assignments.

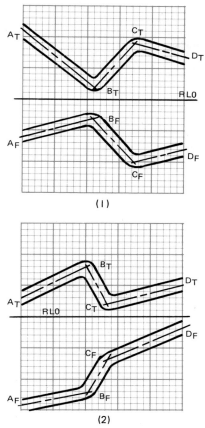

(1)

(2)

FIG. 7-10-A   Assignments.

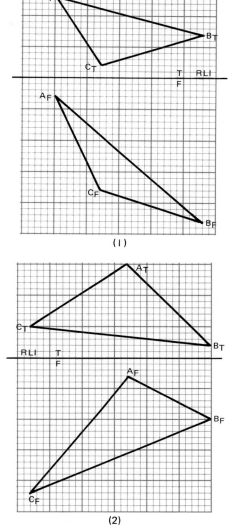

(1)

(2)

FIG. 7-10-B   Assignments.

ASSIGNMENTS FOR UNIT 7-11, ANGLES BETWEEN LINES AND PLANES

17. *The Angle a Line Makes with a Plane.* With the use of a grid, reference lines, and auxiliary views, find the angle the line makes with a plane for the layout shown in Fig. 7-11-A. Scale to suit.

FIG. 7-11-A   Assignments.

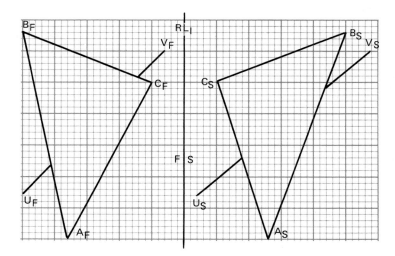

18. *The Angle a Line Makes with a Plane.* With the use of a grid, reference lines, and auxiliary views, find the angle the line makes with a plane for the layout shown in Fig. 7-11-B. Scale to suit.

19. *Edge Line of Two Planes.* With the use of a grid, reference lines, and auxiliary views, find the edge line of the two planes shown in the two problems in Fig. 7-11-C. Scale to suit.

FIG. 7-11-B   Assignments.

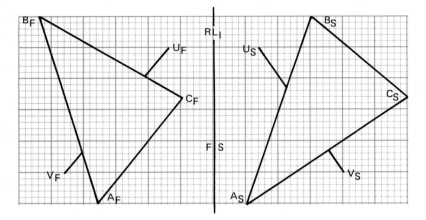

FIG. 7-11-C   Assignments.

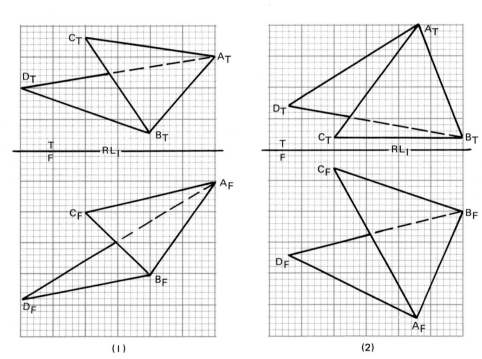

(1)   (2)

CHAPTER 8

BASIC DIMENSIONING

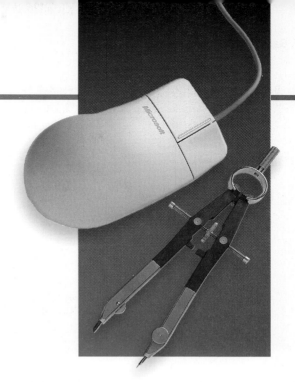

Definitions

Chamfer To cut away the inside or outside piece to facilitate assembly.

Counterbore A flat-bottomed, cylindrical recess that permits the head of a fastening device to lie recessed into the part.

Countersink An angular-sided recess to accommodate the head of flathead screws, rivets, and similar items.

Datum dimensioning The method in which several dimensions emanate from a common reference point or line.

Dimensions Lines and numerical values used to define geometric characteristics, such as lengths, diameters, angles, and locations.

Extension (projection) lines Lines used to indicate the point or line on the drawing to which the dimension applies.

Fits The clearance or interference between two mating parts.

Leaders A straight, inclined line used to direct notes, dimensions, symbols, item numbers, or part numbers to features on the drawing.

Mass production Production in which parts are produced in quantity, requiring special tools and gages.

Notes Written information used to simplify or complement dimensions; they may be general or local (specific).

Slope The slant of a line representing an inclined surface.

Spotface An area where the surface is machined just enough to provide smooth, level seating for a bolt head, nut, or washer.

Symmetrical The quality in which features on each side of the center line or median plane are identical in size, shape, and location.

Taper The ratio of the difference in the diameters of two sections.

Tolerances The permissible variations in the specified form, size, or location of individual features of a part.

Undercutting or **necking** The process of cutting a recess in a diameter to permit two parts to come together.

Unit production Production in which each part is made separately using general-purpose tools and machines.

8-1 BASIC DIMENSIONING

A working drawing is one from which a part can be produced. The drawing must be a complete set of instructions, so that it will not be necessary to give further information to the people fabricating the object. A *working drawing,* then, consists of the views necessary to explain the shape, the dimensions needed for manufacture, and required specifications, such as material and quantity needed. The latter information may be found in the notes on the drawing, or it may be located in the title block.

Dimensioning

Dimensions are given on drawings by extension lines, dimension lines, leaders, arrowheads, figures, notes, and symbols. They define geometric characteristics, such as lengths, diameters, angles, and locations (Fig. 8-1-1, pg. 172). The lines used in dimensioning are thin in contrast to the outline of the object. The dimensions must be clear and concise and permit only one interpretation. In general, each surface, line, or point is located by only one set of dimensions. These dimensions are not duplicated in other views. Deviations from the approved rules for dimensioning should be made only in exceptional cases, when they will improve the clarity of the dimensions. An exception to these rules is for arrowless and tabular dimensioning, which is discussed in Unit 8-4.

Drawings for industry require some form of tolerancing on dimensions so that components can be properly assembled and manufacturing and production requirements can be met. This

FIG. 8-1-1 Basic dimensioning elements.

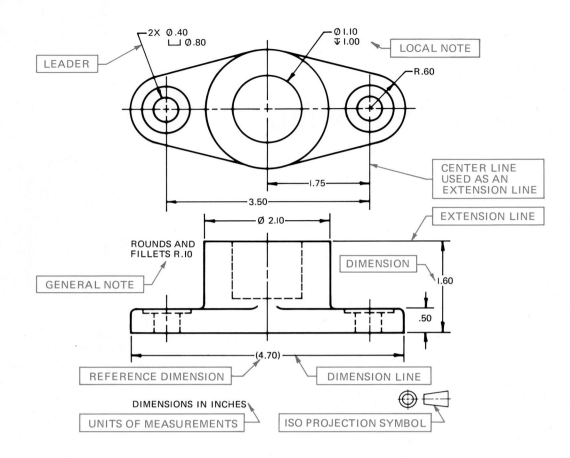

chapter deals only with basic dimensioning and tolerancing techniques. Geometric tolerancing, such as true positioning and tolerance of form, is covered in detail in Chap. 16.

Dimension and Extension Lines

Dimension lines are used to determine the extent and direction of dimensions, and they are normally terminated by uniform arrowheads, as shown in Fig. 8-1-2. Using an oblique line in lieu of an arrowhead is a common method used in architectural drafting. The recommended length and width of arrowheads should be in a ratio of 3:1 (Fig. 8-1-3B). The length of the arrowhead should be equal to the height of the dimension numerals.

A single style of arrowhead should be used throughout the drawing. An industrial practice is to use a small filled-in circle in lieu of two arrowheads where space is limited (Fig. 8-1-3D).

Preferably, dimension lines should be broken for insertion of the dimension that indicates the distance between the extension lines. Where dimension lines are not broken, the dimension is placed above the dimension line.

When several dimension lines are directly above or next to one another, it is good practice to stagger the dimensions in order to improve the clarity of the drawing. The spacing suitable for most drawings between parallel dimension lines is .38 in. (8mm), and the spacing between the outline of the object and the nearest dimension line should be approximately

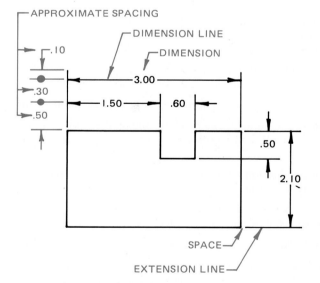

FIG. 8-1-2 Dimension and extension lines.

.50 in. (10mm). When the space between the extension lines is too small to permit the placing of the dimension line complete with arrowheads and dimension, the alternate method of placing the dimension line, dimension, or both outside the extension lines is used (Fig. 8-1-3D). Center lines should never be

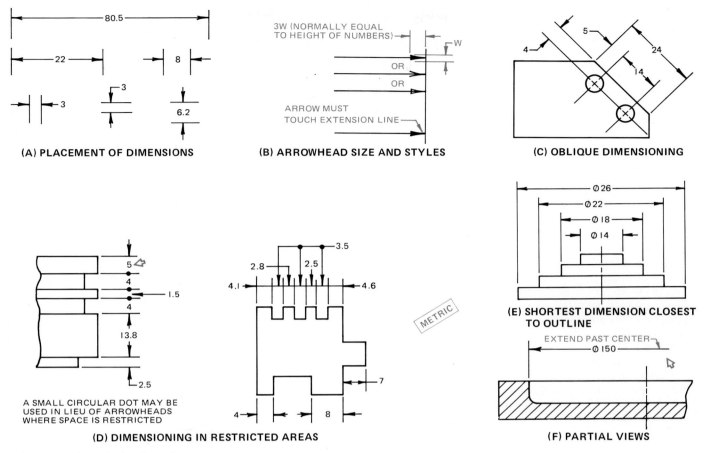

(A) PLACEMENT OF DIMENSIONS

(B) ARROWHEAD SIZE AND STYLES

(C) OBLIQUE DIMENSIONING

A SMALL CIRCULAR DOT MAY BE USED IN LIEU OF ARROWHEADS WHERE SPACE IS RESTRICTED

(D) DIMENSIONING IN RESTRICTED AREAS

(E) SHORTEST DIMENSION CLOSEST TO OUTLINE

(F) PARTIAL VIEWS

FIG. 8-1-3   Dimensioning linear features.

used for dimension lines. Every effort should be made to avoid crossing dimension lines by placing the shortest dimension closest to the outline (Fig. 8-1-3E).

Avoid dimensioning to hidden lines. In order to do so, it may be necessary to use a sectional view or a broken-out section. When the termination for a dimension is not included, as when used on partial or sectional views, the dimension line should extend beyond the center of the feature being dimensioned and shown with only one arrowhead (Fig. 8-1-3F).

Dimension lines should be placed outside the view where possible and should extend to extension lines rather than visible lines. However, when readability is improved by avoiding either extra long extension lines (Fig. 8-1-4) or the crowding of dimensions, placing of dimensions on views is permissible.

Extension (projection) lines are used to indicate the point or line on the drawing to which the dimension applies (Fig. 8-1-5, pg. 174). A small gap is left between the extension line and the outline to which it refers, and the extension line extends about .12 in. (3mm) beyond the outermost dimension line. However, when extension lines refer to points, as in Fig. 8-1-5E, they should extend through the points. Extension lines are usually drawn perpendicular to dimension lines. However, to improve clarity or when there is overcrowding, extension lines may be drawn at an oblique angle as long as clarity is maintained.

Center lines may be used as extension lines in dimensioning. The portion of the center line extending past the circle is not broken, as in Fig. 8-1-5B.

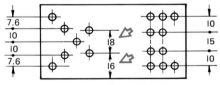

(A) IMPROVING READABILITY OF DRAWING

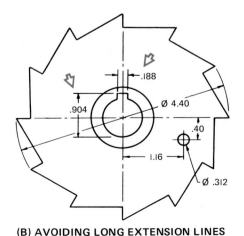

(B) AVOIDING LONG EXTENSION LINES

FIG. 8-1-4   Placing dimensions on views.

Where extension lines cross other extension lines, dimension lines, or visible lines, they are not broken. However, if extension lines cross arrowheads or dimension lines close to arrowheads, a break in the extension line is recommended (Fig. 8-1-5C).

Leaders

Leaders are used to direct notes, dimensions, symbols, item numbers, or part numbers to features on the drawing. See Fig. 8-1-6. A leader should generally be a single straight inclined line (not vertical or horizontal) except for a short horizontal portion extending to the center of the height of the first or last letter or digit of the note. The leader is terminated by an arrowhead or a dot of at least .06 in. (1.5 mm) in diameter. Arrowheads should always terminate on a line; dots should be used within the outline of the object and rest on a surface. Leaders should not be bent in any way unless it is unavoidable. Leaders should not cross one another, and two or more leaders adjacent to one another should be drawn parallel if practicable. It is better to repeat dimensions or references than to use long leaders.

Where a leader is directed to a circle or circular arc, its direction should point to the center of the arc or circle. Regardless of the reading direction used, aligned or unidirectional, all notes and dimensions used with leaders are placed in a horizontal position.

Notes

Notes are used to simplify or complement dimensioning by giving information on a drawing in a condensed and systematic manner. They may be general or local notes, and should be in the present or future tense.

General Notes   These refer to the part or the drawing as a whole. They should be shown in a central position below the view to which they apply or placed in a general note column. Typical examples of this type of note are:

- FINISH ALL OVER
- ROUNDS AND FILLETS R .06
- REMOVE ALL SHARP EDGES

Local Notes   These apply to local requirements only and are connected by a leader to the point to which the note applies.

Repetitive features and dimensions may be specified in the local note by the use of an X in conjunction with the numeral to indicate the "number of times" or "places" they are required (see Figs. 8-1-1 and 8-1-6). A full space is left between the X and the feature dimension. For additional information refer to Unit 8-3.

Typical examples are:

- $4 \times \emptyset 6$
- $2 \times 45°$
- $\emptyset 3$
  $\vee \emptyset 11.5 \times 86°$
- $M12 \times 1.25$

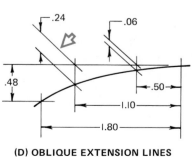

(A) USE OF EXTENSION LINES

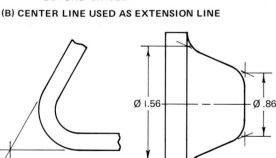

(B) CENTER LINE USED AS EXTENSION LINE

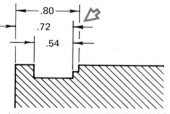

(C) BREAK IN EXTENSION LINES

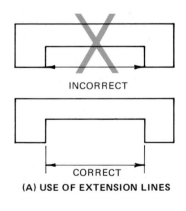

(D) OBLIQUE EXTENSION LINES

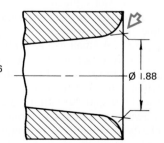

(E) EXTENSION LINE FROM POINTS

FIG. 8-1-5   Extension (projection) lines.

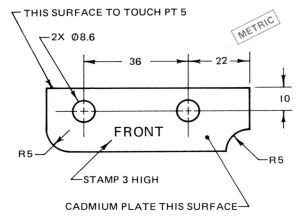

FIG. 8-1-6   Leaders.

Units of Measurement

Although the metric system of dimensioning is becoming the official international standard of measurement, most drawings in the United States are still dimensioned in inches or feet and inches. For this reason, drafters should be familiar with all the dimensioning systems that they may encounter. The dimensions used in this book are primarily decimal inch. However, metric and dual dimensions shown in the problems in this text are also used.

On drawings where all dimensions are either inches or millimeters, individual identification of linear units is not required. However, the drawing should contain a note stating the units of measurement.

Where some inch dimensions, such as nominal pipe sizes, are shown on a millimeter-dimensioned drawing, the abbreviation IN. must follow the inch values.

Inch Units of Measurement

Decimal-Inch System   (U.S. customary linear units)   Parts are designed in basic decimal increments, preferably .02 in., and are expressed with a minimum of two figures to the right of the decimal point (Fig. 8-1-7). Using the .02 in. module, the second decimal place (hundredths) is an even number or zero. By using the design modules having an even number for the last digit, dimensions can be halved for center distances without

increasing the number of decimal places. Decimal dimensions that are not multiples of .02, such as .01, .03, and .15, should be used only when it is essential to meet design requirements, such as to provide clearance, strength, smooth curves, etc. When greater accuracy is required, sizes are expressed as three- or four-place decimal numbers, for example, 1.875.

Whole dimensions will show a minimum of two zeros to the right of the decimal point.

$$24.00 \quad not \quad 24$$

An inch value of less than 1 is shown without a zero to the left of the decimal point.

$$.44 \quad not \quad 0.44$$

In cases where parts have to be aligned with existing parts or commercial products, which are dimensioned in fractions, it may be necessary to use decimal equivalents of fractional dimensions.

Fractional-Inch System   This system of dimensioning has not been recommended for use by ANSI for many years. It is shown in this text only for reference purposes or when making changes to drawings that may still be in use. In this system, parts are designed in basic units of common fractions down to 1/64 in. Decimal dimensions are used when finer divisions than 1/64 in. must be made. Common fractions may be used for specifying the size of holes that are produced by drills ordinarily stocked in fraction sizes and for the sizes of standard screw threads.

When common fractions are used on drawings, the fraction bar must not be omitted and should be horizontal except when applied with a typewriting machine that does not have a horizontal fraction bar.

When a dimension intermediate between 1/64 increments is necessary, it is expressed in decimals, such as .30, .257, or .2575 in.

The inch marks (") are not shown with dimensions. A note such as

DIMENSIONS ARE IN INCHES

should be clearly shown on the drawing. The exception is when the dimension "1 in." is shown on the drawing. The 1 should then be followed by the inch marks—1", *not* 1.

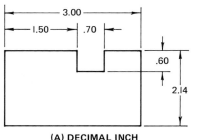

(A) DECIMAL INCH

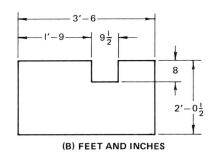

(B) FEET AND INCHES

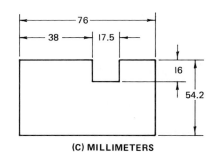

(C) MILLIMETERS

FIG. 8-1-7   Dimensioning units.

Foot-and-Inch System   Feet and inches are often used for installation drawings, drawings of large objects, and floor plans associated with architectural work. In this system all dimensions 12 in. or greater are specified in feet and inches. For example, 24 in. is expressed as 2'–0, and 27 in. is expressed as 2'–3. Parts of an inch are usually expressed as common fractions, rather than as decimals.

The inch marks (") are not shown on drawings. The drawing should carry a note such as

DIMENSIONS ARE IN FEET AND INCHES UNLESS OTHERWISE SPECIFIED

A dash should be placed between the foot and inch values. For example, 1'–3, not 1'3.

SI Metric Units of Measurement

The standard metric units on engineering drawings are the millimeter (mm) for linear measure and micrometer (μm) for surface roughness (Fig. 8-1-7C). For architectural drawings, meter and millimeter units are used.

Whole numbers from 1 to 9 are shown without a zero to the left of the number or a zero to the right of the decimal point.

$$2 \quad not \quad 02 \text{ or } 2.0$$

A millimeter value of less than 1 is shown with a zero to the left of the decimal point.

$$0.2 \quad not \quad .2 \text{ or } .20$$
$$0.26 \quad not \quad .26$$

Decimal points should be uniform and large enough to be clearly visible on reduced-size prints. They should be placed in line with the bottom of the associated digits and be given adequate space.

Neither commas nor spaces are used to separate digits into groups in specifying millimeter dimensions on drawings.

$$32545 \quad not \quad 32\ 545$$

Identification   A metric drawing should include a general note, such as

UNLESS OTHERWISE SPECIFIED DIMENSIONS ARE IN MILLIMETERS

and be identified by the word METRIC prominently displayed near the title block.

Units Common to Either System

Some measurements can be stated so that the callout will satisfy the units of both systems. For example, tapers such as .006 in. per inch and 0.006 mm per millimeter can both be expressed simply as the ratio 0.006:1 or in a note such as TAPER 0.006:1. Angular dimensions are also specified the same in both inch and metric systems.

Standard Items

Fasteners and Threads   Either inch or metric fasteners and threads may be used. Refer to the Appendix and Chap. 10 for additional information.

Hole Sizes   Tables showing standard inch and metric drill sizes are shown in the Appendix.

Dual Dimensioning

With the great exchange of drawings taking place between the United States and the rest of the world, at one time it became advantageous to show drawings in both inches and millimeters. As a result, many internationally operated companies adopted a dual system of dimensioning. Today, however, this type of dimensioning should be avoided if possible. When it may be necessary or desirable to give dimensions in both inches and millimeters on the same drawing, as shown in Figs. 8-1-8 and 8-1-9, a note or illustration should be located near the title block or strip to identify the inch and millimeter dimensions. Examples are

$$\frac{\text{MILLIMETER}}{\text{INCH}} \quad and/or \quad \text{MILLIMETER [INCH]}$$

Angular Units

Angles are measured in degrees. The decimal degree is now preferred over the use of degrees, minutes, and seconds. For example, the use of 60.5° is preferred to the use of 60° 30'. Where only minutes or seconds are specified, the number of minutes or seconds is preceded by 0°, or 0°0', as applicable. Some examples follow.

Decimal Degree	Degrees, Minutes, and Seconds
10° ± 0.5°	10° ± 0°30'
0.75°	0°45'
0.004°	0°0'15"
90° ± 1.0°	90° ± 1°
25.6 ± 0.2°	25°36' ± 0°12'
25.51°	25° 30'36"

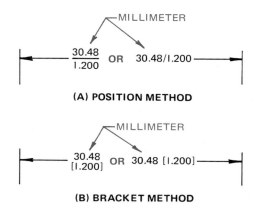

(A) POSITION METHOD

(B) BRACKET METHOD

FIG. 8-1-8   Dual dimensioning.

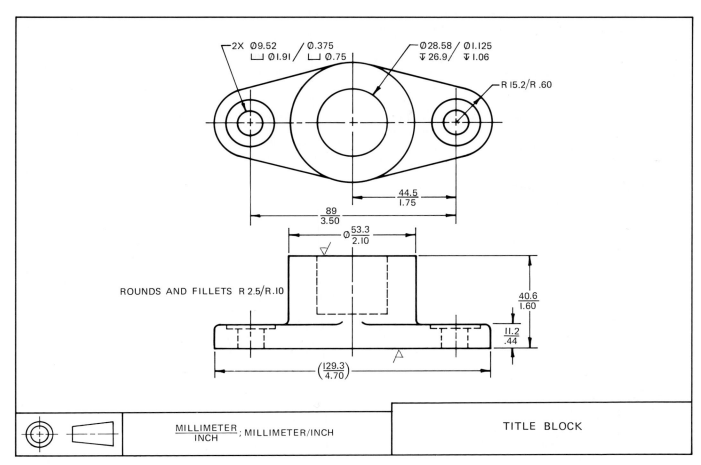

FIG. 8-1-9   Dual-dimensioned drawing.

FIG. 8-1-10   Angular units.

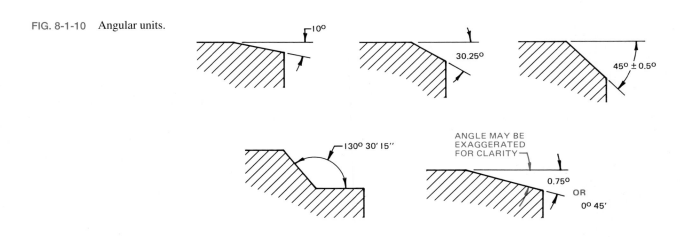

The dimension line of an angle is an arc drawn with the apex of the angle as the center point for the arc, wherever practicable. The position of the dimension varies according to the size of the angle and appears in a horizontal position. Refer to the recommended arrangements as shown in Fig. 8-1-10.

Reading Direction

Dimensions and notes should be placed to be read from the bottom of the drawing for engineering drawings (unidirectional system). For architectural and structural drawings the aligned system of dimensioning is used (Fig. 8-1-11, pg. 178).

In both methods angular dimensions and dimensions and notes shown with leaders should be aligned with the bottom of the drawing.

Basic Rules for Dimensioning

Refer to Fig. 8-1-12.

- Place dimensions between the views when possible.
- Place the dimension line for the shortest length, width, or height nearest the outline of the object. Parallel dimension lines are placed in order of their size, making the longest dimension line the outermost.

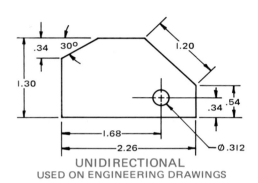

UNIDIRECTIONAL
USED ON ENGINEERING DRAWINGS

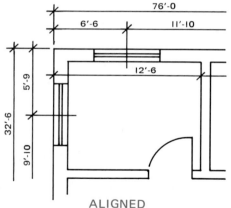

ALIGNED
USED ON ARCHITECTURAL AND STRUCTURAL DRAWINGS

FIG. 8-1-11   Reading direction of dimensions.

- Place dimensions with the view that best shows the characteristic contour or shape of the object. When this rule is applied, dimensions will not always be between views.
- On large views, dimensions can be placed on the view to improve clarity.
- Use only one system of dimensions, either the unidirectional or the aligned, on any one drawing.
- Dimensions should not be duplicated in other views.
- Dimensions should be selected so that it will not be necessary to add or subtract dimensions in order to define or locate a feature.

Symmetrical Outlines

A part is said to be *symmetrical* when the features on each side of the center line or median plane are identical in size, shape, and location. Partial views are often drawn for the sake of economy or space. With CAD duplicating, the other half of the view requires little effort. However, space constraints may preclude this possibility. When only one-half the outline of a symmetrically shaped part is drawn, symmetry is indicated by applying the symmetry symbol to the center line on both sides of the part (Fig. 8-1-13). In such cases, the outline of the part should extend slightly beyond the center line and terminate with a break line. Note the dimensioning method of extending the dimension lines to act as extension lines for the perpendicular dimensions.

Reference Dimensions

A reference dimension is shown for information only, and it is not required for manufacturing or inspection purposes. It is enclosed in parentheses, as shown in Fig. 8-1-14. Formerly the abbreviation REF was used to indicate a reference dimension.

Not-to-Scale Dimensions

When a dimension on a drawing is altered, making it not to scale, it should be underlined (underscored) with a straight, thick line (Fig. 8-1-15), except when the condition is clearly shown by break lines.

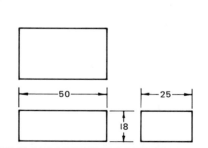

(A) PLACE DIMENSIONS BETWEEN VIEWS

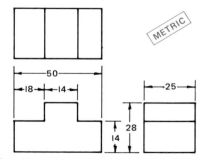

(B) PLACE SMALLEST DIMENSION NEAREST THE VIEW BEING DIMENSIONED

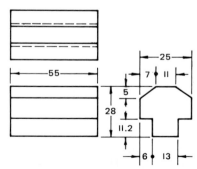

(C) DIMENSION THE VIEW THAT BEST SHOWS THE SHAPE

FIG. 8-1-12   Basic dimensioning rules.

FIG. 8-1-13   Dimensioning symmetrical outlines or features.

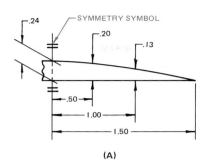

(A)

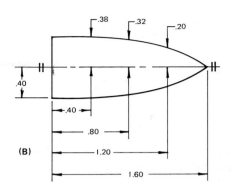

(B)

Operational Names

The use of operational names, such as turn, bore, grind, ream, tap, and thread, with dimensions should be avoided. While the drafter should be aware of the methods by which a part can be produced, the method of manufacture is better left to the producer. If the completed part is adequately dimensioned and has surface texture symbols showing finish quality desired, it remains a shop problem to meet the drawing specifications.

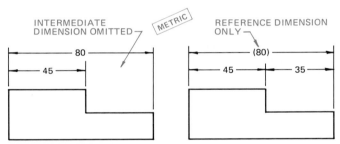

FIG. 8-1-14   Reference dimensions.

FIG. 8-1-15   Not-to-scale dimensions.

FIG. 8-2-1   Dimensioning diameters.

Abbreviations

Abbreviations and symbols are used on drawings to conserve space and time, but only where their meanings are quite clear. See the Appendix for commonly accepted abbreviations.

REFERENCES AND SOURCE MATERIAL

1. ASME Y14.5M–1994, *Dimensioning and Tolerancing.*
2. CAN/CSA B78.2–M91, *Dimensioning and Tolerancing of Technical Drawings.*

ASSIGNMENTS

See Assignments 1 through 3 for Unit 8-1 on pages 209–211.

8-2 DIMENSIONING CIRCULAR FEATURES

Diameters

Where the diameter of a single feature or the diameters of a number of concentric cylindrical features are to be specified, it is recommended that they be shown on the longitudinal view (Fig. 8-2-1).

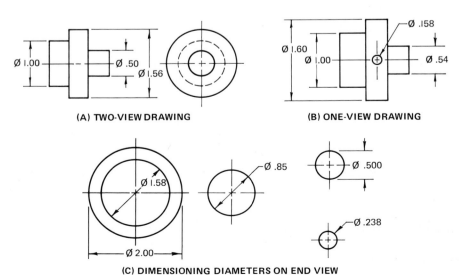

(A) TWO-VIEW DRAWING

(B) ONE-VIEW DRAWING

(C) DIMENSIONING DIAMETERS ON END VIEW

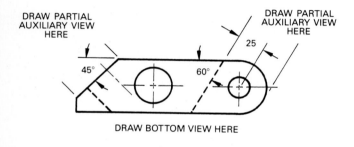

DRAW PARTIAL
AUXILIARY VIEW
HERE

DRAW PARTIAL
AUXILIARY VIEW
HERE

25

45°   60°

DRAW BOTTOM VIEW HERE

4. Make a working drawing of one of the parts shown in Figs. 7-3-D through 7-3-F. Draw partial auxiliary views for Fig. 7-3-D; partial auxiliary and side views for Fig. 7-3-E; and full top and front views, and partial auxiliary views for Fig. 7-3-F.

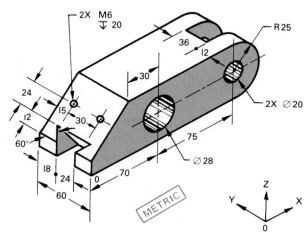

FIG. 7-3-D   Dovetail bracket.

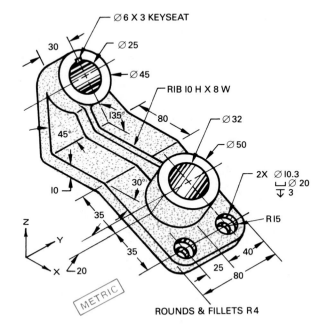

FIG. 7-3-F   Offset guide.

ROUNDS & FILLETS R 4

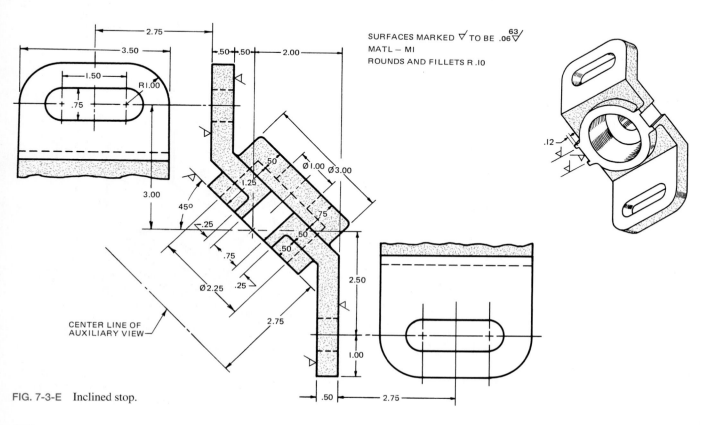

SURFACES MARKED ∇ TO BE .06 63/∇
MATL – MI
ROUNDS AND FILLETS R .10

FIG. 7-3-E   Inclined stop.

Where space is restricted or when only a partial view is used, diameters may be dimensioned as illustrated in Fig. 8-2-2. Regardless of where the diameter dimension is shown, the numerical value is preceded by the diameter symbol Ø for both customary and metric dimensions.

Radii

The general method of dimensioning a circular arc is by giving its radius. A radius dimension line passes through, or is in line with, the radius center and terminates with an arrowhead touching the arc (Fig. 8-2-3). An arrowhead is never used at the radius center. The size of the dimension is preceded by the abbreviation R for both customary and metric dimensioning. Where space is limited, as for a small radius, the radial dimension line may extend through the radius center. Where it is inconvenient to place the arrowhead between the radius center and the arc, it may be placed outside the arc, or a leader may be used (Fig. 8-2-3A).

Where a dimension is given to the center of the radius, a small cross should be drawn at the center (Fig. 8-2-3B). Extension lines and dimension lines are used to locate the

center. Where the location of the center is unimportant, a radial arc may be located by tangent lines (Fig. 8-2-3E).

Where the center of a radius is outside the drawing or interferes with another view, the radius dimension line may be foreshortened (Fig. 8-2-3D). The portion of the dimension line next to the arrowhead should be radial relative to the curved line. Where the radius dimension line is foreshortened and the center is located by coordinate dimensions, the dimensions locating the center should be shown as foreshortened or the dimension shown as not to scale.

Simple fillet and corner radii may also be dimensioned by a general note, such as

ALL ROUNDS AND FILLETS UNLESS OTHERWISE SPECIFIED R.20
or
ALL RADII R5

Where a radius is dimensioned in a view that does not show the true shape of the radius, TRUE R is added before the radius dimension, as illustrated in Fig. 8-2-4.

Rounded Ends

Overall dimensions should be used for parts or features having rounded ends. For fully rounded ends, the radius (R) is shown but not dimensioned (Fig. 8-2-5A). For parts with partially rounded ends, the radius is dimensioned (Fig. 8-2-5B). Where a hole and radius have the same center and the hole location is more critical than the location of a radius, either the radius or the overall length should be shown as a reference dimension (Fig. 8-2-5C).

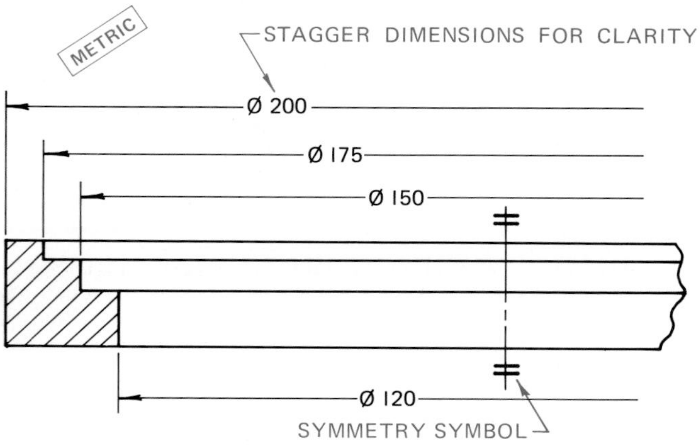

FIG. 8-2-2   Dimensioning diameters where space is restricted.

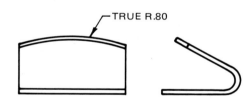

FIG. 8-2-4   Indicating true radius.

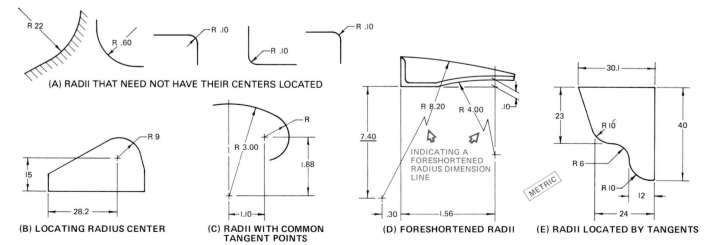

FIG. 8-2-3   Dimensioning radii.

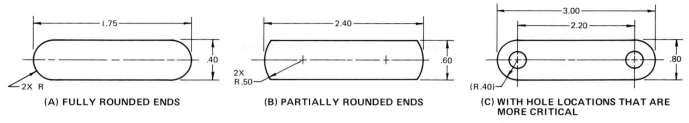

FIG. 8-2-5   Dimensioning external surfaces with rounded ends.

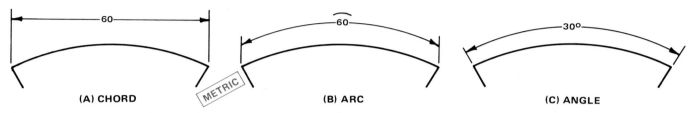

FIG. 8-2-6   Dimensioning chords, arcs, and angles.

Dimensioning Chords, Arcs, and Angles

The difference in dimensioning chords, arcs, and angles is shown in Fig. 8-2-6.

Spherical Features

Spherical surfaces may be dimensioned as diameters or radii, but the dimension should be used with the abbreviations SR or SØ (Fig. 8-2-7).

Cylindrical Holes

Plain, round holes are dimensioned in various ways, depending upon design and manufacturing requirements (Fig. 8-2-8). However, the leader is the method most commonly used. When a *leader* is used to specify diameter sizes, as with small holes, the dimension is identified as a diameter by preceding the numerical value with the diameter symbol Ø.

The size, quantity, and depth may be shown on a single line, or on several lines if preferable. For through holes, the abbreviation THRU should follow the dimension if the drawing does not make this clear. The depth dimension of a blind hole is the depth of the full diameter and is normally included as part of the dimensioning note.

When more than one hole of a size is required, the number of holes should be specified. However, care must be taken to avoid placing the hole size and quantity values together without adequate spacing. It may be better to show the note on two or more lines than to use a single line note that might be misread (Fig. 8-2-8D).

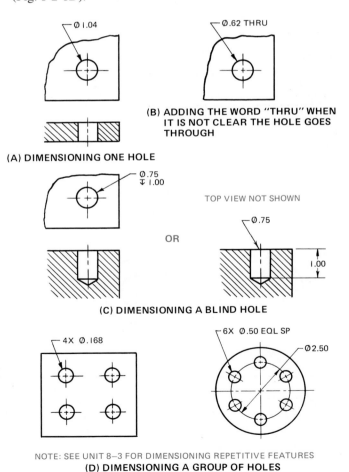

FIG. 8-2-8   Dimensioning cylindrical holes.

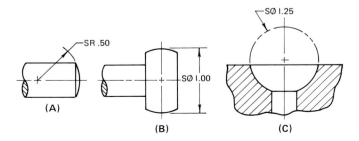

FIG. 8-2-7   Dimensioning spherical surfaces.

Minimizing Leaders   If too many leaders would impair the legibility of a drawing, letters or symbols as shown in Fig. 8-2-9 should be used to identify the features.

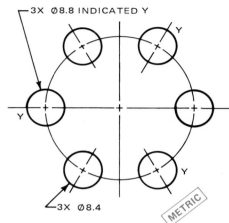

FIG. 8-2-9   Minimizing leaders.

Slotted Holes

Elongated holes and slots are used to compensate for inaccuracies in manufacturing and to provide for adjustment. See Fig. 8-2-10. The method selected to locate the slot would depend on how the slot was made. The method shown in Fig. 8-2-10B is used when the slot is punched out and the location of the punch is given. Figure 8-2-10A shows the dimensioning method used when the slot is machined out.

Countersinks, Counterbores, and Spotfaces

Counterbores, spotfaces, and countersinks are specified on drawings by means of dimension symbols or abbreviations, the symbols being preferred. The symbols or abbreviations indicate the form of the surface only and do not restrict the methods used to produce that form. The dimensions for them are usually given as a note, preceded by the size of the through hole (Figs. 8-2-11 and 8-2-12).

A *countersink* is an angular-sided recess to accommodate the head of flathead screws, rivets, and similar items. The diameter at the surface and the included angle are given. When the depth of the tapered section of the countersink is critical, it

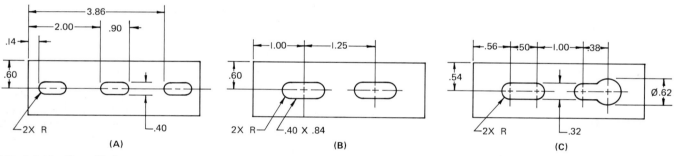

FIG. 8-2-10   Slotted holes.

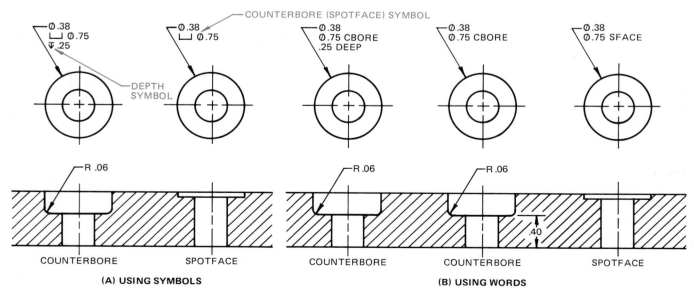

FIG. 8-2-11   Counterbored and spotfaced holes.

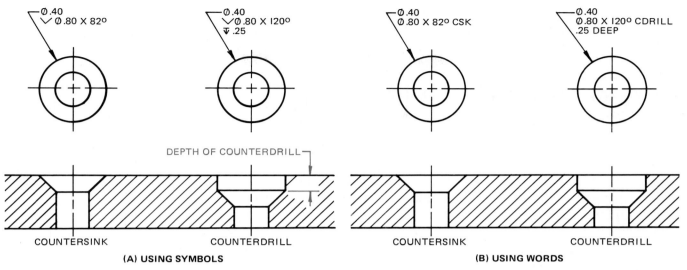

FIG. 8-2-12   Countersunk and counterdrilled holes.

is specified in the note or by a dimension. For counterdrilled holes, the diameter, depth, and included angle of the counterdrill are given.

A *counterbore* is a flat-bottomed, cylindrical recess that permits the head of a fastening device, such as a bolt, to lie recessed into the part. The diameter, depth, and corner radius are specified in a note. In some cases, the thickness of the remaining stock may be dimensioned rather than the depth of the counterbore.

A *spotface* is an area where the surface is machined just enough to provide smooth, level seating for a bolt head, nut, or washer. The diameter of the faced area and either the depth or the remaining thickness are given. A spotface may be specified by a general note and not delineated on the drawing. If no depth or remaining thickness is specified, it is implied that the spotfacing is the minimum depth necessary to clean up the surface to the specified diameter.

The symbols for counterbore or spotface, countersink, and depth are shown in Figs. 8-2-11 and 8-2-12. In each case the symbol precedes the dimension.

REFERENCE AND SOURCE MATERIAL

1. ASME Y14.5M–1994, *Dimensioning and Tolerancing.*
2. CAN/CSA B78.2–M91, *Dimensioning and Tolerancing of Technical Drawings.*

ASSIGNMENTS

See Assignments 4 through 6 for Unit 8-2 on pages 212–214.

8-3 DIMENSIONING COMMON FEATURES

Repetitive Features and Dimensions

Repetitive features and dimensions may be specified on a drawing by the use of an X in conjunction with the numeral to indicate the "number of times" or "places" they are required. A space is shown between the X and the dimension.

An X that means "by" is often used between coordinate dimensions specified in note form. Where both are used on a drawing, care must be taken to ensure each is clear (Fig. 8-3-1, pg. 184).

Chamfers

The process of *chamfering,* that is, cutting away the inside or outside piece, is done to facilitate assembly. Chamfers are normally dimensioned by giving their angle and linear length (Fig. 8-3-2, pg. 184). When the chamfer is 45°, it may be specified as a note.

When a very small chamfer is permissible, primarily to break a sharp corner, it may be dimensioned but not drawn, as in Fig. 8-3-2C. If not otherwise specified, an angle of 45° is understood.

Internal chamfers may be dimensioned in the same manner, but it is often desirable to give the diameter over the chamfer. The angle may also be given as the included angle if this is a design requirement. This type of dimensioning is generally necessary for larger diameters, especially those over 2 in. (50mm), whereas chamfers on small holes are usually expressed as countersinks. Chamfers are never measured along the angular surface.

Slopes and Tapers

Slope

A *slope* is the slant of a line representing an inclined surface. It is expressed as a ratio of the difference in the heights at right angles to the base line, at a specified distance apart (Fig. 8-3-3, pg. 185). Figure 8-3-3D is the preferred method of dimensioning slopes on architectural and structural drawings.

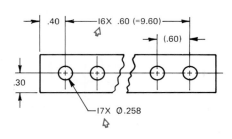

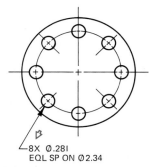

(A) USING "NUMBER OF TIMES" SYMBOL

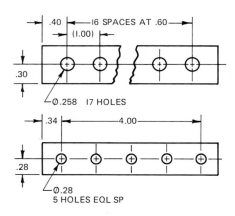

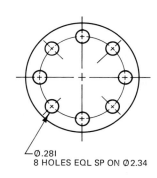

(B) USING DESCRIPTIVE NOTES

FIG. 8-3-1   Dimensioning repetitive detail.

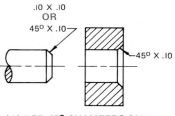

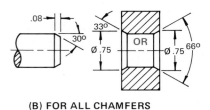

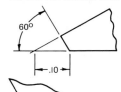

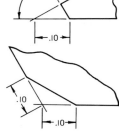

(A) FOR 45° CHAMFERS ONLY

(B) FOR ALL CHAMFERS

(C) SMALL CHAMFERS

(D) CHAMFERS BETWEEN SURFACES AT OTHER THAN 90°

FIG. 8-3-2   Dimensioning chamfers.

The following dimensions and symbol may be used, in different combinations, to define the slope of a line or flat surface:

- The slope specified as a ratio combined with the slope symbol (Fig. 8-3-3A).
- The slope specified by an angle (Fig. 8-3-3B).

- The dimensions showing the difference in the heights of two points from the base line and the distance between them (Fig. 8-3-3C).

Taper

A *taper* is the ratio of the difference in the diameters of two sections (perpendicular to the axis of a cone to the distance between these two sections), as shown in Fig. 8-3-4. When the taper symbol is used, the vertical leg is always shown to the left and precedes the ratio figures. The following dimensions may be used, in suitable combinations, to define the size and form of tapered features:

- The diameter (or width) at one end of the tapered feature
- The length of the tapered feature
- The rate of taper
- The included angle
- The taper ratio
- The diameter at a selected cross section
- The dimension locating the cross section

Knurls

Knurling is specified in terms of type, pitch, and diameter before and after knurling (Fig. 8-3-5). The letter P precedes the pitch number. Where control is not required, the diameter after knurling is omitted. Where only portions of a feature require

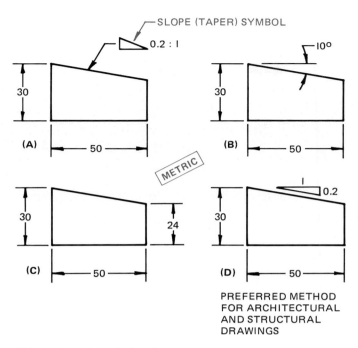

FIG. 8-3-3   Dimensioning slopes.

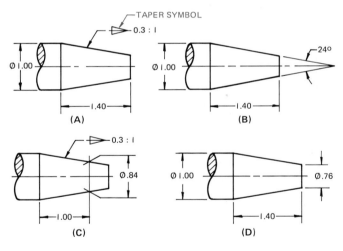

FIG. 8-3-4   Dimensioning tapers.

Formed Parts

In dimensioning formed parts, the inside radius is usually specified, rather than the outside radius, but all forming dimensions should be shown on the same side if possible. Dimensions apply to the side on which the dimensions are shown unless otherwise specified (Fig. 8-3-6).

Undercuts

The operation of *undercutting* or *necking,* that is, cutting a recess in a diameter, is done to permit two parts to come together, as illustrated in Fig. 8-3-7A (pg. 186). It is indicated on the drawing by a note listing the width first and then the diameter. If the radius is shown at the bottom of the undercut, it will be assumed that the radius is equal to one-half the width

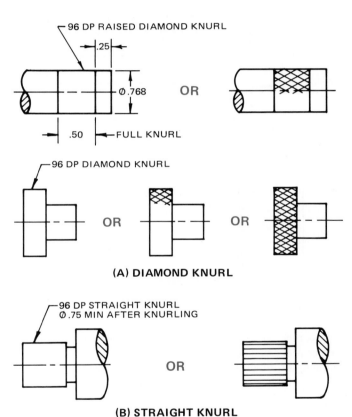

FIG. 8-3-5   Dimensioning knurls.

knurling, axial dimensions must be provided. Where required to provide a press fit between parts, knurling is specified by a note on the drawing that includes the type of knurl required, the pitch, the toleranced diameter of the feature prior to knurling, and the minimum acceptable diameter after knurling. Commonly used types are straight, diagonal, spiral, convex, raised diamond, depressed diamond, and radial. The pitch is usually expressed in terms of teeth per inch or millimeter and may be the straight pitch, circular pitch, or diametral pitch. For cylindrical surfaces, the latter is preferred.

The knurling symbol is optional and is used only to improve clarity on working drawings.

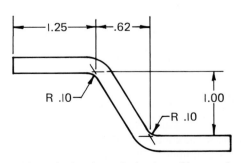

FIG. 8-3-6   Dimensioning theoretical points of intersection.

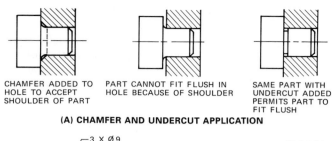

(A) CHAMFER AND UNDERCUT APPLICATION

CHAMFER ADDED TO HOLE TO ACCEPT SHOULDER OF PART

PART CANNOT FIT FLUSH IN HOLE BECAUSE OF SHOULDER

SAME PART WITH UNDERCUT ADDED PERMITS PART TO FIT FLUSH

(B) PLAIN UNDERCUT   METRIC   (C) UNDERCUT WITH RADIUS

FIG. 8-3-7   Dimensioning undercuts.

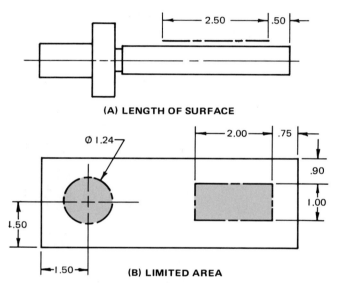

(A) LENGTH OF SURFACE

(B) LIMITED AREA

FIG. 8-3-8   Dimensioning limited lengths and areas.

unless otherwise specified, and the diameter will apply to the center of the undercut. When the size of the undercut is unimportant, the dimension may be left off the drawing.

Limited Lengths and Areas

Sometimes it is necessary to dimension a limited length or area of a surface to indicate a special condition. In such instances, the area or length is indicated by a chain line (Fig. 8-3-8A). When indicating a length of surface, the chain line is drawn parallel and adjacent to the surface. When indicating an area of surface, the area is cross-hatched within the chain line boundary (Fig. 8-3-8B).

Wire, Sheet Metal, and Drill Rod

Wire, sheet metal, and drill rod, which are manufactured to gage or code sizes, should be shown by their decimal dimensions; but gage numbers, drill letters, etc., may be shown in parentheses following those dimensions.

EXAMPLES

Sheet —.141 (NO. 10 USS GA)
    —.081 (NO. 12 B & S GA)

REFERENCES AND SOURCE MATERIAL

1. ASME Y14.5M–1994, *Dimensioning and Tolerancing.*
2. CAN/CSA B78.2–M91, *Dimensioning and Tolerancing of Technical Drawings.*

ASSIGNMENTS

See Assignments 7 through 12 for Unit 8-3 on pages 215 to 216.

8-4   DIMENSIONING METHODS

The choice of the most suitable dimensions and dimensioning methods will depend, to some extent, on how the part will be produced and whether the drawings are intended for unit or mass production. *Unit production* refers to cases where each part is to be made separately, using general-purpose tools and machines. *Mass production* refers to parts produced in quantity, where special tools and gages are usually provided.

Either linear or angular dimensions may locate features with respect to one another (point-to-point) or from a datum. Point-to-point dimensions may be adequate for describing simple parts. Dimensions from a datum may be necessary if a part with more than one critical dimension must mate with another part.

The following systems of dimensioning are used more commonly for engineering drawings.

Rectangular Coordinate Dimensioning

This is a method for indicating distance, location, and size by means of linear dimensions measured parallel or perpendicular to reference axes or datum planes that are perpendicular to one another. Coordinate dimensioning with dimension lines must clearly identify the datum features from which the dimensions originate (Fig. 8-4-1).

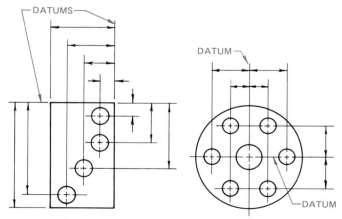

FIG. 8-4-1   Rectangular coordinate dimensioning.

Rectangular Coordinates for Arbitrary Points   Coordinates for arbitrary points of reference without a grid appear adjacent to each point (Fig. 8-4-2) or in tabular form (Fig. 8-4-3). CAD systems will automatically display any point coordinate when it is picked.

Rectangular Coordinate Dimensioning Without Dimension Lines   Dimensions may be shown on extension lines without the use of dimension lines or arrowheads. The base lines may be zero coordinates or they may be labeled as *X*, *Y*, and *Z* (Fig. 8-4-4).

Tabular Dimensioning   Tabular dimensioning is a type of coordinate dimensioning in which dimensions from mutually perpendicular planes are listed in a table on the drawing rather than on the pictorial delineation. This method is used on drawings that require the location of a large number of similarly shaped features when dimensioning parts for numerical control (Fig. 8-4-5).

It may be advantageous to have a part dimensioned symmetrically about its center, as shown in Fig. 8-4-6 (pg. 188). When the center lines are designated as the base (zero) lines, positive and negative values will occur. These values are shown with the dimensions locating the holes.

FIG. 8-4-2   Coordinates for arbitrary points.

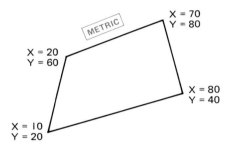

POINT	X	Y
I	10	20
2	80	40
3	70	80
4	20	60

FIG. 8-4-3   Coordinates for arbitrary points in tabular form.

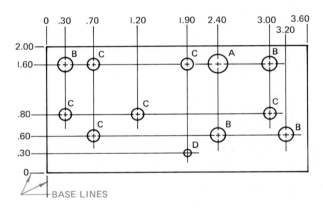

HOLE SYMBOL	HOLE SIZE
A	.246
B	.189
C	.154
D	.125

FIG. 8-4-4   Rectangular coordinate dimensioning without dimension lines (arrowless dimensioning).

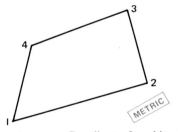

FIG. 8-4-5   Tabular dimensioning.

HOLE DIA	HOLE SYMBOL	LOCATION		
		X	Y	Z
5.6	A₁	60	40	18
4.8	B₁	10	40	THRU
	B₂	75	40	THRU
	B₃	60	16	THRU
	B₄	80	16	THRU
4	C₁	18	40	THRU
	C₂	55	40	THRU
	C₃	10	20	THRU
	C₄	30	20	THRU
	C₅	75	20	THRU
	C₆	18	16	THRU
3.2	D₁	55	8	12
8.1	E₁	42	20	12

FIG. 8-4-6   Tabular dimensioning with origin (0, 0) for *X* and *Y* axes located at the center of the part.

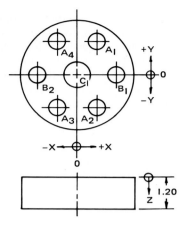

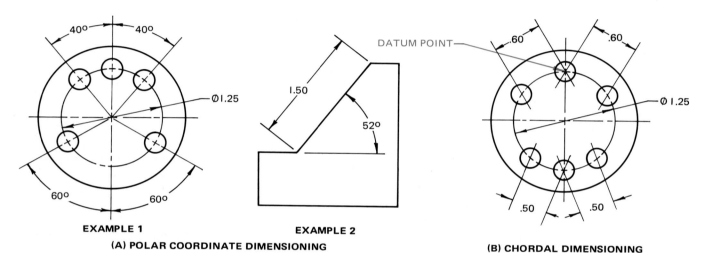

HOLE DIA	HOLE SYMBOL	LOCATION		
		X	Y	Z
.375	A_1	.50	.75	THRU
	A_2	.50	−.75	THRU
	A_3	−.50	−.75	THRU
	A_4	−.50	.75	THRU
.250	B_1	1.00	0	.60
	B_2	−1.00	0	.60
.500	C_1	0	0	THRU

EXAMPLE 1   EXAMPLE 2

(A) POLAR COORDINATE DIMENSIONING

(B) CHORDAL DIMENSIONING

FIG. 8-4-7   Polar coordinate and chordal dimensioning.

Polar Coordinate Dimensioning

Polar coordinate dimensioning is commonly used in circular planes or circular configurations of features. It is a method of indicating the position of a point, line, or surface by means of a linear dimension and an angle, other than 90°, that is implied by the vertical and horizontal center lines (Fig. 8-4-7A).

Chordal Dimensioning

The chordal dimensioning system may also be used for the spacing of points on the circumference of a circle relative to a datum, where manufacturing methods indicate that this will be convenient (Fig. 8-4-7B).

True-Position Dimensioning

True-position dimensioning has many advantages over the coordinate dimensioning system (Fig. 8-4-8). Because of its scope, it is covered as a complete topic in Chap. 16.

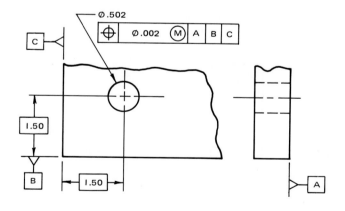

FIG. 8-4-8   True-position dimensioning.

Chain Dimensioning

When a series of dimensions is applied on a point-to-point basis, it is called chain dimensioning (Fig. 8-4-9). A possible disadvantage of this system is that it may result in an undesirable accumulation of tolerances between individual features. See Unit 8-5.

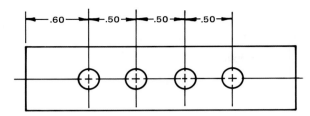

FIG. 8-4-9   Chain dimensioning.

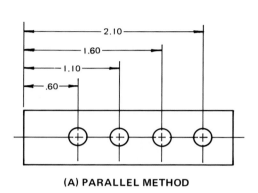

(A) PARALLEL METHOD

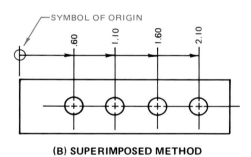

(B) SUPERIMPOSED METHOD

FIG. 8-4-10   Common-point (baseline) dimensioning.

Datum or Common-Point Dimensioning

When several dimensions emanate from a common reference point or line, the method is called *common-point* or *datum dimensioning*. Dimensioning from reference lines may be executed as parallel dimensioning or as a superimposed running dimensioning (Fig. 8-4-10).

Superimposed running dimensioning is simplified parallel dimensioning and may be used where there are space problems. Dimensions should be placed near the arrowhead, in line with the corresponding extension line, as shown in Fig. 8-4-10B. The origin is indicated by a circle and the opposite end of each dimension is terminated with an arrowhead.

It may be advantageous to use superimposed running dimensions in two directions. In such cases, the origins may be shown as in Fig. 8-4-11, or at the center of a hole or other feature.

REFERENCES AND SOURCE MATERIAL

1. ASME Y14.5M–1994, *Dimensioning and Tolerancing.*
2. CAN/CSA B78.2–M91, *Dimensioning and Tolerancing of Technical Drawings.*

ASSIGNMENTS

See Assignments 13 through 18 for Unit 8-4 on pages 217 through 219.

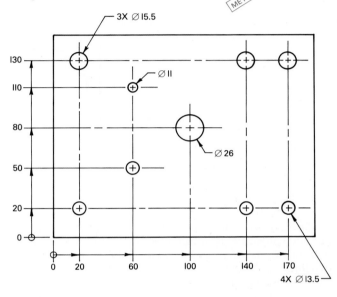

FIG. 8-4-11   Superimposed running dimensions in two directions.

8-5   LIMITS AND TOLERANCES

In the 6000 years of the history of technical drawing as a means for the communication of engineering information, it seems inconceivable that such an elementary practice as the tolerancing of dimensions, which we take so much for granted today, was introduced for the first time about 80 years ago.

Apparently, engineers and fabricators came in a very gradual manner to the realization that exact dimensions and shapes could not be attained in the manufacture of materials and products.

The skilled tradespeople of old prided themselves on being able to work to exact dimensions. What they really meant was that they dimensioned objects with a degree of accuracy greater than that with which they could measure. The use of modern measuring instruments would easily have shown the deviations from the sizes that they called exact.

As soon as it was realized that variations in the sizes of parts had always been present, that such variations could be restricted but not avoided, and also that slight variations in the size that a part was originally intended to have could be tolerated without its correct functioning being impaired, it was evident that interchangeable parts need not be identical parts, but that it would be sufficient if the significant sizes that controlled

FIG. 8-5-1   A working drawing.

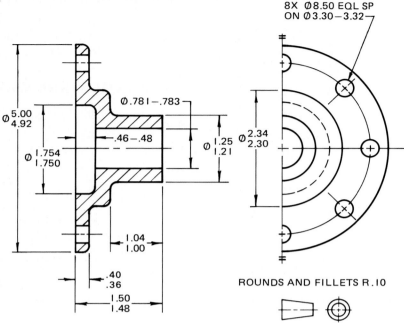

ROUNDS AND FILLETS R .10

their fits lay between definite limits. Accordingly, the problem of interchangeable manufacture evolved from the making of parts to a would-be exact size, to the holding of parts between two limiting sizes lying so closely together that any intermediate size would be acceptable.

Tolerances are the permissible variations in the specified form, size, or location of individual features of a part from that shown on the drawing. The finished form and size into which material is to be fabricated are defined on a drawing by various geometric shapes and dimensions.

As mentioned previously, the manufacturer cannot be expected to produce the exact size of parts as indicated by the dimensions on a drawing, so a certain amount of variation on each dimension must be tolerated. For example, a dimension given as $1.500 \pm .004$ in. means that the manufactured part can be anywhere between 1.496 and 1.504 in. and that the tolerance permitted on this dimension is .008 in. The largest and smallest permissible sizes (1.504 and 1.496 in., respectively) are known as the *limits*.

Greater accuracy costs more money, and since economy in manufacturing would not permit all dimensions to be held to the same accuracy, a system for dimensioning must be used (Fig. 8-5-1). Generally, most parts require only a few features to be held to high accuracy.

In order that assembled parts may function properly and to allow for interchangeable manufacturing, it is necessary to permit only a certain amount of tolerance on each of the mating parts and a certain amount of allowance between them.

Key Concepts

In order to calculate limit dimensions, the following concepts should be clearly understood (refer to Fig. 8-5-2).

Actual Size   The actual size is the measured size.

Basic Size   The basic size of a dimension is the theoretical size from which the limits for that dimension are derived by the application of the allowance and tolerance.

Design Size   Design size refers to the size from which the limits of size are derived by the application of tolerances.

Limits of Size   These limits are the maximum and minimum sizes permissible for a specific dimension.

Nominal Size   The nominal size is the designation used for the purpose of general identification.

Tolerance   The tolerance on a dimension is the total permissible variation in the size of a dimension. The tolerance is the difference between the limits of size.

Bilateral Tolerance   With bilateral tolerance, variation is permitted in both directions from the specified dimension.

TERMINOLOGY	EXAMPLE	EXPLANATION
BASIC SIZE	1.500	
BASIC SIZE WITH TOLERANCE ADDED	1.500 ±.004 ◀— HALF OF	TOTAL TOLERANCE
LIMITS OF SIZE	1.504 1.496	LARGEST AND SMALLEST SIZES PERMITTED
TOLERANCE	.008	DIFFERENCE BETWEEN LIMITS OF SIZE

FIG. 8-5-2   Limit and tolerance terminology.

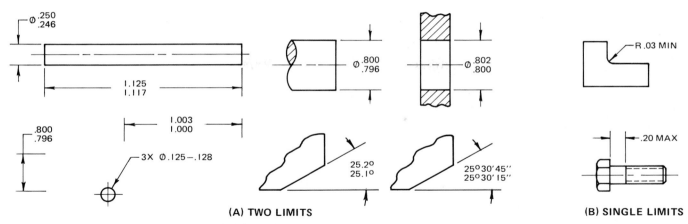

FIG. 8-5-3   Methods of indicating tolerances on drawings.

Unilateral Tolerance

Unilateral Tolerance   With unilateral tolerance, variation is permitted in only one direction from the specified dimension.

Maximum Material Size   The maximum material size is that limit of size of a feature that results in the part containing the maximum amount of material. Thus it is the maximum limit of size for a shaft or an external feature, or the minimum limit of size for a hole or internal feature.

Tolerancing

All dimensions required in the manufacture of a product have a tolerance, except those identified as reference, maximum, minimum, or stock. Tolerances may be expressed in one of the following ways:

- As specified limits of tolerances shown directly on the drawing for a specified dimension (Fig. 8-5-3).
- As plus-and-minus tolerancing.
- Combining a dimension with a tolerance symbol. (See "Symbols" in Unit 8-6.)
- In a general tolerance note, referring to all dimensions on the drawing for which tolerances are not otherwise specified.
- In the form of a note referring to specific dimensions.
- Tolerances on dimensions that locate features may be applied directly to the locating dimensions or by the positional tolerancing method described in Chap. 16.
- Tolerancing applicable to the control of form and runout, referred to as *geometric tolerancing,* is also covered in detail in Chap. 16.

Direct Tolerancing Methods

A tolerance applied directly to a dimension may be expressed in two ways.

Limit Dimensioning   For this method, the high limit (maximum value) is placed above the low limit (minimum value). When it is expressed in a single line, the low limit precedes the high limit and they are separated by a dash (Figs. 8-5-3 and 8-5-4).

Where limit dimensions are used and where either the maximum or minimum dimension has digits to the right of the decimal point, the other value should have the zeros added so that both the limits of size are expressed to the same number of decimal places. This applies to both U.S. customary and metric drawings. For example:

$$\begin{array}{ccccccc} 30.75 & & 30.75 & & .750 & & .75 \\ & not & & and & & not & \\ 30.00 & & 30 & & .748 & & .748 \end{array}$$

Plus-and-Minus Tolerancing   (Refer to Fig. 8-5-5, pg. 192.) For this method the dimension of the specified size is given first and is followed by a plus or minus expression of tolerancing. The plus value should be placed above the minus value. This type of tolerancing can be broken down into bilateral and

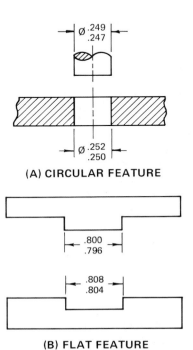

FIG. 8-5-4   Limit dimensioning application.

unilateral tolerancing. In a bilateral tolerance, the plus-and-minus tolerances should normally be equal, but special design considerations may sometimes dictate unequal values (Fig. 8-5-6). The specified size is the design size, and the tolerance represents the desired control of quality and appearance.

Metric Tolerancing   In the metric system the dimension need not be shown to the same number of decimal places as its tolerance. For example:

$$1.5 \pm 0.04 \quad not \quad 1.50 \pm 0.04$$
$$10 \pm 0.1 \quad not \quad 10.0 \pm 0.1$$

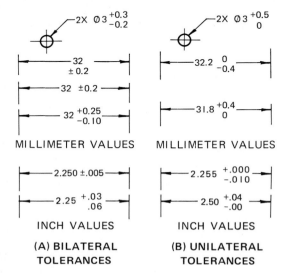

MILLIMETER VALUES

MILLIMETER VALUES

INCH VALUES

INCH VALUES

(A) BILATERAL TOLERANCES

(B) UNILATERAL TOLERANCES

FIG. 8-5-5   Plus-and-minus tolerancing.

Where bilateral tolerancing is used, both the plus and minus values have the same number of decimal places, using zeros where necessary. For example:

$$30^{+0.15}_{-0.10} \quad not \quad 30^{+0.15}_{-0.1}$$

Where unilateral tolerancing is used and either the plus or minus value is nil, a single zero is shown without a plus or minus sign. For example:

$$40^{\ 0}_{-0.15} \quad or \quad 40^{+0.15}_{\ 0}$$

An application of unilateral tolerancing is shown in Fig. 8-5-6B.

Inch Tolerancing   In the inch system the dimension is given to the same number or decimal places as its tolerance. For example:

Bilateral:

$$.500 \pm .004 \quad not \quad .50 \pm .004$$

Unilateral:

$$.750^{+.005}_{-.000} \quad not \quad .750^{+.005}_{-0}.$$
$$30.0° \pm .2° \quad not \quad 30° \pm .2°$$

Conversion charts for tolerances are shown in Fig. 8-5-7.

General Tolerance Notes   The use of general tolerance notes greatly simplifies the drawing and saves considerable layout in

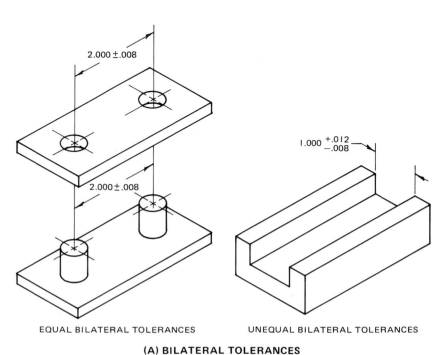

EQUAL BILATERAL TOLERANCES

UNEQUAL BILATERAL TOLERANCES

(A) BILATERAL TOLERANCES

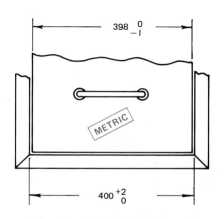

(B) UNILATERAL TOLERANCES

FIG. 8-5-6   Application of tolerances.

its preparation. The following examples illustrate the wide field of application of this system. The values given in the examples are typical.

EXAMPLE 1

EXCEPT WHERE STATED OTHERWISE, TOLERANCES ON FINISHED DECIMAL DIMENSIONS ±0.1.

TOTAL TOLERANCE IN INCHES		MILLIMETER CONVERSION ROUNDED TO
AT LEAST	LESS THAN	
.00004	.0004	4 DECIMAL PLACES
.0004	.004	3 DECIMAL PLACES
.004	.04	2 DECIMAL PLACES
.04	.4	1 DECIMAL PLACE
.4 AND OVER		WHOLE mm

TOTAL TOLERANCE IN MILLIMETERS		INCH CONVERSION ROUNDED TO
AT LEAST	LESS THAN	
0.002	0.02	5 DECIMAL PLACES
0.02	0.2	4 DECIMAL PLACES
0.2	2	3 DECIMAL PLACES
2 AND OVER		2 DECIMAL PLACES

FIG. 8-5-7   Conversion charts for tolerances.

EXAMPLE 2

EXCEPT WHERE STATED OTHERWISE, TOLERANCES ON FINISHED DIMENSIONS TO BE AS FOLLOWS:

Dimension (in.)	Tolerance
UP TO 4.00	± .004
FROM 4.01 TO 12.00	± .003
FROM 12.01 TO 24.00	± .02
OVER 24.00	± .04

A comparison between the tolerancing methods described is shown in Fig. 8-5-8.

Tolerance Accumulation

It is necessary also to consider the effect of each tolerance with respect to other tolerances, and not to permit a chain of tolerances to build up a cumulative tolerance between surfaces or points that have an important relation to one another. Where the position of a surface in any one direction is controlled by more than one tolerance, the tolerances are cumulative. Figure 8-5-9 (pg. 194) compares the tolerance accumulation resulting from three different methods of dimensioning.

Chain Dimensioning   The maximum variation between any two features is equal to the sum of the tolerances on the intermediate distances. This results in the greatest tolerance accumulation, as illustrated by the ± .08 variation between holes X and Y, as shown in Fig. 8-5-9A.

Datum Dimensioning   The maximum variation between any two features is equal to the sum of the tolerances on the two dimensions from the datum to the feature. This reduces the

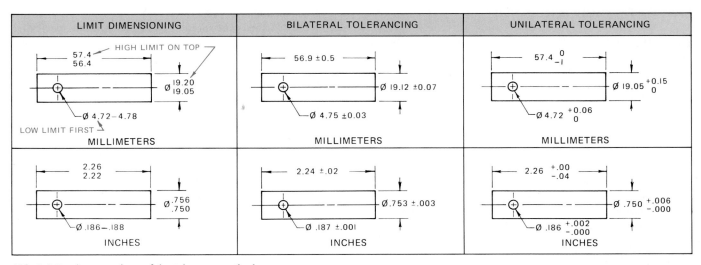

FIG. 8-5-8   A comparison of the tolerance methods.

FIG. 8-5-9   Dimensioning method comparison.

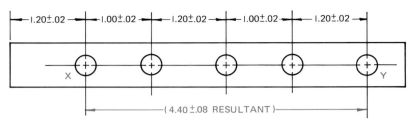

(A) CHAIN DIMENSIONING (GREATEST TOLERANCE ACCUMULATION)

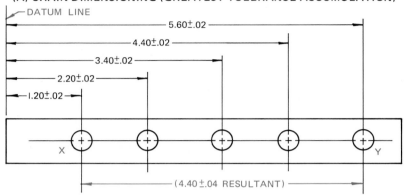

(B) DATUM DIMENSIONING (LESSER TOLERANCE ACCUMULATION)

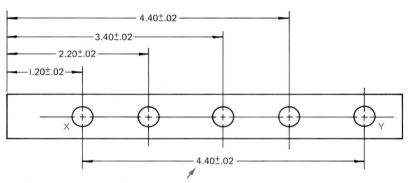

(C) DIRECT DIMENSIONING (LEAST TOLERANCE ACCUMULATION)

tolerance accumulation, as illustrated by the ±.04 variation between holes X and Y in Fig. 8-5-9B.

Direct Dimensioning   The maximum variation between any two features is controlled by the tolerance on the dimension between the features. This results in the least tolerance accumulation, as illustrated by the ± .02 variation between holes X and Y in Fig. 8-5-9C.

Additional Rules for Dimensioning

- The engineering intent must be clearly defined.
- Dimensions must be complete enough to describe the total geometry of each feature. Determining a shape by measuring its size on a drawing or by assuming a distance or size is not acceptable.
- Dimensions should be selected and arranged to avoid unsatisfactory accumulation of tolerances to preclude

more than one interpretation, and to ensure a proper fit between mating parts.

- The finished part should be defined without specifying manufacturing methods. Thus only the diameter of a hole is given, without indicating if it is to be drilled, reamed, punched, or made by any other operation.
- Dimensions must be selected to give required information directly. Dimensions should preferably be shown in true profile views and refer to visible outlines rather than to hidden lines. A common exception to this general rule is a diametral dimension on a section view.
- Drawings that illustrate part surfaces or center lines at right angles to each other, but without an angular dimension, are interpreted as being 90° between these surfaces or center lines. Actual surfaces, axes, and center planes may vary within their specified tolerance of perpendicularity.
- Dimension lines are placed outside the outline of the part and between the views unless the drawing may be simplified or clarified by doing otherwise.

- Dimension lines should be aligned, if practicable, and should be grouped for uniform appearance.

REFERENCES AND SOURCE MATERIAL

1. ASME Y14.5M–1994, *Dimensioning and Tolerancing*.
2. CAN/CSA B78.2–M91, *Dimensioning and Tolerancing of Technical Drawings*.

ASSIGNMENT

See Assignment 19 for Unit 8-5 on page 220.

8-6 FITS AND ALLOWANCES

In order that assembled parts may function properly and to allow for interchangeable manufacturing, it is necessary to permit only a certain amount of tolerance on each of the mating parts and a certain amount of allowance between them.

Fits

The fit between two mating parts is the relationship between them with respect to the amount of clearance or interference present when they are assembled. There are three basic types of fits: clearance, interference, and transition.

Clearance Fit   A fit between mating parts having limits of size so prescribed that a clearance always results in assembly.

Interference Fit   A fit between mating parts having limits of size so prescribed that an interference always results in assembly.

Transition Fit   A fit between mating parts having limits of size so prescribed as to partially or wholly overlap, so that either a clearance or interference may result in assembly.

Allowance

An *allowance* is an intentional difference between the maximum material limits of mating parts. It is the minimum clearance (positive allowance) or maximum interference (negative allowance) between such parts.

The most important terms relating to limits and fits are shown in Fig. 8-6-1. The terms are defined as follows:

Basic Size   The size to which limits or deviations are assigned. The basic size is the same for both members of a fit.

Deviation   The algebraic difference between a size and the corresponding basic size.

Upper Deviation   The algebraic difference between the maximum limit of size and the corresponding basic size.

Lower Deviation   The algebraic difference between the minimum limit of size and the corresponding basic size.

Tolerance   The difference between the maximum and minimum size limits on a part.

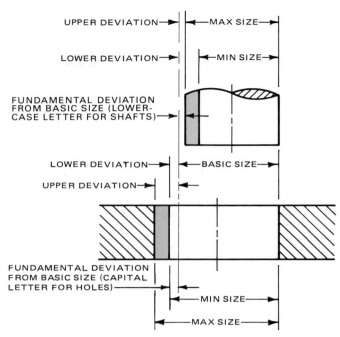

FIG. 8-6-1   Illustration of definitions.

Tolerance Zone   A zone representing the tolerance and its position in relation to the basic size.

Fundamental Deviation   The deviation closest to the basic size.

Description of Fits

Running and Sliding Fits

Running and sliding fits, for which tolerances and clearances are given in the Appendix, represent a special type of clearance fit. These are intended to provide a similar running performance, with suitable lubrication allowance, throughout the range of sizes.

Locational Fits

Locational fits are intended to determine only the location of the mating parts; they may provide rigid or accurate location, as with interference fits, or some freedom of location, as with clearance fits. Accordingly, they are divided into three groups: clearance fits, transition fits, and interference fits.

Locational clearance fits are intended for parts that are normally stationary but that can be freely assembled or disassembled. They run from snug fits for parts requiring accuracy of location, through the medium clearance fits for parts such as ball, race, and housing, to the looser fastener fits where freedom of assembly is of prime importance.

Locational transition fits are a compromise between clearance and interference fits for application where accuracy of location is important but a small amount of either clearance or interference is permissible.

Locational interference fits are used where accuracy of location is of prime importance and for parts requiring rigidity and alignment with no special requirements for bore pressure. Such fits are not intended for parts designed to transmit frictional loads from one part to another by virtue of the tightness of fit; these conditions are covered by force fits.

Drive and Force Fits

Drive and force fits constitute a special type of interference fit, normally characterized by maintenance of constant bore pressures throughout the range of sizes. The interference therefore varies almost directly with diameter, and the difference between its minimum and maximum values is small, to maintain the resulting pressures within reasonable limits.

Interchangeability of Parts

Increased demand for manufactured products led to the development of new production techniques. Interchangeability of parts became the basis for mass-production, low-cost manufacturing, and it brought about the refinement of machinery, machine tools, and measuring devices. Today it is possible and generally practical to design for 100 percent interchangeability.

No part can be manufactured to exact dimensions. Tool wear, machine variations, and the human factor all contribute to some degree of deviation from perfection. It is therefore necessary to determine the deviation and permissible clearance, or interference, to produce the desired fit between parts.

Modern industry has adopted three basic approaches to manufacturing:

1. *The completely interchangeable assembly.* Any and all mating parts of a design are toleranced to permit them to assemble and function properly without the need for machining or fitting at assembly.
2. *The fitted assembly.* Mating features of a design are fabricated either simultaneously or with respect to one another. Individual members of mating features are not interchangeable.
3. *The selected assembly.* All parts are mass-produced, but members of mating features are individually selected to provide the required relationship with one another.

Standard Inch Fits

Standard fits are designated for design purposes in specifications and on design sketches by means of symbols, as shown in Fig. 8-6-2. These symbols, however, are not intended to be shown directly on shop drawings; instead the actual limits of size are determined and specified on the drawings.

The letter symbols used are as follows:

RC   Running and sliding fit
LC   Locational clearance fit
LT   Locational transition fit
LN   Locational interference fit
FN   Force or shrink fit

These letter symbols are used in conjunction with numbers representing the class of fit; for example, FN4 represents a class 4 force fit.

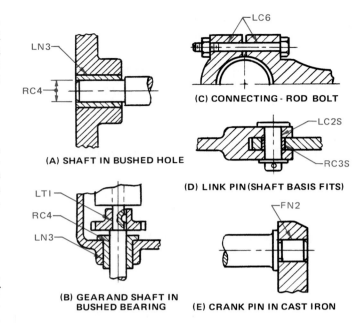

FIG. 8-6-2   Typical design sketches showing classes of fits.

Each of these symbols (two letters and a number) represents a complete fit, for which the minimum and maximum clearance or interference, and the limits of size for the mating parts, are given directly in Appendix tables 43 through 47. See Fig. 8-6-3.

Running and Sliding Fits

RC1 Precision Sliding Fit   This fit is intended for the accurate location of parts that must assemble without perceptible play, for high-precision work such as gages.

RC2 Sliding Fit   This fit is intended for accurate location, but with greater maximum clearance than class RC1. Parts made to this fit move and turn easily but are not intended to run freely, and in the larger sizes may seize with small temperature changes.

Note: LC1 and LC2 locational clearance fits may also be used as sliding fits with greater tolerances.

RC3 Precision Running Fit   This fit is about the closest fit that can be expected to run freely and is intended for precision work for oil-lubricated bearings at slow speeds and light journal pressures, but is not suitable where appreciable temperature differences are likely to be encountered.

RC4 Close Running Fit   This fit is intended chiefly as a running fit for grease- or oil-lubricated bearings on accurate machinery with moderate surface speeds and journal pressures, where accurate location and minimum play are desired.

RC5 and RC6 Medium Running Fits   These fits are intended for higher running speeds and/or where temperature variations are likely to be encountered.

RC7 Free Running Fit   This fit is intended for use where accuracy is not essential and/or where large temperature variations are likely to be encountered.

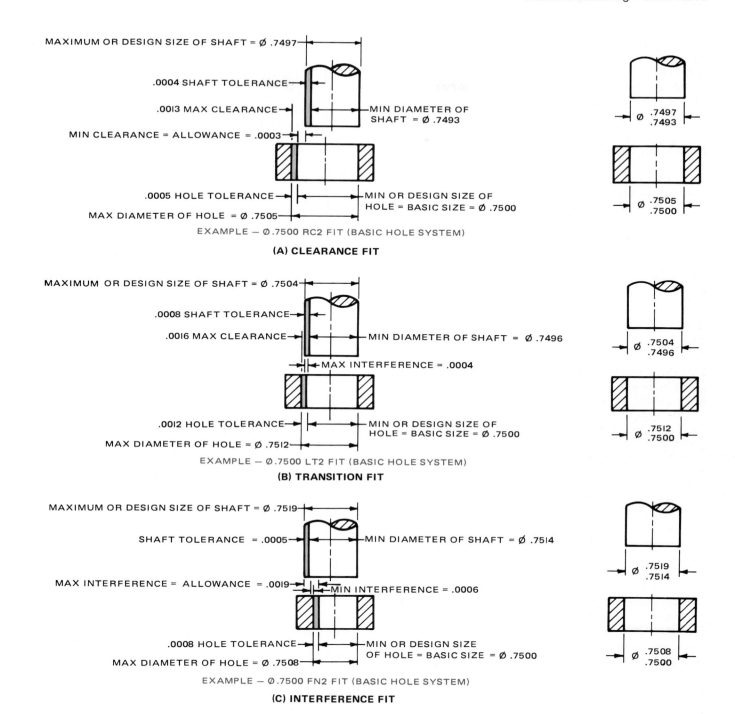

EXAMPLE — Ø.7500 RC2 FIT (BASIC HOLE SYSTEM)

(A) CLEARANCE FIT

EXAMPLE — Ø.7500 LT2 FIT (BASIC HOLE SYSTEM)

(B) TRANSITION FIT

EXAMPLE — Ø.7500 FN2 FIT (BASIC HOLE SYSTEM)

(C) INTERFERENCE FIT

FIG. 8-6-3   Types of inch fits.

RC8 and RC9 Loose Running Fits   These fits are intended for use where materials made to commercial tolerances, such as cold-rolled shafting, tubing, etc., are involved.

Locational Clearance Fits

Locational clearance fits are intended for parts that are normally stationary but that can be freely assembled or disassembled. They run from snug fits for parts requiring accuracy of location,

through the medium-clearance fits for parts such as spigots, etc., to the looser fastener fits where freedom of assembly is of prime importance.

These are classified as follows:

LC1 to LC4   These fits have a minimum zero clearance, but in practice the probability is that the fit will always have a clearance. These fits are suitable for location of nonrunning parts and spigots, although classes LC1 and LC2 may also be used for sliding fits.

LC5 and LC6   These fits have a small minimum clearance, intended for close location fits for nonrunning parts. LC5 can also be used in place of RC2 as a free-slide fit, and LC6 may be used as a medium running fit having greater tolerances than RC5 and RC6.

LC7 and LC11   These fits have progressively larger clearances and tolerances and are useful for various loose clearances for assembly of bolts and similar parts.

Locational Transition Fits

Locational transition fits are a compromise between clearance and interference fits for application where accuracy of location is important, but either a small amount of clearance or interference is permissible.

These are classified as follows:

LT1 and LT2   These fits average a slight clearance, giving a light push fit, and are intended for use where the maximum clearance must be less than for the LC1 to LC3 fits, and where slight interference can be tolerated for assembly by pressure or light hammer blows.

LT3 and LT4   These fits average virtually no clearance and are for use where some interference can be tolerated, for example, to eliminate vibration. These are sometimes referred to as an *easy keying fit* and are used for shaft keys and ball race fits. Assembly is generally by pressure or hammer blows.

LT5 and LT6   These fits average a slight interference, although appreciable assembly force will be required when extreme limits are encountered, and selective assembly may be desirable. These fits are useful for heavy keying, for ball race fits subject to heavy duty and vibration, and as light press fits for steel parts.

Locational Interference Fits

Locational interference fits are used where accuracy of location is of prime importance, and for parts requiring rigidity and alignment with no special requirements for bore pressure. Such fits are not intended for parts designed to transmit frictional loads from one part to another by virtue of the tightness of fit, as these conditions are covered by force fits.

These are classified as follows:

LN1 and LN2   These are light press fits, with very small minimum interference, suitable for parts such as dowel pins, which are assembled with an arbor press in steel, cast iron, or brass. Parts can normally be dismantled and reassembled, as the interference is not likely to overstrain the parts, but the interference is too small for satisfactory fits in elastic materials or light alloys.

LN3   This is suitable as a heavy press fit in steel and brass, or a light press fit in more elastic materials and light alloys.

LN4 to LN6   While LN4 can be used for permanent assembly of steel parts, these fits are primarily intended as press fits for more elastic or soft materials, such as light alloys and the more rigid plastics.

Force or Shrink Fits

Force or shrink fits constitute a special type of interference fit, normally characterized by maintenance of constant bore pressures throughout the range of sizes. The interference therefore varies almost directly with diameter, and the difference between its minimum and maximum values is small to maintain the resulting pressures within reasonable limits.

These fits may be described briefly as follows:

FN1 Light Drive Fit   Requires light assembly pressure and produces more or less permanent assemblies. It is suitable for thin sections or long fits, or in cast iron external members.

FN2 Medium Drive Fit   Suitable for ordinary steel parts or as a shrink fit on light sections. It is about the tightest fit that can be used with high-grade cast iron external members.

FN3 Heavy Drive Fit   Suitable for heavier steel parts or as a shrink fit in medium sections.

FN4 and FN5 Force Fits   Suitable for parts that can be highly stressed and/or for shrink fits where the heavy pressing forces required are impractical.

Basic Hole System

In the basic hole system, which is recommended for general use, the basic size will be the design size for the hole, and the tolerance will be plus. The design size for the shaft will be the basic size minus the minimum clearance, or plus the maximum interference, and the tolerance will be minus, as given in the tables in the Appendix. For example (see Table 43 in the Appendix), for a 1 in. RC7 fit, values of +.0020, .0025, and −.0012 are given; hence limits will be:

$$\text{Hole } \varnothing \; 1.0000 \; {}^{+.0020}_{-.0000}$$

$$\text{Shaft } \varnothing \; .9975 \; {}^{+.0000}_{-.0012}$$

Basic Shaft System

Fits are sometimes required on a basic shaft system, especially in cases where two or more fits are required on the same shaft. This is designated for design purposes by a letter S following the fit symbol; for example, RC7S.

Tolerances for holes and shaft are identical with those for a basic hole system, but the basic size becomes the design size for the shaft and the design size for the hole is found by adding the minimum clearance or subtracting the maximum interference from the basic size.

For example, for a 1 in. RC7S fit, values of +.0020, .0025, and −.0012 are given; therefore, limits will be:

$$\text{Hole } \varnothing \; 1.0025 \; {}^{+.0020}_{-.0000}$$

$$\text{Shaft } \varnothing \; 1.0000 \; {}^{+.0000}_{-.0012}$$

Preferred Metric Limits and Fits

The ISO system of limits and fits for mating parts is approved and adopted for general use in the United States. It establishes the designation symbols used to define specific dimensional limits on drawings.

The general terms "hole" and "shaft" can also be taken as referring to the space containing or contained by two parallel faces of any part, such as the width of a slot, or the thickness of a key.

An "International Tolerance grade" establishes the magnitude of the tolerance zone or the amount of part size variation allowed for internal and external dimensions alike (Table 40, Appendix). There are 18 tolerance grades, which are identified by the prefix IT, such a IT6, IT11, etc. The smaller the grade number, the smaller the tolerance zone. For general applications of IT grades, see Fig. 8-6-4.

Grades 1 to 4 are very precise grades intended primarily for gage making and similar precision work, although grade 4 can also be used for very precise production work.

Grades 5 to 16 represent a progressive series suitable for cutting operations, such as turning, boring, grinding, milling, and sawing. Grade 5 is the most precise grade, obtainable by fine grinding and lapping, while 16 is the coarsest grade for rough sawing and machining.

Grades 12 to 16 are intended for manufacturing operations such as cold heading, pressing, rolling, and other forming operations.

As a guide to the selection of tolerances, Fig. 8-6-4B has been prepared to show grades that may be expected to be held by various manufacturing processes for work in metals. For work in other materials, such as plastics, it may be necessary to use coarser tolerance grades for the same process.

A fundamental deviation establishes the position of the tolerance zone with respect to the basic size. Fundamental deviations are expressed by *tolerance position letters.* Capital letters are used for internal dimensions, and lowercase letters for external dimensions.

Tolerance Symbol

For metric application of limits and fits, the tolerance may be indicated by a basic size and tolerance symbol. By combining the IT grade number and the tolerance position letter, the tolerance symbol is established that identifies the actual maximum and minimum limits of the part. The toleranced sizes are thus defined by the basic size of the part followed by the symbol composed of a letter and a number (Fig. 8-6-5, pg. 200).

Preferred Tolerance Grades

The preferred tolerance grades are shown in Table 40 of the Appendix. The encircled tolerance grades (13 each) are first choice, the framed tolerance grades are second choice, and the open tolerance grades are third choice.

Hole Basis Fits System

In the hole basis fits system (see Table 48 of the Appendix) the basic size will be the minimum size of the hole. For example,

FIG. 8-6-4   International Tolerance (IT) grades.

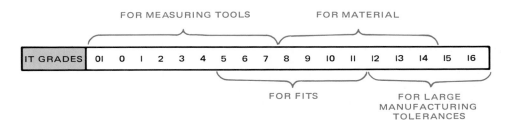

(A) APPLICATIONS

MACHINING PROCESSES	TOLERANCE GRADES									
	4	5	6	7	8	9	10	11	12	13
LAPPING & HONING										
CYLINDRICAL GRINDING										
SURFACE GRINDING										
DIAMOND TURNING										
DIAMOND BORING										
BROACHING										
REAMING										
TURNING										
BORING										
MILLING										
PLANING & SHAPING										
DRILLING										

(B) APPLICATIONS FOR MACHINING PROCESSES

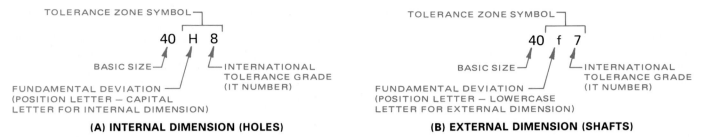

FIG. 8-6-5   Metric tolerance symbol.

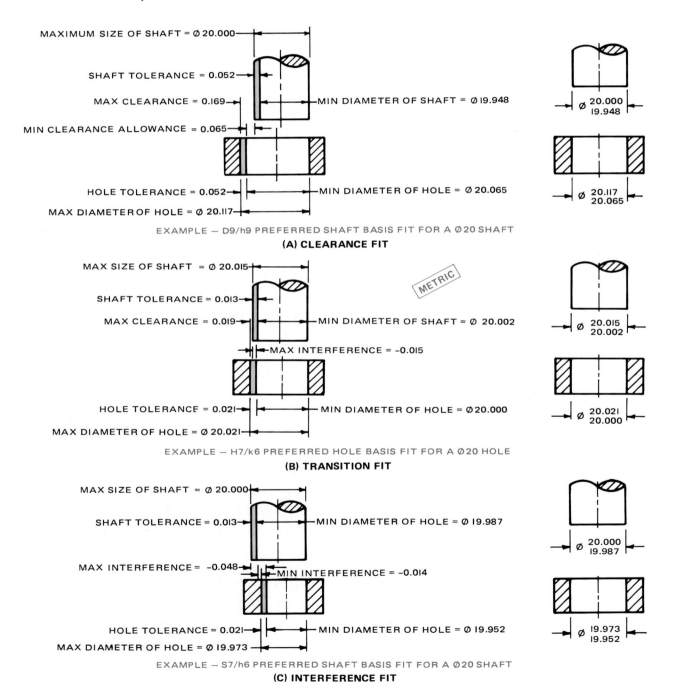

FIG. 8-6-6   Types of metric fits.

for a Ø25H8/f7 fit, which is a preferred hole basis clearance fit, the limits for the hole and shaft will be as follows:

Hole limits = Ø 25.000 – Ø 25.033
Shaft limits = Ø 24.959 – Ø 24.980
Minimum interference = - 0.020
Maximum interference = - 0.074

If a Ø25H7/s6 preferred hole basis interference fit is required, the limits for the hole and shaft will be as follows:

Hole limits = Ø 25.000 – Ø 25.021
Shaft limits = Ø 25.035 – Ø 25.048
Minimum interference = –0.014
Maximum interference = –0.048

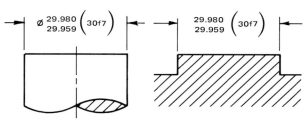

FIG. 8-6-7   Metric fit symbol.

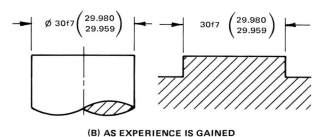

(A) WHEN SYSTEM IS FIRST INTRODUCED

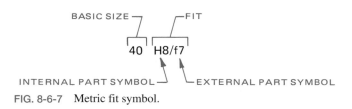

(B) AS EXPERIENCE IS GAINED

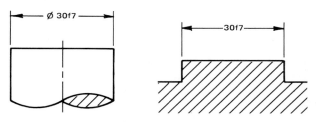

(C) WHEN SYSTEM IS ESTABLISHED

FIG. 8-6-8   Metric tolerance symbol application.

Shaft Basis Fits System

Where more than two fits are required on the same shaft, the shaft basis fits system is recommended. Tolerances for holes and shaft are identical with those for a basic hole system. However, the basic size becomes the maximum shaft size. For example, for a Ø16 C11/h11 fit, which is a preferred shaft basis clearance fit, the limits for the hole and shaft will be as follows (refer to Table 49 of the Appendix):

Hole limits = Ø 16.095 – Ø 16.205
Shaft limits = Ø 15.890 – Ø 16.000
Minimum clearance = 0.095
Maximum clearance = 0.315

Preferred Fits

First-choice tolerance zones are shown to relative scale in Tables 41 and 42 of the Appendix. Hole basis fits have a fundamental deviation of "H" on the hole, and shaft basis fits have a fundamental deviation of "h" on the shaft. Normally, the hole basis system is preferred. Figure 8-6-6 shows examples of three common fits.

Fit Symbol   A fit is indicated by the basic size common to both components, followed by a symbol corresponding to each component, with the internal part symbol preceding the external part symbol (Fig. 8-6-7).

The limits of size for a hole having a tolerance symbol 40H8 (see Table 41) is:

Ø 40.039 Maximum limit
Ø 40.000 Minimum limit

The limits of size for the shaft having a tolerance symbol 40f7 (see Table 42) is:

Ø 39.975 Maximum limit
Ø 39.950 Minimum limit

The method shown in Fig. 8-6-8A is recommended when the system is first introduced. In this case limit dimensions are specified, and the basic size and tolerance symbol are identified as reference.

As experience is gained, the method shown in Fig. 8-6-8B can be used. When the system is established and standard tools, gages, and stock materials are available with size and symbol identification, the method shown in Fig. 8-6-8C may be used.

This would result in a clearance fit of 0.025 to 0.089 mm. A description of the preferred metric fits is shown in Fig. 8-6-9 (pg. 202).

REFERENCES AND SOURCE MATERIAL

1. ANSI B4.2, *Preferred Metric Limits and Fits.*

ASSIGNMENTS

See Assignments 20 through 23 for Unit 8-6 on pages 221 through 224.

FIG. 8-6-9   Description of preferred metric fits.

ISO SYMBOL		DESCRIPTION
HOLE BASIS	SHAFT BASIS	
H11/c11	C11/h11	LOOSE RUNNING FIT FOR WIDE COMMERCIAL TOLERANCES OR ALLOWANCES ON EXTERNAL MEMBERS.
H9/d9	D9/h9	FREE RUNNING FIT NOT FOR USE WHERE ACCURACY IS ESSENTIAL, BUT GOOD FOR LARGE TEMPERATURE VARIATIONS, HIGH RUNNING SPEEDS, OR HEAVY JOURNAL PRESSURES.
H8/f7	F8/h7	CLOSE RUNNING FIT FOR RUNNING ON ACCURATE MACHINES AND FOR ACCURATE LOCATION AT MODERATE SPEEDS AND JOURNAL PRESSURES.
H7/g6	G7/h6	SLIDING FIT NOT INTENDED TO RUN FREELY, BUT TO MOVE AND TURN FREELY AND LOCATE ACCURATELY.
H7/h6	H7/h6	LOCATIONAL CLEARANCE FIT PROVIDES SNUG FIT FOR LOCATING STATIONARY PARTS; BUT CAN BE FREELY ASSEMBLED AND DISASSEMBLED.
H7/k6	K7/h6	LOCATIONAL TRANSITION FIT FOR ACCURATE LOCATION, A COMPROMISE BETWEEN CLEARANCE AND INTERFERENCE.
H7/n6	N7/h6	LOCATIONAL TRANSITION FIT FOR MORE ACCURATE LOCATION WHERE GREATER INTERFERENCE IS PERMISSIBLE.
H7/p6	P7/h6	LOCATIONAL INTERFERENCE FIT FOR PARTS REQUIRING RIGIDITY AND ALIGNMENT WITH PRIME ACCURACY OF LOCATION BUT WITHOUT SPECIAL BORE PRESSURE REQUIREMENTS.
H7/s6	S7/h6	MEDIUM DRIVE FIT FOR ORDINARY STEEL PARTS OR SHRINK FITS ON LIGHT SECTIONS, THE TIGHTEST FIT USABLE WITH CAST IRON.
H7/u6	U7/h6	FORCE FIT SUITABLE FOR PARTS WHICH CAN BE HIGHLY STRESSED OR FOR SHRINK FITS WHERE THE HEAVY PRESSING FORCES REQUIRED ARE IMPRACTICAL.

(Left axis: CLEARANCE FITS, TRANSITION FITS, INTERFERENCE FITS. Right axis: MORE CLEARANCE, MORE INTERFERENCE.)

8-7 SURFACE TEXTURE

Modern development of high-speed machines has resulted in higher loadings and increased speeds of moving parts. To withstand these more severe operating conditions with minimum friction and wear, a particular surface finish is often essential, making it necessary for the designer to accurately describe the required finish to the persons who are actually making the parts.

For accurate machines it is no longer sufficient to indicate the surface finish by various grind marks, such as "g," "f," or "fg." It becomes necessary to define surface finish and take it out of the opinion or guesswork class.

All surface finish control starts in the drafting room. The designer has the responsibility of specifying the right surface to give maximum performance and service life at the lowest cost. In selecting the required surface finish for any particular part, designers base decisions on experience with similar parts, on field service data, or on engineering tests. Such factors as size and function of the parts, type of loading, speed and direction of movement, operating conditions, physical characteristics of both materials on contact, whether they are subjected to stress reversals, type and amount of lubricant, contaminants, temperature, etc., influence the choice.

There are two principal reasons for surface finish control:

1. To reduce friction
2. To control wear

Whenever a film of lubricant must be maintained between two moving parts, the surface irregularities must be small enough so they will not penetrate the oil film under the most severe operating conditions. Bearings, journals, cylinder boxes, piston pins, bushings, pad bearings, helical and worm gears, seal surfaces, machine ways, and so forth, are examples where this condition must be fulfilled.

Surface finish is also important to the wear of certain pieces that are subject to dry friction, such as machine tool bits, threading dies, stamping dies, rolls, clutch plates, brake drums, etc.

Smooth finishes are essential on certain high-precision pieces. In mechanisms such as injectors and high-pressure cylinders, smoothness and lack of waviness are essential to accuracy and pressure-retaining ability.

Surfaces, in general, are very complex in character. Only the height, width, and direction of surface irregularities will be covered in this section since these are of practical importance in specific applications.

Surface Texture Characteristics

Refer to Fig. 8-7-1.

Microinch   A microinch is one-millionth of an inch (.000 001 in.). For written specifications or reference to surface roughness requirements, microinch may be abbreviated as μin.

Micrometer   A micrometer is one-millionth of a meter (0.000 001 m). For written specifications or reference to surface roughness requirements, micrometer may be abbreviated as μm.

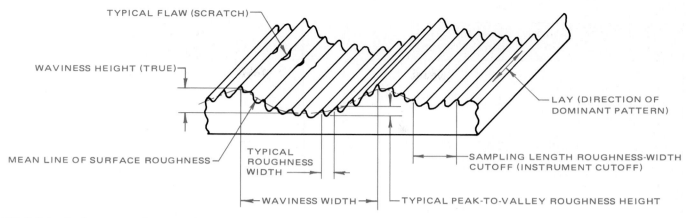

TYPICAL FLAW (SCRATCH)

WAVINESS HEIGHT (TRUE)

LAY (DIRECTION OF DOMINANT PATTERN)

MEAN LINE OF SURFACE ROUGHNESS

TYPICAL ROUGHNESS WIDTH

SAMPLING LENGTH ROUGHNESS-WIDTH CUTOFF (INSTRUMENT CUTOFF)

WAVINESS WIDTH

TYPICAL PEAK-TO-VALLEY ROUGHNESS HEIGHT

FIG. 8-7-1   Surface texture characteristics.

Roughness   Roughness consists of the finer irregularities in the surface texture, usually including those that result from the inherent action of the production process. These include traverse feed marks and other irregularities within the limits of the roughness-width cutoff.

Roughness-Height Value   Roughness-height value is rated as the arithmetic average (AA) deviation expressed in microinches or micrometers measured normal to the center line. ISO and many European countries use the term CLA (center line average) in lieu of AA. Both have the same meaning.

Roughness Spacing   Roughness spacing is the distance parallel to the nominal surface between successive peaks or ridges that constitute the predominant pattern of the roughness. Roughness spacing is rated in inches or millimeters.

Roughness-Width Cutoff   The greatest spacing of repetitive surface irregularities is included in the measurement of average roughness height. Roughness-width cutoff is rated in inches or millimeters and must always be greater than the roughness width in order to obtain the total roughness-height rating.

Waviness   Waviness is usually the most widely spaced of the surface texture components and normally is wider than the roughness-width cutoff. Waviness may result from such factors as machine or work deflections, vibration, chatter, heat treatment, or warping strains. Roughness may be considered as superimposed on a "wavy" surface. Although waviness is not currently in ISO Standards, it is included as part of the surface texture symbol to follow present industrial practices in the United States.

Lay   The direction of the predominant surface pattern, ordinarily determined by the production method used, is the lay.

Flaws   Flaws are irregularities that occur at one place or at relatively infrequent or widely varying intervals in a surface. Flaws include such defects as cracks, blow holes, checks, ridges, and scratches. Unless otherwise specified, the effect of flaws is not included in the roughness-height measurements.

Surface Texture Symbol

Surface characteristics of roughness, waviness, and lay may be controlled by applying the desired values to the surface texture symbol, shown in Figs. 8-7-2 and 8-7-3 (pg. 204), in a general note, or both. Where only the roughness value is indicated, the horizontal extension line on the symbol may be omitted. The horizontal bar is used whenever any surface characteristics are placed above the bar or to the right of the symbol. The point of the symbol should be located on the line indicating the surface, on an extension line from the surface, or on a leader pointing to the surface or extension line. If necessary, the symbol may be connected to the surface by a leader line terminating in an arrow. The symbol applies to the entire surface, unless otherwise specified. The symbol for the same surface should not be duplicated on other views.

When numerical values accompany the symbol, the symbol should be in an upright position in order to be readable from the bottom. This means that the long leg and extension line are always on the right. When no numerical values are shown on the symbol, the symbol may also be positioned to be readable from the right side.

Application

Plain (Unplated or Uncoated) Surfaces   Surface texture values specified on plain surfaces apply to the completed surface unless otherwise noted.

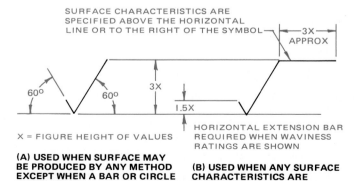

SURFACE CHARACTERISTICS ARE SPECIFIED ABOVE THE HORIZONTAL LINE OR TO THE RIGHT OF THE SYMBOL

3X APPROX

60°   60°

3X

1.5X

X = FIGURE HEIGHT OF VALUES

HORIZONTAL EXTENSION BAR REQUIRED WHEN WAVINESS RATINGS ARE SHOWN

(A) USED WHEN SURFACE MAY BE PRODUCED BY ANY METHOD EXCEPT WHEN A BAR OR CIRCLE IS SPECIFIED

(B) USED WHEN ANY SURFACE CHARACTERISTICS ARE SPECIFIED

FIG. 8-7-2   Basic surface texture symbol.

PRESENT SYMBOLS VALUES SHOWN IN CUSTOMARY OR METRIC			FORMER SYMBOLS VALUES SHOWN IN MICROINCHES AND INCHES	
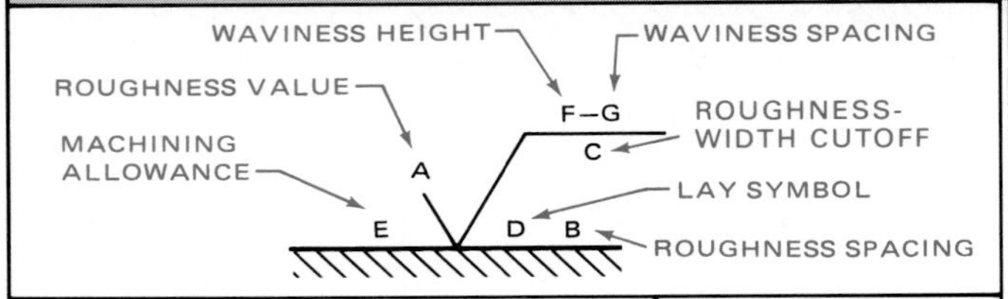 WAVINESS HEIGHT — WAVINESS SPACING / ROUGHNESS VALUE / MACHINING ALLOWANCE — F—G, C ROUGHNESS-WIDTH CUTOFF / A, LAY SYMBOL / E D B ROUGHNESS SPACING			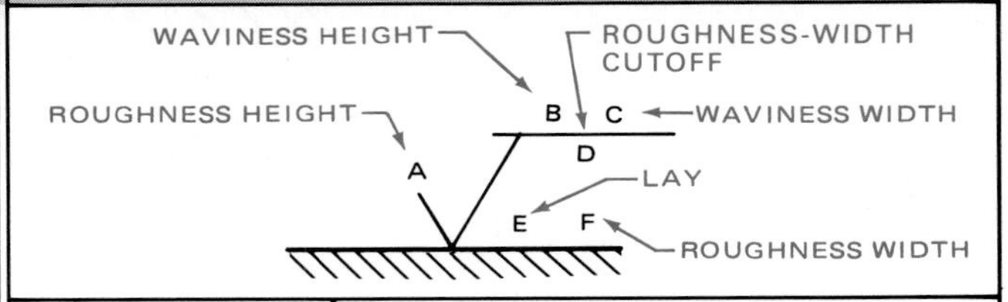 WAVINESS HEIGHT — ROUGHNESS-WIDTH CUTOFF / ROUGHNESS HEIGHT — B C WAVINESS WIDTH / A D LAY / E F ROUGHNESS WIDTH	
BASIC SURFACE TEXTURE SYMBOL	✓		✓	BASIC SURFACE TEXTURE SYMBOL
ROUGHNESS-HEIGHT RATING IN MICROINCHES OR MICROMETERS AND N SERIES ROUGHNESS NUMBERS	63 ✓	N8 ✓	63 ✓	ROUGHNESS-HEIGHT RATING IN MICROINCHES
MAXIMUM AND MINIMUM ROUGHNESS HEIGHT IN MICROINCHES OR MICROMETERS	63 32 ✓		63 32 ✓	MAXIMUM AND MINIMUM ROUGHNESS-HEIGHT RATINGS IN MICROINCHES
WAVINESS HEIGHT IN INCHES OR MILLIMETERS (F)	63 32 ✓ F		63 32 ✓ .002	WAVINESS HEIGHT IN INCHES
WAVINESS SPACING IN INCHES OR MILLIMETERS (G)	63 32 ✓ F—G		63 32 ✓ .002—1	WAVINESS WIDTH IN INCHES
LAY SYMBOL (D)	63 32 ✓ ⊥		63 32 ✓ ⊥	LAY SYMBOL
MAXIMUM ROUGHNESS SPACING IN INCHES OR MILLIMETERS (B)	63 32 ✓ ⊥ B		✓ .002—1 ⊥ .008	SURFACE ROUGHNESS WIDTH IN INCHES
ROUGHNESS SAMPLING LENGTH OR CUTOFF RATING IN INCHES OR MILLIMETERS (C)	63 32 ✓ C ⊥		✓ .002—1 .030 ⊥ .008	ROUGHNESS WIDTH CUTOFF IN INCHES

FIG. 8-7-3   Location of notes and symbols on surface texture symbol.

Plated or Coated Surfaces   Drawings or specifications for plated or coated parts must indicate whether the surface texture value applies before, after, or both before and after plating or coating.

Surface Texture Ratings   The roughness value rating is indicated at the left of the long leg of the symbol (Fig. 8-7-3). The specification of only one rating indicates the maximum value, and any lesser value is acceptable. The specification of two ratings indicates the minimum and maximum values, and anything lying within that range is acceptable. The maximum value is placed over the minimum.

Waviness-height rating is specified in inches or millimeters and is located above the horizontal extension of the symbol. Any lesser value is acceptable.

Waviness spacing is indicated in inches or millimeters and is located above the horizontal extension and to the right, separated from the waviness-height rating by a dash. Any lesser value is acceptable. If the waviness value is a minimum, the abbreviation MIN should be placed after the value.

The surface roughness range for common production methods is shown in Fig. 8-7-4.

Typical surface roughness-height applications are shown in Figs. 8-7-5 and 8-7-6 (pp. 206–207).

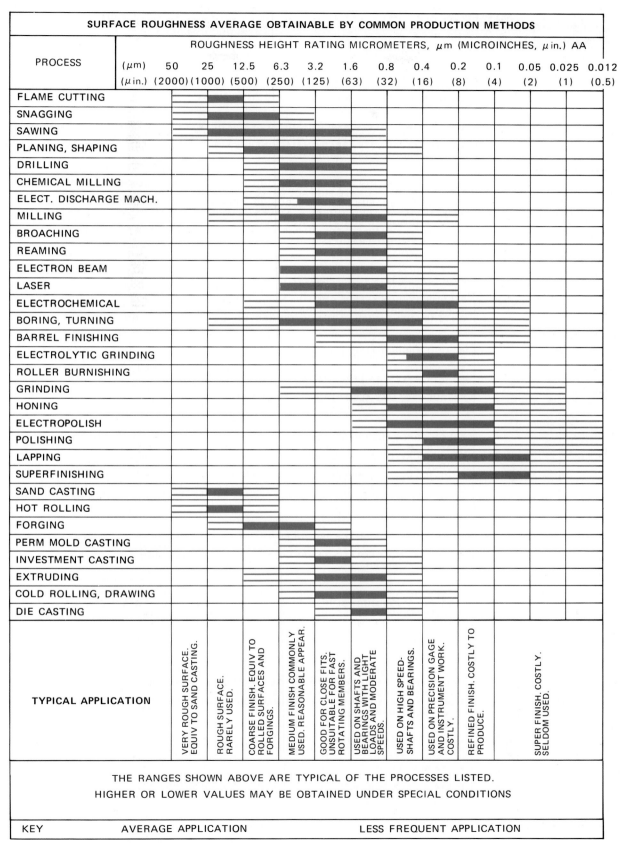

FIG. 8-7-4   Surface roughness range for common production.

MICROINCHES AA RATING	MICROMETERS AA RATING	APPLICATION
1000	25.2	ROUGH, LOW-GRADE SURFACE RESULTING FROM SAND CASTING, TORCH OR SAW CUTTING, CHIPPING, OR ROUGH FORGING. MACHINE OPERATIONS ARE NOT REQUIRED AS APPEARANCE IS NOT OBJECTIONABLE. THIS SURFACE, RARELY SPECIFIED, IS SUITABLE FOR UNMACHINED CLEARANCE AREAS ON ROUGH CONSTRUCTION ITEMS.
500	12.5	ROUGH, LOW-GRADE SURFACE RESULTING FROM HEAVY CUTS AND COARSE FEEDS IN MILLING, TURNING, SHAPING, BORING, AND ROUGH FILING, DISC GRINDING, AND SNAGGING. IT IS SUITABLE FOR CLEARANCE AREAS ON MACHINERY, JIGS, AND FIXTURES. SAND CASTING OR ROUGH FORGING PRODUCES THIS SURFACE.
250	6.3	COARSE PRODUCTION SURFACES, FOR UNIMPORTANT CLEARANCE AND CLEAN-UP OPERATIONS, RESULTING FROM COARSE SURFACE GRIND, ROUGH FILE, DISC GRIND, RAPID FEEDS IN TURNING, MILLING, SHAPING, DRILLING, BORING, GRINDING, ETC., WHERE TOOL MARKS ARE NOT OBJECTIONABLE. THE NATURAL SURFACES OF FORGINGS, PERMANENT MOLD CASTINGS, EXTRUSIONS, AND ROLLED SURFACES ALSO PRODUCE THIS ROUGHNESS. IT CAN BE PRODUCED ECONOMICALLY AND IS USED ON PARTS WHERE STRESS REQUIREMENTS, APPEARANCE, AND CONDITIONS OF OPERATIONS AND DESIGN PERMIT.
125	3.2	THE ROUGHEST SURFACE RECOMMENDED FOR PARTS SUBJECT TO LOADS, VIBRATION, AND HIGH STRESS. IT IS ALSO PERMITTED FOR BEARING SURFACES WHEN MOTION IS SLOW AND LOADS LIGHT OR INFREQUENT. IT IS A MEDIUM COMMERCIAL MACHINE FINISH PRODUCED BY RELATIVELY HIGH SPEEDS AND FINE FEEDS TAKING LIGHT CUTS WITH SHARP TOOLS. IT MAY BE ECONOMICALLY PRODUCED ON LATHES, MILLING MACHINES, SHAPERS, GRINDERS, ETC., OR ON PERMANENT MOLD CASTINGS, DIE CASTINGS, EXTRUSIONS, AND ROLLED SURFACES.
63	1.6	A GOOD MACHINE FINISH PRODUCED UNDER CONTROLLED CONDITIONS USING RELATIVELY HIGH SPEEDS AND FINE FEEDS TO TAKE LIGHT CUTS WITH SHARP CUTTERS. IT MAY BE SPECIFIED FOR CLOSE FITS AND USED FOR ALL STRESSED PARTS, EXCEPT FAST-ROTATING SHAFTS, AXLES, AND PARTS SUBJECT TO SEVERE VIBRATION OR EXTREME TENSION. IT IS SATISFACTORY FOR BEARING SURFACES WHEN MOTION IS SLOW AND LOADS LIGHT OR INFREQUENT. IT MAY ALSO BE OBTAINED ON EXTRUSIONS, ROLLED SURFACES, DIE CASTINGS, AND PERMANENT MOLD CASTINGS WHEN RIGIDLY CONTROLLED.
32	0.8	A HIGH-GRADE MACHINE FINISH REQUIRING CLOSE CONTROL WHEN PRODUCED BY LATHES, SHAPERS, MILLING MACHINES, ETC., BUT RELATIVELY EASY TO PRODUCE BY CENTERLESS, CYLINDRICAL, OR SURFACE GRINDERS. ALSO, EXTRUDING, ROLLING, OR DIE CASTING MAY PRODUCE A COMPARABLE SURFACE WHEN RIGIDLY CONTROLLED. THIS SURFACE MAY BE SPECIFIED IN PARTS WHERE STRESS CONCENTRATION IS PRESENT. IT IS USED FOR BEARINGS WHEN MOTION IS NOT CONTINUOUS AND LOADS ARE LIGHT. WHEN FINER FINISHES ARE SPECIFIED, PRODUCTION COSTS RISE RAPIDLY; THEREFORE, SUCH FINISHES MUST BE ANALYZED CAREFULLY.
16	0.4	A HIGH-QUALITY SURFACE PRODUCED BY FINE CYLINDRICAL GRINDING, EMERY BUFFING, COARSE HONING, OR LAPPING. IT IS SPECIFIED WHERE SMOOTHNESS IS OF PRIMARY IMPORTANCE, SUCH AS RAPIDLY ROTATING SHAFT BEARINGS, HEAVILY LOADED BEARINGS, AND EXTREME TENSION MEMBERS.
8	0.2	A FINE SURFACE PRODUCED BY HONING, LAPPING, OR BUFFING. IT IS SPECIFIED WHERE PACKINGS AND RINGS MUST SLIDE ACROSS THE DIRECTION OF THE SURFACE GRAIN, MAINTAINING OR WITHSTANDING PRESSURES, OR FOR INTERIOR HONED SURFACES OF HYDRAULIC CYLINDERS. IT MAY ALSO BE REQUIRED IN PRECISION GAGES AND INSTRUMENT WORK, OR SENSITIVE-VALUE SURFACES, OR ON RAPIDLY ROTATING SHAFTS AND ON BEARINGS WHERE LUBRICATION IS NOT DEPENDABLE.
4	0.1	A COSTLY REFINED SURFACE PRODUCED BY HONING, LAPPING, AND BUFFING. IT IS SPECIFIED ONLY WHEN THE REQUIREMENTS OF DESIGN MAKE IT MANDATORY. IT IS REQUIRED IN INSTRUMENT WORK, GAGE WORK, AND WHERE PACKINGS AND RINGS MUST SLIDE ACROSS THE DIRECTION OF SURFACE GRAIN, SUCH AS ON CHROME-PLATED PISTON RODS, ETC., WHERE LUBRICATION IS NOT DEPENDABLE.
2 1	0.05 0.025	COSTLY REFINED SURFACES PRODUCED ONLY BY THE FINEST OF MODERN HONING, LAPPING, BUFFING, AND SUPERFINISHING EQUIPMENT. THESE SURFACES MAY HAVE A SATIN OR HIGHLY POLISHED APPEARANCE DEPENDING ON THE FINISHING OPERATION AND MATERIAL. THESE SURFACES ARE SPECIFIED ONLY WHEN DESIGN REQUIREMENTS MAKE IT MANDATORY. THEY ARE SPECIFIED ON FINE OR SENSITIVE INSTRUMENT PARTS OR OTHER LABORATORY ITEMS, AND CERTAIN GAGE SURFACES, SUCH AS PRECISION GAGE BLOCKS.

FIG. 8-7-5   Typical surface roughness-height applications.

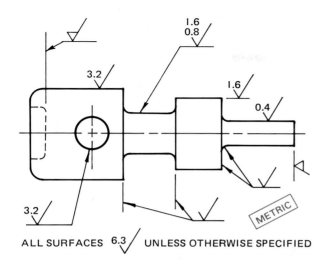

NOTE: VALUES SHOWN ARE IN MICROMETERS.

FIG. 8-7-6   Application of surface texture symbols and notes.

SPECIFYING MAXIMUM ROUGHNESS

SPECIFYING MINIMUM AND MAXIMUM ROUGHNESS

VALUES SHOWN ARE IN MICROINCHES

RECOMMENDED ROUGHNESS-HEIGHT VALUES		N SERIES OF ROUGHNESS GRADE NUMBERS
MICROINCHES μ in.	MICROMETERS μm	
2000	50	N 12
1000	25	N 11
500	12.5	N 10
250	6.3	N 9
125	3.2	N 8
63	1.6	N 7
32	0.8	N 6
16	0.4	N 5
8	0.2	N 4
4	0.1	N 3
2	0.05	N 2
1	0.025	N 1

FIG. 8-7-7   Roughness-height ratings and their equivalent N series grade numbers.

Roughness height ratings and their equivalent N series grade numbers are shown in Fig. 8-7-7.

Lay symbols, which indicate the directional pattern of the surface texture, are shown in Fig. 8-7-8. The symbol is located to the right of the long leg of the symbol. On surfaces having parallel or perpendicular lay designated, the lead resulting from machine feeds may be objectionable. In these cases, the symbol should be supplemented by the words NO LEAD.

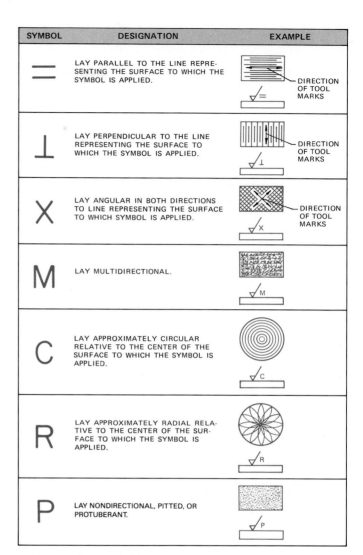

SYMBOL	DESIGNATION	EXAMPLE
=	LAY PARALLEL TO THE LINE REPRESENTING THE SURFACE TO WHICH THE SYMBOL IS APPLIED.	DIRECTION OF TOOL MARKS
⊥	LAY PERPENDICULAR TO THE LINE REPRESENTING THE SURFACE TO WHICH THE SYMBOL IS APPLIED.	DIRECTION OF TOOL MARKS
X	LAY ANGULAR IN BOTH DIRECTIONS TO LINE REPRESENTING THE SURFACE TO WHICH SYMBOL IS APPLIED.	DIRECTION OF TOOL MARKS
M	LAY MULTIDIRECTIONAL.	
C	LAY APPROXIMATELY CIRCULAR RELATIVE TO THE CENTER OF THE SURFACE TO WHICH THE SYMBOL IS APPLIED.	
R	LAY APPROXIMATELY RADIAL RELATIVE TO THE CENTER OF THE SURFACE TO WHICH THE SYMBOL IS APPLIED.	
P	LAY NONDIRECTIONAL, PITTED, OR PROTUBERANT.	

FIG. 8-7-8   Lay symbols.

PRESENT SYMBOL	FORMER SYMBOL
LAY SYMBOLS	
STANDARD ROUGHNESS SAMPLING LENGTH VALUES	
INCHES	MILLIMETERS
.003	0.08
.010	0.25
.030	0.8
.100	2.54
.300	8
1.000	25.4

FIG. 8-7-9   Lay and roughness sampling length applications.

Roughness sampling length or cutoff rating is in inches or millimeters and is located below the horizontal extension (Fig. 8-7-3). Unless otherwise specified, roughness sampling length is .03 in. (0.08 mm). See Fig. 8-7-9.

Notes

Notes relating to surface roughness can be local or general. Normally, a general note is used where a given roughness requirement applies to the whole part or the major portion. Any exceptions to the general note are given in a local note (Fig. 8-7-10).

Machined Surfaces

In preparing working drawings or parts to be cast, molded, or forged, the drafter must indicate the surfaces on the drawing that will require machining or finishing. The symbol $\sqrt{}$ identifies those surfaces that are produced by machining operations (Fig. 8-7-11). It indicates that material is to be provided for removal by machining. Where all the surfaces are to be machined, a general note, such as FINISH ALL OVER, may be used, and the symbols on the drawing may be omitted. Where space is restricted, the machining symbol may be placed on an extension line.

Machining symbols, like dimensions, are not normally duplicated. They should be used on the same view as the dimensions that give the size or location of the surfaces concerned. The symbol is placed on the line representing the surface or, where desirable, on the extension line locating the surface. Figures 8-7-12 and 8-7-13 show examples of the use of machining symbols.

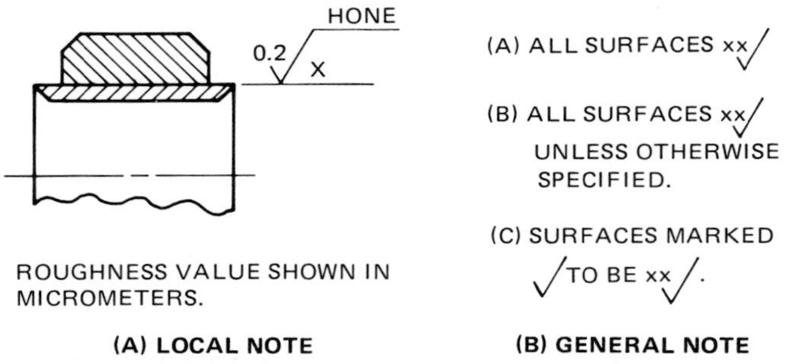

Material Removal Allowance

When it is desirable to indicate the amount of material to be removed, the amount of material in inches or millimeters is shown to the left of the symbol. Illustrations showing material removal allowance are shown in Figs. 8-7-14 and 8-7-15.

Material Removal Prohibited

When it is necessary to indicate that a surface must be produced without material removal, the machining prohibited symbol shown in Fig. 8-7-16 must be used.

Former Machining Symbols

Former machining symbols, as shown in Fig. 8-7-17, may be found on many drawings in use today. When called upon to make changes or revisions to a drawing already in existence, a drafter must adhere to the drawing conventions shown on that drawing.

REFERENCES AND SOURCE MATERIALS

1. ANSI Y14.36, *Surface Texture Symbols*.
2. GAR.
3. General Motors.

ASSIGNMENTS

See Assignments 24 through 28 for Unit 8-7 on pages 225 to 226.

HONE

0.2

ROUGHNESS VALUE SHOWN IN MICROMETERS.

(A) LOCAL NOTE

(A) ALL SURFACES xx$\sqrt{}$

(B) ALL SURFACES xx$\sqrt{}$
UNLESS OTHERWISE SPECIFIED.

(C) SURFACES MARKED
$\sqrt{}$ TO BE xx $\sqrt{}$.

(B) GENERAL NOTE

FIG. 8-7-10   Surface texture notes.

REMOVAL OF MATERIAL BY MACHINING IS

OPTIONAL   OBLIGATORY

FIG. 8-7-11   Indicating the removal of material on the surface texture symbol.

FIG. 8-7-12   Application of surface texture symbol when machining of surface is required.

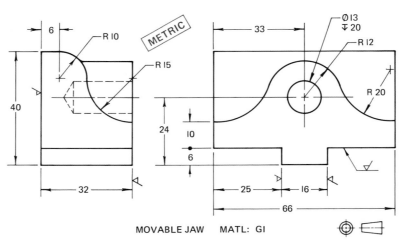

MOVABLE JAW   MATL: GI

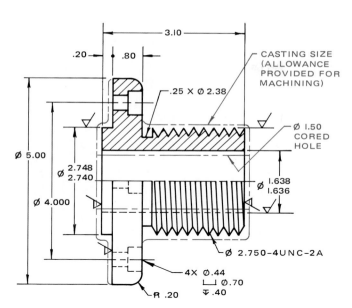

FIG. 8-7-13   Extra metal allowance for machined surfaces.

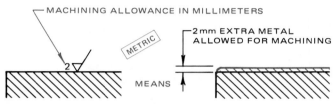

FIG. 8-7-14   Indication of machining allowance.

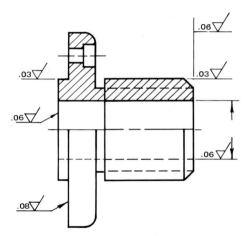

FIG. 8-7-15   Indicating machining allowance on drawings.

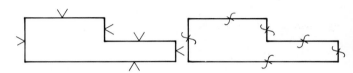

FIG. 8-7-16   Symbol to indicate the removal of material is not permitted.

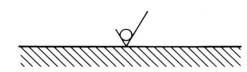

FIG. 8-7-17   Former machining symbols.

ASSIGNMENTS FOR CHAPTER 8

ASSIGNMENTS FOR UNIT 8-1, BASIC DIMENSIONING

1. Select one of the template drawings (Fig. 8-1-A or 8-1-B, pg. 210) and make a one-view drawing, complete with dimensions, of the part.

2. Select one of the parts shown in Figs. 8-1-C through 8-1-F (pg. 210) and make a three-view drawing, complete with dimensions, of the part.

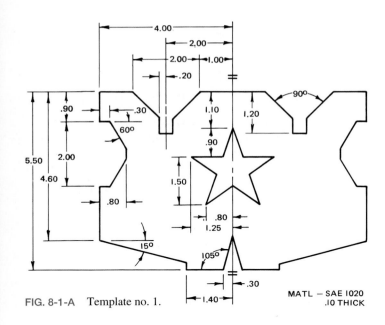

FIG. 8-1-A   Template no. 1.

MATL — SAE 1020
.10 THICK

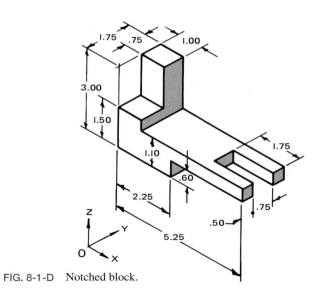

FIG. 8-1-D   Notched block.

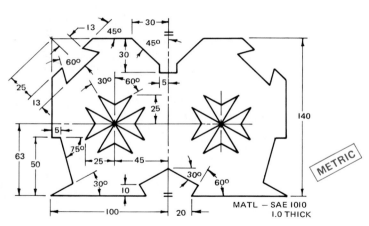

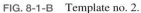

MATL — SAE 1010
1.0 THICK

METRIC

FIG. 8-1-B   Template no. 2.

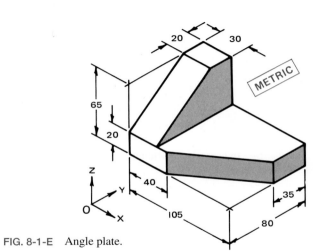

METRIC

FIG. 8-1-E   Angle plate.

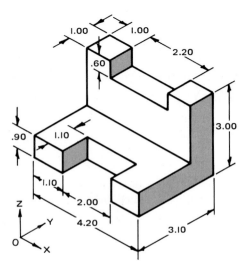

FIG. 8-1-C   Cross slide.

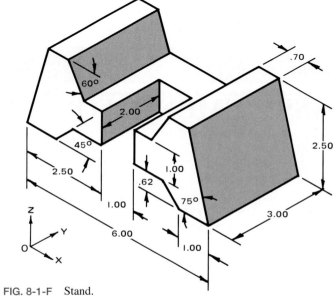

FIG. 8-1-F   Stand.

3. Select one of the parts shown in Figs. 8-1-G through 8-1-L and make a three-view drawing, complete with dimensions, of the part. Show the dimensions with the view that best shows the shape of the part or feature.

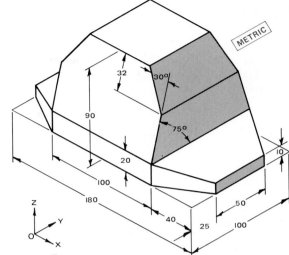

FIG. 8-1-J   Base.

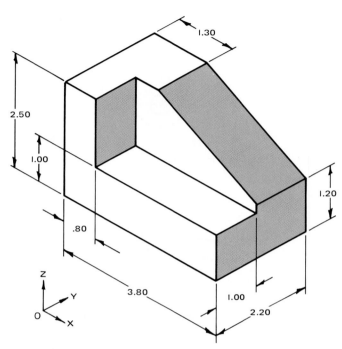

FIG. 8-1-G   Guide stop.

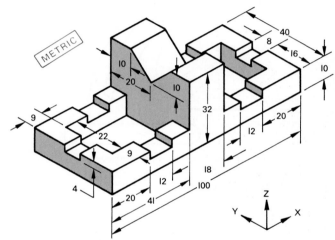

FIG. 8-1-K   Guide support.

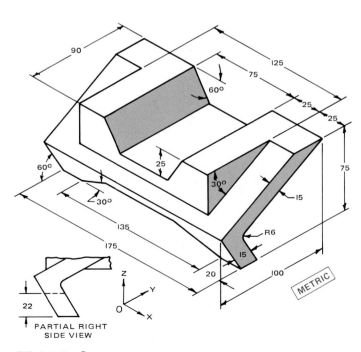

FIG. 8-1-H   Separator.

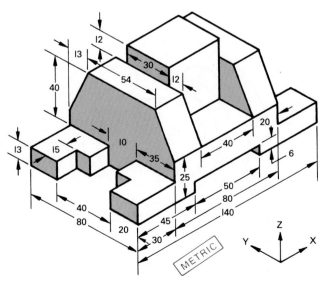

FIG. 8-1-L   Vertical guide.

ASSIGNMENTS FOR UNIT 8-2, DIMENSIONING CIRCULAR FEATURES

4. Select one of the problems shown in Figs. 8-2-A through 8-2-E and make a one-view drawing, complete with dimensions, of the part.

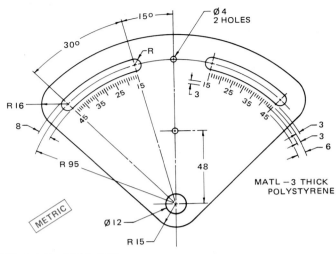

FIG. 8-2-C   Dial indicator.

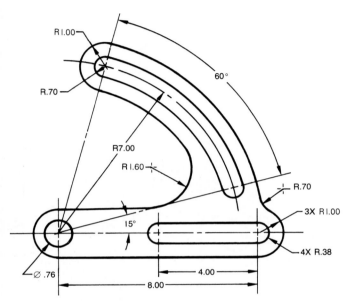

FIG. 8-2-A   Adjustable table support.

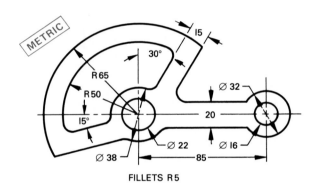

FIG. 8-2-D   Adjustable sector.

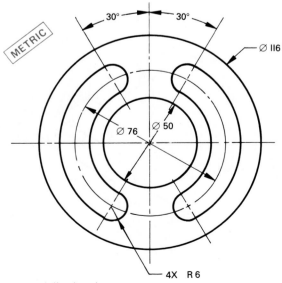

FIG. 8-2-B   Adjusting ring.

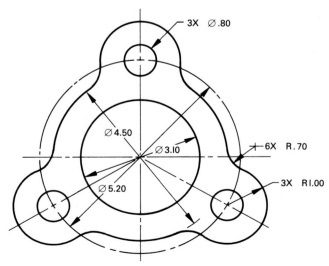

FIG. 8-2-E   Gasket.

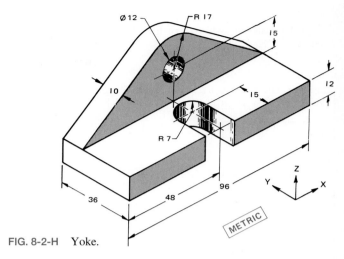

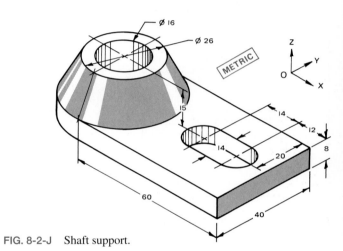

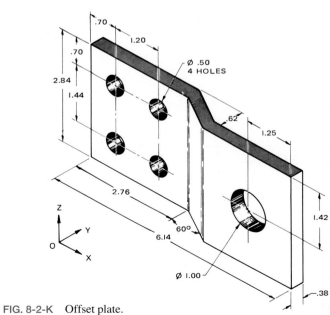

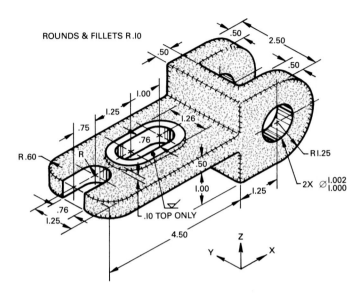

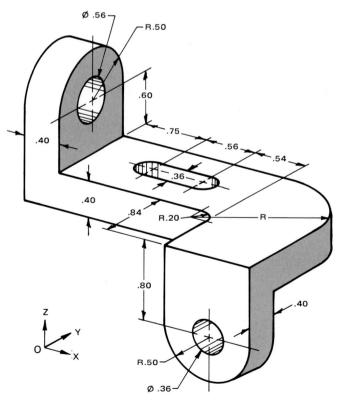

5. Select one of the problems shown in Figs. 8-2-F through 8-2-K and make a three-view drawing, complete with dimensions, of the part.

ROUNDS & FILLETS R.10

FIG. 8-2-F   Swing bracket.

FIG. 8-2-G   Bracket.

FIG. 8-2-H   Yoke.

METRIC

FIG. 8-2-J   Shaft support.

METRIC

FIG. 8-2-K   Offset plate.

6. Select one of the parts shown in Fig. 8-2-L, and using one of the scales shown, redraw the part and add dimensions.

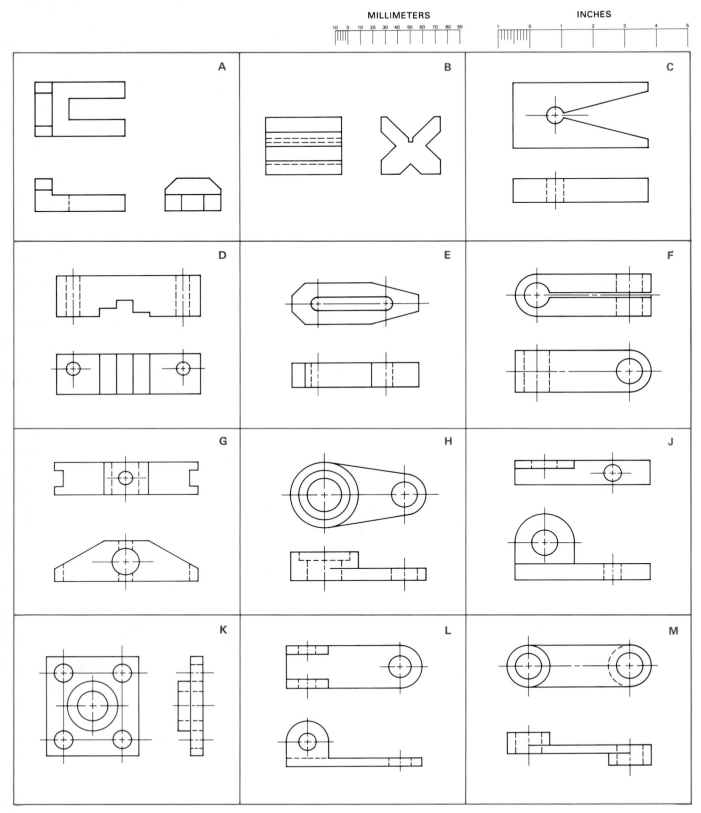

FIG. 8-2-L   Problems in dimensioning practice.

ASSIGNMENTS FOR UNIT 8-3, DIMENSIONING COMMON FEATURES

7. Redraw the handle shown in Fig. 8-3-A. The following features are to be added and dimensioned:
   (a) 45° × .10 chamfer
   (b) 33DP diamond knurl for 1.20 in. starting .80 in. from left end
   (c) 1:8 circular taper for 1.20 in. length on right end of Ø1.25

(d) .16 × Ø.54 in. undercut on Ø.75
(e) Ø.189 × .25 in. deep, 4 holes equally spaced
(f) 30° × .10 chamfer, the .10 in. dimension taken horizontally along the shaft.

8. Redraw the selector shaft shown in Fig. 8-3-B and dimension. Scale the drawing for sizes.
9. Make a one-view drawing (plus a partial view of the blade), with dimensions, of the screwdriver shown in Fig. 8-3-C.
10. Make a one-view drawing with dimensions of the indicator rod shown in Fig. 8-3-D.

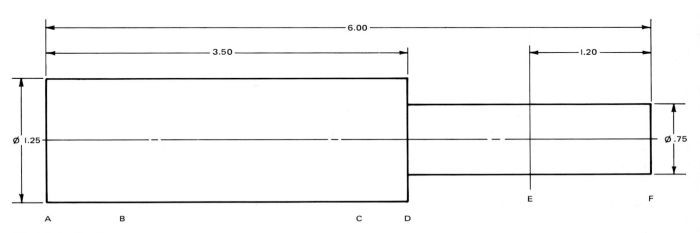

FIG. 8-3-A   Handle.

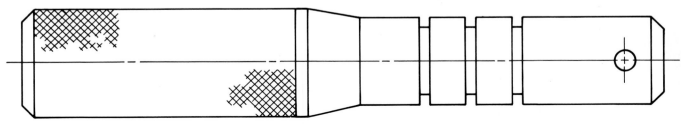

FIG. 8-3-B   Selector shaft.

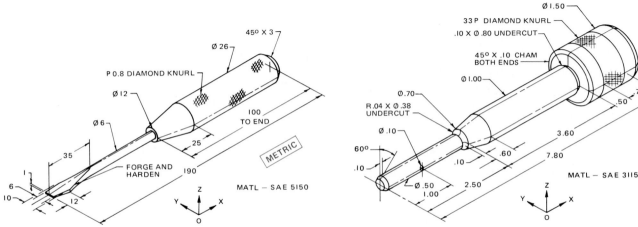

FIG. 8-3-C   Screwdriver.

FIG. 8-3-D   Indicator rod.

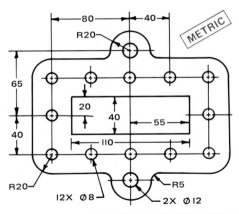

MATL—2mm FLUOROCARBON PLASTIC (MYLAR)

FIG. 8-3-E   Gasket.

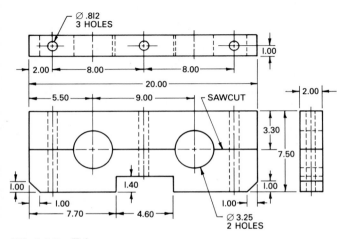

FIG. 8-3-F   Tube support.

11. Make a half-view drawing of one of the parts shown in Figs. 8-3-E through 8-3-G. Add the symmetry symbol to the drawing and dimension using symbolic dimensioning wherever possible. Use the MIRROR command to create the view if CAD is used.
12. Make a one-view drawing of the adjusting locking plate shown in Fig. 8-3-H. If manually drawn, show only two or three holes and teeth. Scale 10:1.

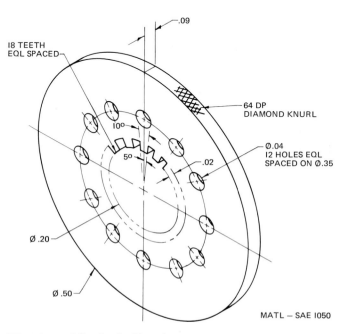

MATL – SAE 1050

FIG. 8-3-H   Adjusting locking plate.

FIG. 8-3-G   Gasket.

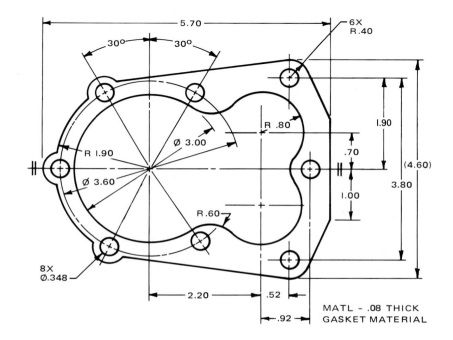

MATL - .08 THICK
GASKET MATERIAL

ASSIGNMENTS FOR UNIT 8-4, DIMENSIONING METHODS

13. Select one of the problems shown in Figs. 8-4-A and 8-4-B, and make a working drawing of the part. The arrowless dimensioning shown is to be replaced with rectangular coordinate dimensioning and has the following dimensioning changes.

For Fig. 8-4-A:

■ Holes A, E, and D are located from the zero coordinates.

■ Holes B are located from center of hole E.
■ Holes C are located from center of hole D.

For Fig. 8-4-B:

■ Holes E and D are located from left and bottom edges.
■ Holes A and C are located from center of hole D.
■ Holes B are located from center of hole E.

14. Redraw the terminal board shown in Fig. 8-4-C using tabular dimensioning. Use the bottom and left-hand edge for the datum surfaces to locate the holes.

15. Divide a sheet into four quadrants by bisecting the vertical and horizontal sides. In each quadrant draw the adaptor plate shown in Fig. 8-4-D. Different methods of dimensioning are to be used for each drawing. The methods are rectangular coordinate, chordal, arrowless, and tabular.

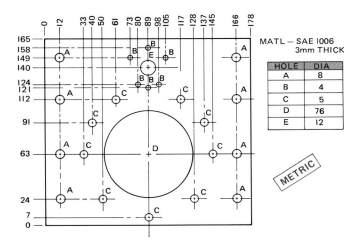

HOLE	DIA
A	8
B	4
C	5
D	76
E	12

MATL — SAE 1006
3mm THICK

FIG. 8-4-A   Cover plate.

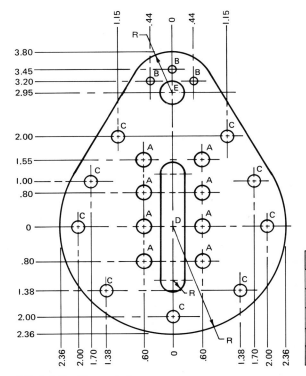

MATL—SAE 1008
12 THICK

HOLE	SIZE
A	.30
B	.16
C	.24
D	.40 X 2.80
E	.50

FIG. 8-4-B   Transmission cover.

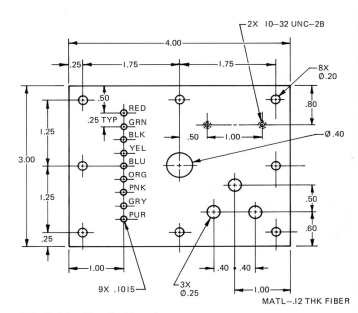

MATL—.12 THK FIBER

FIG. 8-4-C   Terminal board.

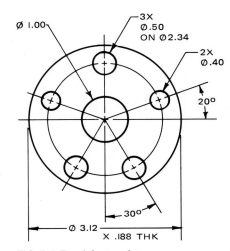

FIG. 8-4-D   Adaptor plate.

217

16. Redraw the oil chute shown in Fig. 8-4-E using tabular dimensioning to locate the holes from datum surfaces *X*, *Y*, and *Z*.

17. Redraw one of the parts shown in Figs. 8-4-F and 8-4-G. Use arrowless or tabular dimensioning. For Fig. 8-4-F use the bottom and left-hand edge for the datum surfaces to locate the holes. For Fig. 8-4-G use the bottom and the center of the part to locate the features. Use the MIRROR command to create the views if CAD is used.

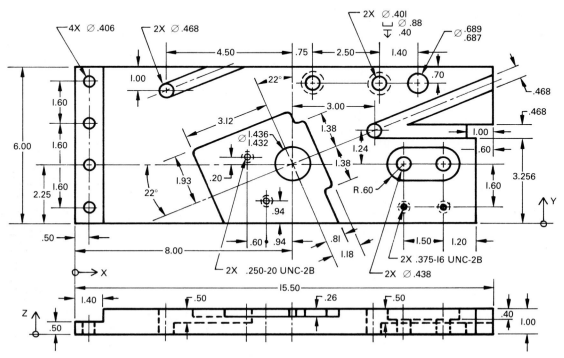

FIG. 8-4-E   Oil chute.

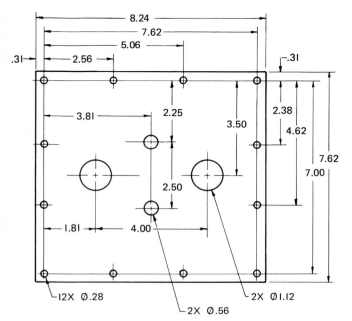

FIG. 8-4-F   Cover plate.

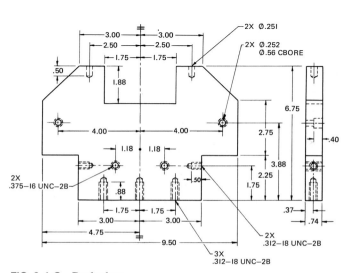

FIG. 8-4-G   Back plate.

18. Redraw the part shown in Fig. 8-4-H and use tabular
    dimensioning for locating the holes. Use the *X*, *Y*, and *Z*
    datums to locate the holes.

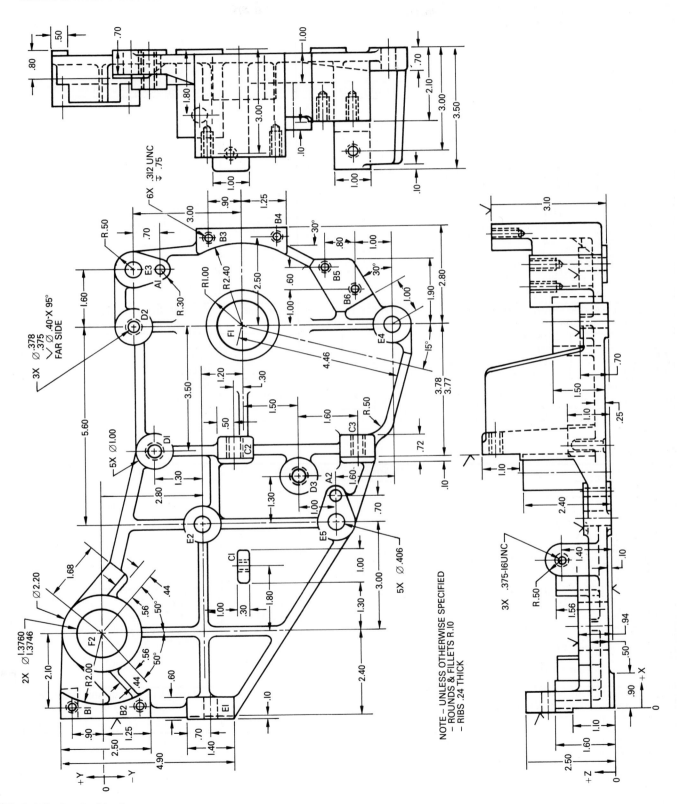

FIG. 8-4-H   Interlocking base.

ASSIGNMENT FOR UNIT 8-5, LIMITS AND TOLERANCES

19. Calculate the sizes and tolerances for one of the drawings shown in Fig. 8-5-A or 8-5-B.

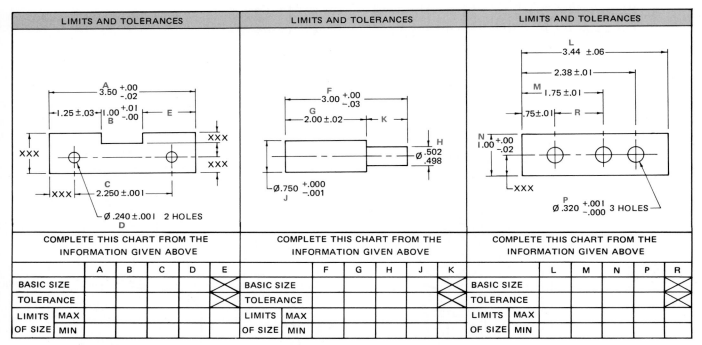

FIG. 8-5-A   Inch limits and tolerances.

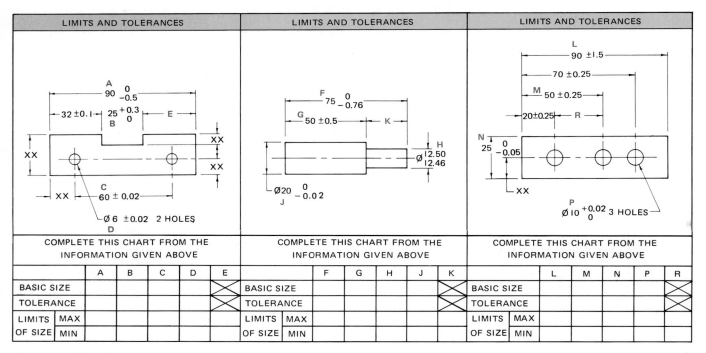

FIG. 8-5-B   Metric limits and tolerances.

ASSIGNMENTS FOR UNIT 8-6, FITS AND ALLOWANCES

20. Using the tables of fits located in the Appendix, calculate the missing dimensions in any of the four charts shown in Figs. 8-6-A through 8-6-D (here and on pg. 222).

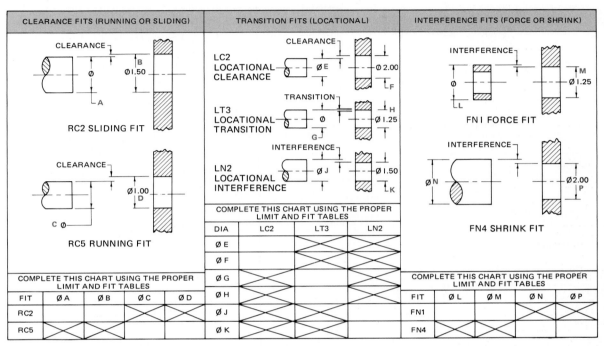

FIG. 8-6-A   Inch fits.

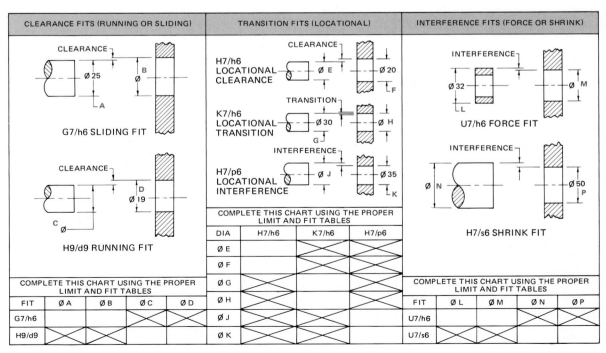

FIG. 8-6-B   Metric fits.

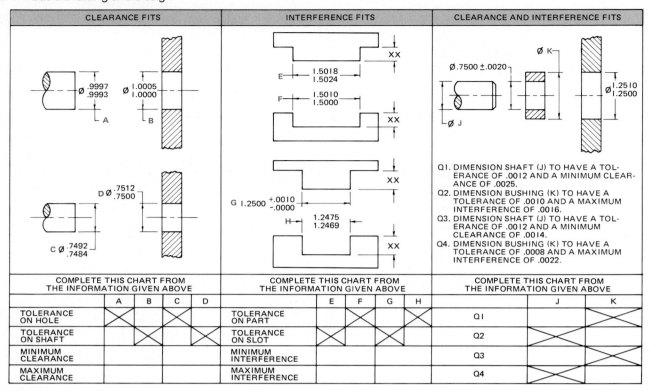

FIG. 8-6-C   Inch fits.

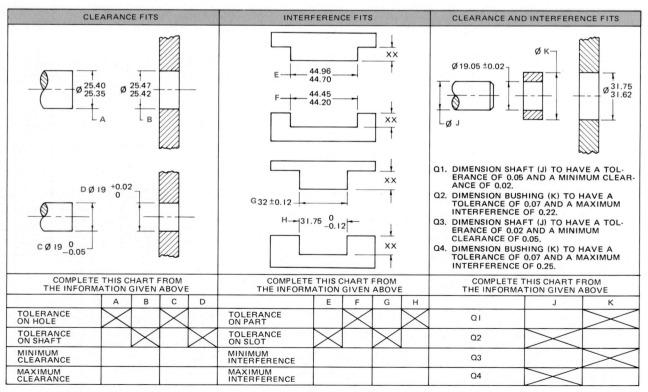

FIG. 8-6-D   Metric fits.

21. Using the fit tables in the Appendix, complete the table shown in Fig. 8-6-E using either U.S. customary or metric sizes.

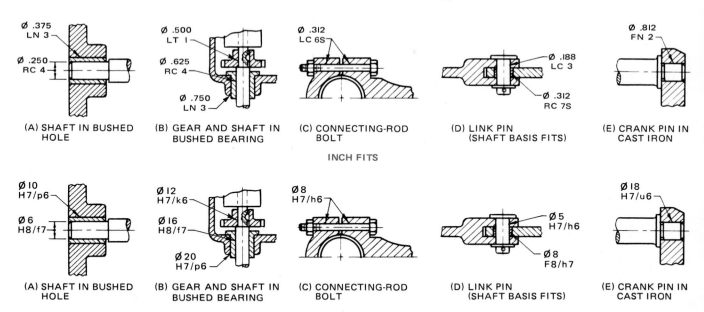

INCH FITS

(A) SHAFT IN BUSHED HOLE

(B) GEAR AND SHAFT IN BUSHED BEARING

(C) CONNECTING-ROD BOLT

(D) LINK PIN (SHAFT BASIS FITS)

(E) CRANK PIN IN CAST IRON

METRIC FITS

DESIGN SKETCH	BASIC DIAMETER SIZE IN. [mm]	SYMBOL	BASIS	FEATURE	LIMITS OF SIZE		CLEARANCE OR INTERFERENCE	
					MAX	MIN	MAX	MIN
A	.375 [10]		HOLE	HOLE				
				SHAFT				
A	.250 [6]		HOLE	HOLE				
				SHAFT				
B	.500 [12]		HOLE	HOLE				
				SHAFT				
B	.625 [16]		HOLE	HOLE				
				SHAFT				
B	.750 [20]		HOLE	HOLE				
				SHAFT				
C	.312 [8]		SHAFT	HOLE				
				SHAFT				
D	.188 [5]		HOLE	HOLE				
				SHAFT				
D	.312 [8]		SHAFT	HOLE				
				SHAFT				
E	.812 [18]		HOLE	HOLE				
				SHAFT				

FIG. 8-6-E   Fit problems.

22. Make a detail drawing of the spindle shown in Fig. 8-6-F. Scale the part for sizes using one of the scales shown with the drawing. Other considerations are:
    (a) "A" diameter to have an LC3 (inch) or H7/h6 (metric) fit.
    (b) "B" diameter requires a 96 diamond knurl or its equivalent.
    (c) "C" diameter to have an LT3 (inch) or H7/k6 (metric) fit.
    (d) "D" diameter to be a minimum relief (undercut).
    (e) "E" to be a standard No. 807 Woodruff key in center of segment and the diameter to be controlled by an RC3 (inch) or H7/g6 (metric) fit.

(f) "F" to be undercut for a standard external retaining ring and dimensioned to manufacturer's specifications.
(g) Dimension in decimal inch or metric.

23. Make a detail drawing of the roller guide base shown in Fig. 8-6-G. Use one of the scales shown on the drawing to scale the part for sizes. Other considerations are:
    (a) Keyseat for a standard square key and limits on the hole controlled by either an H9/d9 (metric) or RC6 (inch) fit.
    (b) Control critical machine surfaces to 0.8 μm or 32 μin.
    (c) Dimension in metric or decimal inch.

FIG. 8-6-F   Spindle.

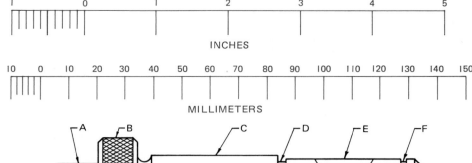

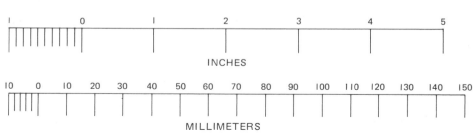

FIG. 8-6-G   Roller guide base.

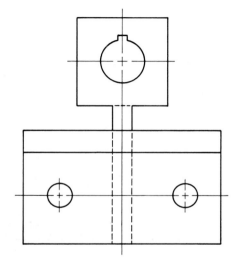

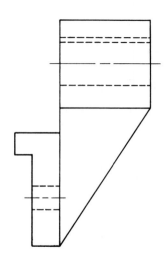

ASSIGNMENTS FOR UNIT 8-7, SURFACE TEXTURE

24. Make a working drawing of the link shown in Fig. 8-7-A. The amount of material to be removed from the end surfaces of the hub is .09 in. and .06 in. on the bosses and bottom of the vertical hub. The two large holes are to have an LN3 fit for journal bearings. Scale 1:1.

25. Make a working drawing of the cross slide shown in Fig. 8-7-B. Scale 1:1. The following information is to be added to the drawing:
    - The dovetail slot is to have a maximum roughness value of 3.2 μm and a machining allowance of 2mm.
    - The ends of the shaft support are to have maximum and minimum roughness values of 1.6 and 0.8 μm and a machining allowance of 2 mm.
    - The hole is to have an H8 tolerance.

26. Make a working drawing of the column bracket shown in Fig. 8-7-C. Scale 1:1. The following information is to be added to the drawing:
    - The bottom of the base is to have a maximum roughness value of 125 μin. and a machining allowance of .06 in.
    - The tops of the bosses are to have a maximum roughness value of 250 min. and a machining allowance of .04 in.
    - The end surfaces of the hubs supporting the shafts are to have maximum and minimum roughness values of 63 and 32 μin. and a machining allowance of .04 in.
    - The large hole is to be dimensioned for an RC4 fit. The small hole is to be dimensioned for an LN3 fit for plain bearings.

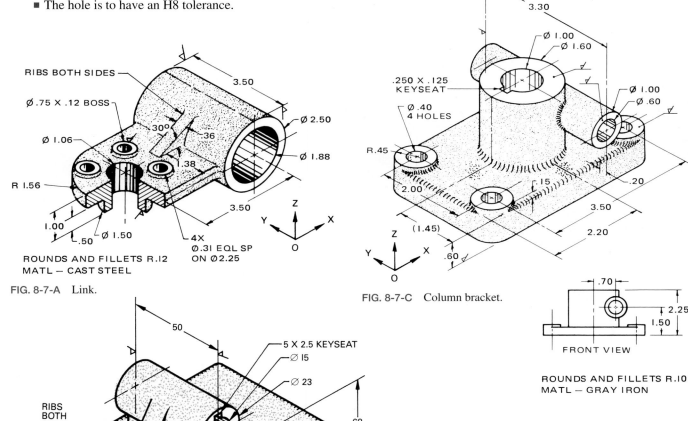

RIBS BOTH SIDES

Ø .75 X .12 BOSS

Ø 1.06

30°

36

Ø 2.50

Ø 1.88

R 1.56

1.38

1.00

.50

Ø 1.50

3.50

3.50

4X
Ø .31 EQL SP
ON Ø2.25

ROUNDS AND FILLETS R.12
MATL — CAST STEEL

FIG. 8-7-A   Link.

FIG. 8-7-C   Column bracket.

3.30

Ø 1.00

Ø 1.60

.250 X .125
KEYSEAT

Ø .40
4 HOLES

R.45

2.00

Ø 1.00

Ø .60

.15

.20

3.50

2.20

(1.45)

.60

.70

2.25

1.50

FRONT VIEW

ROUNDS AND FILLETS R.10
MATL — GRAY IRON

50

5 X 2.5 KEYSEAT

Ø 15

Ø 23

RIBS BOTH SIDES

10

60°

12

68

25

60°

10

30

45

120

50

12

70

34

18

METRIC

ROUNDS AND FILLETS R5

MATL — MALLEABLE IRON

FIG. 8-7-B   Cross slide.

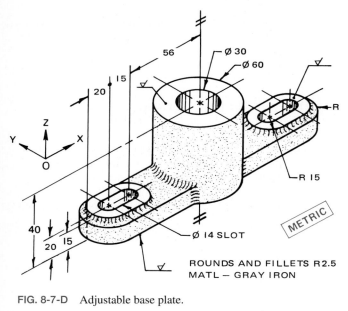

FIG. 8-7-D   Adjustable base plate.

27. Make a working drawing of the adjustable base plate shown in Fig. 8-7-D. The amount of material to be removed on the surfaces requiring machining is 2 mm. The center hole is to be dimensioned having an H8 tolerance. Scale 1:1.

28. Make a working drawing of one of the parts shown in Figs. 8-7-E through 8-7-G. Show limit dimensions for the holes showing fit symbols. Unless otherwise specified, surface finish to be 63 μin (1.6 μm) with a machining allowance of .06 in. (2 mm).

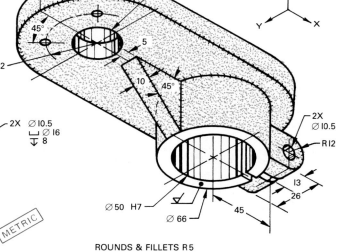

ROUNDS & FILLETS R5

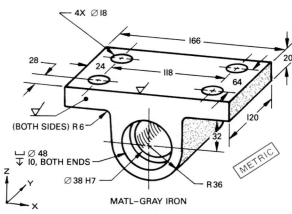

FIG. 8-7-E   Swing bracket.

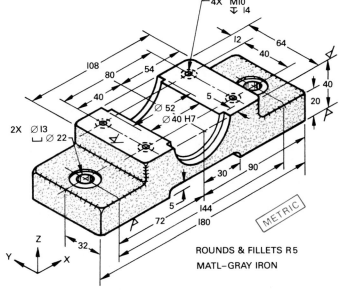

FIG. 8-7-G   Bearing base.

FIG. 8-7-F   Bearing housing.

226

CHAPTER 9

SECTIONS

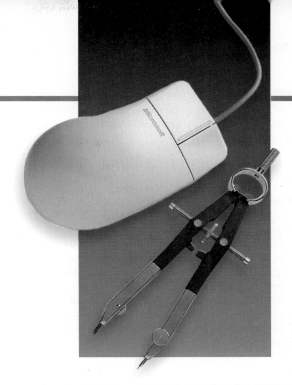

Definitions

Cutting plane The plane at which the exterior view is cut away to reveal the interior view.

Cutting-plane line A line to indicate the location of the cutting plane.

Half-section A view of an assembly or object, usually symmetrical, that shows one-half of the full view in section.

Phantom section A sectional view superimposed on the regular view, used to show the typical interior shape and/or mating part within the object.

Section lining or **cross-hatching** Lines used to indicate either the surface that has been theoretically cut or the material from which the object is to be made.

Sectional views or sections Drawings used to show interior details of objects that are too complicated to be shown clearly in regular views.

9-1 SECTIONAL VIEWS

Sectional views, commonly called *sections,* are used to show interior detail that is too complicated to be shown clearly by regular views containing many hidden lines. For some assembly drawings, they show a difference in materials. A sectional view is obtained by supposing that the nearest part of the object to be cut or broken away is on an imaginary cutting plane. The exposed or cut surfaces are identified by section lining or cross-hatching. Hidden lines and details behind the cutting-plane line are usually omitted unless they are required for clarity or dimensioning. It should be understood that only in the sectional view is any part of the object shown as having been removed.

A sectional view frequently replaces one of the regular views. For example, a regular front view is replaced by a front view in section, as shown in Fig. 9-1-1.

Whenever practical, except for revolved sections, sectional views should be projected perpendicularly to the cutting plane and be placed in the normal position for third-angle projection.

FIG. 9-1-1 A full-section drawing.

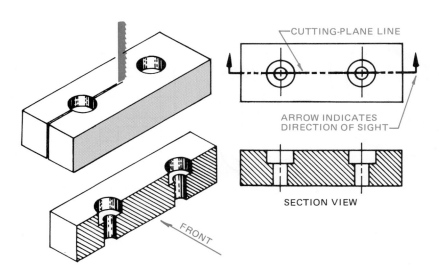

CUTTING-PLANE LINE

ARROW INDICATES DIRECTION OF SIGHT

SECTION VIEW

FRONT

When the preferred placement is not practical, the sectional view may be moved to some other convenient position on the drawing, but it must be clearly identified, usually by two upper-case letters, and labeled.

Cutting-Plane Lines

Cutting-plane lines (Fig. 9-1-2) are used to show the location of cutting planes for sectional views. Two forms of cutting-plane lines are approved for general use.

The first form consists of evenly spaced, thick dashes with arrowheads. The second form consists of alternating long dashes and pairs of short dashes. The long dashes may vary in length, depending on the size of the drawing.

Both forms of lines should be drawn to stand out clearly on the drawing. The ends of the lines are bent at 90° and terminated by bold arrowheads to indicate the direction of sight for viewing the section.

The cutting-plane line can be omitted when it corresponds to the center line of the part and it is obvious where the cutting

plane lies. On drawings with a high density of line work and for offset sections (see Unit 9-6), cutting-plane lines may be modified by omitting the dashes between the line ends for the purpose of obtaining clarity, as shown in Fig. 9-1-2B.

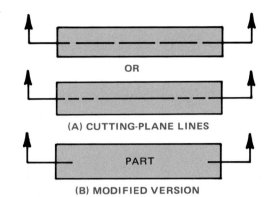

FIG. 9-1-2   Cutting-plane lines.

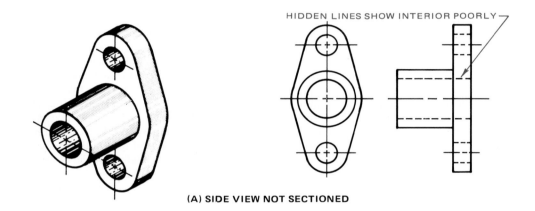

(A) SIDE VIEW NOT SECTIONED

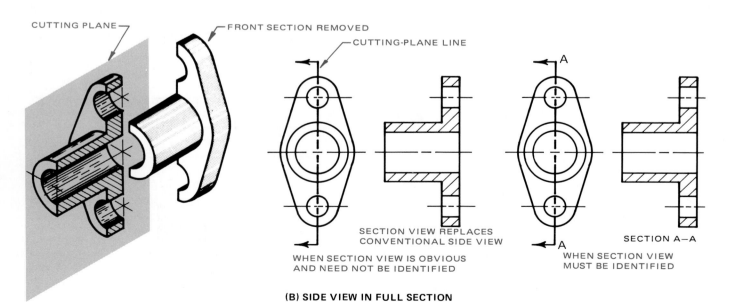

(B) SIDE VIEW IN FULL SECTION

FIG. 9-1-3   Full-section view.

FIG. 9-1-4   Visible and hidden lines on section views.

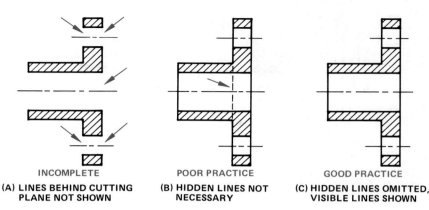

(A) **LINES BEHIND CUTTING PLANE NOT SHOWN**
INCOMPLETE

(B) **HIDDEN LINES NOT NECESSARY**
POOR PRACTICE

(C) **HIDDEN LINES OMITTED, VISIBLE LINES SHOWN**
GOOD PRACTICE

Full Sections

When the cutting plane extends entirely through the object in a straight line and the front half of the object is theoretically removed, a *full section* is obtained (Figs. 9-1-3 and 9-1-4). This type of section is used for both detail and assembly drawings. When the section is on an axis of symmetry, it is not necessary to indicate its location (Fig. 9-1-5). However, it may be identified and indicated in the normal manner to increase clarity, if so desired.

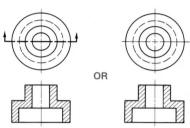

FIG. 9-1-5   Cutting-plane line may be omitted when it corresponds with a center line.

Section Lining

Section lining, sometimes referred to as *cross-hatching,* can serve a double purpose. It indicates the surface that has been theoretically cut and makes it stand out clearly, thus helping the observer to understand the shape of the object. Section lining may also indicate the material from which the object is to be made, when the lining symbols shown in Fig. 9-1-6 are used.

Section Lining for Detail Drawings

Since the exact material specifications for a part are usually given elsewhere on the drawing, the general-purpose section lining symbol is recommended for most detail drawings. An exception may be made for wood when it is desirable to show the direction of the grain.

The lines for section lining are thin and are usually drawn at an angle of 45° to the major outline of the object. The same angle is used for the whole "cut" surface of the object. If the part shape would cause section lines to be parallel, or nearly so, to one of the sides of the part, some angle other than 45° should be chosen. See Fig. 9-1-7 (pg. 230). The spacing of the hatching lines should be reasonably uniform to give a good appearance to the drawing. The pitch, or distance between lines, normally varies between .03 and .12 in. (1 and 3 mm), depending on the size of the area to be sectioned.

As a cost reduction in manual drafting, large areas need not be entirely section-lined (Fig. 9-1-8, pg. 230). Section lining around the outline will usually be sufficient, providing clarity is not sacrificed.

Dimensions or other lettering should not be placed in sectioned areas. When this is unavoidable, the section lining should be omitted for the numerals or lettering (Fig. 9-1-9, pg. 230).

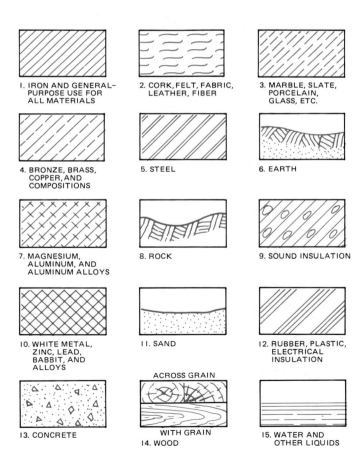

1. IRON AND GENERAL-PURPOSE USE FOR ALL MATERIALS

2. CORK, FELT, FABRIC, LEATHER, FIBER

3. MARBLE, SLATE, PORCELAIN, GLASS, ETC.

4. BRONZE, BRASS, COPPER, AND COMPOSITIONS

5. STEEL

6. EARTH

7. MAGNESIUM, ALUMINUM, AND ALUMINUM ALLOYS

8. ROCK

9. SOUND INSULATION

10. WHITE METAL, ZINC, LEAD, BABBIT, AND ALLOYS

11. SAND

12. RUBBER, PLASTIC, ELECTRICAL INSULATION

13. CONCRETE

ACROSS GRAIN
WITH GRAIN
14. WOOD

15. WATER AND OTHER LIQUIDS

FIG. 9-1-6   Symbolic section lining.

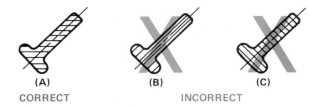

FIG. 9-1-7   Direction of section lining.

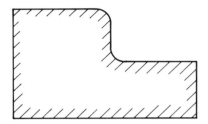

FIG. 9-1-8   Outline section lining.

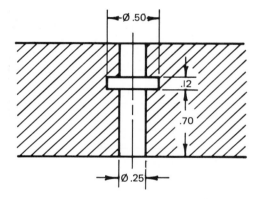

FIG. 9-1-9   Section lining omitted to accommodate dimensions.

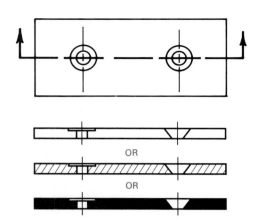

FIG. 9-1-10   Thin parts in section.

Sections that are too thin for effective section lining, such as sheet-metal items, packing, and gaskets, may be shown without section lining, or the area may be filled in completely (Fig. 9-1-10).

ASSIGNMENTS

See Assignments 1 and 2 for Unit 9-1 on pages 240–241.

9-2 TWO OR MORE SECTIONAL VIEWS ON ONE DRAWING

If two or more sections appear on the same drawing, the cutting-plane lines are identified by two identical large Gothic letters, one at each end of the line, placed behind the arrowhead so that the arrow points away from the letter. Normally, begin alphabetically with A-A, then B-B, and so on (Fig. 9-2-1). The identification letters should not include I, O, Q, or Z.

FIG. 9-2-1   Detail drawing having two sectional views.

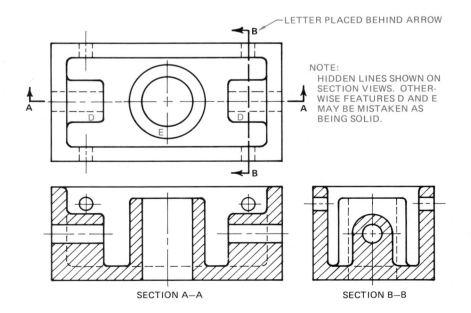

Sectional-view subtitles are given when identification letters are used and appear directly below the view, incorporating the letters at each end of the cutting-plane line thus: SECTION A-A, or abbreviated, SECT. B-B. When the scale is different from the main view, it is stated below the subtitle thus:

SECTION A-A
SCALE 1:10

ASSIGNMENTS

See Assignments 3 and 4 for Unit 9-2 on pages 241–242.

9-3 HALF-SECTIONS

A *half-section* is a view of an assembly or object, usually symmetrical, showing one-half of the view in section (Figs. 9-3-1 and 9-3-2). Two cutting-plane lines, perpendicular to each other, extend halfway through the view, and one-quarter of the view is considered removed with the interior exposed to view.

Similar to the practice followed for full-section drawings, the cutting-plane line need not be drawn for half-sections when it is obvious where the cutting took place. Instead, center lines may be used. When a cutting-plane is used, the common

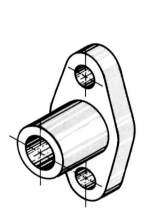
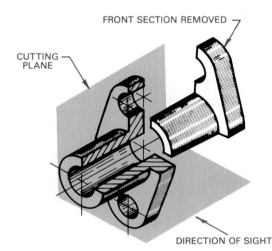
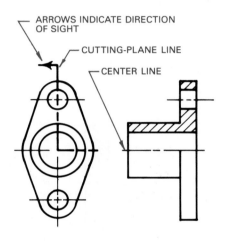

FIG. 9-3-1   Half-section drawing.

FIG. 9-3-2   Half-section views.

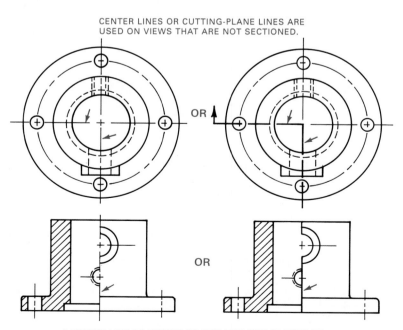

A CENTER LINE OR VISIBLE OBJECT LINE MAY BE USED TO DIVIDE THE SECTIONED HALF FROM THE UNSECTIONED HALF.

practice is to show only one end of the cutting-plane line, terminating with an arrow to show the direction of sight for viewing the section.

On the sectional view a center line or a visible object line may be used to divide the sectioned half from the unsectioned half of the drawing. This type of sectional drawing is best suited for assembly drawings where both internal and external construction is shown on one view and where only overall and center-to-center dimensions are required. The main disadvantage of using this type of sectional drawing for detail drawings is the difficulty in dimensioning internal features without adding hidden lines. However, hidden lines may be added for dimensioning, as shown in Fig. 9-3-3.

ASSIGNMENT

See Assignment 5 for Unit 9-3 on pages 242–243.

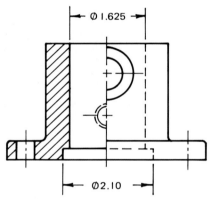

HIDDEN LINES ADDED FOR DIMENSIONING

FIG. 9-3-3   Dimensioning a half-section view.

9-4 THREADS IN SECTION

True representation of a screw thread is seldom provided on working drawings because it would require very laborious and accurate drawing involving repetitious development of the helix curve of the thread. A symbolic representation of threads is now standard practice.

Three types of conventions are in general use for screw-thread representation (Fig. 9-4-1). These are known as detailed, schematic, and simplified representations. Simplified representation should be used whenever it will clearly portray the requirements. Schematic and detailed representations require more drafting time but are sometimes necessary to avoid confusion with other parallel lines or to more clearly portray particular aspects of the threads.

Threaded Assemblies

Any of the thread conventions shown here may be used for assemblies of threaded parts, and two or more methods may be used on the same drawing, as shown in Fig. 9-4-2. In sectional views, the externally threaded part is always shown covering the internally threaded part (Fig. 9-4-3).

ASSIGNMENTS

See Assignments 6 and 7 for Unit 9-4 on pages 243–244.

9-5 ASSEMBLIES IN SECTION

Section Lining on Assembly Drawings

General-purpose section lining is recommended for most assembly drawings, especially if the detail is small. Symbolic

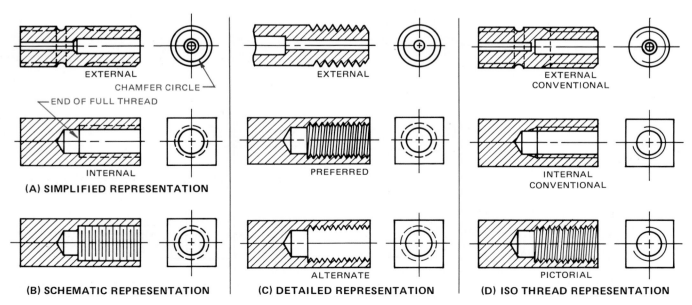

FIG. 9-4-1   Threads in section.

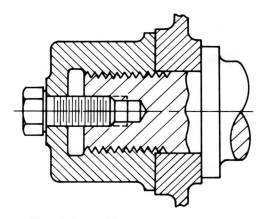

FIG. 9-4-2   Threaded assembly.

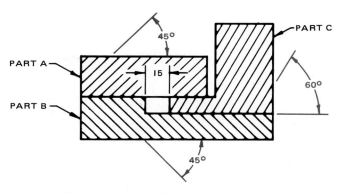

FIG. 9-5-1   Direction of section lining.

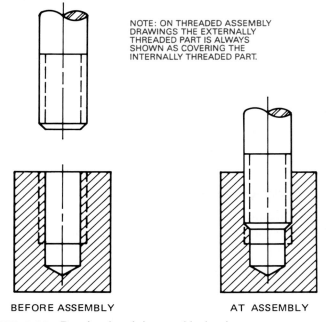

NOTE: ON THREADED ASSEMBLY DRAWINGS THE EXTERNALLY THREADED PART IS ALWAYS SHOWN AS COVERING THE INTERNALLY THREADED PART.

BEFORE ASSEMBLY   AT ASSEMBLY

FIG. 9-4-3   Drawing threads in assembly drawings.

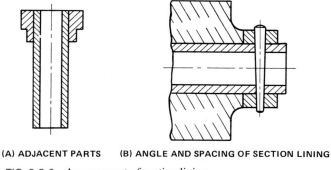

(A) ADJACENT PARTS   (B) ANGLE AND SPACING OF SECTION LINING

FIG. 9-5-2   Arrangement of section lining.

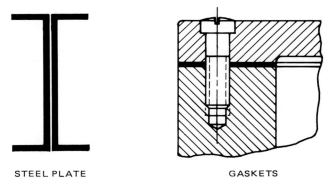

STEEL PLATE   GASKETS

FIG. 9-5-3   Assembly of thin parts in section.

section lining is generally not recommended for drawings that will be microformed.

General-purpose section lining should be drawn at an angle of 45° with the main outlines of the view. On adjacent parts, the section lines should be drawn in the opposite direction, as shown in Figs. 9-5-1 and 9-5-2.

For additional adjacent parts, any suitable angle may be used to make each part stand out separately and clearly. Section lines should not be purposely drawn to meet at common boundaries.

When two or more thin adjacent parts are filled in, a space is left between them, as shown in Fig. 9-5-3.

Symbolic section lining is used on special-purpose assembly drawings, such as illustrations for parts catalogs, display assemblies, and promotional materials, when it is desirable to distinguish between different materials (Fig. 9-1-6).

All assemblies and subassemblies pertaining to one particular set of drawings should use the same symbolic conventions.

Shafts, Bolts, Pins, Keyseats, and Similar Solid Parts, in Section   Shafts, bolts, nuts, rods, rivets, keys, pins, and similar solid parts, the axes of which lie in the cutting plane, should not be sectioned except that a broken-out section of the shaft may be used to describe more clearly the key, keyseat, or pin (Fig. 9-5-4, pg. 234).

ASSIGNMENTS

See Assignments 8 through 10 for Unit 9-5 on pages 244–246.

233

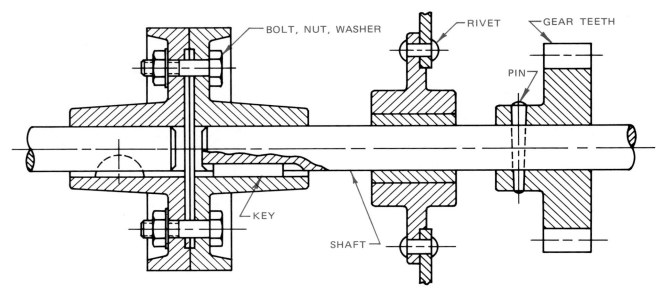

FIG. 9-5-4   Parts that are not section-lined even though the cutting plane passes through them.

9-6 OFFSET SECTIONS

In order to include features that are not in a straight line, the cutting plane may be offset or bent, so as to include several planes or curved surfaces (Figs. 9-6-1 and 9-6-2).

An *offset section* is similar to a full section in that the cutting-plane line extends through the object from one side to the other. The change in direction of the cutting-plane line is not shown in the sectional view.

ASSIGNMENT

See Assignment 11 for Unit 9-6 on pages 246–247.

9-7 RIBS, HOLES, AND LUGS IN SECTION

Ribs in Sections

A true-projection sectional view of a part, such as shown in Fig. 9-7-1, would be misleading when the cutting plane passes longitudinally through the center of the rib. To avoid this impression of solidity, a section not showing the ribs section-lined is preferred. When there is an odd number of ribs, such as those shown in Fig. 9-7-1B, the top rib is aligned with the bottom rib to show its true relationship with the hub and flange. If the rib is not aligned or revolved, it would appear distorted on the sectional view and would therefore be misleading.

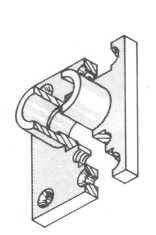

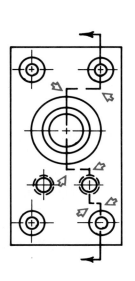

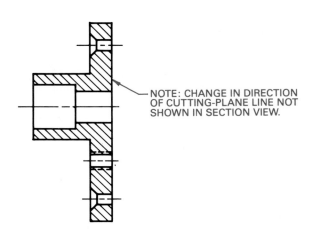

NOTE: CHANGE IN DIRECTION OF CUTTING-PLANE LINE NOT SHOWN IN SECTION VIEW.

FIG. 9-6-1   An offset section.

FIG. 9-6-2   Positioning offset
              sections.

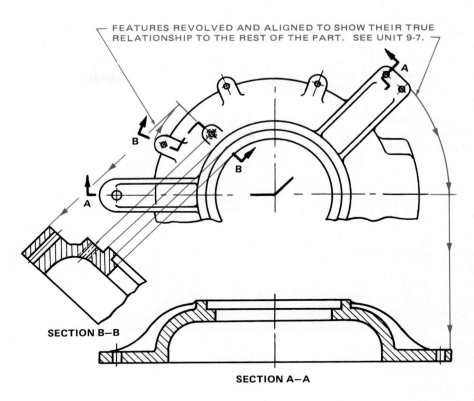

FEATURES REVOLVED AND ALIGNED TO SHOW THEIR TRUE
RELATIONSHIP TO THE REST OF THE PART. SEE UNIT 9-7.

SECTION B—B

SECTION A—A

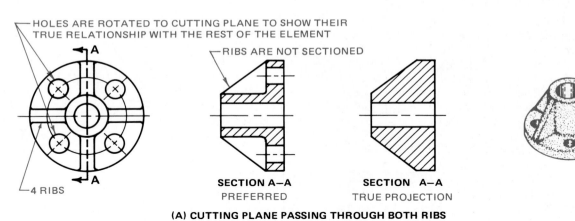

HOLES ARE ROTATED TO CUTTING PLANE TO SHOW THEIR
TRUE RELATIONSHIP WITH THE REST OF THE ELEMENT

RIBS ARE NOT SECTIONED

4 RIBS

SECTION A—A
PREFERRED

SECTION   A—A
TRUE PROJECTION

(A) CUTTING PLANE PASSING THROUGH BOTH RIBS

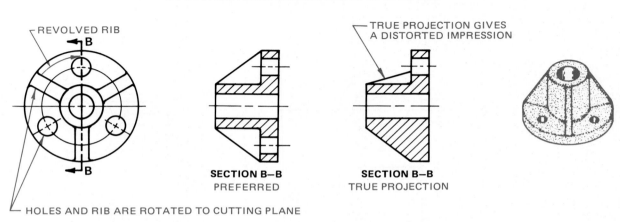

REVOLVED RIB

TRUE PROJECTION GIVES
A DISTORTED IMPRESSION

SECTION B—B
PREFERRED

SECTION B—B
TRUE PROJECTION

HOLES AND RIB ARE ROTATED TO CUTTING PLANE

(B) CUTTING PLANE PASSING THROUGH ONE RIB AND ONE HOLE

FIG. 9-7-1   Preferred and true projection through ribs and holes.

At times it may be necessary to use an alternative method of identifying ribs in a sectional view. Figure 9-7-2 shows a base and a pulley in section. If rib A of the base was not sectioned as previously mentioned, it would appear exactly like rib B in the sectional view and would be misleading. Similarly, ribs C shown on the pulley may be overlooked. To clearly show the relationship of the ribs with the other solid features on the base and pulley, alternate section lining on the ribs is used. The line between the rib and solid portions is shown as a broken line.

FIG. 9-7-2 Alternate method of showing ribs in section.

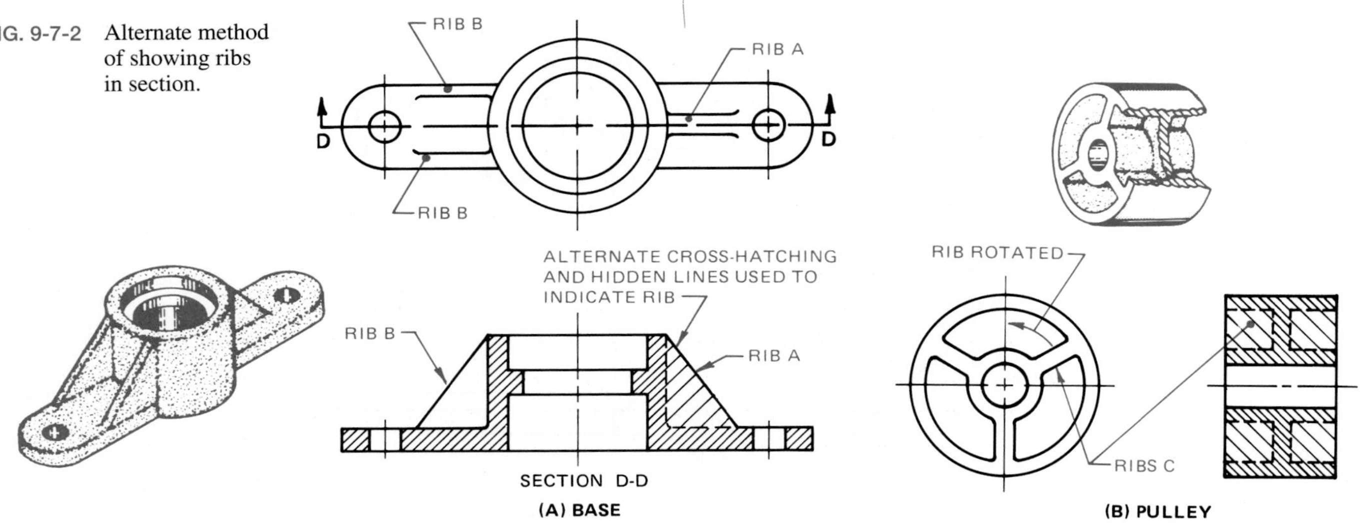

(A) BASE

(B) PULLEY

FIG. 9-7-3 Lugs in section.

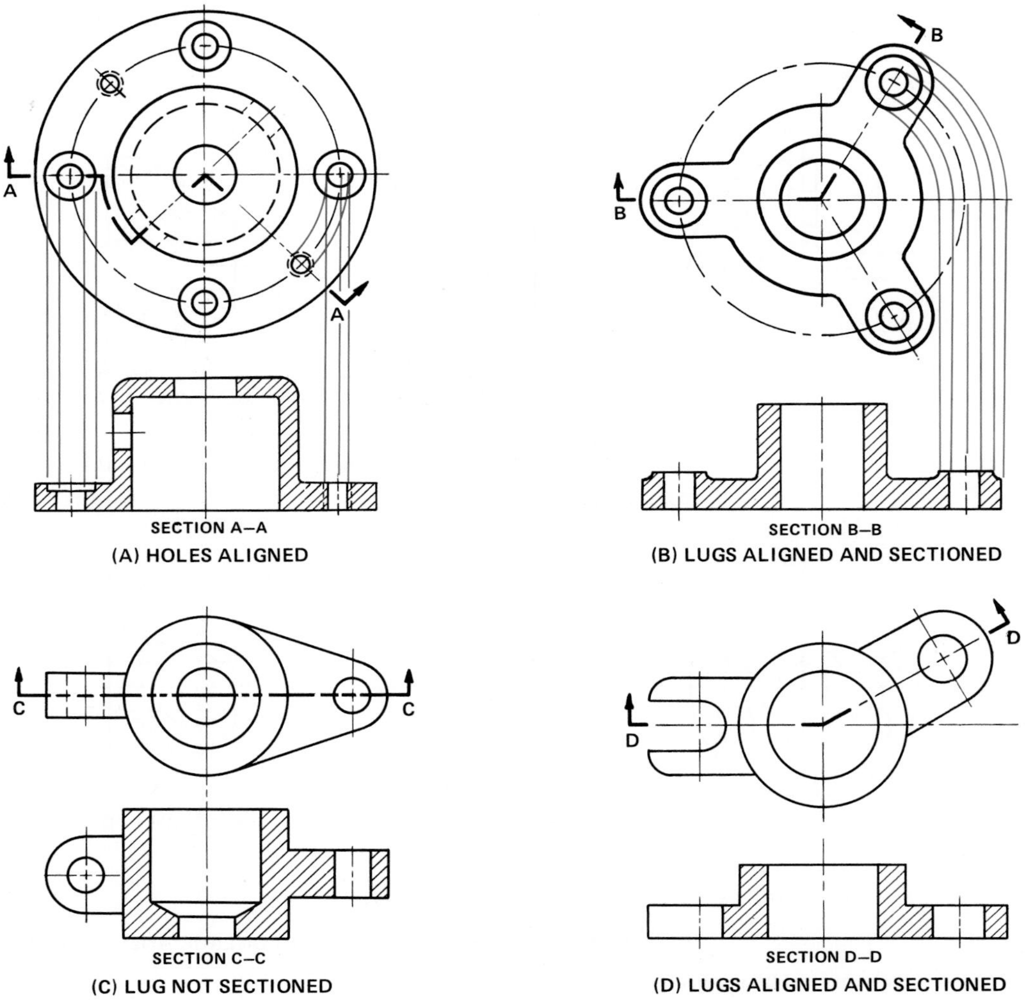

SECTION A—A
(A) HOLES ALIGNED

SECTION B—B
(B) LUGS ALIGNED AND SECTIONED

SECTION C—C
(C) LUG NOT SECTIONED

SECTION D—D
(D) LUGS ALIGNED AND SECTIONED

Holes in Sections

Holes, like ribs, are aligned as shown in Fig. 9-7-1 to show their true relationship to the rest of the part.

Lugs in Section

Lugs, like ribs and spokes, are also aligned to show their true relationship to the rest of the part, because true projection may be misleading. Figure 9-7-3 shows several examples of lugs in section. Note how the cutting-plane line is bent or offset so that the features may be clearly shown in the sectional view.

Some lugs are shown in section, and some are not. When the cutting plane passes through the lug crosswise, the lug is sectioned; otherwise, the lugs are treated in the same manner as ribs.

ASSIGNMENT ▪▪▪▪▪▪▪▪▪▪▪▪▪▪▪▪▪▪▪▪▪▪▪▪▪▪▪

See Assignment 12 for Unit 9-7 on pages 247–248.

9-8 REVOLVED AND REMOVED SECTIONS

Revolved and removed sections are used to show the cross-sectional shape of ribs, spokes, or arms when the shape is not obvious in the regular views (Figs. 9-8-1 through 9-8-3, here and on pg. 238). Often end views are not needed when a revolved section is used. For a revolved section, draw a center line through the shape on the plane to be described, imagine the part to be rotated 90°, and superimposed on the view of the shape that would be seen when rotated (Figs. 9-8-1 and 9-8-2). If the revolved section does not interfere with the view on which it is revolved, the view is not broken unless it would provide for clearer dimensioning. When the revolved section interferes or passes through lines on the view on which it is revolved, the general practice is to break the view (Fig. 9-8-2). Often the break is used to shorten the length of the object. In no circumstances should the lines on the view pass through the section. When superimposed on the view, the outline of the revolved section is a thin, continuous line.

The removed section differs in that the section, instead of being drawn right on the view, is removed to an open area on the drawing (Fig. 9-8-3). Frequently the removed section is drawn to an enlarged scale for clarification and easier dimensioning. Removed sections of symmetrical parts should be placed, whenever possible, on the extension of the center line (Fig. 9-8-3B).

On complicated drawings where the placement of the removed view may be some distance from the cutting plane, auxiliary information, such as the reference zone location (Fig. 9-8-4, pg. 238), may be helpful.

Placement of Sectional Views

Whenever practical, except for revolved sections, sectional views should be projected perpendicularly to the cutting plane and be placed in the normal position for third-angle projection (Fig. 9-8-5, pg. 238).

When the preferred placement is not practical, the sectional view may be removed to some other convenient position on the drawing, but it must be clearly identified, usually by two capital letters, and be labeled.

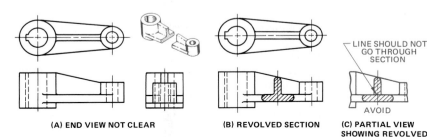
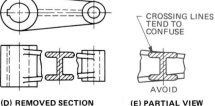

(A) END VIEW NOT CLEAR

(B) REVOLVED SECTION

(C) PARTIAL VIEW SHOWING REVOLVED SECTION

— LINE SHOULD NOT GO THROUGH SECTION

AVOID

(D) REMOVED SECTION WITH MAIN VIEW BROKEN FOR CLARITY

(E) PARTIAL VIEW SHOWING REVOLVED SECTION

CROSSING LINES TEND TO CONFUSE

AVOID

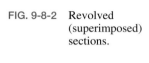

FIG. 9-8-1   Revolved sections.

FIG. 9-8-2   Revolved (superimposed) sections.

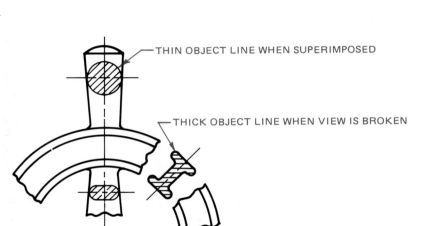

— THIN OBJECT LINE WHEN SUPERIMPOSED

— THICK OBJECT LINE WHEN VIEW IS BROKEN

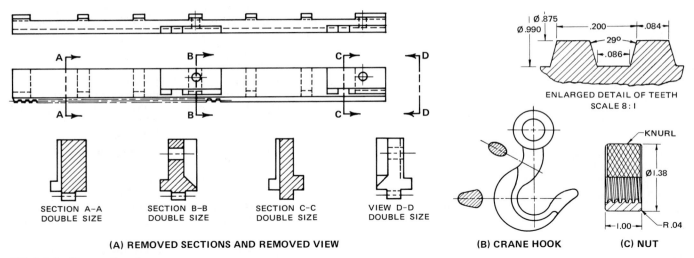

ENLARGED DETAIL OF TEETH
SCALE 8:1

(A) REMOVED SECTIONS AND REMOVED VIEW

SECTION A-A
DOUBLE SIZE

SECTION B-B
DOUBLE SIZE

SECTION C-C
DOUBLE SIZE

VIEW D-D
DOUBLE SIZE

(B) CRANE HOOK   **(C) NUT**

FIG. 9-8-3   Removed sections.

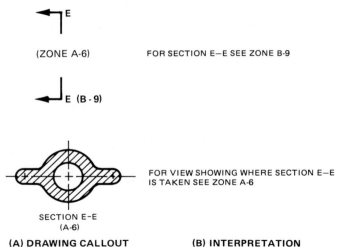

(ZONE A-6)   FOR SECTION E—E SEE ZONE B-9

FOR VIEW SHOWING WHERE SECTION E—E
IS TAKEN SEE ZONE A-6

SECTION E-E
(A-6)

(A) DRAWING CALLOUT   **(B) INTERPRETATION**

FIG. 9-8-4   Reference zone location.

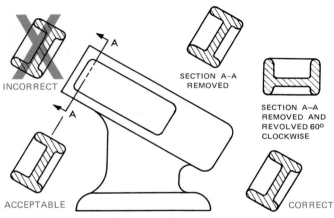

FIG. 9-8-5   Placement of sectional views.

ASSIGNMENTS

See Assignments 13 through 15 for Unit 9-8 on pages 249 to 250.

9-9   SPOKES AND ARMS IN SECTION

A comparison of the true projection of a wheel with spokes and a wheel with a web is made in Figs. 9-9-1A and B. This comparison shows that a preferred section for the wheel and spoke is desirable so that it will not appear to be a wheel with a solid web. In preferred sectioning, any part that is not solid or continuous around the hub is drawn without the section lining, even though the cutting plane passes through it. When there is an odd number of spokes, as shown in Fig. 9-9-1C, the bottom spoke is aligned with the top spoke to show its true relationship to the wheel and to the hub. If the spoke was not revolved or aligned, it would appear distorted in the sectional view.

ASSIGNMENT

See Assignment 16 for Unit 9-9 on page 250.

9-10   PARTIAL OR BROKEN-OUT SECTIONS

Where a sectional view of only a portion of the object is needed, partial sections may be used (Fig. 9-10-1). An irregular break line is used to show the extent of the section. With this type of section, a cutting-plane line is not required.

ASSIGNMENT

See Assignment 17 for Unit 9-10 on pages 250–251.

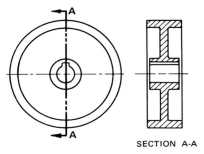

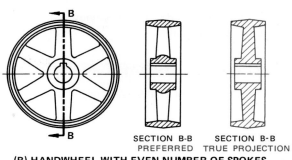

(A) FLAT PULLEY WITH WEB

SECTION A-A

(B) HANDWHEEL WITH EVEN NUMBER OF SPOKES

SECTION B-B
PREFERRED

SECTION B-B
TRUE PROJECTION

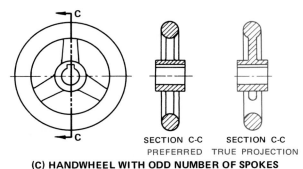

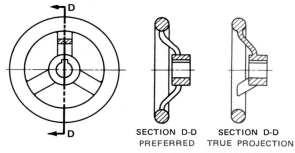

SECTION C-C
PREFERRED

SECTION C-C
TRUE PROJECTION

(C) HANDWHEEL WITH ODD NUMBER OF SPOKES

SECTION D-D
PREFERRED

SECTION D-D
TRUE PROJECTION

(D) HANDWHEEL WITH ODD NUMBER OF OFFSET SPOKES

FIG. 9-9-1   Preferred and true projection of spokes.

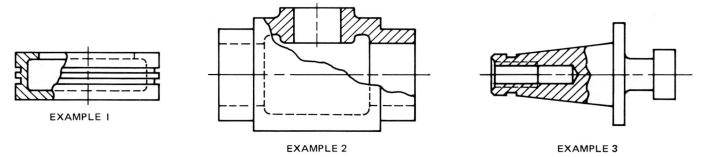

EXAMPLE I

EXAMPLE 2

EXAMPLE 3

FIG. 9-10-1   Broken-out or partial sections.

9-11 PHANTOM OR HIDDEN SECTIONS

A *phantom section* is used to show the typical interior shapes of an object in one view when the part is not truly symmetrical in shape, as well as to show mating parts in an assembly drawing (Fig. 9-11-1). It is a sectional view superimposed on the regular view without the removal of the front portion of the object. The section lining used for phantom sections consists of thin, evenly spaced, broken lines.

ASSIGNMENT

See Assignment 18 for Unit 9-11 on pages 251–252.

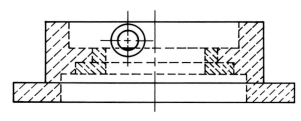

FIG. 9-11-1   Phantom or hidden sections.

9-12 SECTIONAL DRAWING REVIEW

In Units 9-1 through 9-11 the different types of sectional views have been explained and drawing problems have been assigned with each type of section drawing.

In the drafting office it is the drafter who must decide which views are required to fully explain the part to be made. In addition, the drafter must select the proper scale(s) which will show the features clearly.

This unit has been designed to review the sectional-view options open to the drafter.

ASSIGNMENT

See Assignment 19 for Unit 9-12 on pages 253–257.

ASSIGNMENTS FOR CHAPTER 9

ASSIGNMENTS FOR UNIT 9-1, SECTIONAL VIEWS

1. Select one of the problems shown in Fig. 9-1-A or 9-1-B and make a working drawing of the part. Surfaces shown with √ should have a surface texture rating of 125 μin. or 3.2 μm and a machining allowance of .06 in. or 1.5 mm. Use symbolic dimensioning wherever possible. For Fig. 9-1-A draw the top and a full-section front view. For Fig. 9-1-B draw the right side and a full-section front view.

2. Select one of the problems shown in Fig. 9-1-C or 9-1-D and make a three-view working drawing of the part. Surfaces shown with √ should have a surface texture rating of 63 μin. or 1.6 μm and a machining allowance of .06 in. or 2 mm. Use limit dimensions for the holes showing a fit. For Fig. 9-1-C draw the front view in full section. For Fig. 9-1-D draw the right side view in full section through the Ø16 hole.

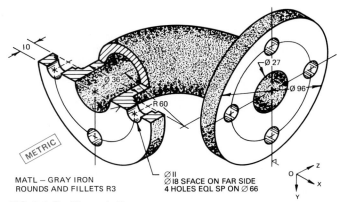

MATL — GRAY IRON
ROUNDS AND FILLETS R3

Ø11
Ø18 SFACE ON FAR SIDE
4 HOLES EQL SP ON Ø66

FIG. 9-1-B   Flanged elbow.

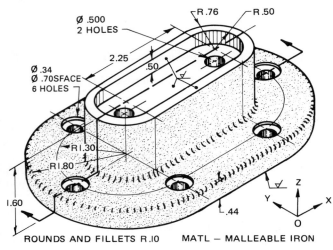

FIG. 9-1-A   Shaft base.

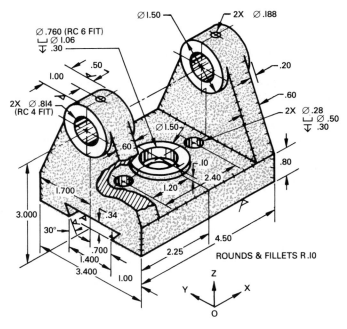

FIG. 9-1-C   Slide bracket.

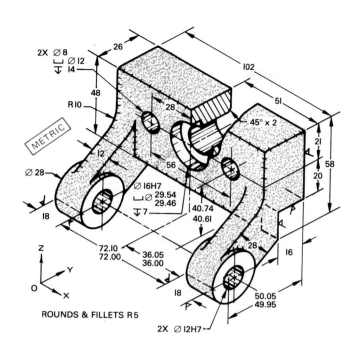

FIG. 9-1-D   Bracket.

ASSIGNMENTS FOR UNIT 9-2, TWO OR MORE SECTIONAL VIEWS ON ONE DRAWING

3. Select one of the problems shown in Fig. 9-2-A or 9-2-B and make a working drawing of the part showing the appropriate views in sections. Refer to the Appendix for taper sizes. Use symbolic dimensioning wherever possible.

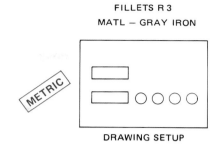

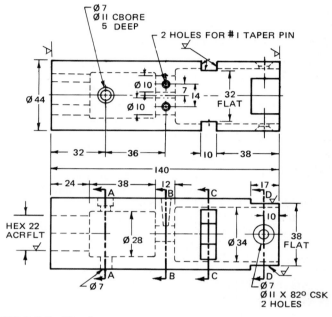

FIG. 9-2-B   Housing.

FIG. 9-2-A   Casing.

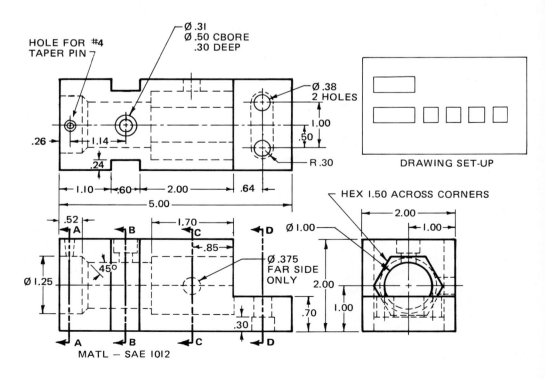

MATL – SAE 1012

FIG. 9-2-C   Guide block.

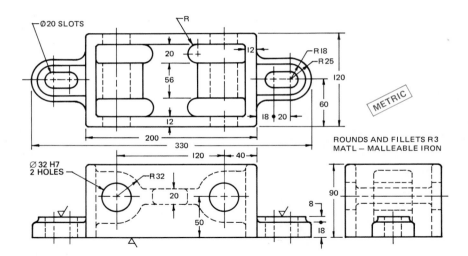

4. Make a three-view working drawing of the guide block shown in Fig. 9-2-C showing the front and side views in full section. The cutting plane for the side view should be taken through the right Ø32 hole. The surface finish for the bottom should have a surface texture rating of 1.6 and a machining allowance of 2 mm. The surface finish for the two bosses should have a surface texture rating of 0.8 and a machining allowance of 1 mm. Show the limits for the Ø32 holes.

ASSIGNMENT FOR UNIT 9-3, HALF-SECTIONS

5. Select one of the parts shown in Figs. 9-3-A through 9-3-D and make a two-view working drawing of the part showing the side view in half-section. Dimension the keyseat as per Chap. 11.

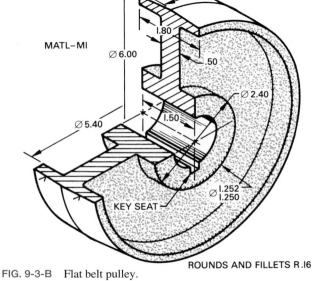

FIG. 9-3-B   Flat belt pulley.

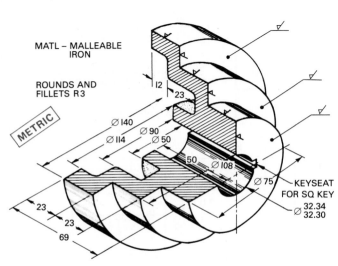

FIG. 9-3-A   Step pulley.

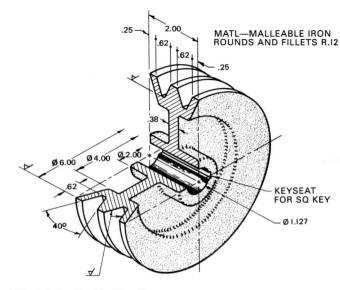

FIG. 9-3-C   Double-V pulley.

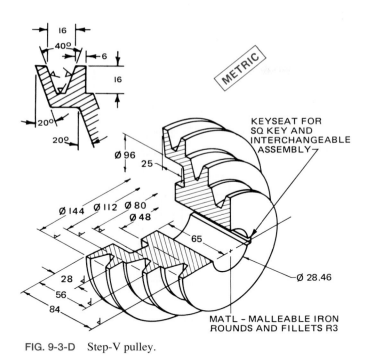

FIG. 9-3-D   Step-V pulley.

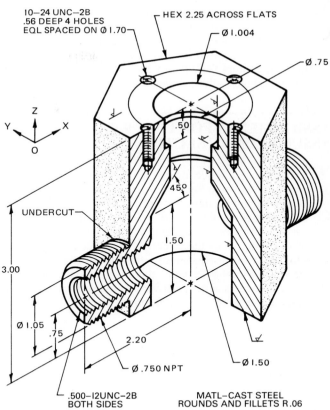

FIG. 9-4-B   Valve body.

ASSIGNMENTS FOR UNIT 9-4, THREADS IN SECTION

6. Select one of the problems shown in Figs. 9-4-A through 9-4-C and make a working drawing of the part. Determine the number of views and the best type of section that will clearly describe the part. Use symbolic dimensioning wherever possible and add undercut sizes.

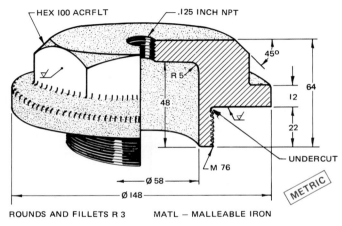

FIG. 9-4-A   Pipe plug.

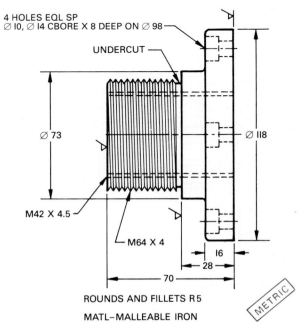

FIG. 9-4-C   End plate.

243

7. Make a three-view working drawing of the control arm shown in Fig. 9-4-D. Draw the front view in full section. Tabular dimensioning is to be used to locate the holes. The origin is located at the bottom left side of the part. Locate the two centers for each of the slotted holes. The two Ø6.5-6.6 holes will be drilled as one operation.

ASSIGNMENTS FOR UNIT 9-5, ASSEMBLIES IN SECTION

8. Make a one-view section assembly drawing of one of the problems shown in Figs. 9-5-A through 9-5-C. Include on your drawing an item list and identify the parts on the assembly. Assuming that this drawing will be used in a catalog, place on the drawing the dimensions and information required by the potential buyer. Scale 1:1.
9. Make a two-view assembly drawing of the cam slide shown in Fig. 9-5-D. Show the top view with the cover plate removed and the front view in full section. Use your judgment for dimensions not shown.

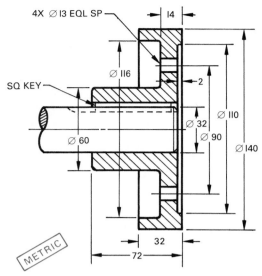

FLANGES HELD TOGETHER BY MI2 X 1.75 X 45 LG HEX HD BOLTS WITH LOCKWASHERS

2mm NEOPRENE GASKET BETWEEN FLANGES

FIG. 9-5-A   Flanged connection.

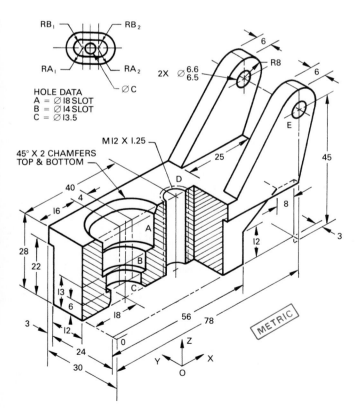

FIG. 9-4-D   Control arm.

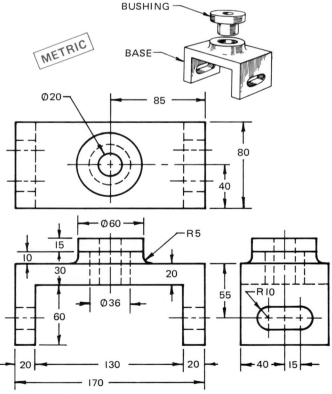

FIG. 9-5-B   Bushing holder.

FIG. 9-5-C   Caster.

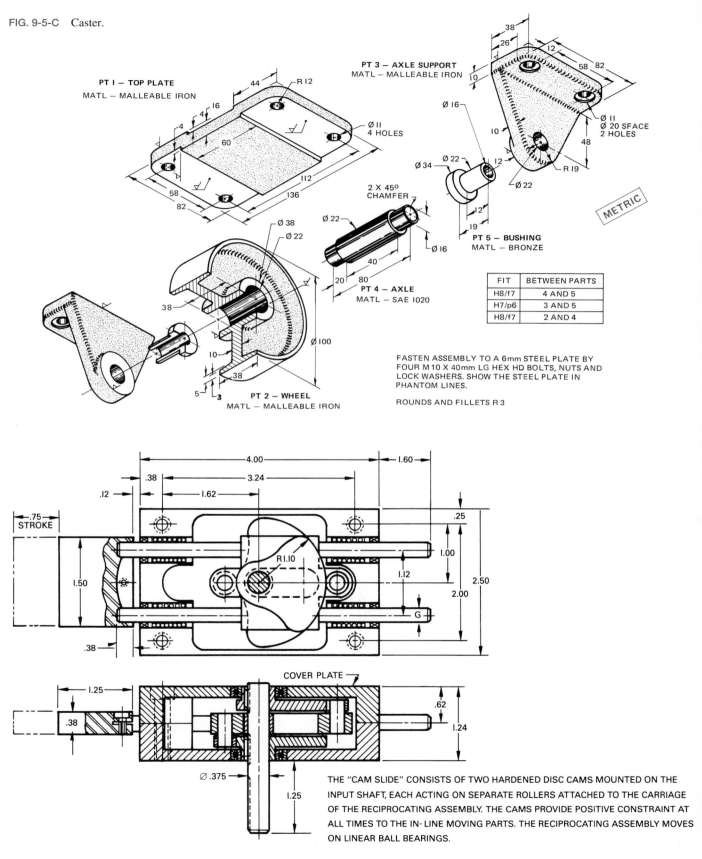

PT 1 — TOP PLATE
MATL — MALLEABLE IRON

PT 3 — AXLE SUPPORT
MATL — MALLEABLE IRON

PT 5 — BUSHING
MATL — BRONZE

PT 4 — AXLE
MATL — SAE 1020

PT 2 — WHEEL
MATL — MALLEABLE IRON

METRIC

FIT	BETWEEN PARTS
H8/f7	4 AND 5
H7/p6	3 AND 5
H8/f7	2 AND 4

FASTEN ASSEMBLY TO A 6mm STEEL PLATE BY
FOUR M 10 X 40mm LG HEX HD BOLTS, NUTS AND
LOCK WASHERS. SHOW THE STEEL PLATE IN
PHANTOM LINES.

ROUNDS AND FILLETS R 3

COVER PLATE

THE "CAM SLIDE" CONSISTS OF TWO HARDENED DISC CAMS MOUNTED ON THE
INPUT SHAFT, EACH ACTING ON SEPARATE ROLLERS ATTACHED TO THE CARRIAGE
OF THE RECIPROCATING ASSEMBLY. THE CAMS PROVIDE POSITIVE CONSTRAINT AT
ALL TIMES TO THE IN- LINE MOVING PARTS. THE RECIPROCATING ASSEMBLY MOVES
ON LINEAR BALL BEARINGS.

FIG. 9-5-D   Cam slide (protected by patent) courtesy Stelron Cam Co.

10. Make a two-view assembly drawing of the connecting link shown in Fig. 9-5-E. Show the top and front views with the front view in full section.

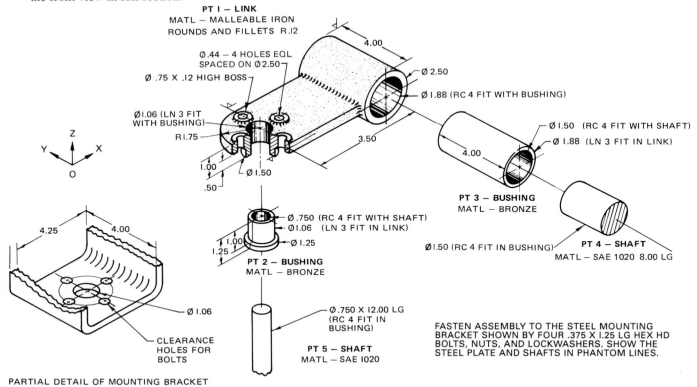

PT I — LINK
MATL — MALLEABLE IRON
ROUNDS AND FILLETS R .12

Ø .44 — 4 HOLES EQL. SPACED ON Ø 2.50

Ø .75 X .12 HIGH BOSS

Ø 1.06 (LN 3 FIT WITH BUSHING)

R 1.75

Ø 1.50

4.00

Ø 2.50

Ø 1.88 (RC 4 FIT WITH BUSHING)

3.50

1.00

.50

Ø .750 (RC 4 FIT WITH SHAFT)
Ø 1.06 (LN 3 FIT IN LINK)
Ø 1.25

1.00
1.25

PT 2 — BUSHING
MATL — BRONZE

Ø .750 X 12.00 LG (RC 4 FIT IN BUSHING)

PT 5 — SHAFT
MATL — SAE 1020

Ø 1.50 (RC 4 FIT WITH SHAFT)
Ø 1.88 (LN 3 FIT IN LINK)

4.00

PT 3 — BUSHING
MATL — BRONZE

Ø 1.50 (RC 4 FIT IN BUSHING)

PT 4 — SHAFT
MATL — SAE 1020 8.00 LG

FASTEN ASSEMBLY TO THE STEEL MOUNTING BRACKET SHOWN BY FOUR .375 X 1.25 LG HEX HD BOLTS, NUTS, AND LOCKWASHERS. SHOW THE STEEL PLATE AND SHAFTS IN PHANTOM LINES.

Z
Y ↑ X
O

4.25   4.00

Ø 1.06

CLEARANCE HOLES FOR BOLTS

PARTIAL DETAIL OF MOUNTING BRACKET

FIG. 9-5-E   Connecting link.

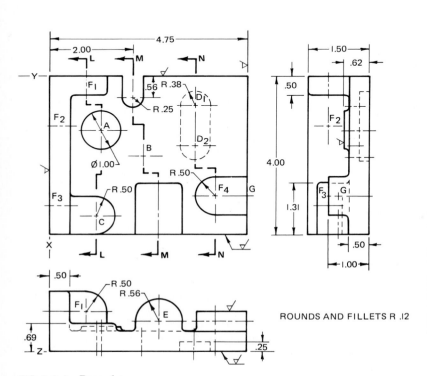

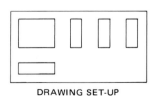

DRAWING SET-UP

REPLACE RIGHT SIDE VIEW WITH SECTIONS G-G, H-H, AND J-J.

4.75
2.00

—Y—
L   M   N

F₁   .56   R .38
R .25   D₁

F₂
A

Ø 1.00   B   D₂

R .50
F₃   R .50   F₄   G
C

X—
L   M   N

1.50
.62
.50

F₂

4.00

F₃ G

1.31

.50
1.00

ROUNDS AND FILLETS R .12

.50
R .50
R .56

F₁   E

.69
Z—
.25

FIG. 9-6-A   Base plate.

HOLE	HOLE SIZE	LOCATION		
		X	Y	Z
A	.500–13UNC–2B	1.25	1.38	
B	Ø .281 CSK Ø .50 X 82°	2.25	1.94	
C	Ø .281 CBORE Ø .50 X .25 DEEP	1.12	3.50	
D₁	Ø .31	3.50	.75	
D₂	Ø .31	3.50	1.75	
E	.500–13UNC–2B X .75 DEEP	2.62		.75
F₁	Ø .50	.88		1.00
F₂	Ø .50		1.25	1.00
F₃	Ø .50		3.25	1.00
F₄	Ø .50	4.00	3.00	
G	Ø .12 THROUGH		3.00	.75

DRAW TOP, FRONT AND 3 SECTION VIEWS
MATL — MALLEABLE IRON

ASSIGNMENT FOR UNIT 9-6, OFFSET SECTIONS

11. Select one of the problems shown in Fig. 9-6-A or 9-6-B and make a working drawing of the part. Scale 1:1.

ASSIGNMENT FOR UNIT 9-7, RIBS, HOLES, AND LUGS IN SECTION

12. Select one of the problems shown in Figs. 9-7-A through 9-7-E (here and on pg. 248) and make a three-view working drawing of the part showing the front and side views in section. For Fig. 9-7-C draw only the front and side views.

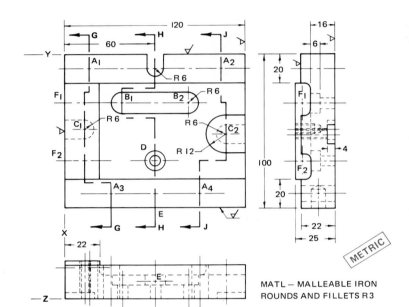

METRIC

MATL — MALLEABLE IRON
ROUNDS AND FILLETS R3

FIG. 9-6-B   Mounting plate.

HOLE	HOLE SIZE	LOCATION		
		X	Y	Z
A1	Ø 12	16	9	
A2	Ø 12	100	9	
A3	Ø 12	30	92	
A4	Ø 12	87	92	
B1	Ø 8	38	32	
B2	Ø 8	80	32	
C1	M6 X 12 DEEP	12	50	
C2	M6	104	52	
D	Ø 6 CBORE Ø 12 X 6 DEEP	58	70	
E	Ø 10 X 12 DEEP	58		11
F1	Ø 6		32	20
F2	Ø 6		70	20

REPLACES RIGHT SIDE VIEW WITH
SECTIONS L-L, M-M, AND N-N

FIG. 9-7-A   Shaft support.

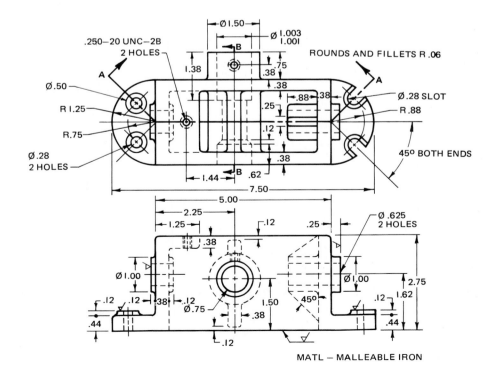

MATL — MALLEABLE IRON

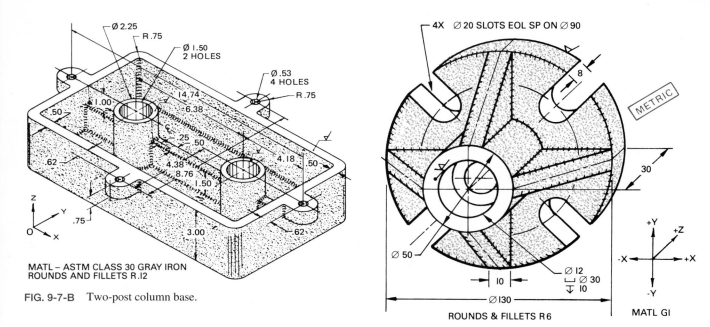

FIG. 9-7-B Two-post column base.

MATL – ASTM CLASS 30 GRAY IRON
ROUNDS AND FILLETS R.12

FIG. 9-7-C Flanged support.

4X Ø 20 SLOTS EOL SP ON Ø 90

METRIC

ROUNDS & FILLETS R 6

MATL GI

FIG. 9-7-D Bracket bearing.

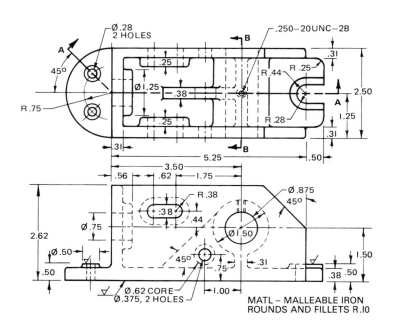

MATL – MALLEABLE IRON
ROUNDS AND FILLETS R.10

FIG. 9-7-E Shaft support base.

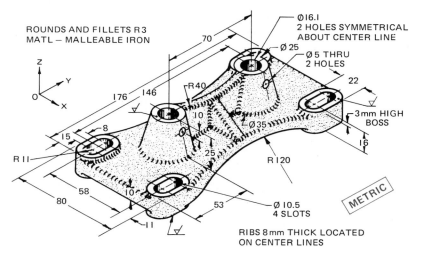

ROUNDS AND FILLETS R 3
MATL – MALLEABLE IRON

Ø16.1
2 HOLES SYMMETRICAL
ABOUT CENTER LINE

Ø 25

Ø 5 THRU
2 HOLES

3mm HIGH
BOSS

Ø 10.5
4 SLOTS

METRIC

RIBS 8mm THICK LOCATED
ON CENTER LINES

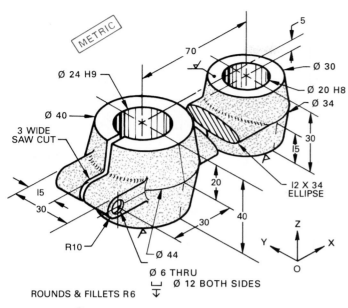

METRIC

70
5
Ø 24 H9
Ø 30
Ø 40
Ø 20 H8
3 WIDE
SAW CUT
Ø 34
30
15
15
20
30
40
30
12 X 34
ELLIPSE
R10
Ø 44
Z
Y
X
Ø 6 THRU
O
Ø 12 BOTH SIDES
ROUNDS & FILLETS R6

FIG. 9-8-A   Connector.

FIG. 9-8-B   Chisel.

ASSIGNMENTS FOR UNIT 9-8, REVOLVED AND REMOVED SECTIONS

13. Make a two-view working drawing of the connector shown in Fig. 9-8-A. Show a revolved section of the arm on the top view. The machined surfaces are to have a surface texture rating of 1.6 and a machining allowance of 2 mm.
14. Make a working drawing of the chisel shown in Fig. 9-8-B. Show either revolved or removed sections taken at planes A through D.
15. Select one of the problems shown in Fig. 9-8-C or 9-8-D (pg. 250) and make a working drawing of the part. For clarity it is recommended that an enlarged removed view be used to show the detail of the inclined hole. Use symbolic dimensioning wherever possible. Scale 1:1.

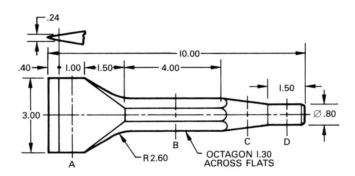

.24
.40
1.00
1.50
4.00
10.00
1.50
3.00
Ø .80
A
R 2.60
B
OCTAGON 1.30
ACROSS FLATS
C
D

FIG. 9-8-C   Shaft support.

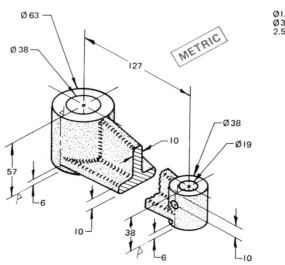

Ø 63
Ø 38
127
METRIC
10
Ø 38
Ø 19
57
6
10
38
6
10

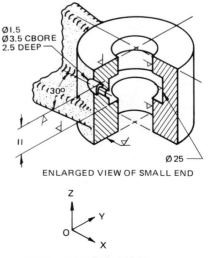

Ø1.5
Ø3.5 CBORE
2.5 DEEP
30°
11
Ø 25
ENLARGED VIEW OF SMALL END
Z
Y
O
X
MATL – MALLEABLE IRON
ROUNDS AND FILLETS R3

FIG. 9-8-D Idler support.

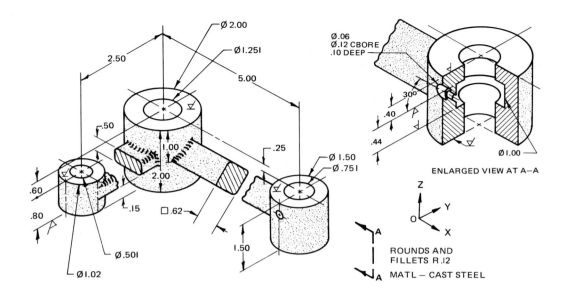

ENLARGED VIEW AT A—A

ROUNDS AND FILLETS R.12

MATL — CAST STEEL

ASSIGNMENT FOR UNIT 9-9, SPOKES AND ARMS IN SECTION

16. Select one of the problems shown in Fig. 9-9-A or 9-9-B and make a two-view working drawing of the part. Draw the side view in full section, and show a revolved section of the spoke in the front view. Scale 1:1.

ASSIGNMENT FOR UNIT 9-10, PARTIAL OR BROKEN-OUT SECTIONS

17. Select one of the problems shown in Fig. 9-10-A or 9-10-B and make a two-view working drawing of the part. Use partial sections where clarity of drawing can be achieved. Scale 1:1.

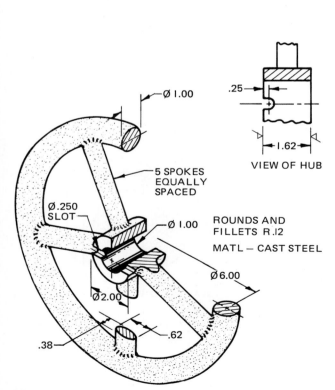

VIEW OF HUB

ROUNDS AND FILLETS R.12

MATL — CAST STEEL

FIG. 9-9-A Handwheel.

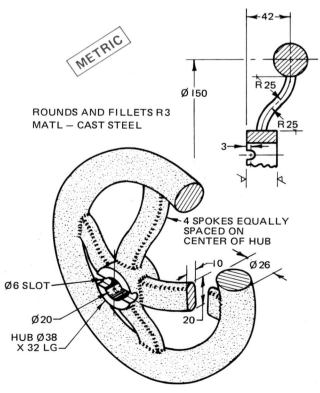

METRIC

ROUNDS AND FILLETS R3
MATL — CAST STEEL

FIG. 9-9-B Offset handwheel.

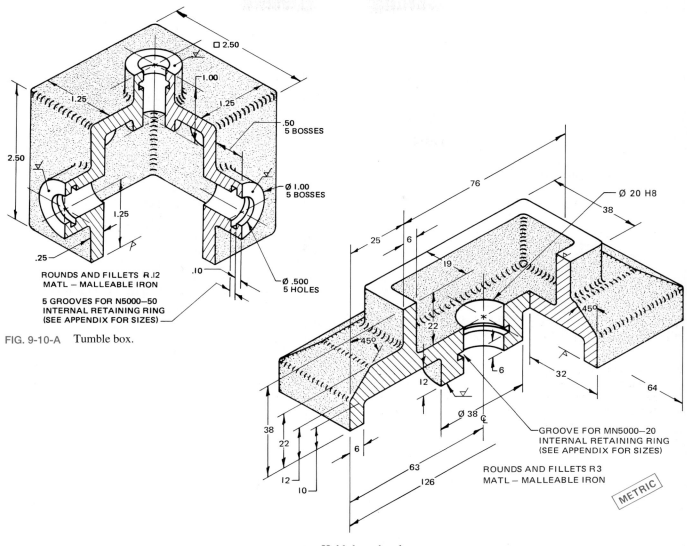

FIG. 9-10-A   Tumble box.

ROUNDS AND FILLETS R.12
MATL — MALLEABLE IRON

5 GROOVES FOR N5000—50
INTERNAL RETAINING RING
(SEE APPENDIX FOR SIZES)

□ 2.50

.50
5 BOSSES

Ø 1.00
5 BOSSES

Ø .500
5 HOLES

GROOVE FOR MN5000—20
INTERNAL RETAINING RING
(SEE APPENDIX FOR SIZES)

ROUNDS AND FILLETS R 3
MATL — MALLEABLE IRON

METRIC

FIG. 9-10-B   Hold-down bracket.

ASSIGNMENT FOR UNIT 9-11, PHANTOM OR HIDDEN SECTIONS

18. Make a two-view drawing of the part or one of the assemblies shown in Figs. 9-11-A through 9-11-C (here and on pg. 252). One view is to be drawn as a phantom section drawing. Show only the hole and shaft sizes for the fits shown. Scale 1:1.

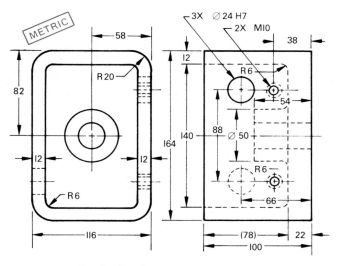

FIG. 9-11-A   Bearing housing.

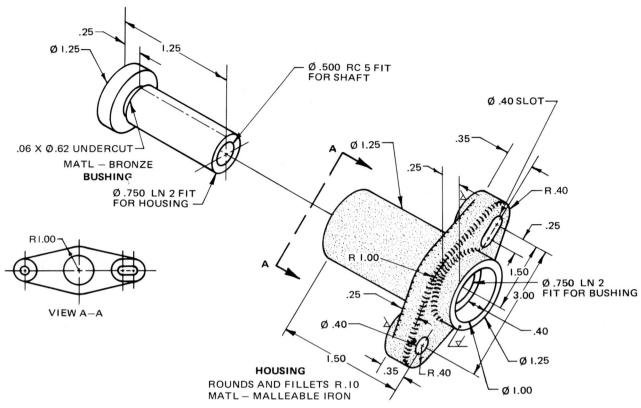

FIG. 9-11-B   Drill jig assembly.

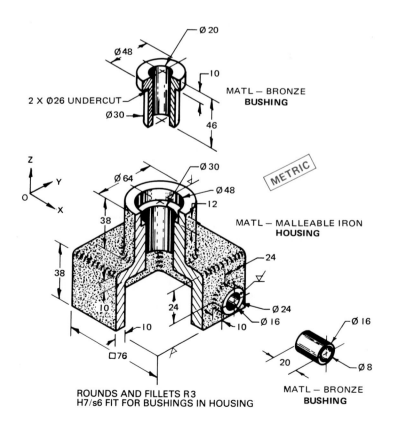

FIG. 9-11-C   Housing.

ASSIGNMENT FOR UNIT 9-12, SECTIONAL DRAWING REVIEW

information on section drawings found in Units 9-1 through 9-11, select appropriate sectional views that will improve the clarity of the drawing.

19. Make a working drawing of one of the parts shown in Figs. 9-12-A through 9-12-H (pp. 253–257). From the

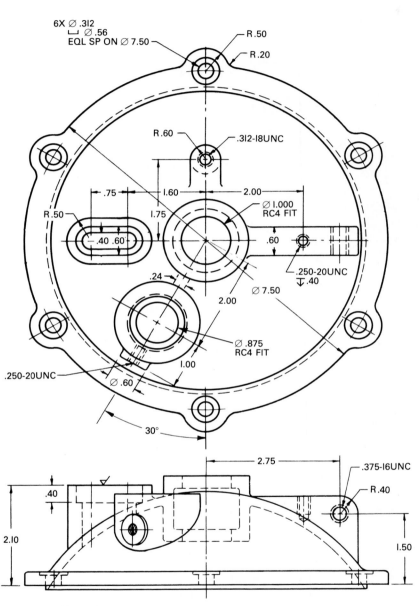

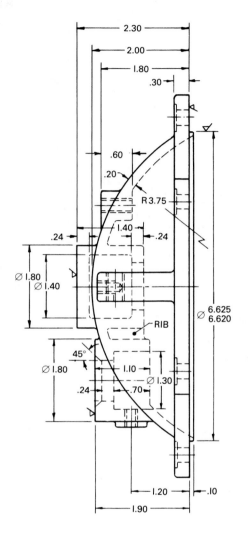

FIG. 9-12-A Domed cover.

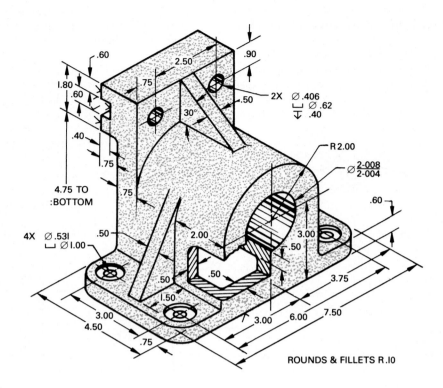

FIG. 9-12-B Slide support.

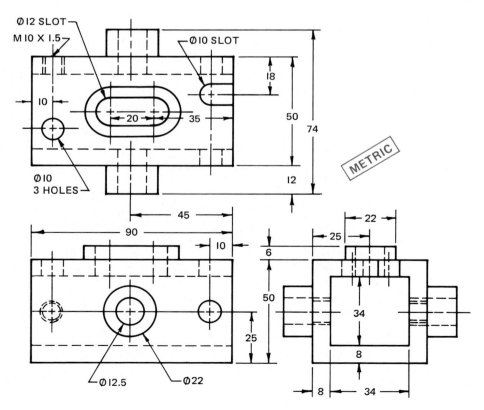

FIG. 9-12-C Jacket.

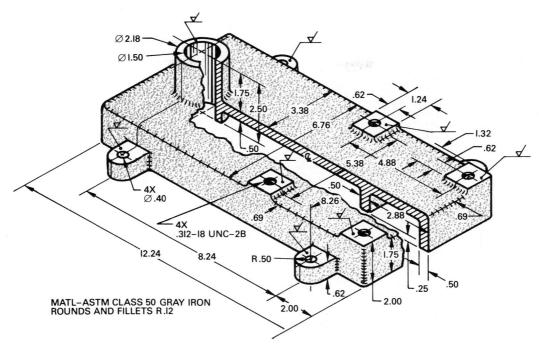

FIG. 9-12-D   Drill press base.

Ø 2.18
Ø 1.50
1.75
2.50
3.38
.62
1.24
6.76
1.32
.62
5.38   4.88
4X
Ø .40
.50
8.26
.50
2.88
.69
4X
.312–18 UNC–2B
.69
12.24   8.24
R .50
1.75
2.00
.62
1.75
.25   .50
2.00

MATL–ASTM CLASS 50 GRAY IRON
ROUNDS AND FILLETS R .12

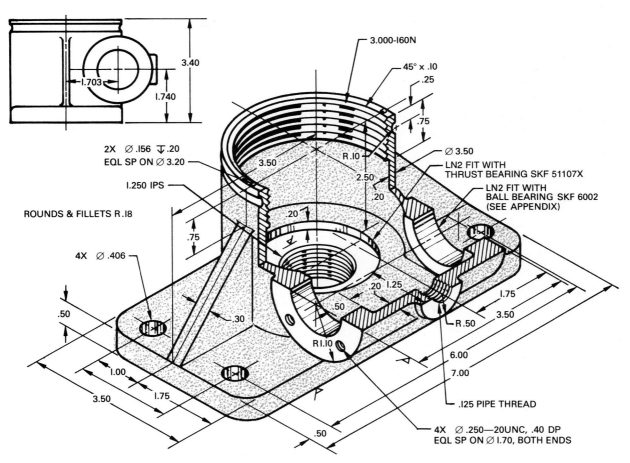

FIG. 9-12-E   Base.

3.40
1.703
1.740

3.000-160N
45° x .10
.25
.75

2X   Ø .156   ↧ .20
EQL SP ON Ø 3.20
3.50
R .10
2.50
.20
Ø 3.50
LN2 FIT WITH
THRUST BEARING SKF 51107X

1.250 IPS
LN2 FIT WITH
BALL BEARING SKF 6002
(SEE APPENDIX)

ROUNDS & FILLETS R .18
.20
.75

4X   Ø .406
.20   1.25
1.75
.30
.50
3.50
R .50
.50
R 1.10
6.00
7.00

.125 PIPE THREAD

4X   Ø .250—20UNC, .40 DP
EQL SP ON Ø 1.70, BOTH ENDS

.50
1.00
1.75
3.50
.50

FIG. 9-12-F   Swivel base.

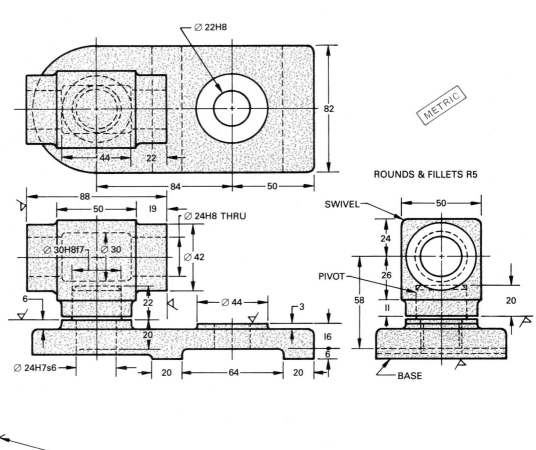

FIG. 9-12-G   Housing.

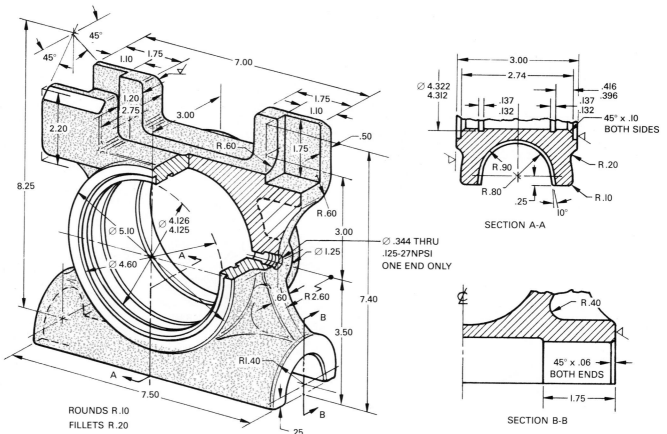

FIG. 9-12-H   Pump base.

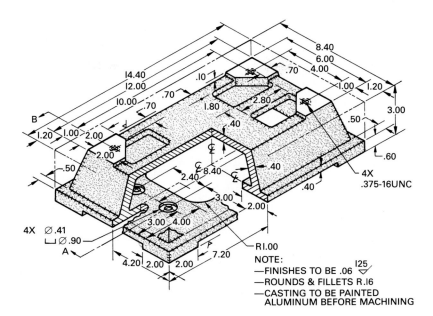

NOTE:
—FINISHES TO BE .06
—ROUNDS & FILLETS R.16
—CASTING TO BE PAINTED
   ALUMINUM BEFORE MACHINING

FASTENERS, MATERIALS, AND FORMING PROCESSES

CHAPTER 10

THREADED FASTENERS

Definitions

Chemical locking Fastening achieved by means of an adhesive.

Clearance drill size A diameter slightly larger than the major diameter of the bolt which permits the free passage of the bolt.

Counterbored hole A circular, flat-bottomed recess that permits the head of a bolt or cap screw to rest below the surface of the part.

Countersunk hole An angular-sided recess that accommodates the shape of a flat-head cap screw or machine screw or an oval-head machine screw.

Detailed representation A method of representation used to show the detail of a screw thread.

Free-spinning devices Fasteners that spin free in the clamping direction, which makes them easy to assemble, and have break-loose torque greater than the seating torque.

Lead The distance the threaded part would move parallel to the axis during one complete rotation in relation to a fixed mating part.

Pitch The distance from a point on the thread form to the corresponding point on the next form.

Prevailing-torque methods Fasteners that make use of increased friction between nut and bolt.

Screw thread A ridge of uniform section in the form of a helix on the external or internal surface of a cylinder.

Series The number of threads per inch, set for different diameters.

Simplified representation A method of representation used to clearly portray the requirements of a thread.

Spotfacing A machine operation that provides a smooth, flat surface where a bolt head or a nut will rest.

Standardization The manufacture and use of parts of similar types and sizes to reduce cost and simplify inventory and quality control.

Tap drill size A diameter equal to the minor diameter of the thread for a tapped hole.

10-1 SIMPLIFIED THREAD REPRESENTATION

Fastening devices are important in the construction of manufactured products, in the machines and devices used in manufacturing processes, and in the construction of all types of buildings. Fastening devices are used in the smallest watch to the largest ocean liner (Fig. 10-1-1).

FIG. 10-1-1 Fasteners. *(Industrial Fasteners Institute)*

There are two basic kinds of fasteners: permanent and removable. Rivets and welds are permanent fasteners. Bolts, screws, studs, nuts, pins, rings, and keys are removable fasteners. As industry progressed, fastening devices became standardized, and they developed definite characteristics and names. A thorough knowledge of the design and graphic representation of the more common fasteners is an essential part of drafting.

The cost of fastening, once considered only incidental, is fast becoming recognized as a critical factor in total product cost. "It's the in-place cost that counts, not the fastener cost" is an old saying of fastener design. The art of holding down fastener cost is not learned simply by scanning a parts catalog. More subtly, it entails weighing such factors as standardization, automatic assembly, tailored fasteners, and joint preparation.

A favorite cost-reducing method, *standardization,* not only cuts the cost of parts but reduces paperwork and simplifies inventory and quality control. By standardizing on type and size, it may be possible to reach the level of usage required to make power tools or automatic assembly feasible.

Screw Threads

A *screw thread* is a ridge of uniform section in the form of a helix on the external or internal surface of a cylinder (Fig. 10-1-2). The helix of a square thread is shown in Fig. 10-1-3.

The *pitch* of a thread, *P,* is the distance from a point on the thread form to the corresponding point on the next form, measured parallel to the axis (Fig. 10-1-4, pg. 262). The *lead, L,* is the distance the threaded part would move parallel to the axis during one complete rotation in relation to a fixed mating part (the distance a screw would enter a threaded hole in one turn).

Thread Forms

Figure 10-1-5 (pg. 262) shows some of the more common thread forms in use today. The ISO metric thread will eventually replace all the V-shaped metric and inch threads. As for the other thread forms shown, the proportions will be the same for both metric- and inch-size threads.

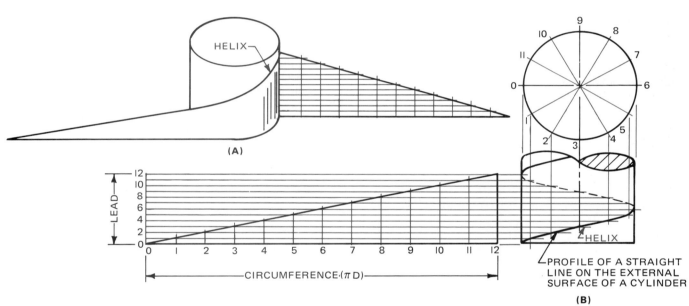

FIG. 10-1-2   The helix.

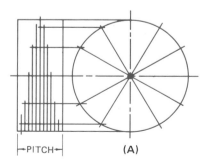
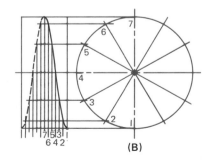
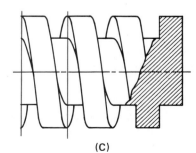

FIG. 10-1-3   The helix of a square thread.

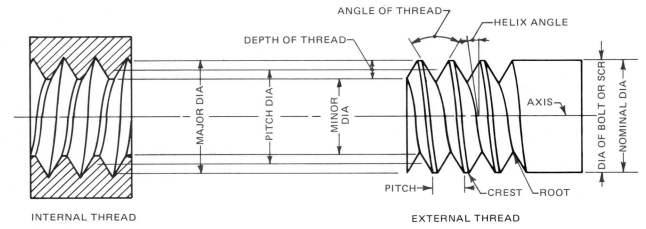

FIG. 10-1-4   Screw thread terms.

FIG. 10-1-5   Common thread forms and proportions.

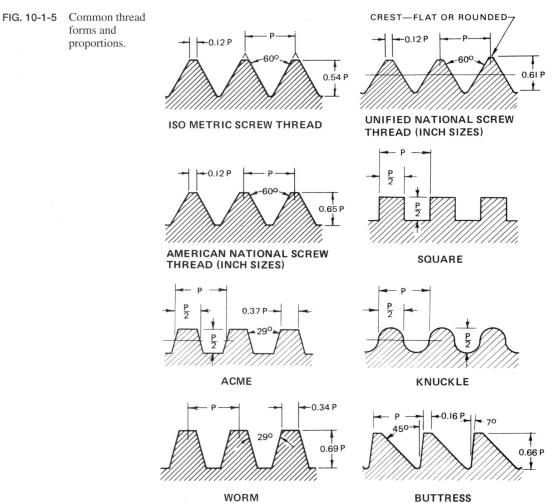

The knuckle thread is usually rolled or cast. A familiar example of this form is seen on electric light bulbs and sockets (Fig. 10-1-6). The square and acme forms are designed to transmit motion or power, as on the lead screw of a lathe. The buttress thread takes pressure in only one direction—against the surface perpendicular to the axis.

Thread Representation

True representation of a screw thread is seldom used on working drawings. Symbolic representation of threads is now standard practice. There are three types of conventions in general use for screw thread representation. These are known as

FIG. 10-1-6   Application of a knuckle thread. *(STUDIOHIO)*

simplified, detailed, and *schematic* (Fig. 10-1-7). *Simplified* representation should be used whenever it will clearly portray the requirements. *Detailed* representation is used to show the detail of a screw thread, especially for dimensioning in enlarged views, layouts, and assemblies. The *schematic* representation is nearly as effective as the detailed representation and is much easier to draw when manual drafting is used. This representation has given way to the simplified representation,

and as such, has been discarded as a thread symbol by most countries.

Right- and Left-Hand Threads

Unless designated otherwise, threads are assumed to be right-hand. A bolt being threaded into a tapped hole would be turned in a right-hand (clockwise) direction (Fig. 10-1-8). For some special applications, such as turnbuckles, left-hand threads are required. When such a thread is necessary, the letters LH are added after the thread designation.

Single and Multiple Threads

Most screws have single threads. It is understood that unless the thread is designated otherwise, it is a single thread. The single thread has a single ridge in the form of a helix (Fig. 10-1-9). The lead of a thread is the distance traveled parallel to the axis in one rotation of a part in relation to a fixed mating part (the distance a nut would travel along the axis of a bolt with one rotation of the nut). In single threads, the lead is equal to the pitch. A double thread has two ridges, started 180° apart, in the form of helices, and the lead is twice the pitch. A triple thread has three ridges, started 120° apart, in the form of helices, and the lead is three times the pitch. Multiple threads are used where fast movement is desired with a minimum

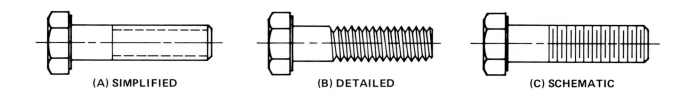

(A) SIMPLIFIED   **(B) DETAILED**   **(C) SCHEMATIC**

FIG. 10-1-7   Symbolic thread representation.

FIG. 10-1-8   Right- and left-hand threads.

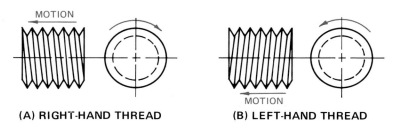

(A) RIGHT-HAND THREAD   **(B) LEFT-HAND THREAD**

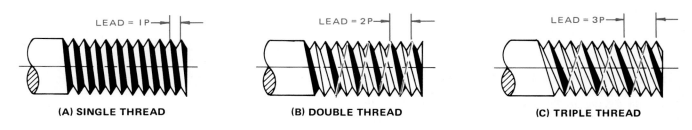

(A) SINGLE THREAD   **(B) DOUBLE THREAD**   **(C) TRIPLE THREAD**

FIG. 10-1-9   Single and multiple threads.

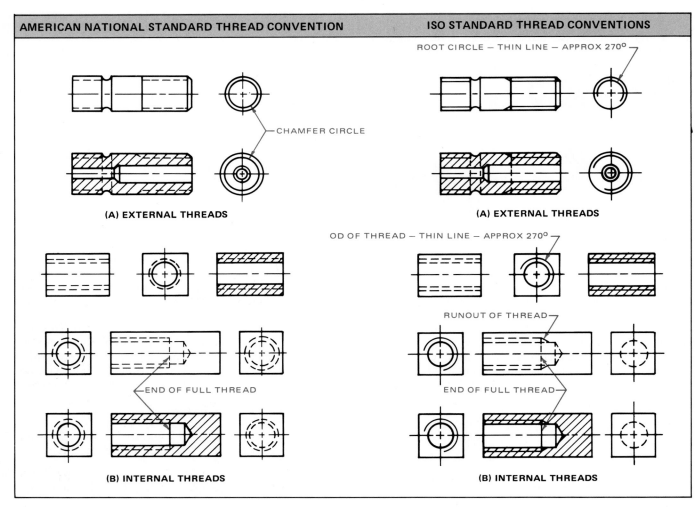

FIG. 10-1-10   Simplified thread representation.

number of rotations, such as on threaded mechanisms for opening and closing windows.

Simplified Thread Representation

In this system the thread crests, except in hidden views, are represented by a thick outline and the thread roots by a thin broken line (Fig. 10-1-10). The end of the full-form thread is indicated by a thick line across the part, and imperfect or runout threads are shown beyond this line by running the root line at an angle to meet the crest line. If the length of runout threads is unimportant, this portion of the convention may be omitted.

Threaded Assemblies

For general use, the simplified representation of threaded parts is recommended for assemblies (Fig. 10-1-11). In sectional views, the externally threaded part is always shown covering the internally threaded part.

Inch Threads

In the United States and Canada a great number of threaded assemblies are still designed using inch-sized threads. In this system the pitch is equal to

$$\frac{1}{\text{Number of threads per inch}}$$

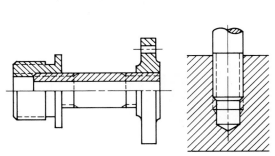

FIG. 10-1-11   Simplified representation of threads in assembly drawings.

The number of threads per inch is set for different diameters in what is called a thread *series*. For the Unified National system, there is the coarse-thread series (UNC) and the fine-thread series (UNF). See Table 8 of the Appendix.

In addition, there is an extra-fine-thread series (UNEF) for use where a small pitch is desirable, such as on thin-walled tubing. For special work and for diameters larger than those specified in the coarse and fine series, the Unified National thread system has three series that provide for the same number of threads per inch regardless of the diameter. These are the 8-thread series, the 12-thread series, and the 16-thread series. These are called *constant-pitch* threads.

Thread Class

Three classes of external thread (classes 1A, 2A, and 3A) and three classes of internal thread (classes 1B, 2B, and 3B) are provided. These classes differ in the amount of allowances and tolerances provided in each class.

The general characteristics and uses of the various classes are as follows.

Classes 1A and 1B   These classes produce the loosest fit, that is, the greatest amount of play (free motion) in assembly. They are useful for work where ease of assembly and disassembly is essential, such as for stove bolts and other rough bolts and nuts.

Classes 2A and 2B   These classes are designed for the ordinary good grade of commercial products, such as machine screws and fasteners, and for most interchangeable parts.

Classes 3A and 3B   These classes are intended for exceptionally high-grade commercial products, where a particularly close or snug fit is essential and the high cost of precision tools and machines is warranted.

Thread Designation

Thread designation for inch threads, whether external or internal, is expressed in this order: diameter (nominal or major diameter in decimal form with a minimum of three or maximum of four decimal places), number of threads per inch, thread form and series, and class of fit (number and letter). See Fig. 10-1-12.

Metric Threads

Metric threads are grouped into diameter-pitch combinations distinguished from one another by the pitch applied to specific diameters (Fig. 10-1-13, pg. 266).

The pitch for metric threads is the distance between corresponding points on adjacent teeth. In addition to a coarse- and fine-pitch series, a series of constant pitches is available. See Table 9 of the Appendix.

Coarse-Thread Series   This series is intended for use in general engineering work and commercial applications.

Fine-Thread Series   The fine-thread series is for general use where a finer thread than the coarse-thread series is desirable. In comparison with a coarse-thread screw, the fine-thread screw is stronger in both tensile and torsional strength and is less likely to loosen under vibration.

Thread Grades and Classes

The *fit* of a screw thread is the amount of clearance between the internal and external threads when they are assembled.

For each of the two main thread elements—pitch diameter and crest diameter—a number of tolerance grades have been established. The number of the tolerance grades reflects the size of the tolerance. For example, grade 4 tolerances are smaller than grade 6 tolerances, and grade 8 tolerances are larger than grade 6 tolerances.

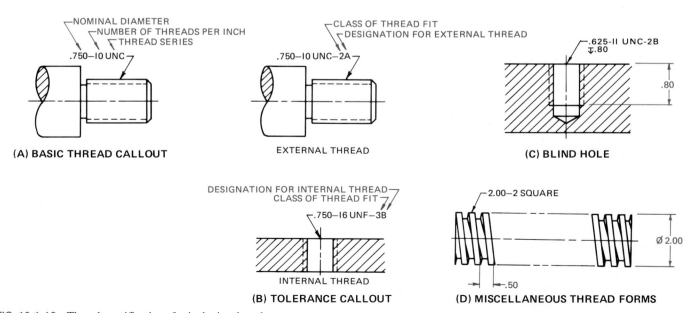

FIG. 10-1-12   Thread specifications for inch-size threads.

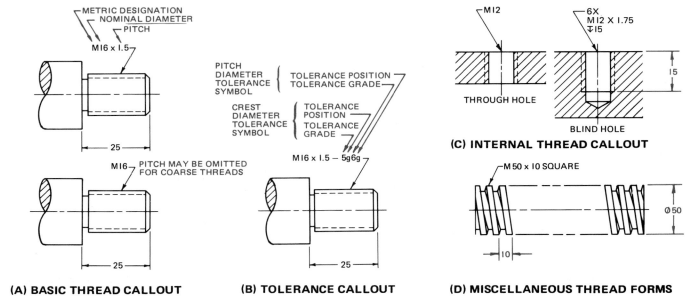

FIG. 10-1-13   Thread specifications for metric threads.

Grade 6 tolerances should be used for medium-quality length-of-engagement applications. The tolerance grades below grade 6 are intended for applications involving fine quality and/or short lengths of engagement. Tolerance grades above grade 6 are intended for coarse quality and/or long lengths of engagement.

In addition to the tolerance grade, a positional tolerance is required. This tolerance defines the maximum-material limits of the pitch and crest diameters of the external and internal threads and indicates their relationship to the basic profile.

In conformance with current coating (or plating) thickness requirements and the demand for ease of assembly, a series of tolerance positions reflecting the application of varying amounts of allowance has been established as follows.

For external threads:

- Tolerance position e (large allowance)
- Tolerance position g (small allowance)
- Tolerance position h (no allowance)

For internal threads:

- Tolerance position G (small allowance)
- Tolerance position H (no allowance)

ISO Metric Screw Thread Designation

ISO metric screw threads are defined by the nominal size (basic major diameter) and pitch, both expressed in millimeters. An M specifying an ISO metric screw thread precedes the nominal size, and an × separates the nominal size from the pitch (Fig. 10-1-13). For the coarse-thread series only, the pitch is not shown unless the dimension for the length of the thread is required. In specifying the length of thread, an × is used to separate the length of thread from the rest of the designations.

For external threads, the length or depth of thread may be given as a dimension on the drawing.

For example, a 10 mm diameter, 1.25 pitch, fine-thread series is expressed as M10 × 1.25. A 10 mm diameter, 1.5 pitch, coarse-thread series is expressed as M10; the pitch is not shown unless the length of thread is required. If the latter thread were 25 mm long and this information was required on the drawing, the thread callout would be M10 × 1.5 × 25.

A complete designation for an ISO metric screw thread comprises, in addition to the basic designation, an identification for the tolerance class. The tolerance class designation is separated from the basic designation by a dash and includes the symbol for the pitch diameter tolerance followed immediately by the symbol for crest diameter tolerance. Each of these symbols consists of a numeral indicating the grade tolerance followed by a letter representing the tolerance position (a capital letter for internal threads and a lowercase letter for external threads). Where the pitch and crest diameter symbols are identical, the symbol needs to be given only once. The complete designation for an ISO metric screw is used only when design requirements warrant it.

For external threads, the length of thread may be given as a dimension on the drawing. The length given is to be the minimum length of full thread. For threaded holes that go all the way through the part, the term THRU is sometimes added to the note. If no depth is given, the hole is assumed to go all the way through. For threaded holes that do not go all the way through, the depth (in conjunction with the depth symbol or word) is given in the note, for example, M12 × 1.75 × 20 DEEP. The depth given is the minimum depth of full thread.

Neither the chamfer shown at the beginning of a thread, nor the undercut at the end of a thread where a small diameter meets a larger diameter is required to be dimensioned, as shown in Fig. 10-1-14. A comparison of customary and metric thread sizes is shown in Fig. 10-1-15.

Pipe Threads

The pipe universally used is the inch-sized pipe. When pipe is ordered, the nominal diameter and wall thickness (in inches or millimeters) are given. In calling for the size of thread, the note used is similar to that for screw threads. When calling for a pipe thread on a metric drawing, the abbreviation IN follows the pipe size (Fig. 10-1-16).

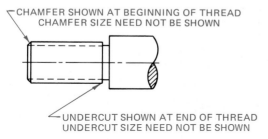

CHAMFER SHOWN AT BEGINNING OF THREAD
CHAMFER SIZE NEED NOT BE SHOWN

UNDERCUT SHOWN AT END OF THREAD
UNDERCUT SIZE NEED NOT BE SHOWN

FIG. 10-1-14   Omission of thread information on detail drawings.

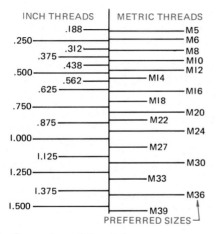

INCH THREADS | METRIC THREADS

FIG. 10-1-15   Comparison of thread sizes.

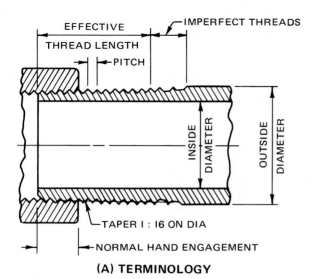

(A) TERMINOLOGY

FIG. 10-1-16   Pipe thread terminology and conventions.

EXAMPLE 1

4 × 8NPT

EXAMPLE 2

4 × 8NPS
     where 4 = nominal diameter of pipe, in inches
           8 = number of threads per inch
          N = American Standard
          P = pipe
          S = straight pipe thread
          T = taper pipe thread

REFERENCES AND SOURCE MATERIAL

1. ANSI Y14.6, *Screw Thread Representation*.

ASSIGNMENTS

See Assignments 1 through 8 for Unit 10-1 on pages 283–285.

10-2 DETAILED AND SCHEMATIC THREAD REPRESENTATION

Detailed Thread Representation

Detailed representation of threads is a close approximation of the actual appearance of a screw thread. The form of the thread is simplified by showing the helices as straight lines and the truncated crests and roots as sharp Vs. It is used when a more realistic thread representation is required (Fig. 10-2-1, pg. 268).

Detailed Representation of V Threads   The detailed representation for V-shaped threads uses the sharp-V profile.

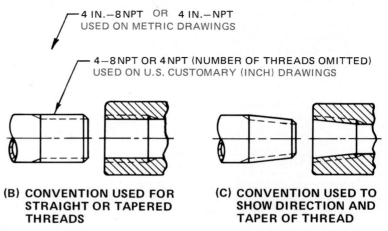

4 IN.–8NPT OR 4 IN.–NPT
USED ON METRIC DRAWINGS

4–8NPT OR 4NPT (NUMBER OF THREADS OMITTED)
USED ON U.S. CUSTOMARY (INCH) DRAWINGS

(B) CONVENTION USED FOR STRAIGHT OR TAPERED THREADS

(C) CONVENTION USED TO SHOW DIRECTION AND TAPER OF THREAD

The order of drawing the screw threads is shown in Fig. 10-2-1. The pitch is seldom drawn to scale; generally it is approximated. Lay off (establish) the pitch P and the half-pitch $P/2$, as shown in step 1. Add the crest lines. In step 2 add the V profile for one thread, top and bottom, locating the root diameter. Add construction lines for the root diameter. In step 3 add one side of the remaining Vs (thread profile), then add the other side of the Vs, completing the thread profile. In step 4, add the root lines which complete the detailed representation of the threads.

Detailed Representation of Square Threads   The depth of the square thread is one-half the pitch. In Fig. 10-2-2A, lay off spaces equal to $P/2$ along the diameter and add construction lines to locate the depth (root dia.) of thread. At B add the crest lines. At C add the root lines, as shown. At D the internal square thread is shown in section. Note the reverse direction of the crest and root lines.

Detailed Representation of Acme Threads   The depth of the acme thread is one-half the pitch (Fig. 10-2-2E through G). The stages in drawing acme threads are shown at E. For drawing purposes locate the pitch diameter midway between the outside diameter and the root diameter. The pitch diameter locates the pitch line. On the pitch line, lay off half-pitch spaces and the root lines to complete the view. The construction shown at F is enlarged.

Sectional views of an internal acme thread are shown at G. Showing the root and crest lines beyond the cutting plane on sectional views is optional.

Threaded Assemblies

It is often desirable to show threaded assembly drawings in detailed form, e.g., in presentation or catalog drawings. Hidden lines are normally omitted on these drawings, as they do nothing

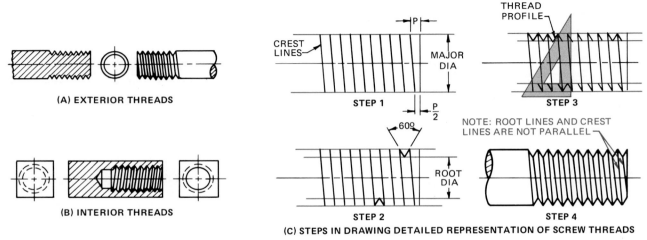

FIG. 10-2-1   Detailed representation of threads.

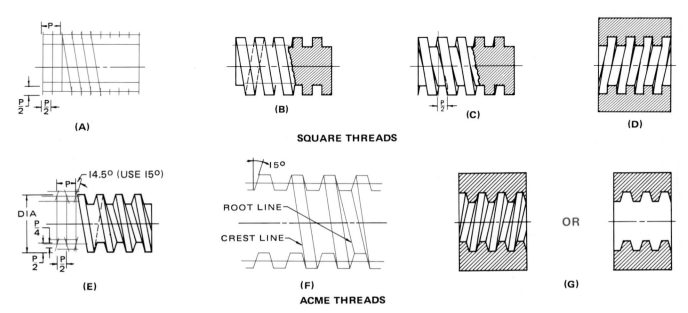

FIG. 10-2-2   Steps in drawing detailed representation of square and acme threads.

FIG. 10-2-3   Detailed threaded assembly.

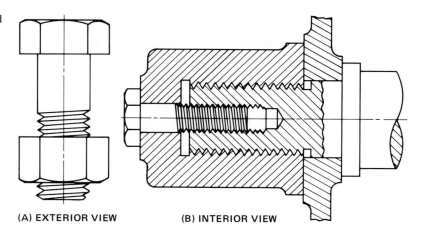

(A) EXTERIOR VIEW   (B) INTERIOR VIEW

to add to the clarity of the drawing (Fig. 10-2-3). One type of thread representation is generally used within any one drawing. When required, however, all three methods may be used.

Schematic Thread Representation

The staggered lines, symbolic of the thread root and crests, normally are perpendicular to the axis of the thread. The spacing between the root and crest lines and the length of the root lines are drawn to any convenient size (Fig. 10-2-4). At one time the root line was shown as a thick line.

REFERENCES AND SOURCE MATERIAL

1. ANSI Y14.6, *Screw Threads Representation.*

ASSIGNMENTS

See Assignments 9 through 11 for Unit 10-2 on pages 286 to 287.

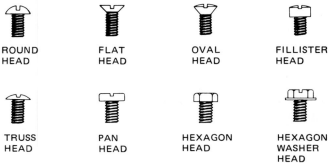

CHAMFERED END OF THREAD

FIG. 10-2-4   Schematic representation of threads.

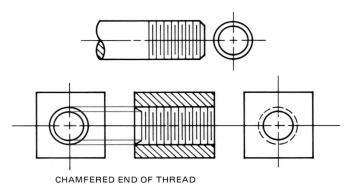

ROUND HEAD   FLAT HEAD   OVAL HEAD   FILLISTER HEAD

TRUSS HEAD   PAN HEAD   HEXAGON HEAD   HEXAGON WASHER HEAD

(A) SCREWS

10-3   COMMON THREADED FASTENERS

Fastener Selection

Fastener manufacturers agree that product selection must begin at the design stage. For it is here, when a product is still a figment of someone's imagination, that the best interests of the designer, production manager, and purchasing agent can be served. Designers, naturally, want optimum performance; production people are interested in the ease and economics of assembly; purchasing agents attempt to minimize initial costs and stocking costs.

The answer, pure and simple, is to determine the objectives of the particular fastening job and then consult fastener suppliers. These technical experts can often shed light on the situation and then recommend the right item at the best in-place cost.

Machine screws are among the most common fasteners in industry (Figs. 10-3-1 and 10-3-2, pg. 270). They are the easiest

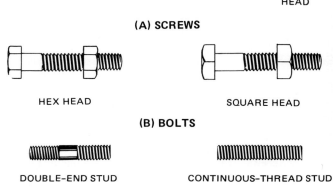

HEX HEAD   SQUARE HEAD

(B) BOLTS

DOUBLE-END STUD   CONTINUOUS-THREAD STUD

(C) STUDS

FIG. 10-3-1   Common threaded fasteners.

FIG. 10-3-2   Fastener applications.

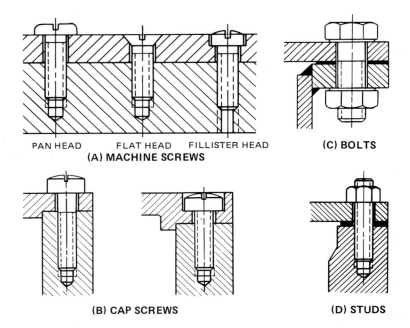

(A) MACHINE SCREWS — PAN HEAD   FLAT HEAD   FILLISTER HEAD

(C) BOLTS

(B) CAP SCREWS

(D) STUDS

to install and remove. They are also among the least understood. To obtain maximum machine-screw efficiency, thorough knowledge of the properties of both the screw and the materials to be fastened together is required.

For a given application, a designer should know the load that the screw must withstand, whether the load is one of tension or shear, and whether the assembly will be subject to impact shock or vibration. Once these factors have been determined, the size, strength, head shape, and thread type can be selected.

Fastener Definitions

Machine Screws   Machine screws have either fine or coarse threads and are available in a variety of heads. They may be used in tapped holes as shown in Fig. 10-3-2A, or with nuts.

Cap Screws   A cap screw is a threaded fastener that joins two or more parts by passing through a clearance hole in one part and screwing into a tapped hole in the other, as in Fig. 10-3-2B. A cap screw is tightened or released by torquing the head. Cap screws range in size from .25 in. (6 mm) in diameter and are available in five basic types of head.

Captive Screws   Captive screws are those that remain attached to the panel or parent material even when disengaged from the mating part. They are used to meet military requirements, to prevent screws from being lost, to speed assembly and disassembly operations, and to prevent damage from loose screws falling into moving parts or electrical circuits.

Tapping Screws   Tapping screws cut or form a mating thread when driven into preformed holes.

Bolts   A bolt is a threaded fastener that passes through clearance holes in assembled parts and threads into a nut (Fig. 10-3-2C).

Bolts and nuts are available in a variety of shapes and sizes. The square and hexagon head are the two most popular designs.

Studs   Studs are shafts threaded at both ends, and they are used in assemblies. One end of the stud is threaded into one of the parts being assembled; and the other assembly parts, such as washers and covers, are guided over the studs through clearance holes and are held together by means of a nut that is threaded over the exposed end of the stud (Fig. 10-3-2D).

Explanatory Data

A bolt is designed for assembly with a nut. A screw is designed to be used in a tapped or other preformed hole in the work. However, because of basic design, it is possible to use certain types of screws in combination with a nut.

The Change to Metric Fasteners

In the United States, the Industrial Fasteners Institute (IFI) has undertaken a major compilation of standards in its *Metric Fastener Standards* book.

Fastener Configuration

Head Styles

Which of the various head configurations to specify depends on the type of driving equipment used (screwdriver, socket wrench, etc.), the type of joint load, and the external appearance desired. The head styles shown in Fig. 10-3-3 can be used for both bolts and screws but are most commonly identified with the fastener category called machine screw or cap screw.

Hex and Square   The hex head is the most commonly used head style. The hex head design offers greater strength, ease of torque input, and area than the square head.

Pan   This head combines the qualities of the truss, binding, and round head types.

Binding   This type of head is commonly used in electrical connections because its undercut prevents fraying of stranded wire.

Washer (flanged)   This configuration eliminates the need for a separate assembly step when a washer is required, increases the bearing areas of the head, and protects the material finish during assembly.

Oval   Characteristics of this head type are similar to those of the flat head but it is sometimes preferred because of its neat appearance.

Flat   Available with various head angles, this fastener centers well and provides a flush surface.

Fillister   The deep slot and small head allow a high torque to be applied during assembly.

Truss   This head covers a large area. It is used where extra holding power is required, holes are oversize, or the material is soft.

12-Point   This 12-sided head is normally used on aircraft-grade fasteners. Multiple sides allow for a very sure grip and high torque during assembly.

Drive Configurations

Figure 10-3-4 shows 16 different driving designs.

Shoulders and Necks

The shoulder of a fastener is the enlarged portion of the body of a threaded fastener or the shank of an unthreaded fastener (Fig. 10-3-5, pg. 272).

Point Styles

The *point* of a fastener is the configuration of the end of the shank of a headed or headless fastener. Standard point styles are shown in Fig. 10-3-6 (pg. 272).

Cup   Most widely used when the cutting-in action of point is not objectionable.

Flat   Used when frequent resetting of a part is required. Particularly suited for use against hardened steel shafts. This point is preferred where walls are thin or the threaded member is a soft material.

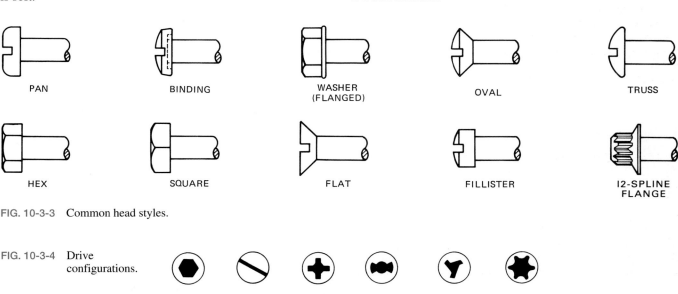

FIG. 10-3-3   Common head styles.

FIG. 10-3-4   Drive configurations.

FIG. 10-3-5   Shoulder and necks.

OVAL SHOULDER   ROUND SHOULDER   FIN NECK   SQUARE (CARRIAGE) NECK

CUP   FLAT   CONE   HALF DOG   OVAL

FIG. 10-3-6   Point styles.

Cone   Used for permanent location of parts. Usually spotted in a hole to half its length.

Oval   Used when frequent adjustment is necessary or for seating against angular surfaces.

Half Dog   Normally applied where permanent location of one part in relation to another is desired.

Property Classes of Fasteners

Inch Fasteners

The strength of customary fasteners for most common uses is determined by the size of the fastener and the material from which it is made. Property classes are defined by the Society of Automotive Engineers (SAE) or the American Society for Testing and Materials (ASTM).

Figure 10-3-7 lists the mechanical requirements of inch-sized fasteners and their identification patterns.

Metric Fasteners

For mechanical and material requirements, metric fasteners are classified under a number of property classes. Bolts, screws, and studs have seven property classes of steel suitable for general engineering applications. The property classes are designated by numbers where increasing numbers generally represent increasing tensile strengths. The designation symbol consists of two parts: the first numeral of a two-digit symbol or the first two numerals of a three-digit symbol approximates one-hundredth of the minimum tensile strength in megapascals (MPa) and the last numeral approximates one-tenth of the ratio expressed as a percentage of minimum yield strength and minimum tensile strength.

EXAMPLE 1

A property class 4.8 fastener (see Fig. 10-3-8) has a minimum tensile strength of 420 MPa and a minimum yield strength of 340 MPa. One percent of 420 is 4.2. The first digit is 4. The minimum yield strength of 340 MPa is equal to approximately 80 percent of the minimum tensile strength of 420 MPa. One-tenth of 80 percent is 8. The last digit of the property class is 8.

Property Class (Equal to or Less Than)	Nominal Diameter	Minimum Tensile Strength MPa	Minimum Yield Strength MPa
4.6	M5 thru M36	400	240
4.8	M1.6 thru M16	420	340
5.8	M5 thru M24	520	420
8.8	M16 thru M36	830	660
9.8	M1.6 thru M16	900	720
10.9	M5 thru M36	1040	940
12.9	M1.6 thru M36	1220	1100

FIG. 10-3-8   Mechanical requirements for metric bolts, screws, and studs.

FIG. 10-3-7   Mechanical requirements for inch-size threaded fasteners.

HEAD DESIGNATION	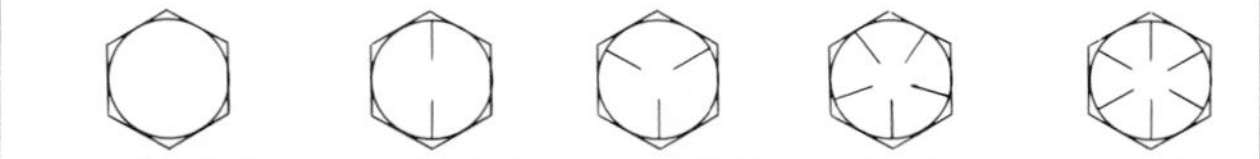				
GRADE	GRADES 0, I, 2	GRADE 3	GRADE 5	GRADE 7	GRADE 8
MINIMUM TENSILE STRENGTH KIPS	0—NO REQUIRE-MENTS 1—55 2—69 64 55	110 100	120 115 105	133	150

EXAMPLE 2

A property class 10.9 fastener (see Fig. 10-3-8) has a minimum tensile strength of 1040 MPa and a minimum yield strength of 940 MPa. One percent of 1040 is 10.4. The first two numerals of the three-digit symbol are 10. The minimum yield strength of 940 MPa is equal to approximately 90 percent of the minimum tensile strength of 1040 MPa. One-tenth of 90 percent is 9. The last digit of the property class is 9.

Machine screws are normally available only in classes 4.8 and 9.8; other bolts, screws, and studs are available in all classes within the specified product size limitations given in Fig. 10-3-7.

For guidance purposes only, to assist designers in selecting a property class, the following information may be used:

- Class 4.6 is approximately equivalent to SAE grade 1 and ASTM A 307, grade A.
- Class 5.8 is approximately equivalent to SAE grade 2.
- Class 8.8 is approximately equivalent to SAE grade 5 and ASTM A 449.
- Class 9.8 has properties approximately 9 percent stronger than SAE grade 5 and ASTM A 449.
- Class 10.9 is approximately equivalent to SAE grade 8 and ASTM A 354 grade BD.

Fastener Markings

Slotted and crossed recessed screws of all sizes and other screws and bolts of sizes .25 in. or M4 and smaller need not be marked. All other bolts and screws of sizes .25 in. or M5 and larger are marked to identify their strength. The property class symbols for metric fasteners are shown in Fig. 10-3-9. The symbol is located on the top of the bolt head or screw. Alternatively, for hex-head products, the markings may be indented on the side of the head.

All studs of size .25 in. or M5 and larger are identified by the property class symbol. The marking is located on the

extreme end of the stud. For studs with an interference-fit thread the markings are located at the nut end. Studs smaller than .50 in. or M12 use different identification symbols.

Nuts

The customary terms *regular* and *thick* for describing nut thicknesses have been replaced by the terms *style 1* and *style 2* for metric nuts. The design of style 1 and 2 steel nuts shown in Fig. 10-3-10 is based on providing sufficient nut strength to reduce the possibility of thread stripping. There are three property classes of steel nuts available (Fig. 10-3-11).

Hex-Flanged Nuts   These nuts are intended for general use in applications requiring a large bearing contact area. The two styles of flanged hex nuts differ dimensionally in thickness

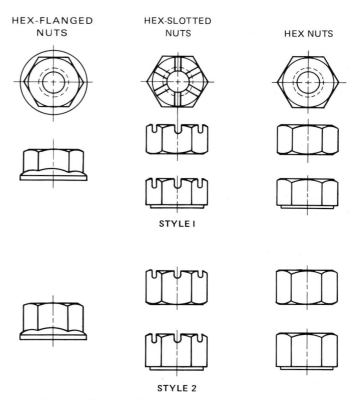

FIG. 10-3-10   Hex-nut styles.

Property Class	Identification Symbol	
	Bolts, Screws and Studs	Studs Smaller Than M12
4.6	4.6	—
4.8	4.8	—
5.8	5.8	—
8.8 (I)	8.8	○
9.8 (I)	9.8	+
10.9 (I)	10.9	□
12.9	12.9	△

Note I: Products made of low-carbon martensite steel shall be additionally identified by underlining the numerals.

FIG. 10-3-9   Metric property class identification symbols for bolts, screws, and studs.

Property Class	Nominal Nut Size	Suggested Property Class of Mating Bolt, Screw, or Stud
5	M5 thru M36	4.6, 4.8, 5.8
9	M5 thru M16 M20 thru M36	5.8, 9.8 5.8, 8.8
10	M6.3 thru M36	10.9

FIG. 10-3-11   Metric nut selection for bolts, screws, and studs.

FIG. 10-3-12   Approximate head proportions for hex-head cap screws, bolts, and nuts.

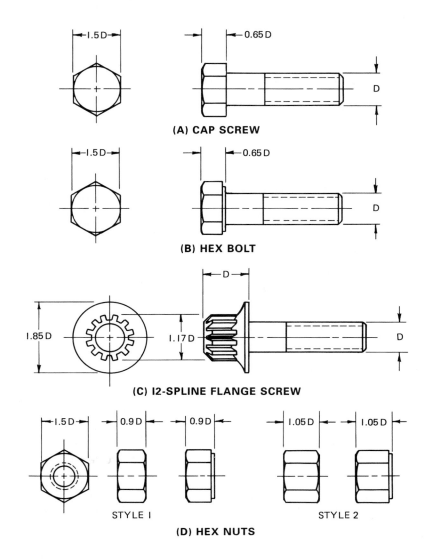

(A) CAP SCREW

(B) HEX BOLT

(C) I2-SPLINE FLANGE SCREW

STYLE I

STYLE 2

(D) HEX NUTS

only. The standard property classes for hex-flanged nuts are identical to the hex nuts. All metric nuts are marked to identify their property class.

Drawing a Bolt and Nut

Bolts and nuts are not normally drawn on detail drawings unless they are of a special size or have been modified. On some assembly drawings it may be necessary to show a nut and bolt. Approximate nut and bolt sizes are shown in Fig. 10-3-12. Actual sizes are found in Table 11 of the Appendix. Nut and bolt templates are also available and are recommended as a cost-saving device for manual drafting. Conventional drawing practice is to show the nuts and bolt heads in the across-corners position in all views.

Studs

Studs, as shown in Fig. 10-3-13, are still used in large quantities to best fulfill the needs of certain design functions and for overall economy.

Double-End Studs   These studs are designated in the following sequence: type and name; nominal size; thread information; stud length; material, including grade identification; and finish (plating or coating) if required.

EXAMPLE

TYPE 2 DOUBLE-END STUD
.500—13 UNC—2A × 4.00
CADMIUM PLATED

Continuous-Thread Studs   These studs are designated in the following sequence: product name, nominal size, thread

(A) DOUBLE-END   (B) CONTINUOUS-THREAD

FIG. 10-3-13   Studs.

information, stud length, material, and finish (plating or coating) if required.

EXAMPLE

TYPE 3 CONTINUOUS-THREAD STUD, M24 × 3 × 200, STEEL CLASS 8.8, ZINC PHOSPHATE AND OIL

Washers

Washers are one of the most common forms of hardware and perform many varied functions in mechanically fastened assemblies. They may only be required to span an oversize clearance hole, to give better bearing for nuts or screw faces, or to distribute loads over a greater area. Often, they serve as locking devices for threaded fasteners. They are also used to maintain a spring-resistance pressure, to guard surfaces against marring, and to provide a seal.

Classification of Washers

Washers are commonly the elements that are added to screw systems to keep them tight, but not all washers are locking types. Many washers serve other functions, such as surface protection, insulation, sealing, electrical connection, and spring-tension take-up devices.

Flat Washers   Plain, or flat, washers are used primarily to provide a bearing surface for a nut or a screw head, to cover large clearance holes, and to distribute fastener loads over a large area—particularly on soft materials such as aluminum or wood (Fig. 10-3-14).

Conical Washers   These washers are used with screws to effectively add spring take-up to the screw elongation.

Helical Spring Washers   These washers are made of slightly trapezoidal wire formed into a helix of one coil so that the free height is approximately twice the thickness of the washer section (Fig. 10-3-15).

Tooth Lock Washers   Made of hardened carbon steel, a tooth lock washer has teeth that are twisted or bent out of the plane of the washer face so that sharp cutting edges are presented to both the workpiece and the bearing face of the screw head or nut (Fig. 10-3-16).

Spring Washers   There are no standard designs for spring washers (Fig. 10-3-17). They are made in a great variety of sizes and shapes and are usually selected from a manufacturer's catalog for some specific purpose.

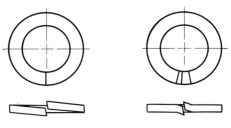

(A) PLAIN	(B) NONLINK POSITIVE

FIG. 10-3-15   Helical spring washers.

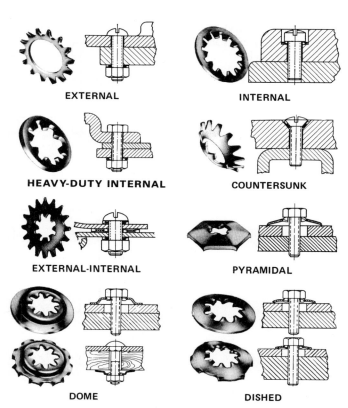

FIG. 10-3-16   Tooth lock washers.

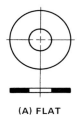

FIG. 10-3-14   Flat and conical washers.

(A) FLAT	(B) CONICAL	(C) RAMP CONICAL

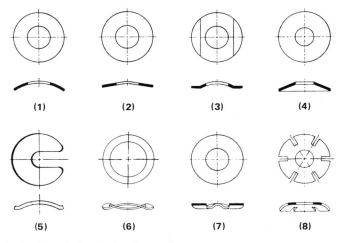

FIG. 10-3-17   Typical spring washers.

Special-Purpose Washers   Molded or stamped nonmetallic washers are available in many materials and may be used as seals, electrical insulators, or for protection of the surface of assembled parts.

Many plain, cone, or tooth washers are available with special mastic sealing compounds firmly attached to the washer. These washers are used for sealing and vibration isolation in high-production industries.

Terms Related to Threaded Fasteners

The *tap drill size* for a threaded (tapped) hole is a diameter equal to the minor diameter of the thread. The *clearance drill size*, which permits the free passage of a bolt, is a diameter slightly greater than the major diameter of the bolt (Fig. 10-3-18). A *counterbored hole* is a circular, flat-bottomed recess that permits the head of a bolt or cap screw to rest below the surface of

FIG. 10-3-18   Specifying threaded fasteners and holes.

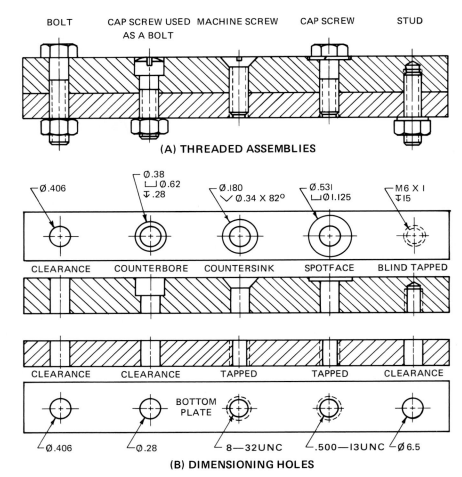

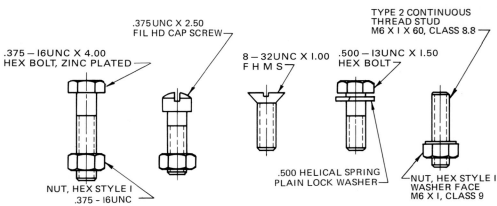

the part. A *countersunk hole* is an angular-sided recess that accommodates the shape of a flat-head cap screw or machine screw or an oval-head machine screw. *Spotfacing* is a machine operation that provides a smooth, flat surface where a bolt head or a nut will rest.

Specifying Fasteners

In order for the purchasing department to properly order the fastening device that has been selected in the design, the following information is required. (*Note:* The information listed will not apply to all types of fasteners.)

1. Type of fastener
2. Thread specifications
3. Fastener length
4. Material
5. Head style
6. Type of driving recess
7. Point type (setscrews only)
8. Property class
9. Finish

EXAMPLES

.375—16 UNC—2A × 4.00 HEX BOLT, ZINC PLATED

M10 × 1.5 × 50, 9.8 12-SPLINE FLANGE SCREW, CADMIUM PLATED

TYPE 2 DOUBLE-END STUD, M10 × 1.5 × 100, STEEL CLASS 9.8, CADMIUM PLATED

NUT, HEX, STYLE 1, .500 UNC STEEL

MACH SCREW, PHILLIPS ROUND HD, 8—32 UNC × 1.00, BRASS

WASHER, FLAT 8.4 ID × 17 OD × 2 THK, STEEL HELICAL SPRING

REFERENCES AND SOURCE MATERIAL

1. *Machine Design,* Fastening and joining reference issue.
2. *Design Engineering* and Staff of Stelco's "Fastener Facts."
3. ANSI Y14.6, *Screw Thread Representation.*
4. "Metric Fastener Standards Handbook" by Industrial Fasteners Institute.

ASSIGNMENTS

See Assignments 12 through 17 for Unit 10-3 on pages 287 through 289.

10-4 SPECIAL FASTENERS

Setscrews

Setscrews are used as semipermanent fasteners to hold a collar, sheave, or gear on a shaft against rotational or translational forces. In contrast to most fastening devices, the setscrew is essentially a compression device. Forces developed by the screw point on tightening produce a strong clamping action that resists relative motion between assembled parts. The basic problem in setscrew selection is to find the best combination of setscrew form, size, and point style that provides the required holding power.

Setscrews can be categorized in two ways: by their head style and by the point style desired (Fig. 10-4-1). Each setscrew style is available in any one of five point styles.

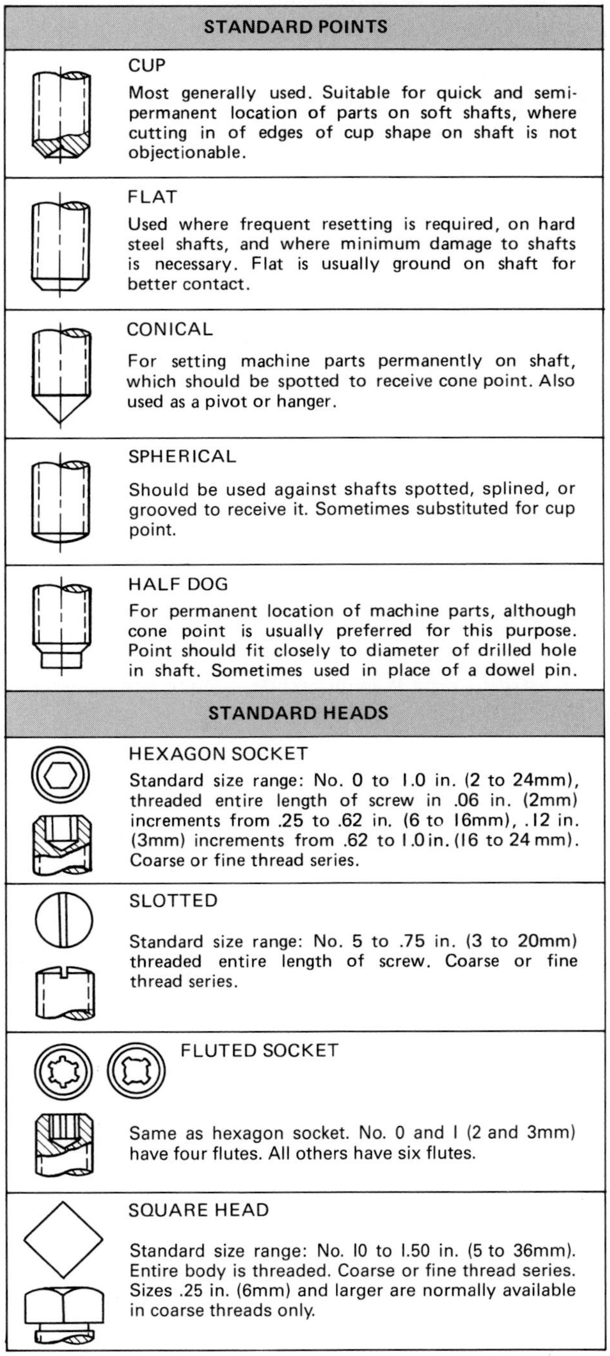

STANDARD POINTS	
CUP	Most generally used. Suitable for quick and semipermanent location of parts on soft shafts, where cutting in of edges of cup shape on shaft is not objectionable.
FLAT	Used where frequent resetting is required, on hard steel shafts, and where minimum damage to shafts is necessary. Flat is usually ground on shaft for better contact.
CONICAL	For setting machine parts permanently on shaft, which should be spotted to receive cone point. Also used as a pivot or hanger.
SPHERICAL	Should be used against shafts spotted, splined, or grooved to receive it. Sometimes substituted for cup point.
HALF DOG	For permanent location of machine parts, although cone point is usually preferred for this purpose. Point should fit closely to diameter of drilled hole in shaft. Sometimes used in place of a dowel pin.
STANDARD HEADS	
HEXAGON SOCKET	Standard size range: No. 0 to 1.0 in. (2 to 24mm), threaded entire length of screw in .06 in. (2mm) increments from .25 to .62 in. (6 to 16mm), .12 in. (3mm) increments from .62 to 1.0 in. (16 to 24 mm). Coarse or fine thread series.
SLOTTED	Standard size range: No. 5 to .75 in. (3 to 20mm) threaded entire length of screw. Coarse or fine thread series.
FLUTED SOCKET	Same as hexagon socket. No. 0 and 1 (2 and 3mm) have four flutes. All others have six flutes.
SQUARE HEAD	Standard size range: No. 10 to 1.50 in. (5 to 36mm). Entire body is threaded. Coarse or fine thread series. Sizes .25 in. (6mm) and larger are normally available in coarse threads only.

FIG. 10-4-1 Setscrews.

The conventional approach to selecting the setscrew diameter is to make it roughly equal to one-half the shaft diameter. This rule of thumb often gives satisfactory results, but its range of usefulness is limited.

Setscrews and Keyseats

When a setscrew is used in combination with a key, the screw diameter should be equal to the width of the key. In this combination the setscrew is locating the parts in an axial direction only. The torsional load on the parts is carried by the key.

The key should be tight-fitting, so that no motion is transmitted to the screw. Key design is covered in Chap. 11.

Keeping Fasteners Tight

Fasteners are inexpensive, but the cost of installing them can be substantial. Probably the simplest way to cut assembly costs is to make sure that, once installed, fasteners stay tight.

The American National Standards Institute has identified three basic locking methods: free-spinning, prevailing-torque, and chemical locking. Each has its own advantages and disadvantages (Fig. 10-4-2).

Free-spinning devices include toothed and spring lockwashers and screws and bolts with washerlike heads. With these arrangements, the fasteners spin free in the clamping direction, which makes them easy to assemble, and the break-loose torque is greater than the seating torque. However, once break-loose torque is exceeded, free-spinning washers have no prevailing torque to prevent further loosening.

Prevailing torque methods make use of increased friction between nut and bolt. Metallic types usually have deformed threads or contoured thread profiles that jam the threads on assembly. Non-metallic types make use of nylon or polyester insert elements that produce interference fits on assembly.

Chemical locking is achieved by coating the fastener with an adhesive.

Locknuts

A *locknut* is a nut with special internal means for gripping a threaded fastener to prevent rotation. It usually has the dimensions, mechanical requirements, and other specifications of a standard nut, but with a locking feature added.

Locknuts are divided into three general classifications: prevailing-torque, free-spinning, and other types. These are shown in Figs. 10-4-3 and 10-4-4.

Prevailing-Torque Locknuts

Prevailing-torque locknuts spin freely for a few turns, and then must be wrenched to final position. The maximum holding and locking power is reached as soon as the threads and the locking feature are engaged. Locking action is maintained until the nut is removed. Prevailing-torque locknuts are classified by basic design principles:

1. Thread deflection causes friction to develop when the threads are mated; thus the nut resists loosening.
2. The out-of-round top portion of the tapped nut grips the bolt threads and resists rotation.

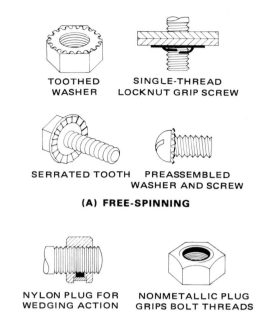

(A) FREE-SPINNING

(B) PREVAILING-TORQUE

FIG. 10-4-2   Basic locking methods for setscrews.

3. The slotted section of the locknut is pressed inward to provide a spring frictional grip on the bolt.
4. Inserts, either nonmetallic or of soft metal, are plastically deformed by the bolt threads to produce a frictional interference fit.
5. A spring wire or pin engages the bolt threads to produce a wedging or ratchet-locking action.

Free-Spinning Locknuts

Free-spinning locknuts are free to spin on the bolt until seated. Additional tightening locks the nut.

Since most free-spinning locknuts depend on clamping force for their locking action, they are usually not recommended for joints that might relax through plastic deformation or for fastening materials that might crack or crumble.

Other Locknut Types

Jam nuts are thin nuts used under full-sized nuts to develop locking action. The large nut has sufficient strength to elastically deform the lead threads of the bolt and jam nut. Thus, a considerable resistance against loosening is built up. The use of jam nuts is decreasing; a one-piece, prevailing-torque locknut usually is used instead at a savings in assembled cost.

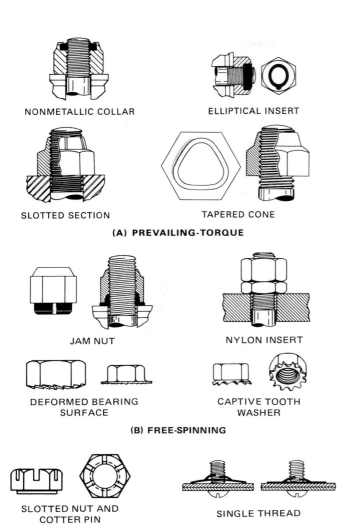

NONMETALLIC COLLAR   ELLIPTICAL INSERT

SLOTTED SECTION   TAPERED CONE

(A) PREVAILING-TORQUE

JAM NUT   NYLON INSERT

DEFORMED BEARING SURFACE   CAPTIVE TOOTH WASHER

(B) FREE-SPINNING

SLOTTED NUT AND COTTER PIN   SINGLE THREAD

(C) OTHER TYPES

FIG. 10-4-3   Locknuts.

FIG. 10-4-4   Single-thread engaging nuts.

Slotted and castle nuts have slots that receive a cotter pin that passes through a drilled hole in the bolt and thus serves as the locking member. Castle nuts differ from slotted nuts in that they have a circular crown of a reduced diameter.

Single-thread locknuts are spring steel fasteners which may be speedily applied. Locking action is provided by the grip of the thread-engaging prongs and the reaction of the arched base. Their use is limited to nonstructural assemblies and usually to screw sizes below 6 mm in diameter (Fig. 10-4-5, pg. 280).

Captive or Self-Retaining Nuts

Captive or self-retaining nuts provide a permanent, strong, multiple-thread fastener for use on thin materials (Fig. 10-4-6, pg. 280). They are especially good where there are blind locations, and they can normally be attached without damaging finishes. Methods of attaching these types of nuts vary and tools required for assembly are generally uncomplicated and inexpensive. The self-retained nuts are grouped according to four means of attachment.

1. Plate or anchor nuts: These nuts have mounting lugs that can be riveted, welded, or screwed to the part.
2. Caged nuts: A spring-steel cage retains a standard nut. The cage snaps into a hole or clips over an edge to hold the nut in position.
3. Clinch nuts: They are specially designed nuts with pilot collars that are clinched or staked into the parent part through a precut hole.
4. Self-piercing nuts: A form of clinch nut that cuts its own hole.

Inserts

Inserts are a special form of nut designed to serve the function of a tapped hole in blind or through-hole locations (Fig. 10-4-7, pg. 280).

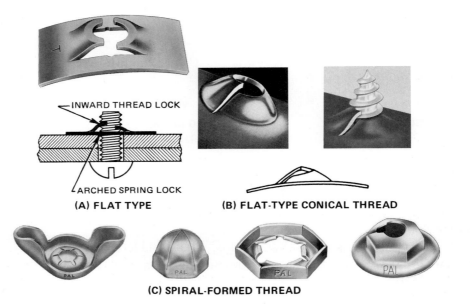

INWARD THREAD LOCK

ARCHED SPRING LOCK

(A) FLAT TYPE   **(B) FLAT-TYPE CONICAL THREAD**

(C) SPIRAL-FORMED THREAD

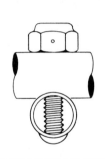

USE OF LOCKNUT FOR
TUBULAR FASTENING

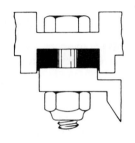

FOR RUBBER-INSULATED
AND CUSHION MOUNT-
INGS WHERE THE NUT
MUST REMAIN STATION-
ARY

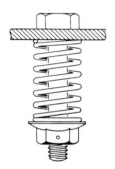

USE OF LOCKNUT ON
A SPRING CLAMP

USE OF LOCKNUT WHERE
ASSEMBLY IS SUBJECTED
TO VIBRATORY OR CYCLIC
MOTIONS THAT COULD
CAUSE LOOSENING

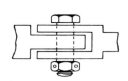

USE OF LOCKNUT ON A
BOLTED CONNECTION
THAT REQUIRES PRE-
DETERMINED PLAY

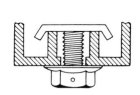

FOR AN EXTRUDED PART
ASSEMBLY

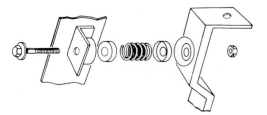

FOR HOLDING A MOTOR MOUNTING
SECURELY IN POSITION

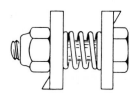

FOR SPRING-MOUNTED
CONNECTIONS WHERE
THE NUT MUST REMAIN
STATIONARY OR IS SUB-
JECT TO ADJUSTMENT

FIG. 10-4-5   Typical locknut applications.

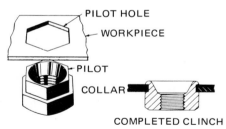

PILOT HOLE

WORKPIECE

PILOT

COLLAR

COMPLETED CLINCH

(I) UNIVERSAL PIERCE NUT

(2) HIGH-STRESS PIERCE NUT

(A) PLATE NUT   **(B) CAGED NUT**   **(C) CLINCH NUT**   **(D) PIERCE NUTS**

FIG. 10-4-6   Captive or self-retaining nuts.

(A) MOLDED-IN INSERT

(B) SELF-TAPPING INSERT

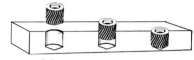

(C) PRESSED-IN INSERT

**(D) EXTERNAL-INTERNAL
THREADED INSERT**

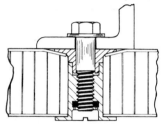

(E) SANDWICH PANEL INSERT

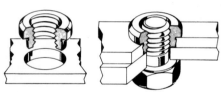

(F) THIN MATERIAL INSERT

FIG. 10-4-7   Inserts.

Sealing Fasteners

Fasteners hold two or more parts together, but they can perform other functions as well. One important auxiliary function is that of sealing gases and liquids against leakage.

Two types of sealed-joint construction are possible with fasteners (Fig. 10-4-8). In one approach, the fasteners enter the sealed medium and are separately sealed. The second approach uses a separate sealing element that is held in place by the clamping forces produced by conventional fasteners, such as rivets or bolts.

There are many methods of obtaining a seal using sealing fasteners, as shown in Fig. 10-4-9.

REFERENCES AND SOURCE MATERIAL

1. *Machine Design,* Fastening and joining reference issue.

ASSIGNMENTS

See Assignments 18 through 20 for Unit 10-4 on page 290.

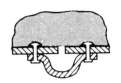 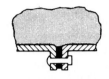

(A) FASTENERS SEPARATELY SEALED   **(B) SEALING ELEMENT CLAMPED IN PLACE**

FIG. 10-4-8   Types of sealed-joint construction.

BRONZE SLEEVE   LEAD WASHER   LIQUID PLASTIC COATING   MOLDED RUBBER RING   MASTIC SEALING COMPOUND

PREASSEMBLED NEOPRENE WASHER   PREASSEMBLED METAL WASHER AND O-RING   PREASSEMBLED METAL AND NEOPRENE WASHER   PREASSEMBLED NYLON WASHER   O-RING   O-RING WITH TEFLON WASHER

(A) SEALING SCREWS

MOLDED RUBBER RING   SOFT-ALUMINUM WASHER   PLASTIC JACKET   O-RING   O-RING   INTERFERENCE FIT

(B) SEALING RIVETS

NYLON PELLET   COPPER INSERT   NYLON COLLAR   NYLON BODY   FLOWED-IN SEALANT   MOLDED RUBBER GASKET OR O-RING

(C) SEALING NUTS

MOLDED NYLON-SEAL RING   MOLDED RUBBER TOROID   LAMINATED NEOPRENE TO METAL   NYLON SLEEVE   O-RING   FLOWED-IN SEALANT

(D) SEALING WASHERS

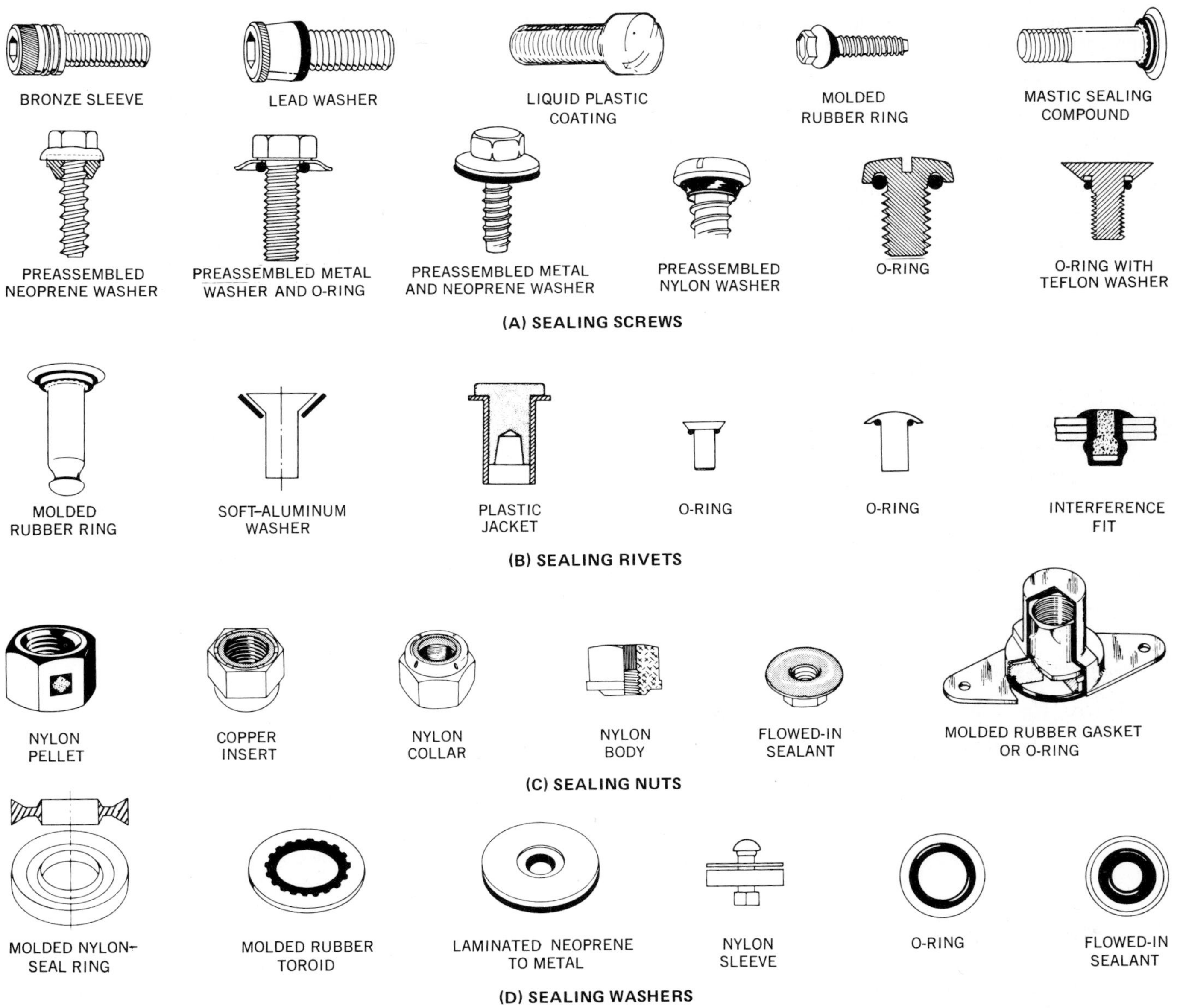

FIG. 10-4-9   Sealing fasteners.

10-5 FASTENERS FOR LIGHT-GAGE METAL, PLASTIC, AND WOOD

Tapping Screws

Tapping screws cut or form a mating thread when driven into drilled or cored holes. These one-piece fasteners permit rapid installation, since nuts are not used and access is required from only one side of the joint. The mating thread produced by the tapping screw fits the screw threads closely, and no clearance is necessary. This close fit usually keeps the screws tight, even under vibrating conditions (Fig. 10-5-1).

Tapping screws are practically all case-hardened and, therefore, can be driven tight and have a relatively high ultimate torsional strength. The screws are used in steel, aluminum (cast, extruded, rolled, or die-formed) die castings, cast iron, forgings, plastics, reinforced plastics, asbestos, and resin-impregnated plywood (Fig. 10-5-2). Coarse threads should be used with weak materials.

Self-drilling tapping screws have special points for drilling and then tapping their own holes (Fig. 10-5-3). These eliminate drilling or punching, but they must be driven by a power screwdriver.

Special Tapping Screws

Typical special tapping screws are the self-captive screws and double-thread combinations for limited drive. Self-captive screws combine a coarse-pitched starting thread (similar to type B) with a finer pitch (machine-screw thread) farther along the screw shank.

Sealing tapping screws, with pre-assembled washers or O-rings (Fig. 10-5-4B) are available in a variety of styles.

TYPE AB   TYPE B   TYPE F   TYPE U

FIG. 10-5-1   Self-tapping screws.

HEAVY GAGE SHEET METAL AND STRUCTURAL STEEL USE TYPES B, U, F.	LIGHT GAGE SHEET METAL USE TYPES AB, B.

Holes may be drilled or clean-punched.

Two parts may have pierced holes to nest burrs. This results in a stronger joint.

Use a pierced hole in workpiece if clearance hole is needed in part to be fastened.

Extruded hole may also be used in workpiece if clearance hole is needed in fastened part.

Holes may be drilled or clean-punched the same size in both sheet metal parts. For thicker sheet metal and structural steel, a clearance hole should be provided in the part to be fastened. Hole size depends on thickness of the workpiece.

Notes: I. Use hex-head on type B screws.
　　　 2. With type U screws, material should be thick enough to permit sufficient thread engagement—at least one screw diameter.

PLASTICS USE TYPES B, U, F.	CASTINGS AND FORGINGS USE TYPES B, U, F.

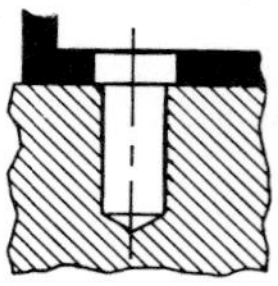

Screw holes may be molded or drilled. If material is brittle or friable, molded holes should be formed with a rounded chamfer, and drilled holes should be machine-chamfered. Provide a clearance in the part to be fastened. Depth of penetration should be held within the "minimum and maximum" limits recommended. The hole should be deeper than the screw penetration to allow for chip clearance.

Holes may be cored if it is practical to maintain close tolerances. Otherwise blind-drill holes to recommended hole size. Provide a clearance hole for screw in the part to be fastened. The hole in the casting, if it is a blind hole, should be deeper than the screw penetration to allow for chip clearance.

Notes: I. Hole in fastened part may be the same size as workpiece hole for type U screws.
　　　 2. Type B is suitable for use only in nonferrous castings.

FIG. 10-5-2   Tapping-screw application chart.

REFERENCES AND SOURCE MATERIAL

1. *Machine Design,* Fastening and joining reference issue.

ASSIGNMENTS

See Assignments 21 through 23 for Unit 10-5 on page 291.

.09

.02

.07

FIG. 10-5-3   Self-drilling tapping screws.

FIG. 10-5-4   Special tapping screws.

(A) TAPPING SCREWS WITH PREASSEMBLED WASHERS

(B) TAPPING SCREWS WITH PREASSEMBLED SEALING WASHERS OR COMPOUNDS

ASSIGNMENTS FOR CHAPTER 10

ASSIGNMENTS FOR UNIT 10-1, SIMPLIFIED THREAD REPRESENTATION

1. Make a working drawing of the gear box shown in Fig. 10-1-A. Draw the top, front, a full section right-side view, and an auxiliary view showing the surface with the tapped holes. Scale 1:1.

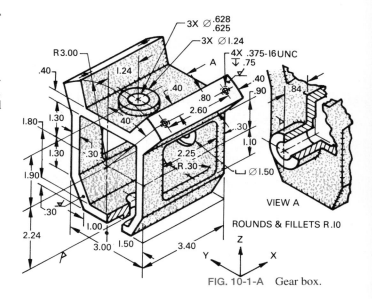

FIG. 10-1-A   Gear box.

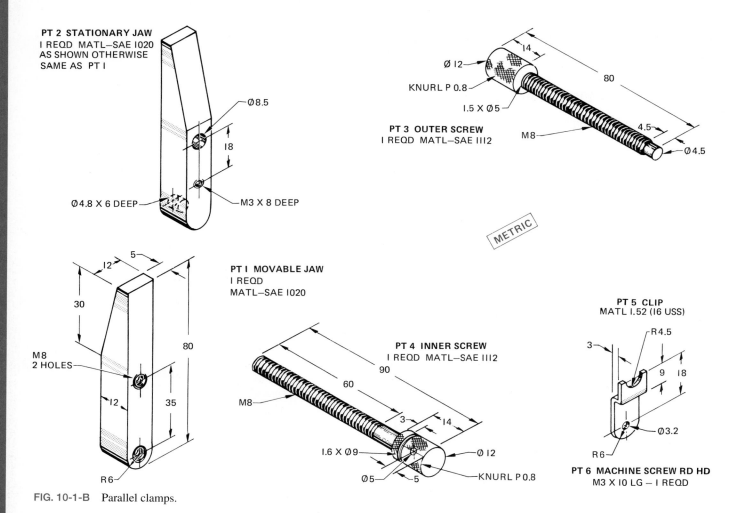

PT 2 STATIONARY JAW
I REQD MATL—SAE 1020
AS SHOWN OTHERWISE
SAME AS PT I

Ø8.5

18

Ø4.8 X 6 DEEP

M3 X 8 DEEP

Ø 12

KNURL P 0.8

1.5 X Ø5

PT 3 OUTER SCREW
I REQD MATL—SAE 1112

14

80

4.5

M8

Ø4.5

METRIC

12   5

30

80

M8
2 HOLES

12   35

R 6

PT I MOVABLE JAW
I REQD
MATL—SAE 1020

PT 4 INNER SCREW
I REQD MATL—SAE 1112

90

60

M8

3   14

1.6 X Ø9

Ø5   5

Ø 12

KNURL P 0.8

PT 5 CLIP
MATL I.52 (16 USS)

R4.5

3

9   18

Ø3.2

R6

PT 6 MACHINE SCREW RD HD
M3 X 10 LG — I REQD

FIG. 10-1-B   Parallel clamps.

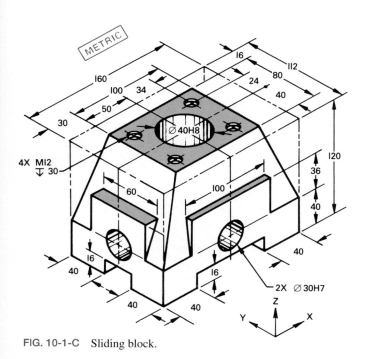

FIG. 10-1-C   Sliding block.

2. Make a two-view assembly drawing of the parallel clamps shown in Fig. 10-1-B. Use simplified thread conventions and include an item list calling for all the parts. The only dimension required on the drawing is the maximum opening of the jaws. Identify the parts on the assembly. Scale 1:1.

3. Make detail drawings of the parts shown in Fig. 10-1-B. Scale 1:1. Use your judgment for the number of views required for each part.

4. Make a working drawing of the sliding block shown in Fig. 10-1-C. Show the top, front, and right-side views. Use simplified thread representation. Use limit dimensions where fit symbols are shown. Scale 1:2.

5. Make a three-view detail drawing of the terminal block shown in Fig. 10-1-D. Use tabular dimensioning for locating the holes from the X, Y, and Z axes. The origin for the X and Y axes will be the center of the Ø4.80 hole. The origin for the Z axis will be the bottom of the part. Scale 1:2.

6. Make a detail drawing of the guide block shown in Fig. 10-1-E. Draw the top, front, and left-side views. Scale 1:1.
7. Make a one-view assembly drawing of the turnbuckle shown in Fig. 10-1-F. Show the assembly in its shortest length and also include the maximum position shown in

phantom lines. The only dimensions required are the minimum and maximum distances between the eye centers. Scale 1:1.
8. Make detail drawings of the parts shown in Fig. 10-1-F. Scale 1:1.

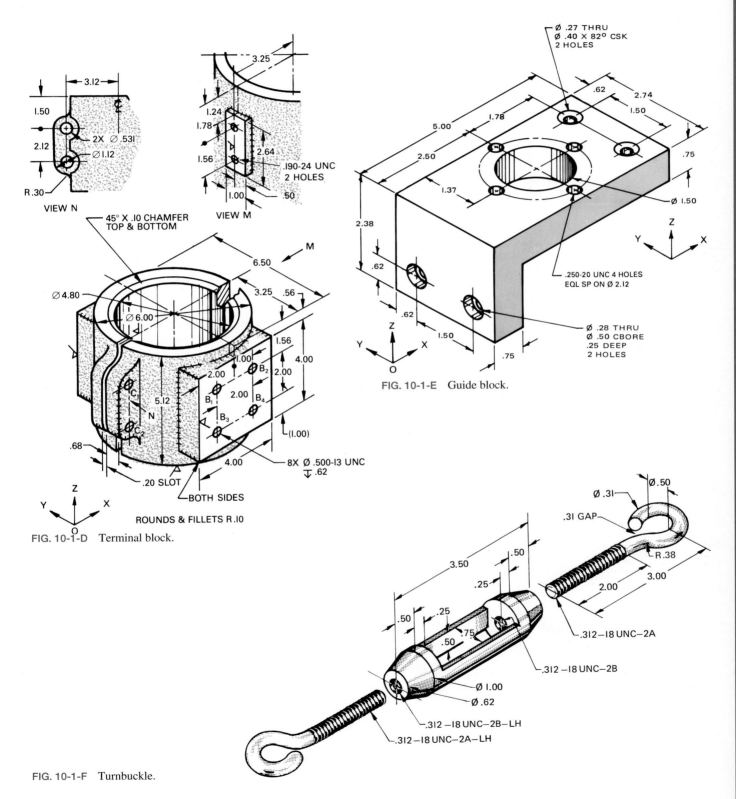

VIEW N

VIEW M

FIG. 10-1-E   Guide block.

FIG. 10-1-D   Terminal block.

FIG. 10-1-F   Turnbuckle.

ASSIGNMENTS FOR UNIT 10-2, DETAILED AND SCHEMATIC THREAD REPRESENTATION

9. Lay out the parts as shown in either Fig. 10-2-A or 10-2-B and draw the threads in detailed representation. The end rods are to be drawn in section. Scale 1:1.

10. Make one-view drawings of the parts shown in Figs. 10-2-C and 10-2-D using detailed thread representation. Use a conventional break to shorten the length of each part. Scale 1:1.

11. Make a one-view drawing of one of the parts shown in Fig. 10-2-E or 10-2-F using detailed thread representation. Scale 1:1 for Fig. 10-2-E and scale 2:1 for Fig. 10-2-F.

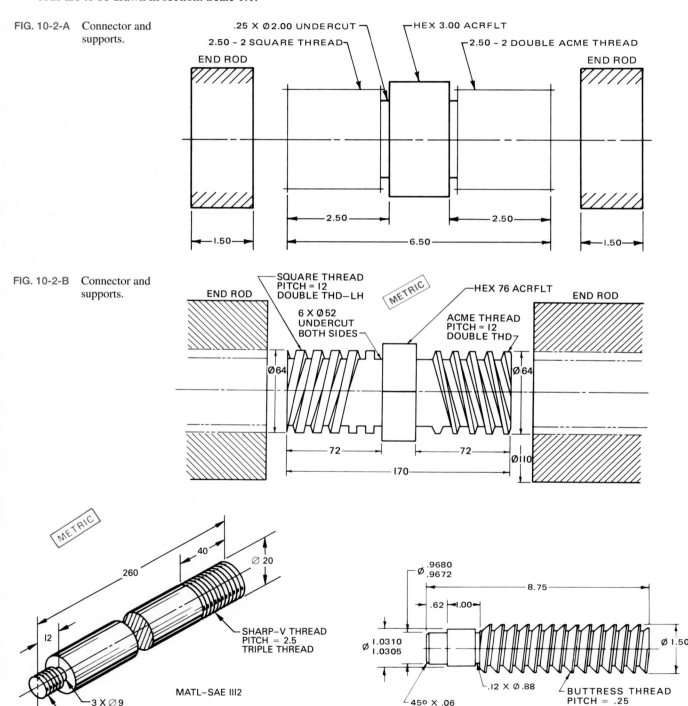

FIG. 10-2-A   Connector and supports.

FIG. 10-2-B   Connector and supports.

FIG. 10-2-C   Guide rod.

FIG. 10-2-D   Jack screw.

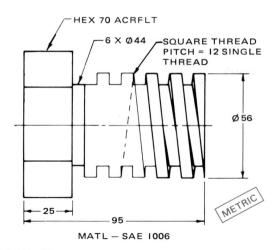

HEX 70 ACRFLT

6 X Ø44

SQUARE THREAD
PITCH = 12 SINGLE
THREAD

Ø56

25

95

MATL — SAE 1006

FIG. 10-2-E   Plug.

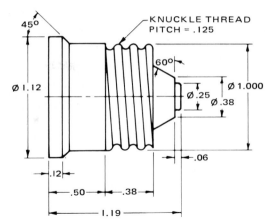

45°

KNUCKLE THREAD
PITCH = .125

60°

Ø1.12

Ø.25

Ø1.000

Ø.38

.06

.12

.50

.38

1.19

FIG. 10-2-F   Fuse.

ASSIGNMENTS FOR UNIT 10-3, COMMON THREADED FASTENERS

12. Prepare a full-section assembly drawing of the four fastener assemblies shown in Fig. 10-3-A. Dimension both the clearance and threaded holes. A top view may be shown if required. Scale 1:1.

13. Prepare full-section assembly drawings of the four fastener assemblies shown in Fig. 10-3-B. Dimension both the clearance and threaded holes. A top view of the fastener may be shown if required. Scale 1:1.

FIG. 10-3-A   Threaded fasteners.

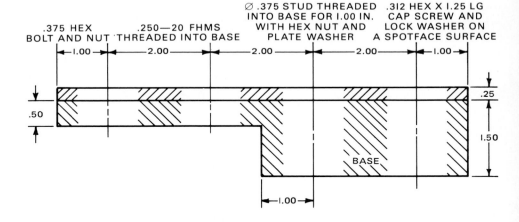

.375 HEX
BOLT AND NUT

.250—20 FHMS
THREADED INTO BASE

Ø.375 STUD THREADED
INTO BASE FOR 1.00 IN.
WITH HEX NUT AND
PLATE WASHER

.312 HEX X 1.25 LG
CAP SCREW AND
LOCK WASHER ON
A SPOTFACE SURFACE

1.00   2.00   2.00   2.00   1.00

.50   .25   1.50

BASE

1.00

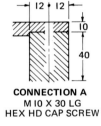

12 12

10

40

CONNECTION A
M 10 X 30 LG
HEX HD CAP SCREW

12

CONNECTION B
M 10 X 40 LG STUD
THREAD EACH END 20 LG
HEX NUT STYLE I AND
SPRING LOCK WASHER

CONNECTION C
M 10 X 30 LG
FL HD CAP SCREW

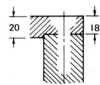

20   18

CONNECTION D
M 10 X 1.25 X 25 LG
SOCKET HEAD CAP SCREW
AND SPRING LOCK WASHER

FIG. 10-3-B   Threaded fasteners.

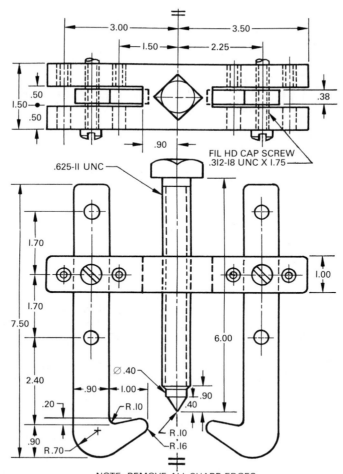

FIG. 10-3-C   Wheel-puller assembly.

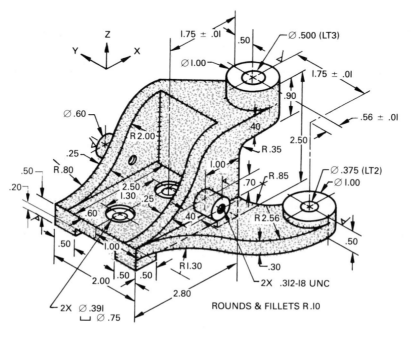

FIG. 10-3-D   Shaft intermediate support.

14. Make a two-view assembly drawing of the wheel-puller shown in Fig. 10-3-C. Use simplified thread representation and include an item list calling for all the parts. Scale 1:1.
15. Make detail drawings of the parts shown in Fig. 10-3-C. Use your judgment for the selection and number of views required for each part. Scale 1:1.
16. Make a working drawing of the shaft intermediate support shown in Fig. 10-3-D. Show the top, front, and left-side views. Surfaces shown ✓ to have a maximum roughness value of 250 μin. and a machining allowance of .04 in. Show the limits of size for the Ø.500 and Ø.375 holes. Scale 1:1.
17. Make a working drawing of the base in Fig. 10-3-E. Show the limits of size for the Ø15 and Ø18 holes. Surfaces shown ✓ to have a maximum roughness value of 3.2 μm and a machining allowance of 2 mm. Scale 1:1.

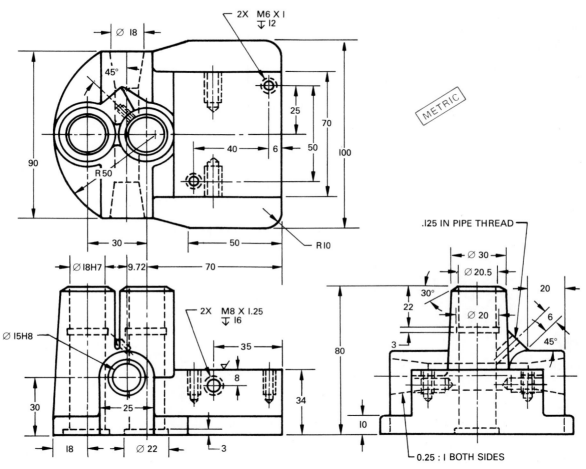

FIG. 10-3-E   Base.

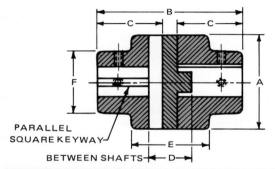

PARALLEL
SQUARE KEYWAY

BETWEEN SHAFTS

MAXIMUM BORE	A	B	C	D	E	F
.9375	3.00	3.75	1.75	.88	1.50	2.38
1.1875	3.50	4.69	2.19	1.06	1.81	2.75
1.4375	4.00	5.62	2.62	1.25	2.12	3.12
1.6875	5.00	6.56	3.06	1.44	2.44	3.50
1.9375	5.50	7.50	3.50	1.50	2.50	4.00
2.1875	6.00	8.44	3.94	1.81	3.06	4.38

DIMENSIONS SHOWN ARE IN INCHES

ASSIGNMENTS FOR UNIT 10-4, SPECIAL FASTENERS

18. Make a one-view assembly drawing of the flexible coupling shown in Fig. 10-4-A. The shafts, which are coupled, are 1.50 in. in diameter and are to be shown in the assembly. They are to extend beyond the coupling for approximately 2.00 in. and end with a conventional break. Show the setscrews and keys in position. Scale 1:1.

FIG. 10-4-A   Flexible coupling.

PT 3 YOKE
MATL—CI I REQD
ROUNDS AND
FILLETS R 3

R 10

M 10
3 HOLES

PT 6 SETSCREW
M 10 X 30 LG 2 REQD

R 12

Ø 38

PT 7 SETSCREW
M 10 X 10 LG
HEX SOCKET
DOG POINT
2 REQD

Ø 20 H8f7 FIT
WITH PT 2

PT 8 HEX HD JAM NUT
M 10 2 REQD

M 10

Ø 20 H8f7 FIT WITH PT 2

Ø 38

ROUNDS AND FILLETS R 3

Ø 8 SLOTS

PT I BASE
MATL—CI I REQD

Courtesy Boston Gear Works

FIG. 10-4-B   Adjustable shaft support. *(Boston Gear Works)*

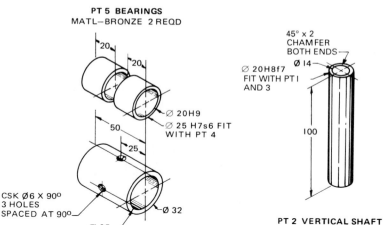

PT 5 BEARINGS
MATL—BRONZE 2 REQD

Ø 20H9
Ø 25 H7s6 FIT
WITH PT 4

CSK Ø6 X 90°
3 HOLES
SPACED AT 90°

Ø 25
H7s6 FIT WITH PT 5

Ø 32

PT 4 BEARING HOUSING
MATL—STEEL I REQD

45° x 2
CHAMFER
BOTH ENDS

Ø 14

Ø 20H8f7
FIT WITH PT I
AND 3

100

PT 2 VERTICAL SHAFT
MATL—STEEL I REQD

METRIC

19. Make a one-view assembly drawing of the adjustable shaft support shown in Fig. 10-4-B. Show the base in full section. A broken-out section is recommended to clearly show the setscrews in the yoke. Add part numbers to the assembly drawing and include an item list. Do not dimension. Scale 1:1.

20. Make detail drawings for the parts shown in Fig. 10-4-B. Use your judgment for the number of views required for each part. Show the limits of size for the holes. Scale 1:1.

ASSIGNMENTS FOR UNIT 10-5, FASTENERS FOR LIGHT-GAGE METAL, PLASTIC, AND WOOD

21. Make a drawing of the assemblies shown in Fig. 10-5-A. Either inch or metric fasteners may be used. The steel post is fastened to the panel by two rows of tapping screws. The steel strap is held to the post by a single tapping screw which has the equivalent strength (body area) of at least three of the other tapping screws. Dimension the holes and fastener sizes. Scale to suit.

22. Make a two-view assembly drawing of the woodworking vise shown in Fig. 10-5-B. Have the opening between jaws 1.50 in. Include on the drawing an item list calling for all the parts. Scale 1:2.

23. Make detail drawings of the parts shown in Fig. 10-5-B. Use your judgment for the selection and number of views required for each part. Show the limits of size where fits are given. Scale 1:2.

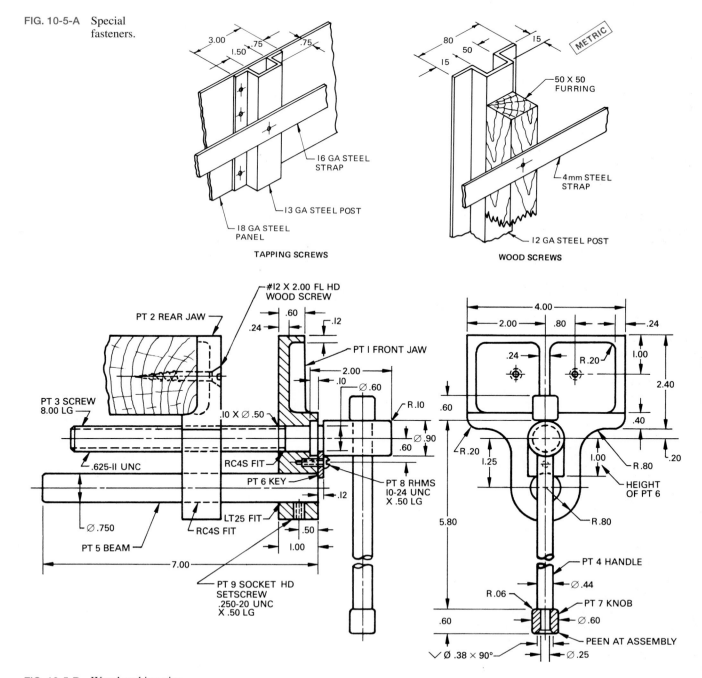

FIG. 10-5-A   Special fasteners.

FIG. 10-5-B   Woodworking vise.

CHAPTER 11

MISCELLANEOUS TYPES OF FASTENERS

Definitions

Adhesion   The force that holds materials together.

Compression spring   An open-coiled, helical spring that offers resistance to a compressive force.

Edge distance   The interval between the edge of the part and the center line of the fastener (rivet, bolt, etc.).

Extension spring   A close-coiled, helical spring that offers resistance to a pulling force.

Key   A piece of steel lying partly in a groove in the shaft and extending into another groove in the hub.

Keyseat   The groove that holds the key in the shaft.

Keyway   The groove in the hub or surrounding part that holds the key.

Pitch distance   The interval between center lines of adjacent fasteners (rivets, bolts, etc.).

Quick-release pins   Fasteners designed to release quickly and easily.

1. RETAINING COMPOUND JOINT	2. PRESS FIT	3. KNURLED JOINT
4. TAPERED SHAFT	5. SLIDING FIT	6. DRIVEN KEY
7. SPLINE	8. SLIP FIT WITH KEY	9. BRAZED JOINT
10. SETSCREW	11. PINS	12. SPLIT HUB

FIG. 11-1-1   Miscellaneous types of fasteners.

Resistance-welded fastener   A threaded metal part designed to be fused permanently in place by standard production welding equipment.

Retaining or **snap rings**   Fasteners used to provide a removable shoulder to accurately locate, retain, or lock components on shafts and in bores of housings.

Rivet   A ductile metal pin that is inserted through holes in two or more parts, and having the ends formed over to securely hold the parts.

Semipermanent pins   Fasteners that require application of pressure or the aid of tools for installation or removal.

Serrations   Shallow, involute splines with 45° pressure angles.

Splined shaft   A shaft having multiple grooves, or keyseats, cut around its circumference.

Spring clip   A self-retaining clip, requiring only a flange, panel edge, or mounting hole to clip to.

Stress   The force pulling materials apart.

Torsion spring, flat coil spring, and **flat spring**   Types of springs that exert pressure in a circular arc.

11-1 KEYS, SPLINES, AND SERRATIONS

Keys

A *key* is a piece of steel lying partly in a groove in the shaft and extending into another groove in the hub. The groove in the shaft is referred to as a *keyseat,* while the groove in the hub or surrounding part is referred to as a *keyway* (see boxes 6 and 8 in Fig. 11-1-1). A key is used to secure gears, pulleys, cranks, handles, and similar machine parts to shafts, so that the motion of the part is transmitted to the shaft, or the motion of the shaft to the part, without slippage. The key may also act in a safety capacity; its size is generally calculated so that when overloading takes place, the key will shear or break before the part or shaft breaks or deforms.

There are many kinds of keys. The most common types are shown in Fig. 11-1-2. Square and flat keys are widely used in industry. The width of the square and flat key should be approximately one-quarter the shaft diameter, but for proper key selection refer to Table 21 in the Appendix. These keys are also available with a 1:100 taper on their top surfaces and are then known as *squared-taper* or *flat-tapered* keys. The keyway in the hub is tapered to accommodate the taper on the key.

The gib-head key is the same as the squared or flat-tapered key but has a head added for easy removal.

The Pratt and Whitney key is rectangular with rounded ends. Two-thirds of this key sits in the shaft, one-third sits in the hub.

The Woodruff key is semicircular and fits into a semicircular keyseat in the shaft and a rectangular keyway in the hub. The width of the key should be approximately one-quarter

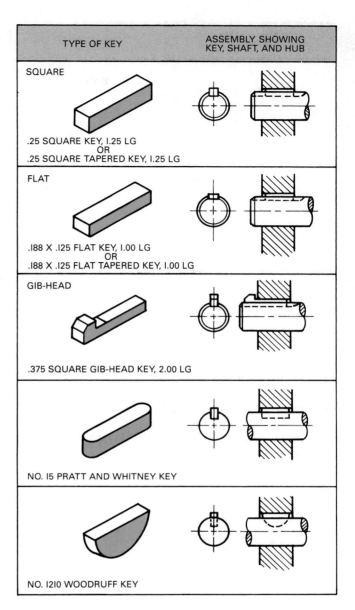

TYPE OF KEY	ASSEMBLY SHOWING KEY, SHAFT, AND HUB
SQUARE	
.25 SQUARE KEY, 1.25 LG OR .25 SQUARE TAPERED KEY, 1.25 LG	
FLAT	
.188 X .125 FLAT KEY, 1.00 LG OR .188 X .125 FLAT TAPERED KEY, 1.00 LG	
GIB-HEAD	
.375 SQUARE GIB-HEAD KEY, 2.00 LG	
NO. 15 PRATT AND WHITNEY KEY	
NO. 1210 WOODRUFF KEY	

FIG. 11-1-2   Common keys.

the diameter of the shaft, and its diameter should approximate the diameter of the shaft. Half the width of the key extends above the shaft and into the hub. Refer to the Appendix for exact sizes. Woodruff keys are identified by a number that gives the nominal dimensions of the key. The numbering system, which originated many years ago, is identified with the fractional-inch system of measurement. The last two digits of the number give the normal diameter in eighths of an inch, and the digits preceding the last two give the nominal width in thirty-seconds of an inch. For example, a No. 1210 Woodruff key describes a key $\frac{12}{32} \times \frac{10}{8}$ in., or a $\frac{3}{8} \times 1\frac{1}{4}$ in. key.

In calling up keys on the item list, only the information shown in the callout in Fig. 11-1-2 need be given.

Dimensioning of Keyseats

Keyseats and keyways are dimensioned by width, depth, location, and if required, length. The depth is dimensioned from the opposite side of the shaft or hole (Fig. 11-1-3).

Tapered Keyseats   The depth of tapered keyways in hubs, which is shown on the drawing, is the nominal depth $H/2$ minus an allowance. This is always the depth at the large end of the tapered keyseat and is indicated on the drawing by the abbreviation LE.

The radii of fillets, when required, must be dimensioned on the drawing.

Since standard milling cutters for Woodruff keys have the same appropriate number, it is possible to call for a Woodruff keyseat by the number only. Where it is desirable to detail Woodruff keyseats on a drawing, all dimensions are given in the form of a note in the following order: width, depth, and radius of cutter. Woodruff keyseats may alternately be dimensioned in the same manner as for square and flat keys, specifying first the width and then the depth (Fig. 11-1-4).

Splines and Serrations

A *splined shaft* is a shaft having multiple grooves, or keyseats, cut around its circumference for a portion of its length, in order that a sliding engagement may be made with corresponding internal grooves of a mating part.

Splines are capable of carrying heavier loads than keys, permit lateral movement of a part while maintaining positive rotation, and allow the attached part to be indexed or changed to another angular position.

Splines have either straight-sided teeth or curved-sided teeth. The latter type is known as an *involute spline*.

Involute Splines   The splines are similar in shape to involute gear teeth but have pressure angles of 30, 37.5 or 45°. There are two types of fits, the *side fit* and the *major-diameter fit* (Fig. 11-1-5).

Straight-Side Splines   The most popular are the SAE straight-side splines, as shown in Fig. 11-1-6. They have been

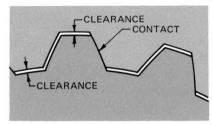

(A) SIDE FIT

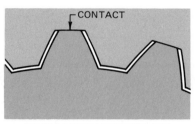

(B) MAJOR DIAMETER FIT

FIG. 11-1-5   Involute splines.

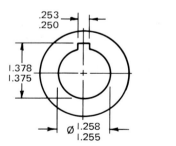

FIG. 11-1-3   Dimensioning keyseats.

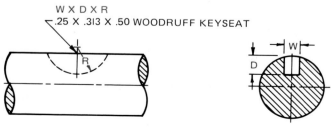

W X D X R
.25 X .313 X .50 WOODRUFF KEYSEAT

FIG. 11-1-4   Alternate method of detailing a Woodruff keyseat.

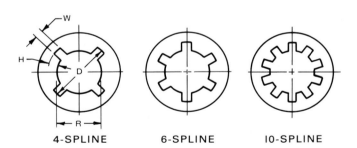

4-SPLINE   6-SPLINE   10-SPLINE

NUMBER OF SPLINES	W FOR ALL FITS	PERMANENT FIT		TO SLIDE WITHOUT LOAD		TO SLIDE UNDER LOAD	
		H	R	H	R	H	R
4	0.241 D	0.075 D	0.85 D	0.125 D	0.75 D		
6	0.250 D	0.050 D	0.90 D	0.075 D	0.85 D	0.100 D	0.80 D
10	0.156 D	0.045 D	0.91 D	0.070 D	0.86 D	0.095 D	0.81 D
16	0.098 D	0.045 D	0.91 D	0.070 D	0.86 D	0.095 D	0.81 D

FIG. 11-1-6   Sizes of SAE parallel-side splines.

used in many applications in the automotive and machine industries.

Serrations   *Serrations* are shallow, involute splines with 45° pressure angles. They are primarily used for holding parts, such as plastic knobs, on steel shafts.

Drawing Data

It is essential that a uniform system of drawing and specifying splines and serrations be used on drawings. The conventional method of showing and calling out splines on a drawing is shown in Fig. 11-1-7. Distance *L* does not include the cutter

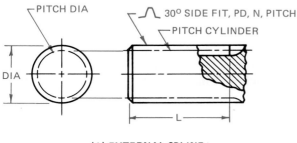

(A) EXTERNAL SPLINE

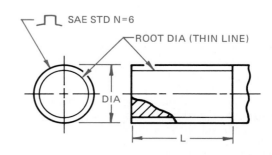

(A) EXTERNAL SPLINE

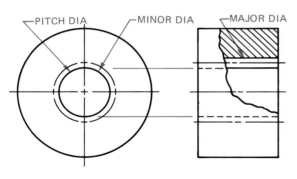

(B) INTERNAL SPLINE

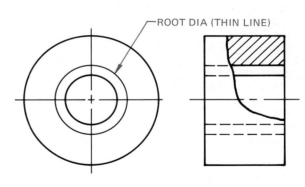

(B) INTERNAL SPLINE

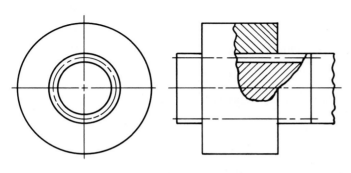

(C) ASSEMBLY DRAWING

INVOLUTE SPLINES

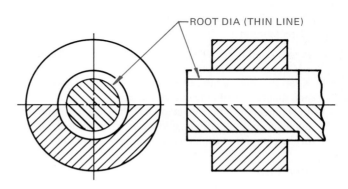

(C) ASSEMBLY DRAWING

STRAIGHT-SIDED SPLINES

FIG. 11-1-7   Callout and representation of splines.

HARDENED AND GROUND DOWEL PIN	TAPER PIN	CLEVIS PIN	COTTER PIN
Standardized in nominal diameters ranging from .12 to .88 (3 to 22mm). 1. Holding laminated sections together with surfaces either drawn up tight or separated in some fixed relationship. 2. Fastening machine parts where accuracy of alignment is a primary consideration. 3. Locking components on shafts, in the form of transverse pin key.	Standard pins have a taper of 1:48 measured on the diameter. Basic dimension is the diameter of the large end. Used for light-duty service in the attachment of wheels, levers, and similar components to shafts. Torque capacity is determined on the basis of double shear, using the average diameter along the tapered section in the shaft for area calculations.	Standard nominal diameters for clevis pins range from .19 to 1.00 (5 to 25mm). Basic function of the clevis pin is to connect mating yoke, or fork, and eye members in knuckle-joint assemblies. Held in place by a small cotter pin or other fastening means, it provides a mobile joint construction, which can be readily disconnected for adjustment or maintenance.	Sizes have been standardized in nominal diameters ranging from .03 to .75 (1 to 20mm). Locking device for other fasteners. Used with a castle or slotted nut on bolts, screws, or studs, it provides a convenient, low-cost locknut assembly. Hold standard clevis pins in place. Can be used with or without a plain washer as an artificial shoulder to lock parts in position on shafts.

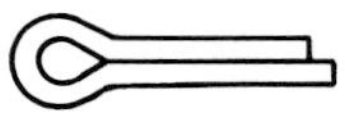

FIG. 11-2-1   Machine pins.

runout. The drawing callout shows the symbol indicating the type of spline followed by the type of fit, the pitch diameter, number of teeth and pitch for involute splines, and number of teeth and outside diameter for straight-sided teeth.

ASSIGNMENTS

See Assignments 1 and 2 for Unit 11-1 on pages 315–316.

11-2 PIN FASTENERS

Pin fasteners are an inexpensive and effective method of assembly where loading is primarily in shear. They can be separated into two groups: semipermanent and quick-release.

Semipermanent Pins

Semipermanent pin fasteners require application of pressure or the aid of tools for installation or removal. The two basic types are machine pins and radial locking pins.

The following general design rules apply to all types of semipermanent pins:

- Avoid conditions where the direction of vibration parallels the axis of the pin.
- Keep the shear plane of the pin a minimum distance of one diameter from the end of the pin.
- In applications where engaged length is at a minimum and appearance is not critical, allow pins to protrude the length of the chamfer at each end for maximum locking effect.

Machine Pins

Four types are generally considered to be most important: hardened and ground dowel pins and commercial straight pins, taper pins, clevis pins, and standard cotter pins. Descriptive data and recommended assembly practices for these four traditional types of machine pins are presented in Fig. 11-2-1. For proper size selection of cotter pins, refer to Fig. 11-2-2.

Radial Locking Pins

Two basic pin forms are employed: solid with grooved surfaces and hollow spring pins, which may be either slotted or spiral-wrapped.

Grooved Straight Pins   Locking action of the grooved pin is provided by parallel, longitudinal grooves uniformly spaced around the pin surface. Rolled or pressed into solid pin stock, the grooves expand the effective pin diameter. When the pin is driven into a drilled hole corresponding in size to nominal pin diameter, elastic deformation of the raised groove edges produces a secure force-fit with the hole wall. Figure 11-2-3 shows

NOMINAL THREAD SIZE	NOMINAL COTTER PIN SIZE	COTTER PIN HOLE	END CLEARANCE*
.250 (6)	.062 (1.5)	.078 (1.9)	.11 (3)
.312 (8)	.078 (2)	.094 (2.4)	.11 (3)
.375 (10)	.094 (2.5)	.109 (2.8)	.14 (4)
.500 (12)	.125 (3)	.141 (3.4)	.17 (6)
.625 (14)	.156 (3)	.172 (3.4)	.23 (5)
.750 (20)	.156 (4)	.172 (4.5)	.27 (7)
1.000 (24)	.188 (5)	.203 (5.6)	.31 (8)
1.125 (27)	.188 (5)	.203 (5.6)	.39 (8)
1.250 (30)	.219 (6)	.234 (6.3)	.41 (10)
1.375 (36)	.219 (6)	.234 (6.3)	.44 (11)
1.500 (42)	.250 (6)	.266 (6.3)	.48 (12)
1.750 (48)	.312 (8)	.328 (8.5)	.55 (14)

*DISTANCE FROM EXTREME POINT OF BOLT OR SCREW TO
   CENTER OF COTTER PIN HOLE.   Inch (mm)

FIG. 11-2-2   Recommended cotter pin sizes.

FIG. 11-2-3   Grooved radial locking pins.

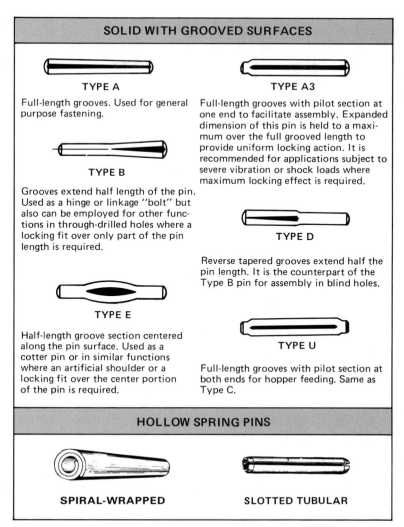

SOLID WITH GROOVED SURFACES

TYPE A

Full-length grooves. Used for general purpose fastening.

TYPE B

Grooves extend half length of the pin. Used as a hinge or linkage "bolt" but also can be employed for other functions in through-drilled holes where a locking fit over only part of the pin length is required.

TYPE E

Half-length groove section centered along the pin surface. Used as a cotter pin or in similar functions where an artificial shoulder or a locking fit over the center portion of the pin is required.

TYPE A3

Full-length grooves with pilot section at one end to facilitate assembly. Expanded dimension of this pin is held to a maximum over the full grooved length to provide uniform locking action. It is recommended for applications subject to severe vibration or shock loads where maximum locking effect is required.

TYPE D

Reverse tapered grooves extend half the pin length. It is the counterpart of the Type B pin for assembly in blind holes.

TYPE U

Full-length grooves with pilot section at both ends for hopper feeding. Same as Type C.

HOLLOW SPRING PINS

SPIRAL-WRAPPED

SLOTTED TUBULAR

six of the grooved-pin constructions that have been standardized. For typical grooved pin applications and size selection, refer to Figs. 11-2-4 and 11-2-5 (pg. 298).

Hollow Spring Pins   Resilience of hollow cylinder walls under radial compression forces is the principle of spiral-wrapped and slotted tubular pins (Fig. 11-2-3). Both pin forms are made to controlled diameters greater than the holes into which they are pressed. Compressed when driven into the hole, the pins exert spring pressure against the hole wall along their entire engaged length to develop locking action.

Standard slotted tubular pins are designed so that several sizes can be used inside one another. In such combinations, shear strengths of the individual pins are additive. For spring pin applications, refer to Fig. 11-2-6 (pg. 299).

Quick-Release Pins

Commercially available quick-release pins vary widely in head styles, types of locking and release mechanisms, and range of pin lengths (Fig. 11-2-7, pg. 299).

Quick-release pins may be divided into two basic types: push-pull and positive-locking pins. The positive-locking pins can be further divided into three categories: heavy-duty cotter pins, single-acting pins, and double-acting pins.

Push-Pull Pins

These pins are made with either a solid or a hollow shank, containing a detent assembly in the form of a locking lug, button, or ball, backed up by some type of resilient core, plug, or spring. The detent member projects from the surface of the pin body until sufficient force is applied in assembly or removal to cause it to retract against the spring action of the resilient core and release the pin for movement.

Positive-Locking Pins

For some quick-release fasteners, the locking action is independent of insertion and removal forces. As in the case of push-pull pins, these pins are primarily suited for shear-load applications. However, some degree of tension loading usually can be tolerated without affecting the pin function.

| SHAFT DIA | | TRANSVERSE KEY | | | LONGITUDINAL KEY | |
| | | PIN DIA | | TAPER PIN | PIN DIA | |
IN.	(MM)	IN.	(MM)	NO.	IN.	(MM)
.375	(10)	.125	(3)	3/0	.094	(2.5)
.438	(12)	.156	(4)	0	.125	(3)
.500	(14)	.156	(5)	0	.125	(4)
.562	(16)	.188	(5)	2	.156	(4)
.625	(18)	.188	(6)	2	.156	(5)
.750	(20)	.250	(6)	4	.156	(5)
.875	(22)	.250	(6)	4	.219	(6)
1.000	(24)	.312	(8)	6	.250	(6)
1.062	(26)	.312	(8)	6	—	—
1.125	(28)	.375	(10)	7	—	—
1.188	(30)	.375	(10)	7	—	—
1.250	(32)	.375	(10)	7	.312	(8)
1.375	(34)	.438	(11)	7	.375	(10)
1.438	(36)	.438	(11)	7	—	—
1.500	(38)	.500	(12)	8	.438	(11)

FIG. 11-2-4   Recommended groove pin sizes.

REFERENCES AND SOURCE MATERIAL

1. *Machine Design*, Fastening and joining reference issue.

ASSIGNMENTS

See Assignments 3 through 5 for Unit 11-2 on pages 316 through 318.

11-3   RETAINING RINGS

Retaining rings, or *snap rings*, are used to provide a removable shoulder to accurately locate, retain, or lock components on shafts and in bores of housings (see Fig. 11-3-1, pg. 300). They are easily installed and removed, and since they are usually made of spring steel, retaining rings have a high shear strength and impact capacity. In addition to fastening and positioning, a number of rings are designed for taking up end play caused by accumulated tolerances or wear in the parts being retained. In general, these devices can be placed into three categories, which describe the type and method of fabrication: stamped retaining rings, wire-formed rings, and spiral-wound retaining rings.

Stamped Retaining Rings

Stamped retaining rings, in contrast to wire-formed rings with their uniform cross-sectional area, have a tapered radial width

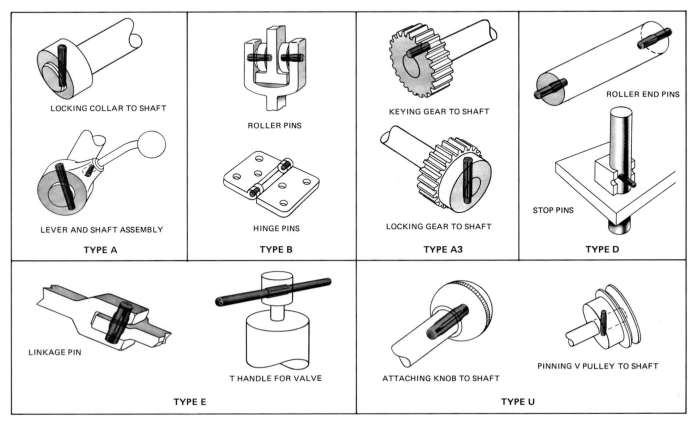

FIG. 11-2-5   Groove pin applications.

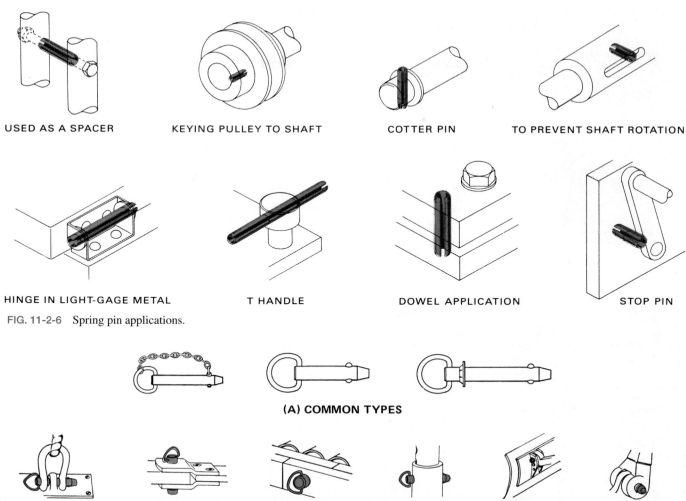

USED AS A SPACER   KEYING PULLEY TO SHAFT   COTTER PIN   TO PREVENT SHAFT ROTATION

HINGE IN LIGHT-GAGE METAL   T HANDLE   DOWEL APPLICATION   STOP PIN

FIG. 11-2-6   Spring pin applications.

(A) COMMON TYPES

CLEVIS-SHACKLE PIN   DRAW-BAR HITCH PIN   RIGID COUPLING PIN   TUBING LOCKPIN   ADJUSTMENT PIN   SWIVEL HINGE PIN

(B) APPLICATIONS

FIG. 11-2-7   Quick-release pins.

that decreases symmetrically from the center section to the free ends. The tapered construction permits the rings to remain circular when they are expanded for assembly over a shaft or contracted for insertion into a bore or housing. This constant circularity ensures maximum contact surface with the bottom of the groove.

Stamped retaining rings can be classified into three groups: axially assembled rings, radially assembled rings, and self-locking rings which do not require grooves. *Axially assembled rings* slip over the ends of shafts or down into bores, while *radially assembled rings* have side openings that permit the rings to be snapped directly into grooves on a shaft.

Commonly used types of stamped retaining rings are illustrated and compared in Fig. 11-3-2 (pg. 300).

Wire-Formed Retaining Rings

The *wire-formed retaining ring* is a split ring formed and cut from spring wire of uniform cross-sectional size and shape. The wire is cold-drawn or rolled into shape from a continuous

coil or bar. Then the gap ends are cut into various configurations for ease of application and removal.

Rings are available in many cross-sectional shapes, but the most commonly used are the rectangular and circular cross sections.

Spiral-Wound Retaining Rings

Spiral-wound retaining rings consist of two or more turns of rectangular material, wound on edge to provide continuous crimped or uncrimped coil.

REFERENCES AND SOURCE MATERIAL

1. *Machine Design*, Fastening and joining reference issue.

ASSIGNMENTS ▐▊▊▊▊▊▊▊▊▊▊▊▊▊▊▊▊▊▊▊▊▊▊▊▊▊▊▊

See Assignments 6 through 8 for Unit 11-3 on pages 318–319.

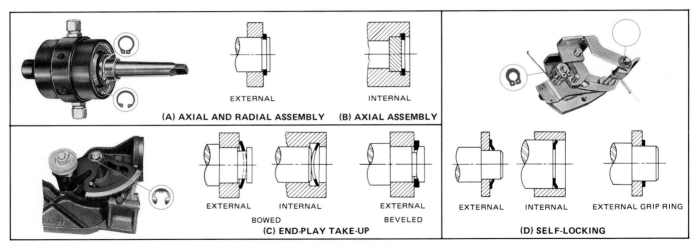

FIG. 11-3-1   Retaining ring applications.

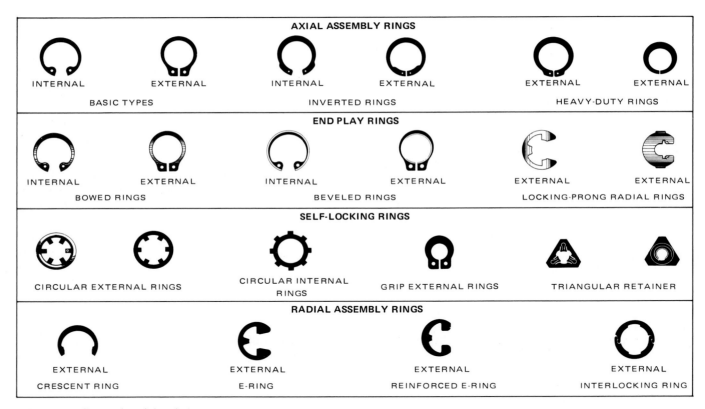

FIG. 11-3-2   Stamped retaining rings.

11-4   SPRINGS

Springs may be classified into three general groups according to their application.

Controlled Action Springs   Controlled action springs have a well-defined function, or a constant range of action for each cycle of operation. Examples are valve, die, and switch springs.

Variable-Action Springs   Variable-action springs have a changing range of action because of the variable conditions imposed upon them. Examples are suspension, clutch, and cushion springs.

Static Springs   Static springs exert a comparatively constant pressure or tension between parts. Examples are packing or bearing pressure, antirattle, and seal springs.

Types of Springs

The type or name of a spring is determined by characteristics such as function, shape of material, application, or design.

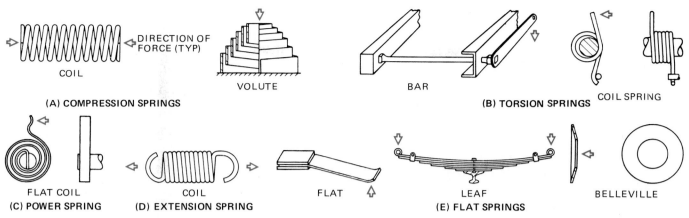

FIG. 11-4-1   Types of springs.

Figure 11-4-1 illustrates common springs in use. Figure 11-4-2 designates spring nomenclature.

Compression Springs

A *compression spring* is an open-coiled helical spring that offers resistance to a compressive force.

Compression Spring Ends   Figure 11-4-3A (pg. 302) shows the ends commonly used on compression springs.

Plain open ends are produced by straight cutoff with no reduction of helix angle (Fig. 11-4-4, pg. 302). The spring should be guided on a rod or in a hole to operate satisfactorily.

Ground open ends are produced by parallel grinding of open-end coil springs. Advantages of this type of end are improved stability and a larger number of total coils.

Plain closed ends are produced with a straight cutoff and with reduction of helix angle to obtain closed-end coils, resulting in a more stable spring.

Ground closed ends are produced by parallel grinding of closed-end coil springs, resulting in maximum stability.

Extension Springs

An *extension spring* is a close-coiled, helical spring that offers resistance to a pulling force. It is made from round or square wire.

Extension Spring Ends   The end of an extension spring is usually the most highly stressed part. Thus, proper consideration should be given to its selection. The types of ends shown in Fig. 11-4-3B are most commonly used on extension springs. Different types of ends can be used on the same spring.

Torsion Springs

Springs exerting pressure along a path that is a circular arc, or in other words, providing a torque, are called torsion springs, motor springs, power springs, etc. The term *torsion spring* is usually applied to a helical spring of round, square, or rectangular wire, loaded by torque.

The variation in ends used is almost limitless, but a few of the more common types are illustrated in Fig. 11-4-3C.

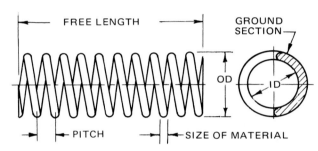

FIG. 11-4-2   Spring nomenclature.

Torsion Bar Springs   A *torsion bar spring* is a relatively straight bar anchored at one end, on which a torque may be exerted at the other end, thus tending to twist it about its axis.

Power Springs

Clock or Motor Type   A *flat coil spring*, also known as a clock or motor spring, consists of a strip of tempered steel wound on an arbor and usually confined in a case or drum.

Flat Springs

Flat springs are made of flat material formed in such a manner as to apply force in the desired direction when deflected in the opposite direction.

Leaf Springs   A *leaf spring* is composed of a series of flat springs nested together and arranged to provide approximately uniform distribution of stress throughout its length. Springs may be used in multiple arrangements, as shown in Fig. 11-4-5A (pg. 302).

Belleville Springs   *Belleville springs* are washer-shaped, made in the form of a short, truncated cone.

Belleville washers may be assembled in series to accommodate greater deflections, in parallel to resist greater forces, or in combination of series and parallel, as shown in Fig. 11-4-5B.

Spring Drawings

On working drawings, a schematic drawing of a helical spring is recommended to save drafting time (Fig. 11-4-6, pg. 303).

FIG. 11-4-3   End styles for helical springs.

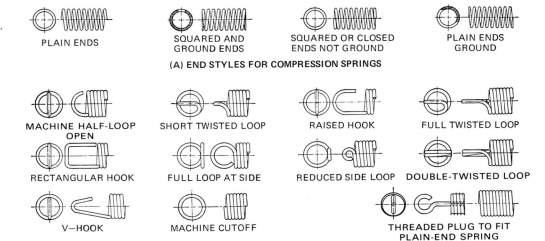

PLAIN ENDS   SQUARED AND GROUND ENDS   SQUARED OR CLOSED ENDS NOT GROUND   PLAIN ENDS GROUND

(A) END STYLES FOR COMPRESSION SPRINGS

MACHINE HALF-LOOP OPEN   SHORT TWISTED LOOP   RAISED HOOK   FULL TWISTED LOOP

RECTANGULAR HOOK   FULL LOOP AT SIDE   REDUCED SIDE LOOP   DOUBLE-TWISTED LOOP

V-HOOK   MACHINE CUTOFF   THREADED PLUG TO FIT PLAIN-END SPRING

(B) END STYLES FOR EXTENSION SPRINGS

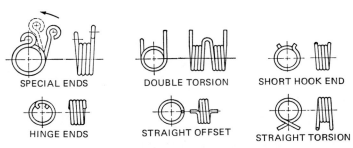

SPECIAL ENDS   DOUBLE TORSION   SHORT HOOK END

HINGE ENDS   STRAIGHT OFFSET   STRAIGHT TORSION

(C) END STYLES FOR TORSION SPRINGS

FIG. 11-4-4   Coil definitions.

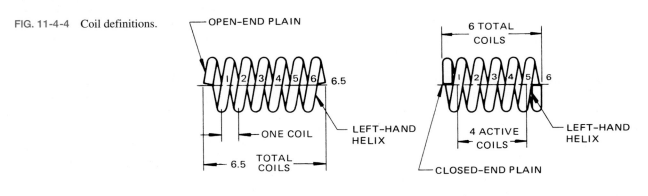

OPEN-END PLAIN

6.5

ONE COIL   LEFT-HAND HELIX

6.5   TOTAL COILS

6 TOTAL COILS

6   LEFT-HAND HELIX

4 ACTIVE COILS

CLOSED-END PLAIN

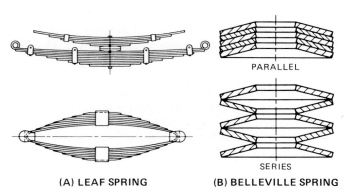

(A) LEAF SPRING   (B) BELLEVILLE SPRING

PARALLEL

SERIES

FIG. 11-4-5   Spring arrangements.

As in screw-thread representation, straight lines are used in place of the helical curves. On assembly drawings, springs are normally shown in section, and either cross-hatching lines or solid black shading is recommended, depending on the size of the wire's diameter (Fig. 11-4-7).

Dimensioning Springs

The following information should be given on a drawing of a spring:

- Size, shape, and kind of material used in the spring
- Diameter (outside or inside)
- Pitch or number of coils

FIG. 11-4-6   Schematic drawing of springs.

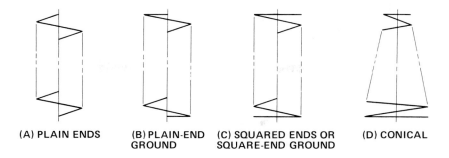

(A) PLAIN ENDS   (B) PLAIN-END GROUND   (C) SQUARED ENDS OR SQUARE-END GROUND   (D) CONICAL

FIG. 11-4-7   Showing helical springs on assembly drawings.

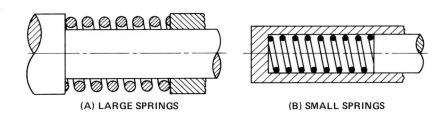

(A) LARGE SPRINGS   (B) SMALL SPRINGS

- Shape of ends
- Length
- Load and rate (not covered in this text)

EXAMPLE

ONE HELICAL TENSION SPRING 3.00 LG (OR NUMBER OF COILS), .50 ID, PITCH .25, 18 B & S GA SPRING BRASS WIRE

In using single-line representation, the dimensions should state the applicable size of material required to ensure correct interpretation on such features as inside diameter, outside diameter, and end loops (Fig. 11-4-8).

Spring Clips

Spring clips are a relatively new class of industrial fasteners. They perform multiple functions, eliminate the handling of several small parts, and thus reduce assembly costs (Fig. 11-4-9, pg. 304).

The *spring clip* is generally self-retaining, requiring only a flange, panel edge, or mounting hole to clip to. Basically, spring clips are light-duty fasteners and serve the same function as small bolts and nuts, self-tapping screws, clamps, spot welding, and formed retaining plates.

Dart-type Spring Clips   Dart-shaped panel retaining elements have hips to engage within panel or component holes. The top of arms of the fastener can be formed in any shape to perform unlimited fastening functions.

Stud Receiver Clips   There are three basic types of stud receivers: push-ons, tubular types, and self-threading fasteners. All are designed to make attachments to unthreaded studs, rivets, pins, or rods of metal or plastic.

Cable, Wire, and Tube Clips   These fasteners incorporate self-retaining elements for engaging panel holes or mounting on panel edges and flanges.

Spring-clip cable, wire, and tubing fasteners are front-mounting devices, requiring no access to the back of the panel.

Spring Molding Clips   Molding retaining clips are formed with legs that hold the clips to a panel and arms that positively engage the flanges of various sizes and shapes of trim molding and pull the molding tightly to the attaching panel.

U-, S-, and C-Shaped Spring Clips   These spring clips get their names from their shapes. The fastening function is accomplished by using inward compressive spring force to secure assembly components or provide self-retention after installation.

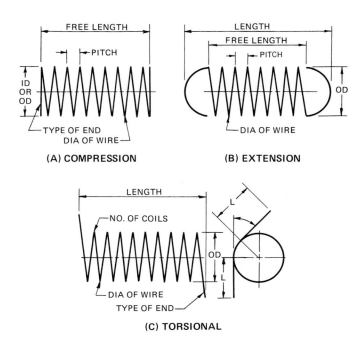

(A) COMPRESSION   (B) EXTENSION

(C) TORSIONAL

FIG. 11-4-8   Dimensioning springs.

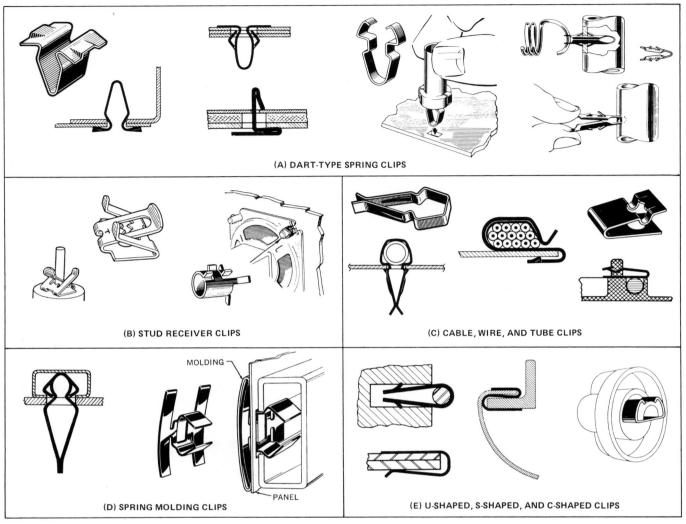

(A) DART-TYPE SPRING CLIPS

(B) STUD RECEIVER CLIPS

(C) CABLE, WIRE, AND TUBE CLIPS

MOLDING

(D) SPRING MOLDING CLIPS

PANEL

(E) U-SHAPED, S-SHAPED, AND C-SHAPED CLIPS

FIG. 11-4-9   Spring clips.

REFERENCES AND SOURCE MATERIAL

1. General Motors Corp.
2. The Wallace Barnes Co. Ltd.
3. *Machine Design*, Fastening and joining reference issue.

ASSIGNMENTS ▓▓▓▓▓▓▓▓▓▓▓▓▓▓▓▓▓▓

See Assignments 9 through 11 for Unit 11-4 on pages 320 to 321.

11-5   RIVETS

Standard Rivets

Riveting is a popular method of fastening and joining, primarily because of its simplicity, dependability, and low cost.

A myriad of manufactured products and structures, both small and large, are held together by these fasteners. Rivets are classified as permanent fastenings, as distinguished from removable fasteners, such as bolts and screws.

Basically, a *rivet* is a ductile metal pin that is inserted through holes in two or more parts, and having the ends formed over to securely hold the parts.

Another important reason for riveting is versatility, with respect to both the properties of rivets as fasteners and the method of clinching.

- *Part materials:* Rivets can be used to join dissimilar materials, metallic or nonmetallic, in various thicknesses.
- *Multiple functions:* Rivets can serve as fasteners, pivot shafts, spacers, electric contacts, stops, or inserts.
- *Fastening finished parts:* Rivets can be used to fasten parts that have already received a final painting or other finishing.

Riveted joints are neither watertight nor airtight, although such a joint may be attained at some added cost by using a sealing compound. The riveted parts cannot be disassembled for maintenance or replacement without knocking the rivet out and clinching a new one in place for reassembly. Common riveted joints are shown in Fig. 11-5-1.

Large Rivets

Large rivets are used in structural work of buildings and bridges. Today, however, high-strength bolts have almost completely replaced rivets in field connections because of cost, strength, and the noise factor. Rivet joints are of two types: butt and lapped. The more common types of large rivets are shown in Fig. 11-5-2. In order to show the difference between *shop rivets* (rivets that are put in the structure at the shop) and *field*

rivets (rivets that are used on the site), two types of symbols are used. In drawing shop rivets, the diameter of the rivet head is shown on the drawings. For field rivets, the shaft diameter is used. Figure 11-5-3 shows the conventional rivet symbols adopted by the American and Canadian Institutes of Steel Construction.

Rivets for Aerospace Equipment

The following representation of rivets on drawings for aerospace equipment is also recommended for other fields of work involving rivets.

The symbolic representation for a set (installed) rivet consists of a cross indicating its position. This representation is

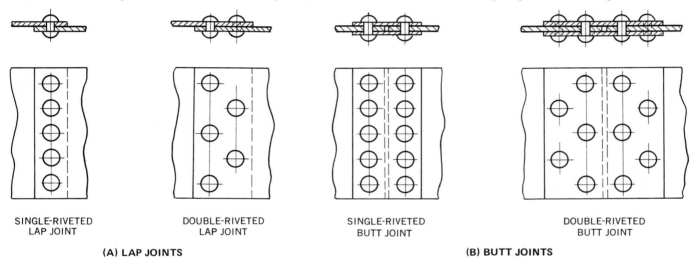

FIG. 11-5-1 Common riveted joints.

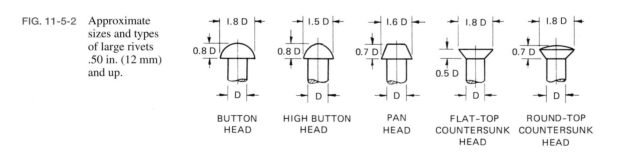

FIG. 11-5-2 Approximate sizes and types of large rivets .50 in. (12 mm) and up.

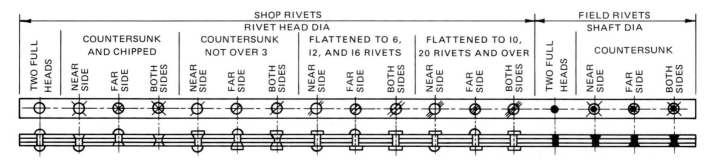

FIG. 11-5-3 Conventional rivet symbols.

supplemented by the relevant information regarding rivet and rivet assembly (Fig. 11-5-4).

The upper left-hand quadrant of the symbol shows the part number for the rivet used in the item list on the drawing or in a table on the drawing that clearly defines the part. This number is preceded by the capital letter R (Fig. 11-5-4B). Where a composite rivet is used (rivet plus sleeve), the item reference numbers for both rivet and sleeve are shown (Fig. 11-5-4C).

SYMBOLIC REPRESENTATION	DESCRIPTION AND MEANING
A	POSITION OF RIVET
B R 17	SOLID RIVET R 17 = RIVET, ITEM REFERENCE 17 SHOWN ON ITEM LIST OR TABLE ON THE DRAWING 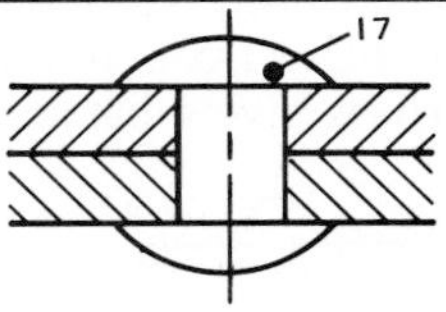
C R 32 35	COMPOSITE RIVET R 32 = RIVET, ITEM REFERENCE 32 SHOWN ON ITEM LIST OR TABLE ON THE DRAWING 35 = SLEEVE, ITEM REFERENCE 35 SHOWN ON ITEM LIST OR TABLE ON THE DRAWING 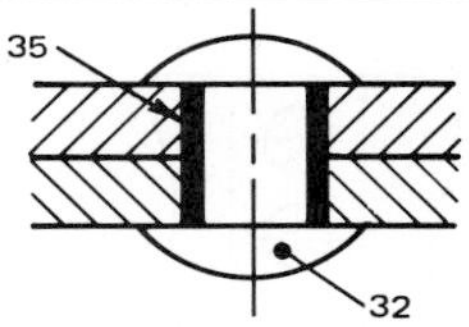
D N OR F	N = PREFORMED HEAD OF THE RIVET ON NEAR SIDE F = PREFORMED HEAD OF THE RIVET ON FAR SIDE
E	100° COUNTERSINK ON NEAR SIDE 2 ▽ 82 82° COUNTERSINK ON FAR SIDE 100° COUNTERSINK ON BOTH SIDES
F	100° DIMPLING ON NEAR SIDE 2 ∧ 82 TWO SHEETS 82° DIMPLED ON FAR SIDE
G V△	{ FIRST SHEET DIMPLED 100° ON NEAR SIDE { SECOND SHEET COUNTERSINK 100° ON FAR SIDE V 82 △ 82° { FIRST SHEET DIMPLED 82° ON NEAR SIDE { SECOND SHEET DIMPLED 82° ON FAR SIDE

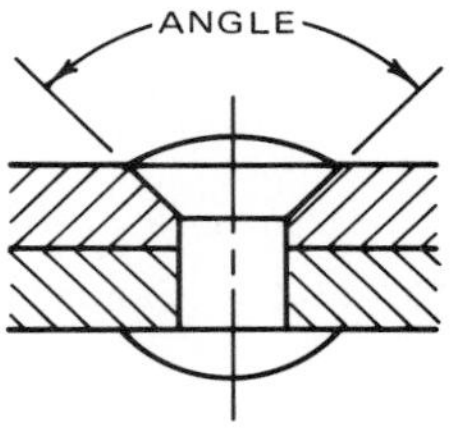

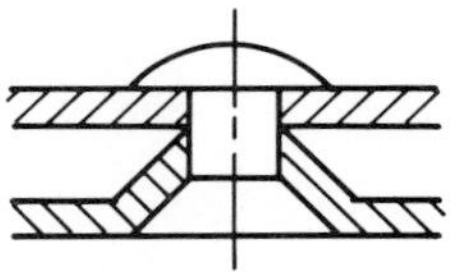

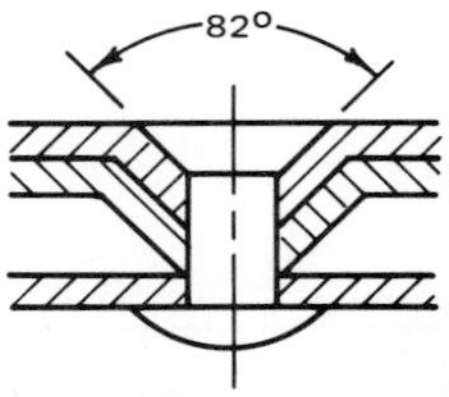

FIG. 11-5-4   Symbolic representation for a set (installed) rivet used on aerospace equipment.

The upper right-hand quadrant of the symbol contains a capital letter giving the position of the preformed head (Fig. 11-5-4D).

The lower left-hand quadrant of the symbol contains information on the position of either a countersink or a dimpling, or a combination of both. The countersink to be made on the parts to be riveted is indicated by an equilateral triangle oriented to indicate either near or far side (Fig. 11-5-4E). If other than 100°, the value of the angle in degrees is placed on the right of the countersink symbol. Where dimpling of the sheets to be riveted is required, it is indicated by an open isosceles triangle oriented to indicate either near or far side (Fig. 11-5-4F). If other than 100°, the value of the angle in degrees is placed on the right of the dimpling symbol. Where the combination of a countersink on one part and a dimpling on the other part is required, it is indicated by showing both the countersink and dimpling symbols. If other than 100°, the value of the angle in degrees is placed to the right of the countersink and dimpling symbol (Fig. 11-5-4G). The lower right-hand quadrant of the symbol is left blank.

Symbolic Representation of a Line of Rivets   The crosses (symbol representing the fixed rivet) are aligned along the axes of the drawing, and the number of places for rivets are indicated.

The supplemental information is placed directly on the drawing if space is available or with a leader line indicating the corresponding rivet assembly (Fig. 11-5-5A).

When the rivets are aligned, identical, and equidistant, the symbols should be shown in the first and last positions, together with the total number of pitches and distance (Fig. 11-5-5B).

Small Rivets

Design of small rivet assemblies is influenced by two major considerations:

1. The joint itself, its strength, appearance, and configuration
2. The final riveting operation, in terms of equipment capabilities and production sequence

Types of Small Rivets

Four types of small rivets are illustrated in Fig. 11-5-6 and described as follows.

Semitubular   This is the most widely used type of small rivet. The depth of the hole in the rivet, measured along the wall, does not exceed 112 percent of the mean shank diameter.

FIG. 11-5-5   Drawing callout for rivets used on aerospace equipment.

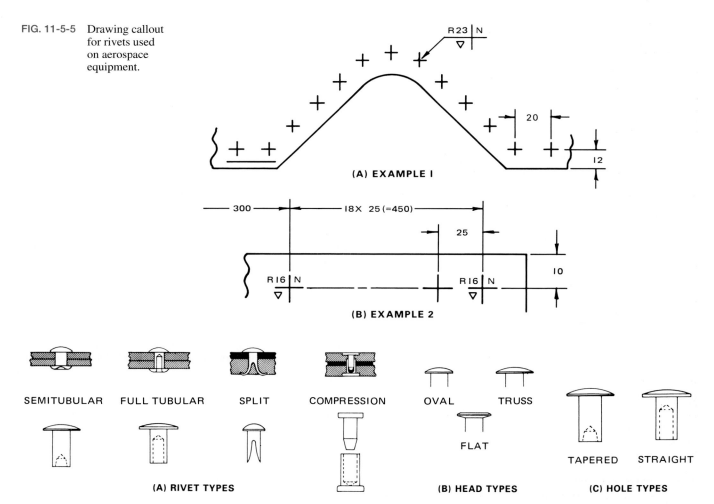

(A) EXAMPLE I

(B) EXAMPLE 2

SEMITUBULAR   FULL TUBULAR   SPLIT   COMPRESSION   OVAL   TRUSS   FLAT   TAPERED   STRAIGHT

(A) RIVET TYPES

(B) HEAD TYPES

(C) HOLE TYPES

FIG. 11-5-6   Basic types of small rivets.

The hole may be extruded (straight or tapered) or drilled (straight), depending on the manufacturer and/or rivet size.

Full Tubular   This rivet has a drilled shank with a hole depth more than 112 percent of the mean shank diameter. It can be used to punch its own hole in fabric, some plastic sheets, and other soft materials, eliminating a preliminary punching or drilling operation.

Bifurcated (Split)   The rivet body is sawed or punched to produce a pronged shank that punches its own hole through fiber, wood, or plastic.

Compression   This rivet consists of two elements: the solid or blank rivet and the deep-drilled tubular member. Pressed together, these form an interference fit.

Design Recommendations

Figure 11-5-7 shows the preferred methods of drawing small-rivet connections for various types of joints, materials, clearances, and so forth. The following are considerations that should be taken into account when small rivets are to be used as fasteners.

Select the Right Rivets   Basic types are covered in Fig. 11-5-6. Rivet standards for all types but compression rivets have been published by the Tubular and Split Rivet Council.

Rivet Diameters   The optimum rivet diameter is determined, not by performance requirements, but by economics—the costs of the rivet and the labor to install it. The rivet length-to-diameter ratio should not exceed 6:1.

Rivet Positioning   The location of the rivet in the assembled product influences both joint strength and clinching requirements. The important dimensions are edge distance and pitch distance.
   Edge distance is the interval between the edge of the part and the center line of the rivet. The recommended edge distance for plastic materials, either solid or laminated, is between two and three diameters, depending on the thickness and inherent strength of the material.
   Pitch distance—the interval between center lines of adjacent rivets—should not be too small. Unnecessarily high stress concentrations in the riveted material and buckling at adjacent empty holes can result if the pitch distance is less than three times the diameter of the largest rivet in the assembly (metal parts) or five times the diameter (plastic parts).

Blind Rivets

Blind riveting is a technique for setting a rivet without access to the reverse side of the joint. However, blind rivets may also be used in applications in which both sides of the joint are actually accessible.
   Blind rivets are classified according to the methods with which they are set: pull-mandrel, drive-pin, and chemically expanded (Fig. 11-5-8).

Design Considerations

Blind-rivet design data are illustrated in Fig. 11-5-9 (pg. 310).

Type of Rivet   Selection depends on a number of factors, such as speed of assembly, clamping capacity, available sizes, adaptability to the assembly, ease of removal, cost, and structural integrity of the joint.

Joint Design   Factors that must be known include allowable tolerances of rivet length versus assembly thickness, hole clearance, joint configuration, and type of loading.

Speed of Installation   The fastest, most efficient installation is done with power tools—air, hydraulic, or electric. Manual tools, such as special pliers, can be used efficiently with practically no training.

In-Place Costs   Blind rivets often have lower in-place costs than solid rivets or tapping screws.

Loading   A blind-rivet joint is usually in compression or shear.

Material Thickness   Some rivets can be set in materials as thin as .02 in. (0.5 mm). Also, if one component is of compressible material, rivets with extra-large head diameter should be used.

Edge Distance   The average recommended edge distance is twice the diameter of the rivet.

Spacing   Rivet pitch should be three times the diameter of the rivet.

Length   The amount of length needed for clinching action varies greatly. Most rivet manufacturers provide data on grip ranges of their rivets.

Backup Clearance   Full entry of the rivet is essential for tightly clinched joints. Sufficient backup clearance must be provided to accommodate the full length of the unclinched rivet.

Blind Holes or Slots   A useful application of a blind rivet is in fastening members in a blind hole. At A in Fig. 11-5-9, the formed head bears against the side of the hole only. This joint is not as strong as the other two (B and C).

Riveted Joints   Riveted cleat or batten holds a butt joint, A. The simple lap joint, B, must have sufficient material beyond the hole for strength. Excessive material beyond rivet hole C may curl up or vibrate or cause interference problems, depending on the installation.

Flush Joints   Generally, flush joints are made by countersinking one of the sections and using a rivet with a countersunk head, A.

Weatherproof Joints   A hollow-core rivet can be sealed by capping it, A; by plugging it, B; or by using both a cap and a plug, C. To obtain a true seal, however, a gasket or mastic should be used between the sections and perhaps under the rivet head. An ideal solution is to use a closed-end rivet.

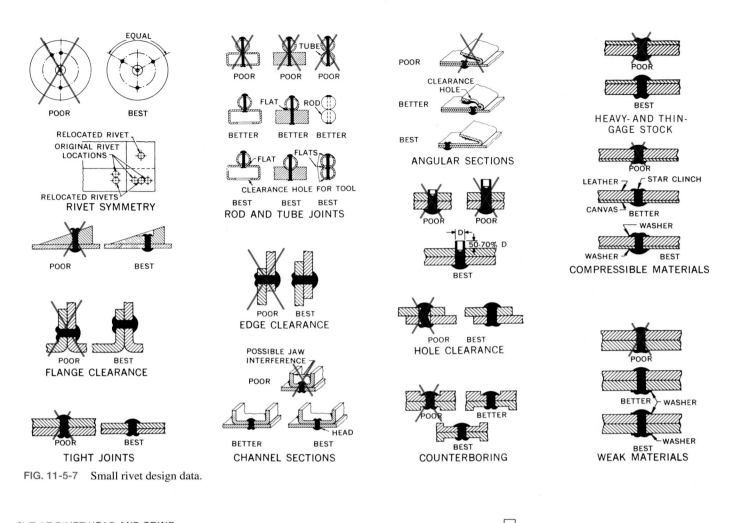

FIG. 11-5-7   Small rivet design data.

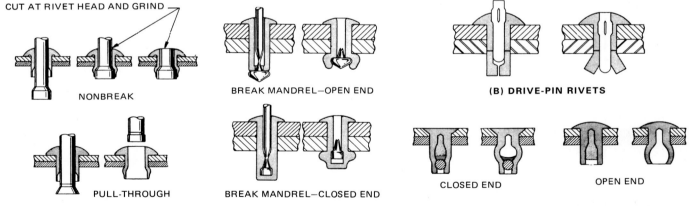

FIG. 11-5-8   Basic types of blind rivets and methods of setting.
(Machine Design, Vol. 53, No. 26*)*

Rubber, Plastic, and Fabric Joints   Some plastics, such as reinforced molded fiberglass or polystyrene, which are reasonably rigid, present no problem for most small rivets. However, when the material is very flexible or is a fabric, set the rivet as shown at A or B, with the upset head against the solid member. If this practice is not possible, use a backup strip as shown at C.

Pivoted Joints   There are a number of ways of producing a pivoted assembly. Three are shown.

Attaching Solid Rod   When attaching a rod to other members, the usual practice is to pass the rivet completely through the rod.

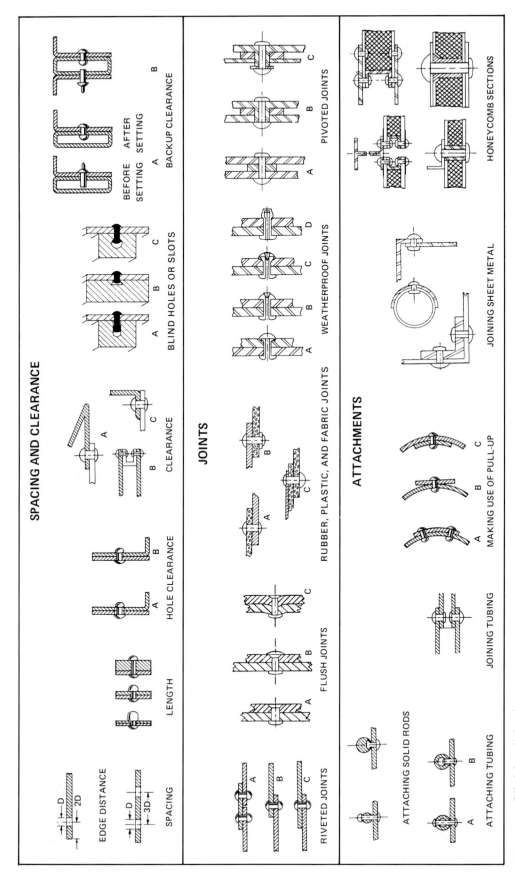

FIG. 11-5-9   Blind rivet design data.

Attaching Tubing   Attaching tubing is an application for which the blind rivet is ideally suited.

Joining Tubing   This tubing joint is a common form of blind riveting, used for both structural and low-cost power transmission assemblies.

Making Use of Pull-up   By judicious positioning of rivets and parts that are to be assembled with rivets, the setting force can sometimes be used to pull together unlike parts.

Honeycomb Sections   Inserts should be employed to strengthen the section and provide a strong joint.

REFERENCES AND SOURCE MATERIAL

1. *Machine Design,* Fastening and joining reference issue.

ASSIGNMENTS

See Assignments 12 and 13 for Unit 11-5 on pages 322–323.

11-6   WELDED FASTENERS

The most common forms of welded fasteners are screws and nuts. In this unit, welded fasteners are grouped into resistance-welded threaded fasteners and arc-welded studs.

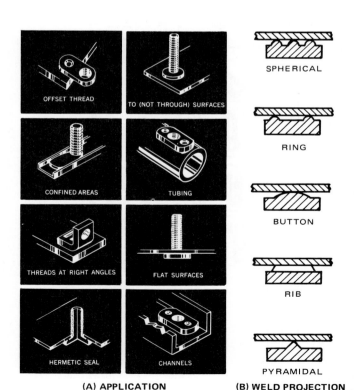

(A) APPLICATION   (B) WELD PROJECTION

FIG. 11-6-1   Resistance-welded fasteners.

Resistance-Welded Fasteners

Simply defined, a *resistance-welded fastener* is an externally or internally threaded metal part designed to be fused permanently in place by standard production welding equipment. Two methods of resistance welding are used to attach these fasteners: projection welding and spot welding.

Design Considerations

Before fasteners can be used, three basic requirements must be met (Figs. 11-6-1 to 11-6-3, here and on pg. 312).

1. The materials to be joined, both part and fastener, must be suitable for resistance welding.
2. The parts to be welded must be portable enough to be carried to the welder.
3. Production volume should be great enough to justify tooling costs.

Figure 11-6-1 shows typical resistance-welded fasteners.

Arc-Welded Studs

There are two basic stud welding processes: electric-arc and capacitor-discharge.

Use Projection-Weld Fasteners When
■ Suitable projection welding equipment is available.
■ Appearance is an important consideration. Projection welding does not mark the surface on the opposite side of the weld.
■ Simultaneous welding of multiple fasteners is required.
■ Spacing between fasteners must be kept close.
■ Fasteners must be welded to part sections of varying thicknesses.
■ Fasteners must be welded to parts of unusual shape or a watertight weld joint is required.
■ Welding fixtures can be used for easier locating or automatic feeding.
■ Length of production run without maintenance is critical.
Use Spot-Weld Fasteners When
■ Suitable rocker-arm welding equipment is available.
■ Appearance of the part surface opposite the weld is not critical. Spot welding leaves a slight indentation from the electrode tips.
■ Other spot welds are being performed on parts of the assembly.
■ Length of production run without maintenance is not too important. Spot-welding electrode tips will mushroom to some extent in production welding. Shorter runs before refacing or redressing must be expected.
■ Dissimilar materials, such as aluminum, copper, or magnesium, are being welded.
■ Shape, size, or space requirements do not permit use of projection-welded fasteners.

FIG. 11-6-2   Guide to weld fastener selection.

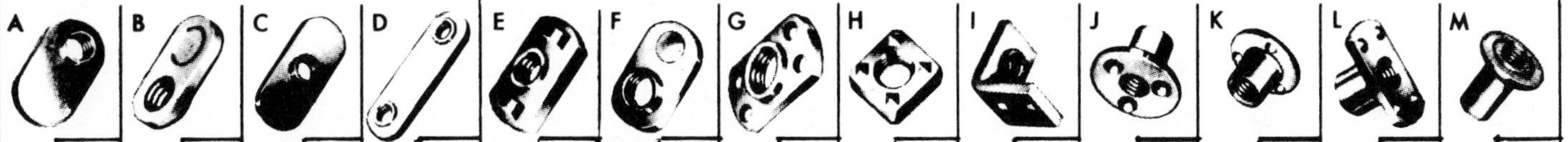

Application Factors

─ Spot-Weld Nuts ─ / ─ Projection-Weld Nuts ─

Application Factors	A	B	C	D	E	F	G	H	I	J	K	L	M
Flat Surfaces	◄	▼	◄	▼	◄	◄	◄	◄	◄	▼	▼	▼	◄
Curved Surfaces (concave)	◄	▼	◄		◄	▼	◄	◄	◄	▼	▼	▼	
Round Surfaces (convex)	◄	▼	◄	▼	◄	▼	◄	▼	▼	▼	▼	▼	
Tubing	◄	▼	◄	▼	▼	▼		◄	▼			◄	
Channels	◄	◄	◄	◄	◄	▼	◄	◄	◄			◄	
Narrow Flanges	◄	◄	◄	◄	◄	▼	◄	◄	◄			◄	
Offset	◄	▼	◄		◄	▼		◄	▼			▼	
Wall Corners	◄	▼	▼	▼	◄	◄		◄	◄				
Blind Hole								◄	◄	◄		◄	◄
Wire	◄	◄	▼	▼			▼		▼				◄
Through Hole	◄	▼	◄	◄	◄	◄	▼	◄	◄	▼	▼	▼	◄
Tension Against Weld	◄	◄	◄		◄	◄	◄	◄	◄		◄	◄	◄
Hermetic Seal							▼		▼				◄
Right Angle	▼	▼		▼		▼			◄				◄
Extra Thread			◄		◄		◄			◄	◄		◄
Bridging				◄	◄				◄			◄	◄
Dual Tapped				◄									
Self-Locating Pilot	◄	◄		▼	◄	◄	◄	◄	◄		◄	◄	◄
No Hole Required in Sheet				▼		▼							
Used with Keyhole Slot													
Pilotless								◄					

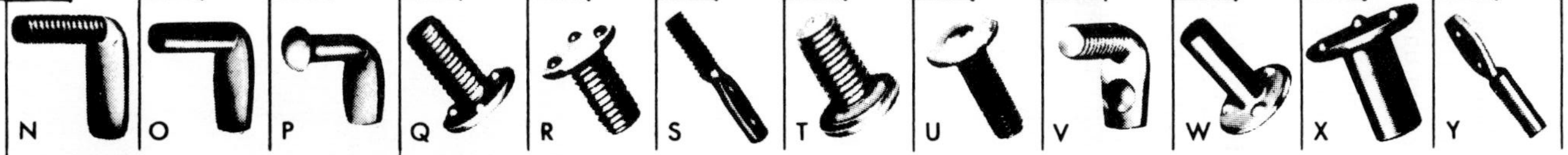

─ Spot-Weld Screws and Pins ─ / ─ Projection-Weld Screws and Pins ─

N	O	P	Q	R	S	T	U	V	W	X	Y

LEGEND

Spot-Weld Screws and Pins
A Single Tab
B Targeted
C Double Tab
D Dual Tapped
E Dual Projection
F Button Projection
G Four-Button Projection
H Pilotless
I Right-Angle Bracket
J Blind-Hole Flange
K Through Hole
L Tee-Shape
M Hermetic Seal
N Right-Angle Spade

Projection-Weld Screws and Pins
O Right-Angle Spade Pin
P Keyhole-Slot, Right-Angle Spade Pin
Q Through Hole
R Blind Hole
S Spade
T Hermetic Seal
U Button-Projection, Blind Hole
V Button, Right-Angle Spade
W Through-Hole Pin
X Blind-Hole Pin
Y Spade Pin

FIG. 11-6-3   Resistance-welded fastener guide.

Electric-Arc Stud Welding   The more widely used stud welding process is a semiautomatic electric-arc process. To avoid burn-through, the plate thickness should be at least one-fifth the weld base diameter.

Capacitor-Discharge Stud Welding   This stud welding process derives its heat from an arc produced by a rapid discharge of stored electrical energy.

Design Considerations

In most instances, the thickness of the plate for stud attachment will determine the stud welding process. Electric-arc stud welding is generally used for fasteners .32 in. (8 mm) and larger.

REFERENCES AND SOURCE MATERIAL

1. *Machine Design,* Fastening and joining reference issue.

ASSIGNMENT

See Assignment 14 for Unit 11-6 on page 324.

11-7   ADHESIVE FASTENINGS

Industrial designers and manufacturers are relying on adhesives more than ever before. They allow greater versatility in design, styling, and materials. They can also cut costs. However, as with any engineering tool, there are limitations as well as advantages. For physical properties and application data of typical adhesives, refer to Table 51 of the Appendix.

Adhesion Versus Stress

Adhesion is the force that holds materials together. *Stress* on the other hand, is the force pulling materials apart (Fig. 11-7-1). The basic types of stress in adhesives are:

1. *Tensile.* Pull is exerted equally over the entire joint. Pull direction is straight and away from the adhesive bond. All adhesive contributes to bond strength.

2. *Shear.* Pull direction is across the adhesive bond. The bonded materials are being forced to slide over one another.
3. *Clearance.* Pull is concentrated at one edge of the joint and exerts a prying force on the bond. The other edge of the joint is theoretically under zero stress.
4. *Peel.* One surface must be flexible. Stress is concentrated along a thin line at the edge of the bond.

Resistance to stress is one reason for the rapid increase in the use of adhesives for product assembly. The following points elaborate on stress resistance and the other advantages of adhesives.

Advantages

1. Adhesives allow uniform distribution of stress over the entire bond area. (Fig. 11-7-1). This eliminates stress concentration caused by rivets, bolts, spot welds, and similar fastening techniques. Lighter, thinner materials can be used without sacrificing strength.
2. Adhesives can effectively bond dissimilar materials.
3. Continuous contact between mating surfaces effectively bonds and seals against many environmental conditions.
4. Adhesives eliminate holes needed for mechanical fasteners and surface marks resulting from spot welding, brazing, etc.

Limitations

1. Adhesive bonding can be slow or require critical processing. This is particularly true in mass production. Some adhesives require heat and pressure or special jigs and fixtures to establish the bond.
2. Adhesives are sensitive to surface conditions. Special surface preparation may be required.
3. Some adhesive solvents present hazards. Special ventilation may be required to protect employees from toxic vapors.
4. Environmental conditions can reduce bond strength of some adhesives. Some do not hold well when exposed to low temperatures, high humidity, severe heat, chemicals, water, etc.

FIG. 11-7-1   Stresses in bonded joints.

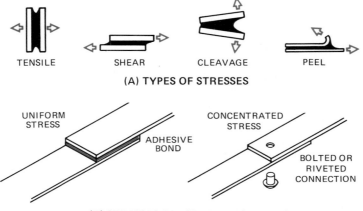

TENSILE   SHEAR   CLEAVAGE   PEEL

(A) TYPES OF STRESSES

UNIFORM STRESS — ADHESIVE BOND

CONCENTRATED STRESS — BOLTED OR RIVETED CONNECTION

(B) STRESSES CAUSED BY FASTENERS

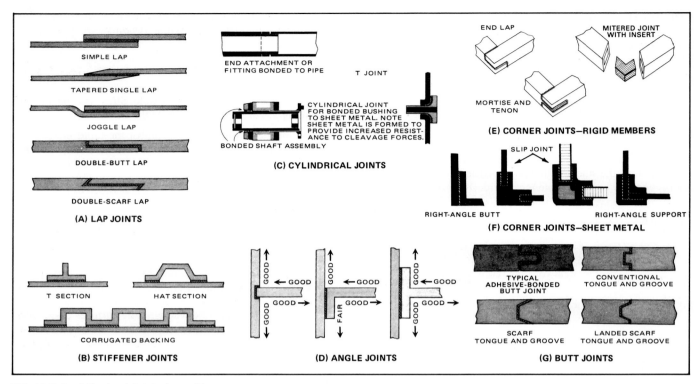

FIG. 11-7-2   Adhesive joint design guide.

Joint Design

Joints should be specifically designed for use with structural adhesives. First, the joint should be designed so that all the bonded area shares the load equally. Second, the joint configuration should be designed so that basic stress is primarily in shear or tensile, with cleavage and peel minimized or eliminated.

The following structural joints and their advantages and disadvantages illustrate some typical design alternatives (Fig. 11-7-2).

Lap Joints   Lap joints are most practical and applicable in bonding thin materials. The simple lap joint is offset. This can result in cleavage and peel stress under load when thin materials are used. A tapered single lap joint is more efficient than a simple lap joint. The tapered edge allows bending of the joint edge under stress. The joggle lap joint gives more uniform stress distribution than either the simple or tapered lap joint.

The double-butt lap joint gives more uniform stress distribution in the load-bearing area than the above joints. This type of joint, however, requires machining, which is not always feasible with thinner-gage metals. Double-scarf lap joints have better resistance to bending forces than double-butt joints.

Angle Joints   Angle joints give rise to either peel or cleavage stress depending on the gage of the metal. Typical approaches to the reduction of cleavage are illustrated.

Butt Joints   The following recessed butt joints are recommended: landed scarf tongue and groove, conventional tongue and groove, and scarf tongue and groove.

Cylindrical Joints   The T joint and overlap slip joint are typical for bonding cylindrical parts such as tubing, bushings, and shafts.

Corner Joints—Sheet Metal   Corner joints can be assembled with adhesives by using simple supplementary attachments. This permits joining and sealing in a single operation. Typical designs are right-angle butt joints, slip joints, and right-angle support joints.

Corner Joints—Rigid Members   Corner joints, as in storm doors or decorative frames, can be adhesive-bonded. End lap joints are the simplest design type, although they require machining. Mortise and tenon joints are excellent from a design standpoint, but they also require machining. The mitered joint with an insert is best if both members are hollow extrusions.

Stiffener Joints   Deflection and flutter of thin metal sheets can be minimized with adhesive-bonded stiffeners.

REFERENCES AND SOURCE MATERIAL

1. 3M Co.

ASSIGNMENT

See Assignment 15 for Unit 11-7 on page 325.

11-8 FASTENER REVIEW FOR CHAPTERS 10 AND 11

In chapters 10 and 11 the more common types of fasteners were explained and drawing problems were assigned for each type of fastener. In this unit selected assignments, which incorporate a variety of fasteners, were chosen to provide a thorough review of the numerous types of fasteners available to the designer.

ASSIGNMENTS

See Assignments 16 and 17 for Unit 11-8 on pages 325–326.

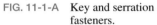

ASSIGNMENTS FOR CHAPTER 11

ASSIGNMENTS FOR UNIT 11-1, KEYS, SPLINES, AND SERRATIONS

1. Lay out the two fastener assemblies shown in Fig. 11-1-A or 11-1-B (pg. 316). The following fasteners are used:
   For 11-1-A
   - *Assembly A:* flat key
   - *Assembly B:* serrations

FIG. 11-1-A   Key and serration fasteners.

For 11-1-B
- *Assembly A:* square key
- *Assembly B:* Woodruff key

Refer to the Appendix and manufacturers' catalogs for sizes and use your judgment for dimensions not shown. Show the dimensions for the keyseats and serrations. Scale 1:1.

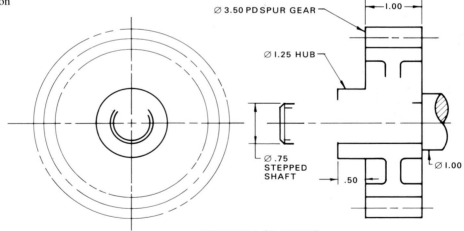

ASSEMBLY A (FLAT KEY)

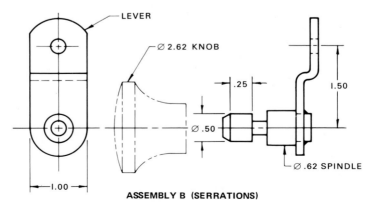

ASSEMBLY B (SERRATIONS)

2. Make a working drawing of the axle shown in Fig. 11-1-C. Dimension the keyseats according to Fig. 11-1-3. Refer to the Appendix.

ASSIGNMENTS FOR UNIT 11-2, PIN FASTENERS

3. Complete the pin assemblies shown in Fig. 11-2-A or 11-2-B, given the following information:

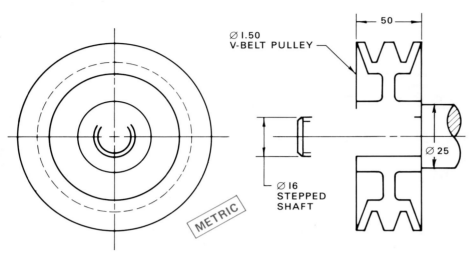

ASSEMBLY A (SQUARE KEY)

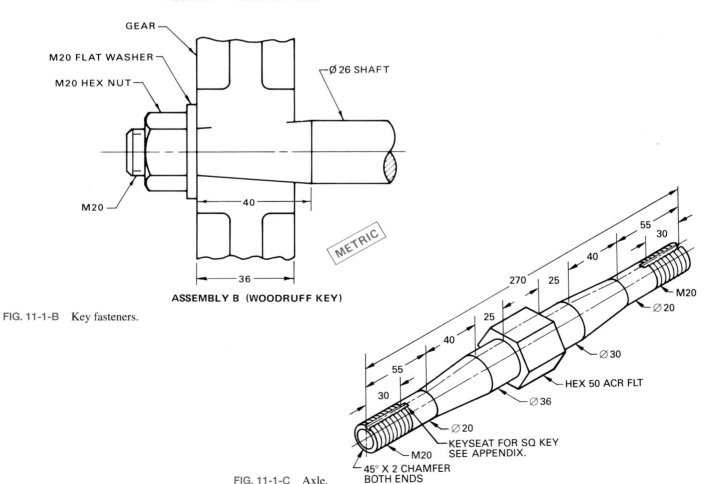

ASSEMBLY B (WOODRUFF KEY)

FIG. 11-1-B   Key fasteners.

FIG. 11-1-C   Axle.

For Fig. 11-2-A
- *Assembly A.* Slotted tubular spring pins are used to fasten the cap and handle to the shaft. Scale 1:2.
- *Assembly B.* A clevis pin whose area is equal to the four rivets is used to fasten the trailer hitch to the tractor draw bar. Scale 1:2.

For Fig. 11-2-B
- *Assembly A.* A type E grooved pin holds the roller to the bracket. A washer and cotter pin are used to fasten the bracket to the push rod. Scale 1:1.
- *Assembly B.* A type A3 grooved pin holds the V-belt pulley to the shaft. Scale 1:1.

Refer to manufacturers' catalogs for pin sizes and provide the complete information to order each fastener.

FIG. 11-2-A   Pin fasteners.

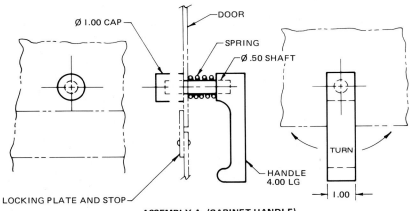

ASSEMBLY A (CABINET HANDLE)

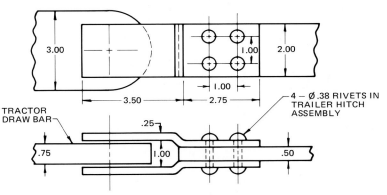

ASSEMBLY B (DRAW BAR HITCH)

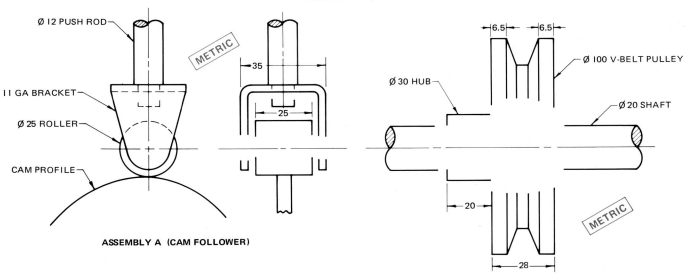

ASSEMBLY A (CAM FOLLOWER)

ASSEMBLY B (V-BELT PULLEY)

FIG. 11-2-B   Pin fasteners.

4. Make a two-view assembly drawing of the crane hook shown in Fig. 11-2-C. The hook is to be held to the U-frame with a slotted locknut. A spring pin is inserted through the locknut slots to prevent the nut from turning. A clevis pin with washer and cotter pin holds the pulley to the frame. Include on the drawing an item list. Scale 1:1.

5. Prepare detail drawings of the parts in Assignment 4. Use your judgment for the scale and selection of views.

ASSIGNMENTS FOR UNIT 11-3, RETAINING RINGS

6. Complete the assemblies shown in Fig. 11-3-A or 11-3-B by adding suitable retaining rings as per the information supplied below. Refer to the Appendix and manufacturers'

FIG. 11-2-C   Crane hook.

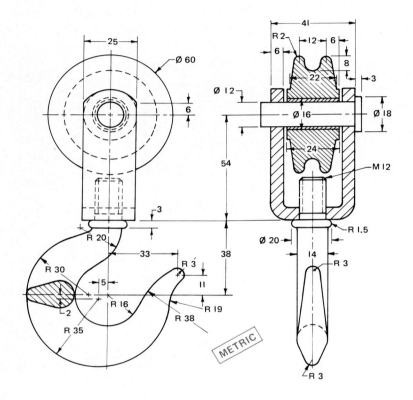

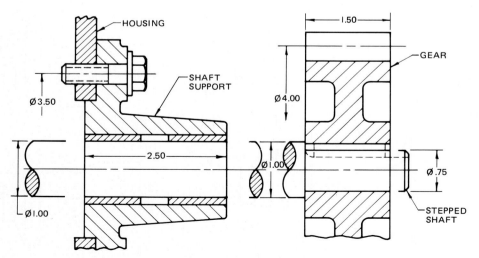

FIG. 11-3-A   Retaining ring fasteners.

ASSEMBLY A (EXTERNAL RETAINING RINGS)

catalogs and show on the drawing the catalog number for the retaining ring. Add ring and groove sizes. Scale 1:1. Use your judgment for dimensions not shown.

For Fig. 11-3-A an external radial retaining ring mounted on the shaft is to act as a shoulder for the shaft support. An external axial retaining ring is required to hold the gear on the shaft.

For Fig. 11-3-B

- *Assembly A.* External self-locking retaining rings hold the roller shaft in position on the bracket.
- *Assembly B.* An external self-locking ring holds the plastic housing to the viewer case. An internal self-locking ring holds the lens in position.

7. Complete the power drive assembly shown in Fig. 11-3-C given the following information. The shaft is positioned in the housing by an SKF #6005 bearing. The end cap and a retaining ring hold the bearing in place and a retaining ring holds the cap in the housing. Two retaining rings position the bearing on the shaft. The gear is positioned on the clutch by a retaining ring and a square key. The clutch is locked to the shaft by a square key held in position by a setscrew. The pulley drive is positioned and held to the shaft by a square key and two retaining rings. Include an item list on the drawing calling out the purchased parts.

8. Make detail drawings of the end cap and the partial view of the shaft in Assignment 7.

FIG. 11-3-B   Retaining ring fasteners.

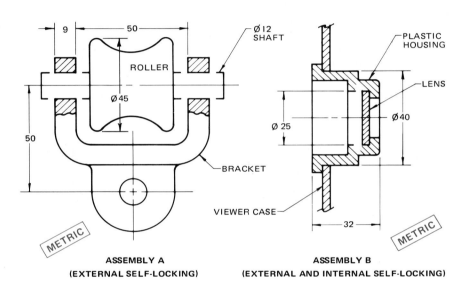

ASSEMBLY A
(EXTERNAL SELF-LOCKING)

ASSEMBLY B
(EXTERNAL AND INTERNAL SELF-LOCKING)

FIG. 11-3-C   Power drive assembly.

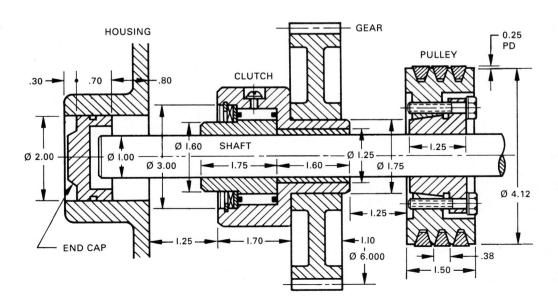

ASSIGNMENTS FOR UNIT 11-4, SPRINGS

9. Lay out the two assembly drawings as shown in Figs. 11-4-A or 11-4-B. Complete the drawings from the information supplied below, and make detail drawings of the springs. Use your judgment for sizes not given.
   For Fig. 11-4-A
   - *Assembly A.* The license plate holder is held to the frame of the car by a hinge. A torsion spring is required to keep the plate holder in position. The torsion spring is slipped over the hinge pin during assembly, and one end of the spring passes through the hole in the bumper. The other end of the spring is locked into the spring-retaining notch in the license plate holder. Scale 1:2.
   - *Assembly B.* Flat springs are positioned in openings C and D in the tape deck player. These springs hold the cassette against the bottom and the locating pin positioned in the left side of the tape deck. Scale 1:2.
   For Fig. 11-4-B
   - *Assembly A.* An extension spring controls the lever. The spring is fastened to the neck in the pin and through the hole in the lever. Scale 1:1.

FIG. 11-4-A   Spring fasteners.

- *Assembly B.* A compression spring mounted on the shaft of the handle provides sufficient pressure to hold the lever in position, thus maintaining the door against the panel. To open the door, the handle is pushed in and turned. This action compresses the spring and forces the lever away from the notch in the panel edge, thus permitting the lever to turn. Scale 1:1.

10. Complete the punch holder assembly shown in Fig. 11-4-C given the following information. The two helical springs have plain closed ends and are Ø.06 and have a pitch of .10. The plunger and punch are held in the punch holder by retaining rings. An RC3 fit is required for the Ø.30 shaft. Include on the drawing an item list.

11. Make working drawings of the parts from the completed assembly drawing in Assignment 10. Use your judgment for dimensions not shown.

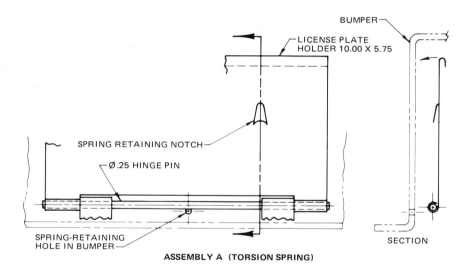

ASSEMBLY A (TORSION SPRING)

ASSEMBLY B (FLAT SPRINGS)

FIG. 11-4-B   Spring fasteners.

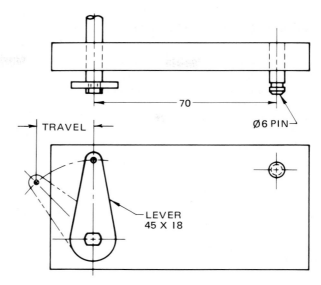

ASSEMBLY A (EXTENSION SPRING)

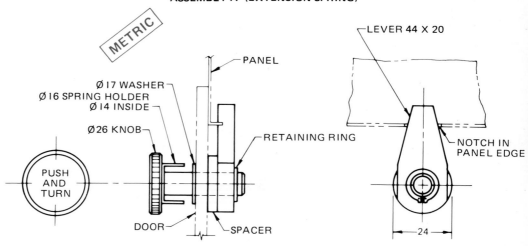

ASSEMBLY B (COMPRESSION SPRING)

FIG. 11-4-C   Punch holder
assembly.

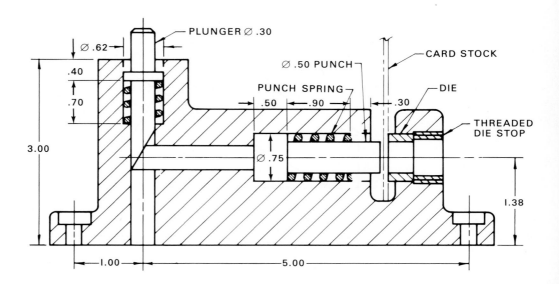

ASSIGNMENTS FOR UNIT 11-5, RIVETS

12. Complete the two assembly drawings shown in Fig. 11-5-A or 11-5-B from the information supplied below. Refer to manufacturers' catalogs for rivet type and sizes, and on each assembly show the callout for the rivets. Use your judgment for sizes not given.

    For Fig. 11-5-A
    - *Assembly A.* Padlock brackets are riveted to the locker door and door frame with two blind rivets in each bracket. Scale 1:1.
    - *Assembly B.* The roof truss is assembled in the shop with five evenly spaced Ø.50 in. (12 mm) rivets in each angle. Scale 1:4.

    For Fig. 11-5-B
    - *Assembly A.* The grill is held to the panel by four truss-head full tubular rivets. Scale 1:1.

- *Assembly B.* The support is held to the plywood panel by drive rivets uniformly spaced on the gage lines. Two rivets hold the bracket to the support. Scale 1:1.

13. Complete the assembly shown in Fig. 11-5-C using the graphical symbols of rivets for aerospace equipment and given the following information:
    - *Assembly A.* Ø8 rivets equally spaced at 55 OC; item reference 22; 100° countersunk both sides; preformed head near side.
    - *Assembly B.* Ø6 combined rivets equally spaced at 50 OC; item reference 19, sleeve item reference 21; preformed head far side.
    - *Assembly C.* Ø4 rivets equally spaced at 40 OC (4 sides); item reference 16; preformed head far side; 82° dimple near side.

FIG. 11-5-A   Rivet fasteners.

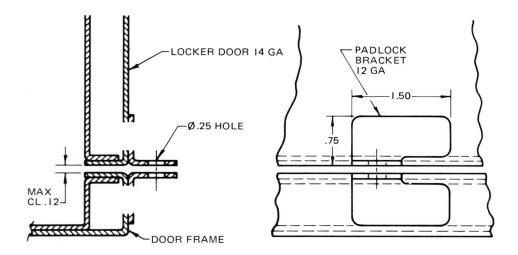

ASSEMBLY A (BLIND RIVETS)

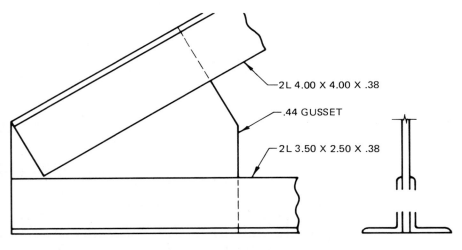

ASSEMBLY B (LARGE STRUCTURAL RIVETS)

FIG. 11-5-B   Rivet fasteners.

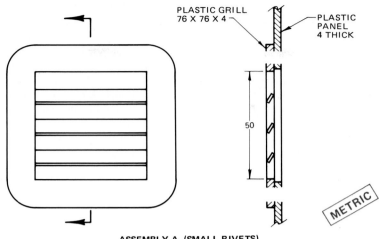

ASSEMBLY A (SMALL RIVETS)

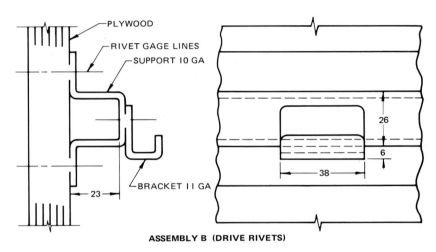

ASSEMBLY B (DRIVE RIVETS)

FIG. 11-5-C   Rivets for aerospace equipment.

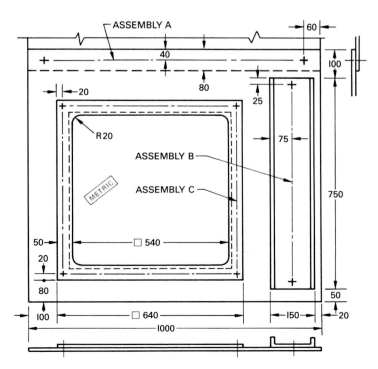

ASSIGNMENT FOR UNIT 11-6, WELDED FASTENERS

14. Complete the two assemblies shown in Fig. 11-6-A or 11-6-B. Refer to manufacturers' catalogs and the Appendix for standard fastener components. Complete the drawings from the information supplied below. Use your judgment for sizes not given. Scale 1:1.

    For Fig. 11-6-A

    ■ *Assembly A.* Two resistance-welded threaded fasteners, one on each side of the pipe, are required. The bracket drops over the fasteners, and lock washers and nuts secure the bracket to the pipe.

    ■ *Assembly B.* A leakproof attaching method (stud welding) is required to hold the adaptor to the panel.

    For Fig. 11-6-B

    ■ *Assembly A.* A spot-weld nut is to be attached to the panel. A hole in the clamp permits a machine screw to fasten the pipe clamp to the nut.

    ■ *Assembly B.* A right-angle bracket is to be fastened (projection welding) to the bottom plate. The vertical plate is secured to the bracket by a machine screw and lock washer.

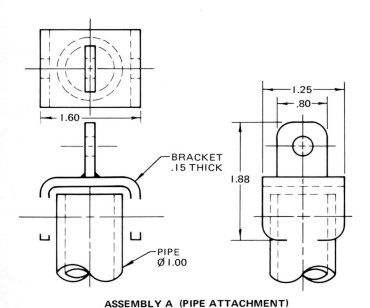

ASSEMBLY A (PIPE ATTACHMENT)

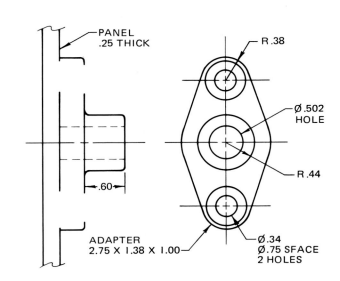

ASSEMBLY B (LEAKPROOF ATTACHMENT)

FIG. 11-6-A   Welded fasteners.

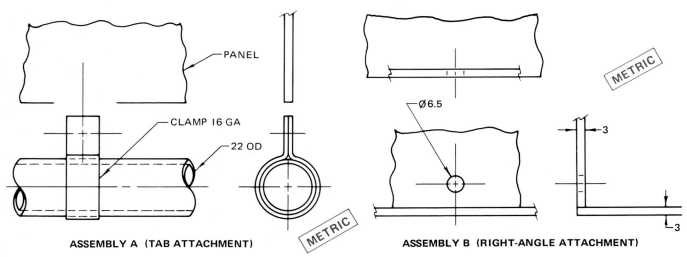

ASSEMBLY A (TAB ATTACHMENT)

ASSEMBLY B (RIGHT-ANGLE ATTACHMENT)

FIG. 11-6-B   Welded fasteners.

ASSIGNMENT FOR UNIT 11-7, ADHESIVE FASTENINGS

15. Complete the two adhesive-bonded assemblies shown in Fig. 11-7-A or 11-7-B from the information supplied below and the adhesive chart in the Appendix. List the adhesive product number and state the method of application you would recommend. Use your judgment for sizes not shown, and dimension the joint. Scale is to suit.

For Fig. 11-7-A
- *Assembly A.* The riveted joint shown is to be replaced by a joggle lap joint. It must be fast-drying.
- *Assembly B.* The sheet-metal corner joint shown is to be replaced by a slip joint. It must be water-resistant.

For Fig. 11-7-B
- *Assembly A.* Three pieces of wood are to be assembled into the shape shown. Joint design has not been shown.
- *Assembly B.* The riveted joint shown is to be replaced by a joggle lap joint. See Table 51 of the Appendix for more information on military (MMM) specs.

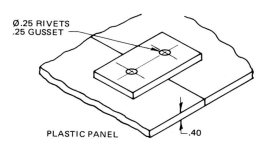

Ø.25 RIVETS
.25 GUSSET

PLASTIC PANEL   .40

ASSEMBLY A (BUTT JOINT)

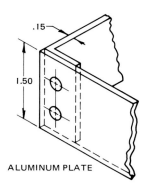

.15

1.50

ALUMINUM PLATE

ASSEMBLY B (SLIP JOINT)

FIG. 11-7-A   Adhesive fastenings.

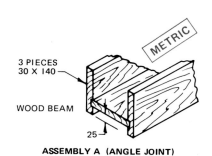

METRIC

3 PIECES
30 X 140

WOOD BEAM

25

ASSEMBLY A (ANGLE JOINT)

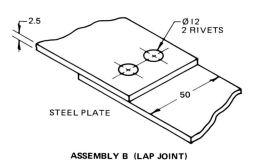

2.5

Ø12
2 RIVETS

STEEL PLATE

50

ASSEMBLY B (LAP JOINT)

FIG. 11-7-B   Adhesive fastenings.

ASSIGNMENTS FOR UNIT 11-8, FASTENER REVIEW FOR CHAPTERS 10 AND 11

16. Prepare detail drawings of the parts shown in Fig. 11-8-A. Include on the drawing an item list. The shaft is to have an RC4 fit with the bushing and the bushing an LN3 fit in the body. Use your judgment for the selection and number of views for each part.

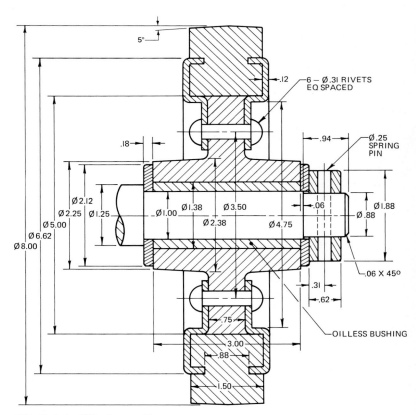

5°

.12

6 – Ø.31 RIVETS
EQ SPACED

.94

Ø.25
SPRING
PIN

.18

Ø2.12
Ø2.25   Ø1.25   Ø1.00   Ø1.38   Ø3.50
Ø2.38   Ø4.75

.06

Ø.88   Ø1.88

Ø5.00

Ø6.62
Ø8.00

.06 X 45°

.31

.62

OILLESS BUSHING

.75

3.00

.88

1.50

FIG. 11-8-A   Wheel assembly.

17. Make a one-view assembly drawing of the universal joint
    shown in Fig. 11-8-B. Include on the drawing an item list.

FIG. 11-8-B   Universal joint.

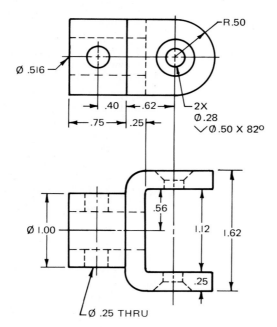

PT I – FORK – 2 REQD

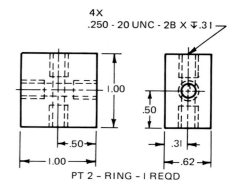

PT 2 – RING – I REQD

PT 3 – Ø .25 SPRING PIN – 2 REQD
PT 4 – .250 – 20 FHMS – .62 LG – 4 REQD

CHAPTER 12

MANUFACTURING MATERIALS

Definitions

Chilling   A process that produces white iron.

High alloy   Steel castings that contain a minimum of eight percent nickel and/or chromium.

Plastics   Nonmetallic materials capable of being formed or molded with the aid of heat, pressure, chemical reactions, or a combination of these.

Thermoplastics   Materials that soften, or liquefy, and flow when heat is applied.

Thermosetting plastics   Materials which undergo an irreversible chemical change when heat is applied or when a catalyst or reactant is added.

12-1   CAST IRONS AND FERROUS METALS

This chapter is an up-to-date reference on manufacturing materials. It provides the drafter and designer with basic information on materials and their properties to ensure the proper selection of the product material.

Ferrous Metals

Iron and the large family of iron alloys called steel are the most frequently specified metals. Iron is abundant (iron ore constitutes about five percent of the earth's crust), easy to convert from ore to a useful form, and iron and steel are sufficiently strong and stable for most engineering applications.

All commercial forms of iron and steel contain carbon, which is an integral part of the metallurgy of iron and steel.

Cast Iron

Because of its low cost, cast iron is often considered a simple metal to produce and to specify. Actually, the metallurgy of cast iron is more complex than that of steel and other familiar design materials. Whereas most other metals are usually specified by a standard chemical analysis, the same analysis of cast iron can produce several entirely different types of iron, depending upon rate of cooling, thickness of the casting, and how long the casting remains in the mold. By controlling these variables, the foundry can produce a variety of irons for heat- or wear-resistant uses, or for high-strength components (Fig. 12-1-1, pg. 328).

Types of Cast Iron

Ductile (Nodular) Iron   Ductile iron, sometimes called nodular iron, is not as available as gray iron, and it is more difficult to control in production. However, ductile iron can be used where higher ductility or strength is required than is available in gray iron (Fig. 12-1-2, pg. 328).

Ductile iron is used in applications such as crankshafts because of its good machinability, fatigue strength, and high modulus of elasticity; heavy-duty gears because of its high yield strength and wear resistance; and automobile door hinges because of its ductility.

Gray Iron   Gray iron is a supersaturated solution of carbon in an iron matrix. The excess carbon precipitates out in the form of graphite flakes. Typical applications of gray iron include automotive blocks, flywheels, brake disks and drums, machine bases, and gears. Gray iron normally serves well in any machinery application because of its fatigue resistance.

White Iron   White iron is produced by a process called *chilling,* which prevents graphite carbon from precipitating out. Either gray or ductile iron can be chilled to produce a surface of white iron. In castings that are white iron throughout, however,

FIG. 12-1-1   Schematic diagram of a blast furnace, hot blast stone, and skiploader. *(American Iron and Steel Institute)*

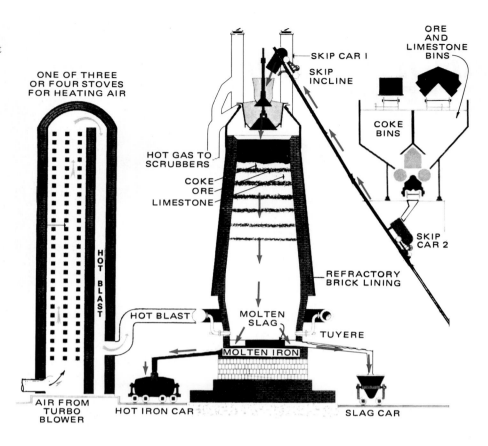

FIG. 12-1-2   Mechanical properties of cast iron.

MECHANICAL PROPERTY		DUCTILE					GRAY						MALLEABLE						
		50-55-06	60-40-18	100-70-03	120-90-02	WHITE	20	25	30	40	50	60	32510	35018	40010	45006	50005	70003	90001
Yield strength	10^3 lb/in.^{2}	60–75	45–60	75–90	90–125								32	35	40	45	50	70	90
	MPa	410–520	310–410	520–620	620–860		*	*	*	*	*	*	220	240	275	310	345	485	620
Tensile strength	10^3 lb/in.^{2}	90–110	60–80	100–120	125–150	20–50	20–25	25–30	30–35	40–48	50–57	60–66	50	53	60	65	70	85	105
	MPa	620–760	410–550	690–825	860–1035	140–345	140–170	170–205	205–240	275–330	345–390	415–455	345	365	415	450	480	585	725
Elongation in 2.00 in. (50 mm)	%	3–10	10–25	6–10	2–7	—	1	1	1	0.8	0.5	0.5	10	18	10	6	5	3	1
Modulus of elasticity	10^3 lb/in.^{2}	22–25	22–25	22–25	22–25	—	12	13	15	17	19	20	25	25	26	26	26	26–28	26–28
	10^3 MPa	150–170	150–170	150–170	150–170	8 —	83	90	103	117	131	138	172	172	180	180	180	180–193	180–193

*Yield strength usually about 65–80% of tensile strength.

the composition of iron is selected according to part size to ensure that the volume of metal involved can chill rapidly enough to produce white iron.

Because of their extreme hardness, white irons are used primarily for applications requiring wear and abrasion resistance, such as mill liners and shot-blasting nozzles. Other uses include railroad brake shoes, rolling-mill rolls, clay-mixing and brick-making equipment, and crushers and pulverizers. Plain (unalloyed) white iron usually costs less than other cast irons.

The principal disadvantage of white iron is that it is very brittle.

High-Alloy Irons   High-alloy irons are ductile, gray, or white irons that contain over three percent alloy content. These irons have properties that are significantly different from the unalloyed irons and are usually produced by specialized foundries.

Malleable Iron   Malleable iron is white iron that has been converted to a malleable condition by a two-stage heat-treating process.

It is a commercial cast material that is similar to steel in many respects. It is strong and ductile, has good impact and fatigue properties, and has excellent machining characteristics.

The two basic types of malleable iron are ferritic and pearlitic. Ferritic grades are more machinable and ductile, whereas the pearlite grades are stronger and harder.

Forming Process

For design information on the preparation of metal castings, see Chap. 13, Units 13-1 and 13-3.

REFERENCES AND SOURCE MATERIAL

1. *Machine Design,* Materials reference issue.

ASSIGNMENT

See Assignment 1 for Unit 12-1 on page 344.

12-2 CARBON STEEL

Carbon steel is essentially an iron-carbon alloy with small amounts of other elements (either intentionally added or unavoidably present), such as silicon, magnesium, copper, and sulfur. Steels can be either cast to shape or wrought into various mill forms from which finished parts can be machined, forged, formed, stamped, or otherwise generated.

Wrought steel is either poured into ingots or is sand-cast. After solidification the metal is reheated and hot-rolled—often in several steps—into the finished wrought form. Hot-rolled steel is characterized by a scaled surface and a decarburized skin.

Carbon and Low-Alloy Cast Steels

Carbon and low-alloy cast steels lend themselves to the formation of streamlined, intricate parts with high strength and rigidity. A number of advantages favor steel casting as a method of construction:

1. The metallographic structure of steel castings is uniform in all directions. It is free from the directional variations in properties of wrought-steel products.
2. Cast steels are available in a wide range of mechanical properties depending on the compositions and heat treatments.
3. Steel castings can be annealed, normalized, tempered, hardened, or carburized.
4. Steel castings are as easy to machine as wrought steels.
5. Most compositions of carbon and low-alloy cast steels are easily welded because their carbon content is under 0.45 percent.

The making of steel is illustrated in Fig. 12-2-1 (pg. 330).

High-Alloy Cast Steels

The term *high alloy* is applied arbitrarily to steel castings containing a minimum of eight percent nickel and/or chromium. Such castings are used mostly to resist corrosion or provide strength at temperatures above 1200°F (560°C).

Carbon Steels

Carbon steels are the workhorse of product design. They account for over 90 percent of total steel production. More carbon steels are used in product manufacturing than all other metals combined.

A thorough understanding of the selection and specification criteria for all types of steel requires knowledge of what is implied by carbon-steel mill forms, qualities, grades, tempers, finishes, edges, and heat treatments; also how and where these terms relate to dimensions, tolerances, physical and mechanical properties, and manufacturing requirements.

The designer's specification job really begins the instant that molten steel hits the mold. The conditions under which steel solidifies have a significant effect on production and on performance of subsequent mill products.

Steel Specification

Several ways are used to identify a specific steel: by chemical or mechanical properties, by its ability to meet a standard specification or industry-accepted practice, or by its ability to be fabricated into an identified part.

Chemical Composition

The steel producer can be instructed to produce a desired composition in one of three ways:

1. By a maximum limit
2. By a minimum limit
3. By an acceptable range

The following are some commonly specified elements.

Carbon   Carbon is the principal hardening element in steel. As carbon content is increased to about 0.85 percent, hardness and tensile strength increase, but ductility and weldability decrease.

Manganese   Manganese is a lesser contributor to hardness and strength. Properties depend on carbon content. Increasing manganese increases the rate of carbon penetration during carburizing but decreases weldability.

Phosphorus   Large amounts of phosphorus increase strength and hardness but reduce ductility and impact toughness, particularly in the higher-carbon grades. Phosphorus in low-carbon, free-machining steels improves machinability.

Silicon   A principal deoxidizer in the steel industry, silicon increases strength and hardness but to a lesser extent than manganese. However, it reduces machinability.

FIG. 12-2-1 Flowchart for steelmaking.

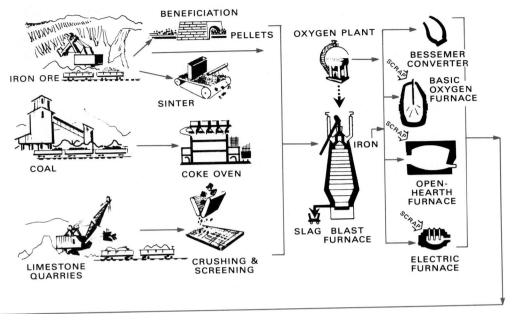

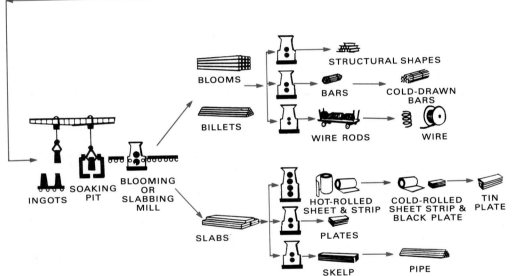

Sulfur Increased sulfur content reduces transverse ductility, notch-impact toughness, and weldability. Sulfur is added to improve machinability of steel.

Copper Copper improves atmospheric corrosion resistance when present in excess of 0.15 percent.

Lead Lead improves the machinability of steel.

Classification Bodies

The specifications covering the composition of iron and steel have been issued by various classification bodies. These specifications serve as a selection guide and provide a means for the buyer to conveniently specify certain known and recognized requirements. The main classification bodies are:

SAE—Society of Automotive Engineers

AISI—American Iron and Steel Institute This is an association of steel producers that issues steel specifications for the steelmaking industry and cooperates with the SAE in using the same numbers for the same steel.

ASTM—American Society for Testing and Materials This group is interested in materials of all kinds and writes specifications. The ASTM steel specifications for steel plate and structural shapes are used by all steelmakers in North America.

The ASTM has several specifications covering structural steel. Both the AISA and AISC (American Institute of Steel Construction) refer to ASTM specifications.

ASME—American Society of Mechanical Engineers This group is interested in the steel used in pressure vessels and other mechanical equipment.

SAE and AISI—Systems of Steel Identification

The specifications for steel bar are based on a code that specifies the composition of each type of steel covered. They include both plain carbon and alloy steels. The code is a four-number system (Figs. 12-2-2 and 12-2-3). Each figure in the number has the following specific function: the first or left-side figure represents the major class of steel, the second figure represents a subdivision of the major class. For example, the series having *one* (1) as the left-hand figure covers the carbon steels. The second figure breaks this class up into normal low-sulfur steels, the high-sulfur free-machining grades, and another grade having higher than normal manganese.

Originally the second figure represented the percentage of the major alloying element present, and this is true of many of the alloy steels. However, this had to be varied in order to account for all the steels that are available.

The third and fourth figures represent carbon content in hundredths of one percent, thus the figure xx15 means 0.15 of one percent carbon.

EXAMPLE

SAE 2335 is a nickel steel containing 3.5 percent nickel and 0.35 of one percent carbon.

Carbon-Steel Sheets

Flat-rolled carbon-steel sheets are made from heated slabs that are progressively reduced in size as they move through a series of rolls. Typical properties of rolled carbon steels are shown in Fig. 12-2-4 (pg. 332).

Hot-Rolled Sheets   Hot-rolled sheets are produced in three principal qualities: commercial, drawing, and physical.

Cold-Rolled Sheets   Cold-rolled sheets are made from hot-rolled coils that are pickled, then cold-reduced to the desired thickness. The commercial quality of cold-rolled sheets is normally produced with a matte finish suitable for painting or enameling but not suitable for electroplating.

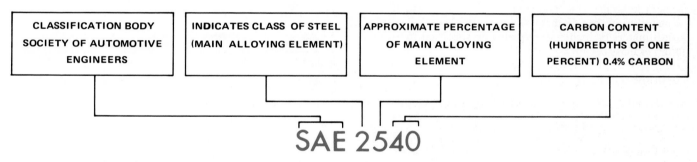

| CLASSIFICATION BODY SOCIETY OF AUTOMOTIVE ENGINEERS | INDICATES CLASS OF STEEL (MAIN ALLOYING ELEMENT) | APPROXIMATE PERCENTAGE OF MAIN ALLOYING ELEMENT | CARBON CONTENT (HUNDREDTHS OF ONE PERCENT) 0.4% CARBON |

SAE 2540

FIG. 12-2-2   Steel designation system.

FIG. 12-2-3   Carbon steel designations, properties and uses.

TYPE OF CARBON STEEL	NUMBER SYMBOL	PRINCIPAL PROPERTIES	COMMON USES
Plain carbon	10XX		
Low-carbon steel (0.06 to 0.20% carbon)	1006 to 1020	Toughness and less strength	Chains, rivets, shafts, and pressed steel products
Medium-carbon steel (0.20 to 0.50% carbon)	1020 to 1050	Toughness and strength	Gears, axles, machine parts, forgings, bolts, and nuts
High-carbon steel (over 0.50% carbon)	1050 and over	Less toughness and greater hardness	Saws, drills, knives, razors, finishing tools, and music wire
Sulfurized (free-cutting)	11XX	Improves machinability	Threads, splines, and machined parts
Phosphorized	12XX	Increases strength and hardness but reduces ductility	
Manganese steels	13XX	Improves surface finish	

MECHANICAL PROPERTY		AISI STEEL											
		1015/1020/1022			1035/1040			1045/1050			1095		
		HOT-ROLLED	COLD DRAWN	ANNEALED	HOT-ROLLED	COLD-DRAWN	QUENCHED AND TEMPERED	HOT-ROLLED	COLD-DRAWN	QUENCHED AND TEMPERED	HOT-ROLLED	COLD-DRAWN AND ANNEALED	QUENCHED AND TEMPERED
Yield strength	10^3 lb/in.^{2}	40	51	42	42	71	63–96	49	84	68–117	66	76	80–152
	MPa	270	350	295	290	440	435–660	335	580	470–800	455	525	580–1050
Tensile strength	10^3 lb/in.^{2}	65	61	60	76	85	96–130	90	100	105–137	130	99	130–216
	MPa	450	420	415	525	585	660–895	620	690	725–945	895	680	895–1490
% Elongation in 2.00 in. (50 mm)		25	15	38	18	12	17–24	15	10	25–15	9	13	10–84

FIG. 12-2-4   Typical mechanical properties of rolled carbon steel.

Carbon-Steel Plates

Carbon-steel plates are produced (in rectangular plates or in coils) by hot rolling directly from the ingot or slab. Plate thickness ranges from .19 in. (4 mm) and thicker for plates up to 48 in. (1200 mm) wide, and from .25 in. (6 mm) and thicker for plates wider than 48 in. (1200 mm). Thickness is specified in millimeters or inches. It can also be specified by weight (lb/ft^2) or mass (kg/m^2).

Carbon-Steel Bars

Hot-Rolled Bars   Hot-rolled carbon-steel bars are produced from blooms or billets in a variety of cross sections and sizes (Figs. 12-2-5 and 12-2-6).

Cold-Finished Bars   Cold-finished carbon-steel bars are produced from hot-rolled steel by a cold-finishing process which improves surface finish, dimensional accuracy, and alignment. Cold drawing and cold rolling also increase the yield and tensile strength. For machinability ratings of cold-drawn carbon steel, see Fig. 12-2-7.

Steel Wire

Steel wire is made from hot-rolled rods produced in continuous-length coils. Most wire is drawn, but some special shapes are rolled.

Pipe and Tubing

Pipe and tubing range from the familiar plumber's black pipe to high-precision mechanical tubing for bearing races. Pipe and tubing may contain fluids, support structures, or be a primary shape from which products are fabricated.

Welded Tubular Products   Welded tubular products are made from hot-rolled or cold-rolled flat steel coils.

Pipe   Pipe is produced from carbon or alloy steel to nominal dimensions. Nominal pipe sizes are expressed in inch sizes, but in the metric system the outside diameter and the wall thickness are expressed in millimeters. The outside diameter is often much larger than the nominal size. For example, a .75-in. standard-weight pipe has an outside diameter of 1.050 in. (26.7 mm).

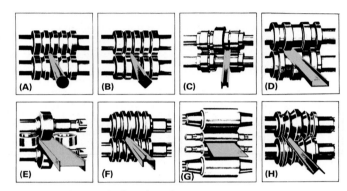

FIG. 12-2-5   Standard steel stock. *(American Iron and Steel Institute)*

The outside diameter of nominal-size pipe always remains the same and the mass or wall thickness changes. ANSI B36 has developed 10 different wall thicknesses (schedules) of pipe. (See Table 57 in the Appendix.)

Nominal pipe-size designation stops at 12 in. Pipe 14 in. and over is listed on the basis of outside diameter and wall thickness.

Tubing   Tubing is usually specified by a combination of either outside diameter, inside diameter, or wall thickness. Sizes range from approximately .25 to 5.00 in. (6 to 125 mm), in increments of .12 in. (3 mm). Wall thickness is usually specified in inches, millimeters, or by gage numbers.

Structural-Steel Shapes

A large tonnage of structural-steel shapes goes into manufactured products rather than buildings. The frame of a truck, railroad car, or earth-moving equipment is a structural design problem, just as is a high-rise building.

Size Designation   Several ways are used to describe a structural section in a specification, depending primarily on its shape.

1. Beams and channels are measured by the depth of the section in inches (millimeters) and by weight (lb/ft) or mass (kg/m).

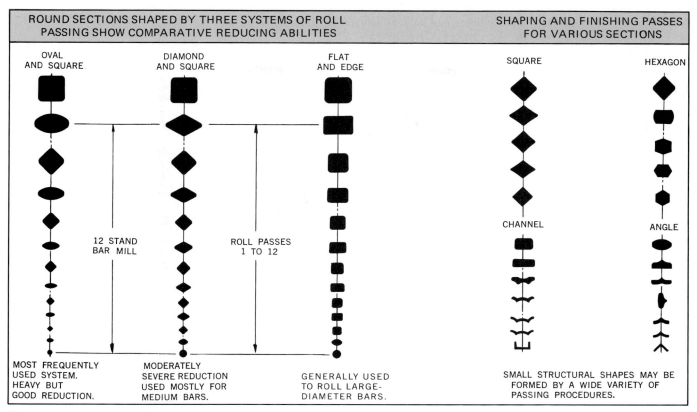

ROUND SECTIONS SHAPED BY THREE SYSTEMS OF ROLL PASSING SHOW COMPARATIVE REDUCING ABILITIES

SHAPING AND FINISHING PASSES FOR VARIOUS SECTIONS

OVAL AND SQUARE

DIAMOND AND SQUARE

FLAT AND EDGE

SQUARE

HEXAGON

12 STAND BAR MILL

ROLL PASSES 1 TO 12

CHANNEL

ANGLE

MOST FREQUENTLY USED SYSTEM. HEAVY BUT GOOD REDUCTION.

MODERATELY SEVERE REDUCTION USED MOSTLY FOR MEDIUM BARS.

GENERALLY USED TO ROLL LARGE-DIAMETER BARS.

SMALL STRUCTURAL SHAPES MAY BE FORMED BY A WIDE VARIETY OF PASSING PROCEDURES.

FIG. 12-2-6   Bar-mill roll passes. *(American Iron and Steel Institute)*

AISI NO.	RATING*	AISI NO.	RATING*
12L14	195	1114	85
1213	137	1137	72
1215	137	1141	69
1212	100	1018	66
1211	91	1045	55
1117	89		

*Based on 1212 = 100%

FIG. 12-2-7   Machinability rating of cold-drawn carbon steel.

2. Angles are described by length of legs and thickness in inches (millimeters), or more commonly, by length of legs and weight (lb/ft) or mass (kg/m). The longest leg is always stated first.
3. Tees are specified by width of flange, overall depth of stem, and pounds per foot (kilograms per meter) in that order.
4. Zees are specified by width of flange and thickness in inches (millimeters), or by depth, width across flange, and pounds per foot (kilograms per meter).
5. Wide-flange sections are described by depth, width across flange, and pounds per foot (kilograms per meter).

High-Strength Low-Alloy Steels

The properties of high-strength low-alloy (HSLA) steels generally exceed those of conventional carbon structural steels. These low-alloy steels are usually chosen for their high ratios of yield to tensile strength, resistance to puncturing, abrasion resistance, corrosion resistance, and toughness.

ASTM Specifications

ASTM has six specifications covering high-strength low-alloy steels. These are:

ASTM A94   Used primarily for riveted and bolted structures and for special structural purposes.

ASTM A242   Used primarily for structural members where light weight or low mass and durability are important.

ASTM A374   Used where high strength is required and where resistance to atmospheric corrosion must be at least equal to that of plain copper-bearing steel.

ASTM A375   This specification differs slightly from ASTM A374 in that material can be specified in the annealed or normalized condition.

ASTM A440   This covers high-strength intermediate-manganese steels for nonwelded applications.

ASTM A441   This covers the intermediate-manganese HSLA steels, which are readily weldable when proper welding procedures are used.

Low- and Medium-Alloy Steels

There are two basic types of alloy steel: *through hardenable* and *surface hardenable*. Each type contains a broad family of steels whose chemical, physical, and mechanical properties make them suitable for specific product applications (Fig. 12-2-8).

Stainless Steels

Stainless steels have many industrial uses because of their desirable corrosion-resistance and strength properties.

Free-Machining Steels

A whole family of free-machining steels has been developed for fast and economical machining (Fig. 12-2-9). These steels are available in bar stock in various compositions, some standard

TYPE OF STEEL	ALLOY SERIES	APPROXIMATE ALLOY CONTENT (%)	PRINCIPAL PROPERTIES	COMMON USES
Manganese steel	13xx	Mn 1.6–1.9	Improve surface finish	
Molybdenum steels	40xx 41xx 43xx 44xx 46xx 47xx 48xx	Mo 0.15–0.3 Cr 0.4–1.1; Mo 0.08–0.35 Ni 1.65–2; Cr 0.4–0.9; Mo 0.2–0.3 Mo 0.45–0.6 Ni 0.7–2; Mo 0.15–0.3 Ni 0.9–1.2; Cr 0.35–0.55; Mo 0.15–0.4 Ni 3.25–3.75; Mo 0.2–0.3	High strength	Axles, forgings, gears, cams, mechanical parts
Chromium steels	50xx 51xx E51100 E52100	Cr 0.3–0.5 Cr 0.7–1.15 C 1.0; Cr 0.9–1.15 C 1.0; Cr 0.9–1.15	Hardness, great strength and toughness	Gears, shafts, bearings, springs, connecting rods
Chromium-vanadium steel	61xx	Cr 0.5–1.1; V0.1–0.15	Hardness and strength	Punches and dies, piston rods, gears, axles
Nickel-chromium-molybdenum steels	86xx 87xx 88xx	Ni 0.4–0.7; Cr 0.4–0.6; Mo 0.15–0.25 Ni 0.4–0.7; Cr 0.4–0.6; Mo 0.2–0.3 Ni 0.4–0.7; Cr 0.4–0.6; Mo 0.3–0.4	Rust resistance, hardness, and strength	Food containers, surgical equipment
Silicon-manganese steel	92xx	Si 1.8–2.2	Springiness and elasticity	Springs

FIG. 12-2-8   AISI designation system for alloy steel.

DESIGNATION	PHOSPHORIZED				SULFURIZED								
	12L13/12L14/12L15		1211/1212/1213		1117/1118/1119			1137			1141/1144		
MECHANICAL PROPERTY	HOT-ROLLED	COLD-DRAWN	HOT-ROLLED	COLD-DRAWN	HOT-ROLLED	COLD-DRAWN	QUENCHED AND TEMPERED	HOT-ROLLED	COLD-DRAWN	QUENCHED AND TEMPERED	HOT-ROLLED	COLD-DRAWN	QUENCHED AND TEMPERED
Yield strength 10^3 lb/in.^{2}	34	60-80	33	58	34-46	51-68	50-76	48	82	136	51	90	163
MPa	235	416-550	225	400	235-315	350-470	345-525	330	565	335	350	620	1120
Tensile strength 10^3 lb/in.^{2}	57	70-90	55	75	62-76	69-78	89-113	88	98	157	94	100	190
MPa	390	480-620	380	517	425-525	475-535	615-780	605	675	1080	650	690	1310
% Elongation in 2.00 in. (50 mm)	22	10-18	25	10	23-33	15-20	17-22	15	10	5	15	10	9
Machinability (B1212 = 100)	195–296		91–137		89–100			71			69		

FIG. 12-2-9   Typical mechanical properties of free-machining carbon steels.

and some proprietary. When utilized properly, they lower the cost of machining by reducing metal removal time.

REFERENCES AND SOURCE MATERIAL

1. *Machine Design,* Materials reference issue.

ASSIGNMENT

See Assignment 2 for Unit 12-2 on page 344.

12-3 NONFERROUS METALS

Although ferrous alloys are specified for more engineering applications than all nonferrous metals combined, the large family of nonferrous metals offers a wider variety of characteristics and mechanical properties. For example, the lightest metal is lithium, .02 lb/in.^{3} (0.53 g/cm^3); the heaviest is osmium with a weight of .81 lb/in.^{3} (mass of 22.5 g/cm^3)—nearly twice the weight of lead. Mercury melts at around −38°F (−30°C), while tungsten, the highest-melting metal, liquefies at 6170°F (3410°C).

Availability, abundance, and the cost to convert the metal into useful forms all play an important part in selecting a nonferrous metal. Although nearly 80 percent of all elements are called "metals," only about two dozen of these are used as structural engineering materials. Of the balance, however, many are used as coatings, in electronic devices, as nuclear materials, and as minor constituents in other systems.

One of the most important aspects in selecting a material for a mechanical or structural application is how easily the material can be shaped into the finished part—and how its properties can be either intentionally or inadvertently altered in the process (Fig. 12-3-1). Frequently, metals are simply cast into the finished part. In other cases, metals are cast into an intermediate form (such as an ingot), then worked or "wrought" by rolling, forging, extruding, or other deformation processes.

Manufacturing with Metals

Machining   Most metals can be machined. Machinability is best for metals that allow easy chip removal with minimum tool wear.

Powder Metallurgy (PM) Compacting   Parts can be made from most metals and alloys by PM compacting, although only a few are economically justified. Iron and iron-copper alloys are most commonly used.

Casting   Theoretically, any metal that can be melted and poured can be cast. However, economic limitations usually narrow down the number of ways metals are cast commercially.

Extruding and Forging   Metals to be forged or extruded must be ductile and not work-harden at working temperature. Some metals show these characteristics at room temperature and can be cold-worked; others must be heated.

Stamping and Forming   Most metals, except brittle alloys, can be press-worked.

FORMING METHOD	ALUMINUM	COPPER	IRON	LEAD	MAGNESIUM	NICKEL	SILVER, GOLD, PLATINUM	MOLYBDENUM, COPPER, TANTALUM, TUNGSTEN	STEEL	TIN	TITANIUM	ZINC
Casting												
Centrifugal	✔	✔	✔			✔			✔			
Continuous	✔	✔	✔	✔					✔			
Ceramic mold	✔	✔	✔		✔	✔			✔			✔
Investment	✔	✔			✔	✔	✔		✔			
Permanent mold	✔	✔	✔	✔	✔	✔			✔	✔		✔
Sand	✔	✔	✔	✔	✔	✔			✔	✔		✔
Shell mold	✔	✔	✔		✔	✔			✔			
Die casting	✔	✔		✔	✔					✔		✔
Cold heading	✔	✔		✔		✔	✔		✔			
Deep drawing	✔	✔			✔	✔			✔		✔	
Extruding	✔	✔	✔		✔	✔		✔	✔	✔	✔	
Forging	✔	✔			✔	✔		✔	✔		✔	
Machining	✔	✔	✔		✔	✔	✔	✔	✔		✔	✔
PM compacting	✔	✔	✔			✔	✔	✔	✔		✔	
Stamping and forming	✔	✔			✔	✔	✔	✔	✔		✔	✔

FIG. 12-3-1   Common methods of forming metals.

Cold Heading   Metals must be ductile and should not work-harden rapidly. Annealing should restore ductility and softness in cold-heading alloys.

Deep Drawing   Deep drawing involves severe deformation and the metal is usually stretched over the die.

Aluminum

The density of aluminum is about one-third that of steel, brass, nickel, or copper. Yet some alloys of aluminum are stronger than structural steel. Under most service conditions, aluminum has high resistance to corrosion and forms no colored salts that might stain or discolor adjacent components (Fig. 12-3-2).

Copper

Copper alloys, approximately 250 of them, are fabricated in rod, sheet, tube, and wire form. Each of these alloys has some property or combination of properties that makes it unique. They can be grouped into several general headings, such as coppers, brasses, leaded brasses, phosphor bronzes, aluminum bronzes, silicon bronzes, beryllium coppers, cupronickels, and nickel silvers (Fig. 12-3-3).

Copper alloys are used where one or more of the following properties is needed: thermal or electrical conductivity, corrosion resistance, strength, ease of forming, ease of joining, and color.

The major alloy usages are:

1. Copper in pure form as a conductor in the electrical industry
2. Copper or alloy tubing for water, drainage, air conditioning, and refrigeration lines
3. Brasses, phosphor bronzes, and nickel silvers as springs or in construction of equipment if corrosive conditions are too severe for iron or steel

An advantage of copper and its alloys, offered by no other metals, is the wide range of colors available.

Nickel

Commercially pure wrought nickel is a grayish-white metal capable of taking a high polish. Because of its combination of attractive mechanical properties, corrosion resistance, and formability, nickel or its alloys is used in a variety of structural applications usually requiring specific corrosion resistance.

Magnesium

Magnesium, with density of only .06 lb/in.^{3} (1.74 g/cm^3), is the world's lightest structural metal. The combination of low density and good mechanical strength makes possible alloys with a high strength-to-weight ratio.

Zinc

Zinc is a relatively inexpensive metal that has moderate strength and toughness and outstanding corrosion resistance in many types of service.

The principal characteristics that influence the selection of zinc alloys for die castings include the dimensional accuracy obtainable, castability of thin sections, smooth surface, dimensional stability, and adaptability to a wide variety of finishes.

Titanium

Titanium is a light metal at .16 lb/in.^{3} (4.43 g/cm^3); it is 60 percent heavier than aluminum but 45 percent lighter than alloy steel. It is the fourth most abundant metallic element in the earth's crust and the ninth most common element.

Titanium-based alloys are much stronger than aluminum alloys and superior in many respects to most alloy steels.

Beryllium

Beryllium has a strength-to-weight ratio comparable to high-strength steel, yet it is lighter than aluminum. Its melting point

MAJOR ALLOYING ELEMENT	DESIGNATION
Aluminum (99% or more)	1xxx
Copper	2xxx
Manganese	3xxx
Silicon	4xxx
Magnesium	5xxx
Magnesium and silicon	6xxx
Zinc	7xxx
Other elements	8xxx
Unused series	9xxx

FIG. 12-3-2   Wrought aluminum alloy designations.

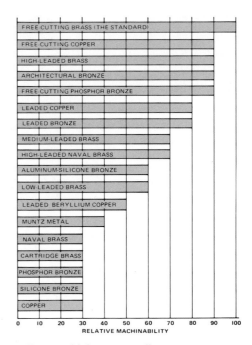

FIG. 12-3-3   Free-machining copper alloys.

is 2345°F (1285°C) and it has excellent thermal conductivity. It is nonmagnetic and a good conductor of electricity.

Refractory Metals

Refractory metals are those metals with melting points above 3600°F (2000°C). Among these, the best known and most extensively used are tungsten, tantalum, molybdenum and niobium. Refractory metals are characterized by high-temperature strength, corrosion resistance, and high melting points.

Tantalum and Niobium

Tantalum and niobium are usually discussed together, since most of their working operations are identical. Unlike molybdenum and tungsten, tantalum and niobium can be worked at room temperatures. The major differences between tantalum and niobium are in density, nuclear cross section, and corrosion resistance. The density of tantalum is almost twice that of niobium.

Molybdenum

Molybdenum is widely used in missiles, aircraft, industrial furnaces, and nuclear projects. Its melting point is lower than that of tantalum and tungsten. Molybdenum has a high strength-to-weight ratio and a low vapor pressure, is a good conductor of heat and electricity, and has a high modulus of elasticity and a low coefficient of expansion.

Tungsten

Tungsten is the only refractory metal that has the combination of excellent corrosion resistance, good electrical and thermal conductivity, a low coefficient of expansion, and high strength at elevated temperatures.

Precious Metals

Gold costs over 8,000 times more than an equal amount of iron; rhodium costs nearly 32,000 times more than copper. With prices such as these, why are precious metals ever specified?

In some cases, precious metals are used for their unique surface characteristics. They reflect light better than other metals. Gold, for example, is specified as a surface for heat reflectors, insulators, and collectors because of its outstanding ability to reflect ultraviolet radiation.

The family of metals called precious metals can be divided into three sub-groups: silver and silver alloys; gold and gold alloys; and the so-called platinum metals, which are platinum, palladium, rhodium, ruthenium, iridium, and osmium.

REFERENCES AND SOURCE MATERIAL

1. *Machine Design*, Materials reference issue.

ASSIGNMENTS ▨▨▨▨▨▨▨▨▨▨▨▨▨▨▨▨▨▨▨▨▨

See Assignments 3 and 4 for Unit 12-3 on pages 344–345.

12-4 PLASTICS

This unit will acquaint drafters with the general characteristics of commercially available plastics so that they can make proper use of plastics in products.

Plastics may be defined as nonmetallic materials capable of being formed or molded with the aid of heat, pressure, chemical reactions, or a combination of these (Fig. 12-4-1).

Plastics are strong, tough, durable materials that solve many problems in machine and equipment design. Metals, it is true, are hard and rigid. This means that they can be machined, to very close tolerances, into cams, bearings, bushings, and gears, which will work smoothly under heavy loads for long periods. Although some come close, no plastic has the hardness and creep resistance of steel, for example. However, metals have many weaknesses that engineering plastics do not. Metals corrode or rust, they must be lubricated, their working surfaces wear readily, they cannot be used as electrical or thermal insulators, they are opaque and noisy, and where they must flex, metals fatigue rapidly.

Plastics can resolve these weaknesses, though not necessarily all with one material. The engineering plastics are resistant to most chemicals; fluorocarbon is one of the most chemically inert substances known. None of the engineering plastics corrode or rust; acetal resin and fluorocarbon are unaffected even when continuously immersed in water. Engineering plastics can be run at low speeds and loads and, without lubrication, are among the world's slipperiest solids, being comparable to ice. Engineering plastics are resilient; therefore, they run more quietly and smoothly than equivalent metal products, and they are able to stand periodic overloads without harmful effects.

Plastics are a family of materials, not a single material. Each material has its special advantages. Being manufactured, plastics raw materials are capable of being variously combined to give almost any property desired in an end product. But these are controlled variations unlike those of nature's products. Some thermoplastics can be sterilized.

The widespread and growing use of plastics in almost every phase of modern living can be credited in large part to their unique combinations of advantages. These advantages are light

FIG. 12-4-1   A variety of plastics parts.

THERMOPLASTICS			
NAME OF PLASTIC	PROPERTIES	FORMS AND METHODS OF FORMING	USES
ABS (Acrylonitrile Butadiene-Styrene)	Strong, tough, good electrical properties.	Available in powder or granules for injection molding, extrusion, and calendering and as sheet for vacuum forming.	Pipe, wheels, football helmets, battery cases, radio cases, children's skates, tote boxes.
Acetal Resin	Rigid without being brittle, tough, resistant to extreme temperatures, good electrical properties.	Produced in powder form for molding and extrusion, available in rod, bar, tube, strip, slab.	Automobile instrument clusters, gears, bearings, bushings, door handles, plumbing fixtures, threaded fasteners, cams.
Acrylics	Exceptional clarity and good light transmission. Strong, rigid, and resistant to sharp blows. Excellent insulator. Colorless or full range of transparent, translucent, or opaque colors.	Available in sheet, rod, tube, and molding powders. Plastic products can be produced by fabricating of sheets, rods, and tubes, hot forming of sheets, injection and compression molding of powder, extrusion, casting.	Airplane canopies and windows, television and camera viewing lenses, combs, costume jewelry, salad bowls, trays, lamp bases, scale models, automobile tail lights, outdoor signs.
Cellulosics (A) Cellulose Acetate		Available in pellets, sheets, film, rods, tubes, strips, coated cord. Can be made into products by injection, compression molding, extrusion, blow molding, and vacuum forming, or sheets and coating.	Eyeglass frames, toys, lamp shades, combs, shoe heels.
(B) Cellulose Acetate Butyrate	Among the toughest of plastics. Retains a lustrous finish under normal wear. Transparent, translucent, or opaque in wide variety of colors and in clear transparent. Good insulators.	Available in pellets, sheets, rods, tubes, strips and as a coating. Can be made into products by injection, compression molding, extrusion, blowing and drawing of sheet, laminating, coating.	Steering wheels, radio cases, pipe and tubing, tool handles, playing cards.
(C) Cellulose Propionate		Available in pellets for injection extrusion or compression molding.	Appliance housing, telephone handsets, pens and pencils.
(D) Ethyl Cellulose		Available in granules, flake, sheet, rod, tube, film, or foil. Can be made into finished products by injection, compression molding, extrusion, drawing.	Edge moldings, flashlights, electrical parts.
(E) Cellulose Nitrate		Available in rods, tubes, sheets for machining and as a coating.	Shoe heel covers, fabric coating.
Fluorocarbons	Low coefficient of friction, resistant to extreme heat and cold. Strong, hard, and good insulators.	Available as powder and granules in resin form. Sheet, rod, tube, film, tape, and dispersions. Molded, extruded, and machined.	Valve seats, gaskets, coatings, linings, tubings.

FIG. 12-4-2   Thermoplastics. *(The Society of Plastics Industry)*

weight, range of color, good physical properties, adaptability to mass-production methods, and often, lower cost.

Aside from the range of uses attributable to the special qualities of different plastics, these materials achieve still greater variety through the many forms in which they can be produced.

They may be made into definite shapes like dinnerware and electric switchboxes. They may be made into flexible film and sheeting such as shower curtains and upholstery. Plastics may be made into sheets, rods, and tubes that are later shaped or machined into internally lighted signs or airplane blisters. They may be made into filaments for use in household screening, industrial strainers, and sieves. Plastics may be used as a coating on textiles and paper. They may be used to bind such materials as fibers of glass and sheets of paper or wood to form boat hulls, airplane wing tips, and tabletops.

Plastics are usually classified as either thermoplastic or thermosetting.

Thermoplastics

Thermoplastics soften, or liquefy, and flow when heat is applied. Removal of the heat causes these materials to set or solidify. They may be reheated and reformed or reused. In this group fall the acrylics, the cellulosics, nylons (polyamides),

THERMOPLASTICS			
NAME OF PLASTIC	PROPERTIES	FORMS AND METHODS OF FORMING	USES
Nylon (Polyamides)	Resistant to extreme temperatures. Strong and long-wearing range of soft colors.	Available as a molding powder, in sheets, rods, tubes, and filaments. Injection, compression, blow molding, and extrusion.	Tumblers, faucet washers, gears. As a filament, it is used as brush bristles, fishing line.
Polycarbonate	High impact strength, resistant to weather, transparent.	Primarily a molding material, may take form of film, extrusion, coatings, fibers, or elastomers.	Parts for aircraft, automobiles, business machines, gages, safety-glass lenses.
Polyethylene	Excellent insulating properties, moisture proof, clear transparent, translucent.	Available in pellet, powder, sheet, film, filament, rod, tube, and foamed. Injection, compression, blow molding, extrusion, coating, and casting.	Ice cube trays, tumblers, dishes, bottles, bags, balloons, toys, moisture barriers.
Polystyrene	Clear, transparent, translucent, or opaque. All colors. Water and weather resistant, resistance to heat or cold.	Available in molding powders or granules, sheets, rods, foamed blocks, liquid solution, coatings, and adhesives. Injection, compression molding, extrusion, laminating, machining.	Kitchen items, food containers, wall tile, toys, instrument panels.
Polypropylenes	Good heat resistance. High resistance to cracking. Wide range of colors.	Processed by injection molding, blow molding, and extrusion.	Thermal dishware, washing machine agitators, pipe and pipe fittings, wire and cable insulation, battery boxes, packaging film and sheets.
Urethanes	Tough and shock-resistant for solid materials. Flexible for foamed material, can be foamed in place.	Solid type—starting two reactants, final article can be extruded, molded, calendered, or cast. Foamed type—can be made by either a prepolymer or one-shot process. In either slab stock or molded form.	Mattresses, cushioning, padding, toys, rug underlays, crash-pads, sponges, mats, adhesion, thermal insulation, industrial tires.
Vinyls	Strong and abrasion-resisting. Resistant to heat and cold. Wide color range.	Available in molding powder, sheet, rod, tube, granules, powder. It can be formed by extrusion, casting, calendering, compression, and injection molding.	Raincoats, garment bags, inflatable toys, hose, records, floor and wall tile, shower curtains, draperies, pipe, paneling.

FIG. 12-4-2   Thermoplastics. (continued)

polyethylene, polystyrene, polyfluorocarbons, the vinyls, polyvinylidene, ABS, acetal resin, polypropylene, and polycarbonates (Fig. 12-4-2).

Thermosetting Plastics

Thermosetting plastics undergo an irreversible chemical change when heat is applied or when a catalyst or reactant is added. They become hard, insoluble, and infusible, and they do not soften upon reapplication of heat. Thermosetting plastics include phenolics, amino plastics (melamine and urea), cold-molded polyesters, epoxies, silicones, alkyds, allylics, and casein (Fig. 12-4-3, pg. 340).

Machining

Practically all thermoplastics and thermosets can be satisfactorily machined on standard equipment with adequate tooling.

The nature of the plastic will determine whether heat should be applied, as in some laminates, or avoided, as in buffing some thermoplastics. Standard machining operations can be used, such as turning, drilling, tapping, milling, blanking, and punching.

Material Selection

One of the first decisions a designer makes is the choice of materials. The choice is influenced by many factors, such as the end use of the product and the properties of the selected material (see Fig. 12-4-4, pg. 341). No attempt is made at this point to discuss the engineering approach to selection of materials.

However, a basic examination and selection of a plastic material at this time will help acquaint the drafter with the wide range of plastics available.

THERMOSETTING PLASTICS			
NAME OF PLASTIC	PROPERTIES	FORMS AND METHODS OF FORMING	USES
Alkyds	Excellent dielectric strength, heat resistance, and resistance to moisture.	Available in molding powder and liquid resin. Finished molded products are produced by compression molding.	Light switches, electric motor insulator and mounting cases, television tuning devices and tube supports. Enamels and lacquers for automobiles, refrigerators, and stoves are typical uses for the liquid form.
Allylics	Excellent dielectric strength and insulation resistance. No moisture absorption; stain resistance. Full range of opaque and transparent colors.	Available in the form of monomers, prepolymers, and powders. Finished articles may be made by transfer or compression molding, lamination, coating, or impregnation.	Electrical connectors, appliance handles, knobs, etc. Laminated overlays or coatings for plywood, hardboard, and other laminated materials needing protection from moisture.
Amino (Melamine and Urea)	Full range of translucent and opaque colors. Very hard, strong, but not unbreakable. Good electrical qualities.	Available as molding powder or granules, as a foamed material in solution, and as resins. Finished products can be made by compression, transfer, plunger molding, and laminating with wood, paper, etc.	Melamine—Tablewear, buttons, distributor cases, tabletops, plywood adhesive, and as a paper and textile treatment. Urea—Scale housing, radio cabinets, electrical devices, appliance housings, stove knobs in resin form as baking enamel coatings, plywood adhesive and as a paper and textile treatment.
Casein	Excellent surface polish. Wide range of near transparent and opaque colors. Strong, rigid, affected by humidity and temperature changes.	Available in rigid sheets, rods, and tubes, as a powder and liquid. Finished products are made by machining of the sheets, rods, and tubes.	Buttons, buckles, beads, game counters, knitting needles, toys, and adhesives.

Cold-Molded 3 Types:
  Bitumin
  Phenolic
  Cement-Asbestos
	Resistance to high heat, solvents, water, and oil.	Available in compounds. Finished articles produced by molding and curing.	Switch bases and plugs, insulators, small gears, handles and knobs, tiles, jigs and dies, toy building blocks.

Epoxy	Good electrical properties; water and weather resistance.	Available as molding compounds, resins, foamed blocks, liquid solutions, adhesives, coatings, sealants.	Protective coating for appliances, cans, drums, gymnasium floors, and other hard-to-protect surfaces. They firmly bond metals, glass, ceramics, hard rubber and plastics, printed circuits, laminated tools and jigs, and liquid storage tanks.
Phenolics	Strong and hard. Heat and cold resistant; excellent insulators.	Cast and molded.	Radio and tv cabinets, washing machine agitators, jukebox housings, jewelry, pulleys, electrical insulation.
Polyesters (Fiberglass)	Strong and tough, bright and pastel colors. High dielectric qualities.	Produced as liquids, dry powders, premix molding compounds, and as cast sheets, rods, and tubes. They are formed by molding, casting, impregnating, and premixing.	Used to impregnate cloth or mats of glass fibers, paper, cotton, and other fibers in the making of reinforced plastic for use in boats, automobile bodies, luggage.
Silicones	Heat resistant, good dielectric properties.	Available as molding compounds, resins, coatings, greases, fluids, and silicon rubber. Finished by compression and transfer molding, extrusion, coating, calendering, casting, foaming, and impregnating.	Coil forms, switch parts, insulation for motors, and generator coils.

FIG. 12-4-3   Thermosetting plastics. *(The Society of Plastics Industry)*

MATERIALS PROPERTY	Thermoplastics											Thermosets				
	ABS	ACETAL	ACRYLIC	CELLULOSIC	FLUOROCARBON	POLYAMIDE	POLYCARBONATE	POLYETHYLENE	POLYPROPYLENE	POLYVINYL CHLORIDE	POLYVINYL	EPOXY	PHENOLICS	POLYESTER	SILICONE	UREA & MELAMINE
Tensile Strength	–	2	2	4	–	1	1	–	–	3	–	3	3	2	1	2
Flexural Strength	4	4	3	4	1	2	3	–	1	3	3	3	2	1	3	3
Impact Strength	2	–	–	2	1	–	2	–	3	2	2	–	1	2	3	3
Hardness	2	2	1	–	–	2	2	–	–	3	–	3	1	3	4	2
Continuous Heat Resistance	–	4	–	–	1	2	–	–	3	–	–	4	2	3	1	4
Weather Resistance	–	2	1	–	1	2	–	2	2	–	2	3	3	1	2	4
Resistance to Heat Expansion	2	–	3	–	–	1	2	–	–	2	3	4	2	2	1	3
Electrical Properties	–	–	4	–	2	–	1	3	3	2	1	1	1	1	1	1
Maximum Volume per Kilogram	3	–	–	4	–	–	–	2	1	3	–	4	1	1	3	2
Chemical Resistance	–	2	–	–	1	3	–	4	4	–	–	1	1	1	1	1
Transparency	–	–	1	3	–	–	4	–	–	2	4	–	–	–	–	–
Resistance to Cold Flow	2	–	3	–	–	–	3	–	–	2	1	–	–	–	–	–
Dimensional Stability to Moisture	2	2	3	–	1	–	2	1	1	1	2	2	–	1	3	2
Colorability	1	3	1	2	4	3	2	3	3	1	3	–	–	1	–	2

Note: Materials rated number 1 are best of those listed for property indicated. Dashes indicate material is not considered for that particular property. Do not compare properties between thermoplastic and thermosetting materials.

FIG. 12-4-4   Property comparison chart for plastics *(General Motors Corp.)*

PRODUCTION REPORT					
Part Name	Material		Reason for Selection	Machining Required	Color
	1st Choice	2nd Choice			
Telephone Case	ABS	Cellulosics	Good impact strength Good range of colors Excellent surface finish Good electrical properties Variety of forming methods	None	Green Blue White Tan Red Black

FIG. 12-4-5   Selection of material. *(STUDIOHIO)*

For instance, the preliminary production report for the material selection of the telephone case shown in Fig. 12-4-5 is an example of the type of research required in selecting a material.

Forming Processes

For design information on the preparation of molded plastics, see Chap. 13, Unit 13-4.

REFERENCES AND SOURCE MATERIAL

1. The Society of the Plastics Industry, Inc.
2. Crystaplex Plastics.
3. General Motors Corp.

ASSIGNMENTS

See Assignments 5 through 8 for Unit 12-4 on page 345–346.

12-5 RUBBER

This unit is to acquaint the drafter with the general characteristics of rubber, both natural and synthetic.

The use of rubber is advantageous when design considerations involve one or more of the following factors:

- Electrical insulation
- Vibration isolation
- Sealing surfaces
- Chemical resistance
- Flexibility

Material and Characteristics

Elastomers (rubber-like substances) are derived from either natural or synthetic sources. Rubber can be formed into useful rigid or flexible shapes, usually with the aid of heat, pressure, or both. The most outstanding characteristics of vulcanized rubber are its low modulus of elasticity and its ability to withstand large deformations and to quickly recover its shape when released. Vulcanized rubber is most compressible. In general, natural rubber has good flex life and low temperature flexibility. Certain synthetic rubbers have characteristics that offer improved performance under conditions that involve such deteriorating effects as heat, oil, and weather.

The cost of each type of rubber and ease of processing are factors to be considered when selecting materials for any application. The use of rubber varies from cements and coatings to soft or hard mechanical goods. Typical formed part are tires, tubes, battery cases, drive belts, machinery mounts, hoses, seals, floormats, gaskets, and weather strips.

Kinds of Rubber

Rubber parts are produced in either mechanical (solid) or cellular form, depending upon the desired performance of the part.They are categorized into two kinds of rubber, natural and synthetic. The synthetic rubbers are divided into several kinds.

Mechanical Rubber

Mechanical rubber is used in either pressure-molded, cast, or extruded forms. Typical parts produced by these methods are tires, belts, and bumpers. Mechanical rubber should be used in preference to sponge rubber because of its superior physical properties.

Cellular Rubber

Cellular rubber can be produced with "open" or "closed" cells. Open-cell sponge rubber is made by the inclusion of a gas-forming chemical compound in the mixture before vulcanization. The heat of the vulcanizing process causes a gas to form in the rubber, making a cellular structure. If compressed, the air is expelled from the cells. When the pressure is released, air is absorbed, allowing the part to quickly recover its shape. Typical applications are pads and weather stripping. Foam rubber is a specialized type of open cell.

Closed-cell sponge rubber is made by an inert gas solution method which produces innumerable ball-shaped cells with continuous walls. When closed-cell sponge rubber is deformed, the cells are displaced rather than deflated. Closed-cell rubber is very springy when squeezed.

Both open- and closed-cell sponge rubber is available in block or sheet form that can be cut to size and shape. This characteristic can sometimes provide a low-cost method of producing relatively simple parts.

Assembly Methods

Several methods of fastening rubber parts to other components of an assembly can be used. When selecting the method of attachment, the designer should consider the hardness of the rubber, operating conditions, and disassembly requirements.

Fastener Inserts

Rubber can be molded to various metallic inserts, as illustrated in Fig. 12-5-1 (pg. 343). Some of the advantages of this practice are the elimination of loose attaching parts, simplification of assembly operations, and reduction of assembly equipment.

Inserts should be designed with holes, undercuts, or such shape that the rubber can overhang an edge. This design provides a mechanical anchor and additional adhesive bond of the rubber to the metal. Sharp edges that cause stress concentrations should be avoided.

Grip Fit

Many molded and extruded soft rubber parts and shapes are designed to take advantage of their gripping action to hold them in place at assembly. This action derives from the characteristic of rubber that permits it to be stretched or extended (Fig. 12-5-2). The grip fit can also serve as a seal for most elements. In applications containing liquids under pressure, additional fasteners should be used to ensure retention and a positive seal.

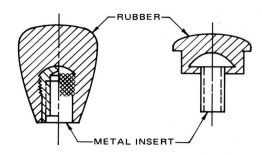

FIG. 12-5-1   Fastener inserts.

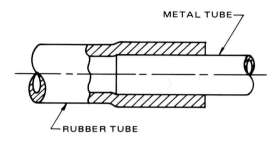

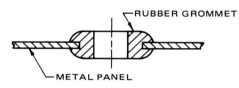

FIG. 12-5-2   Grip fits.

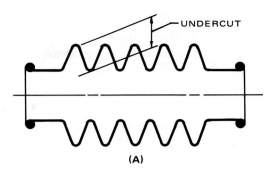

(A)

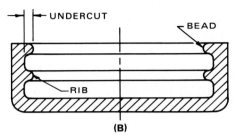

(B)

FIG. 12-5-3   Ribs, undercuts, and beads.

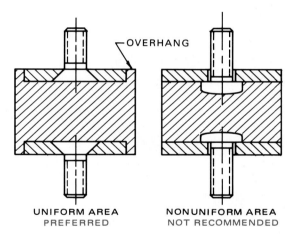

UNIFORM AREA
PREFERRED

NONUNIFORM AREA
NOT RECOMMENDED

FIG. 12-5-4   Wall and section thickness.

Design Considerations

Hard rubber molded parts present problems similar to those of plastics, which are described in Unit 12-4. The following points should be considered in the designing of soft rubber molded parts.

- Reenforcing ribs generally do not represent molding problems. When the inside size is relatively large and the undercut is not too deep, the part may easily be stripped from the mold because of its elasticity (Fig. 12-5-3).
- The thickness of walls and sections depends upon the loading requirements and the hardness of the rubber. Because of the resilience of soft rubber, sections should be of uniform cross section (Fig. 12-5-4).
- Due to the flexibility of rubber and the size and shape of the part, many items do not require draft. However, draft, or taper, usually facilitates molding. The amount of draft depends upon the hardness of the rubber, length of surface, and types of inserts. Generally, at least 0.5° draft per side should be provided.
- Fillets and radii improve the flow of rubber to the various sections. Where rubber is bonded to inserts, the bond will be less likely to fail if a certain amount of rubber is permitted to overhang the edges of the inserts (Fig. 12-5-4).

Specifying Rubber on Drawings

Rubber specifications should always be determined in consultation with a material engineer or rubber parts supplier. Since the broad spectrum of properties of rubber are easily varied by

the ingredients and conditioning of processing, rubber materials should be specified on the basis of performance rather than chemical composition. Specifications normally cover tensile strength, elongation, hardness, compression, set, and various aging and weather tests.

In parts formed of soft rubber compounds, the hardness is usually specified because it is quickly and easily measured and is related to modulus.

REFERENCES AND SOURCE MATERIAL

1. General Motors Corp.

ASSIGNMENTS

See Assignments 9 through 11 for Unit 12-5 on page 346.

ASSIGNMENTS FOR CHAPTER 12

ASSIGNMENT FOR UNIT 12-1, CAST IRONS

1. Make a two-view working drawing of one of the parts shown in Fig. 12-1-A or 12-1-B. Use a revolved section to show the center section of the arm. Select a suitable cast iron for the part. Scale 1:1.

ASSIGNMENT FOR UNIT 12-2, CARBON STEEL

2. Make a working drawing of one of the parts shown in Fig. 12-2-A or 12-2-B. Show the worm threads in pictorial form. Scale 2:1. Select a suitable steel for the part. Conventional breaks may be used to shorten the length of the view.

ASSIGNMENTS FOR UNIT 12-3, NONFERROUS METALS

3. Make a three-view working drawing of the outboard motor clamp shown in Fig. 12-3-A or 12-3-B. Add a full section top view having the cutting plane located at line *AD*. Use lines or surfaces marked *A*, *B*, and *C* as the zero lines and use arrowless dimensioning. Scale 1:2. Select a suitable material noting that the part must be water-resistant, have a painted finish, have moderate strength, and have a light weight or mass.

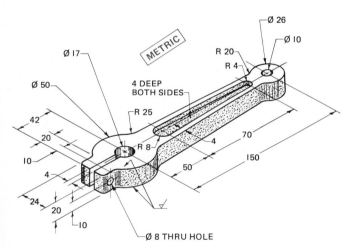

FIG. 12-1-A   Plug wrench.

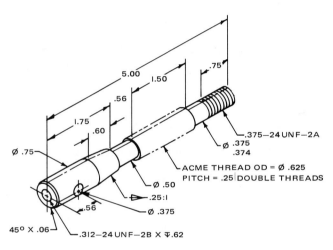

FIG. 12-2-A   Raising bar.

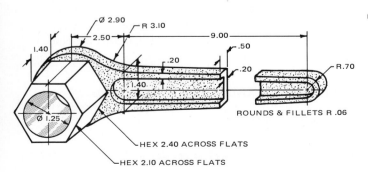

FIG. 12-1-B   Door closer arm.

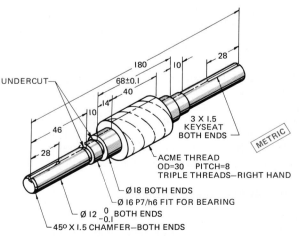

FIG. 12-2-B   Worm for gear jack.

4. Make a two-view assembly drawing, complete with an item list, of the coupling assembly shown in Fig. 12-3-C. The coupling is bolted to an 8 mm steel plate. Using phantom lines, show the plate and the shafts extending a short distance beyond the parts. Select suitable material for the parts. Scale 1:1.

ASSIGNMENTS FOR UNIT 12-4, PLASTICS

5. Design and prepare working drawings for a plastic tee and rubber grommet used for golfing. The design can be standard or novel. The tee is held on to a #20 Am. St. gage aluminum plate by the grommet.
6. Design and prepare a working drawing of a gearshift handle to screw on to a Ø.375 in. (or Ø10 mm) shaft. A threaded insert (see Unit 13-4) is recommended. Selection of material and color to be included on the drawing.
7. Make a one-view detailed assembly drawing of the shaft coupling shown in Fig. 12-4-A. The metal hubs are joined by an elastomer. Assembly sizes: overall length 2.90; shaft diameter .750 (RC4 hole basis fit); hub Ø1.50. Use your judgment for dimensions not shown. Include on the drawing an item list. Scale 1:1.
8. Make a two-view exploded orthographic assembly drawing of the connecting link shown in Fig. 12-4-B (pg. 346). Include on the drawing a material list. The student is to select the material. Show only overall dimensions and shaft sizes.

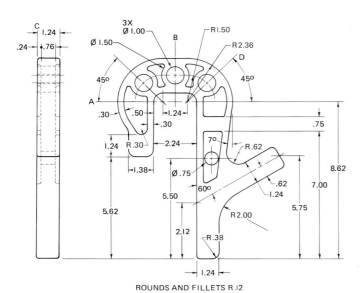

FIG. 12-3-A   Outboard motor clamp.

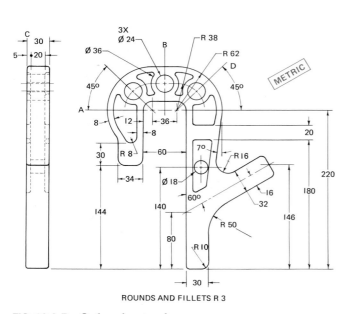

FIG. 12-3-B   Outboard motor clamp.

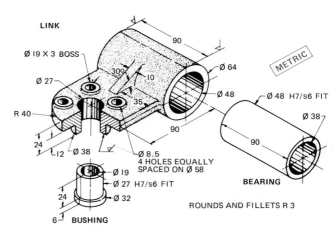

FIG. 12-3-C   Coupling.

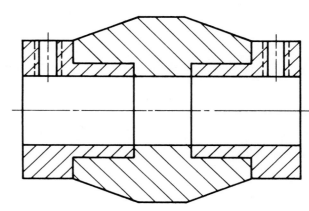

FIG. 12-4-A   Shaft coupling.

ASSIGNMENTS FOR UNIT 12-5, RUBBER

9. Design a rubber boot, similar to that shown in Fig. 12-5-3, to fit on the universal joint shown in Fig. 12-5-A. The purpose of the boot is to prevent dirt and other contaminants from forming around the joint.

10. Make a detail drawing of the boot in Assignment 9.
11. Make a one-view full section assembly drawing of the caster assembly shown in Fig. 12-5-B. Include on the drawing an item list and overall assembly sizes. Select a suitable material for part 1. Scale 1:1.

FIG. 12-4-B   Connecting link.

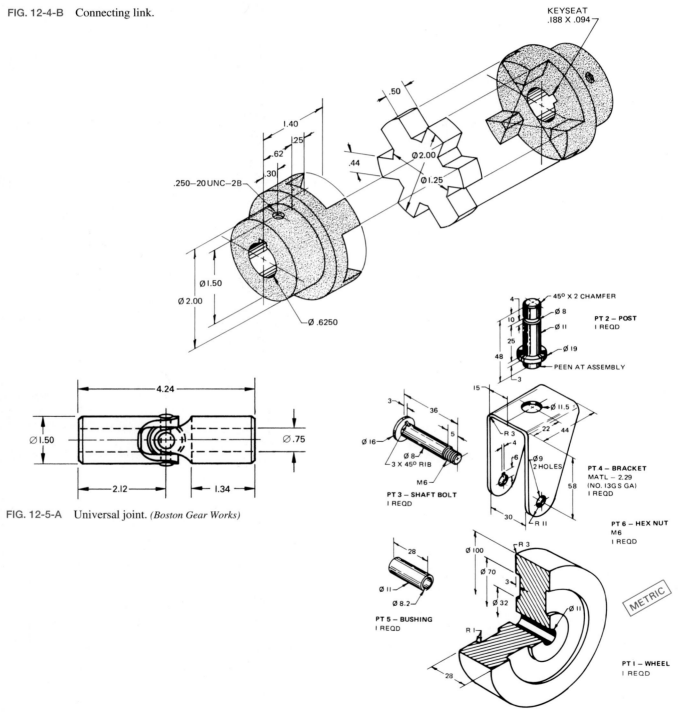

FIG. 12-5-A   Universal joint. *(Boston Gear Works)*

FIG. 12-5-B   Caster assembly.

CHAPTER 13

FORMING PROCESSES

Definitions

Briquetting machines   Machines used to compress powders into finished shapes.

Casting   The process whereby parts are produced by pouring molten metal into a mold.

Closed-die forging   All forging operations involving three-dimensional control.

Cold chamber   A type of die-casting machine used for high-melting nonferrous alloys.

Cold shut   A lap where two surfaces of metal have folded against each other.

Datum surfaces   The surfaces that provide a common reference for measuring, machining, and assembly.

Datum-locating dimension   The dimension between each casting datum surface and the corresponding machining datum surface.

Draft   The slope given to the side walls of a pattern to facilitate easy removal from a mold, or a casting from a die.

Ferrous   Metals that contain iron.

Flash space   The space provided between the die surfaces for the escape of the excess metal, called flash.

Forging   Plastically deforming a cast or sintered ingot, a wrought bar or billet, or a powder-metal shape to produce a desired shape with good mechanical properties.

Investment   The refractory material used to encase a wax pattern.

Nonferrous   Alloys that contain metals such as aluminum, magnesium, and copper but contain no iron.

Parting line   A line along which the pattern is divided for molding, or along which the sections of a mold separate.

Powder metallurgy   The process of making parts by compressing and sintering various metallic and nonmetallic powders into shape.

Shrinkage   The difference between dimensions of the mold and the corresponding dimensions of the molded part.

Slurry   A mixture of plaster of paris and fillers with water and setting-control agents.

Submerged plunger   A type of die-casting machine used for low-melting alloys.

13-1 METAL CASTINGS

Forming Processes

When a component of a machine takes shape on the drawing board or CAD monitor of the designer, the method of its manufacture may still be entirely open. The number of possible manufacturing processes is increasing day by day, and the optimum process is found only by carefully weighing technological advantages and drawbacks in relation to the economics of production.

The choice of the manufacturing process depends on the size, shape, and quantity of the component. Manufacturing processes are therefore important to the engineer and drafter in order to properly design a part. They must be familiar with the advantages, disadvantages, costs, and machines necessary for manufacturing. Since the cost of the part is influenced by the production method, such as welding or casting, the designer must be able to choose wisely the method that will reduce the cost. In some cases it may be necessary to recommend the purchase of a new or different machine in order to produce the part at a competitive price. This means the designer should design the part for the process as well as for the function. Most of all, unnecessarily close tolerances on nonfunctional dimensions should be avoided.

This chapter covers the following manufacturing processes: casting, forging, and powder metallurgy. Forming by means of welding is covered in Chap. 18.

Casting Processes

Casting is the process whereby parts are produced by pouring molten metal into a mold. A typical cast part is shown in Fig. 13-1-1. Casting processes for metals can be classified by either the type of mold or pattern or the pressure or force used to fill the mold. Conventional sand, shell, and plaster molds use a permanent pattern, but the mold is used only once. Permanent molds and die-casting dies are machined in metal or graphite sections and are employed for a large number of castings. Investment casting and the relatively new full mold process involve both an expendable mold and an expendable pattern.

Casting metals are usually alloys or compounds of two or more metals. They are generally classed as ferrous or nonferrous metals. *Ferrous metals* are those that contain iron, the most common being gray iron, steel, and malleable iron. *Nonferrous alloys,* which contain no iron, are those containing metals such as aluminum, magnesium, and copper.

Sand Mold Casting

The most widely employed casting process for metals uses a permanent pattern of metal or wood that shapes the mold cavity when loose molding material is compacted around the pattern. This material consists of a relatively fine sand, which serves as the refractory aggregate, plus a binder.

A typical sand mold, with the various provisions for pouring the molten metal and compensating for contraction of the solidifying metal, and a sand core for forming a cavity in the casting are shown in Fig. 13-1-2. Sand molds consist of two or more sections: bottom (*drag*), top (*cope*), and intermediate sections (*cheeks*) when required. The sand is contained in flasks equipped with pins and plates to ensure the alignment of the cope and drag.

Molten metal is poured into the sprue, and connecting runners provide flow channels for the metal to enter the mold cavity through gates. Riser cavities are located over the heavier sections of the casting. A vent is usually added to permit the escape of gases that are formed during the pouring of metal.

When a hollow casting is required, a form called a *core* is usually used. Cores occupy that part of the mold that is intended to be hollow in the casting. Cores, like molds, are formed of sand and placed in the supporting impressions, or *core prints,* in the molds. The core prints ensure positive location of the core in the mold and, as such, should be placed so that they support the mass of the core uniformly to prevent shifting or sagging. Metal core supports called *chaplets,* which

are used in the mold cavity and which fuse into the casting, are sometimes used by the foundry in addition to core prints (Fig. 13-1-3A, pg. 350). Chaplets and their locations are not usually specified on drawings.

In producing sand molds, a metal or wooden pattern must first be made. The pattern, normally made in two parts, is slightly larger in every dimension than the part to be cast, to allow for shrinkage when the casting cools. This is known as *shrinkage allowance,* and the pattern maker allows for it by using a shrink rule for each of the cast metals.

Drafts, or slight tapers, are also placed on the pattern to allow for easy withdrawal from the sand mold. The parting line location and amount of draft are very important considerations in the design process.

In the construction of patterns for castings in which various points on the surface of the casting must be machined, sufficient excess metal should be provided for all machined surfaces. Allowance depends on the metal used, the shape and size of the part, the tendency to warp, the machining method, and set-up.

After a sand mold has been used, the sand is broken and the casting removed. Next the excess metal, gates, and risers are removed and remelted.

Shell Mold Casting

The refractory sand used in shell molding is bonded by a thermostable resin that forms a relatively thin shell mold. A heated, reusable metal pattern plate (Fig. 13-1-3B) is used to form each half of the mold by either dumping a sand-resin mixture on top of the heated pattern or by blowing resin-coated sand under air pressure against the pattern.

Plaster Mold Casting

Plaster of paris and fillers are mixed with water and setting-control agents to form a *slurry.* This slurry is poured around a reusable metal or rubber pattern and sets to form a gypsum mold (Fig. 13-1-3C). The molds are then dried, assembled, and filled with molten (nonferrous) metals. Plaster mold casting is ideal for producing thin, sound walls. As in sand mold casting, a new mold is required for each casting. Castings made by this process have smoother finish, finer detail, and greater dimensional accuracy than sand castings.

Permanent Mold Casting

Permanent mold casting makes use of a metal mold, similar to a die, which is utilized to produce many castings from each mold (Fig. 13-1-4, pg. 350). It is used to produce some ferrous alloy castings, but due to rapid deterioration of the mold caused by the high pouring temperatures of these alloys and the high mold cost, the process is confined largely to production of nonferrous alloy castings.

Investment Mold Casting

Investment castings have been better known in the past by the term *lost wax castings.* The term *investment* refers to the refractory material used to encase the wax patterns.

This process uses both an expendable pattern and an expendable mold. Patterns of wax, plaster, or frozen mercury

FIG. 13-1-1   Typical cast part.

FIG. 13-1-2   Sequence in preparing a sand casting.

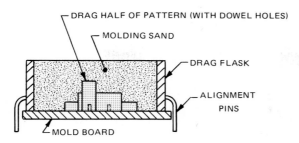

DRAG HALF OF PATTERN (WITH DOWEL HOLES)
MOLDING SAND
DRAG FLASK
ALIGNMENT PINS
MOLD BOARD

(A) STARTING TO MAKE THE SAND MOLD

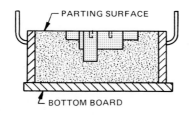

PARTING SURFACE
BOTTOM BOARD

(B) AFTER ROLLING OVER THE DRAG

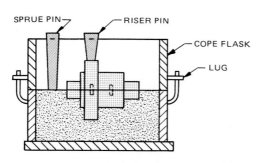

SPRUE PIN
RISER PIN
COPE FLASK
LUG

(C) PREPARING TO RAM MOLDING SAND IN COPE

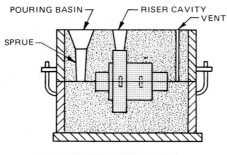

POURING BASIN
RISER CAVITY
VENT
SPRUE

(D) REMOVING RISER AND SPRUE PINS AND ADDING POURING BASIN

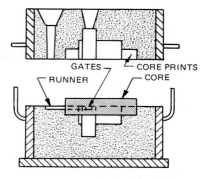

GATES
CORE PRINTS
RUNNER
CORE

(E) PARTING FLASKS TO REMOVE PATTERN AND TO ADD CORE AND RUNNER

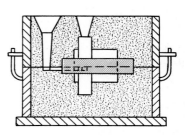

(F) SAND MOLD READY FOR POURING

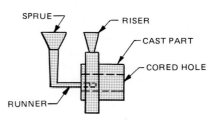

SPRUE
RISER
CAST PART
CORED HOLE
RUNNER

SPRUE, RISER, AND RUNNER TO BE REMOVED FROM CASTING.

(G) CASTING AS REMOVED FROM THE MOLD

FIG. 13-1-3   Mold casting techniques.

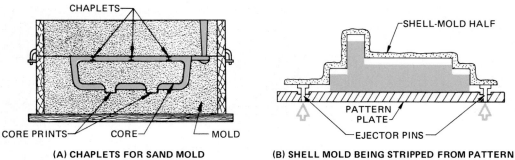

(A) CHAPLETS FOR SAND MOLD

(B) SHELL MOLD BEING STRIPPED FROM PATTERN

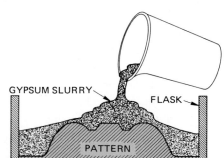

(C) POURING SLURRY OVER A PLASTER MOLD PATTERN

FIG. 13-1-4   Permanent mold casting. *(General Motors Corp.)*

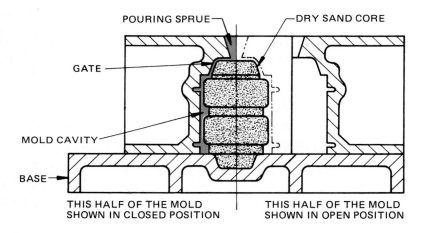

THIS HALF OF THE MOLD SHOWN IN CLOSED POSITION

THIS HALF OF THE MOLD SHOWN IN OPEN POSITION

are cast in metal dies. The molds are formed either by pouring a slurry of a refractory material around the pattern positioned in a flask or by building a thick layer of shell refractory on the pattern by repeated dipping into slurries and drying. The arrangement of the wax patterns in the flask method is shown in Fig. 13-1-5.

Full Mold Casting

The characteristic feature of the full mold process is the use of consumable patterns made of foamed plastic. These are not extracted from the mold but are vaporized by the molten metal.

The full mold process is suitable for individual castings. The advantages it offers are obvious: it is very economical and reduces the delivery time required for prototypes, articles urgently needed for repair jobs, or individual large machine parts.

Centrifugal Casting

In the centrifugal casting process, commonly applied to cylindrical casting of either ferrous or nonferrous alloys, a permanent mold is rotated rapidly about the axis of the casting while a measured amount of molten metal is poured into the mold cavity (Fig. 13-1-6A). The centrifugal force is used to hold the metal against the outer walls of the mold with the volume of metal poured determining the wall thickness of the casting. Rotation speed is rapid enough to form the central hole without a core. Castings made by this method are smooth, sound, and clean on the outside.

Continuous Casting

Continuous casting produces semifinished shapes, such as uniform section rounds, ovals, squares, rectangles, and plates.

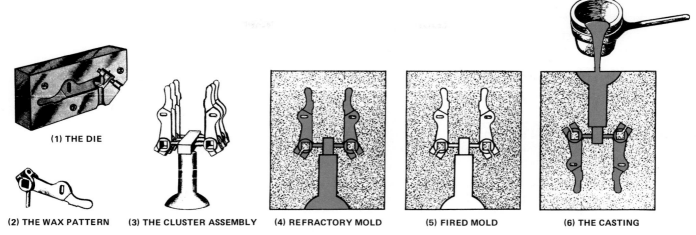

FIG. 13-1-5   Investment mold casting.

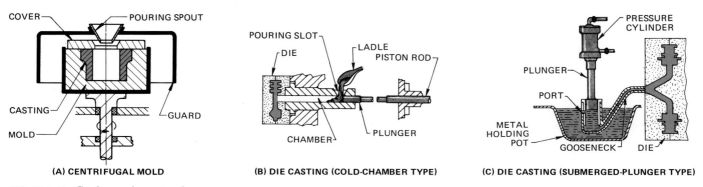

FIG. 13-1-6   Casting equipment and processes.

These shapes are cast from nearly all ferrous and nonferrous metals by continuously pouring the molten metal into a water-jacketed mold. The metal solidifies in the mold, and the solid billet exits continuously into a water spray. These sections are processed further by rolling, drawing, or extruding into smaller, more intricate shapes. Iron bars cast by this process are finished by machining.

Die Casting

One of the least expensive, fastest, and most efficient processes used in the production of metal parts is die casting. Die castings are made by forcing molten metal into a die or mold. Large quantities, accurately cast, can be produced with a die-casting die, thus eliminating or reducing machining costs. Many parts are completely finished when taken from a die. Since die castings can be accurate to within .001 in. (0.02 mm) of size, internal and external threads, gear teeth, and lugs can readily be cast.

Die casting has its limitations. Only nonferrous alloys can be die-cast economically because of the lack of a suitable die material to withstand the higher temperatures required for steel and iron.

Die-casting machines are of two types: the *submerged-plunger* type for low-melting alloys containing zinc, tin, lead, etc., and the *cold-chamber* type for high-melting nonferrous alloys containing aluminum and magnesium (Figs. 13-1-6B and C).

Selection of Process

Selection of the most feasible casting process for a given part requires an evaluation of the type of metal, the number of castings required, their shape and size, the dimensional accuracy required, and the casting finish required. When the casting can be produced by a number of methods, selection of the process is based on the most economical production of the total requirement. Since final cost of the part, rather than price of the rough casting, is the significant factor, the number of finishing operations necessary on the casting is also considered. Those processes that provide the closest dimensions, the best surface finish, and the most intricate detail usually require the smallest number of finishing operations.

A direct comparison of the capabilities, production characteristics, and limitations of several processes is provided in Fig. 13-1-7 (pg. 352).

Design Considerations

The advantages of using castings for engineering components are well appreciated by designers. Of major importance is the

PROCESS	METALS CAST	USUAL WEIGHT (MASS) RANGE	MINIMUM PRODUCTION QUANTITIES	RELATIVE SET-UP COST	CASTING DETAIL FEASIBLE	MINIMUM THICKNESS IN. (MM)	DIMENSIONAL TOLERANCES IN. (MM)	SURFACE FINISH, RMS (μIN.)
SAND (Green, Dry, and Core) CO_2 Sand	All ferrous and nonferrous	Less than 1 lb. (0.5 kg) to several tons	3, without mechanization	Very low to high depending on mechanization	Fair	.12 to .25 (3 to 6) .10 to .25 (2.5 to 6)	± .03 (0.8) ± .02 (0.5)	350 250
SHELL	All ferrous and nonferrous	0.5 to 30 lb. (0.2 to 15 kg)	50	Moderate to high depending on mechanization	Fair to good	.03 to .10 (0.8 to 2.5)	± .015 (0.4)	200
PLASTER	Al, Mg, Cu, and Zn alloys	Less than 1 lb. to 3000 lb. (0.5 to 1350 kg)	1	Moderate	Excellent	.03 to .08 (0.8 to 2)	± .01 (0.2)	100
INVESTMENT	All ferrous and nonferrous	Less than 1 oz. to 50 lb. (30 g to 25 kg)	25	Moderate	Excellent	0.2 to .06 (0.5 to 1.5)	± .005 (0.1)	80
PERMANENT MOLD Metal Mold Graphite Mold	Nonferrous and cast iron Steel	1 to 40 lb. (0.5 to 20 kg) 5 to 300 lb. (2 to 150 kg)	100 100	Moderate to high	Poor	.18 to .25 (4.5 to 6) .25 (6)	± .02 (0.5) ± .03 (0.8)	200 200
DIE	Sn, Pb, Zn, Al, Mg, and Cu alloys	Less than 1 lb. to 20 lb. (0.5 to 10 kg)	1000	High	Excellent	.05 to .08 (1.2 to 2)	± .002 (0.5)	60

* Values listed are primarily for aluminum alloys, but data applies generally to other metals also.

‡ Depends on surface area. Double if dimension is across parting line.

FIG. 13-1-7   General characteristics of casting processes.

fact that they can produce shapes of any degree of complexity and of virtually any size.

Solidification of Metal in a Mold

While this is not the first step in the sequence of events, it is of such fundamental importance that it forms the most logical point to begin understanding the making of a casting. Consider a few simple shapes transformed into mold cavities and filled with molten metal.

In a sphere, heat dissipates from the surface through the mold while solidification commences from the outside and proceeds progressively inward, in a series of layers (Fig. 13-1-8A). As liquid metal solidifies, it contracts in volume, and unless feed metal is supplied, a shrinkage cavity may form in the center.

The designer must realize that a shrinkage problem exists and that the foundry worker must attach risers to the casting or resort to other means to overcome it.

When the simple sphere has solidified further, it continues to contract in volume, so that the final casting is smaller than the mold cavity.

Consider a shape with a square cross section, such as the one shown in Fig. 13-1-8B. Here again, cooling proceeds at

right angles to the surface and is necessarily faster at the corners of the casting. Thus solidification proceeds more rapidly at the corners.

The resulting hot spot prolongs solidification, promoting solidification shrinkage and lack of density in this area. The only logical solution, from the designer's viewpoint, is the provision of very generous fillets or radii at the corners. Additionally, the relative size or shape of the two sections forming the corner is of importance. If they are materially different, as in Fig. 13-1-8E, contraction in the lighter member will occur at a different rate from that in the heavier member. Differential contraction is the major cause of casting stress, warping, and cracking.

General Design Rules

Design for Casting Soundness   Most metals and alloys shrink when they solidify. Therefore, the design must be such that all members of the parts increase in dimension progressively to one or more suitable locations where feeder heads can be placed to offset liquid shrinkage (Fig. 13-1-9).

Fillet or Round All Sharp Angles   Fillets have three functional purposes: to reduce stress concentration in the casting in service; to eliminate cracks, tears, and draws at reentry angles;

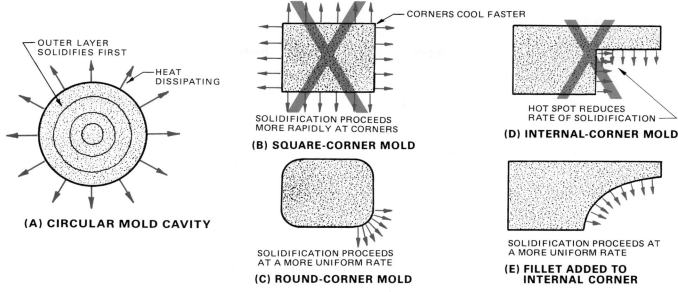

FIG. 13-1-8   Cooling effect on mold cavities filled with molten metal.

FIG. 13-1-9   Design members so that all parts increase progressively to feeder risers. *(Meehanite Metal Corp.)*

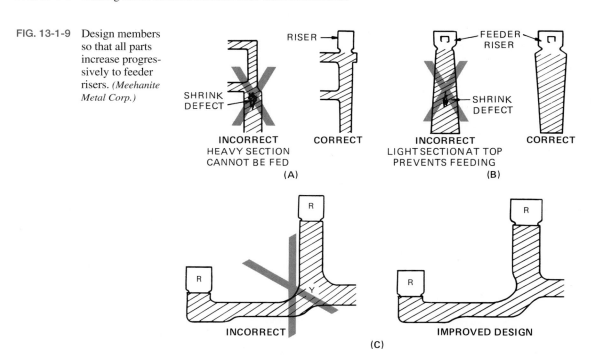

and to make corners more moldable to eliminate hot spots (Fig. 13-1-10, pg. 354).

Bring the Minimum Number of Adjoining Sections Together
A well-designed casting brings the minimum number of sections together and avoids acute angles (Fig. 13-1-11, pg. 354).

Design All Sections as Nearly Uniform in Thickness as Possible   Shrink defects and casting strains existed in the casting illustrated in Fig. 13-1-12 (pg. 354). Redesigning eliminated excessive metal and resulted in a casting that was free from defects, was lighter in weight (mass), and prevented the development of casting strains in the light radial veins.

Avoid Abrupt Section Changes—Eliminate Sharp Corners at Adjoining Sections   The difference in the relative thickness of adjoining sections should be minimum and not exceed a 2:1 ratio (Fig. 13-1-13, pg. 355).

When a change of thickness must be less than 2:1, it may take the form of a fillet; where the difference must be greater, the form recommended is that of a wedge.

Wedge-shaped changes in wall thickness are to be designated with a taper not exceeding 1 in 4.

Design Ribs for Maximum Effectiveness   Ribs have two functions: to increase stiffness and to reduce the mass. If too

FIG. 13-1-10   Fillet all sharp angles.

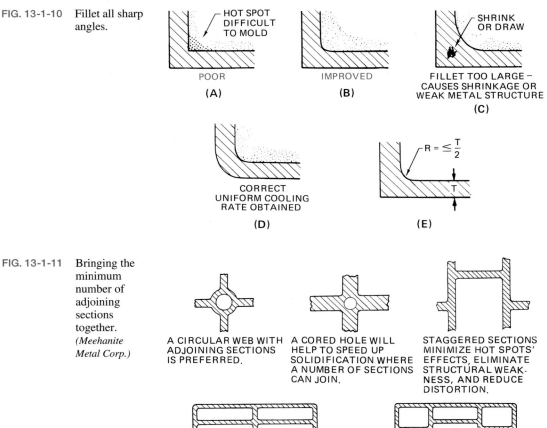

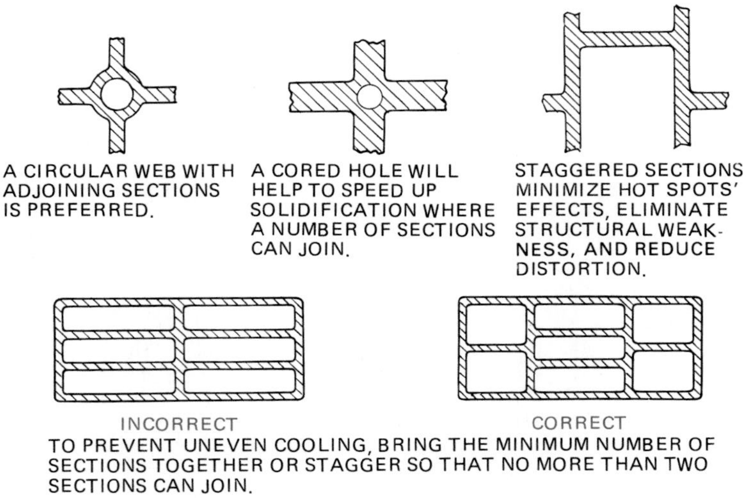

FIG. 13-1-11   Bringing the minimum number of adjoining sections together. *(Meehanite Metal Corp.)*

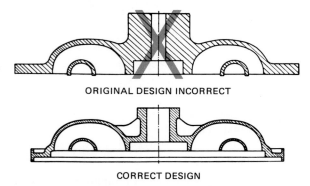

FIG. 13-1-12   Design all sections as nearly uniform in thickness as possible. *(Meehanite Metal Corp.)*

shallow in depth or too widely spaced, they are ineffectual (Fig. 13-1-14).

Avoid Bosses and Pads Unless Absolutely Necessary
Bosses and pads increase metal thickness, create hot spots, and cause open grain or draws. Blend these into the casting by tapering or flattening the fillets. Bosses should not be included

in casting design when the surface to support bolts, etc., may be obtained by milling or countersinking.

Use Curved Spokes   In spoked wheels, a curved spoke is preferred to a straight one. It will tend to straighten slightly, thereby offsetting the dangers of cracking (Fig. 13-1-15).

Use an Odd Number of Spokes   A wheel having an odd number of spokes will not have the same direct tensile stress along the arms as one having an even number and will have more resiliency to casting stresses.

Consider Wall Thicknesses   Walls should be of minimum thickness, consistent with good foundry practice, and should provide adequate strength and stiffness. Wall thicknesses for different materials are as follows:

1. Walls of gray-iron castings and aluminum sand castings should not be less than .16 in. (4 mm) thick.
2. Walls of malleable iron and steel castings should not be less than .18 in. (5 mm) thick.
3. Walls of bronze, brass, or magnesium castings should not be less than .10 in. (2.4 mm) thick.

Select Parting Lines   A *parting line* is a line along which the pattern is divided for molding, or along which the sections of a

FIG. 13-1-13   Avoid abrupt changes. *(Meehanite Metal Corp.)*

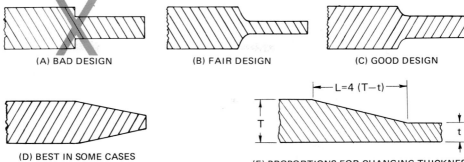

(A) BAD DESIGN   (B) FAIR DESIGN   (C) GOOD DESIGN

(D) BEST IN SOME CASES

$L = 4\,(T - t)$

(E) PROPORTIONS FOR CHANGING THICKNESS

FIG. 13-1-14   Design ribs for maximum effectiveness. *(Meehanite Metal Corp.)*

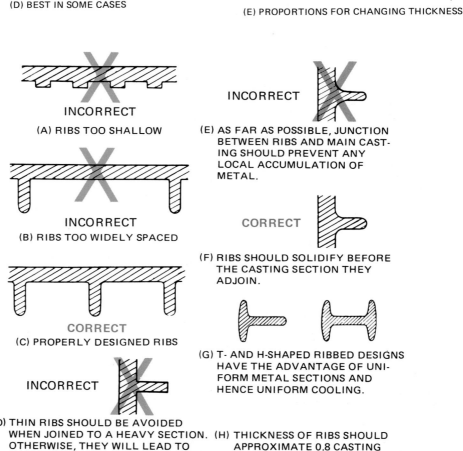

INCORRECT
(A) RIBS TOO SHALLOW

INCORRECT
(B) RIBS TOO WIDELY SPACED

CORRECT
(C) PROPERLY DESIGNED RIBS

INCORRECT
(D) THIN RIBS SHOULD BE AVOIDED WHEN JOINED TO A HEAVY SECTION. OTHERWISE, THEY WILL LEAD TO HIGH STRESSES AND CRACKING.

INCORRECT
(E) AS FAR AS POSSIBLE, JUNCTION BETWEEN RIBS AND MAIN CASTING SHOULD PREVENT ANY LOCAL ACCUMULATION OF METAL.

CORRECT
(F) RIBS SHOULD SOLIDIFY BEFORE THE CASTING SECTION THEY ADJOIN.

(G) T- AND H-SHAPED RIBBED DESIGNS HAVE THE ADVANTAGE OF UNIFORM METAL SECTIONS AND HENCE UNIFORM COOLING.

(H) THICKNESS OF RIBS SHOULD APPROXIMATE 0.8 CASTING THICKNESS

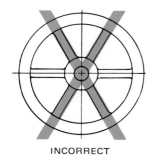

INCORRECT

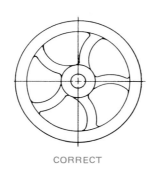

CORRECT

(A) USE AN ODD NUMBER OF CURVED SPOKES

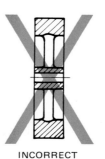

INCORRECT

CAREFULLY BLEND SECTIONS

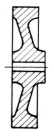

CORRECT

(B) AVOID EXCESSIVE SECTION VARIATION

FIG. 13-1-15   Spoked-wheel design. *(Meehanite Metal Corp.)*

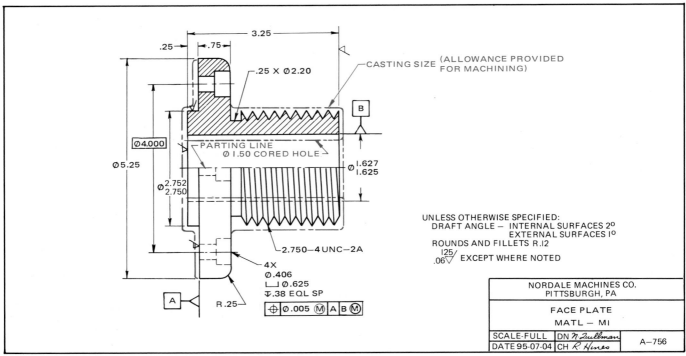

(A) WORKING DRAWING OF A CAST PART

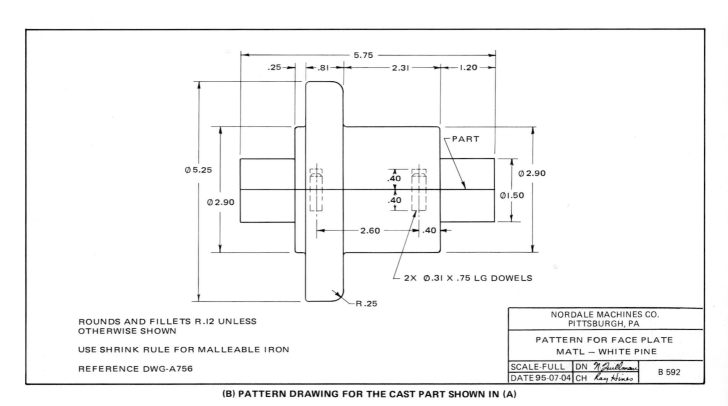

(B) PATTERN DRAWING FOR THE CAST PART SHOWN IN (A)

FIG. 13-1-16   Cast-part drawings.

mold separate. Selection of a parting line depends on a number of factors:

- Shape of the casting
- Elimination of machining on draft surfaces
- Method of supporting cores
- Location of gates and feeders

Drill Holes in Castings   Small holes usually are drilled and not cored.

Drafting Practices

It is important that a detail drawing give complete information on all cast parts, e.g.:

- Machining allowances
- Surface texture
- Draft angles
- Limits on cast surfaces that must be controlled
- Locating points
- Parting lines

On small, simple parts all casting information is included on the finished drawing (Fig. 13-1-16). On more complicated parts, it may be necessary to show additional casting views and sections to completely illustrate the construction of the casting. These additional views should show the rough casting outline in phantom lines and the finished contour in solid lines.

Material   In the selection of material for any particular application, the designer is influenced primarily by the physical characteristics, such as strength, hardness, density, resistance to wear, mass, antifrictional properties, conductivity, corrosion resistance, shrinkage, and melting point.

Machining Allowance   In the construction of patterns for castings in which various points on the surface of the casting must be machined, sufficient excess metal should be provided for all machined surfaces. Unless otherwise specified, Fig. 13-1-17 may be used as a guide to machine finish allowance.

Fillets and Radii   Generous fillets and radii (rounds) should be provided on cast corners and specified on the drawing.

Casting Tolerances   A great many factors contribute to the dimensional variations of castings. However, the standard drawing tolerances specified in Fig. 13-1-17 can be satisfactorily attained in the production of castings.

Draft   All casting methods require a draft or taper on all surfaces perpendicular to the parting line to facilitate removal of the pattern and ejection of the casting. The permissible draft must be specified on the drawing, in either degrees of taper for each surface, inches of taper per inch of length, or millimeters of taper per millimeter of length.

Suitable draft angles in general use for both sand and die castings are 1° for external surfaces and 2° for internal surfaces, as shown in Fig. 13-1-18.

The drawing must always clearly indicate whether the draft should be added to, or subtracted from, the casting dimensions.

Casting Datums

It is recognized that in many cases a drawing is made of the fully machined end product, and casting dimensions, draft, and machining allowances are left entirely to the pattern maker or foundry worker. However, for mass-production purposes it is generally advisable to make a separate casting drawing, with carefully selected datums, to ensure that parts will fit into machining jigs and fixtures and will meet final requirements after machining. Under these circumstances, dimensioning requires the selection of two sets of datum surfaces, lines, or points—one for the casting and one for the machining—to provide common reference points for measuring, machining, and assembly. To select suitable datums, the designer must know how the casting is to be made, where the parting line or lines are to be, and how the part is going to fit into machining jigs and fixtures.

The first step in dimensioning is to select a primary datum surface, sometimes referred to as the *base surface* for the casting, and to identify it as datum *A* (Fig. 13-1-19, pg. 358).

CASTING ALLOY	DIMENSIONS WITHIN THIS RANGE	CASTING ALLOWANCE	STANDARD DRAWING TOLERANCE (±)
Cast Iron, Aluminum, Bronze, Etc. Sand Castings	Up to 8.00	.06	.03
	8.00 to 16.00	.09	.06
	16.00 to 24.00	.12	.07
	24.00 to 32.00	.18	.09
	Over 32.00	.25	.12
Pearlitic, Malleable, and Steel Sand Castings	Up to 8.00	.06	.03
	8.00 to 16.00	.09	.06
	16.00 to 24.00	.18	.09
	Over 24.00	.25	.12
Permanent and Semipermanent Mold Castings	Up to 12.00	.06	.03
	12.00 to 24.00	.09	.06
	Over 24.00	.18	.09
Plaster Mold Castings	Up to 8.00	.03	.02
	8.00 to 12.00	.06	.03
	Over 12.00	.10	.06

FIG. 13-1-17   Guide to machining and tolerance allowance in inches for castings.

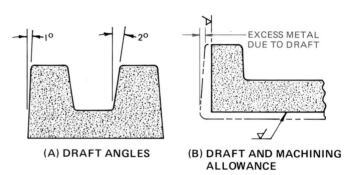

(A) DRAFT ANGLES   (B) DRAFT AND MACHINING ALLOWANCE

FIG. 13-1-18   Draft for removing of casting from mold.

This primary datum should be a surface that meets the following criteria as closely as possible:

1. It must be a surface, or datum targets on a surface (see Fig. 13-1-20), that can be used as the basis for measuring the casting and that can later be used for mounting and locating the part in a jig or fixture, for the purpose of machining the finished part.
2. It should be a surface that will not be removed by machining, so that control of material to be removed is not lost, and can be checked at final inspection.
3. It should be parallel with the top of the mold, or parting line, that is, a surface that has no draft or taper.
4. It should be integral with the main body of the casting, so that measurements from it to the main surfaces of the casting will be least affected by cored surfaces, parting lines, or gated surfaces.
5. It should be a surface, or target areas on a surface, on which the part can be clamped without causing any distortion, so that the casting will not be under a distortional stress for the first machining operation.

6. It should be a surface that will provide locating points as far apart as possible, so that the effect of any flatness error will be minimized.

The second step is to select two other planes to serve as secondary and tertiary surfaces. These planes should be at right angles to one another and to the primary datum surface. They probably will not coincide with actual surfaces, because of taper or draft, except at one point, usually a point adjacent to the primary datum surface. These are identified as datum *B* and datum *C*, respectively, as shown in Fig. 13-1-21.

In the case of a circular part, the endview center lines may be selected as secondary and tertiary datums, as shown in Fig. 13-1-22B. In this case, unless otherwise specified, the center lines represent the center of the outside or overall diameter of the part.

Machining Datums

The first step in dimensioning the machined or finished part is to select a primary datum surface for machining and to identify it as datum *D* (Fig. 13-1-22A). This surface is the first surface on the casting to be machined and is thereafter used as the datum surface for all other machining operations. It should be selected to meet the following criteria:

1. It is generally preferable, though not essential, that it be a surface that is parallel to the primary casting datum surface.
2. It may be a large, flat, machined surface or several small areas of surfaces in the same or parallel planes.

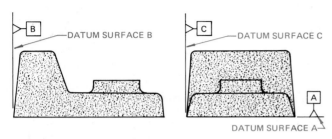

FIG. 13-1-19   Casting datums.

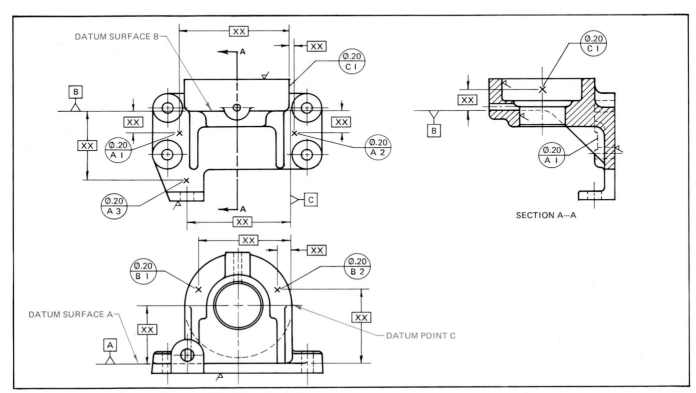

FIG. 13-1-20   Machined cast drawing illustrating datum lines, set-up points, and surface finish.

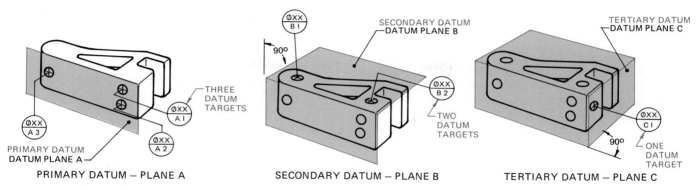

FIG. 13-1-21   Datum planes and datum targets.

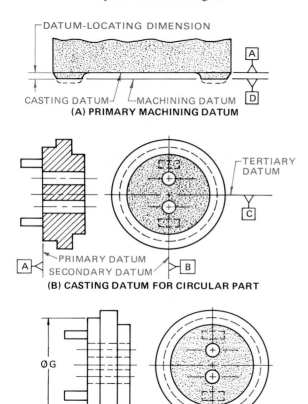

FIG. 13-1-22   Casting and machining datums.

3. If the primary casting datum surface is smooth and does not require machining, as in die castings, or if suitable target areas have been selected, the same surface may be used as the machining datum surface.

4. If the primary casting datum surface of sand castings appears to be the only suitable surface, it is recommended that three or four pads be provided, which can be machined to form the machining datum surface, as shown in Fig. 13-1-22A.

5. When pads or small target areas are selected, they should be placed as far apart as possible and located where the part can be readily clamped in jigs or fixtures without distorting it or interfering with other machining operations.

The second step is to select two other surfaces to serve as secondary and tertiary datums. If these datum surfaces are required only for locating and dimensioning purposes, and not for clamping in a jig or fixture, some suitable datums other than flat, machined surfaces may be chosen. These could be the same datums as used for casting, if the locating point in each case is clearly defined and is not removed in machining. For circular parts, a hole drilled in the center, or a turned diameter other than the outside diameter, may provide a suitable center line for use as secondary datum surfaces (Fig. 13-1-22C).

The third step is to specify the *datum-locating dimension*, that is, the dimension between each casting datum surface and the corresponding machining datum surface (Fig. 13-1-22A). There is never more than one such dimension from each casting datum surface.

Dimensions

When suitable datum surfaces have been selected, with datum-locating dimensions for the machined-casting drawing, dimensioning may proceed, with dimensions being specified directly from the datums to all main surfaces. However, where it is necessary to maintain a particular relationship between two or more surfaces or features, regular point-to-point dimensioning is usually the preferred method. This will normally include all such items as thickness of ribs, height of bosses, projections, depth of grooves, most diameters and radii, and center distances between holes and similar features. Whenever possible, specify dimensions to surfaces or surface intersections, rather than to radii centers or nonexistent center lines. Dimensions given on the casting drawing should not be repeated, except as reference dimensions, on the machined-part drawing.

REFERENCES AND SOURCE MATERIAL

1. American Iron and Steel Institute, *Principles of Forging Design.*
2. General Motors Corp.
3. Meehanite Metal Corp.

ASSIGNMENTS ▓▓▓▓▓▓▓▓▓▓▓▓▓▓▓▓▓▓▓▓▓▓▓▓▓

See Assignments 1 through 6 for Unit 13-1 on pages 371 through 373.

13-2 FORGINGS

Forging consists of plastically deforming, either by a squeezing pressure or sharp blows, a cast or sintered ingot, a wrought bar or billet, or a powder-metal shape, to produce a desired shape with good mechanical properties. Practically all ductile metals can be forged (Fig. 13-2-1).

Closed-Die Forging

Impression Dies

Closed-die forgings are made by hammering or pressing metal until it conforms closely to the shape of the enclosing dies. Grain flow in the closed-die-forged parts can be oriented in the direction requiring greatest strength. In practice, *closed-die forging* has become the term applied to all forging operations involving three-dimensional control.

Three-dimensional control of the material to be forged requires a closed die, a simple and common form of which is the impression die.

In the simplest example of impression-die forging (Fig. 13-2-2) the workpiece is cylindrical and is placed in the bottom-half die. On closing of the top-half die, the cylinder undergoes elastic compression until its enlarged sides touch the side walls of the die impression. At this point, a small amount of excess material begins to form the flash between the two die faces.

The forging impression die gives control over all three directions, except when the die is similar to that shown in Fig. 13-2-2, and the deforming forging machine tool has an unlimited stroke (e.g., a hammer or hydraulic press). In the latter case, the die must be shaped to allow complete closing of the striking faces at the end of the stroke.

Forging dies can be divided into three main classes: single-impression, double-impression, and interlocking (Fig. 13-2-3). Single-impression dies have the impression of the desired forging entirely in one half of the die. Double-impression dies have part of the impression of the desired forging sunk in each die in such a manner that no part of the die projects past the parting line into the other die. This type is the most common class of forging.

Trimming Dies

Because the quantity of forging metal is generally in excess of the space in the die cavity, space is provided between the die surfaces for the escape of the excess metal. This space is called the *flash space*, and the excess metal that flows into it is called *flash*. The flash thickness is proportionate to the mass of the forging.

The flash is removed from forgings by trimming dies, which are formed to the outline of the part (Fig. 13-2-4).

General Design Rules

Corner and Fillet Radii   It is important in forging design to use correct radii where two surfaces meet. Corner and fillet radii on forgings should be sufficient to facilitate the flow of metal.

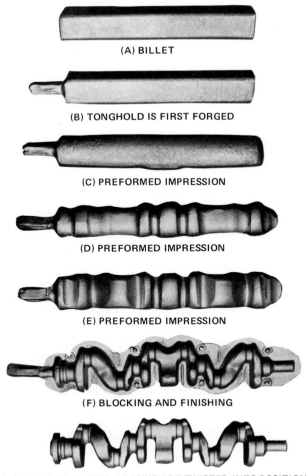

(A) BILLET

(B) TONGHOLD IS FIRST FORGED

(C) PREFORMED IMPRESSION

(D) PREFORMED IMPRESSION

(E) PREFORMED IMPRESSION

(F) BLOCKING AND FINISHING

(G) AFTER TRIMMING, CRANKS ARE TWISTED INTO POSITION

FIG. 13-2-1   The forging of a crankshaft. *(Wyman-Gordon Co.)*

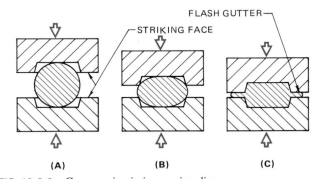

(A)   (B)   (C)

FIG. 13-2-2   Compression in impression dies.

Stress concentrations resulting from abrupt changes in section thickness or direction are minimized by corner and fillet radii of correct size. Any radius larger than recommended will increase die life. Any radius smaller than recommended will decrease die life. See Fig. 13-2-5 for recommendations.

Sharp fillets cause the formation of cold shuts. In a forging, a *cold shut* is a lap where two surfaces of metal have folded against each other, forming an undesirable flow of metal. A cold shut causes a weak spot that may be opened into a crack

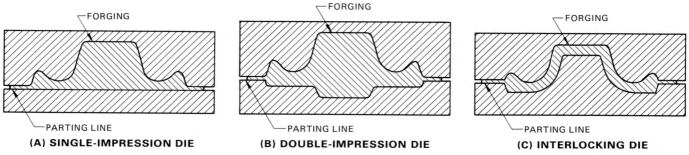

(A) SINGLE-IMPRESSION DIE   **(B) DOUBLE-IMPRESSION DIE**   **(C) INTERLOCKING DIE**

FIG. 13-2-3   Forging dies.

BEFORE TRIMMING   FLASH   AFTER TRIMMING

FIG. 13-2-4   Flash trimming.

by heat treatment. Cold shuts are most likely to form at fillets in deep depressions or in deep sections, especially where the metal is confined (Fig. 13-2-6, pg. 362). In these cases larger fillets are required, as shown in Fig. 13-2-5.

Draft Angle   Draft is one of the first factors to be considered in designing a forged part (Fig. 13-2-7, pg. 362). *Draft* is defined as the slope given to the side walls of the die in order to facilitate removal of the forging. Where little or no draft is allowed, stripper or ejection mechanisms must be used. The usual amount of draft for exterior contours is 7° and for interior contours, 10°.

Die draft equivalent is the amount of offset that results from draft. Figure 13-2-8 (pg. 362) shows the draft equivalents for varying angles and depth of draft.

Parting Line   The surfaces of dies that meet in forgings are the striking surfaces. The line of meeting is the parting line. The parting line of the forging must be established in order to determine the amount of draft and its location.

The location and the type of parting as applied to simple forgings are shown in Fig. 13-2-9 (pg. 362).

Drafting Practices

In preparing forging drawings, it is important to consider drafting practices that may be peculiar to forgings, such as:

- Draft angles and parting lines
- Corner and fillet radii
- Forging tolerances
- Machining allowances

FIG. 13-2-5   Corner and fillet radii.

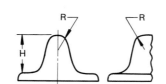

H		R
OVER	TO AND INCL	
0	1.00 (25)	.06 (1.5)
1.00 (25)	1.50 (35)	.09 (2.5)
1.50 (35)	2.00 (50)	.12 (3)
2.00 (50)	3.00 (80)	.18 (4.5)

MIN CORNER RADII

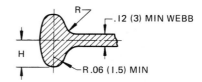

H		R
OVER	TO AND INCL	
0	.30 (8)	R = H
.30 (8)	.50 (13)	$R = \dfrac{3H}{4}$

FILLET RADII FOR SMALL RIBS

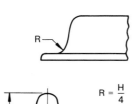

$R = \dfrac{H}{4}$

FILLET RADII WHEN METAL IS CONFINED

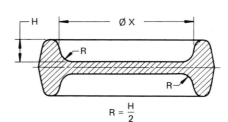

$R = \dfrac{H}{2}$

DEPTH OF A FORGED RECESS SHOULD NOT EXCEED 0.67 X DIA.

FILLET RADII WHEN METAL IS NOT CONFINED

FIG. 13-2-6   Cold shut.

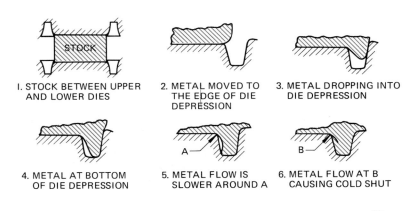

1. STOCK BETWEEN UPPER AND LOWER DIES

2. METAL MOVED TO THE EDGE OF DIE DEPRESSION

3. METAL DROPPING INTO DIE DEPRESSION

4. METAL AT BOTTOM OF DIE DEPRESSION

5. METAL FLOW IS SLOWER AROUND A

6. METAL FLOW AT B CAUSING COLD SHUT

FIG. 13-2-7   Draft application.

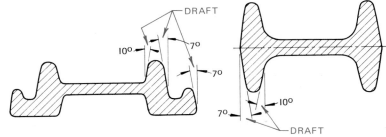

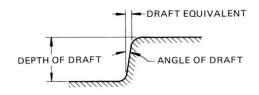

DEPTH OF DRAFT	DRAFT EQUIV FOR ANGLE OF		
	5°	7°	10°
.20 (5)	.018 (0.437)	.024 (0.614)	.035 (0.882)
.40 (10)	.035 (0.875)	.050 (1.228)	.070 (1.763)
.60 (15)	.052 (1.312)	.074 (1.842)	.106 (2.645)
.80 (20)	.070 (1.750)	.100 (2.456)	.140 (3.527)
1.00 (25)	.088 (2.187)	.123 (3.070)	.176 (4.408)

FIG. 13-2-8   Die draft equivalent.

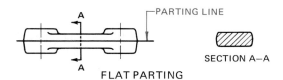

FLAT PARTING

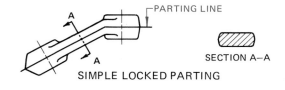

SIMPLE LOCKED PARTING

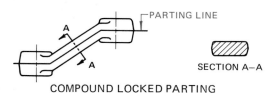

COMPOUND LOCKED PARTING

FIG. 13-2-9   Parting line application.

- Heat treatment
- Location of trademark, part number, and vendor specification

Dimensioning   It is usually desirable to apply dimensions of the depths of the die impressions to the forged part. Draft is additive to these dimensions and should be expressed in degrees or linear dimensions.

When the depth of the die impression is located, only one dimension should originate from the parting line. This surface should then be used to establish other dimensions, as shown in Fig. 13-2-10A.

Allowance for Machining   When a forging is to be machined, allowance must be made for metal to be removed.

Composite Drawings   Generally, a forged part should be shown on one drawing with the forging outline shown in phantom lines, as in Fig. 13-2-10B. Forging outlines for machining allowance should not be dimensioned unless the amount of finish cannot be controlled by the machining symbol.

FIG. 13-2-10   Forged-part drawings.

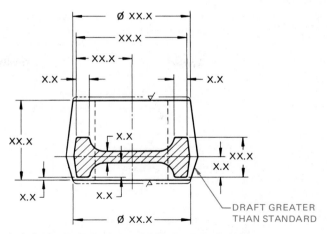

(A) DIMENSIONING A FORGED DRAWING

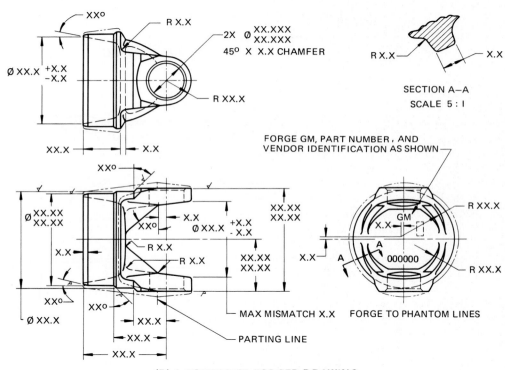

(B) A COMPOSITE FORGED DRAWING

Separate drawings for rough forgings should be made only when the part is complicated and the outline of the rough forging cannot be clearly visualized, or where the outline of the rough forging must be maintained for tooling purposes.

Where both the forging and machining drawings are shown on the same sheet, as in Fig. 13-2-11 (pg. 364), place the headings FORGING DRAWING and MACHINING DRAWING directly under the corresponding views.

REFERENCES AND SOURCE MATERIAL

1. Frank Burbank, "Forging," *Machine Design*, Vol. 37, No. 21.

ASSIGNMENTS

See Assignments 7 and 8 for Unit 13-2 on page 374.

13-3 POWDER METALLURGY

Powder metallurgy is the process of making parts by compressing and sintering various metallic and nonmetallic powders into shape (Fig. 13-3-1, pg. 364).

Dies and presses known as *briquetting machines* are used to compress the powders into shape. These briquets or compacts

FIG. 13-2-11   Separate forging and machining drawings.

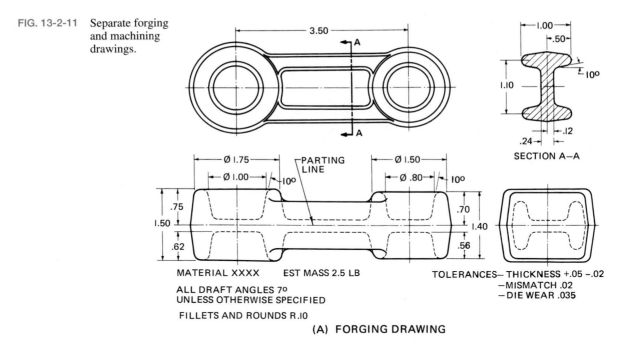

MATERIAL XXXX   EST MASS 2.5 LB

ALL DRAFT ANGLES 7°
UNLESS OTHERWISE SPECIFIED

FILLETS AND ROUNDS R.10

TOLERANCES— THICKNESS +.05 −.02
—MISMATCH .02
—DIE WEAR .035

(A) FORGING DRAWING

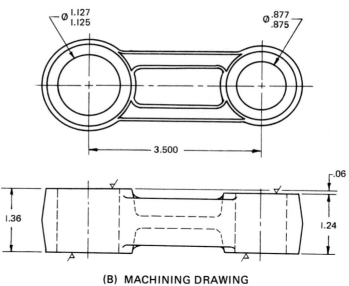

(B) MACHINING DRAWING

FIG. 13-3-1   Compacting sequence for powder metallurgy.

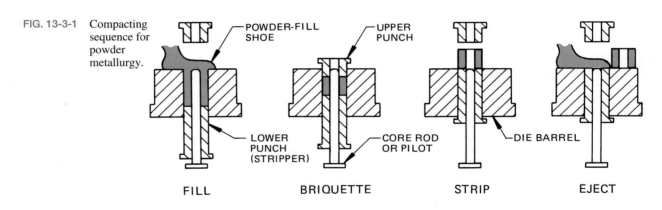

FILL   BRIQUETTE   STRIP   EJECT

are then sintered, or heated, in an atmosphere-controlled furnace, bonding the powdered materials.

Design Considerations

The following considerations should be taken into account when powder-metal parts are designed in order to realize the maximum benefits from the powder metallurgy process (Fig. 13-3-2). This process is most applicable to the production of cylindrical, rectangular, or irregular shapes that do not have large variations in cross-sectional dimensions. Splines, gear teeth, axial holes, counterbores, straight knurls, serrations, slots, and keyseats present few problems.

Ejection from the Die   The shape of the part must permit ejection from the die. The design requirements for some parts can be achieved only by subsequent machining, as in some corner relief designs, reverse tapers, holes at right angles to the direction of pressing, diamond knurls, and undercuts.

Axial Variations   Slots having a depth greater than one-fourth the axial length of the part require multiple-punch action and result in high production costs.

Corner Reliefs   Corner reliefs can be molded or machined. A molded corner relief will save machining.

Reverse Tapers   Reverse tapers cannot be molded. They must be machined.

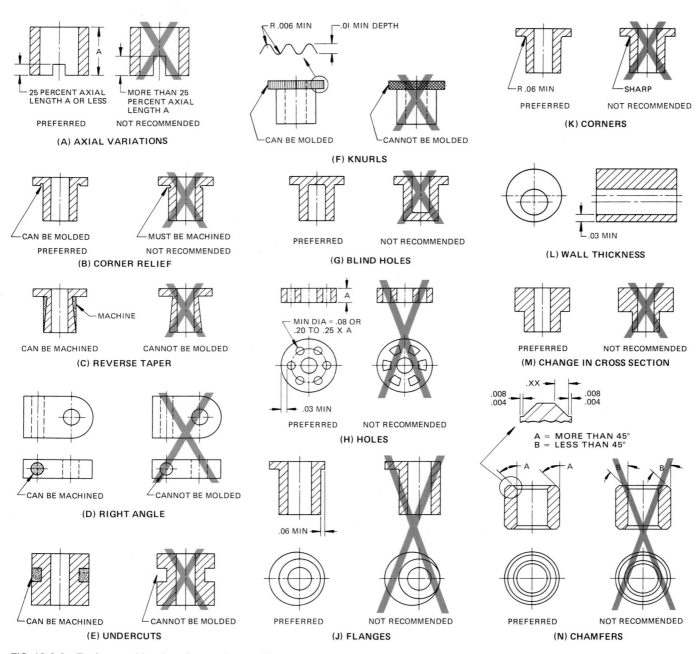

FIG. 13-3-2   Design considerations for powder metallurgy.

Holes at Right Angles to the Direction of Pressing   Right-angle holes must be machined.

Undercuts   Undercuts must be machined.

Knurls   Straight knurls can be molded; diamond knurls cannot.

Blind Holes   If a flange is opposite the blind end of the hole, the part must be modified to allow powder to fill in the die.

Holes   The use of round holes, instead of odd-shaped holes, will simplify tooling, strengthen the part and reduce costs.

Flanges   A .06 in. (1.5 mm) minimum flange overhang is desired to provide longer tool life.

Corners   A fillet radius must be provided under the flange on a flanged part. It allows uniform powder flow in the die and produces a high-strength part.

Wall Thickness   In general, sidewalls bordering a depression or hole should be a minimum of .03 in. (0.8 mm) thick.

Changes in Cross Section   Large changes in cross section should be avoided because they cause density variation. Warping and cracking are likely to occur during sintering.

Chamfers   Care in the design of chamfers minimizes sharp edges on tools and improves tool life.

REFERENCES AND SOURCE MATERIALS

1. General Motors Corp.

ASSIGNMENTS

See Assignments 9 and 10 for Unit 13-3 on pages 375–376.

13-4 PLASTIC MOLDED PARTS

Single Parts

The design of molded parts involves several factors not normally encountered with machine-fabricated and assembled parts. It is important that designers take these factors into consideration.

Shrinkage   *Shrinkage* is defined as the difference between dimensions of the mold and the corresponding dimensions of the molded part. Normally the mold designer is more concerned with shrinkage than the molded-part designer. Shrinkage does, however, affect dimensions, warpage, residual stress, and moldability.

Section Thickness   Solidification is a function of heat transfer from or to the mold for both thermoplastics and thermosets. Each material has a fixed rate of heat transfer. Therefore, where section thickness varies, areas within a molded part will solidify at different rates. The varying rates will cause irregular shrinkage, sink marks, additional strain, and warpage. For these reasons, uniform section thickness is important and may be maintained by adding holes or depressions, as shown in Fig. 13-4-1A and B.

Gates   Gate location should be anticipated during the design stage. Avoid gating into areas subjected to high stress levels, fatigue, or impact. To optimize molding, locate gates in the heaviest section of the part (Fig. 13-4-1C).

Parting or Flash Line   As described earlier, flash is that portion of the molding material that flows or exudes from the mold parting line during molding. Any mold that is made of two or more parts may produce flash at the line of junction of the mold parts. The thickness of flash usually varies between .002 and .016 in. (0.05 and 0.40 mm), depending upon the accuracy of the mold, type of material, and the process used (Fig. 13-4-1D).

Fillets and Radii   The principal functions of fillets and radii (rounds) are to ease the flow of plastic within the mold, to facilitate ejection of the part, and to distribute stress in the part in service. During molding, the material is liquefied, but it is a heavy, viscous liquid that does not easily flow around sharp corners. The liquid tends to bend around corners; therefore, rounded corners permit the liquid plastic to flow smoothly and easily through the mold. For recommended radii, see Fig. 13-4-1E.

Molded Holes   A through hole is more advantageous than a blind hole since it is more accurate and economical. Blind holes should not be more than twice as deep as their diameter, as shown in Fig. 13-4-1F. Avoid placing holes at angles other than perpendicular to the flash line. If such holes are necessary, consider using a drilled hole to maintain simple molding.

Internal and External Draft   Draft is necessary on all rigid molded articles to facilitate removal of the part from the mold. Draft may vary from 0.25 to 4° per side, depending upon the length of the vertical wall, surface area, finish, kind of material, and the mold or method of ejection used.

Threads   External and internal threads can be easily molded by means of loose-piece inserts and rotating core pins. External threads may be formed by placing the cavity so that the threads are formed in the mold pattern.

Ribs and Bosses   Ribs increase rigidity of a molded part without increasing wall thickness and sometimes facilitate flow during molding. Bosses reinforce small, stressed areas, providing sufficient strength for assembly with inserts or screws. Recommended proportions for ribs and bosses are shown in Fig. 13-4-1G.

Undercuts   Parts with undercuts should be avoided. Normally, parts with external undercuts cannot be withdrawn from a one-piece mold. Internal undercuts are considered impractical and should be avoided. If an internal undercut is essential, it may be achieved by machining or by use of a flexible mold core material (Fig. 13-4-1H).

Assemblies

The design of molded parts that are to be assembled with typical fastening methods involves factors different from those normally encountered with metal.

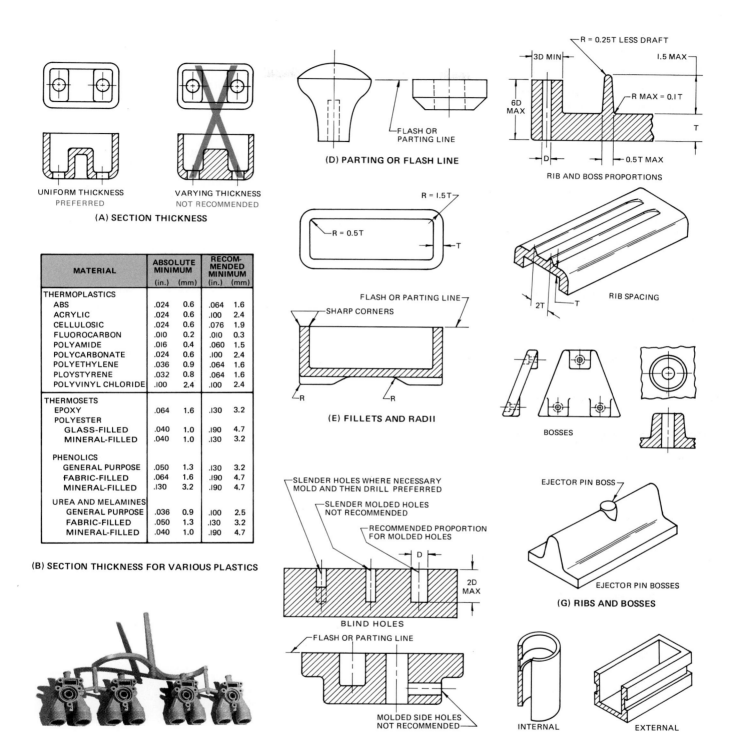

FIG. 13-4-1   Design considerations for plastic molded parts.

Holes and Threads   Mechanical fasteners, in general, depend upon a hole of some type. Holes should be designed and located to provide maximum strength and minimum molding problems. Any straight hole, molded or machined, should have between it and an adjacent hole, or side wall, an amount of material equal to or greater than the diameter or width of the hole. Any threaded hole, molded or tapped, should have

between it and an adjacent hole, or side wall, an amount of material at least three times the outside diameter of the thread. Spacing may be reduced, however, by proper use of bosses.

Drilled holes are often more accurate and easier to produce than molded holes, even though they require a second operation.

Tapped holes provide an economical means of joining a molded part to its assembly. The designer should avoid threads with a pitch of less than .03 in. (0.8 mm). Holes that are to be tapped should be countersunk to prevent chipping when the tap is inserted.

External and internal threads can be molded integrally with the part. Molded threads are usually more expensive to form than other threads because either a method of unscrewing the part from the mold must be provided or a split mold must be used.

Inserts   After the molding material has been determined, the insert should be designed. The molded part should be designed around the insert.

Inserts of round rod stock, coarse diamond-knurled and grooved, provide the strongest anchorage under torque and tension. A large single groove with knurling on each end is superior to two or more grooves with smaller knurled surface areas. See examples of inserts in Fig. 13-4-2.

Press and Shrink Fits   Inserts may be secured by a press fit, or the plastic molding material may be assembled to a larger part by a shrink fit, as shown in Fig. 13-4-3. Both methods rely on shrinkage of the material, which is greatest immediately after removal from the mold.

Heat Forming and Heat Sealing   Most thermoplastics can be re-formed by the application of heat and pressure, as shown in Fig. 13-4-4. This re-forming often eliminates the need for other assembly methods, such as adhesive bonding and mechanical fasteners. This method cannot be used with thermosetting materials.

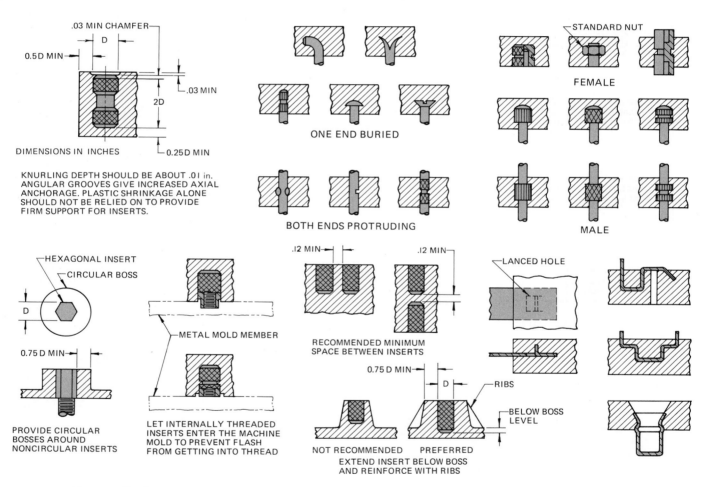

FIG. 13-4-2   Insert applications.

FIG. 13-4-3   Press and shrink fits.

Mechanical Fastening   Various designs of mechanical fasteners are commercially available. Spring-type metal hinges and clips, speed clips or nuts, and expanding rivets are a few of these designs. Design of the parts for assembly requires that molded parts have sufficient sectional strength to withstand the stresses that will be encountered with fasteners. A strengthening of the area that will receive the brunt of these applied stresses is usually required (Fig. 13-4-5).

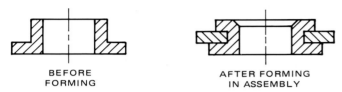

BEFORE
FORMING

AFTER FORMING
IN ASSEMBLY

FIG. 13-4-4   Heat forming.

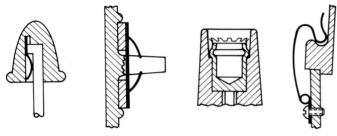

FIG. 13-4-5   Mechanical fasteners.

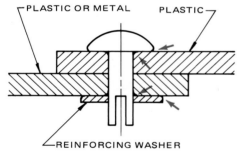

PLASTIC OR METAL   PLASTIC

REINFORCING WASHER

NOTE: BREAK ALL SHARP EDGES ON RIVET, WASHER, AND HOLES.

FIG. 13-4-6   Recommended riveting procedures.

FIG. 13-4-7   Boss cap design.

THREAD SIZE	DIMENSIONS				
	A	B	C	D	R
# 6(.138)	.14	.21	.10	.28	.02
# 8(.164)	.16	.25	.13	.34	.02
# 10(.190)	.18	.29	.17	.40	.02

Rivets   Conventional riveting equipment and procedures can be used with plastics. Care must be exercised to minimize stresses induced during the fastening operation. To do this, the rivet head should be 2.5 to 3 times the shank diameter. Also, rivets should be backed with either plates or washers to avoid high localized stresses (Fig. 13-4-6).

Drilled holes rather than punched holes are preferred for fasteners. If possible, fastener clearance in the hole should be at least .01 in. (0.3 mm) to maintain a plane stress condition at the fastener.

Boss Caps   A *boss cap* is a cup-shaped metal ring that is pressed onto the boss by hand, with an air cylinder, or with a light-duty press. It is designed to reinforce the boss against the expansion force exerted by tapping screws (Fig. 13-4-7).

Adhesive Bonding   When two or more parts are to be joined into an assembly, adhesives permit a strong, durable fastening between similar materials and often are the only fastening method available for joining dissimilar materials. Structural adhesives are made from the same basic resins as many plastics and thus react to their operating environment in a similar manner. In order to provide maximum strength, adhesives must be applied as a liquid to thoroughly wet the surface of the part. The bonding surface must be chemically clean to permit complete wetting. Basic plastics vary in physical properties, so adhesives made from these materials also vary. Fig. 13-4-8 (pg. 370) shows adhesively bonded joints.

Ultrasonic Bonding   Ultrasonic bonding often is used instead of solvent cementing to bond plastic parts. By using this technique, irregularly shaped parts can be bonded in two seconds or less. The bonded parts may be handled and used at reasonable temperatures within minutes after joining.

Only one of the mating parts comes in contact with the horn (Fig. 13-4-9, pg. 370). The part transmits the ultrasonic vibration to small, hidden bonding areas, resulting in fast, perfect welds. Both mating halves remain cool except at the seam, where the energy is quickly dissipated.

This technique is not recommended where high impact strength is required in the bond area.

Ultrasonic Staking   Ultrasonic staking frequently involves the assembly of metal parts. In this technique, a stud molded

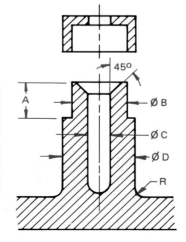

45°

A

Ø B

Ø C

Ø D

R

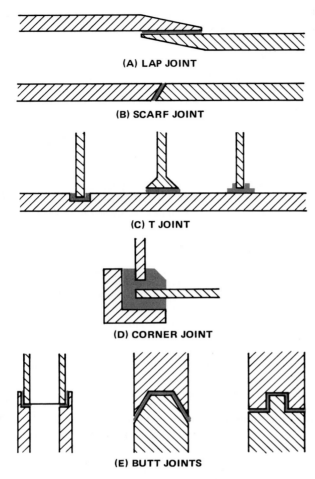

(A) LAP JOINT

(B) SCARF JOINT

(C) T JOINT

(D) CORNER JOINT

(E) BUTT JOINTS

FIG. 13-4-8 Adhesive bonding.

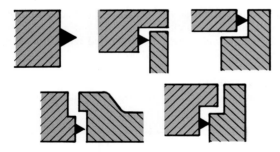

FIG. 13-4-9 Design joints for ultrasonic bonding.

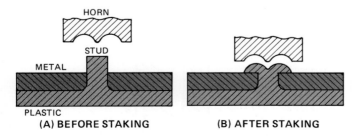

HORN

STUD

METAL

PLASTIC
(A) BEFORE STAKING **(B) AFTER STAKING**

FIG. 13-4-10 Typical ultrasonic staking operation.

into the plastic part protrudes through a hole in the metal part. The surface of the stud is vibrated with a horn having high amplitude and a relatively small contact area. The vibration causes the stud to melt and re-form in the configuration of the horn tip (Fig. 13-4-10).

Friction or Spin Welding This welding technique is limited to parts with circular joints. It is especially useful for large parts where ultrasonic welding or chemical bonding is impractical.

In friction or spin welding, the faces to be joined are pressed together while one part is spun and the other is held fixed. Frictional heat produces a molten zone that becomes a weld when spinning stops (Fig. 13-4-11).

Drawings

In addition to the usual considerations, the following points should be taken into account when a detail drawing of a plastic part is made:

1. Can the part be removed from the mold?
2. Is location of flash line consistent with design requirements?
3. Is section thickness consistent? Are there thick sections? Thin sections? Could greater uniformity of section thickness be maintained?
4. Has the material been correctly specified?
5. Is each feature in accordance with the thinking of competent materials engineers and molders?
6. Have close tolerance requirements been reviewed with responsible engineers?
7. Have marking requirements been specified to inform field service people of the material from which the part is fabricated?

REFERENCES AND SOURCE MATERIAL

1. General Motors Corp.
2. General Electric Co.

ASSIGNMENTS

See Assignments 11 through 13 for Unit 13-4 on pages 376 to 377.

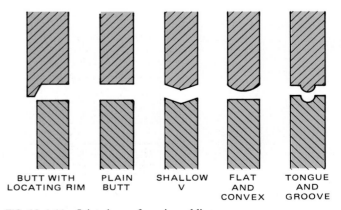

BUTT WITH LOCATING RIM   PLAIN BUTT   SHALLOW V   FLAT AND CONVEX   TONGUE AND GROOVE

FIG. 13-4-11 Joint shapes for spin welding.

ASSIGNMENTS FOR CHAPTER 13

ASSIGNMENTS FOR UNIT 13-1, CASTINGS

1. Complete the assembly drawing of the fork for the hinged pipe vise shown in Fig. 13-1-A. Use your judgment for dimensions not given. Scale 1:1.

2. Complete the detail drawing of the base for the adjustable shaft support assembly shown in Fig. 13-1-B. Cored holes are to be used for the shaft holes. Scale 1:1.

FIG. 13-1-A   Pipe vise.

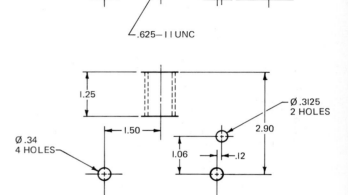

FIG. 13-1-B   Adjustable shaft
support.

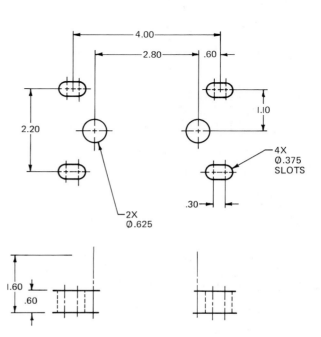

3. Prepare both the casting and the machining drawings for the connector shown in Fig. 13-1-C. Draw a one-view full section, complete with the necessary dimensions for each drawing. Scale 1:1.

4. Prepare both the casting and the machining drawing for the pump bracket shown in Fig. 13-1-D. Cored holes of Ø20 and Ø9 are to be used for the Ø24 and Ø12 holes. The machined surfaces are to have a maximum roughness

FIG. 13-1-C   Connector.

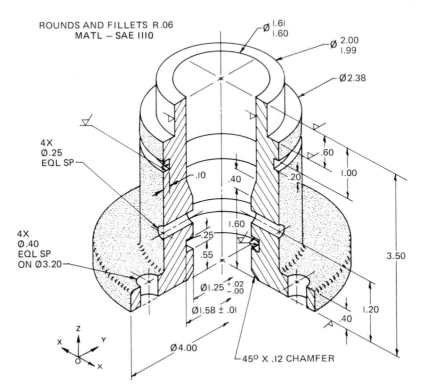

FIG. 13-1-D   Pump bracket.

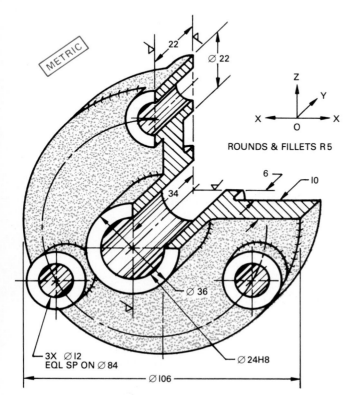

FIG. 13-1-E   Top plate.

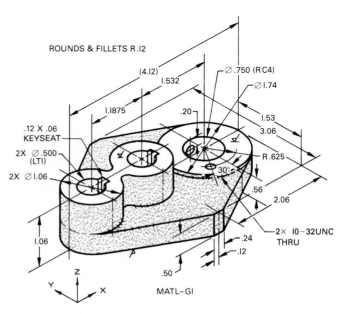

value of 3.2 µm and a machining allowance of 2 mm. Show the limit dimensions for the Ø24 hole.

5. Prepare both the casting and machining drawings of the top plate shown in Fig. 13-1-E. Cored holes are to be used for the three vertical holes. The machined surfaces are to have a maximum roughness value of 63 µin and a machining allowance of .06 in. Show the limit dimensions where fits are indicated. Scale 1:1.

6. Redesign one of the welded parts shown in Figs. 13-1-F through 13-1-H into a cast part. Make a machine drawing given the following information. Show the limit sizes where fits are indicated. Surfaces shown with the letter A are to have a maximum roughness value of 1.6 µm and a machining allowance of 2 mm or its equivalent. Use symbolic dimensioning. For Fig. 13-1-F add a spotface to the Ø.78 hole and increase the top and bottom thickness.

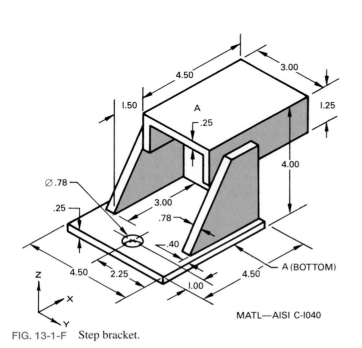

MATL—AISI C-1040

FIG. 13-1-F Step bracket.

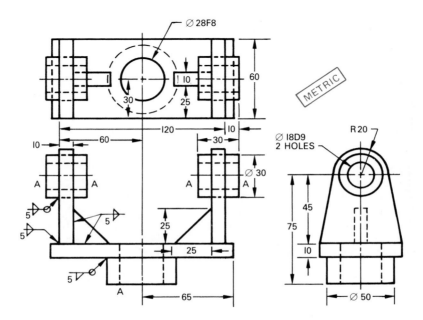

MATL–AISI C–1040

FIG. 13-1-G Swing bracket.

FIG. 13-1-H Shaft support.

ASSIGNMENTS FOR UNIT 13-2, FORGINGS

7. Prepare a forging drawing of one of the parts shown in Fig. 13-2-A or 13-2-B. Scale 1:1.

8. Prepare a forging drawing for the wrench handle shown in Fig. 13-2-C. Scale 1:1.

FIG. 13-2-A   Open-end wrench.

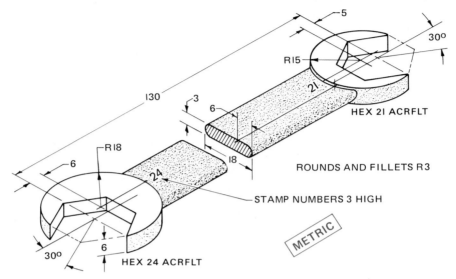

FIG. 13-2-B   Bracket.

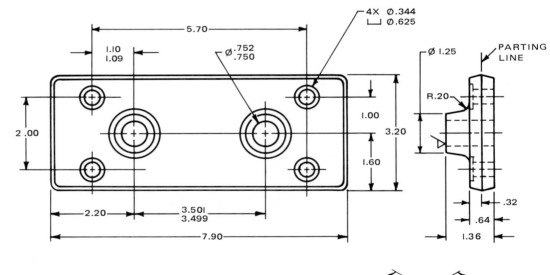

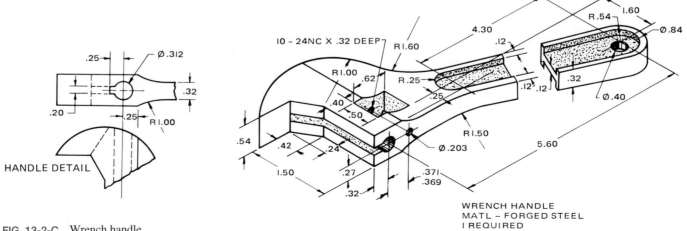

HANDLE DETAIL

FIG. 13-2-C   Wrench handle.

ASSIGNMENTS FOR UNIT 13-3, POWDER METALLURGY

9. Prepare two drawings, one for machining the part, the second for the making of the briquet (powder metallurgy) for one of the parts shown in Fig. 13-1-C, 13-3-A, or 13-3-B. Scale 1:1.

10. Prepare one or two drawings as required, one for machining the part, the second for making the briquet (powder metallurgy) for one of the prefabricated (welded) parts in Figs. 13-3-C through 13-3-E (here and on pg. 376). Scale 1:1.

FIG. 13-3-A   Bracket.

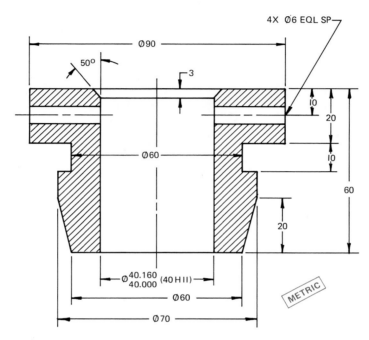

FIG. 13-3-B   Tool holder.

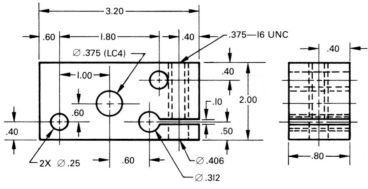

FIG. 13-3-C   Shaft base.

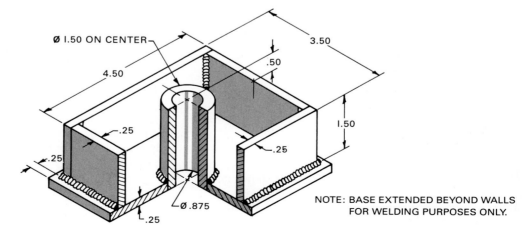

NOTE: BASE EXTENDED BEYOND WALLS
FOR WELDING PURPOSES ONLY.

375

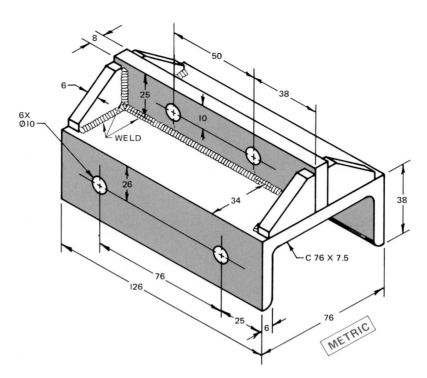

FIG. 13-3-D   Caster frame.

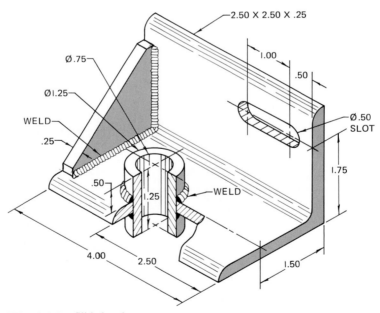

FIG. 13-3-E   Slide bracket

ASSIGNMENTS FOR UNITS 13-4, PLASTIC MOLDED PARTS

11. Redesign for plastic molding one of the parts shown in Figs. 13-3-C through 13-3-E. Where required, prepare two drawings—one for the mold, the other for machining.

Refer to the molding recommendations shown in this unit and indicate the parting line on the drawing. Use your judgment for dimensions not given. Scale 1:1.

12. Using a plastic molding design, add threaded inserts to one of the parts shown in Figs. 13-4-A through 13-4-D. Use your judgment for dimensions not shown and the type and number of views required. Scale 2:1.

13. Make a plastic molding assembly drawing of the parts shown in Fig. 13-4-E. The retaining ring is to be positioned in the center of the part and molded into position. Modification to the retaining ring may be required to prevent the ring from turning in the wheel. Scale 5:1. Show a top view and a full-section view. Dimension the finished assembly.

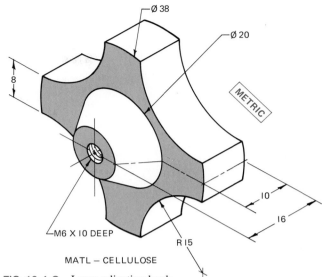

MATL – CELLULOSE

FIG. 13-4-C   Lamp adjusting knob.

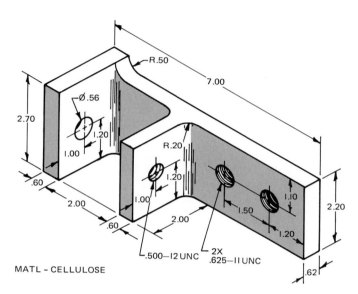

MATL – CELLULOSE

FIG. 13-4-A   Connector.

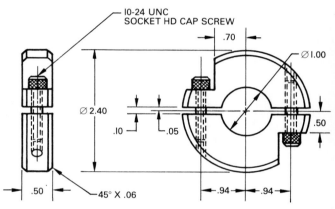

FIG. 13-4-D   Gear clamp.

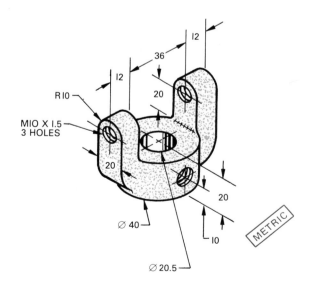

FIG. 13-4-B   Pivot arm.

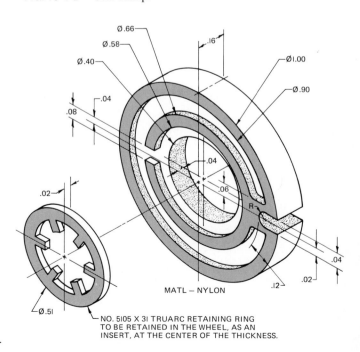

MATL – NYLON

NO. 5105 X 31 TRUARC RETAINING RING TO BE RETAINED IN THE WHEEL, AS AN INSERT, AT THE CENTER OF THE THICKNESS.

FIG. 13-4-E   Cassette-tape drive wheel.

377

WORKING DRAWINGS AND DESIGN

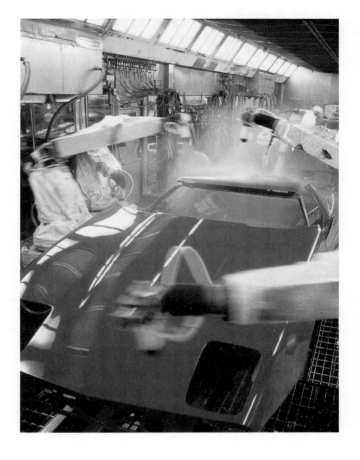

CHAPTER 14

DETAIL AND ASSEMBLY DRAWINGS

Definitions

Appliqués   Pressure-sensitive overlays used to depict common parts, shapes, symbols, surface textures, or notes.

Assembly drawing   A drawing showing the product in its completed state.

Detailer   The draftsperson (drafter) who works from a complete set of instructions and drawings, or who makes working drawings of parts that involve the design of the part.

Item list or **bill of material**   An itemized list of all the components shown on an assembly drawing or a detail drawing.

Subassemblies   Preassembled components and individual parts used to make up a finished product.

Working drawing   A drawing that supplies information and instructions for the manufacture or construction of machines or structures.

14-1 DRAWING QUALITY ASSURANCE

The requirement that engineering drawings be complete, clear, accurate, conform to standards, and ensure proper functional operation is the responsibility of the drafter, checker, and other specialists assigned to review the drawing prior to release. Use of the following recommendations and specific considerations, as applicable, when reviewing drawings is advised as a means of promoting the preparation of quality drawings. The term *drawings* refers to the depiction of details, assemblies, installations, or other types of graphical representations.

Knowledge of the design requirements, the manufacturing process involved, and drafting practices on the part of the drafter, checker, and other reviewers has a definite influence on the accuracy and cost of parts and assemblies. Layouts should be carefully studied and, where necessary, discussed with the designer and responsible engineer to ensure full understanding of the function and application of the design. Suggestions for improvement in design or manufacture should be discussed with accountable personnel. Finished drawings should reflect the objective findings of all responsible reviewers.

Although the review procedure may vary, it is recommended that reference to layouts, proven similar designs, and other pertinent design data be used by the reviewers. Throughout the review it is vital to be constantly on the alert for omitted or incomplete information.

Review Considerations

The following items are typical of those that need to be considered, as applicable, in the preparation and review of drawings.

Applied Surface Finish   Any applied surface finish requirements must be completely defined.

Expansion   Dimensions and tolerances should be adjusted for thermal expansion or contraction during operation. Differences in expansion coefficients of various materials should be kept in mind.

Grain Flow   A part made from a forging or from sheet metal must have direction of grain flow indicated where it is important to the durability of the part.

Inspection Processes   Inspection processes, such as magnetic particle, fluorescent penetrants, and X rays, must be noted on the drawing where required.

Interchangeability   Requirements for interchangeability must be considered.

Locking Feature   Locking features for the retention of parts, such as lockwire holes and tab washer slots, should be shown where required.

Material   Proper material and heat treatment must be specified.

Procurability   Where an item is vendor-supplied, or includes vendor-controlled features, such as material, process, or operational devices, its availability should be considered.

Protective Finish   Protective finish specifications, such as painting or plating, should be specified.

Seizure   Where parts come in contact, material and surface treatments subject to "seizing," galvanic action, or similar effects should normally be avoided.

Service   Accessibility must be provided for servicing, assembling, inspection, and adjustment.

Standard Parts   Standard parts should be used wherever applicable.

Standard Practices   Standards pertaining to design, materials, processes, etc. should be used.

Strength   Design must adequately meet all stress requirements, such as thermal, dynamic, and fatigue stresses. Deterioration (embrittlement, corrosion, and wear) must be considered.

Surface Texture (Roughness)   Surface texture values must be specified for all surfaces requiring control. The values shown should be compatible with overall design requirements.

Tolerances   The tolerances indicated by the linear and angular dimensions and by local, general, or title block notes must ensure the proper assembly and functioning of the parts. Tolerances should be as liberal as the design will permit.

Drawing Considerations

Abbreviations   Abbreviations should conform to the country's, or the individual company's, drawing standards.

Conformance to Drawing Standards   Drawings should conform to the country's, or the individual company's, drawing standards in regard to size of sheet, format, zone marking, microfilm alignment arrowheads, arrangement of views, line characteristics, scale, letter and dimension heights, notes, and general appearance. Lines and lettering must be distinct and dark enough to ensure legible reproduction, including microfilm reduction. Letter (and number) form and size must be compatible with microfilming and reduced-size prints.

Dimensions   The part must be fully dimensioned and the dimensions clearly positioned. True-position relationship should be shown where applicable. Dimensions should not be repeated or shown in a manner that constitutes double dimensioning. Dimensions should not result in objectionable tolerance accumulation. Dimensions should emphasize function of design in preference to production operations or processes and should be such as to minimize shop calculations. Developed lengths and stock size should be specified as applicable.

Draft Angle and Radii   Proper draft angles, fillets, and corner radii should be specified (see Chap. 13).

Geometric Surface Relationship   All requirements covering necessary geometric relationships, such as straightness, runout, squareness, and parallelism, must be shown (see Chap. 16).

Revisions   All revisions must be properly recorded and all lines damaged by erasing during the making of revisions must be restored. All related drawings should be revised to conform.

Scale   The drawing should be to scale and the scale should be identified. Where drawings are to no scale, they should be identified as such.

Surface Texture Symbols   Surface texture symbols and values must be specified for all surfaces requiring control. The values should be compatible with overall design requirements.

Tolerances   The tolerances specified by the linear and angular dimensions and by local, general, or title block notes must ensure the proper assembly and functioning of the part. The selection of positional tolerancing or coordinate tolerancing should be carefully considered. Tolerances should be as liberal as the design will permit.

Symbols   Wherever possible, symbols should be used in lieu of words. The placement and use of symbols should reflect the latest standards.

Symmetrical Opposite Parts   An AS SHOWN and OPPOSITE HAND note with proper identification numbers must be shown for all such parts, unless a separate drawing is made for each hand.

Views   Sufficient full and sectional views must be shown and must be in proper relation to each other if third-angle projection is used, or properly identified if the reference arrow layout method is used (see Chap. 6).

Fabrication Considerations

Adhesives   The drawing must clearly identify the type of joint and adhesive used.

Brazing, Soldering, and Welding   The drawing must include local or general notes or symbols, as applicable, for the method of fabrication used.

Casting   Where the part is made as a casting, sufficient tolerances must be provided for draft, warpage, core shifting, or crossing of the parting line. Can coring be simplified or eliminated? Is the cast part number located in a practical position?

Centers   Where manufacturing can be facilitated by providing machining centers, they should be specified on the drawing.

Economy   Is the design the most economical approach; or would redesign result in a more economical approach without sacrificing quality?

Forged and Molded Parts   For parts made by forging or molding, sufficient tolerances must be allowed for warping, die shift, and die closure.

Holes   Are tolerances adequate to permit economical drilling or reaming? Blind holes must be sufficiently deep to permit threading and reaming.

Machining Lugs   Where a part is cast or forged, manufacturing can often be facilitated by providing clamping lugs and locating pads. Removal of such lugs after machining, where required, should be noted on the drawing.

Numerical Control Machining　Parts to be machined on numerically controlled equipment may be dimensioned to facilitate programming.

Processing Clearance　Design must allow sufficient clearance for drills, cutters, grinding wheels, as well as welding, riveting, and other processing tools.

Special Considerations　Notes for sandblast, vapor blast, and any other special operations should be included where required.

Stamping　All dimensions should be given to the same side of metal, where practical.

Tooling　Dimensions on drawings should reflect the use of standard tooling, such as reamers, cutters, and drills, wherever possible, without specifically calling out the type of tooling to be used, e.g. Ø6.30, not 6.30 DRILL.

Installation Considerations

Assembly　Parts should be designed so there is no possibility of misassembly. Often a dowel, offset bolt hole, or similar feature can be provided to ensure correct, one-way assembly. The design should permit servicing without unreasonable complications.

Clearance　The part must have sufficient clearance with surrounding parts to permit assembly and operation.

Driving Feature　Threaded parts require a slot, hex, or other driving feature.

Puller Feature　Where a part has a tight fit, it may require a puller lip, a jackscrew thread, a knockout hole, or some similar extraction feature.

Tool Clearance　Adequate clearance must be provided for wrenches or other assembly tools.

Torque Values　Required wrench torque values should be specified where items are assembled by means of bolts, cap screws, nuts, or similar features.

REFERENCES AND SOURCE MATERIAL

1. General Motors Corp.

14-2　FUNCTIONAL DRAFTING

Since the basic function of the drafting department is to provide sufficient information to produce or assemble parts, functional drafting must embrace every possible means to communicate this information in the least expensive manner. Functional drafting also applies to any method that would lower the cost of producing the part. New technological developments have provided many new ways of producing drawings at lower costs and/or in less time. This means that drafters must be prepared to discard some of the old, traditional methods in favor of these newer means of communication.

There are many ways to reduce drafting time in preparing a drawing. These drawing shortcuts, when collectively used, are of prime importance in an effective drafting system. These newer techniques cannot be blindly applied, however, but must be carefully evaluated to make certain that the benefits outweigh the potential disadvantages. This evaluation should answer the following questions:

- What is its purpose?
- Is it a personal preference disguised as a project requirement?
- Does it meet contractual requirements?
- Will the shortcut increase costs in other areas, such as manufacturing, purchasing, or inspection?
- Is it an effective communication link?
- How much training or education is required to make effective use of it?
- Are facilities available to implement it?
- Does the shortcut bypass a real bottleneck?

As each of these categories is examined, the advantages of the shortcuts will become apparent.

Procedural Shortcuts

There are a number of procedural shortcuts which, if properly applied and carefully managed, can shorten the drawing preparation cycle and result in savings.

Streamlined Approval Requirements　It is obvious that the more signatures required on a drawing, the greater the delays in releasing data. The decision as to who will approve drawings and drawing changes must be carefully considered to make certain that all necessary functions have been taken into account (checkers, responsible engineers, important technical specialists, etc.) without imposing undue restrictions. Project ground rules and contractual requirements also play an important part in this decision.

Eliminating the Drawing Check from the Preparation Cycle　One of the most common suggested shortcuts, usually proposed when a project is behind schedule or exceeding its budget or when experienced personnel are involved, is to eliminate checking from the drawing preparation cycle.

Using Standard and Existing Drawings　Drawings of parts are constantly being prepared that are repetitions of existing drawings. If the drafter were to incorporate the existing design parts that were already drawn with the new drawings, many drawing hours would be saved. Good drawing application records and an efficient multiple-use drawing system can eliminate a great deal of duplication. Standard tabulated drawings may be used to eliminate hundreds of drawings (Figs. 14-2-1 and 14-2-2).

Standard Drafting Practices　Standard drafting practices are obviously the backbone of efficient drafting room operations. The best way to establish and implement these practices is through a good drafting room manual, with requirements that must be strictly observed by all personnel.

The drafting room manual should contain data on the use and preparation of specific types of drawings, drawing and part number requirements, standard and special drafting practices,

FIG. 14-2-1   Standard tabulated drawings.

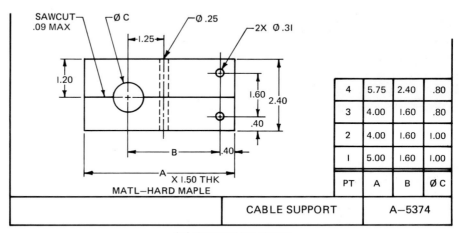

QTY	PART	MATL	DESCRIPTION	PT NO.
2	CABLE SUPPORT	MAPLE	A—5374 PT I	I
2	CABLE SUPPORT	MAPLE	A—5374 PT 2	2
3	CABLE SUPPORT	MAPLE	A—5374 PT 4	3

(A) DRAWING CALLOUT

4	5.75	2.40	.80
3	4.00	1.60	.80
2	4.00	1.60	1.00
I	5.00	1.60	1.00
PT	A	B	Ø C

CABLE SUPPORT   A—5374

(B) STANDARD PART

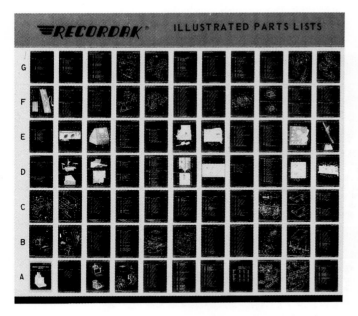

FIG. 14-2-2   Standard parts drawings stored on microfilm. *(Eastman Kodak Co.)*

rules for dimensioning and tolerancing, specifications for associated lists, and company procedures for the preparation, handling, release, and control of drawings.

Team Drafting   Many engineering departments have turned out drawings by the method of one drafter to one drawing. Team drafting involves a number of people producing one drawing. While this may seem uneconomical, it is an expeditious approach, with visible cost savings over the traditional method.

Some firms are using team drafting because it is a better utilization of skill levels. It is a training program through which drafting skills are taught and semiskilled people are given an opportunity to gain experience.

Data Retrieval   The use of microform reader-printers in the drafting room provides quick and ready access to standard drawings and parts. The use of microfiche cards is becoming popular, because they can hold up to 70 pages of information. However, for this method to be effective, a full-time librarian is needed.

Standard Parts and Design-Standard Information Encouraging the use of standard parts and standard approaches to design will not only result in drafting time saved but will cut costs in areas such as purchasing, material control, manufacturing, etc. The odd-size cutout that requires special tooling, the design that calls for nonstandard hardware, and the equipment that uses a wide variety of fasteners when only one or two would suffice are typical cases where properly applied standards would reduce both time and cost.

Copying Machines   One of the most important time-saving devices, which should be available in every drafting area, is a copying machine for reference copies, checking prints of work in preparation, and similar uses (Fig. 14-2-3, pg. 384). When a drafter needs a copy, work is delayed until the copy is made available. Therefore, a good copying machine will soon pay for itself in drawing hours saved.

FIG. 14-2-3   Copying machine. (*Doug Martin*)

Training Programs   To provide drafters with standard procedures and technical information is not enough; they must be trained in their use. New drafters are frequently overwhelmed by a strange environment, while old employees fail to keep up with new requirements or properly use the services available. Training programs for the indoctrination of new personnel and the updating of long-service employees are rewarded by more efficient and versatile operation.

Manual Drafting Equipment and Materials

The quality of the material and supplies used in the preparation of drawings is as important as the quality of the instruments used in fabrication.

Numerous timesaving devices are available for manual drafting: templates for every application, "pens" for easier line work, and more application of tape to artwork, transfer-type lettering, etc. Since drafting applications vary so widely, only the drafting supervisor can determine which devices will increase the drafting production.

Drafting aids are designed to facilitate the making of drawings by removing or reducing some of the more tedious aspects of drafting.

Templates, such as shown in Fig. 14-2-4, play an important part in functional drafting, for they save a great deal of time in drawing common shapes of details such as rounds, squares, hexagons, and ellipses. In addition to common shapes, templates have been made for standard parts, such as nuts and bolt heads, for electrical symbols, outlines of tools and equipment, and many other outlines that are often repeated.

Reducing the Number of Drawings Required

The cost of a project is, to some extent, directly related to the number of drawings that must be prepared. Therefore, careful planning to reduce the number of drawings required can result in significant savings. Some ways to reduce the number of drawings are explained below.

FIG. 14-2-4   Templates are made for many different uses and save a lot of time.

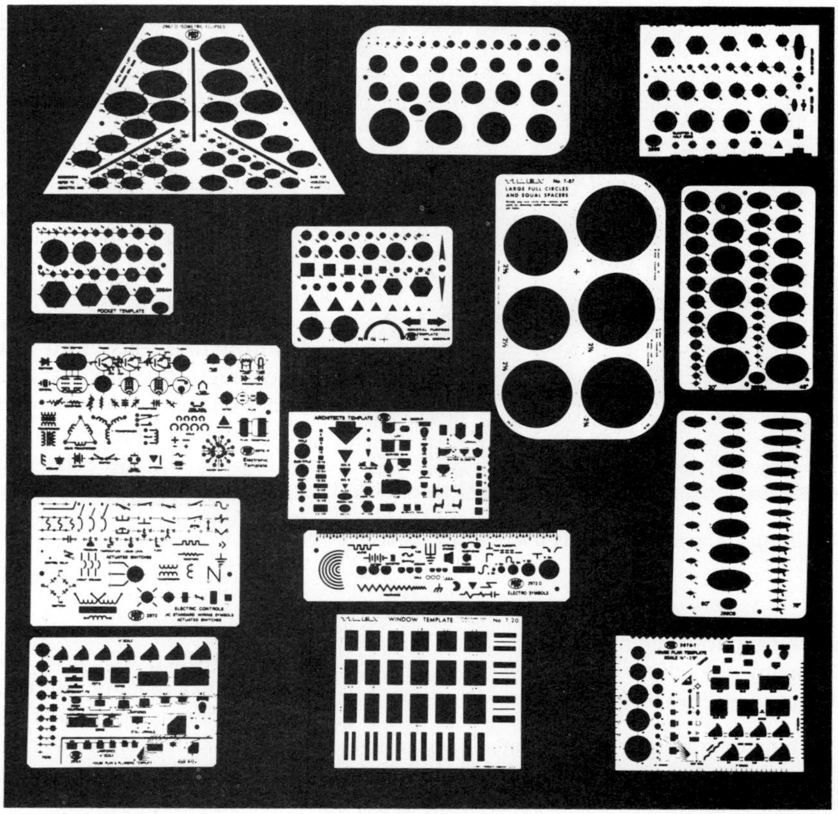

Detail Assembly Drawings   Detail assembly drawings, in which parts are detailed in place on the assembly (Fig. 14-2-5), and multidetail assembly drawings, in which there are separate detail views for the assembly and each of its parts, will reduce the number of drawings required. However, they must be used with extreme care. They can easily become too complicated and confusing to be an effective means of communication (see Unit 14-8).

Selecting the Most Suitable Type of Projection to Describe the Part   The selection of the type of projection (orthographic, isometric, or oblique) can greatly increase the ease with which some drawings can be read and, in many cases, reduce drafting time. For example, a single-line piping drawing drawn in isometric projection simplifies an otherwise difficult drawing problem in orthographic projection (Fig. 14-2-6).

Simplified Representations in Drawings

The steady rise of simplified representation in drawings by various industries has prompted ISO to prepare an international standard that lists together the various methods of simplified representation in general use for detail and assembly drawings.

Simplified representation in drawings is not new. Simplified thread and pipe symbols are two simplified representations that have been used for many years. Promoting and using simplified representation has many advantages; simplified representation:

- Raises the design efficiency.
- Accelerates the course of design.
- Reduces the workload in the drafting office.
- Enhances legibility of the drawing, so as to meet the requirements for drawings in computer graphics and microcopying.

In addition to the following recommendations, simplified representation of features is shown throughout this text where the appropriate drawing practices are explained.

1. Avoid unnecessary views. In many cases one or two views are sufficient to explain the part fully.
2. Use simplified drawing practices, as described throughout this text, especially on threads and common features.
3. The use of the symmetry symbol means that all dimensions are symmetrical about that line.

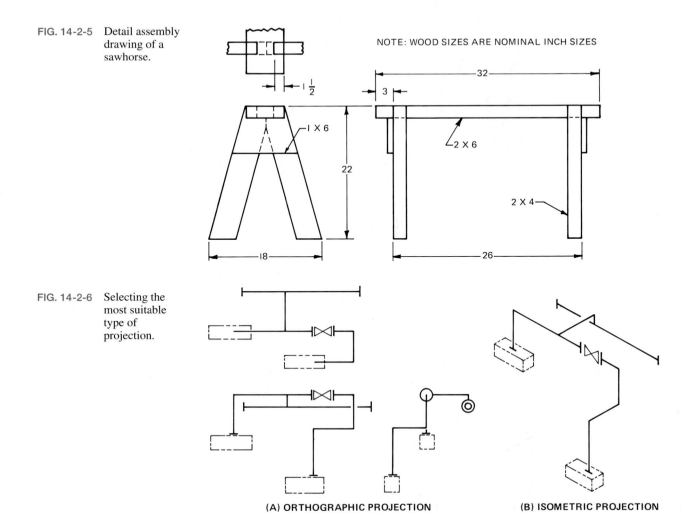

FIG. 14-2-5   Detail assembly drawing of a sawhorse.

NOTE: WOOD SIZES ARE NOMINAL INCH SIZES

FIG. 14-2-6   Selecting the most suitable type of projection.

(A) ORTHOGRAPHIC PROJECTION

(B) ISOMETRIC PROJECTION

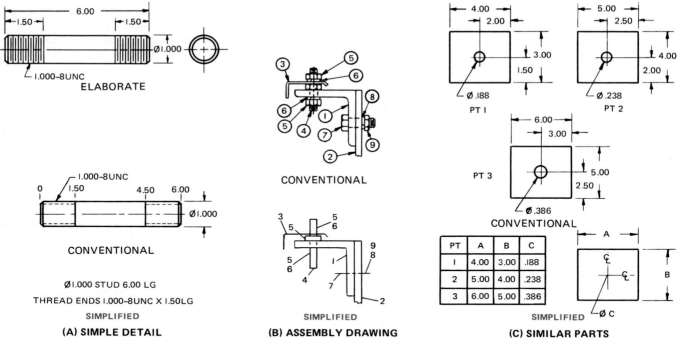

FIG. 14-2-7   Comparison between conventional and simplified representation.

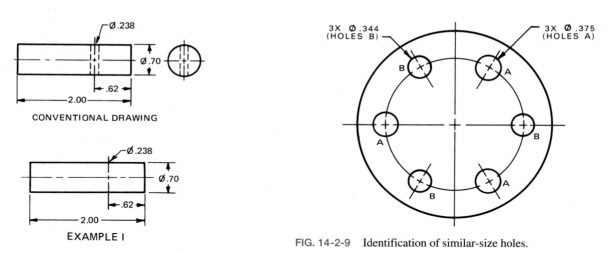

FIG. 14-2-8   Simplified representation for detailed parts.

FIG. 14-2-9   Identification of similar-size holes.

the holes. In such cases, the identification of similar-size holes should be made clear (Fig. 14-2-9).

4. Complicated parts are best described by means of a drawing. However, explanatory notes can complement the drawing, thereby eliminating views that are time-consuming to draw (Figs. 14-2-7 and 14-2-8).

5. When a number of holes of similar size are to be made in a part, there is a chance that the person producing the part may misinterpret the diameter of some of

6. A simplified assembly drawing should be used for assembly purposes only. Some means of simplification are:
   - Standard parts, such as nuts, bolts, and washers, need not be drawn.
   - Small fillets and rounds on cast parts need not be shown.
   - Phantom outlines of complicated details can often be used.

7. Use templates or symbol libraries.

8. Within limits, a small drawing is made more quickly than a large drawing when manual drafting is used.

9. Eliminate hidden lines that do not add clarification.

10. Show only partial views of symmetrical objects (Fig. 14-2-10).
11. Avoid the use of elaborate pictorial and repetitive detail when manual drafting is used.
12. Eliminate repetitive data by use of general notes or phantom lines when manual drafting is used.
13. Eliminate views where the shape or dimension can be given by description, for example, Ø, □, HEX, THK, etc.

Freehand Sketching

Many shops care little whether the drawing is freehand, whether one view is shown, or whether the drawing is to scale, as long as the proportions are approximate. They want the necessary information clearly shown. Freehand sketches and drawings made with instruments can be shown on one sheet. However, it must be clearly understood that the use of freehand sketching does not give the drafter a license to turn out sloppy work.

Savings as high as 30 percent in the preparation of working drawings have been attributed to the use of freehand sketches as opposed to instrument-produced drawings. However, freehand sketching has its limitations. It is highly effective on simple detailed parts, for small radii, such as rounds and fillets, and for small holes. In many cases the term *freehand* is not entirely correct. For instance, templates may be used to draw circles, resistors, or other common features, or a straightedge may be used to produce long lines since it is faster and more accurate than freehand sketching. But short lines are drawn more quickly freehand.

Drawing paper with nonreproducible grid lines is ideal for freehand sketching (Fig. 14-2-11). For this reason many companies have their drawing paper made with nonreproducible grid lines over the entire drawing area. Other advantages of having the grid lines on the paper are that (1) they may serve as guidelines in lettering notes and dimensions and (2) they may be used for measuring distances, thereby reducing the number of times the scale is used for measuring.

Reproduction Shortcuts

In the past few years, a number of reproduction techniques have been developed which, if properly used, can greatly reduce drawing preparation time. An understanding of available techniques and their limitations, supported by the close cooperation of a reproduction group familiar with drafting operations, can help the drafting supervisor make significant cost savings.

New Drawings Made From Existing Drawings

When a new drawing is to be made from an existing drawing with few changes, CAD makes this task easy by simply removing the unwanted material and drawing in the new. With manual drafting, reproducibles will save a great deal of preparation time. This procedure involves making a translucent or transparent print from the original drawing, removing unwanted material from this print, and adding the new information to the drawing. The main drawback to this method is that the existing drawing may not conform to the latest standard drawing practice.

Cut-and-Paste Drafting

No matter how original a design may be, a great number of part features are repetitive. With the aid of modern reproduction methods, drawings can be created by using unchanged portions of existing drawings. When manual drafting is used, transferring them from one drawing to the next is accomplished by scissors and paste-up drafting. It provides a way of using all or parts of existing drawings, notes, charts, and drawing forms to revise existing drawings and to create new drawings. Through the utilization of existing drawings much valuable drafting time is freed for creative design drafting rather than hand copying.

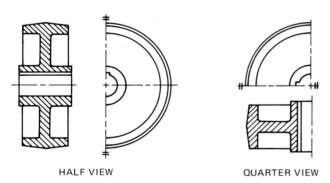

HALF VIEW   QUARTER VIEW

FIG. 14-2-10   Partial views.

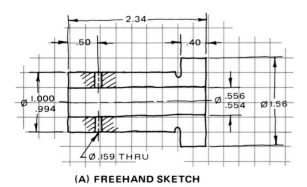

(A) FREEHAND SKETCH

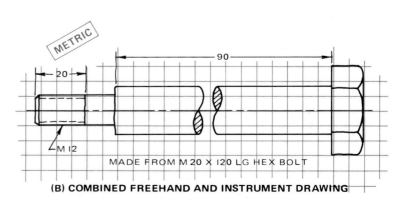

(B) COMBINED FREEHAND AND INSTRUMENT DRAWING

FIG. 14-2-11   Sketching of parts on coordinate paper.

Finished prints can be made on paper, acetate, or vellum. They can be the same size or reduced to different sizes, depending on the reproduction equipment being used.

Another important advantage of cut-and-paste drafting is that materials copied from existing drawings do not have to be minutely rechecked, as must be done with new drawings. Rechecking time is reduced (Fig. 14-2-12).

Appliqués

One of the most successful methods of reducing drawing time, when manual drafting is used, is the use of appliqués. When parts, shapes, symbols, or notes are used repeatedly, *appliqués* should be considered. These pressure-sensitive overlays may be printed on opaque, transparent, or translucent sheets with an adhesive backing.

Appliqués are available in a great variety of standard symbols or patterns (Fig. 14-2-13) and in blank (unprinted) sheets. A matte surface on the blank sheet will accept typewriter copy as well as pencil or ink lines. This material is often used for making corrections on drawings and for adding materials lists or detailed notes that can be typed faster than they can be lettered. Figure 14-2-14 shows a drawing that used many of the appliqués shown in Fig. 14-2-13.

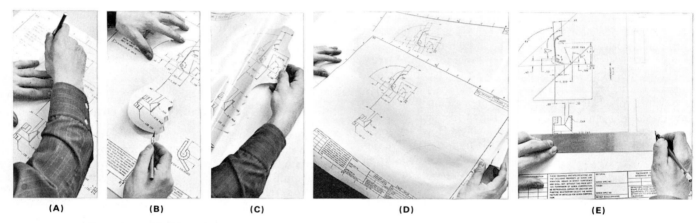

(A)   **(B)**   **(C)**   **(D)**   **(E)**

FIG. 14-2-12   Cut-and-paste drafting. *(Xerox Corp.)*

FIG. 14-2-13   A variety of shapes and sizes of appliqués. *(Graphic Standard Instruments Co.)*

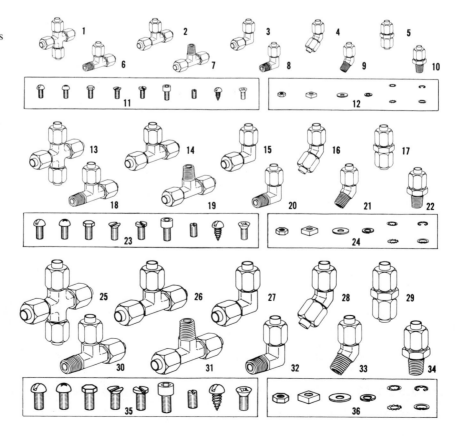

FIG. 14-2-14 Application of appliqués shown in Fig. 14-2-13. *(Graphic Standard Instruments Co.)*

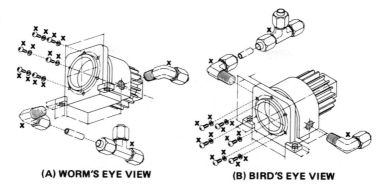

(A) WORM'S EYE VIEW   **(B) BIRD'S EYE VIEW**

Appliqués are available in two basic types: cutout and transfer. *Cutout appliqués* are applied by positioning the desired image in the correct position on the drawing, burnishing (rubbing) the image area, and cutting around it to remove the portion not wanted. The *transfer-type* pressure-sensitive appliqué works on a somewhat different principle. The carrier is removed from the translucent image sheet, and the area to be transferred is placed in position on the drawing. The image to be transferred is then rubbed over the top surface of the transfer sheet with a burnishing stick.

The combined use of cut-and-paste drafting and appliqués for new drawings is found extensively in industry, especially in the electronics and piping fields (Fig. 14-2-15).

Photodrawings

Photodrawings, that is, engineering drawings into which one photograph, or more, is incorporated, have increased in popularity because they can sometimes present a subject even more clearly than conventional drawings. Photodrawings supplement rather than replace conventional engineering drawings by eliminating much of the tedious and time-consuming effort involved when the subject is difficult to draw. They are particularly useful for assembly drawings, piping diagrams, large machine installations, switchboards, etc., provided, of course, that the subject of the drawings exists so that it may be photographed.

Photodrawings are also a comprehensive means of clearly transmitting technical information; they free the drafter from having to draw things that already exist. See Fig. 14-2-16 (pg. 390).

Photodrawings have other advantages. They are easy to make and usually take much less time to prepare than an equivalent amount of conventional drafting.

Background Any photodrawing must begin with a photograph of an object, a part or assembly, a building, a model, or whatever else may be the subject of the drawing.

Photography The best photographic angle usually is one that shows the subject in a flat view with as little perspective as possible. (If the situation calls for a perspective, select the angle that best describes the object.) Make certain that all the parts important to the photodrawing are in view of the camera.

ASSIGNMENTS

See Assignments 1 through 9 for Unit 14-2 on pages 401–406.

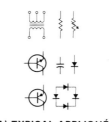

(A) TYPICAL APPLIQUÉS

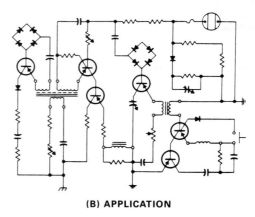

(B) APPLICATION

FIG. 14-2-15   Electronics appliqués. *(Bishop Industries Corp.)*

14-3 DETAIL DRAWINGS

A *working drawing* is a drawing that supplies information and instructions for the manufacture or construction of machines or structures. Generally, working drawings may be classified into two groups: *detail drawings,* which provide the necessary information for the manufacture of the parts, and *assembly drawings,* which supply the necessary information for their assembly.

Since working drawings may be sent to other companies to make or assemble the parts, the drawings should conform with the drawing standards of that company. For this reason, most companies follow the drawing standards of their country. The drawing standards recommended by ANSI and ASME have been adopted by the majority of industries in the United States.

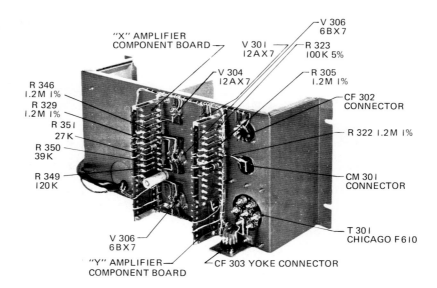

FIG. 14-2-16   Photodrawings.
(Eastman Kodak Co.)

Detail Drawing Requirements

A detail drawing (Fig. 14-3-1) must supply the complete information for the construction of a part. This information may be classified under three headings: shape description, size description, and specifications.

Shape Description   This term refers to the selection and number of views to show or describe the shape of the part. The part may be shown in either pictorial or orthographic projection, the latter being used more frequently. Sectional views, auxiliary views, and enlarged detail views may be added to the drawing in order to provide a clearer image of the part.

Size Description   Dimensions that show the size and location of the shape features are then added to the drawing. The manufacturing process will influence the selection of some dimensions, such as datum features. Tolerances are then selected for each dimension.

Specifications   This term refers to general notes, material, heat treatment, finish, general tolerances, and number required. This information is located on or near the title block or strip.

Additional Drawing Information   In addition to the information pertaining to the part, a detail drawing includes additional information such as drawing number, scale, method of projection, date, name of part or parts, and the drafter's name.

The selection of paper or finished plot size is determined by the number of views selected, the number of general notes required, and the drawing scale used. If the drawing is to be microformed, the lettering size would be another factor to consider. The drawing number may carry a prefix or suffix number or letter to indicate the sheet size, such as A-571 or 4-571; the letter A indicates that it is made on an 8.50 × 11.00 in. sheet, and the number 4 indicates that the drawing is made on a 210 × 297 mm sheet.

Drawing Checklist

As an added precaution against errors occurring on a drawing, many companies have provided checklists for drafters to follow before a drawing is issued to the shop. A typical checklist may be as follows:

1. *Dimensions.* Is the part fully dimensioned, and are the dimensions clearly positioned? Is the drawing dimensioned to avoid unnecessary shop calculations?
2. *Scale.* Is the drawing to scale? Is the scale shown? What will the plot scale be?
3. *Tolerances.* Are the clearances and tolerances specified by the linear and angular dimensions and by local, general, or title block notes suitable for proper functioning? Are they realistic? Can they be liberalized?
4. *Standards.* Have standard parts, design, materials, processes, or other items been used where possible?
5. *Surface texture.* Have surface roughness values been shown where required? Are the values shown compatible with overall design requirements?
6. *Material.* Have proper material and heat treatment been specified?

Qualifications of a Detailer

The detailer should have a thorough understanding of materials, shop processes, and operations in order to properly dimension the part and call for the correct finish and material. In addition, the detailer must have a thorough knowledge of how the part functions in order to provide the correct data and tolerances for each dimension.

The detailer may be called upon to work from a complete set of instructions and drawings, or he or she may be required to make working drawings of parts that involve the design of the part. Design considerations are limited in this unit but are covered in detail in Chap. 19.

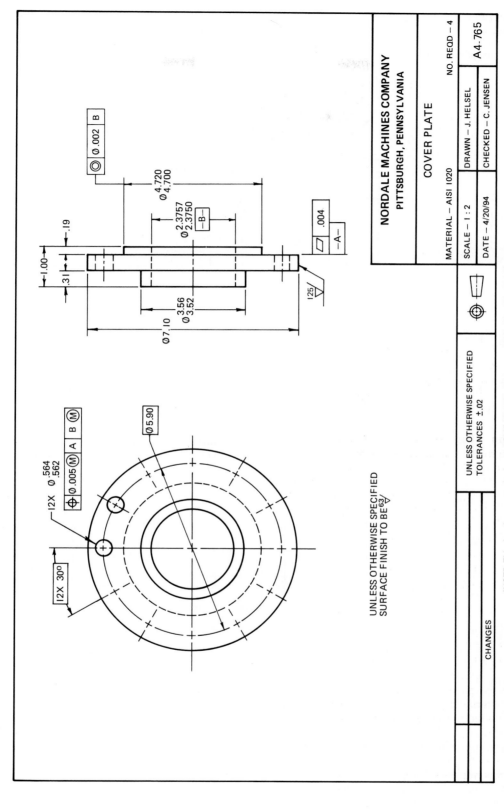

NORDALE MACHINES COMPANY
PITTSBURGH, PENNSYLVANIA

COVER PLATE

MATERIAL — AISI 1020		NO. REQD — 4
SCALE — 1 : 2	DRAWN — J. HELSEL	A4-765
DATE — 4/20/94	CHECKED — C. JENSEN	

UNLESS OTHERWISE SPECIFIED
TOLERANCES ±.02

UNLESS OTHERWISE SPECIFIED
SURFACE FINISH TO BE 63/

CHANGES

FIG. 14-3-1   A simple detail drawing.   [NOTE: This drawing, created prior to ASME Y14.5M-1994, uses ANSI Y14.5M-1982 (R 1988) standard.]

FIG. 14-3-2 Manufacturing process influences the shape of the part.

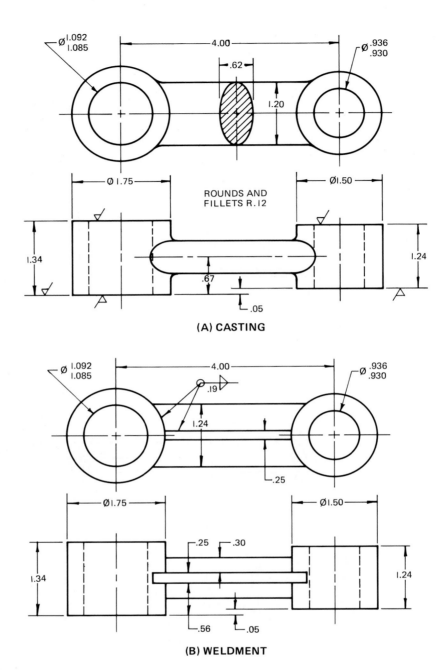

(A) CASTING

(B) WELDMENT

Manufacturing Methods

The type of manufacturing process will influence the selection of material and detailed feature of a part (Fig. 14-3-2). For example, if the part is to be cast, rounds and fillets will be added. Additional material will also be required where surfaces are to be finished.

The more common manufacturing processes are machining from standard stock; prefabrication, which includes welding, riveting, soldering, brazing, and gluing; forming from sheet stock; casting; and forging. The latter two processes can be justified only when large quantities are required for specially designed parts. All these processes are described in detail in other chapters.

Several drawings may be made for the same part, each one giving only the information necessary for a particular step in the manufacture of the part. A part that is to be produced by forging, for example, may have one drawing showing the original rough forged part and one detail of the finished forged part. (Figs. 14-3-2C and 14-3-2D).

ASSIGNMENTS

See Assignments 10 through 20 for Unit 14-3 on pages 407 through 414.

FIG. 14-3-2   Manufacturing process influences the shape of the part. (continued)

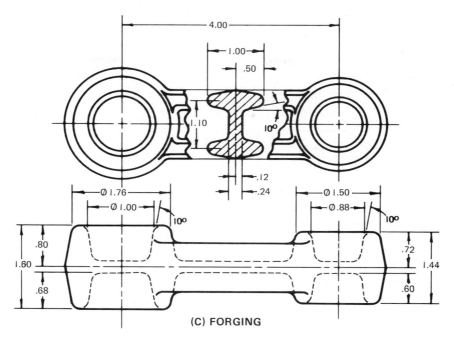

(C) FORGING

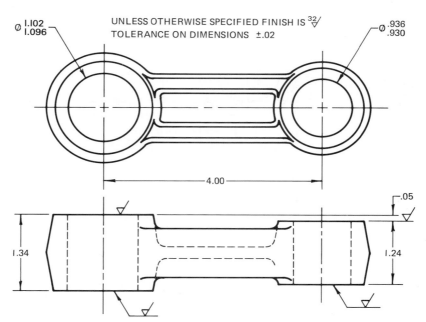

UNLESS OTHERWISE SPECIFIED FINISH IS $\frac{32}{}$ ⩗
TOLERANCE ON DIMENSIONS ±.02

(D) MACHINING DRAWING FOR FORGED PART SHOWN IN (C)

14-4 MULTIPLE DETAIL DRAWINGS

Detail drawings may be shown on separate sheets, or they may be grouped on one or more large sheets.

Often the detailing of parts is grouped according to the department in which they are made. For example, wood, fiber, and metal parts are used in the assembly of a transformer. Three separate detail sheets—one for wood parts, one for fiber parts, and the third for the metal parts—may be drawn. These parts would be made in the different shops and sent to another area for assembly. In order to facilitate assembly, each part is given an identification part number, which is shown on the assembly drawing. A typical detail drawing showing multiple parts is illustrated in Fig. 14-4-1 (pg. 394).

If the details are few, the assembly drawing may appear on the same sheet or sheets.

ASSIGNMENTS

See Assignments 21 through 27 for Unit 14-4 on pages 415 through 421.

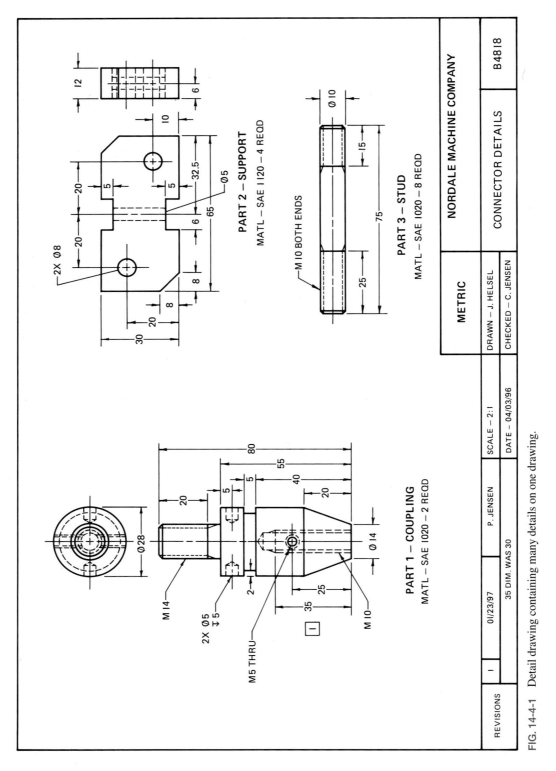

PART 2 – SUPPORT
MATL – SAE 1120 – 4 REQD

2X Ø8

PART 3 – STUD
MATL – SAE 1020 – 8 REQD

M 10 BOTH ENDS

PART 1 – COUPLING
MATL – SAE 1020 – 2 REQD

M 14

2X Ø5 ∓ 5

M5 THRU

M 10

NORDALE MACHINE COMPANY		B4818
	CONNECTOR DETAILS	
METRIC	DRAWN – J. HELSEL	
	CHECKED – C. JENSEN	
	P. JENSEN	SCALE – 2:1
		DATE – 04/03/96
REVISIONS	01/23/97	35 DIM. WAS 30

FIG. 14-4-1   Detail drawing containing many details on one drawing.

14-5 DRAWING REVISIONS

Revisions are made to an existing drawing when manufacturing methods are improved, to reduce cost, to correct errors, and to improve design. A clear record of these revisions must be registered on the drawing.

All drawings must carry a change or revision table, either down the right-hand side or across the bottom of the drawing. In addition to a description of drawing changes, provision may be made for recording a revision symbol, a zone location, an issue number, a date, and the approval of the change. Should the drawing revision cause a dimension or dimensions to be other than the scale indicated on the drawing, the dimensions that are not to scale should be indicated by the method shown in Fig. 8-1-15. Typical revision tables are shown in Fig. 14-5-1.

At times, when there are a large number of revisions to be made, it may be more economical to make a new drawing. When this is done, the words REDRAWN and REVISED should appear in the revision column of the new drawing. A new date is also shown for updating old prints.

REFERENCES AND SOURCE MATERIAL

1. ANSI Y14.5M, 1982, *Dimensions and Tolerancing.*
2. CAN/CSA-B78.2, *Dimensioning and Tolerancing of Technical Drawings.*

ASSIGNMENT

See Assignment 28 for Unit 14-5 on page 421.

14-6 ASSEMBLY DRAWINGS

All machines and mechanisms are composed of numerous parts. A drawing showing the product in its completed state is called an *assembly drawing* (Fig. 14-6-1).

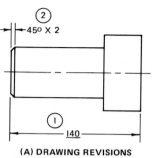
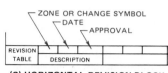

REVISIONS

SYMBOL	DESCRIPTION	DATE & APPROVAL
1	LENGTH WAS 150	J. Helsel 3-4-96
2	CHAMFER ADDED	F. Newman 2-2-97

(B) VERTICAL REVISION BLOCK

(A) DRAWING REVISIONS

ZONE OR CHANGE SYMBOL
DATE
APPROVAL

REVISION TABLE	DESCRIPTION		

(C) HORIZONTAL REVISION BLOCK

FIG. 14-5-1   Drawing revisions.

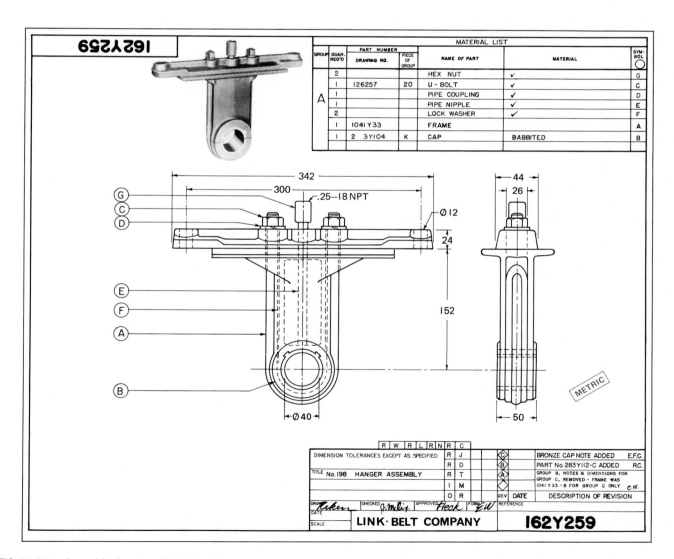

GROUP	QUAN. REQ'D	PART NUMBER		PIECE OF GROUP	NAME OF PART	MATERIAL	SYM-BOL
		DRAWING NO.					
	2				HEX NUT	✓	G
	1	126257		20	U - BOLT	✓	C
A	1				PIPE COUPLING	✓	D
	1				PIPE NIPPLE	✓	E
	2				LOCK WASHER	✓	F
	1	1041Y33			FRAME		A
	1	2 3Y104		K	CAP	BABBITED	B

	R	W	R	L	R	N	R	C

DIMENSION TOLERANCES EXCEPT AS SPECIFIED

R	J	
R	D	
R	T	
I	M	
O	R	

TITLE No.198   HANGER ASSEMBLY

C	BRONZE CAP NOTE ADDED	E.F.C.
B	PART No. 283Y112-C ADDED	R.C.
A	GROUP B, NOTES & DIMENSIONS FOR GROUP C, REMOVED - FRAME WAS 1041 Y 33 - B FOR GROUP C ONLY	C.W.
REV	DATE	DESCRIPTION OF REVISION

REFERENCE

DRAWN   CHECKED O.Melin   APPROVED Heck   FORM EW

SCALE   **LINK·BELT COMPANY**   **162Y259**

FIG. 14-6-1   Assembly drawing. *(Link-Belt Co.)*

Assembly drawings vary greatly in the amount and type of information they give, depending on the nature of the machine or mechanism they depict. The primary functions of the assembly drawing are to show the product in its completed shape, to indicate the relationship of its various components, and to designate these components by a part or detail number. Other information that might be given includes overall dimensions, capacity dimensions, relationship dimensions between parts (necessary information for assembly), operating instructions, and data on design characteristics.

Design Assembly Drawings

When a machine is designed, an assembly drawing or a design layout is first drawn to clearly visualize the performance, shape, and clearances of the various parts. From this assembly drawing, the detail drawings are made and each part is given a part number. To assist in the assembling of the machine, part numbers of the various details are placed on the assembly drawing. The part number is attached to the corresponding part with a leader, as illustrated in Fig. 14-6-2. It is important that the detail drawings not use identical numbering schemes when several item lists are used. Circling the part number is optional.

Installation Assembly Drawings

This type of assembly drawing is used when many unskilled people are employed to mass-assemble parts. Since these people are not normally trained to read technical drawings, simplified pictorial assembly drawings similar to the one shown in Fig. 14-6-3 are used.

Assembly Drawings for Catalogs

Special assembly drawings are prepared for company catalogs. These assembly drawings show only pertinent details and dimensions that would interest the potential buyer. Often one drawing, having letter dimensions accompanied by a chart, is used to cover a range of sizes, such as the pillow block shown in Fig. 14-6-4B.

Item List

An *item list,* often referred to as a *bill of material* (BOM), is an itemized list of all the components shown on an assembly drawing or a detail drawing (Fig. 14-6-5, pg. 398). Often an item list is placed on a separate sheet for ease of handling and duplicating. Since the item list is used by the purchasing department to order the necessary material for the design, the item list should show the raw material size rather than the finished size of the part.

For castings a pattern number should appear in the size column in lieu of the physical size of the part.

Standard components, which are purchased rather than fabricated, such as bolts, nuts, and bearings, should have a part number and appear on the item list. Information in the descriptive column should be sufficient for the purchasing agent to order these parts.

Item lists placed on the bottom of the drawing should read from bottom to top, while item lists placed on the top of the drawing should read from top to bottom. This practice allows additions to be made at a later date.

ASSIGNMENTS

See Assignments 29 through 40 for Unit 14-6 on pages 421 through 434.

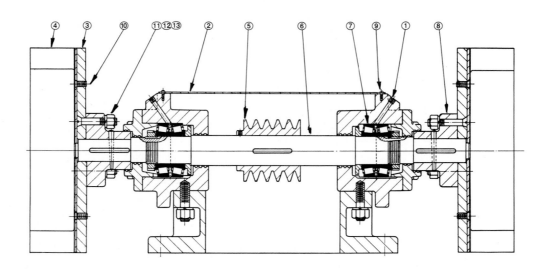

FLOOR STAND GRINDING MACHINE

FIG. 14-6-2   Design assembly drawing. *(Timken Roller Bearing Co.)*

FIG. 14-6-3   Installation assembly drawing.

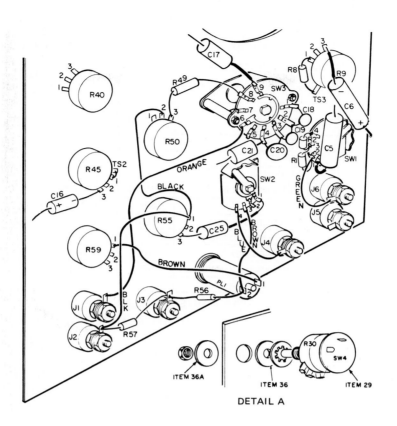

DETAIL A

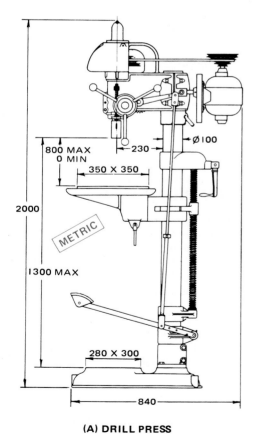

(A) DRILL PRESS

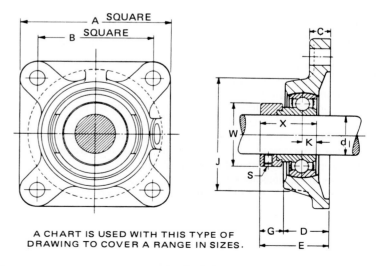

A CHART IS USED WITH THIS TYPE OF
DRAWING TO COVER A RANGE IN SIZES.

(B) PILLOW BLOCK

FIG. 14-6-4   Assembly drawings used in catalogs.

FIG. 14-6-5   Item list.

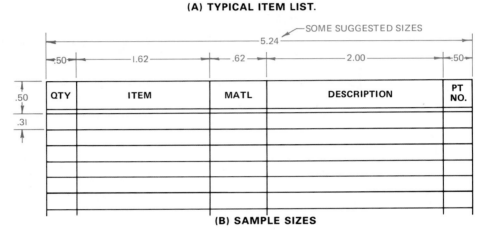

QTY	ITEM	MATL	DESCRIPTION	PT NO.
I	BASE	GI	PATTERN # A3I54	I
I	CAP	GI	PATTERN # B7I56	2
I	SUPPORT	AISI-I2I2	.38 X 2.00 X 4.38	3
I	BRACE	AISI-I2I2	.25 X I.00 X 2.00	4
I	COVER	AISI-I035	.I345 (#I0 GA USS) X 6.00 X 7.50	5
I	SHAFT	AISI-I2I2	ØI.00 X 6.50	6
2	BEARINGS	SKF	RADIAL BALL # 6200Z	7
2	RETAINING CLIP	TRUARC	N5000-725	8
I	KEY	STL	WOODRUFF # 608	9
I	SETSCREW	CUP POINT	HEX SOCKET .25UNC X I.50	I0
4	BOLT—HEX HD—REG	SEMI-FIN	.38UNC X I.50LG	I I
4	NUT—REG HEX	STL	.38UNC	I2
4	LOCK WASHER—SPRING	STL	.38 – MED	I3
				I4

NOTE: PARTS 7 TO I3 ARE PURCHASED ITEMS.

(A) TYPICAL ITEM LIST.

—SOME SUGGESTED SIZES

5.24

.50   1.62   .62   2.00   .50

.50

.31

QTY	ITEM	MATL	DESCRIPTION	PT NO.

(B) SAMPLE SIZES

14-7 EXPLODED ASSEMBLY DRAWINGS

In many instances parts must be identified or assembled by persons unskilled in the reading of engineering drawings. Examples are found in the appliance-repair industry, which relies on assembly drawings for repair work and for reordering parts. Exploded assembly drawings, like that shown in Fig. 14-7-1, are used extensively in these cases, for they are easier to read. This type of assembly drawing is also used frequently by companies that manufacture do-it-yourself assembly kits, such as model-making kits.

For this type of drawing, the parts are aligned in position. Frequently, shading techniques are used to make the drawings appear more realistic.

14-8 DETAIL ASSEMBLY DRAWINGS

Often these are made for fairly simple objects, such as pieces of furniture, when the parts are few in number and are not intricate in shape. All the dimensions and information necessary for the construction of each part and for the assembly of the parts are given directly on the assembly drawing. Separate views of specific parts, in enlargements showing the fitting together of parts, may also be drawn in addition to the regular assembly drawing. Note that in Fig. 14-8-1 (pg. 400) the enlarged views are drawn in picture form, not as regular orthographic views. This method is peculiar to the cabinet-making trade and is not normally used in mechanical drawing.

ASSIGNMENTS ▦▦▦▦▦▦▦▦▦▦▦▦▦▦

See Assignments 41 and 42 for Unit 14-7 on pages 435–436.

ASSIGNMENT ▦▦▦▦▦▦▦▦▦▦▦▦▦▦

See Assignment 43 for Unit 14-8 on pages 436–437.

FIG. 14-7-1   Exploded
assembly
drawings.

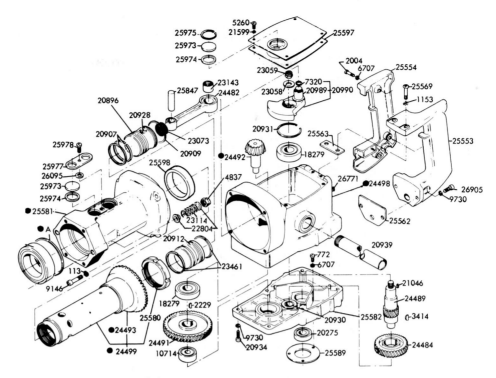

(A) PICTORIAL EXPLODED ASSEMBLY

NOTE:
FRICTION PLATE USES 3 CLUTCH
DISC UNITS WITH 4 CLUTCH
DISCS ON FRICTION PLATE.

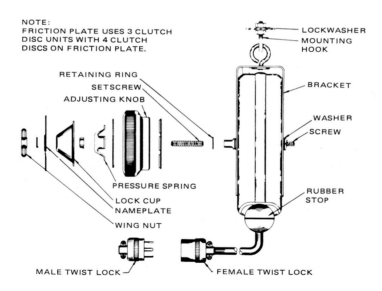

(B) ORTHOGRAPHIC EXPLODED ASSEMBLY

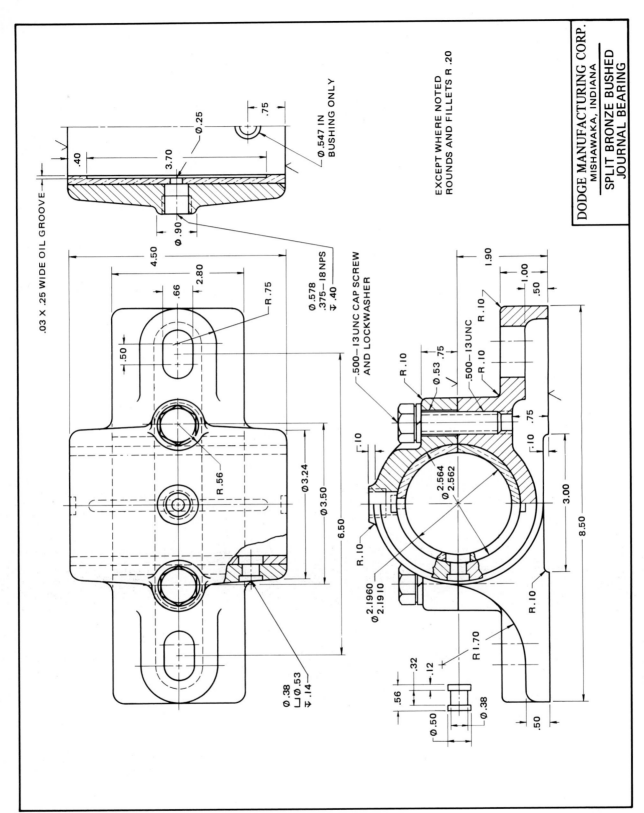

.03 X .25 WIDE OIL GROOVE

.40

Ø .25

3.70

Ø .547 IN
BUSHING ONLY

.75

Ø .90

4.50

.66

2.80

.50

R .75

Ø .578
.375—18 NPS
⌵ .40

.500—13UNC CAP SCREW
AND LOCKWASHER

R .56

Ø 3.24

Ø 3.50

6.50

Ø .38
⌴ Ø .53
⌵ .14

EXCEPT WHERE NOTED
ROUNDS AND FILLETS R .20

DODGE MANUFACTURING CORP.
MISHAWAKA, INDIANA
SPLIT BRONZE BUSHED
JOURNAL BEARING

1.90

R .10

1.00

.50

R .10

.75

Ø .53

.500—13UNC

R .10

.10

Ø 2.564
Ø 2.562

.75

.10

3.00

8.50

R .10

R .10

Ø 2.1960
Ø 2.1910

R 1.70

.32

.12

.56

Ø .50

Ø .38

.50

FIG. 14-8-1 Detail assembly drawing.

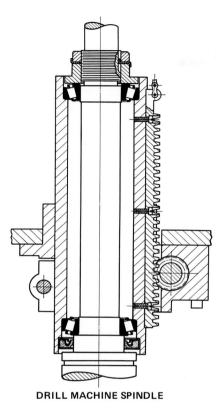

DRILL MACHINE SPINDLE

FIG. 14-9-1   Subassembly drawing. *(Timken Roller Bearing Co.)*

14-9 SUBASSEMBLY DRAWINGS

Many completely assembled items, such as a car and a television set, are assembled with many preassembled components as well as individual parts. These preassembled units are referred to as *subassemblies* (Fig. 14-9-1). The assembly drawings of a transmission for an automobile and the transformer for a television set are typical examples of subassembly drawings.

Subassemblies are designed to simplify final assembly as well as permit the item to be either assembled in a more suitable area or purchased from an outside source. This type of drawing shows only those dimensions that would be required for the completed assembly. Examples are size of the mounting holes and their location, shaft locations, and overall sizes. This type of drawing is found frequently in catalogs. The pillow block shown in Fig. 14-6-4B is a typical subassembly drawing.

ASSIGNMENTS

See Assignments 44 through 47 for Unit 14-9 on pages 438 through 440.

ASSIGNMENTS FOR CHAPTER 14

Note: Convert to symbolic and limit dimensioning, wherever practical, for all drawing assignments in this chapter.

ASSIGNMENTS FOR UNIT 14-2, FUNCTIONAL DRAFTING

1. After the number of drawings made over the last 6 months was reviewed, it was discovered that a great number of cable straps, shown in Fig. 14-2-A (pg. 402), were being made that were similar in design. Prepare a standard tabulated drawing similar to Fig. 14-1-1, reducing the number of standard parts to 4. Scale 1:1.
2. The rod guide shown in Fig. 14-2-B (pg. 402) is to be drawn twice and the drawing time for each recorded. First, on plain paper make an isometric drawing of the part, using a compass to draw the circles and arcs. Next, repeat the drawing, only this time use isometric grid paper and a template for drawing the circles and arcs. From the drawing times recorded, state in percentage the time saved by the use of grid paper and templates. Scale 1:1. Do not dimension.
3. The book rack shown in Fig. 14-2-C (pg. 402) is to be drawn twice and the drawing time for each drawing recorded. The first drawing is to show a three-view orthographic projection of the book rack assembly showing only those dimensions and instructions pertinent to the assembly. On the same drawing prepare detail drawings for the parts required. Scale to suit. On the second drawing make an orthographic detailed assembly drawing of the book rack showing the dimensions and instructions necessary to completely make and assemble the parts. Scale to suit. From the drawing times recorded, state a percentage of time saved by the use of detailed assembly drawings.

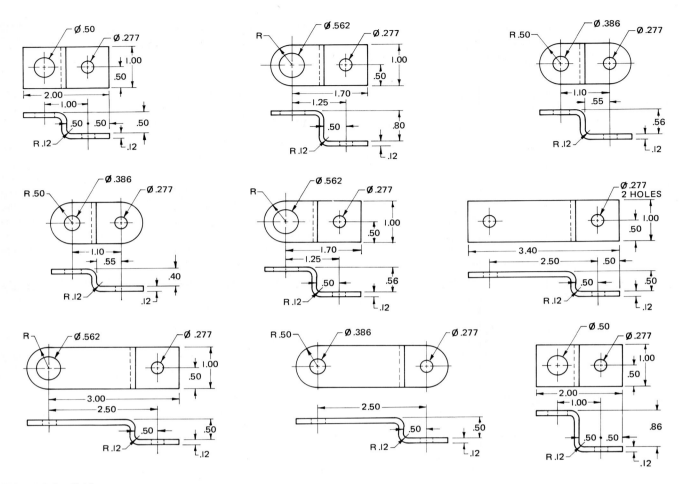

FIG. 14-2-A Cable straps.

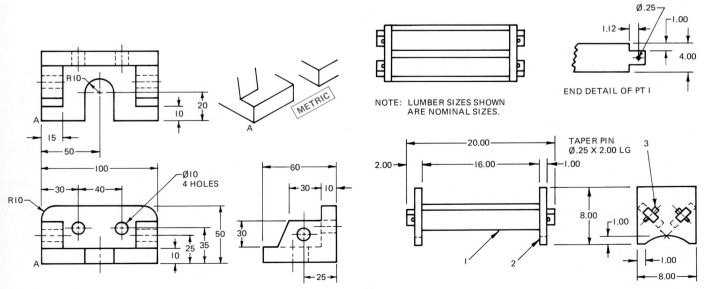

FIG. 14-2-B Rod guide.

FIG. 14-2-C Book rack.

NOTE: LUMBER SIZES SHOWN ARE NOMINAL SIZES.

END DETAIL OF PT I

4. Redraw the part shown in Fig. 14-2-D using arrowless dimensioning and simplified drawing practices. Scale 1:12. Use the bottom and left-hand edge for the datum surfaces.

5. Redraw the two parts shown in Figs. 14-2-E and 14-2-F using partial views and the symmetry symbol. Scale to suit.

FIG. 14-2-D   Cover plate.

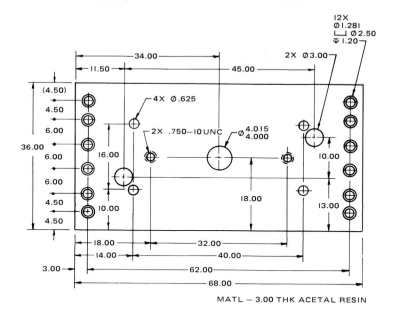

MATL — 3.00 THK ACETAL RESIN

FIG. 14-2-E   Tube support.

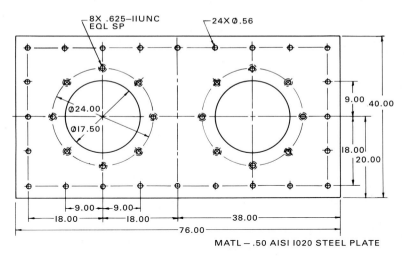

MATL — .50 AISI 1020 STEEL PLATE

FIG. 14-2-F   Gasket.

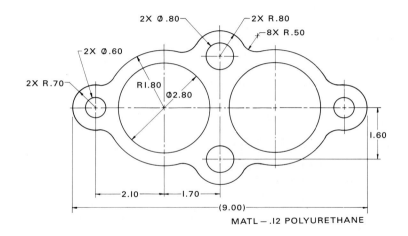

MATL — .12 POLYURETHANE

6. Make simplified drawings of the parts shown in Figs. 14-2-G and 14-2-H. Refer to Fig. 14-2-7. Scale to suit.

7. An exploded orthographic assembly drawing of the wheel-puller shown in Fig. 14-2-J is urgently required. Time does not permit one drafter to do the entire drawing; thus, three drafters will be required to draw the parts. Scale 1:2. Draw all the parts. When all the parts have been completed, make prints of them. Cut out the parts and assemble them in the exploded position on a B (A3) size sheet. Make a suitable print of the exploded assembly on a copying machine.

8. Draw the electronics diagram shown in Fig. 14-2-K using the CAD library (you may have to make your own) or use appliqués and templates if manual drafting is to be used. If appliqués of the electronic components are not available, make your own by photostating Fig. 14-2-15, then cut them out and glue them to your drawing. There is no scale.

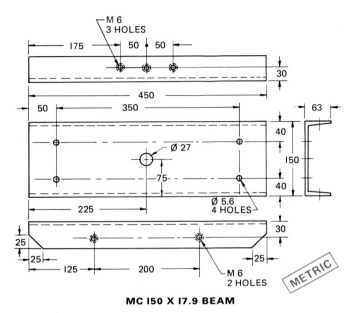

FIG. 14-2-G   Clamp.

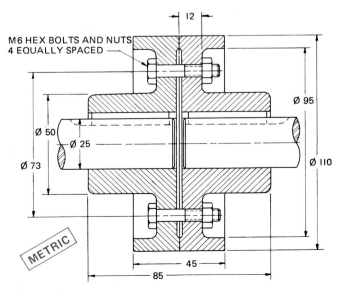

FIG. 14-2-H   Flanged coupling.

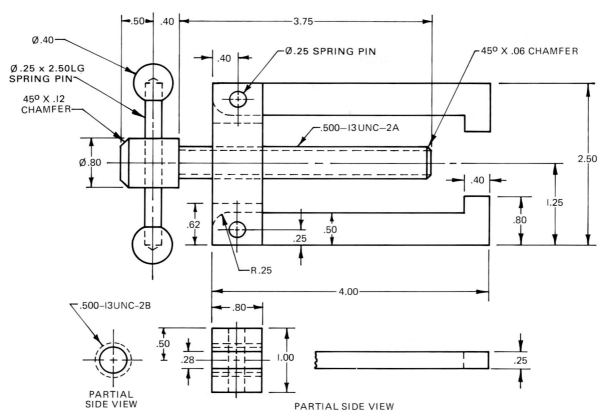

FIG. 14-2-J   Wheel-puller.

FIG. 14-2-K   Electronics diagram.

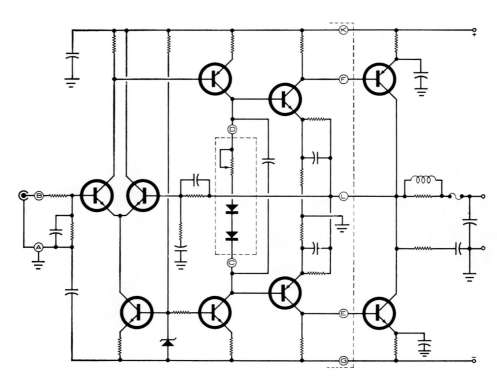

9. Make a photostat of the exploded sprocket assembly shown in Fig. 14-2-L. The photostat, which is to serve as a photodrawing, is to replace the two-view drawing. Make a new chart listing metric sizes to replace the existing chart. Leaders, dimensions, and dimension lines are to be added to the photodrawing.

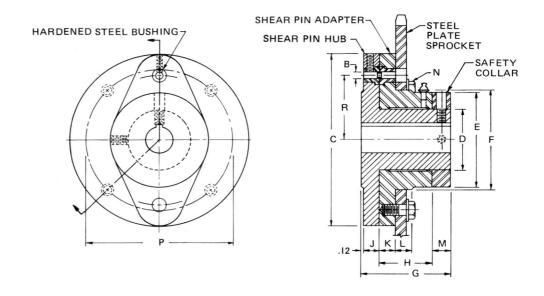

Hub Bore Range	Shear Pin Assembly Number	Shear Pin		Diameters				Length Thru			Hub Flange Thickness	Adapt Flange Thickness	Sprocket Seat Width	Bolts	
		Radius	Pin Dia.	Flange	Shear Pin Hub	Adapt Hub & Collar	Sprocket Seat	Shear Pin Hub	Adapt	Collar				Number & Size	Bolt Circle
		R	B	C	D	E	F	G	H	M	J	K	L	N	P
1.00 & under	SP-17	1.80	.25	5.25	1.75	2.50	2.62	2.44	1.38	.38	.56	.56	.44	4-.38	4.00
1.06-1.25	SP-18	2.18	.25	6.00	2.25	3.25	3.38	2.94	1.75	.50	.56	.56	.56	4-.38	4.75
1.30-1.50	SP-19	2.56	.30	6.75	2.75	4.00	4.12	3.56	2.12	.62	.68	.68	.68	4-.50	5.50
1.56-1.75	SP-20	3.00	.38	7.75	3.25	4.75	4.88	4.18	2.50	.75	.80	.80	.68	4-.50	6.25
1.80-2.00	SP-21	3.30	.45	8.75	3.75	5.25	5.38	4.80	2.88	.88	.94	.94	.94	4-.62	7.00
2.06-2.25	SP-22	3.80	.50	9.75	4.25	6.25	6.38	5.18	3.00	1.00	1.06	1.06	1.18	4-.62	8.00
2.30-2.50	SP-23	4.00	.50	10.00	4.50	6.50	6.62	5.68	3.50	1.00	1.06	1.06	1.38	4-.62	8.25
2.56-2.75	SP-24	4.40	.55	11.50	5.00	7.00	7.12	6.30	3.88	1.12	1.18	1.18	1.38	4-.62	9.25
2.80-3.00	SP-25	4.90	.62	12.50	5.50	8.00	8.12	6.94	4.25	1.25	1.30	1.30	1.38	6-.62	10.25

FIG. 14-2-L   Sprocket assembly.

ASSIGNMENTS FOR UNIT 14-3, DETAIL DRAWINGS

10. Make a detail drawing of one of the parts shown in Figs. 14-3-A through 14-3-G. Select the scale and the number of views required.

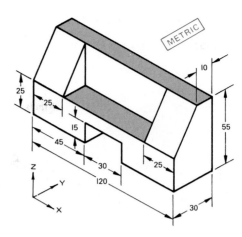

FIG. 14-3-D   Angle block.

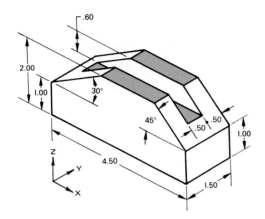

FIG. 14-3-A   Guide block.

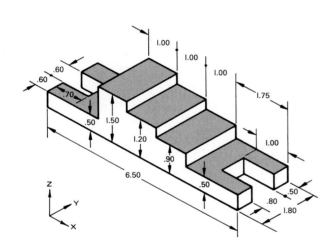

FIG. 14-3-B   Step block.

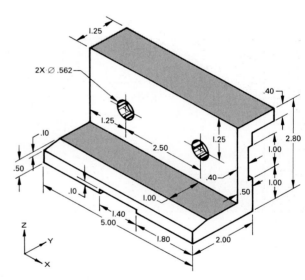

FIG. 14-3-E   Guide bracket.

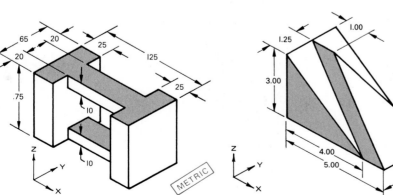

FIG. 14-3-C   Hanger.

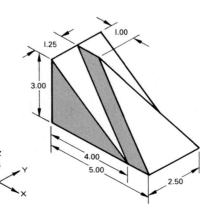

FIG. 14-3-F   Control link.

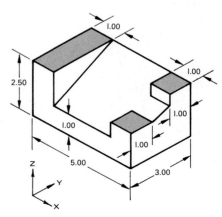

FIG. 14-3-G   End bracket.

11. Make a detail drawing of one of the parts shown in Figs. 14-3-H through 14-3-N. Select the scale and the number of views required.

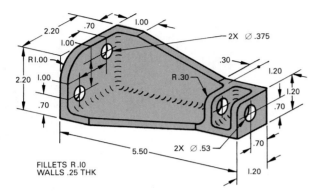

FILLETS R.10
WALLS .25 THK

FIG. 14-3-H   Caster leg.

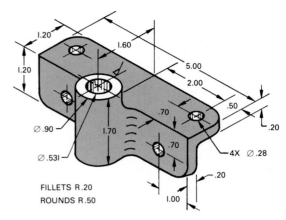

FILLETS R.20
ROUNDS R.50

FIG. 14-3-L   Oarlock socket.

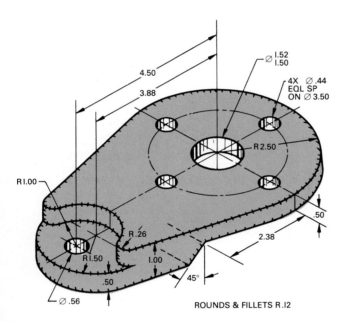

ROUNDS & FILLETS R.12

FIG. 14-3-J   Spacer.

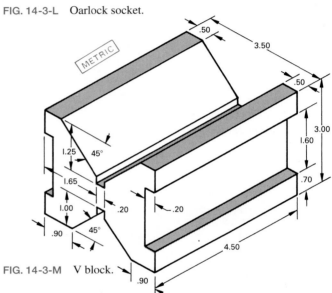

FIG. 14-3-M   V block.

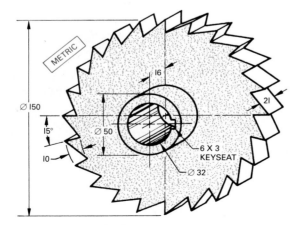

FIG. 14-3-K   Ratchet wheel.

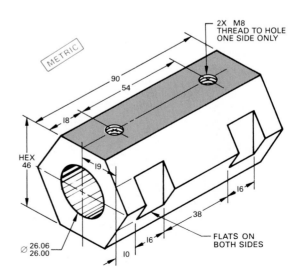

FIG. 14-3-N   Guide rack.

12. Make a detail drawing of one of the parts shown in Figs. 14-3-P through 14-3-T. Select the scale and the number of views required.

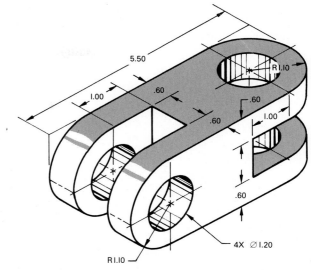

FIG. 14-3-R   Coupling.

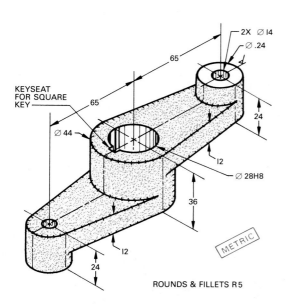

FIG. 14-3-P   Control arm.

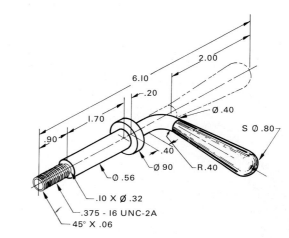

FIG. 14-3-S   Handle.

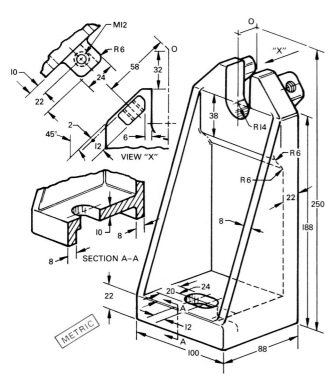

FIG. 14-3-Q   End base.

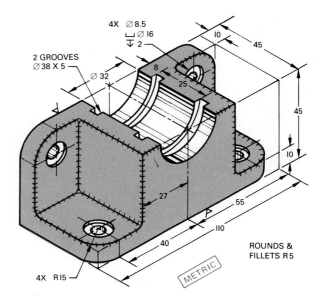

FIG. 14-3-T   Column support.

409

13. Make a detail drawing of one of the parts shown in Figs. 14-3-U through 14-3-Y. Select the scale and the number of views required.

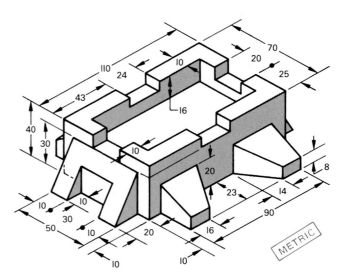

FIG. 14-3-W Base plate.

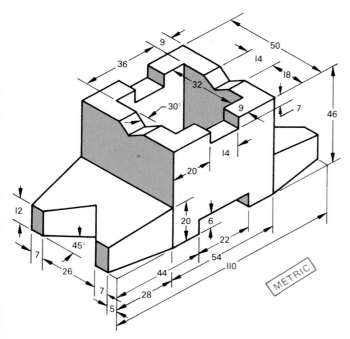

FIG. 14-3-U Trunion.

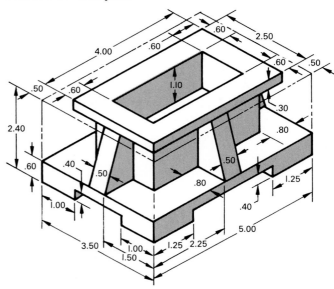

FIG. 14-3-X Cradle bracket.

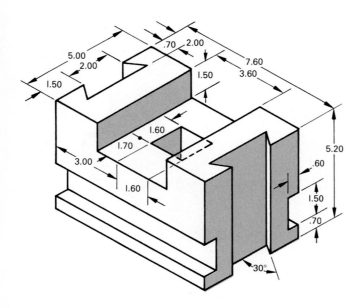

FIG. 14-3-V Base.

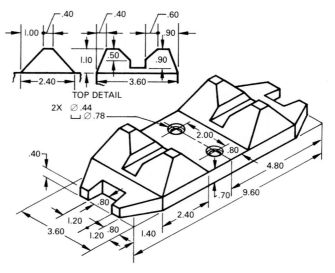

FIG. 14-3-Y Sliding block.

14. Select one of the parts shown in Figs. 14-3-Z and 14-3-AA and make a three-view working drawing. Dimensions are to be converted to millimeters. Only the dovetail and T slot dimensions are critical and must be taken to an accuracy of two points beyond the decimal point. All other dimensions are to be rounded off to whole numbers.

FIG. 14-3-Z   Locating stand.

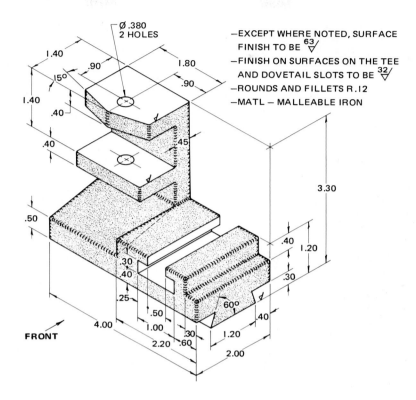

FIG. 14-3-AA   Cross slide.

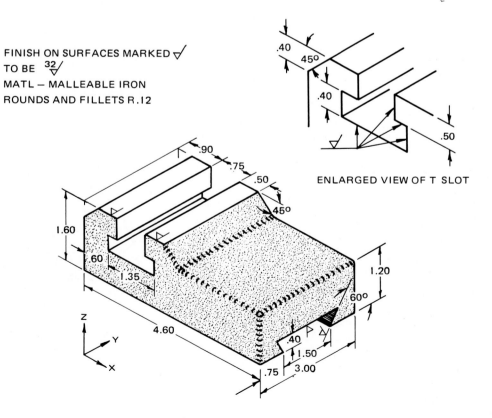

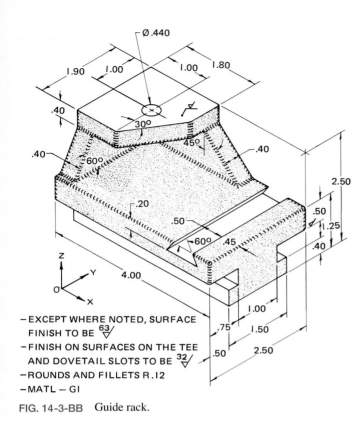

-EXCEPT WHERE NOTED, SURFACE
  FINISH TO BE ⁶³⎺
-FINISH ON SURFACES ON THE TEE
  AND DOVETAIL SLOTS TO BE ³²⎺
-ROUNDS AND FILLETS R.12
-MATL - GI

EXCEPT WHERE NOTED, SURFACE
  FINISH TO BE 63
FINISH ON SURFACES ON THE TEE
  AND DOVETAIL SLOTS TO BE 32
ROUNDS AND FILLETS R.12
MATL - GI

FIG. 14-3-BB   Guide rack.

15. Make a detail drawing of one of the parts shown in Figs. 14-3-BB and 14-3-CC. Select the scale and the number of views required.
16. On a C (A2) size sheet, make a detail drawing of the part shown in Fig. 14-3-DD. To clearly show the features, a section and bottom view should also be drawn. The recommended drawing layout is shown. The slots are to have

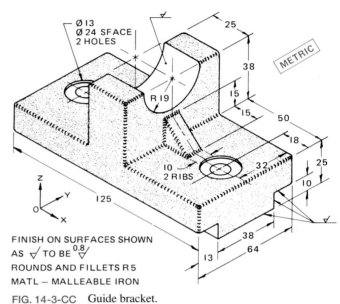

FINISH ON SURFACES SHOWN
AS ⎺ TO BE ^{0.8}⎺
ROUNDS AND FILLETS R 5
MATL - MALLEABLE IRON

FIG. 14-3-CC   Guide bracket.

FIG. 14-3-DD   Pipe vise base.

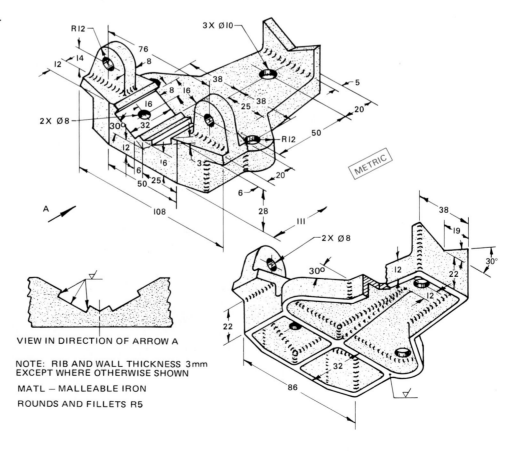

VIEW IN DIRECTION OF ARROW A

NOTE: RIB AND WALL THICKNESS 3mm
EXCEPT WHERE OTHERWISE SHOWN

MATL - MALLEABLE IRON

ROUNDS AND FILLETS R5

a surface finish of 3.2 mm and a machining allowance of 2 mm. The base is to have the same surface finish but with a 3 mm machining allowance.

17. Make a detail drawing of one of the parts shown in Figs. 14-3-EE and 14-3-FF. Select the scale and the number of views required.

18. Make a detail drawing of the part shown in Fig. 14-3-GG. The surfaces shown with a ✓ are to have a surface finish of 63 μin and a machining allowance of .06 in. Select the scale and the number of views required.

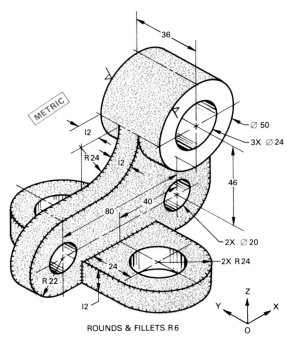

FIG. 14-3-EE   Swivel hanger.

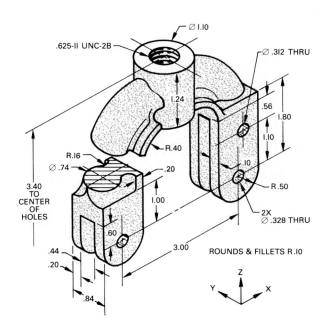

FIG. 14-3-FF   Fork.

FIG. 14-3-GG   Offset bracket.

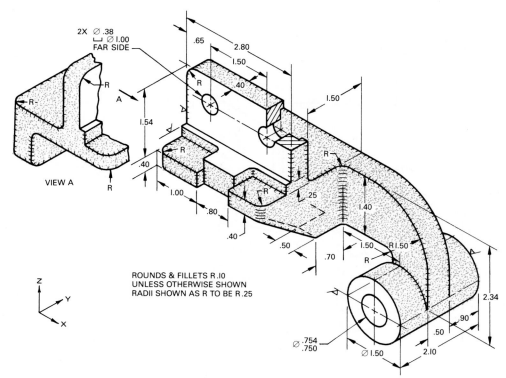

413

19. On a C (A2) size sheet draw the part shown in Fig. 14-3-HH. Draw all six views plus a partial auxiliary view for the .250 threaded hole. Scale 1:2.

20. Make a detail drawing of the part shown in Fig. 14-3-JJ. The surfaces shown with a ✓ are to have a surface finish of 63 μin and a machining allowance of .06 in. Select the scale and the number of views required.

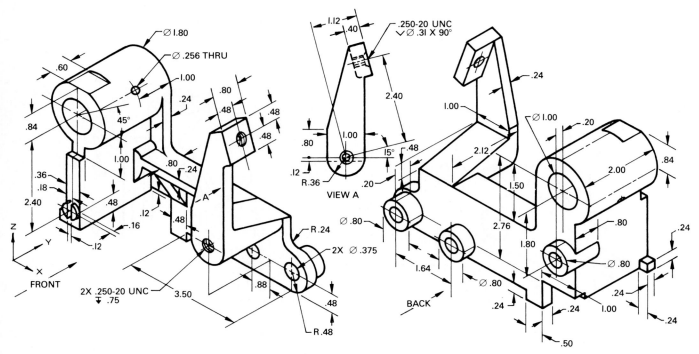

FIG. 14-3-HH   Control bracket.

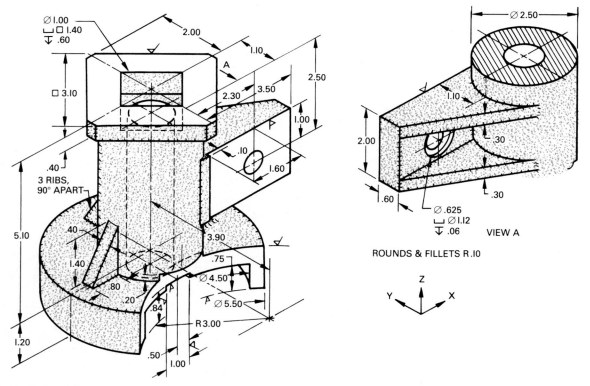

FIG. 14-3-JJ   Pedestal.

ASSIGNMENTS FOR UNIT 14-4, MULTIPLE DETAIL DRAWINGS

21. Make detail drawings of all the parts shown of one of the assemblies in Figs. 14-4-A and 14-4-B. Since time is money, select only the views necessary to describe each part. Below each part show the following information: part number, name of part, material, number required. Scale 1:1.

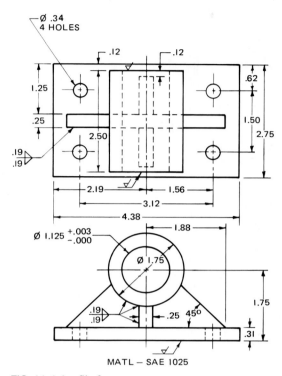

FIG. 14-4-A   Shaft support.

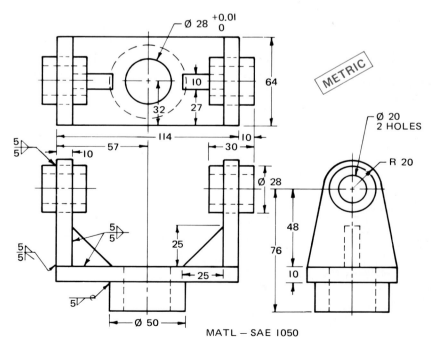

FIG. 14-4-B   Shaft pivot support.

22. Make detail drawings of the nonstandard parts shown on one of the assemblies in Figs. 14-4-C and 14-4-D. Select the scale and the number of views required. Add to the drawing an item list.

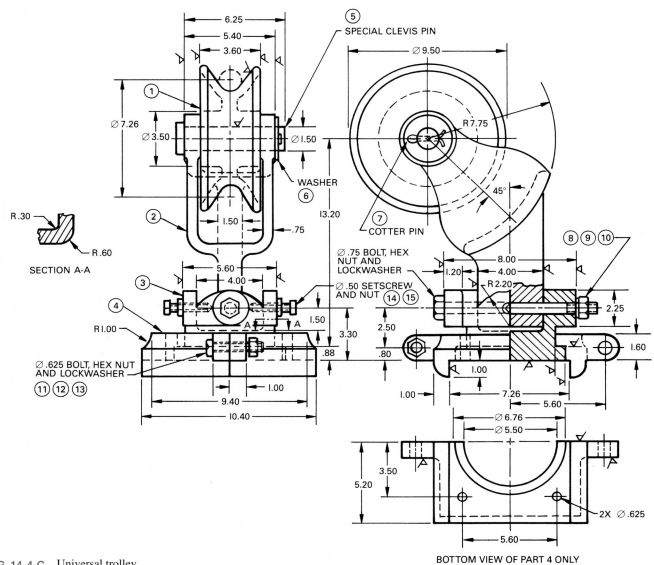

FIG. 14-4-C   Universal trolley.

FIG. 14-4-D   Bearing bracket.

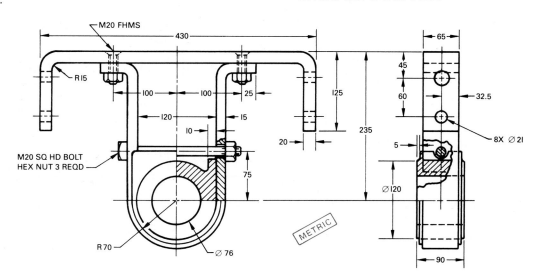

23. Make detail drawings of the nonstandard parts shown in Fig. 14-4-E. Select the scale and the number of views required. Add to the drawing an item list. The surfaces shown with a √ are to have a surface finish of 63 μin and a machining allowance of .06 in.

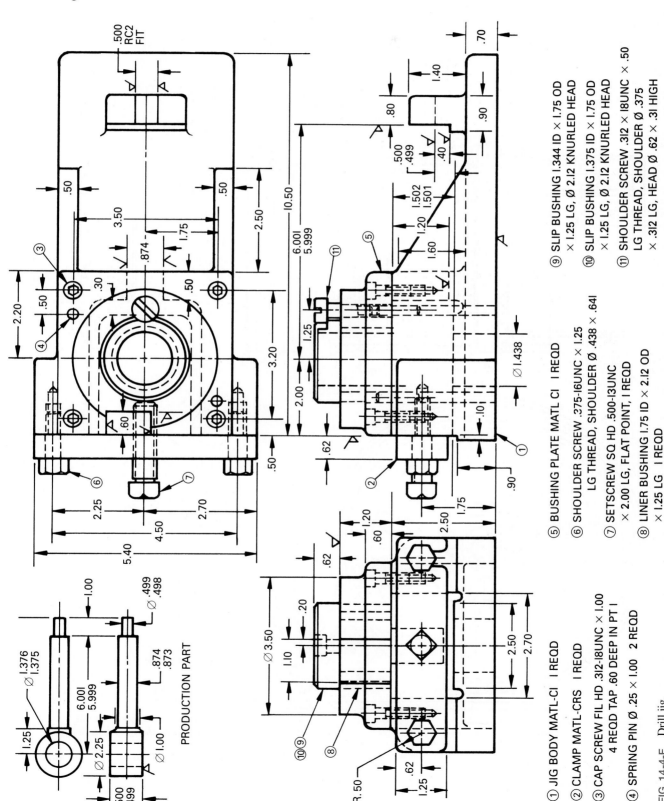

① JIG BODY MATL-CI I REQD
② CLAMP MATL-CRS I REQD
③ CAP SCREW FIL HD .3I2-I8UNC × 1.00
    4 REQD TAP .60 DEEP IN PT I
④ SPRING PIN Ø .25 × 1.00 2 REQD

⑤ BUSHING PLATE MATL CI I REQD
⑥ SHOULDER SCREW .375-I6UNC × 1.25
    LG THREAD, SHOULDER Ø .438 × .641
⑦ SETSCREW SQ HD .500-I3UNC
    × 2.00 LG, FLAT POINT, I REQD
⑧ LINER BUSHING 1.75 ID × 2.12 OD
    × 1.25 LG I REQD

⑨ SLIP BUSHING 1.344 ID × 1.75 OD
    × 1.25 LG, Ø 2.12 KNURLED HEAD
⑩ SLIP BUSHING 1.375 ID × 1.75 OD
    × 1.25 LG, Ø 2.12 KNURLED HEAD
⑪ SHOULDER SCREW .3I2 × I8UNC × .50
    LG THREAD, SHOULDER Ø .375
    × .3I2 LG, HEAD Ø .62 × .3I HIGH

FIG. 14-4-E   Drill jig.

PRODUCTION PART

417

24. On C (A2) size sheets make detail drawings of the parts shown in Fig. 14-4-F. Select the scale and the number of views required. An LN3 fit is required between the bushings and housing, and an RC4 fit between the shafts and bushing. Include an item list for the parts.

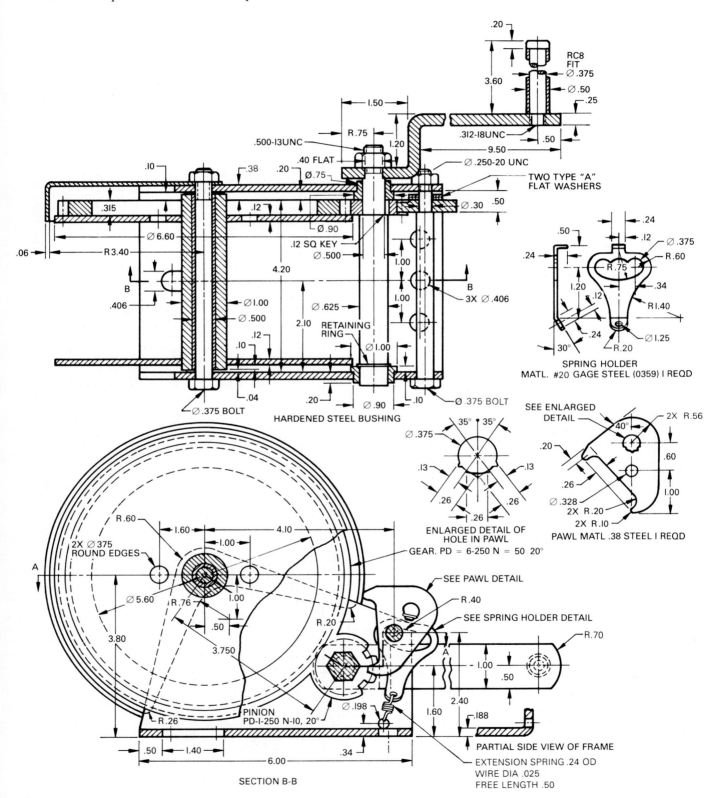

FIG. 14-4-F   Winch. *(Fulton Co.)*

25. Prepare detail drawings of any of the parts assigned by your instructor from the assembly drawings shown in Figs. 14-4-G and 14-4-H. For Fig. 14-4-G the following fits are to be used: Ø28H9/d9; Ø45H7/s6; and Ø35H8/f7.

For Fig. 14-4-H an RC4 fit is required for the Ø1.20 shaft. The scale and selection of views are to be decided by the student. Include an item list for the parts.

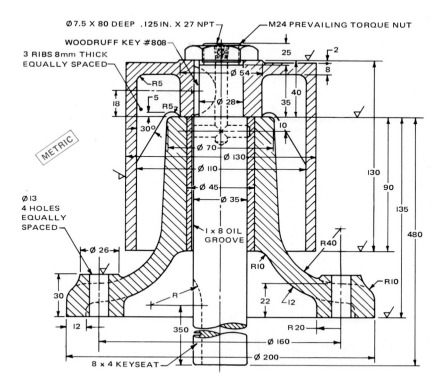

FIG. 14-4-G   Pulley assembly.

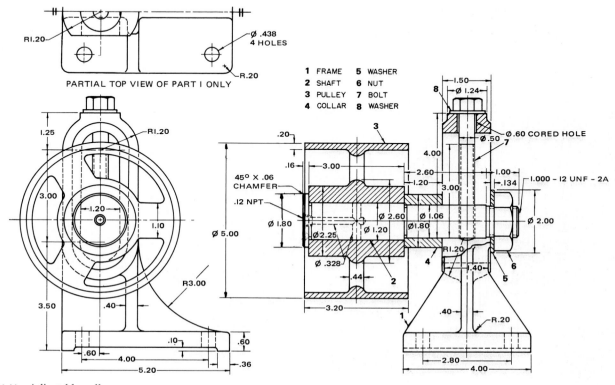

FIG. 14-4-H   Adjustable pulley.

26. Prepare detail drawings for the nonstandard parts for the assemblies shown in Figs. 14-4-J and 14-4-K. Select the scale and number of views required. Include an item list for the parts.

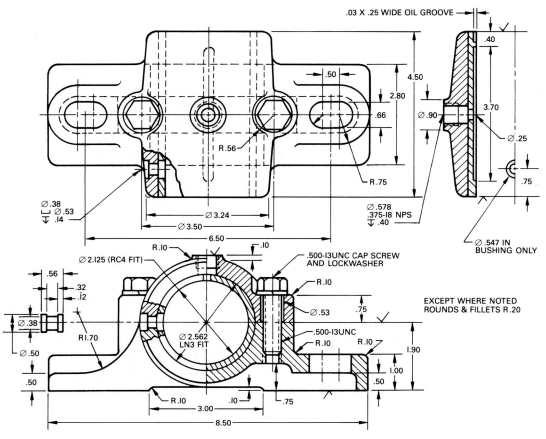

FIG. 14-4-J   Split bushing.

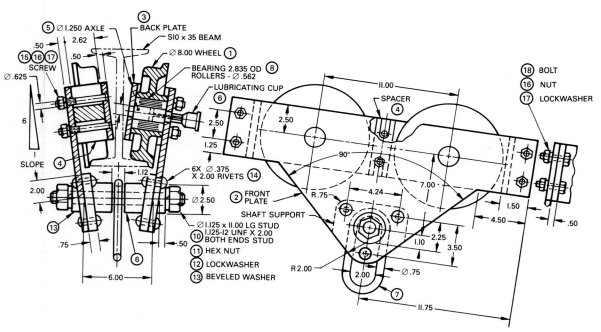

FIG. 14-4-K   Four-wheel trolley.

27. Prepare detail drawings for the nonstandard parts for any of the assemblies shown in Figs. 14-6-A through 14-6-G (here and through pg. 427). The student is to select the scale and the number of views required. Include an item list for the parts.

ASSIGNMENT FOR UNIT 14-5, DRAWING REVISIONS

28. Select one of the drawings shown in Figs. 14-5-A and 14-5-B and make appropriate revisions to these drawings, recording the changes in a drawing revision column and indicating which dimensions are not to scale.

ASSIGNMENTS FOR UNIT 14-6, ASSEMBLY DRAWINGS

29. Make a one-view assembly drawing of one of the assemblies shown in Figs. 14-6-A and 14-6-B. For Fig. 14-6-B show a round bar Ø24 mm in phantom being held in position. Include on the drawing an item list and identification part numbers. Scale 1:1.

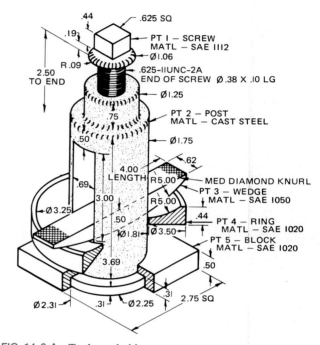

FIG. 14-6-A   Tool post holder.

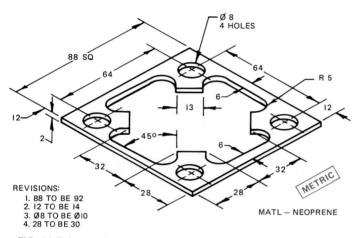

REVISIONS:
1. 88 TO BE 92
2. 12 TO BE 14
3. Ø8 TO BE Ø10
4. 28 TO BE 30

MATL — NEOPRENE

FIG. 14-5-A   Gasket.

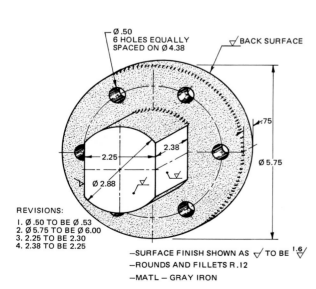

REVISIONS:
1. Ø.50 TO BE Ø.53
2. Ø5.75 TO BE Ø6.00
3. 2.25 TO BE 2.30
4. 2.38 TO BE 2.25

—SURFACE FINISH SHOWN AS ▽ TO BE ¹·⁶/▽
—ROUNDS AND FILLETS R.12
—MATL — GRAY IRON

FIG. 14-5-B   Axle cap.

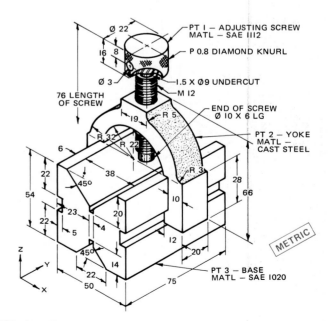

FIG. 14-6-B   V-block clamp.

30. Make a one-view assembly drawing of the bench vise shown in Fig. 14-6-C. Show the vise jaws open 50 mm and place on the drawing only pertinent dimensions. Include on the drawing an item list and identification part numbers. Scale 1:1.

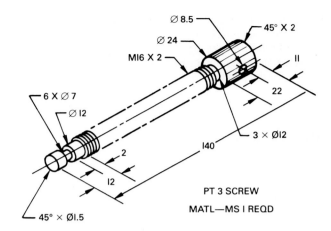

PT 3 SCREW

MATL—MS I REQD

PT 6 HANDLE ∅ 8 X 100 LG THREAD BOTH ENDS

M8 X 1.25 X 10 LG MATL—CRS I REQD

PT 7 FHMS- M6 X I X 20 LG, I REQD

METRIC

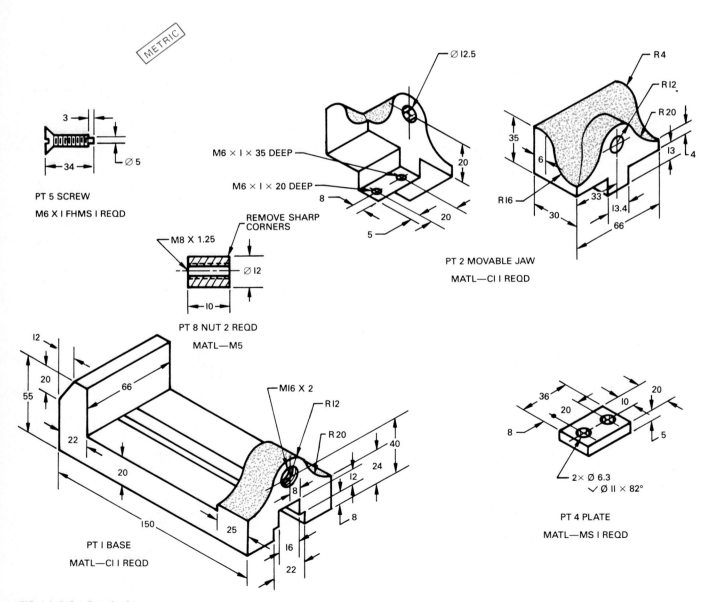

PT 5 SCREW

M6 X I FHMS I REQD

PT 8 NUT 2 REQD

MATL—M5

PT 2 MOVABLE JAW

MATL—CI I REQD

PT 4 PLATE

MATL—MS I REQD

PT I BASE

MATL—CI I REQD

FIG. 14-6-C   Bench vise.

31. Make a one-view assembly drawing of the wrench shown in Fig. 14-6-D. Include on the drawing an item list and identification part numbers. Scale 1:1.

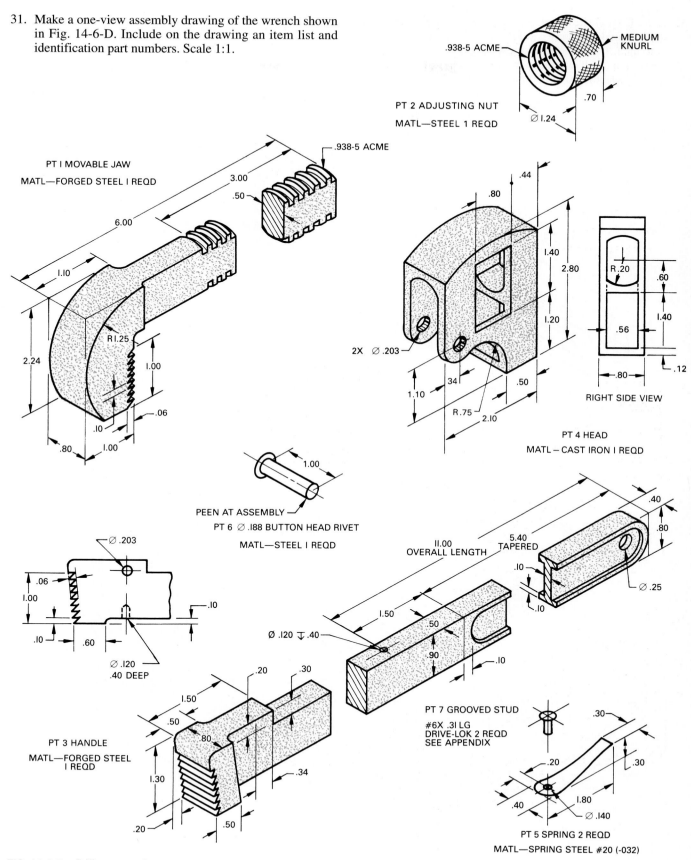

.938-5 ACME — — MEDIUM KNURL

PT 2 ADJUSTING NUT
MATL—STEEL 1 REQD
Ø 1.24
.70

PT I MOVABLE JAW
MATL—FORGED STEEL I REQD
.938-5 ACME
3.00
6.00
.50
1.10
R1.25
2.24
1.00
.06
.10
1.00
.80

2X Ø .203
1.10
34
R .75
2.10
.50
1.20
1.40
2.80
.80
.44

PT 4 HEAD
MATL – CAST IRON I REQD

RIGHT SIDE VIEW
R.20
.60
1.40
.56
.80
.12

PEEN AT ASSEMBLY
PT 6 Ø .188 BUTTON HEAD RIVET
MATL—STEEL I REQD
1.00

Ø .203
.06
1.00
.10
.10   .60
Ø .120
.40 DEEP
.10

PT 3 HANDLE
MATL—FORGED STEEL I REQD
1.50
.50
.80
1.30
.20   .50
.20   .30
.34

Ø .120 ⌴ .40
11.00 OVERALL LENGTH
1.50
.50
.90
.10
5.40 TAPERED
.40
.80
.10
.10
Ø .25

PT 7 GROOVED STUD
#6X .31 LG
DRIVE-LOK 2 REQD
SEE APPENDIX

.30
.20
.40
1.80
.30
Ø .140

PT 5 SPRING 2 REQD
MATL—SPRING STEEL #20 (-032)

FIG. 14-6-D   Stillson wrench.

32. On a C (A2) sheet make a three-view assembly drawing of the wood vise shown in Fig. 14-6-E with the jaws open one inch. Draw the front view in full section, a bottom view, and a half end view. Include on the drawing an item list and identification part numbers. Scale 1:1.

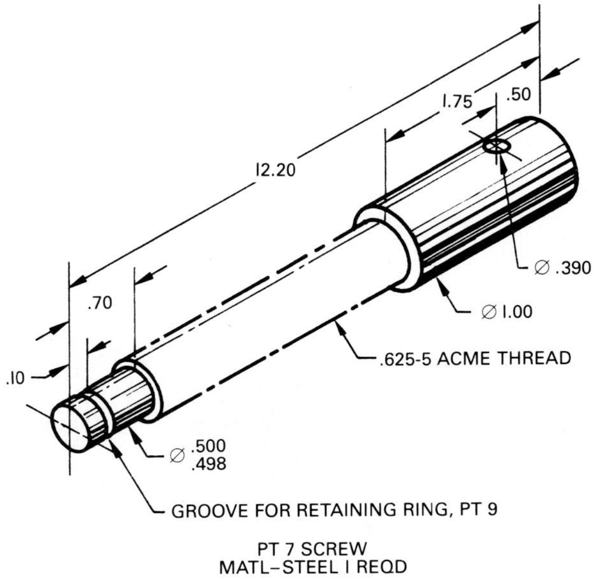

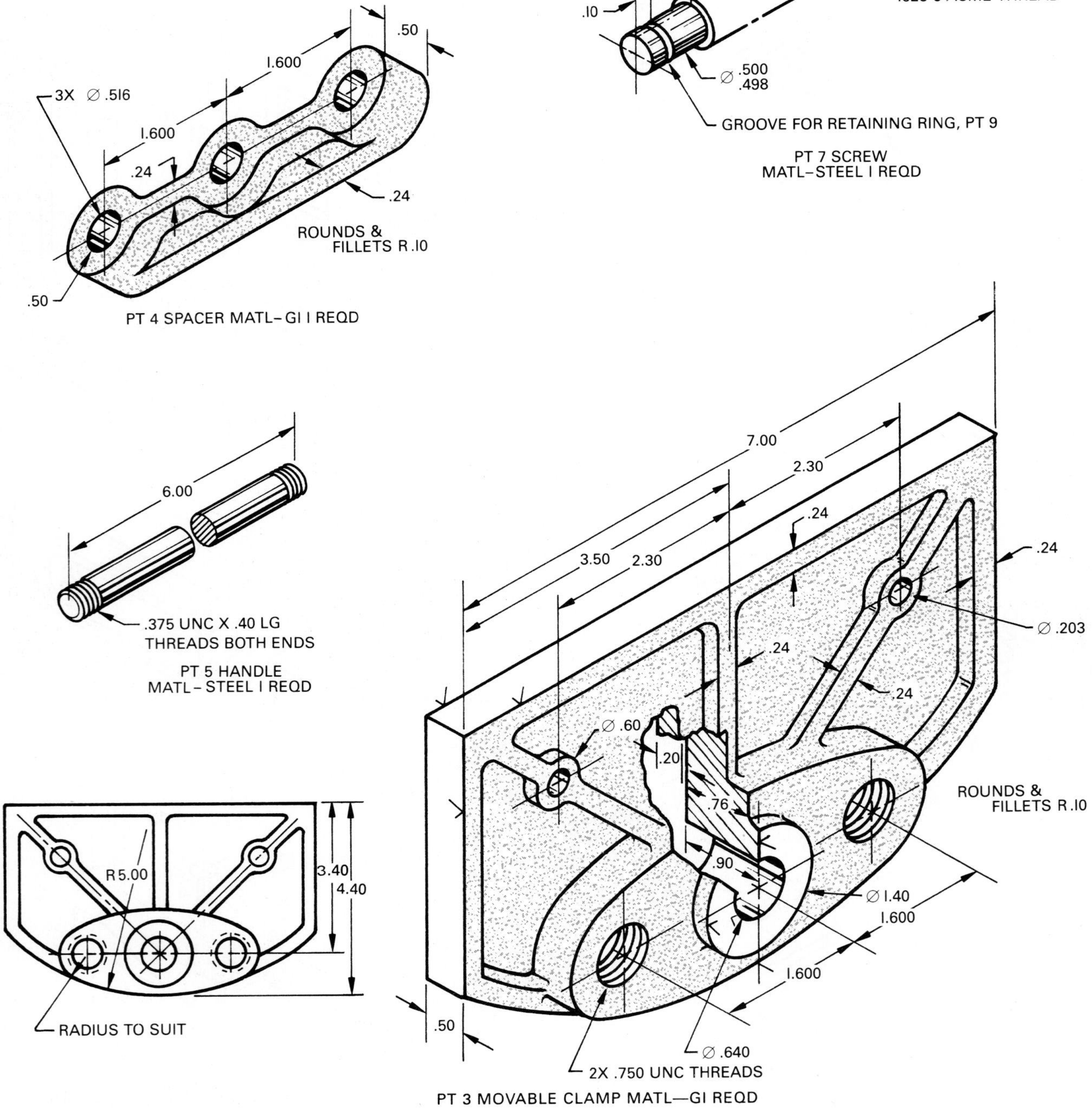

GROOVE FOR RETAINING RING, PT 9

Ø .390
Ø 1.00
.625-5 ACME THREAD
Ø .500 / .498
1.75   .50
12.20
.70
.10

PT 7 SCREW
MATL–STEEL I REQD

3X Ø .516
1.600
1.600
.24
.24
.50
.50
ROUNDS & FILLETS R .10

PT 4 SPACER MATL– GI I REQD

6.00
.375 UNC X .40 LG
THREADS BOTH ENDS

PT 5 HANDLE
MATL– STEEL I REQD

R 5.00
3.40
4.40
RADIUS TO SUIT

7.00
2.30
.24
.24
3.50   2.30
.24
.24
.24
Ø .203
Ø .60
.20
.76
.90
Ø 1.40
1.600
1.600
ROUNDS & FILLETS R .10
Ø .640
2X .750 UNC THREADS
.50

PT 3 MOVABLE CLAMP MATL—GI REQD

FIG. 14-6-E   Wood vise. (Woden Tools)

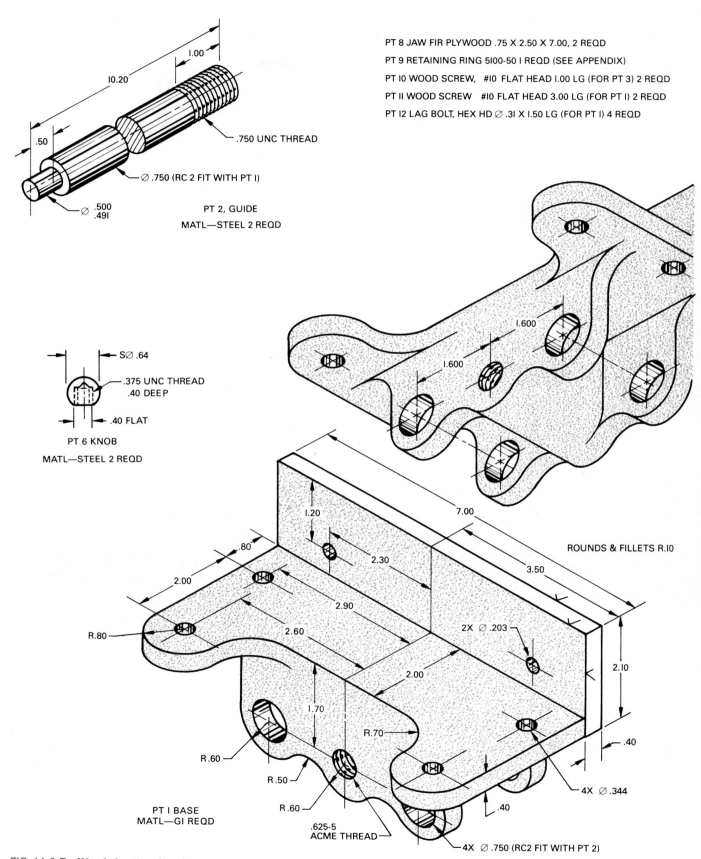

10.20

1.00

.50

.750 UNC THREAD

∅ .750 (RC 2 FIT WITH PT I)

∅ .500
   .491

PT 2, GUIDE
MATL—STEEL 2 REQD

PT 8 JAW FIR PLYWOOD .75 X 2.50 X 7.00, 2 REQD
PT 9 RETAINING RING 5I00-50 I REQD (SEE APPENDIX)
PT I0 WOOD SCREW, #I0 FLAT HEAD I.00 LG (FOR PT 3) 2 REQD
PT II WOOD SCREW #I0 FLAT HEAD 3.00 LG (FOR PT I) 2 REQD
PT I2 LAG BOLT, HEX HD ∅ .3I X I.50 LG (FOR PT I) 4 REQD

S∅ .64

.375 UNC THREAD
.40 DEEP

.40 FLAT

PT 6 KNOB

MATL—STEEL 2 REQD

1.600

1.600

1.20

.80

7.00

ROUNDS & FILLETS R.I0

2.00

2.30

3.50

2.90

2X ∅ .203

2.60

2.00

2.10

R.80

1.70

R.70

R.60

.40

R.50

R.60

.40

4X ∅ .344

PT I BASE
MATL—GI REQD

.625-5
ACME THREAD

4X ∅ .750 (RC2 FIT WITH PT 2)

FIG. 14-6-E   Wood vise. (continued)

33. Make a two-view assembly drawing of the check valve shown in Fig. 14-6-F. Show the front view in full section. Include on the drawing an item list and identification part numbers. Scale 1:1.

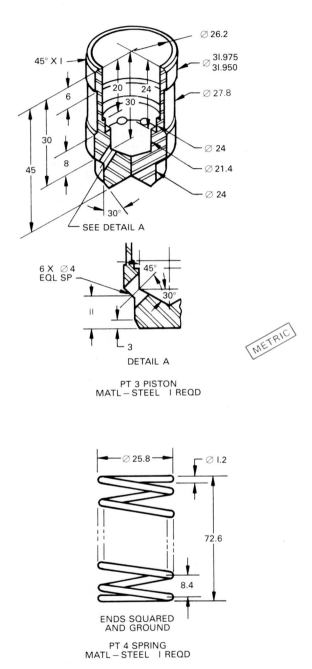

PT 3 PISTON
MATL — STEEL   1 REQD

PT 4 SPRING
MATL — STEEL   1 REQD

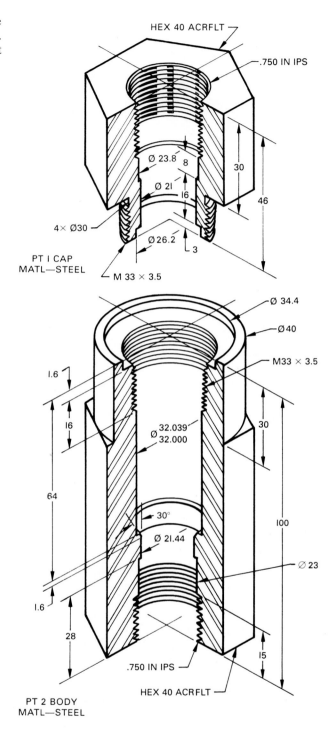

PT 1 CAP
MATL — STEEL

PT 2 BODY
MATL — STEEL

PT 5 O-RING Ø 2 × 30 ID

FIG. 14-6-F   Check valve. *(Bellows-Valvair)*

34. Make a two-view assembly drawing of the check valve shown in Fig. 14-6-G. Show the front view in full section and the top view through part 5. Include on the drawing an item list and identification part numbers. Scale 1:1.

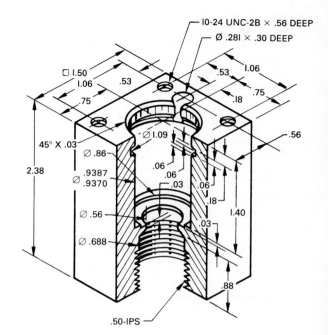

PT 2 BODY MATL—STEEL I REQD

PT 7 CAP SCREW, SOCKET HD
I0-24 UNC X I.75 LONG 4 REQD

PT 8 LOCKWASHER, #I0, 4 REQD

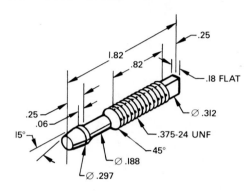

PT 5 VALVE
MATL—STEEL I REQD.

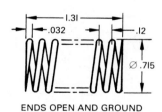

ENDS OPEN AND GROUND
PT 4 SPRING
MATL—STEEL I REQD

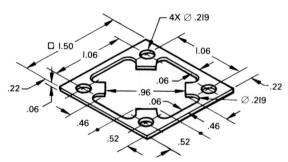

PT 6 GASKET MATL—NEOPRENE I REQD

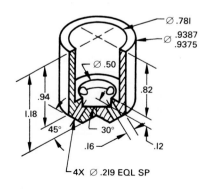

PT 3 PISTON
MATL—ALUMINUM I REQD

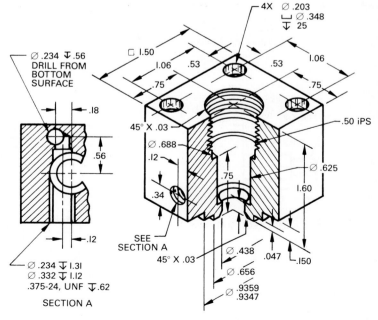

SECTION A

PT I CAP MATL— STEEL I REQD

FIG. 14-6-G   Check valve. *(Bellows-Valvair)*

427

35. On a B (A3) size sheet make a two-view assembly drawing of the trolley shown in Fig. 14-6-H mounted on an S200 × 34 beam. Show the side view in half section and place on the drawing the dimensions suitable for a catalog. Scale 1:2. Refer to Figs. 25-1-7 and 25-1-8 for dimensions of the beam.

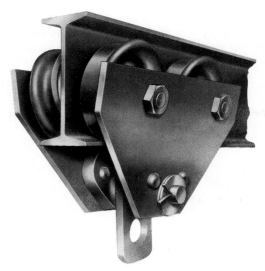

PT 8 ADJUSTING WASHER 26 ID × 44 OD × 4 THK
    12 REQD, MATL—STL
PT 9 RIVET, BUTTON HEAD, Ø10 × 60 LG, 4 REQD
PT I0 WASHER 26 ID × 65 OD × 3 THK, 4 REQD
PT II LOCKNUT MI6 × 2, 4 REQD
    PREVAILING TORQUE INSERT-TYPE
PT I2 COTTER PIN Ø6 × 40 LG, 6 REQD

METRIC

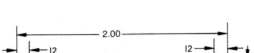

PT 3 AXLE I REQD MATL—CRS

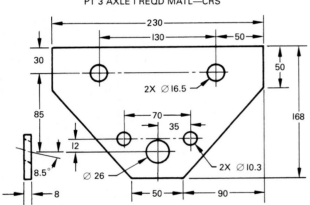

PT I SIDE PLATE 2 REQD MATL—MST

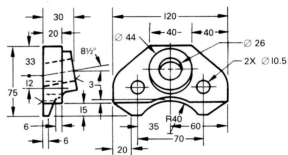

PT 2 GUIDE 2 REQD MATL—CI

PT 5 ROLLER BEARING

44 REQD MATL—CRS
CASE HARDEN!

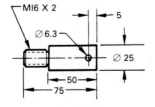

PT 7 AXLE 4 REQD MAT—CRS

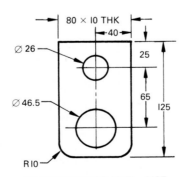

PT 4 HOOK I REQD MATL—MST

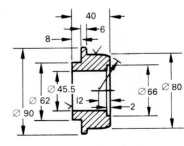

PT 6 WHEEL 4 REQD MAT—CI

FIG. 14-6-H   Trolley.

36. On a C (A2) size sheet make a two-view (one view can be a partial view) assembly drawing of the pipe cutter shown in Fig. 14-6-J. Sizes shown are nominal sizes. Include an item list and identification part numbers on the drawing. Scale 1:1.

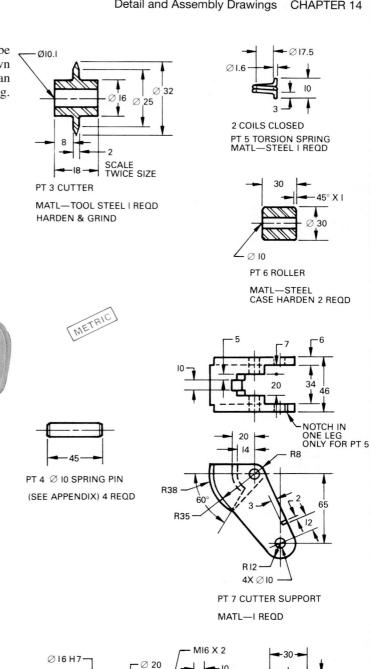

Ø10.1

Ø 16   Ø 25   Ø 32

8   2

18

SCALE TWICE SIZE

PT 3 CUTTER

MATL—TOOL STEEL I REQD
HARDEN & GRIND

Ø1.6   Ø 17.5   10   3

2 COILS CLOSED
PT 5 TORSION SPRING
MATL—STEEL I REQD

30   45° X I   Ø 30   Ø 10

PT 6 ROLLER

MATL—STEEL
CASE HARDEN 2 REQD

METRIC

5   7   6
10   20   34   46

NOTCH IN ONE LEG ONLY FOR PT 5

20   14   R8
R38   60°   3   2   65   12
R35
R 12
4X Ø 10

PT 7 CUTTER SUPPORT

MATL—I REQD

PT 4   Ø 10 SPRING PIN
45

(SEE APPENDIX) 4 REQD

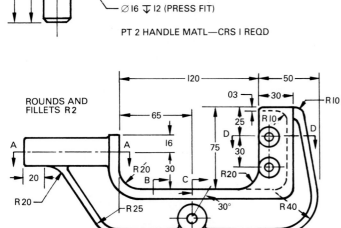

Ø 25   45° X 2   300
6   4√   MI6 X 2   5
110
55
Ø 25   RADIUS TO SUIT AND HARDEN END
Ø 16 ∓ 12 (PRESS FIT)

PT 2 HANDLE MATL—CRS I REQD

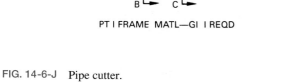

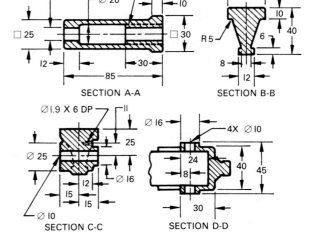

ROUNDS AND FILLETS R 2

120   50
03   30   R 10
65   25   R 10
A   16   D   D
R 20   30   75   R 20
20   R 20   C   30°   R 40
R 20   R 25
B   C

PT I FRAME MATL—GI I REQD

Ø16 H7   Ø 20   MI6 X 2   10
□ 25   □ 30   R 5
12   30
85

SECTION A-A

30   10   40   6   8   12

SECTION B-B

Ø1.9 X 6 DP   II
25
Ø 25   12
Ø 16
15
15
Ø 10

SECTION C-C

Ø 16   4X Ø 10
24   40   45
8
30

SECTION D-D

FIG. 14-6-J   Pipe cutter.

37. On a C (A2) size sheet make a one-view assembly drawing of the two-arm parallel puller shown in Fig. 14-6-K removing the tapered roller bearing from the shaft shown in assembly A. Include on the drawing an item list and identification part numbers. Scale 1:1.

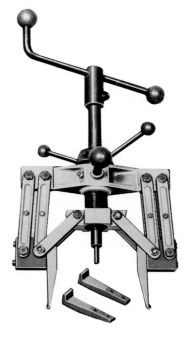

PT 19 BALL BEARING ⌀ .375 MATL—STEEL, 1 REQD

PT 20 GREASE CUP

PT 21 BOLT—HEX HD .312 UNF X 1.50 LG, 6 REQD

PT 22 BOLT—HEX HD .312 UNF X 1.75 LG, 5 REQD

PT 23 BOLT—HEX HD .312 UNF X 2.50 LG, 4 REQD

PT 24 MACH SCREW—HEX HD 8-32 X 1.25 LG, 4 REQD

PT 25 NUT—HEX HD .312 UNF, 10 REQD

PT 26 NUT—HEX HD 8-32, 4 REQD

PT 27 SETSCREW—HEADLESS .375 UNF X .50 LG
        CUP POINT, 2 REQD

PT 28 SETSCREW—HEADLESS 8-32 X .25 LG
        FULL DOG, 2 REQD

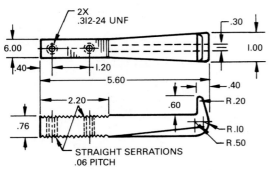

PT 11 FINGERS MATL—CI 2 REQD

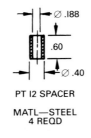

PT 12 SPACER

MATL—STEEL
4 REQD

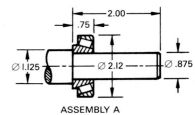

ASSEMBLY A

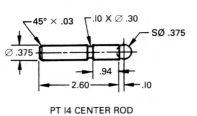

PT 14 CENTER ROD

MATL—STEEL 1 REQD

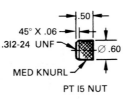

PT 15 NUT

MATL—STEEL 1 REQD

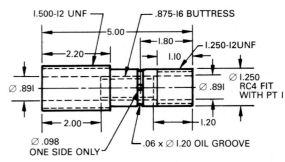

PT 13 ADJUSTING SCREW MATL—STEEL 1 REQD

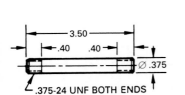

.375-24 UNF BOTH ENDS

PT 16 HANDLE MATL—STEEL 2 REQD

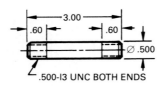

.500-13 UNC BOTH ENDS

PT 17 HANDLE MATL—STEEL 1 REQD

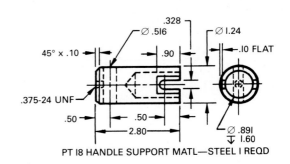

PT 18 HANDLE SUPPORT MATL—STEEL 1 REQD

FIG. 14-6-K Two-arm parallel puller. *(Delro Industries)*

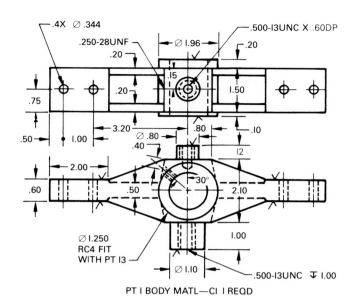

PT I BODY MATL—CI I REQD

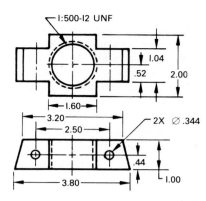

PT 2 SLIDE BAR MATL—STEEL I REQD

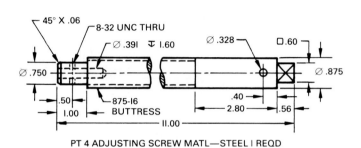

PT 4 ADJUSTING SCREW MATL—STEEL I REQD

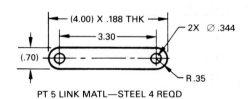

PT 3 JAW MATL—STEEL 2 REQD

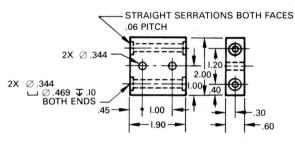

PT 5 LINK MATL—STEEL 4 REQD

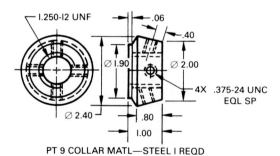

PT 9 COLLAR MATL—STEEL I REQD

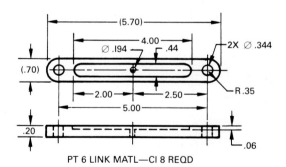

PT 6 LINK MATL—CI 8 REQD

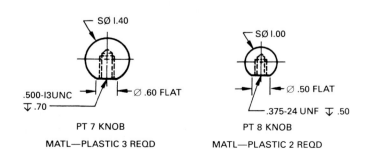

PT 7 KNOB

MATL—PLASTIC 3 REQD

PT 8 KNOB

MATL—PLASTIC 2 REQD

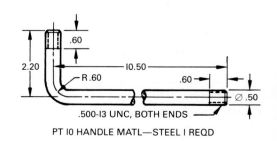

PT I0 HANDLE MATL—STEEL I REQD

FIG. 14-6-K   Two-arm parallel puller. (continued)

38. On a C (A2) size sheet make a front-view, full-section assembly drawing plus a partial side view of the journal jack shown in Fig. 14-6-L. Show the jack in its lowest position and a phantom view of part three 3.00 in. higher. Use this phantom view to indicate the maximum jack height and the regular view to indicate the minimum height. Include an item list and identification part numbers. Scale 1:1.

PT 19 FLAT WASHER MATL—STEEL

    1.19 ID × 2.25 OD × .180 1 REQD

PT 20 PIN MATL—STEEL Ø.188 × 1.00 LG 1 REQD

PT 21 BALL BEARING SØ .625 MATL—STEEL 12 REQD

PT 22 KEY—608 WOODRUFF, 1 REQD

PT 23 PIN MATL—STEEL Ø.25 × .40 LG 1 REQD

PT 24 COTTER PIN Ø.125 × 1.25 LG 1 REQD

PT 25 COTTER PIN Ø.094 × .75 LG 1 REQD

PT 26 HANDLE .875 ID × 1.00 OD × 18.00 LG STL 1 REQD

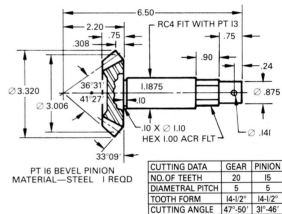

CUTTING DATA	GEAR	PINION
NO. OF TEETH	20	15
DIAMETRAL PITCH	5	5
TOOTH FORM	14-1/2°	14-1/2°
CUTTING ANGLE	47°-50'	31°-46'
WHOLE DEPTH	.431	.431
CHORDAL ADD	.203	.203
CHORDAL THICK	.314	.314

PT 16 BEVEL PINION
MATERIAL—STEEL 1 REQD

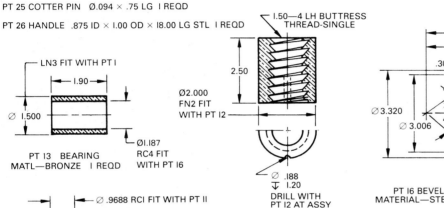

PT 13 BEARING
MATL—BRONZE 1 REQD

PT 17 BEARING
MATL—BRONZE 1 REQD

PT 14 LIFTING NUT
MATL—BRONZE 1 REQD

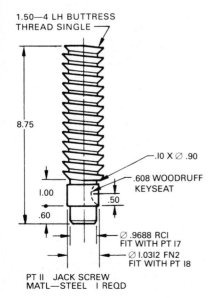

PT 11 JACK SCREW
MATL—STEEL 1 REQD

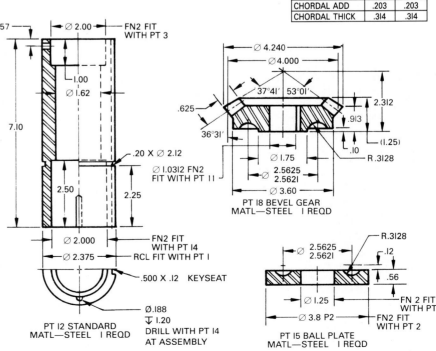

PT 12 STANDARD
MATL—STEEL 1 REQD

PT 18 BEVEL GEAR
MATL—STEEL 1 REQD

PT 15 BALL PLATE
MATL—STEEL 1 REQD

FIG. 14-6-L Journal jack. *(Duff-Norton Co.)*

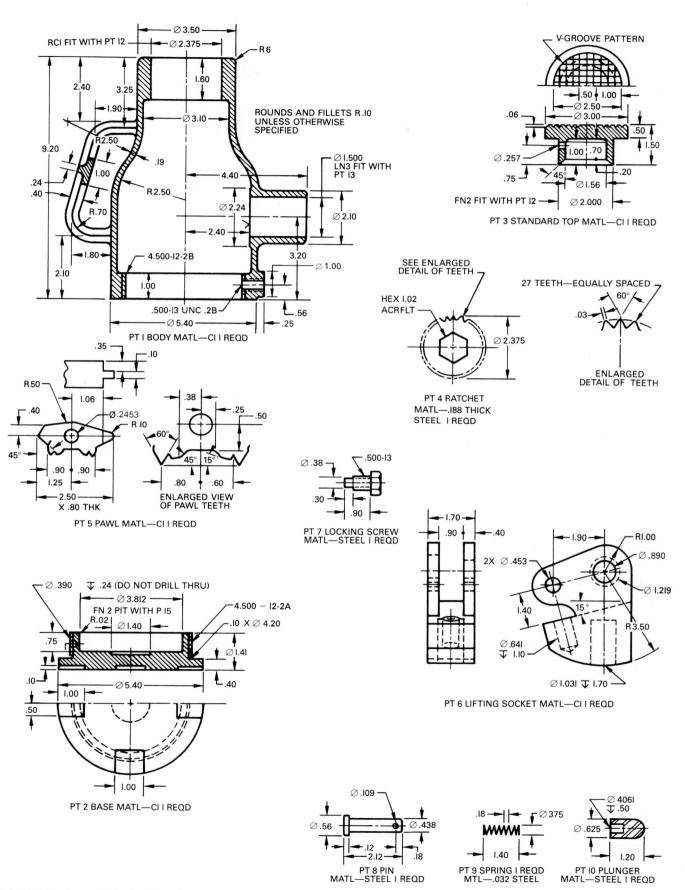

FIG. 14-6-L   Journal jack. (continued)

39. On a C (A2) size sheet make a two-view assembly drawing (front and side) of the grinder shown in Fig. 14-6-M.
Show the front view in half section. Include on the drawing an item list and identification part numbers. Scale 1:1.

40. Make detail drawings of parts 1 and 4 shown in Fig.
14-6-M, replacing the descriptive fit terms with appropriate dimensions.

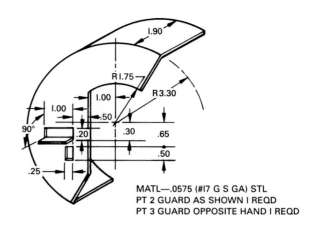

MATL—.0575 (#17 G S GA) STL
PT 2 GUARD AS SHOWN I REQD
PT 3 GUARD OPPOSITE HAND I REQD

PT 9 BEARING SKF 600Z-2Z (SEE APPENDIX) 2 REQD

PT I0 GRINDING WHEEL—FINE 6.00 OD × 1.00 THK × Ø .50I BORE, I REQD

PT II GRINDING WHEEL—MED. 6.00 OD × 1.00 THK × Ø .50I BORE, I REQD

PT I2 CARRIAGE BOLT .250-20 UNC × I.00 LG, 4 REQD

PT I3 WASHER PLAIN TYPE A .28I ID × .625 OD × .065, 4 REQD

PT I4 SETSCREW—SLOTTED HEAD CUP POINT .250—20UNC × .3I LG, I REQD

PT I5 WING NUT .250-20 UNC, 4 REQD

PT I6 NUT—REG HEX .500 UNC, I REQD

PT I7 NUT—REG HEX .500 UNC-LH, I REQD

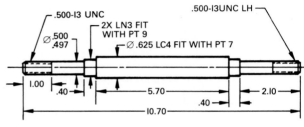

PT 4 SHAFT MATL—CRS I REQD

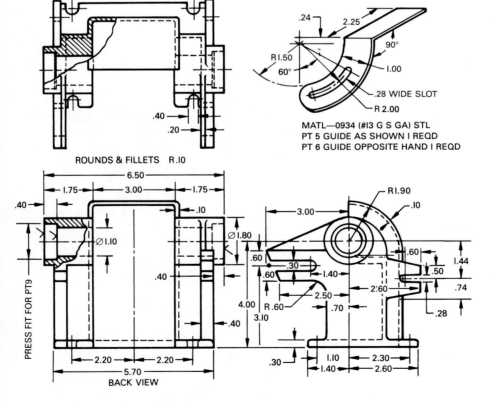

ROUNDS & FILLETS   R.I0

MATL—0934 (#I3 G S GA) STL
PT 5 GUIDE AS SHOWN I REQD
PT 6 GUIDE OPPOSITE HAND I REQD

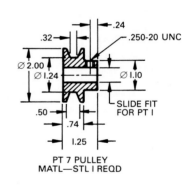

PT 7 PULLEY
MATL—STL I REQD

PT I BASE MATERIAL—CI I REQD

BACK VIEW

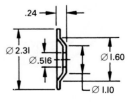

PT 8 SPACER
MATL—.0934 (#I3 G S GA)
4 REQD

FIG. 14-6-M   Bench grinder.

ASSIGNMENTS FOR UNIT 14-7, EXPLODED ASSEMBLY DRAWINGS

41. Make an exploded isometric assembly drawing of one of the assemblies shown in Figs. 14-2-J, 14-7-A, and 14-7-B.

Use center lines to align parts. Include on the drawing an item list and identification part numbers. Scale 1:1.

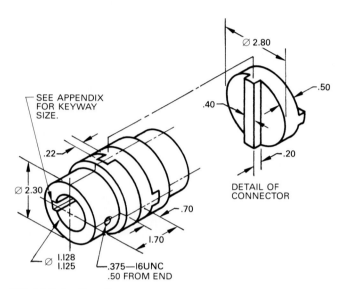

FIG. 14-7-A   Coupling.

FIG. 14-7-B   Universal joint.

ITEM LIST			
PT	ITEM	QTY	MATL	DESCRIPTION
1	FORK	2	WI	
2	RING	1	STL	
3	STUD	4	STL	
4	NO.4 TAPER PIN	2	STL	PURCHASED

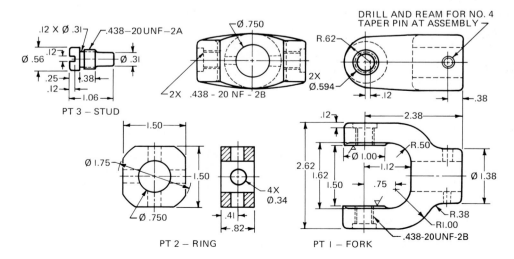

42. Make an exploded assembly drawing in orthographic projection of one of the assemblies shown in Figs. 14-7-B and 14-7-C. Use center lines to align the parts. Include on the drawing an item list and identification part numbers.

ASSIGNMENT FOR UNIT 14-8, DETAILED ASSEMBLY DRAWINGS

43. Make a three-view detailed assembly drawing of any one of the assemblies shown in Figs. 14-8-A, 14-8-C, and

FIG. 14-7-C   Caster.

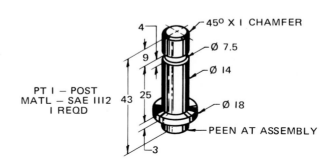

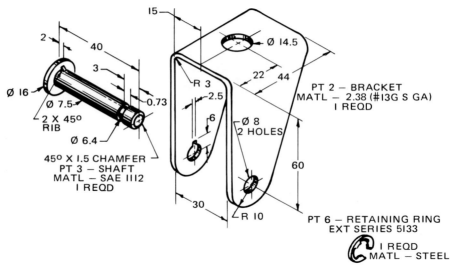

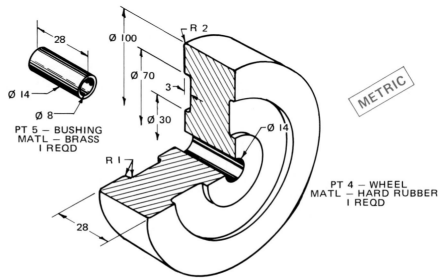

14-8-D. For Fig. 14-8-B only a partial assembly drawing is required. Include on the drawing the method of assembly (i.e., nails, wood screws, dowels, etc.) and an item list and identification part numbers. Include in the item list the assembly parts. The student is to select scale and drawing paper size. For Fig. 14-8-C the basic sizes are given. Design a table of your choice. Show on the drawing how the sides and feet are designed and fastened.

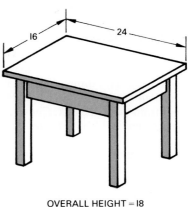

16   24

OVERALL HEIGHT = 18

DIMENSIONS SHOWN ARE IN INCHES.

FIG. 14-8-C   Night table.

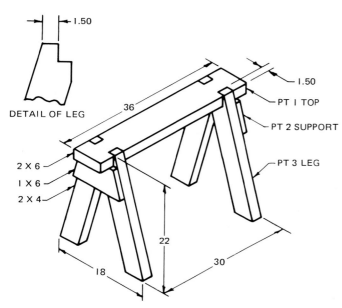

1.50

DETAIL OF LEG

1.50

36

PT 1 TOP

PT 2 SUPPORT

PT 3 LEG

2 X 6
1 X 6
2 X 4

22

18   30

MATL — CONSTRUCTION GRADE SPRUCE

NOTE: WOOD SIZES (THICKNESS AND WIDTH) ARE NOMINAL INCH SIZES.

FIG. 14-8-A   Sawhorse.

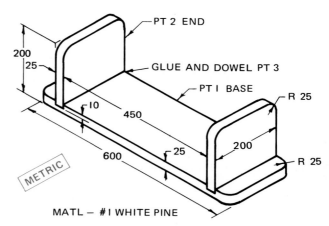

PT 2 END

200
25

GLUE AND DOWEL PT 3

PT 1 BASE

R 25

10
450

600   25   200

R 25

METRIC

MATL — #1 WHITE PINE

FIG. 14-8-D   Book rack.

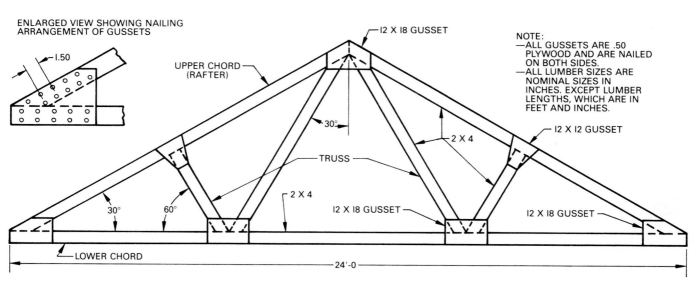

ENLARGED VIEW SHOWING NAILING ARRANGEMENT OF GUSSETS

1.50

12 X 18 GUSSET

UPPER CHORD (RAFTER)

30°

2 X 4

12 X 12 GUSSET

TRUSS

30°   60°

2 X 4

12 X 18 GUSSET

12 X 18 GUSSET

LOWER CHORD

24'-0

NOTE:
—ALL GUSSETS ARE .50 PLYWOOD AND ARE NAILED ON BOTH SIDES.
—ALL LUMBER SIZES ARE NOMINAL SIZES IN INCHES. EXCEPT LUMBER LENGTHS, WHICH ARE IN FEET AND INCHES.

FIG. 14-8-B   Roof truss.

ASSIGNMENTS FOR UNIT 14-9, SUBASSEMBLY DRAWINGS

44. For Fig. 14-9-A make a two-view subassembly drawing of die set B1-361 with dimensions, identification part numbers, and an item list. K = 7.50. Select a plain (journal) bearing from the Appendix. Convert to decimal inch dimensions. Scale 1:2. Identify the hole and shaft sizes for the fits shown.

45. For Fig. 14-9-B make a one-view subassembly drawing of the wheel. A broken-out or partial section view is recommended to show the interior features. Include on the drawing pertinent dimensions, identification part numbers, and an item list. Four Ø10 mm bolts fasten the wheel to an 8 mm plate. Scale 1:1.

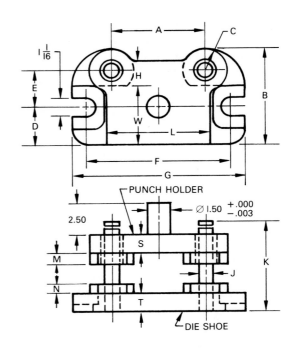

RC4 FIT BETWEEN BUSHING AND SHAFT J

LN2 FIT BETWEEN BUSHING AND PUNCH HOLDER

LN2 FIT BETWEEN SHAFT AND DIE SHOE

Width Inches W	Length Inches L	Thickness Die T	Thickness Punch S	Catalog Number	A	B	C	D	E	F	G	H	J	K	M	N
3	4	1-1/4	1	BI-341	3	5	1-1/8	2	1-7/8	6-1/2	7-1/2	1-1/2	3/4	SPECIFY	5/8	1/2
		1-1/2	1	BI-342												
3	6	1-1/2	1-1/4	BI-361	5	5-1/4	1-1/4	2	2	8-1/2	10	1-11/16	7/8	SPECIFY	3/4	5/8
		1-1/2	1-3/4	BI-362												
		2	1-1/4	BI-363												
		2	1-3/4	BI-364												

FIG. 14-9-A   Die sets. *(E.A. Baumbach Mfg. Co.)*

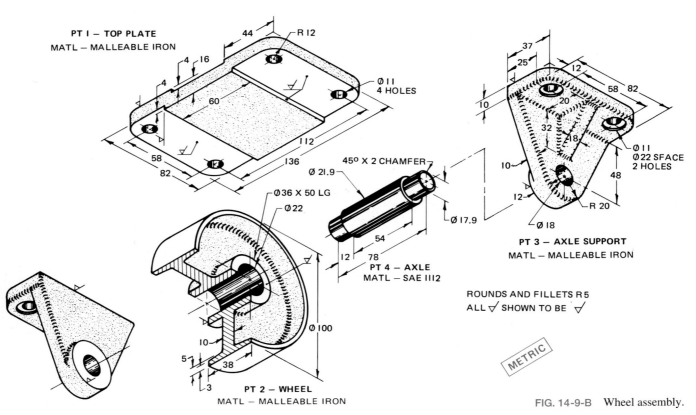

PT 1 – TOP PLATE
MATL – MALLEABLE IRON

Ø11 4 HOLES

45° X 2 CHAMFER

Ø 21.9

Ø36 X 50 LG

Ø 22

Ø 17.9

PT 4 – AXLE
MATL – SAE 1112

PT 3 – AXLE SUPPORT
MATL – MALLEABLE IRON

Ø11
Ø22 SFACE
2 HOLES

R 20

Ø 18

ROUNDS AND FILLETS R 5
ALL ∇ SHOWN TO BE ∇

METRIC

Ø 100

PT 2 – WHEEL
MATL – MALLEABLE IRON

FIG. 14-9-B   Wheel assembly.

46. On a B (A3) size sheet make a partial-section assembly drawing of the idler pulley shown in Fig. 14-9-C. Place on the drawing dimensions suitable for a catalog. Add to the drawing an item list and identification part numbers. Scale 1:1.

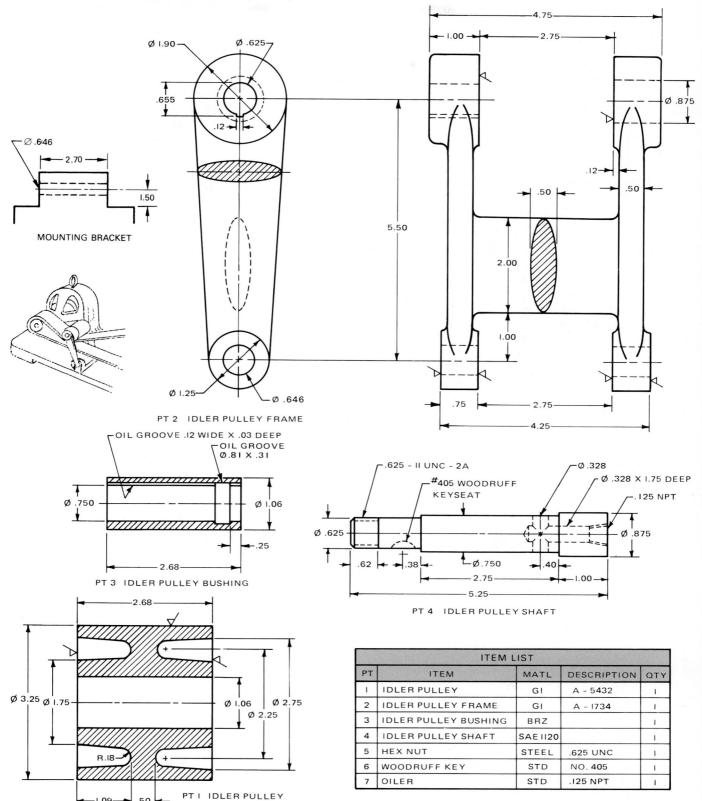

MOUNTING BRACKET

PT 2   IDLER PULLEY FRAME

OIL GROOVE .12 WIDE X .03 DEEP
OIL GROOVE Ø.81 X .31

PT 3   IDLER PULLEY BUSHING

.625 – 11 UNC – 2A
#405 WOODRUFF KEYSEAT
Ø.328
Ø.328 X 1.75 DEEP
.125 NPT

PT 4   IDLER PULLEY SHAFT

PT 1   IDLER PULLEY

ITEM LIST				
PT	ITEM	MATL	DESCRIPTION	QTY
1	IDLER PULLEY	GI	A – 5432	1
2	IDLER PULLEY FRAME	GI	A – 1734	1
3	IDLER PULLEY BUSHING	BRZ		1
4	IDLER PULLEY SHAFT	SAE 1120		1
5	HEX NUT	STEEL	.625 UNC	1
6	WOODRUFF KEY	STD	NO. 405	1
7	OILER	STD	.125 NPT	1

FIG. 14-9-C   Idler pulley.

47. On a B (A3) size sheet make a one-view, full-section drawing of the clutch shown in Fig. 14-9-D. A gear is mounted on the hub of a Formsprag overrunning clutch, Model 12. Shaft dia. 1.375 in. Use your judgment for dimensions not shown. Gear data: 20° spur gear; 6.000 PD; DP = 4; 1.00 tooth face; hub Ø3.50 × 1.340 wide; hub projection one side. Scale 1:1.

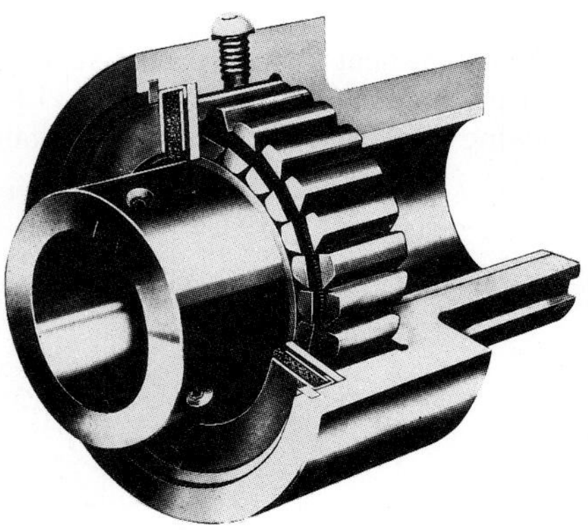

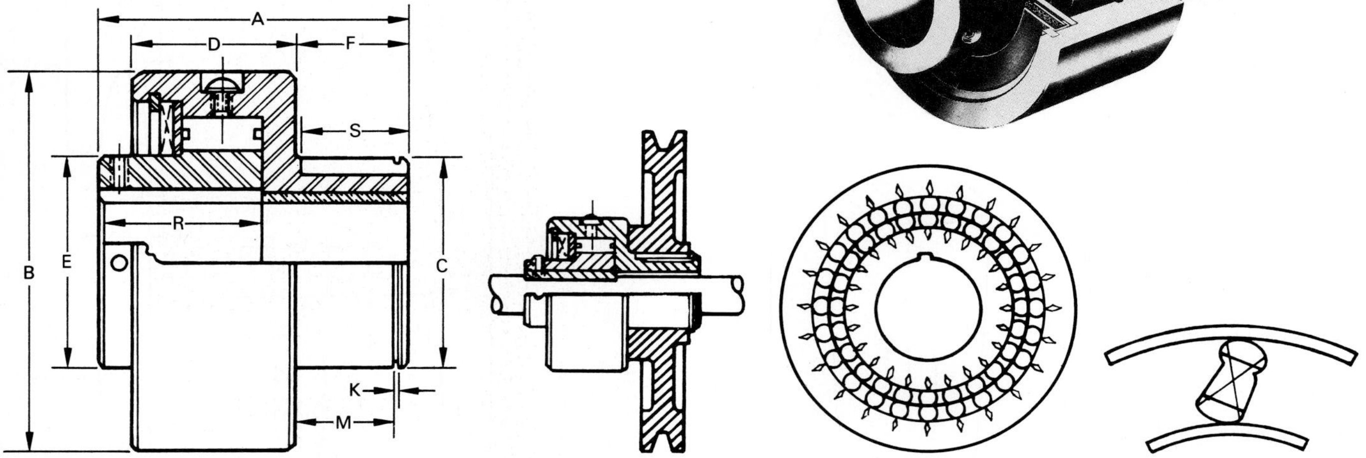

F-S SERIES MODEL NUMBER	3	5	6	8	10	12	14	16
Torque Capacity (Lb.Ft.)	15	40	130	190	420	600	1100	1300
Standard Bore Size	.375	.500 .625	.750	.875 1.000	1.125 1.250	1.375 1.500	1.625 1.750	1.875 2.000
Keyway		1/8 × 1/16 3/16 × 3/32	3/16 × 3/32	1/4 × 1/8	5/16 × 5/32	5/16 × 5/32 3/8 × 3/16	7/16 × 7/32	1/2 × 1/4
Standard Hub Keyway	1/8 × 1/16	3/16 × 3/32	3/16 × 3/32	1/4 × 1/8	5/16 × 5/32	3/8 × 3/16	7/16 × 7/32	1/2 × 1/4
A	2	2 3/4	3 3/16	3 5/16	3 5/8	3 7/8	4 3/8	4.75
B	1 5/8	2 3/16	2 7/8	3 1/4	3 3/4	4 7/16	5 1/4	5.90
C	.875 .874	1.250 1.249	1.375 1.374	1.750 1.749	2.250 2.249	2.500 2.499	2.875 2.874	3.250 3.249
D	13/16	1 1/4	1 15/32	1 5/8	1 3/4	1 15/16	2 1/4	2 1/2
E	1	1 1/16	1 1/4	1 5/8	2 1/32	2 3/8	2 7/8	3 1/4
F	13/16	1	1 5/16	1 5/16	1 7/16	1 7/16	1 5/8	1 3/4
K	.056 .036	.068 .048	.068 .048	.076 .056	.076 .056	.076 .056	.076 .056	.088 .068
M	.720 .715	.905 .900	1.220 1.215	1.220 1.215	1.345 1.340	1.345 1.340	1.530 1.525	1.655 1.650
R	1	1 15/32	1 9/16	1 11/16	1 51/64	2 1/8	2 11/32	2 1/2
S	1/2	9/16	15/16	7/8	15/16	1 3/8	1 9/16	1 11/16
Oil Hole		1/4 – 28	1/4 – 28	1/4 – 28	1/4 – 28	1/4 – 28	1/4 – 28	1/4 – 28

A full complement of sprags between concentric inner and outer races transmits power from one race to the other by wedging action of the sprags when either race is rotated in the driving direction. Rotation in the opposite direction frees the sprags and clutch is disengaged or "overruns."

FIG. 14-9-D Overrunning clutch. *(Formsprag)*

CHAPTER 15

PICTORIAL DRAWINGS

Definitions

Axonometric projection   Projection in which the lines of sight are perpendicular to the plane of projection, but in which the three faces of a rectangular object are all inclined to the plane of projection.

Isometric drawing   Projection based on the procedure of revolving the object at an angle of 45° to the horizontal, so that the front corner is toward the viewer, then tipping the object up or down at an angle of 35°16'.

Non-isometric lines   Lines used to depict sloping surfaces in isometric drawings.

Oblique projection   Projection based on the procedure of placing the object with one face parallel to the frontal plane and placing the other two faces on oblique planes.

Perspective projection   A method of drawing that depicts a three-dimensional object on a flat plane as it appears to the eye.

Unidirectional dimensioning   The preferred method of dimensioning isometric drawings.

15-1 PICTORIAL DRAWINGS

Pictorial drawing is the oldest written method of communication known, but the character of pictorial drawing has continually changed with the advance of civilization. In this text only those kinds of pictorial drawings commonly used by the engineer, designer, and drafter are considered. Pictorial drawings are useful in design, construction or production, erection or assembly, service or repairs, and sales.

There are three general types into which pictorial drawings may be divided: axonometric, oblique, and perspective. These three differ from one another in the fundamental scheme of projection, as shown in Fig. 15-1-1. The type of pictorial drawing used depends on the purpose for which it is drawn. They are used to explain complicated engineering drawings to people who do not have the training or ability to read the conventional multiview drawings; to help the designer work out problems in space, including clearances and interferences; to train new employees in the shop; to speed up and clarify the

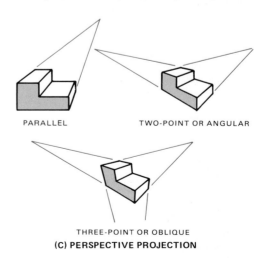

ISOMETRIC   DIMETRIC   TRIMETRIC
(A) AXONOMETRIC PROJECTION

CAVALIER   CABINET
(B) OBLIQUE PROJECTION

PARALLEL   TWO-POINT OR ANGULAR

THREE-POINT OR OBLIQUE
(C) PERSPECTIVE PROJECTION

FIG. 15-1-1   Types of pictorial drawings.

assembly of a machine or the ordering of new parts; to transmit ideas from one person to another, from shop to shop, or from salesperson to purchaser; and as an aid in developing the power of visualization.

Axonometric Projection

A projected view in which the lines of sight are perpendicular to the plane of projection, but in which the three faces of a rectangular object are all inclined to the plane of projection, is called an *axonometric projection* (Fig. 15-1-2). The projections of the three principal axes may make any angle with one another except 90°. Axonometric drawings, as shown in Figs. 15-1-3 and 15-1-4, are classified into three forms: *isometric drawings,* where the three principal faces and axes of the object are equally inclined to the plane of projection; *dimetric drawings,* where two of the three principal faces and axes of the object are equally inclined to the plane of projection; and *trimetric drawings,* where all three faces and axes of the object make different angles with the plane of projection. The most popular form of axonometric projection is the isometric.

Figure 15-1-5 illustrates the three types of axonometric projection, showing the compatible ellipse selection and the

percentage the lines are foreshortened. The 15° angles for dimetric projection and the 11.5° and 30° angles for trimetric projection are shown as these angles are used extensively by industry.

Isometric Drawings

This method is based on a procedure of revolving the object at an angle of 45° to the horizontal, so that the front corner is toward the viewer, then tipping the object up or down at an angle of 35° 16' (Fig. 15-1-6, pg. 444). When this is done to a cube, the three faces visible to the viewer appear equal in shape and size, and the side faces are at an angle of 30° to the horizontal. If the isometric view were actually projected from a view of the object in the tipped position, the lines in the isometric view would be foreshortened and would, therefore, not be seen in their true length. To simplify the drawing of an isometric view, the actual measurements of the object are used. Although the object appears slightly larger without the allowance for shortening, the proportions are not affected.

All isometric drawings are started by constructing the isometric axes, which are a vertical line for height and isometric lines to left and right, at an angle of 30° from the horizontal, for length and width. The three faces seen in the isometric view are the same faces that would be seen in the normal orthographic views: top, front, and side. Figure 15-1-6B illustrates the selection of the front corner *(A)*, the construction of the isometric axes, and the completed isometric view. Note that all lines are drawn to their true length, measured along the isometric axes, and that hidden lines are usually omitted. Vertical

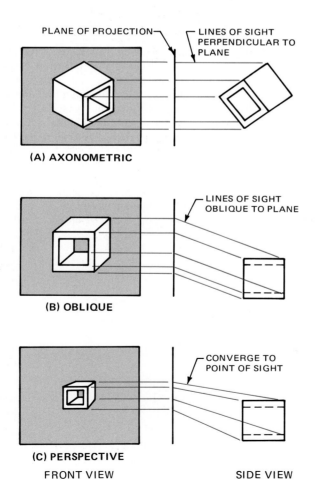

(A) AXONOMETRIC

(B) OBLIQUE

(C) PERSPECTIVE

FRONT VIEW   SIDE VIEW

FIG. 15-1-2   Kinds of projections.

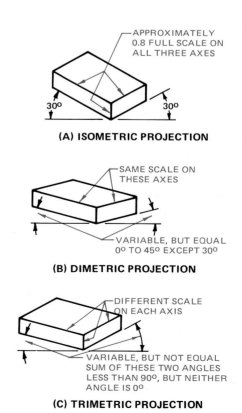

(A) ISOMETRIC PROJECTION

(B) DIMETRIC PROJECTION

(C) TRIMETRIC PROJECTION

FIG. 15-1-3   Types of axonometric drawings.

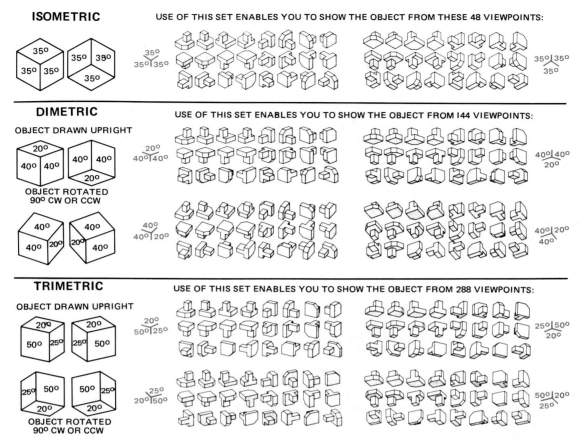

FIG. 15-1-4   Axonometric projection. *(Graphic Standard Instrument Co.)*

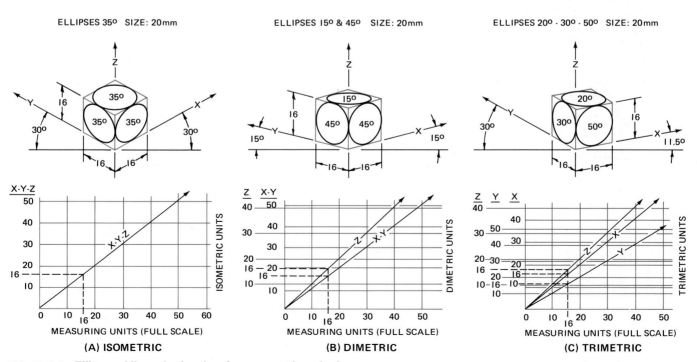

FIG. 15-1-5   Ellipse and line reduction sizes for axonometric projection.
(General Motors Corp.)

FIG. 15-1-6   Isometric axes and projection.

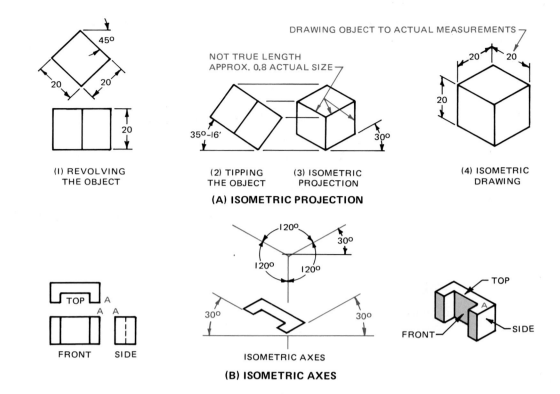

(A) ISOMETRIC PROJECTION

(B) ISOMETRIC AXES

edges are represented by vertical lines, and horizontal edges by lines at 30° to the horizontal.

Two techniques can be used for making an isometric drawing of an irregularly shaped object, as illustrated in Fig. 15-1-7. In one method, the object is divided mentally into a number of sections and the sections are created one at a time in their proper relationship to one another. In the second method, a box is created with the maximum height, width, and depth of the object; then the parts of the box that are not part of the object are removed, leaving the pieces that form the total object.

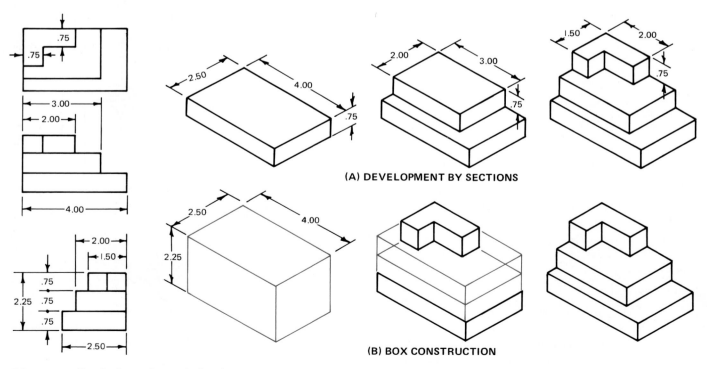

(A) DEVELOPMENT BY SECTIONS

(B) BOX CONSTRUCTION

FIG. 15-1-7   Developing an isometric drawing.

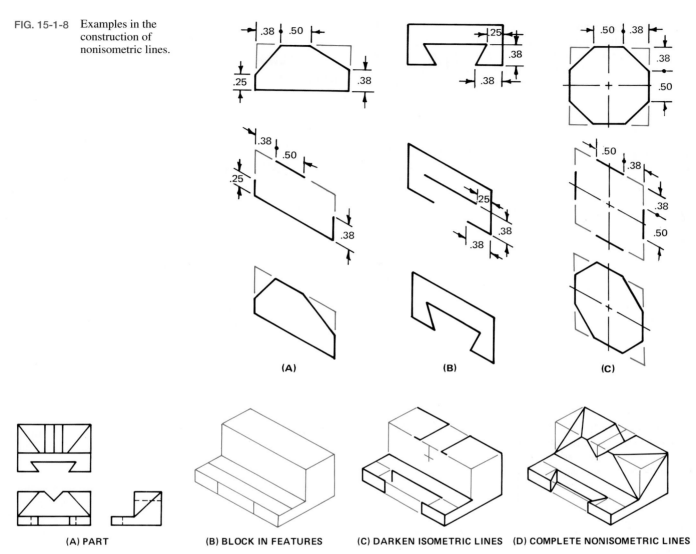

FIG. 15-1-8   Examples in the construction of nonisometric lines.

(A)

(B)

(C)

FIG. 15-1-9   Sequence in drawing an object having nonisometric lines.

(A) PART   (B) BLOCK IN FEATURES   (C) DARKEN ISOMETRIC LINES   (D) COMPLETE NONISOMETRIC LINES

Nonisometric Lines

Many objects have sloping surfaces that are represented by sloping lines in the orthographic views. In isometric drawing, sloping surfaces appear as *nonisometric* lines. To create them, locate their endpoints, found on the ends of isometric lines, and join them with a straight line. Figures 15-1-8 and 15-1-9 illustrate examples in the construction of nonisometric lines.

Dimensioning Isometric Drawings

At times, an isometric drawing of a simple object may serve as a working drawing. In such cases, the necessary dimensions and specifications are placed on the drawing.

Dimension lines, extension lines, and the line being dimensioned are shown in the same plane. Arrowheads should be in the plane of the dimension and extension lines. (Fig. 15-1-10).

Unidirectional dimensioning is the preferred method of dimensioning isometric drawings. The letters and numbers are

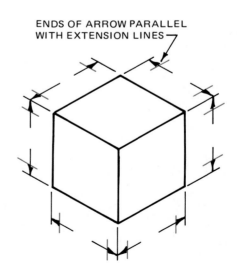

ENDS OF ARROW PARALLEL WITH EXTENSION LINES

FIG. 15-1-10   Orienting the dimension line, arrowhead, and extension line.

vertical and read from the bottom of the sheet. An example of this type of dimensioning is shown in Fig. 15-1-11.

Since the isometric is a one-view drawing, it is not usually possible to avoid placing dimensions on the view or across dimension lines. However, this practice should be avoided whenever possible.

Isometric Grid Paper

Isometric grid sheets are another timesaving device. Designers and engineers frequently use isometric grid paper on which they sketch their ideas and designs (Fig. 15-1-12). Many companies, such as those which prepare pipe drawings, have large drawing sheets made with nonreproducible isometric grid lines.

 CAD

A pictorial drawing may be created by using any CAD two-dimensional (2D) system. Normally, a grid pattern peculiar to the type of pictorial will be employed. Since isometrics are the most popular, an ISOMETRIC GRID option is found on virtually every system.

Automatic generation of a pictorial is common to the axonometric and perspective types only.

CAD systems provide a MODELING option, often referred to as 3D MODELING. With this option the model, isometric

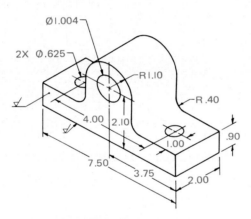

FIG. 15-1-11   Isometric dimensioning.

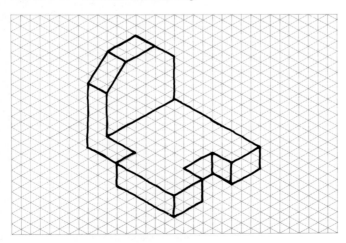

FIG. 15-1-12   Isometric grid paper.

or otherwise, can be automatically generated from the multi-view drawing.

REFERENCES AND SOURCE MATERIAL

1. ANSI Y14.4, *American Drafting Standards Manual, Pictorial Drawing.*
2. General Motors Corp.

ASSIGNMENTS

See Assignments 1 through 6 for Unit 15-1 on pages 465 through 468.

15-2 CURVED SURFACES IN ISOMETRIC

Circles and Arcs in Isometric

A circle on any of the three faces of an object drawn in isometric has the shape of an ellipse (Fig. 15-2-1). Figure 15-2-2 illustrates the steps in drawing circular features on isometric drawings.

1. Draw the center lines and a square, with sides equal to the circle diameter, in isometric.
2. Using the obtuse-angled (120°) corners as centers, draw arcs tangent to the sides forming the obtuse-angled corners, stopping at the points where the center lines cross the sides of the square.
3. Draw construction lines from these same points to the opposite obtuse-angled corners. The points at which these construction lines intersect are the centers for arcs drawn tangent to the sides forming the acute-angled corners, meeting the first arcs.

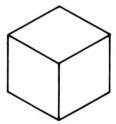

(A) A SQUARE DRAWN IN THE THREE ISOMETRIC POSITIONS

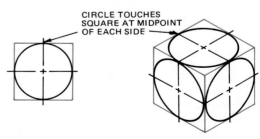

(B) A CIRCLE PLACED INSIDE A SQUARE AND DRAWN IN THE THREE ISOMETRIC POSITIONS

FIG. 15-2-1   Circles in isometric.

FIG. 15-2-2   Sequence in
drawing isometric
circles.

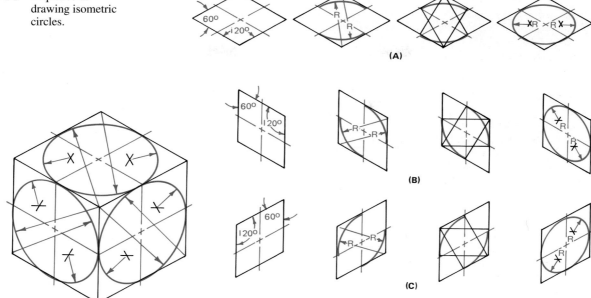

(A)

(B)

(C)

When concentric circles are drawn, each circle must have its own set of centers for the arcs, as shown in Fig. 15-2-3.

The same technique is used for drawing part-circles (arcs), as shown in Fig. 15-2-4. Construct an isometric square with sides equal to twice the radius, and draw that portion of the ellipse necessary to join the two faces. When these faces are

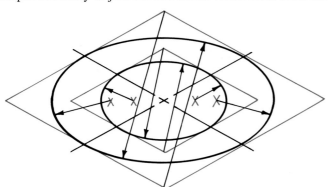

FIG. 15-2-3   Drawing concentric isometric circles.

FIG. 15-2-4   Drawing isometric arcs
and circles.

parallel, draw half of an ellipse (one long radius and one short radius); when they are at an obtuse angle (120°), draw one long radius; and when they are at an acute angle (60°), draw one short radius.

Isometric Templates

For convenience and time saving, isometric ellipse templates should be used whenever possible. A wide variety of elliptical templates are available. The template shown in Fig. 15-2-5 (pg. 448) combines ellipses, scales, and angles. Markings on the ellipses coincide with the center lines of the holes, speeding up the drawing of circles and arcs. Figure 15-2-6 (pg. 448) shows the same part that appears in Fig. 15-2-4 but with the arcs and circles being constructed with a template.

Sketching Circles and Arcs

In sketching circles and arcs on isometric grid paper, locate the center lines first, then lightly sketch in construction boxes

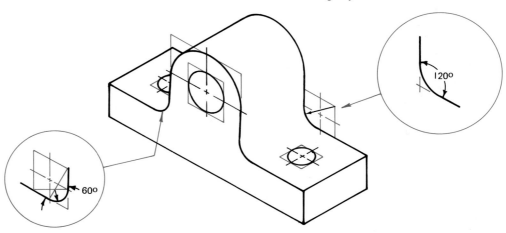

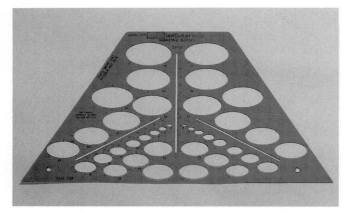

FIG. 15-2-5   Isometric ellipse template. *(STUDIOHIO)*

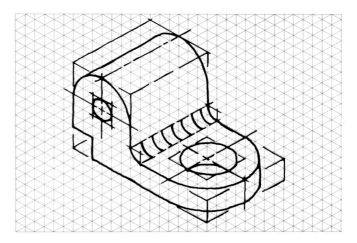

FIG. 15-2-7   Sketching isometric circles and arcs.

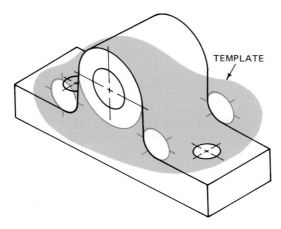

FIG. 15-2-6   Circles and arcs drawn with isometric ellipse template.

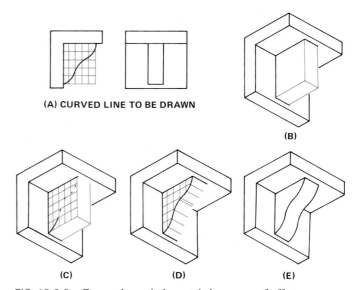

(A) CURVED LINE TO BE DRAWN

(B)

(C)   **(D)**   **(E)**

FIG. 15-2-8   Curves drawn in isometric by means of offset measurements.

(isometric squares) where the circles and arcs should be (Fig. 15-2-7). Sketch the ellipse (isometric circle) just touching the center of each of the four sides of the square.

Drawing Irregular Curves in Isometric

To draw curves other than circles or arcs, the plotting method shown in Fig. 15-2-8 is used.

1. Draw an orthographic view, and divide the area enclosing the curved line into equal squares.
2. Produce an equivalent area on the isometric drawing, showing the offset squares.
3. Take positions relative to the squares from the orthographic view, and plot them on the corresponding squares on the isometric view.
4. Draw a smooth curve through the established points with the aid of an irregular curve.

 CAD

The ELLIPSE command in CAD provides two basic methods for constructing an ellipse. The first method allows for the specification of the major and minor axes or diameters. While this is the most common method of ellipse construction,

specification of the major and minor axis values does not allow for the construction of an ellipse at a known degree of exposure or eccentricity. The second method of ellipse construction in CAD is used to construct an ellipse with a known angle of exposure. The center and major diameter of the ellipse are identified, and then the angle of exposure or rotation in degrees is specified. For curves other than circles or arcs, a grid pattern is employed to allow for offset construction. A series of points are located on the curve using offset construction techniques. A polyline or spline is then generated through those points to create the curve. The larger the number of points used to generate the curve, the more accurate the construction of the curve will be.

ASSIGNMENTS

See Assignments 7 through 9 for Unit 15-2 on pages 468 to 469.

15-3 COMMON FEATURES IN ISOMETRIC

Isometric Sectioning

Isometric drawings are usually made showing exterior views, but sometimes a sectional view is needed. The section is taken on an isometric plane, that is, on a plane parallel to one of the faces of the cube. Figure 15-3-1 shows isometric full sections taken on a different plane for each of three objects. Note the construction lines representing the part that has been cut away. Isometric half-sections are illustrated in Fig. 15-3-2.

When an isometric drawing is sectioned, the section lines are shown at an angle of 60° with the horizontal or in a horizontal position, depending on where the cutting-plane line is located. In half sections, the section lines are sloped in opposite directions, as shown in Fig. 15-3-2.

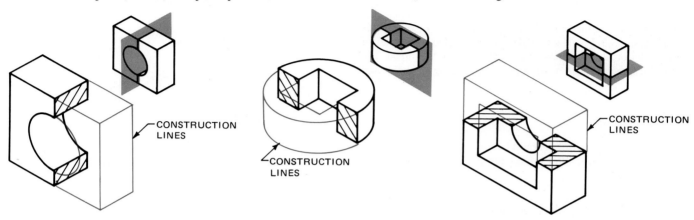

FIG. 15-3-1   Examples of isometric full sections.

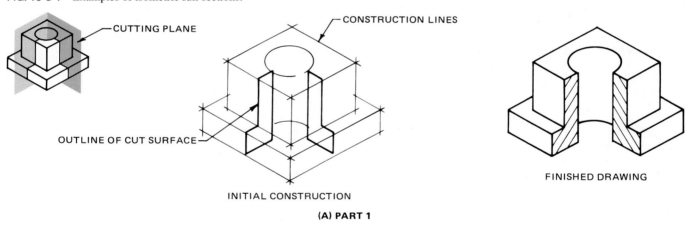

(A) PART 1

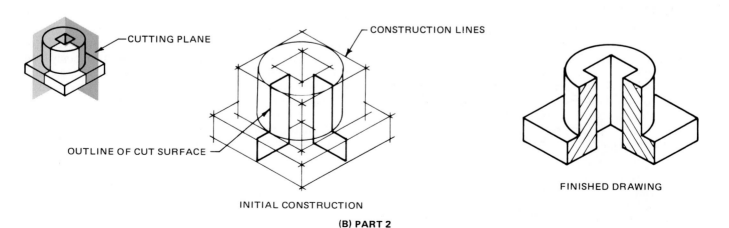

(B) PART 2

FIG. 15-3-2   Examples of isometric half-sections.

Fillets and Rounds

For most isometric drawings of parts having small fillets and rounds, the adopted practice is to draw the corners as sharp features. However, when it is desirable to represent the part, normally a casting, as having a more realistic appearance, either of the methods shown in Fig. 15-3-3 may be used.

Threads

The conventional method for showing threads in isometric is shown in Fig. 15-3-4. The threads are represented by a series of ellipses uniformly spaced along the center line of the thread. The spacing of the ellipses need not be the spacing of the actual pitch.

Break Lines

For long parts, break lines should be used to shorten the length of the drawing. Freehand breaks are preferred, as shown in Fig. 15-3-5.

Isometric Assembly Drawings

Regular or exploded assembly drawings are frequently used in catalogs and sales literature, as illustrated by Fig. 15-3-6.

ASSIGNMENTS

See Assignments 10 through 16 for Unit 15-3 on pages 469 through 472.

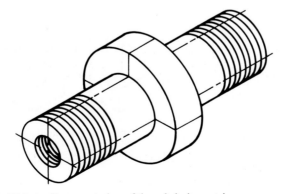

(A) CURVED LINE   **(B) STRAIGHT LINE**

FIG. 15-3-3   Representation of fillets and rounds in isometric.

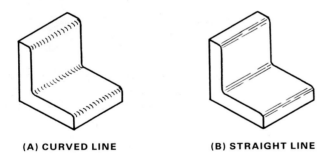

FIG. 15-3-4   Representation of threads in isometric.

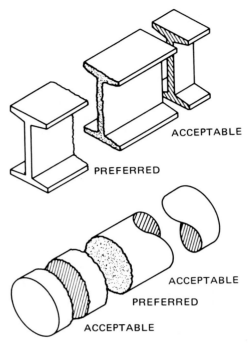

FIG. 15-3-5   Conventional breaks in isometric.

15-4 OBLIQUE PROJECTION

This method of pictorial drawing is based on the procedure of placing the object with one face parallel to the frontal plane and placing the other two faces on oblique (or receding) planes, to left or right, top or bottom, at a convenient angle. The three axes of projection are vertical, horizontal, and receding. Figure 15-4-1 illustrates a cube drawn in typical positions with the receding axis at 60°, 45°, and 30°. This form of projection has the advantage of showing one face of the object without distortion. The face with the greatest irregularity of outline or contour, or the face with the greatest number of circular features, or the face with the longest dimension faces the front (Fig. 15-4-2).

Two types of oblique projection are used extensively. In *cavalier oblique,* all lines are made to their true length, measured on the axes of the projection. In *cabinet oblique,* the lines on the receding axis are shortened by one-half their true length to compensate for distortion and to approximate more closely what the human eye would see. For this reason, and because of the simplicity of projection, cabinet oblique is a commonly used form of pictorial representation, especially when circles and arcs are to be drawn. Figure 15-4-3 (pg. 452) shows a comparison of cavalier and cabinet oblique. Note that hidden lines are omitted unless required for clarity. Many of the drawing techniques for isometric projection apply to oblique projection. Figure 15-4-4 (pg. 452) illustrates the construction of an irregularly shaped object by the box method.

Inclined Surfaces

Angles that are parallel to the picture plane are drawn as their true size. Other angles can be laid off by locating the ends of the inclined line.

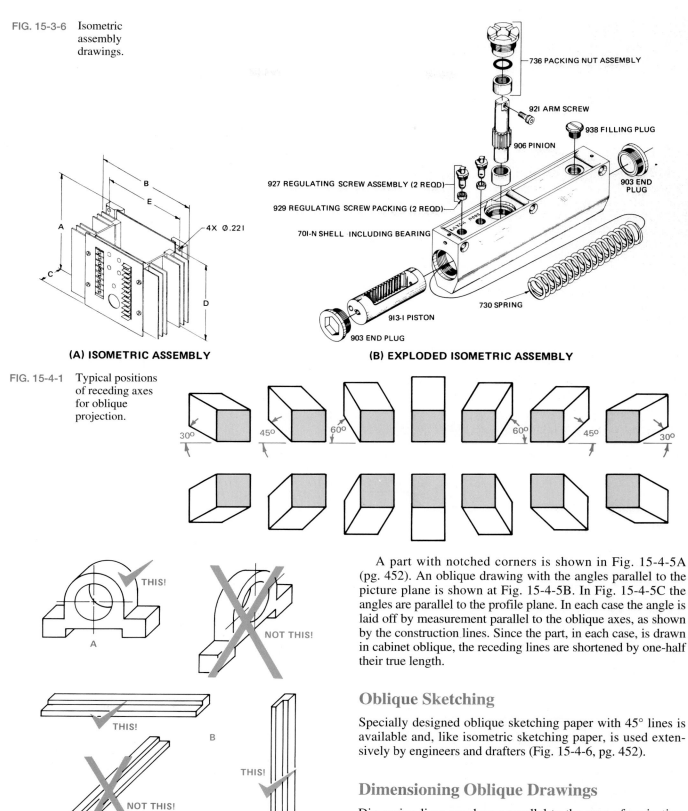

FIG. 15-3-6   Isometric assembly drawings.

736 PACKING NUT ASSEMBLY

921 ARM SCREW

938 FILLING PLUG

906 PINION

927 REGULATING SCREW ASSEMBLY (2 REQD)

929 REGULATING SCREW PACKING (2 REQD)

903 END PLUG

701-N SHELL INCLUDING BEARING

913-1 PISTON

730 SPRING

903 END PLUG

4X Ø.221

(A) ISOMETRIC ASSEMBLY

(B) EXPLODED ISOMETRIC ASSEMBLY

FIG. 15-4-1   Typical positions of receding axes for oblique projection.

30°   45°   60°   60°   45°   30°

THIS!

NOT THIS!

THIS!

B

THIS!

NOT THIS!

A

FIG. 15-4-2   Two general rules for oblique drawings.

A part with notched corners is shown in Fig. 15-4-5A (pg. 452). An oblique drawing with the angles parallel to the picture plane is shown at Fig. 15-4-5B. In Fig. 15-4-5C the angles are parallel to the profile plane. In each case the angle is laid off by measurement parallel to the oblique axes, as shown by the construction lines. Since the part, in each case, is drawn in cabinet oblique, the receding lines are shortened by one-half their true length.

Oblique Sketching

Specially designed oblique sketching paper with 45° lines is available and, like isometric sketching paper, is used extensively by engineers and drafters (Fig. 15-4-6, pg. 452).

Dimensioning Oblique Drawings

Dimension lines are drawn parallel to the axes of projection. Extension lines are projected from the horizontal and vertical object lines whenever possible.

The dimensioning of an oblique drawing is similar to that of an isometric drawing. The recommended method is

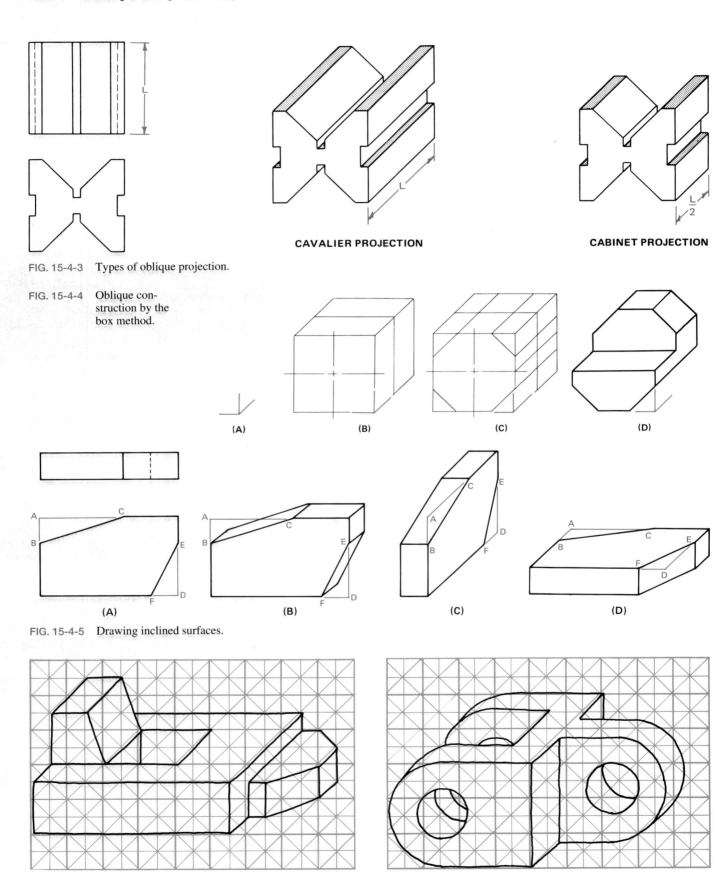

FIG. 15-4-3   Types of oblique projection.

CAVALIER PROJECTION

CABINET PROJECTION

FIG. 15-4-4   Oblique construction by the box method.

(A)   (B)   (C)   (D)

FIG. 15-4-5   Drawing inclined surfaces.

(A)   (B)   (C)   (D)

FIG. 15-4-6   Oblique sketching paper.

FIG. 15-4-7   Dimensioning an
oblique drawing.

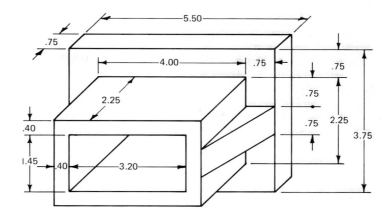

unidirectional dimensioning, which is shown in Fig. 15-4-7.
As in isometric dimensioning, some dimensions must be
placed directly on the view.

CAD

Specially designed oblique lines or grids with horizontal, vertical, and 45° reference lines are available on some CAD systems. If not available, a rectangular pattern may be used to
develop the true-size front face. For the two oblique faces a 45°
or 60° line pattern is used.
    Oblique modeling is not a CAD option.

ASSIGNMENTS

See Assignments 17 through 21 for Unit 15-4 on pages 472
through 474.

FIG. 15-5-1   Application of an
oblique drawing.

15-5 COMMON FEATURES IN OBLIQUE

Circles and Arcs

Whenever possible, the face of the object having circles or
arcs should be selected as the front face, so that such circles
or arcs can be easily drawn in their true shape (Fig. 15-5-1).
When circles or arcs must be drawn on one of the oblique
faces, the *offset measurement method* illustrated in Fig. 15-5-2
(pg. 454) may be used.

1. Draw an oblique square about the center lines, with sides
   equal to the diameter.
2. Draw a true circle within the oblique square, and establish equally spaced points about its circumference.
3. Project these point positions to the edge of the oblique
   square, and draw lines on the oblique axis from these

SCHEMATIC OF A COMPLETELY AUTOMATIC REGISTRATION CONTROL SYSTEM
MAINTAINING THE LOCATION OF CUTOFF ON A CONTINUOUS PRINTED WEB.

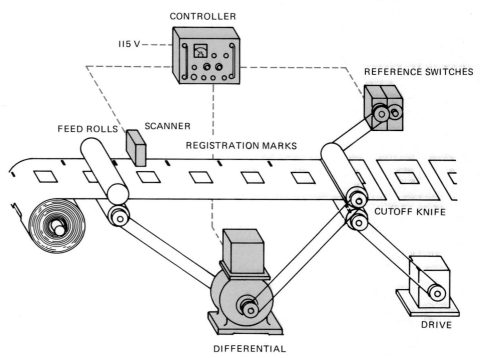

positions. Similarly spaced lines are drawn on the other axis for a cavalier oblique drawing and the spaces halved for a cabinet oblique drawing, forming offset squares and giving intersection points for the oval shape.

Another method used when circles or arcs must be drawn on one of the oblique surfaces is the *four-center method*. In Fig. 15-5-3 a circle is shown as it would be drawn on a front plane, a side plane, and a top plane.

Circles not parallel to the picture plane when drawn by the approximate method are not pleasing but are satisfactory for some purposes. Ellipse templates, when available, should be used because they reduce drawing time and give much better results. If a template is used, the oblique circle should first be blocked in as an oblique square in order to locate the proper position of the circle. Blocking in the circle first also helps the drafter select the proper size and shape of the ellipse. The construction and dimensioning of an oblique part are shown in Fig. 15-5-4.

Oblique Sectioning

Oblique drawings are usually made as outside views, but sometimes a sectional view is necessary. The section is taken on a plane parallel to one of the faces of an oblique cube. Figure 15-5-5 shows an oblique full section and an oblique half-section. Construction lines show the part that has been cut away.

Treatment of Conventional Features

Fillets and Rounds Small fillets and rounds normally are drawn as sharp corners. When it is desirable to show the corners rounded, either of the methods shown in Fig. 15-5-6 is recommended.

Threads The conventional method of showing threads in oblique is shown in Fig. 15-5-7. The threads are represented by a series of circles uniformly spaced along the center line of the thread. The spacing of the circles need not be the spacing of the pitch.

Breaks Figure 15-5-8 shows the conventional method for representing breaks.

🖥 CAD

The CIRCLE, ARC, and FILLET commands are used to create circles and arcs on the front face of an oblique drawing. For the oblique faces use the ELLIPSE command, and specify the degree of exposure (i.e., 45°) to construct circles and arcs.

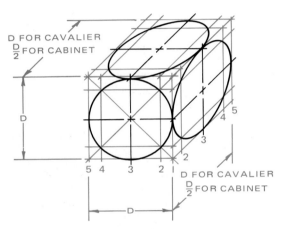

FIG. 15-5-2 Drawing oblique circles by means of offset measurements.

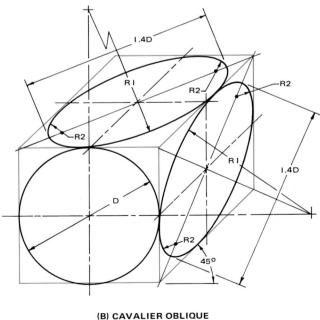

(A) CABINET OBLIQUE

(B) CAVALIER OBLIQUE

FIG. 15-5-3 Approximate ellipse construction for oblique drawings with 45° axis.

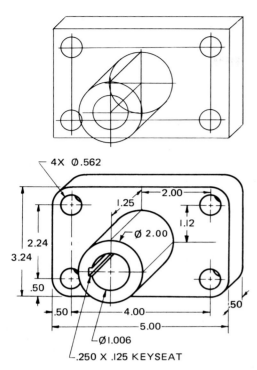

4X Ø.562

2.00

1.25

Ø 2.00

1.12

2.24

3.24

.50

.50   4.00   .50

5.00

Ø1.006

.250 X .125 KEYSEAT

FIG. 15-5-4   Construction and dimensioning of an oblique object.

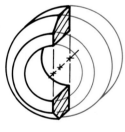

(A) FULL SECTION

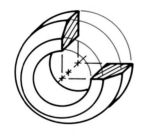

(B) HALF-SECTION

FIG. 15-5-5   Oblique full and half-sections.

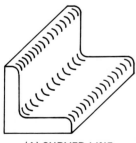

(A) CURVED LINE

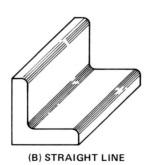

(B) STRAIGHT LINE

FIG. 15-5-6   Representing rounds and fillets.

Thus far, all discussion has been confined to two-dimensional (2D) software. For the vast majority of engineering drawing applications, two-dimensional drafting will suffice. Multiview projection, which uses two or more two-dimensional views to describe a three-dimensional object, is considered the standard operating procedure for many design offices. Occasionally, however, three-dimensional (axonometric)

FIG. 15-5-7   Representation of threads in oblique.

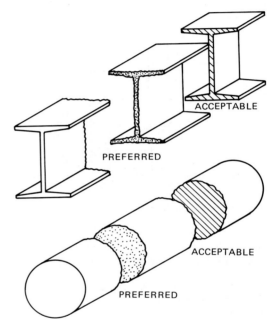

ACCEPTABLE

PREFERRED

ACCEPTABLE

PREFERRED

FIG. 15-5-8   Conventional breaks.

drawing capability is desirable. This can be accomplished as previously shown using a basic two-dimensional system. Lines are drawn inclined both to the right or left at 30° from the horizontal. These, with a vertical line, establish the three major axes of an isometric drawing. This method, however, may become quite cumbersome, requiring a considerable time expenditure. CAD systems provide a modeling option often referred to as *3D modeling*. With this option the model, isometric or otherwise, is automatically generated from the drawing. The method commonly used in modeling is to draw one view (often the plan or top view) and key in the third dimension (thickness or height). All *X, Y,* and *Z* coordinate information is provided to the system. The single-view (plan or top) method to create the object is shown in Fig. 15-5-9 (pg. 456).

ASSIGNMENTS

See Assignments 22 through 25 for Unit 15-5 on pages 474 through 477.

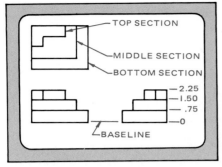

(A) THE PART TO BE MODELED

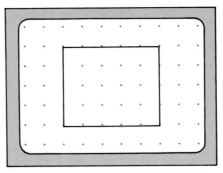

(B) PLAN VIEW OF BOTTOM SECTION

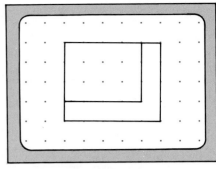

(C) PLAN VIEW WITH CENTER SECTION ADDED

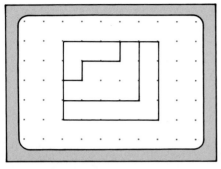

(D) TOP SECTION ADDED

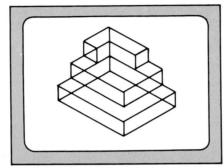

(E) WIRE-FRAME ISOMETRIC ILLUSTRATED WITH ALL LINES

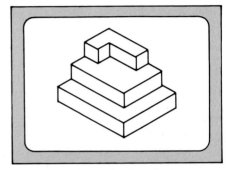

(F) WIRE FRAME ISOMETRIC ILLUSTRATED WITH HIDDEN LINES REMOVED

FIG. 15-5-9   CAD modeling.

15-6 PARALLEL, OR ONE-POINT, PERSPECTIVE

Perspective Projection

Perspective is a method of drawing that depicts a three-dimensional object on a flat plane as it appears to the eye (Fig. 15-6-1). A pictorial drawing made by the intersection of the picture plane with lines of sight converging from points on the object to the point of sight, which is located at a finite distance from the picture plane, is called a perspective (Fig. 15-6-2).

Perspective drawings are more realistic than axonometric or oblique drawings because the object is shown as the eye would see it. Since they are far more difficult to draw than the other types of pictorial drawings, their use in drafting is limited mainly to production or presentation illustrations and illustrations of proposed structures by architects.

FIG. 15-6-1   Application of parallel and angular perspective drawings. *(General Motors Corp.)*

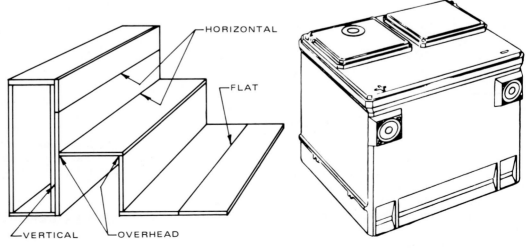

(A) PARALLEL PERSPECTIVE

(B) ANGULAR PERSPECTIVE

The main elements of a perspective drawing are the *picture plane* (plane of projection), the *station point* (the position of the observer's eye when viewing the object), the *horizon* (an imaginary horizontal line taken at eye level), the *vanishing point* or *points* (a point or points on the horizon where all the receding lines converge), and the *ground line* (the base line of the picture plane and object).

To avoid undue distortion in perspective, the point of sight (station point) should be located so that the cone of rays from the observer's eye has an angle at the apex not greater than 30°. This would place the station point a distance away from the outside portion of the object of approximately 2 to 2 ½ times the width of the object being viewed (see Figs. 15-6-2 and 15-6-3).

Types of Perspective Drawings

There are three types of perspective drawings:

1. *Parallel:* one vanishing point
2. *Angular:* two vanishing points
3. *Oblique:* three vanishing points

In industry they are normally referred to as one-point, two-point, and three-point perspectives, respectively (Fig. 15-6-4, pg. 458). Only parallel and angular perspectives are covered in this text.

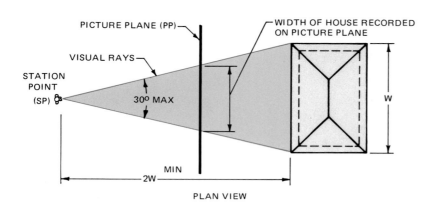

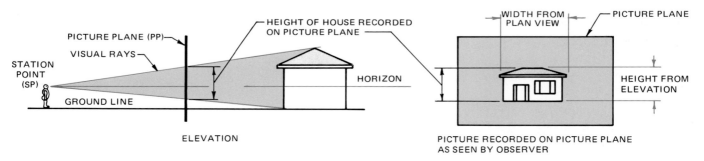

FIG. 15-6-2   Recording the picture on the picture plane.

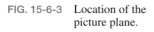
FIG. 15-6-3   Location of the picture plane.

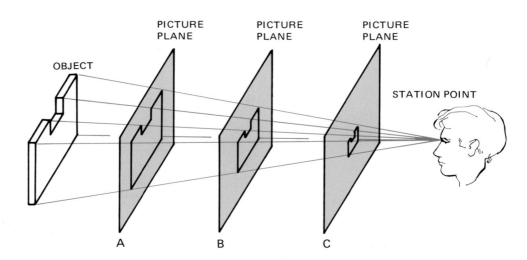

Parallel, or One-Point, Perspective

Parallel-perspective drawings are similar to oblique drawings, except the receding lines all converge at one point on the horizon. In drawing a *parallel-perspective drawing,* one face of the object is placed on the picture-plane line so that it will be drawn in its true size and shape, as shown in Fig. 15-6-5. The *PP* line shown in the top view represents the picture-plane line, and

point *SP* (station point) is the position of the observer. The lines of the object, which are not on the picture-plane, are found by projecting lines down from the top view from the point of intersection of the visual ray and the picture plane, as shown by point *N* in Fig. 15-6-5A(1).

Where the true height of a line or a point does not lie on the picture plane, such as point *P* in Fig. 15-6-5A(2), the true height may be found by extending line *PR* to point *S* on the

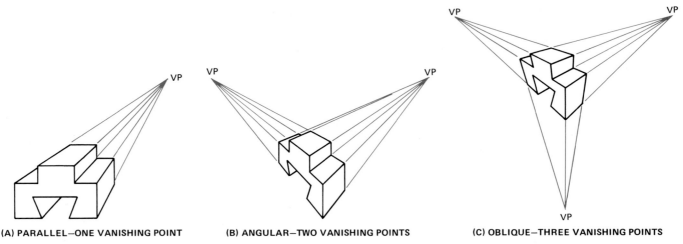

(A) PARALLEL—ONE VANISHING POINT   **(B) ANGULAR—TWO VANISHING POINTS**   **(C) OBLIQUE—THREE VANISHING POINTS**

FIG. 15-6-4   Types of perspective drawings.

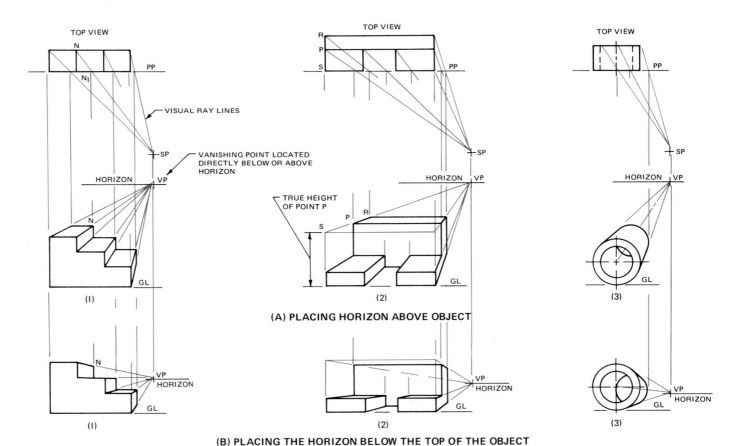

FIG. 15-6-5   Parallel, or one-point, perspective.

picture plane. Since point *S* lies on the picture plane and is the same height as point *P,* it may readily be found on the perspective drawing. Point *P* will lie on the receding line joining point *S* to the line *VP.*

In drawing a one-point perspective, a side or front view and a top view are normally drawn first—the top view to locate the part with respect to the picture plane and the side or front view to obtain the height of the various features. Figure 15-6-6 shows a simple, one-point perspective drawing with construction lines.

One of the most common uses of a parallel-perspective drawing is for representing the interior of a building. With this type of drawing, the vanishing point is located inside the room and is normally at eye level (Fig. 15-6-7).

Parallel-Perspective Grids

A variety of perspective grid sheets are available which enable the drafter to produce perspective drawings in less time than the conventional manner. Using a grid eliminates the tedious

FIG. 15-6-6 Construction of a one-point perspective.

FIG. 15-6-7 Parallel-perspective drawing of an interior of a house.

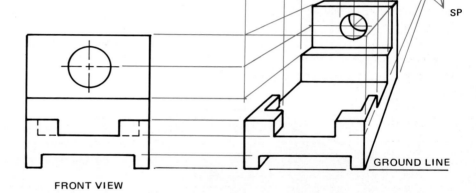

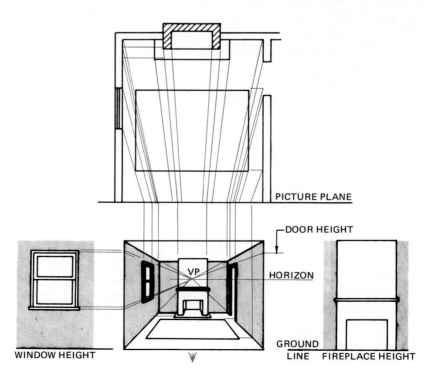

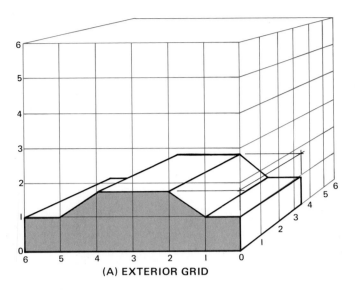

(A) EXTERIOR GRID

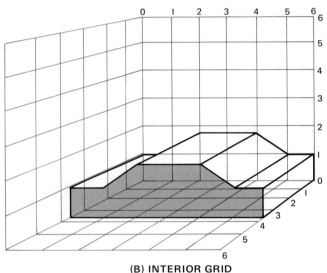

(B) INTERIOR GRID

FIG. 15-6-8   Parallel-perspective grids.

effort of establishing and projecting from the vanishing points for each individual feature. It also eliminates the problem of having the vanishing points located, in many instances, beyond the drawing area.

The cube grid, which is most widely used, has two basic variations: an exterior grid and an interior grid (Fig. 15-6-8). The grid sizes are dependent upon the desired scale of the parts to be drawn. The height and width planes are subdivided into identical increments, each increment representing any convenient size, such as 1.00 in., 1 ft or 10, 100, or 1000 mm. The plane or surface representing the depth is subdivided into increments that are proportionately foreshortened as they recede from the picture plane and thus create the perspective illusion (Fig. 15-6-9).

REFERENCES AND SOURCE MATERIAL

1. ANSI Y14.4, *American Drafting Standards Manual, Pictorial Drawing.*
2. General Motors Corp.

ASSIGNMENTS

See Assignments 26 through 28 for Unit 15-6 on pages 478 to 479.

15-7 ANGULAR, OR TWO-POINT, PERSPECTIVE

Two-point perspective is used quite extensively for architectural and product illustration, as shown in Fig. 15-7-1. *Angular-perspective drawings* are similar to axonometric drawings except that the receding lines converge at two vanishing points located on the horizon. Normally the height, or vertical, lines are parallel to the picture plane, and the length and width lines recede.

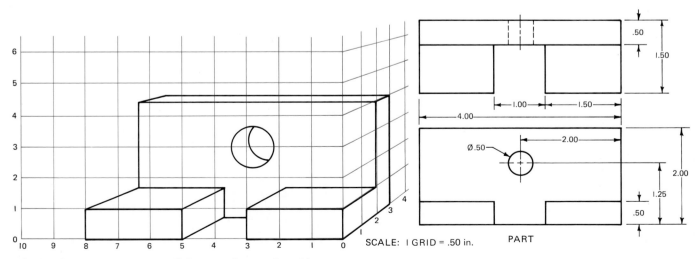

SCALE: 1 GRID = .50 in.   PART

FIG. 15-6-9   Part drawn on a parallel-perspective exterior grid.

The construction for a simple prism is shown in Fig. 15-7-2. Since line 1-2 rests on the picture plane, it will appear as its true height on the perspective drawing and will be located directly below line 1-2 on the top view. The next step is to join points 1 and 2 with light receding lines to both vanishing points. These receding lines represent the width and length lines of the prism; the width lines recede to *VPL* and the length lines recede to *VPR*. Since line 3-4 on the top view does not rest on the picture plane, it will not appear in its true height nor as its true distance from line 1-2 in the perspective. To find its position on the perspective drawing, join line 3-4, which appears as a point in the top view, to *SP* with a visual ray line. Where this visual ray line intersects the picture plane at *C*, project a vertical line down to the perspective view until it intersects the receding lines 1-*VPR* and 2-*VPR* at points 3 and 4, respectively. Line 5-6 may be found in the same manner. Next join point 3 to *VPL* and point 5 to *VPR* with light receding lines. The intersection of these lines is point 7.

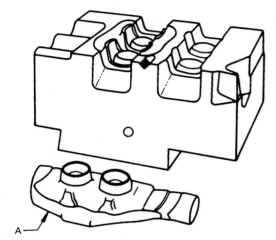

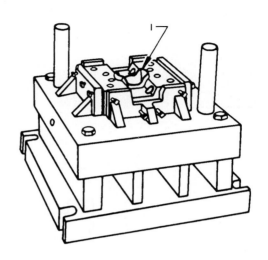

FIG. 15-7-1 Angular-, or two-point, perspective drawings. *(General Motors Corp.)*

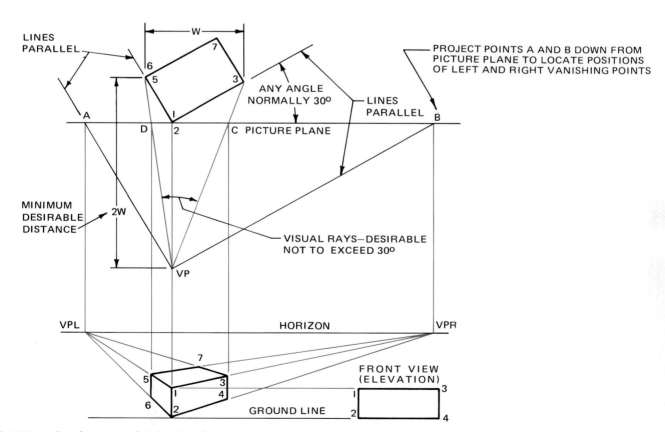

FIG. 15-7-2 Angular-perspective drawing of a prism.

461

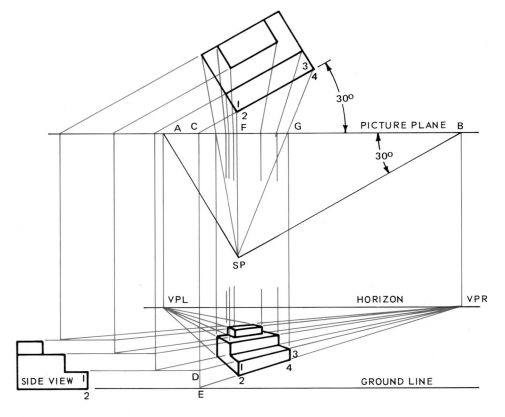

FIG. 15-7-3   Angular-perspective drawing of an object that does not touch the picture plane.

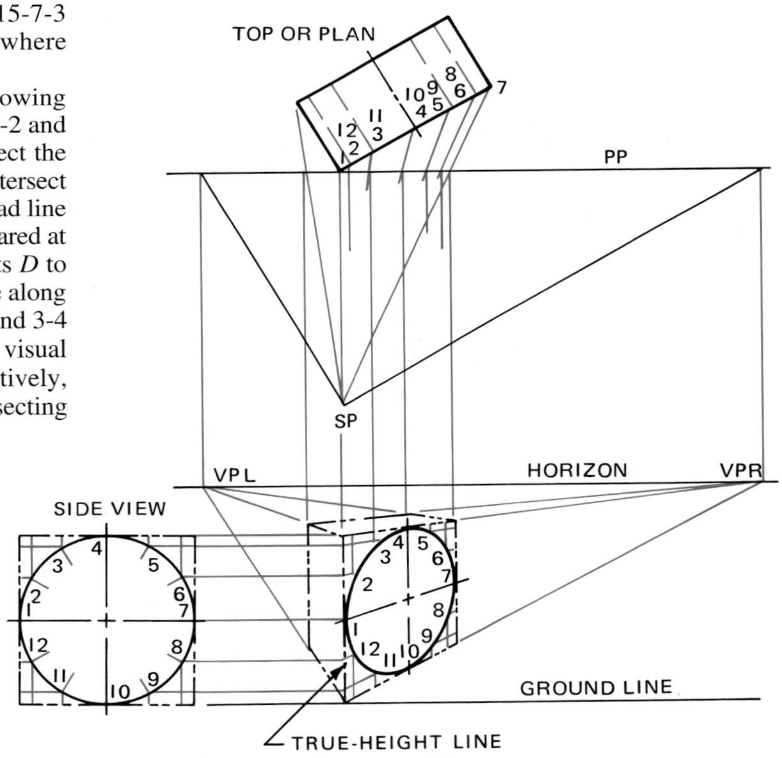

FIG. 15-7-4   Construction of a circle in angular perspective.

Lines Not Touching on the Picture Plane   Figure 15-7-3 illustrates the construction of a perspective drawing where none of the object lines touch the picture plane.

All these lines can be constructed by using the following procedure, which locates the position and size of lines 1-2 and 3-4. Extend line 1-3 (and 2-4) in the top view to intersect the picture plane at point C. Project a line down from C to intersect horizontal lines 1-D and 2-E at D and E, respectively. Had line 1-2 been located at C in the top view, it would have appeared at its true height and at D-E on the perspective. Join points D to VPR and E to VPR with light receding lines. Somewhere along these lines are points 1, 2, 3, and 4. Next join lines 1-2 and 3-4 in the top view to SP with visual ray lines. Where these visual ray lines intersect the picture plane at F and G, respectively, project vertical lines down to the perspective view intersecting line D-VPR at 1 and 3 and E-VPR at 2 and 4.

Construction of Circles and Curves in Perspective   Circles and curves may be constructed in perspective, as illustrated in Fig. 15-7-4. Using orthographic projections, oriented with respect to the subject in the plan and side views, plot and label the desired points (using numbers) on the curved surfaces.

From the plan view project these points to the picture plane, then vertically down to the perspective view. Project the height of the plotting numbers horizontally from the side view to the true-height line in the perspective view. The position of the plotting numbers may now be located on the perspective view.

Locate the points of intersection of the lines projected down from the picture plane with the visual ray lines receding to the right vanishing point from the appropriate numbers on the true-height line.

Horizon Line   Figure 15-7-5 illustrates different effects produced by repositioning the object with respect to the horizon.

Angular-Perspective Grids

Exterior Grid   When the three adjacent exterior planes of the cube are developed, the resultant image is referred to as an *exterior grid*. When this grid is used, the points are projected from the top plane downward and from the picture planes away from the observer (Figs. 15-7-6 and 15-7-7).

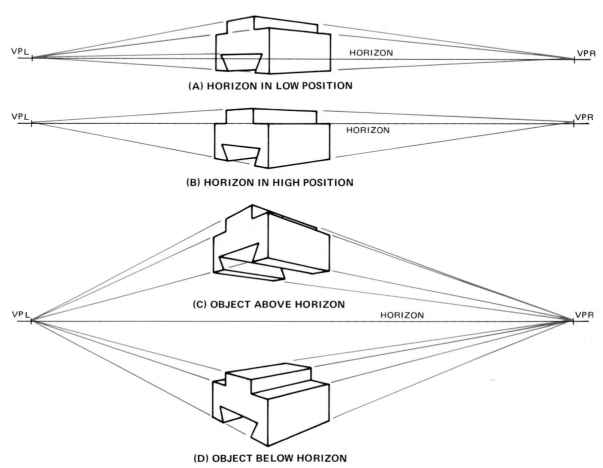

FIG. 15-7-5   Horizon lines.

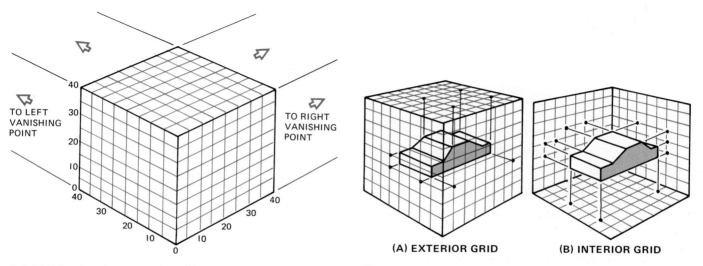

FIG. 15-7-6   Angular-perspective grid.

FIG. 15-7-7   Types of angular-perspective grids.

463

Interior Grid   When the three adjacent interior planes of the cube are exposed and developed, the resultant image is referred to as an *interior grid*. When this grid is used, the points are projected from the base plane upward and from the picture planes toward the observer. Each produces the same results.

Two further variations of both the exterior and interior grids are known as the *bird's-eye* and *worm's-eye grids.* These effects are achieved by rotating the vertical plane of the grid about the horizon line. Objects drawn in the bird's-eye grid appear as if they were being viewed from above the horizon line, as seen in Fig. 15-7-8. Objects drawn in the worm's-eye grid appear as if they were being viewed from below the horizon line.

Grid Increments

The three surfaces or planes of the grid are subdivided into multiple vertical and horizontal increments. Each increment is proportionately foreshortened as it recedes from the picture plane and thus creates the perspective illusion. The grid increments can be any size desired (Fig. 15-7-9).

 CAD

CAD systems that support 3-D modeling usually support the generation perspective drawings. In the construction of the perspective drawing, the position of the "camera" represents the person viewing the object. Several parameters such as zoom, distance, and view clip are then manipulated to create the required perspective view of the object. Only the view is manipulated as the original model of the object remains unaffected.

REFERENCES AND SOURCE MATERIAL

1. General Motors Corp.

ASSIGNMENT

See Assignment 29 for Unit 15-7 on pages 479–480.

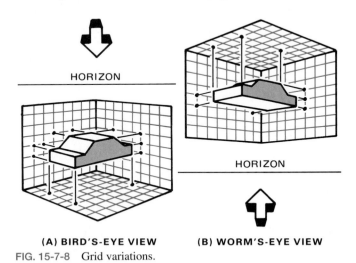

(A) BIRD'S-EYE VIEW   **(B) WORM'S-EYE VIEW**

FIG. 15-7-8   Grid variations.

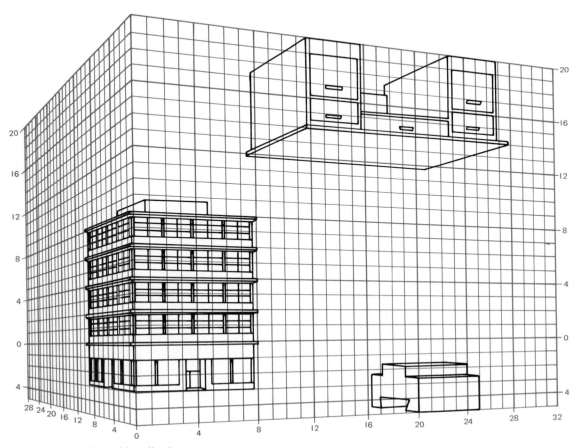

FIG. 15-7-9   Angular-perspective grid applications.

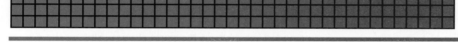

ASSIGNMENTS FOR CHAPTER 15

ASSIGNMENTS FOR UNIT 15-1, PICTORIAL DRAWINGS

1. On isometric grid paper or using the CAD isometric grid, draw the parts shown in Fig. 15-1-A. Do not show hidden lines. Each square shown on the drawing represents one isometric square on the grid.

2. On isometric grid paper or using the CAD isometric grid, draw the parts shown in Fig. 15-1-B. Do not show hidden lines. Each square shown on the drawing represents one isometric square on the grid.

3. On isometric grid paper or using the CAD isometric grid, sketch the parts shown in Fig. 15-1-C. Do not show hidden lines.

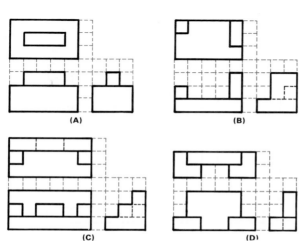

FIG. 15-1-A   Isometric flat-surface assignments.

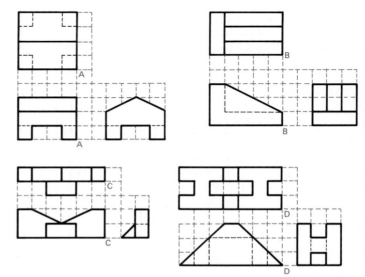

FIG. 15-1-B   Isometric flat-surface assignments.

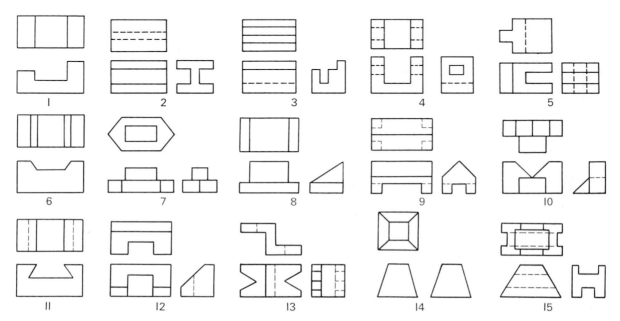

FIG. 15-1-C   Sketching assignments.

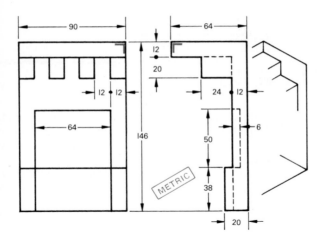

FIG. 15-1-D Tablet.

4. On isometric grid paper or using the CAD isometric grid, make an isometric drawing of one of the parts shown in Figs. 15-1-D through 15-1-F. A partial starting layout is provided for each of the parts shown. Start at the corner indicated by thick lines. Scale 1:1.

5. On isometric grid paper or using the CAD isometric grid, make an isometric drawing of one of the parts shown in Figs. 15-1-G through 15-1-J. For Fig. 15-1-G use the layout shown in Fig. 15-1-E; for Fig. 15-1-H use the layout shown

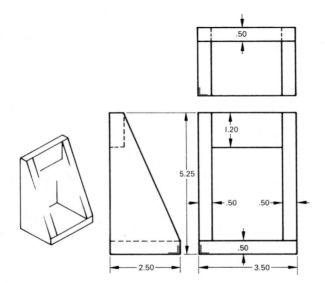

FIG. 15-1-E Stirrup.

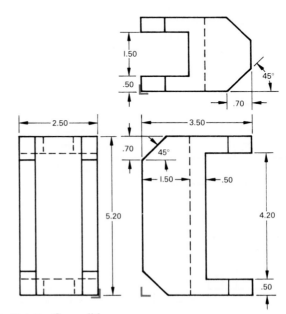

FIG. 15-1-G Cross slide.

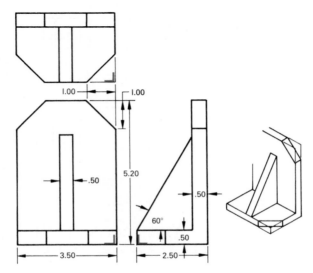

FIG. 15-1-F Brace.

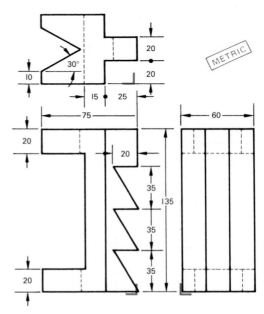

FIG. 15-1-H Ratchet.

in Fig. 15-1-F; and for Fig. 15-1-J use the layout shown in Fig. 15-1-D. Scale 1:1.

6. Make an isometric drawing, complete with dimensions, of one of the parts shown in Figs. 15-1-K through 15-1-Q (here and on pg. 468). Scale 1:1.

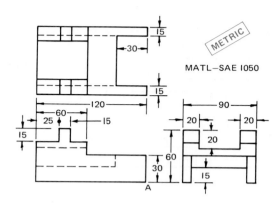

MATL—SAE 1050

FIG. 15-1-M   Step block.

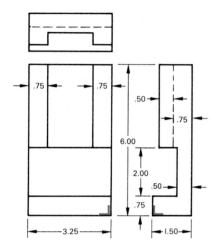

FIG. 15-1-J   Stop.

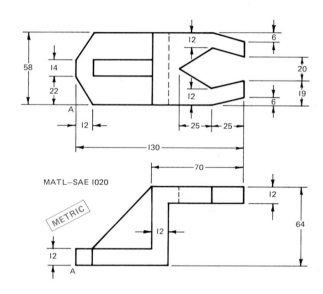

MATL—SAE 1020

FIG. 15-1-N   Support bracket.

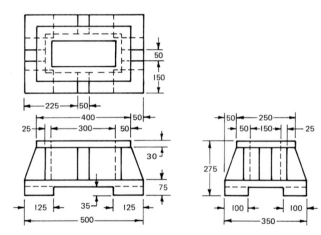

FIG. 15-1-K   Planter box.

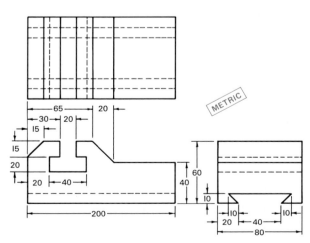

FIG. 15-1-L   Cross slide.

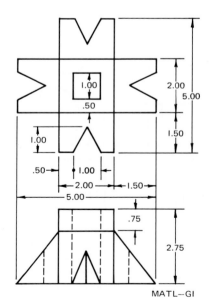

MATL—GI

FIG. 15-1-P   Stand.

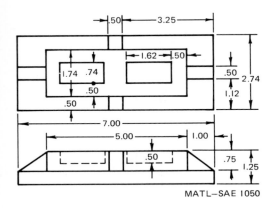

FIG. 15-1-Q   Base plate.

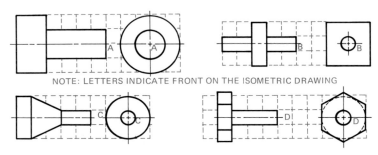

FIG. 15-2-B   Isometric curved-surface assignments.

ASSIGNMENTS FOR UNIT 15-2, CURVED SURFACES IN ISOMETRIC

7. On isometric grid paper or using the CAD isometric grid, draw the parts shown in Fig. 15-2-A. Each square shown on the figure represents one square on the isometric grid. Hidden lines may be omitted for clarity.

8. On isometric grid paper or using the CAD isometric grid, draw the parts shown in Fig. 15-2-B. Each square shown on the figure represents one square on the isometric grid. Hidden lines may be omitted for clarity.

9. Make an isometric drawing, complete with dimensions, of one of the parts shown in Figs. 15-2-C through 15-2-F. Scale 1:2 for Fig. 15-2-F. For all others the scale is 1:1.

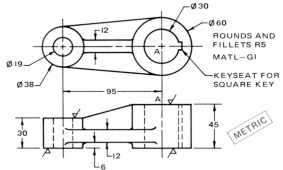

FIG. 15-2-C   Link.

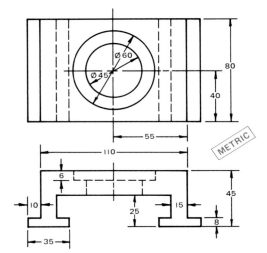

FIG. 15-2-D   T-guide.

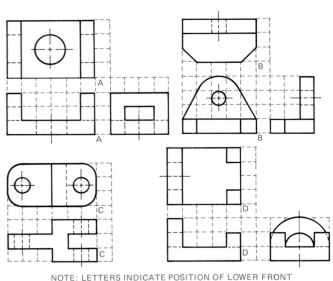

NOTE: LETTERS INDICATE POSITION OF LOWER FRONT CORNER ON THE ISOMETRIC DRAWING

FIG. 15-2-A   Isometric curved-surface assignments.

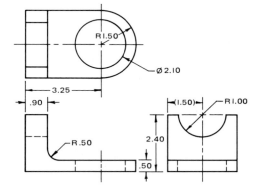

FIG. 15-2-E   Cradle bracket.

FIG. 15-2-F   Base.

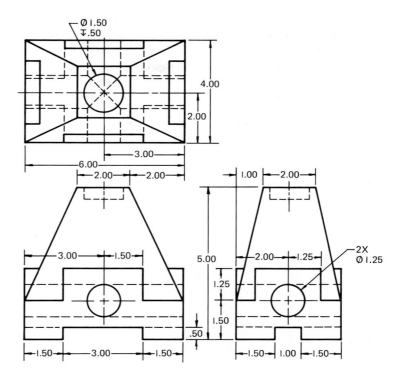

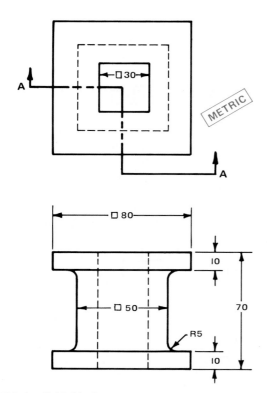

FIG. 15-3-A   Guide block.

ASSIGNMENTS FOR UNIT 15-3, COMMON FEATURES IN ISOMETRIC

10. Make an isometric half-section view, complete with dimensions, of one of the parts shown in Figs. 15-3-A and 15-3-B. Scale 1:1.

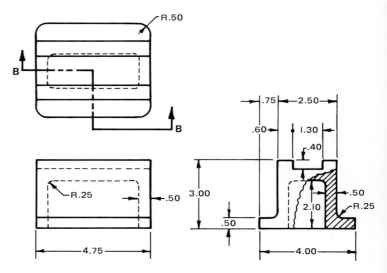

FIG. 15-3-B   Base.

11. Make an isometric full-section drawing, complete with dimensions, of one of the parts shown in Figs. 15-3-C through 15-3-E. Scale 1:1.

12. Make an isometric drawing, complete with dimensions, of one of the parts shown in Figs. 15-3-F and 15-3-G. Use a conventional break to shorten the length for Fig. 15-3-F. Scale 1:1.

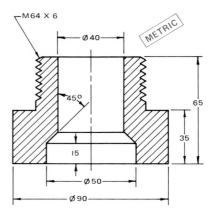

FIG. 15-3-E   Adapter.

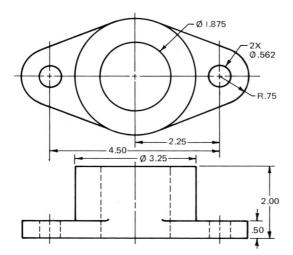

FIG. 15-3-C   Bearing support.

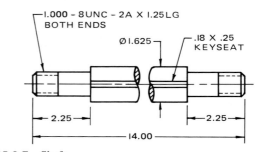

FIG. 15-3-F   Shaft.

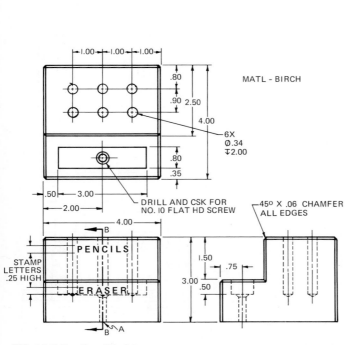

FIG. 15-3-D   Pencil holder.

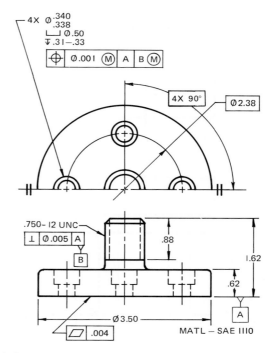

FIG. 15-3-G

13. Make an isometric assembly drawing of the two-post die set, Model 302, shown in Fig. 15-3-H. Allow 2 in. between the top and base. Scale 1:2. Do not dimension. Include on the drawing an item list. Using part numbers, identify the parts on the assembly.

14. Make an isometric exploded assembly drawing of the book rack shown in Fig. 15-3-J. Use a B (A3) sheet. Scale 1:2. Do not dimension. Include on the drawing an item list. Using part numbers, identify the parts on the assembly.

FIG. 15-3-H   Two-post die set.

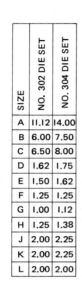

SIZE	NO. 302 DIE SET	NO. 304 DIE SET
A	11.12	14.00
B	6.00	7.50
C	6.50	8.00
D	1.62	1.75
E	1.50	1.62
F	1.25	1.25
G	1.00	1.12
H	1.25	1.38
J	2.00	2.25
K	2.00	2.25
L	2.00	2.00

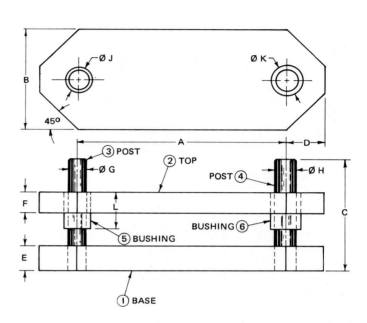

FIG. 15-3-J   Book rack.

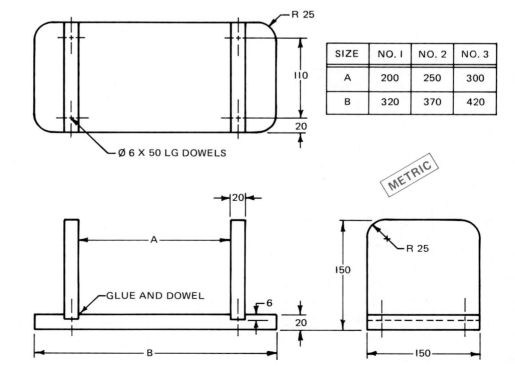

SIZE	NO. 1	NO. 2	NO. 3
A	200	250	300
B	320	370	420

471

15. Make an isometric assembly drawing of the gear clamp shown in Fig. 15-3-K. Add dimensions, identification part numbers, and an item list. Scale 1:1.

16. Make an isometric exploded assembly drawing of the universal joint shown in Fig. 15-3-L. Scale 1:1. Do not dimension. Include on the drawing an item list. Using part numbers, identify the parts on the assembly.

ASSIGNMENTS FOR UNIT 15-4, OBLIQUE PROJECTION

17. On coordinate grid paper or using the CAD grid, make oblique drawings of the three parts shown in Fig. 15-4-A. Each square shown on the figure represents one square on the grid. Hidden lines may be omitted to improve clarity.

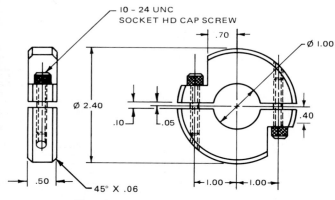

10 – 24 UNC
SOCKET HD CAP SCREW

FIG. 15-3-K   Gear clamp.

PT I – FORK – 2 REQD

4X
.250 - 20 UNC - 2B X ⌄.31

PT 2 – RING – I REQD
PT 3 – Ø .25 SPRING PIN – 2 REQD
PT 4 – .250 – 20 FHMS – .62 LG – 4 REQD

FIG. 15-3-L   Universal joint.

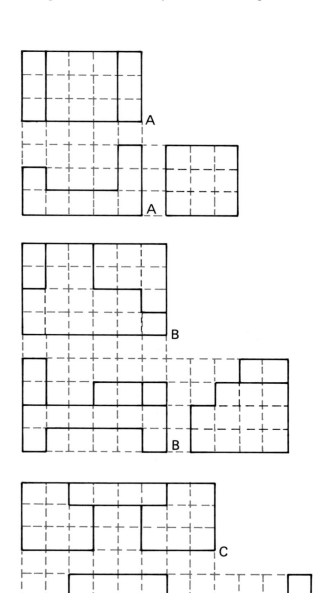

FIG. 15-4-A   Oblique flat-surface assignments.

18. On coordinate grid paper or using the CAD grid, make oblique drawings of the three parts shown in Fig. 15-4-B. Each square shown on the figure represents one square on the grid. Hidden lines may be omitted to improve clarity.

19. On oblique grid paper or using the CAD grid, sketch the parts shown in Fig. 15-4-C. Do not dimension or show hidden lines.

20. Make an oblique drawing of one of the parts shown in Figs. 15-4-D and 15-4-E. A partial starting layout is provided for each of the parts shown. Start at the corner indicated by thick lines. Scale 1:1.

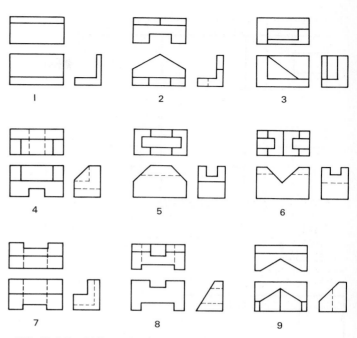

FIG. 15-4-C Oblique sketching assignments.

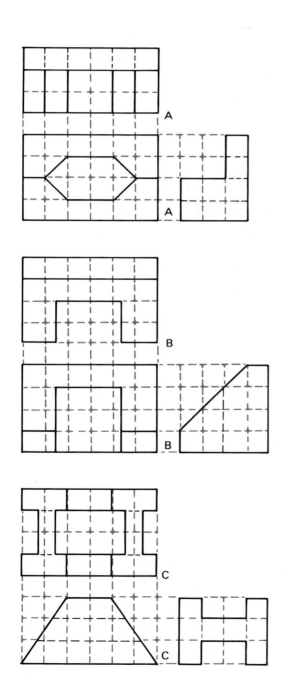

FIG. 15-4-B Oblique flat-surface assignments.

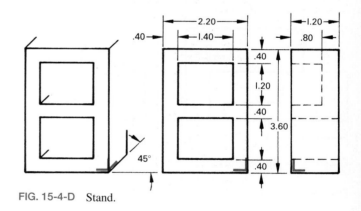

FIG. 15-4-D Stand.

FIG. 15-4-E V-block.

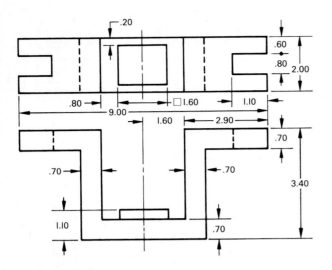

FIG. 15-4-F   Control arm.

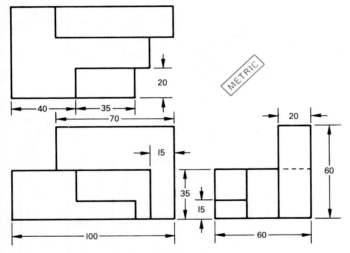

FIG. 15-4-G   Spacer block.

FIG. 15-4-H   V-block rest.

21. Make an oblique drawing, complete with dimensions, of one of the parts shown in Figs. 15-4-F through 15-4-J. Scale 1:1.

ASSIGNMENTS FOR UNIT 15-5, COMMON FEATURES IN OBLIQUE

22. Make an oblique drawing of one of the parts shown in Figs. 15-5-A and 15-5-B. A partial starting layout is provided for each of the parts shown. Add dimensions. Scale 1:1.

23. Make an oblique drawing, complete with dimensions, of one of the parts shown in Figs. 15-5-C through 15-5-G. Scale 1:1. For Fig. 15-5-G show either a broken-out or phantom section to show the Ø.406 hole.

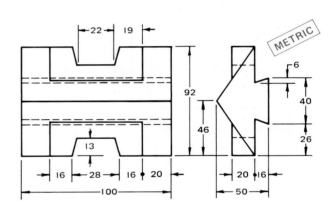

FIG. 15-4-J   Dovetail guide.

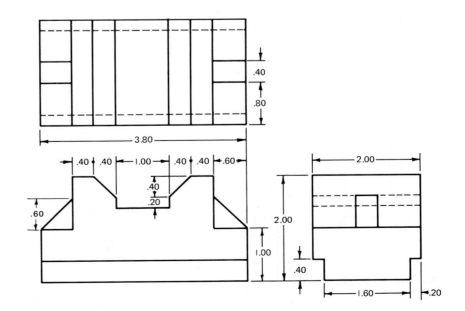

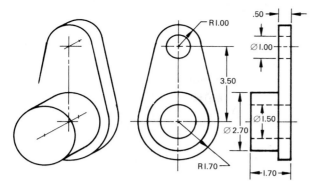

FIG. 15-5-A   Crank.

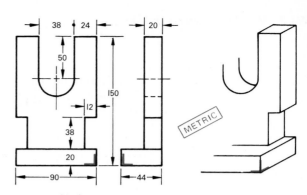

FIG. 15-5-B   Shaft support.

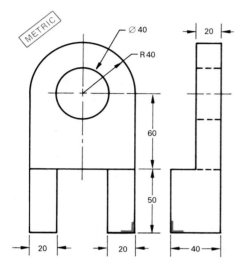

FIG. 15-5-C   Forked guide.

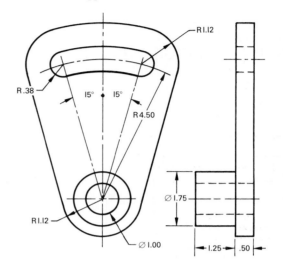

FIG. 15-5-D   Slotted sector.

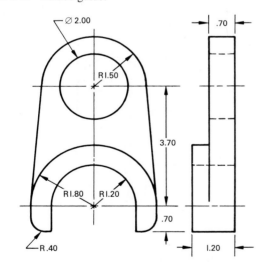

FIG. 15-5-E   Guide link.

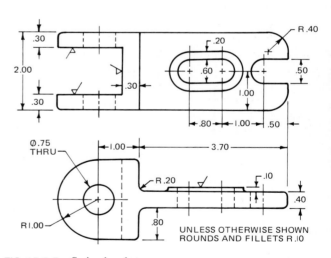

FIG. 15-5-F   Swing bracket.

FIG. 15-5-G   Tool holder.

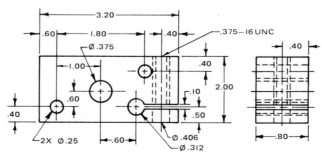

475

24. Make an oblique drawing, complete with dimensions, of one of the parts shown in Figs. 15-5-H through 15-5-M. For Figs. 15-5-J through 15-5-L use the straight line method of showing the rounds and fillets. Scale to suit.

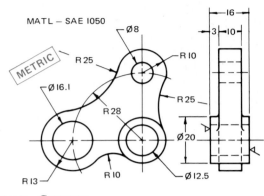

MATL — SAE 1050

FIG. 15-5-K   Connector.

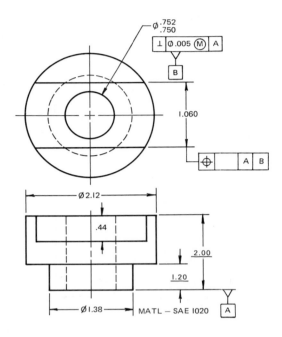

FIG. 15-5-H   Drive link.

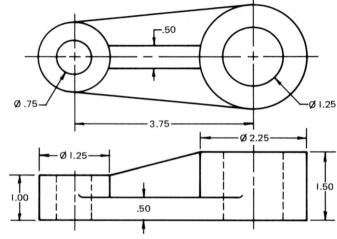

FIG. 15-5-L   Link.

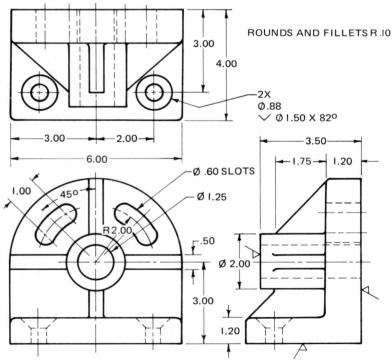

FIG. 15-5-J   End bracket.

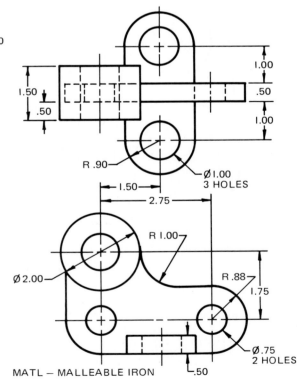

MATL — MALLEABLE IRON

FIG. 15-5-M   Swivel hanger.

25. Make an oblique drawing, complete with dimensions, of one of the parts shown in Figs. 15-5-N through 15-5-U. For Fig. 15-5-P use a conventional break to shorten the length. Show a half-section for Figs. 15-5-S and 15-5-T. Scale to suit.

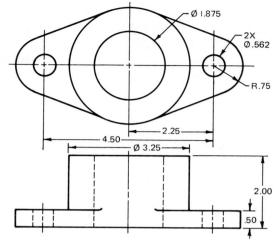

FIG. 15-5-S Bearing support.

FIG. 15-5-T Bushing holder.

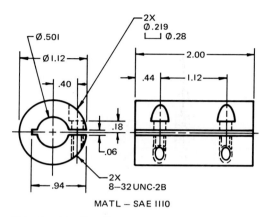

FIG. 15-5-N Coupling.

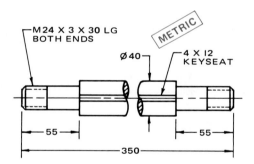

FIG. 15-5-P Shaft.

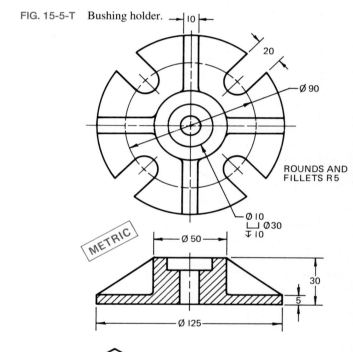

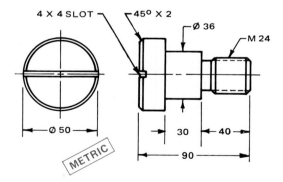

FIG. 15-5-R Stop button.

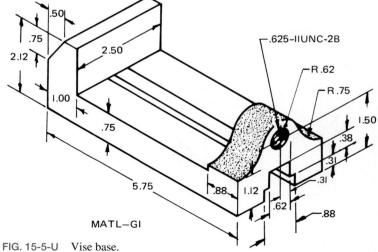

MATL—GI

FIG. 15-5-U Vise base.

477

ASSIGNMENTS FOR UNIT 15-6, PARALLEL, OR ONE-POINT, PERSPECTIVE

Note: If a one-point perspective grid is not available, copy the grid shown in Fig. 15-6-9. For a worm's-eye view rotate the grid 180°. Position the part on the grid in order to best show the part.

26. Using a one-point perspective grid, make a drawing of one of the parts shown in Figs. 15-6-A through 15-6-C. Add dimensions. Scale to suit.
27. Using a one-point perspective grid, make a drawing of one of the parts shown in Figs. 15-6-D through 15-6-F. Add dimensions. Scale to suit.

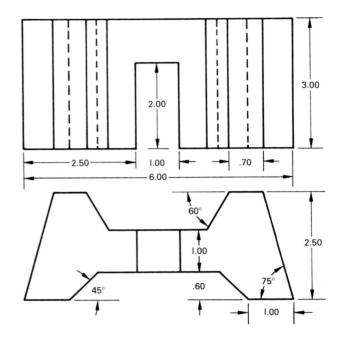

FIG. 15-6-C   Base.

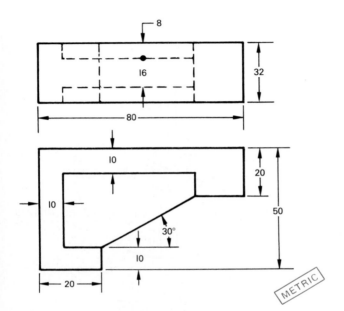

FIG. 15-6-A   Bracket.

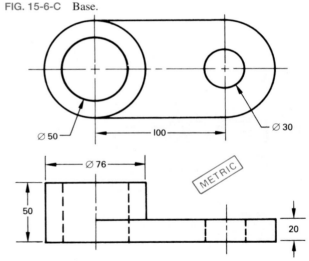

FIG. 15-6-D   Bearing.

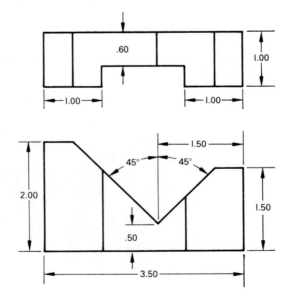

FIG. 15-6-B   V-slide.

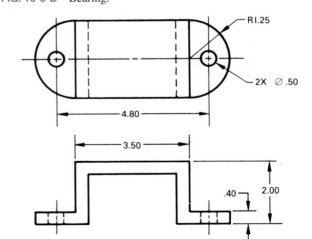

FIG. 15-6-E   Clamp.

478

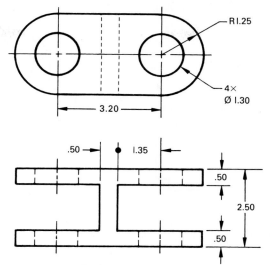

FIG. 15-6-F   Rod spacer.

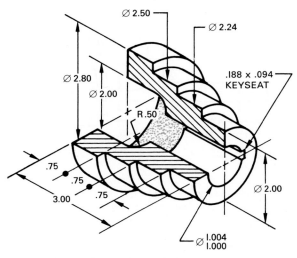

FIG. 15-6-G   Step pulley.

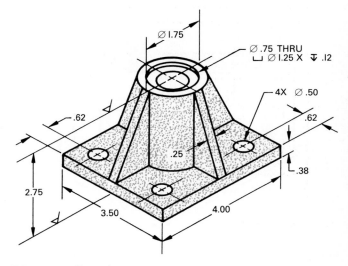

FIG. 15-6-J   Base plate.

28. Using a one-point perspective grid, make a half-section drawing of one of the parts shown in Figs. 15-6-G through 15-6-J. Add dimensions. Scale to suit.

ASSIGNMENT FOR UNIT 15-7, ANGULAR, OR TWO-POINT, PERSPECTIVE

Note: If a two-point perspective grid is not available, copy the grid shown in Fig. 15-7-9. For a bird's-eye view rotate the grid 180°. Position the part on the grid in order to best show the part.

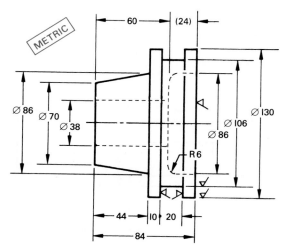

FIG. 15-6-H   Cone spacer.

29. Using a two-point perspective grid make a drawing of one of the parts shown in Figs. 15-7-A through 15-7-G. Add dimensions. Scale to suit.

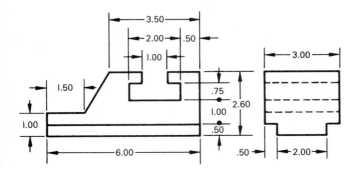

FIG. 15-7-A Tool support.

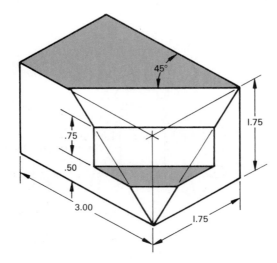

FIG. 15-7-B Corner block.

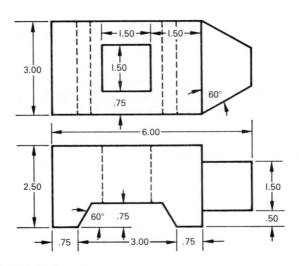

FIG. 15-7-C Locating support.

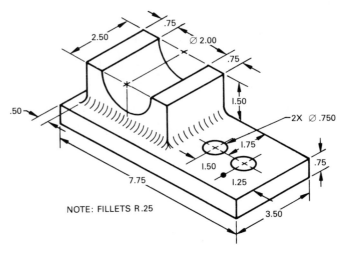

NOTE: FILLETS R.25

FIG. 15-7-D Horizontal guide.

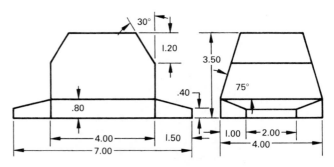

FIG. 15-7-E Base.

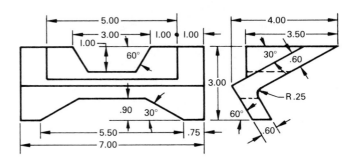

FIG. 15-7-F Separator.

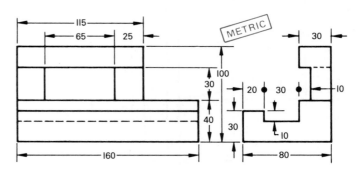

FIG. 15-7-G Support guide.

STANDARD PARTS AND TECHNICAL DATA

ANSI	Y1.1	Abbreviations for Use on Drawings and in Text
ANSI	Y14.1	Drawing Sheet Size and Format
ANSI	Y14.2M	Line Conventions and Lettering
ANSI	Y14.3	Projections
ANSI	Y14.4	Pictorial Drawing
ANSI	Y14.5M	Dimensioning and Tolerancing for Engineering Drawings
ANSI	Y14.6	Screw Thread Representation
ANSI	Y14.7	Gears, Splines, and Serrations
ANSI	Y14.7.	Gear Drawing Standards—Part 1, for Spur, Helical, Double Helical and Rack
ANSI	Y14.9	Forgings
ANSI	Y14.10	Metal Stampings
ANSI	Y14.11	Plastics
ANSI	Y14.14	Mechanical Assemblies
ANSI	Y14.15	Electrical and Electronics Diagrams
ANSI	Y14.15A	Interconnection Diagrams
ANSI	Y14.17	Fluid Power Diagrams
ANSI	Y14.36	Surface Texture Symbols
ANSI	Y32.2	Graphic Symbols for Electrical and Electronics Diagrams
ANSI	Y32.9	Graphic Electrical Wiring Symbols for Architectural and Electrical Layout Drawings
ANSI	B1.1	Unified Screw Threads
ANSI	B4.2	Preferred Metric Limits and Fits
ANSI	B17.1	Keys and Keyseats
ANSI	B17.2	Woodruff Key and Keyslot Dimensions
ANSI	B18.2.1	Square and Hex Bolts and Screws
ANSI	B18.2.2	Square and Hex Nuts
ANSI	B18.3	Socket Cap, Shoulder, and Setscrews
ANSI	B18.6.2	Slotted-Head Cap Screws, Square-Head Setscrews, Slotted-Headless Setscrews
ANSI	B18.6.3	Machine Screws and Machine Screw Nuts
ANSI	B18.21.1	Lock Washers
ANSI	B27.2	Plain Washers
ANSI	B46.1	Surface Texture
ANSI	B94.6	Knurling
ANSI	B94.11M	Twist Drills
ANSI	Z210.1	Metric Practice

TABLE 1   ANSI publications of interest to drafters.

$\frac{1}{64}$	0.015625	$\frac{17}{64}$	0.265625	$\frac{33}{64}$	0.515625	$\frac{49}{64}$	0.765625
$\frac{1}{32}$	0.03125	$\frac{9}{32}$	0.28125	$\frac{17}{32}$	0.53125	$\frac{25}{32}$	0.78125
$\frac{3}{64}$	0.046875	$\frac{19}{64}$	0.296875	$\frac{35}{64}$	0.546875	$\frac{51}{64}$	0.796875
$\frac{1}{16}$	0.0625	$\frac{5}{16}$	0.3125	$\frac{9}{16}$	0.5625	$\frac{13}{16}$	0.8125
$\frac{5}{64}$	0.078125	$\frac{21}{64}$	0.328125	$\frac{37}{64}$	0.578125	$\frac{53}{64}$	0.828125
$\frac{3}{32}$	0.09375	$\frac{11}{32}$	0.34375	$\frac{19}{32}$	0.59375	$\frac{27}{32}$	0.84375
$\frac{7}{64}$	0.109375	$\frac{23}{64}$	0.359375	$\frac{39}{64}$	0.609375	$\frac{55}{64}$	0.859375
$\frac{1}{8}$	0.1250	$\frac{3}{8}$	0.3750	$\frac{5}{8}$	0.6250	$\frac{7}{8}$	0.8750
$\frac{9}{64}$	0.140625	$\frac{25}{64}$	0.390625	$\frac{41}{64}$	0.640625	$\frac{57}{64}$	0.890625
$\frac{5}{32}$	0.15625	$\frac{13}{32}$	0.40625	$\frac{21}{32}$	0.65625	$\frac{29}{32}$	0.90625
$\frac{11}{64}$	0.171875	$\frac{27}{64}$	0.421875	$\frac{43}{64}$	0.671875	$\frac{59}{64}$	0.921875
$\frac{3}{16}$	0.1875	$\frac{7}{16}$	0.4375	$\frac{11}{16}$	0.6875	$\frac{15}{16}$	0.9375
$\frac{13}{64}$	0.203125	$\frac{29}{64}$	0.453125	$\frac{45}{64}$	0.703125	$\frac{61}{64}$	0.953125
$\frac{7}{32}$	0.21875	$\frac{15}{32}$	0.46875	$\frac{23}{32}$	0.71875	$\frac{31}{32}$	0.96875
$\frac{15}{64}$	0.234375	$\frac{31}{64}$	0.484375	$\frac{47}{64}$	0.734375	$\frac{63}{64}$	0.984375
$\frac{1}{4}$	0.2500	$\frac{1}{2}$	0.5000	$\frac{3}{4}$	0.7500	1	1.0000

TABLE 2   Decimal equivalents of common inch fractions.

Quantity	Metric Unit	Symbol	Metric to Inch-Pound Unit	Inch-Pound to Metric Unit
Length	millimeter	mm	1 mm = 0.0394 in.	1 in. = 25.4 mm
	centimeter	cm	1 cm = 0.394 in.	1 ft. = 30.5 cm
	meter	m	1 m = 39.37 in. = 3.28 ft	1 yd. = 0.914 m = 914 mm
	kilometer	km	1 km = 0.62 mile	1 mile = 1.61 km
Area	square millimeter	mm²	1 mm² = 0.001 55 sq. in.	1 sq. in. = 6 452 mm²
	square centimeter	cm²	1 cm² = 0.155 sq. in.	1 sq. ft. = 0.093 m²
	square meter	m²	1 m² = 10.8 sq. ft.	1 sq. yd. = 0.836 m²
			= 1.2 sq. yd.	
Mass	milligram	mg	1 g = 0.035 oz.	1 oz. = 28.3 g
	gram	g	1 kg = 2.205 lb.	1 lb. = 0.454 kg
	kilogram	kg	1 tonne = 1.102 tons	1 ton = 907.2 kg
	tonne	t		= 0.907 tonnes
Volume	cubic centimeter	cm³	1 mm³ = 0.000 061 cu. in.	1 fl. oz. = 28.4 cm³
	cubic meter	m³	1 cm³ = 0.061 cu. in.	1 cu. in. = 16.387 cm³
	milliliter	m	1 m³ = 35.3 cu ft.	1 cu. ft. = 0.028 m³
			= 1.308 cu. yd.	1 cu. yd. = 0.756 m³
			1 mℓ = 0.035 fl. oz.	
Capacity	liter	L	U.S. Measure	U.S. Measure
			1 pt. = 0.473 L	1 L = 2.113 pt.
			1 qt. = 0.946 L	= 1.057 qt.
			1 gal = 3.785 L	= 0.264 gal.
			Imperial Measure	Imperial Measure
			1 pt. = 0.568 L	1 L = 1.76 pt.
			1 qt. = 1.137 L	= 0.88 qt.
			1 gal = 4.546 L	= 0.22 gal.
Temperature	Celsius degree	°C	$°C = \frac{5}{9}(°F-32)$	$°F = \frac{9}{5} \times °C + 32$
Force	newton	N	1 N = 0.225 lb (f)	1 lb (f) = 4.45N
	kilonewton	kN	1 kN = 0.225 kip (f)	= 0.004 448 kN
			= 0.112 ton (f)	
Energy/Work	joule	J	1 J = 0.737 ft · lb	1 ft · lb = 1.355 J
	kilojoule	kJ	1 J = 0.948 Btu	1 Btu = 1.055 J
	megajoule	MJ	1 MJ = 0.278 kWh	1 kWh = 3.6 MJ
Power	kilowatt	kW	1 kW = 1.34 hp	1 hp (550 ft · lb/s) = 0.746 kW
			1 W = 0.0226 ft · lb/min.	1 ft · lb/min = 44.2537 W
Pressure	kilopascal	kPa	1 kPa = 0.145 psi	1 psi = 6.895 kPa
			= 20.885 psf	1 lb-force/sq. ft. = 47.88 Pa
			= 0.01 ton-force per sq. ft.	1 ton-force/sq. ft. = 95.76 kPa
	*kilogram per square centimeter	kg/cm²	1 kg/cm² = 13.780 psi	
Torque	newton meter	N · m	1 N · m = 0.74 lb · ft	1 lb · ft = 1.36 N · m
	*kilogram meter	kg/m	1 kg/m = 7.24 lb · ft	1 lb · ft = 0.14 kg/m
	*kilogram per centimeter	kg/cm	1 kg/cm = 0.86 lb · in	1 lb · in = 1.2 kg/cm
Speed/Velocity	meters per second	m/s	1 m/s = 3.28 ft/s	1 ft/s = 0.305 m/s
	kilometers per hour	km/h	1 km/h = 0.62 mph	1 mph = 1.61 km/h

*Not SI units, but included here because they are employed on some of the gages and indicators currently in use in industry.

TABLE 3 Metric conversion tables.

Across Flats	ACRFLT	Machine Steel	MST
American National Standards Institute	ANSI	Machined	⋎
And	&	Malleable Iron	MI
Angular	ANLR	Material	MATL
Approximate	APPROX	Maximum	MAX
Assembly	ASSY	Maximum Material Condition	Ⓜ or MMC
Between	↔	Meter	m
Bill of Material	B/M	Metric Thread	M
Bolt Circle	BC	Micrometer	μm
Brass	BR	Millimeter	mm
Brown and Sharpe Gage	B&S GA	Minimum	MIN
Bushing	BUSH	Module	MDL
Canada Standards Institute	CSI	Newton	N
Carbon Steel	CS	Nominal	NOM
Casting	CSTG	Not to Scale	xx
Cast Iron	CI	Number	NO
Center Line	CL or ₵	On Center	OC
Center to Center	C to C	Outside Diameter	OD
Centimeter	cm	Parallel	PAR
Chamfer	CHAM	Pascal	Pa
Circularity	CIR	Perpendicular	PERP
Cold-Rolled Steel	CRS	Pitch	P
Concentric	CONC	Pitch Circle	PC
Counterbore	⌴ or CBORE	Pitch Diameter	PD
Counterdrill	CDRILL	Plate	PL
Countersink	⌵ or CSK	Radius	R
Cubic Centimeter	cm³	Reference or Reference Dimension	( ) or REF
Cubic Meter	m³	Regardless of Feature Size	* Ⓢ or RFS
Datum	DAT	Revolutions per Minute	rev/min
Degree (Angle)	° or DEG	Right Hand	RH
Depth	DP or ↧	Root Diameter	RD
Diameter	⌀ or DIA	Second (Arc)	(")
Diametral Pitch	DP	Second (Time)	SEC
Dimension	DIM	Section	SECT
Drawing	DWG	Slotted	SLOT
Eccentric	ECC	Socket	SOCK
Equally Spaced	EQL SP	Spherical	SPHER
Figure	FIG	Spherical Diameter	S⌀
Finish All Over	FAO	Spherical Radius	SR
Flat	FL	Spotface	⌴ or SFACE
Gage	GA	Square	☐ or SQ
Gray Iron	GI	Square Centimeter	cm²
Head	HD	Square Meter	m²
Heat Treat	HT TR	Steel	STL
Heavy	HVY	Straight	STR
Hexagon	HEX	Symmetrical	⊣⊢ or SYM
Hydraulic	HYDR	Taper—Flat	▷
Inside Diameter	ID	—Conical	▷
International Organization for Standardization	ISO	Taper Pipe Thread	NPT
International Pipe Standard	IPS	Thread	THD
Kilogram	kg	Through	THRU
Kilometer	km	Tolerance	TOL
Least Material Condition	Ⓛ or LMC	True Profile	TP
Left Hand	LH	U.S. Gage	USG
Length	LG	Watt	W
Liter	L	Wrought Iron	WI
		Wrought Steel	WS

Datum: A△ or A▲

*Symbol no longer used in ASME Y14.5M-1994 standards.

TABLE 4 Abbreviations and symbols used on technical drawings.
(Note: See Tables 53 and 54 for geometric tolerancing symbols.)

ANGLE	SINE	COSINE	TAN	COTAN	ANGLE
0°	.0000	1.0000	.0000	θ	90°
1°	0.0175	0.9998	0.0175	57.290	89°
2°	0.0349	0.9994	0.0349	28.636	88°
3°	0.0523	0.9986	0.0524	19.081	87°
4°	0.0698	0.9976	0.0699	14.301	86°
5°	0.0872	0.9962	0.0875	11.430	85°
6°	0.1045	0.9945	0.1051	9.5144	84°
7°	0.1219	0.9925	0.1228	8.1443	83°
8°	0.1392	0.9903	0.1405	7.1154	82°
9°	0.1564	0.9877	0.1584	6.3138	81°
10°	0.1736	0.9848	0.1763	5.6713	80°
11°	0.1908	0.9816	0.1944	5.1446	79°
12°	0.2079	0.9781	0.2126	4.7046	78°
13°	0.2250	0.9744	0.2309	4.3315	77°
14°	0.2419	0.9703	0.2493	4.0108	76°
15°	0.2588	0.9659	0.2679	3.7321	75°
16°	0.2756	0.9613	0.2867	3.4874	74°
17°	0.2924	0.9563	0.3057	3.2709	73°
18°	0.3090	0.9511	0.3249	3.0777	72°
19°	0.3256	0.9455	0.3443	2.9042	71°
20°	0.3420	0.9397	0.3640	2.7475	70°
21°	0.3584	0.9336	0.3839	2.6051	69°
22°	0.3746	0.9272	0.4040	2.4751	68°
23°	0.3907	0.9205	0.4245	2.3559	67°
24°	0.4067	0.9135	0.4452	2.2460	66°
25°	0.4226	0.9063	0.4663	2.1445	65°
26°	0.4384	0.8988	0.4877	2.0503	64°
27°	0.4540	0.8910	0.5095	1.9626	63°
28°	0.4695	0.8829	0.5317	1.8807	62°
29°	0.4848	0.8746	0.5543	1.8040	61°
30°	0.5000	0.8660	0.5774	1.7321	60°
31°	0.5150	0.8572	0.6009	1.6643	59°
32°	0.5299	0.8480	0.6249	1.6003	58°
33°	0.5446	0.8387	0.6494	1.5399	57°
34°	0.5592	0.8290	0.6745	1.4826	56°
35°	0.5736	0.8192	0.7002	1.4281	55°
36°	0.5878	0.8090	0.7265	1.3764	54°
37°	0.6018	0.7986	0.7536	1.3270	53°
38°	0.6157	0.7880	0.7813	1.2799	52°
39°	0.6293	0.7771	0.8098	1.2349	51°
40°	0.6428	0.7660	0.8391	1.1918	50°
41°	0.6561	0.7547	0.8693	1.1504	49°
42°	0.6691	0.7431	0.9004	1.1106	48°
43°	0.6820	0.7314	0.9325	1.0724	47°
44°	0.6947	0.7193	0.9657	1.0355	46°
45°	0.7071	0.7071	0.0000	1.0000	45°
ANGLE	COSINE	SINE	COTAN	TAN	ANGLE

TABLE 5 Trigonometric functions.

NUMBER OR LETTER SIZE DRILL	SIZE		NUMBER OR LETTER SIZE DRILL	SIZE		NUMBER OR LETTER SIZE DRILL	SIZE		NUMBER OR LETTER SIZE DRILL	SIZE	
	mm	INCHES		mm	INCHES		mm	INCHES		mm	INCHES
80	0.343	.014	50	1.778	.070	20	4.089	.161	K	7.137	.281
79	0.368	.015	49	1.854	.073	19	4.216	.166	L	7.366	.290
78	0.406	.016	48	1.930	.076	18	4.305	.170	M	7.493	.295
77	0.457	.018	47	1.994	.079	17	4.394	.173	N	7.671	.302
76	0.508	.020	46	2.057	.081	16	4.496	.177	O	8.026	.316
75	0.533	.021	45	2.083	.082	15	4.572	.180	P	8.204	.323
74	0.572	.023	44	2.184	.086	14	4.623	.182	Q	8.433	.332
73	0.610	.024	43	2.261	.089	13	4.700	.185	R	8.611	.339
72	0.635	.025	42	2.375	.094	12	4.800	.189	S	8.839	.348
71	0.660	.026	41	2.438	.096	11	4.851	.191	T	9.093	.358
70	0.711	.028	40	2.489	.098	10	4.915	.194	U	9.347	.368
69	0.742	.029	39	2.527	.100	9	4.978	.196	V	9.576	.377
68	0.787	.031	38	2.578	.102	8	5.080	.199	W	9.804	.386
67	0.813	.032	37	2.642	.104	7	5.105	.201	X	10.084	.397
66	0.838	.033	36	2.705	.107	6	5.182	.204	Y	10.262	.404
65	0.889	.035	35	2.794	.110	5	5.220	.206	Z	10.490	.413
64	0.914	.036	34	2.819	.111	4	5.309	.209			
63	0.940	.037	33	2.870	.113	3	5.410	.213			
62	0.965	.038	32	2.946	.116	2	5.613	.221			
61	0.991	.039	31	3.048	.120	1	5.791	.228			
60	1.016	.040	30	3.264	.129	A	5.944	.234			
59	1.041	.041	29	3.354	.136	B	6.045	.238			
58	1.069	.042	28	3.569	.141	C	6.147	.242			
57	1.092	.043	27	3.658	.144	D	6.248	.246			
56	1.181	.047	26	3.734	.147	E	6.350	.250			
55	1.321	.052	25	3.797	.150	F	6.528	.257			
54	1.397	.055	24	3.861	.152	G	6.629	.261			
53	1.511	.060	23	3.912	.154	H	6.756	.266			
52	1.613	.064	22	3.988	.157	I	6.909	.272			
51	1.702	.067	21	4.039	.159	J	7.036	.277			

TABLE 6   Number and letter-size drills.

METRIC DRILL SIZES		Reference Decimal Equivalent (Inches)	METRIC DRILL SIZES		Reference Decimal Equivalent (Inches)	METRIC DRILL SIZES		Reference Decimal Equivalent (Inches)
Preferred	Available		Preferred	Available		Preferred	Available	
—	0.40	.0157	—	2.7	.1063	14	—	.5512
—	0.42	.0165	2.8	—	.1102	—	14.5	.5709
—	0.45	.0177	—	2.9	.1142	15	—	.5906
—	0.48	.0189	3.0	—	.1181	—	15.5	.6102
0.50	—	.0197	—	3.1	.1220	16	—	.6299
—	0.52	.0205	3.2	—	.1260	—	16.5	.6496
0.55	—	.0217	—	3.3	.1299	17	—	.6693
—	0.58	.0228	3.4	—	.1339	—	17.5	.6890
0.60	—	.0236	—	3.5	.1378	18	—	.7087
—	0.62	.0244	3.6	—	.1417	—	18.5	.7283
0.65	—	.0256	—	3.7	.1457	19	—	.7480
—	0.68	.0268	3.8	—	.1496	—	19.5	.7677
0.70	—	.0276	—	3.9	.1535	20	—	.7874
—	0.72	.0283	4.0	—	.1575	—	20.5	.8071
0.75	—	.0295	—	4.1	.1614	21	—	.8268
—	0.78	.0307	4.2	—	.1654	—	21.5	.8465
0.80	—	.0315	—	4.4	.1732	22	—	.8661
—	0.82	.0323	4.5	—	.1772	—	23	.9055
0.85	—	.0335	—	4.6	.1811	24	—	.9449
—	0.88	.0346	4.8	—	.1890	25	—	.9843
0.90	—	.0354	5.0	—	.1969	26	—	1.0236
—	0.92	.0362	—	5.2	.2047	—	27	1.0630
0.95	—	.0374	5.3	—	.2087	28	—	1.1024
—	0.98	.0386	—	5.4	.2126	—	29	1.1417
1.00	—	.0394	5.6	—	.2205	30	—	1.1811
—	1.03	.0406	—	5.8	.2283	—	31	1.2205
1.05	—	.0413	6.0	—	.2362	32	—	1.2598
—	1.08	.0425	—	6.2	.2441	—	33	1.2992
1.10	—	.0433	6.3	—	.2480	34	—	1.3386
—	1.15	.0453	—	6.5	.2559	—	35	1.3780
1.20	—	.0472	6.7	—	.2638	36	—	1.4173
1.25	—	.0492	—	6.8	.2677	—	37	1.4567
1.3	—	.0512	—	6.9	.2717	38	—	1.4361
—	1.35	.0531	7.1	—	.2795	—	39	1.5354
1.4	—	.0551	—	7.3	.2874	40	—	1.5748
—	1.45	.0571	7.5	—	.2953	—	41	1.6142
1.5	—	.0591	—	7.8	.3071	42	—	1.6535
—	1.55	.0610	8.0	—	.3150	—	43.5	1.7126
1.6	—	.0630	—	8.2	.3228	45	—	1.7717
—	1.65	.0650	8.5	—	.3346	—	46.5	1.8307
1.7	—	.0669	—	8.8	.3465	48	—	1.8898
—	1.75	.0689	9.0	—	.3543	50	—	1.9685
1.8	—	.0709	—	9.2	.3622	—	51.5	2.0276
—	1.85	.0728	9.5	—	.3740	53	—	2.0866
1.9	—	.0748	—	9.8	.3858	—	54	2.1260
—	1.95	.0768	10	—	.3937	56	—	2.2047
2.0	—	.0787	—	10.3	.4055	—	58	2.2835
—	2.05	.0807	10.5	—	.4134	60	—	2.3622
2.1	—	.0827	—	10.8	.4252			
—	2.15	.0846	11	—	.4331			
2.2	—	.0866	—	11.5	.4528			
—	2.3	.0906	12	—	.4724			
2.4	—	.0945	12.5	—	.4921			
2.5	—	.0984	13	—	.5118			
2.6	—	.1024	—	13.5	.5315			

TABLE 7 Metric twist drill sizes.

SIZE INCHES		Graded Pitch Series						Constant Pitch Series					
		Coarse UNC		Fine UNF		Extra Fine UNEF		8 UN		12 UN		16 UN	
Number or Fraction	Decimal	Threads per Inch	Tap Drill Dia.	Threads per Inch	Tap Drill Dia.	Threads per Inch	Tap Drill Dia.	Threads per Inch	Tap Drill Dia.	Threads per Inch	Tap Drill Dia.	Threads per Inch	Tap Drill Dia.
0	.060	—	—	80	$3/64$	—	—	—	—	—	—	—	—
2	.086	56	No. 50	64	No. 49	—	—	—	—	—	—	—	—
4	.112	40	No. 43	48	No. 42	—	—	—	—	—	—	—	—
5	.125	40	No. 38	44	No. 37	—	—	—	—	—	—	—	—
6	.138	32	No. 36	40	No. 33	—	—	—	—	—	—	—	—
8	.164	32	No. 29	36	No. 29	—	—	—	—	—	—	—	—
10	.190	24	No. 25	32	No. 21	—	—	—	—	—	—	—	—
1/4	.250	20	7	28	3	32	.219	—	—	—	—	—	—
5/16	.312	18	F	24	1	32	.281	—	—	—	—	—	—
3/8	.375	16	.312	24	Q	32	.344	—	—	—	—	UNC	—
7/16	.438	14	U	20	.391	28	Y	—	—	—	—	16	V
1/2	.500	13	.422	20	.453	28	.469	—	—	—	—	16	.438
9/16	.562	12	.484	18	.516	24	.516	—	—	UNC	—	16	.500
5/8	.625	11	.531	18	.578	24	.578	—	—	12	.547	16	.562
3/4	.750	10	.656	16	.688	20	.703	—	—	12	.672	UNF	—
7/8	.875	9	.766	14	.812	20	.828	—	—	12	.797	16	.812
1	1.000	8	.875	12	.922	20	.953	UNC	—	UNF	—	16	.938
1 1/8	1.125	7	.984	12	1.047	18	1.078	8	1.000	UNF	—	16	1.062
1 1/4	1.250	7	1.109	12	1.172	18	1.188	8	1.125	UNF	—	16	1.188
1 3/8	1.375	6	1.219	12	1.297	18	1.312	8	1.250	UNF	—	16	1.312
1 1/2	1.500	6	1.344	12	1.422	18	1.438	8	1.375	UNF	—	16	1.438
1 5/8	1.625	—	—	—	—	18	—	8	1.500	12	1.547	16	1.562
1 3/4	1.750	5	1.562	—	—	—	—	8	1.625	12	1.672	16	1.688
1 7/8	1.875	—	—	—	—	—	—	8	1.750	12	1.797	16	1.812
2	2.000	4.5	1.781	—	—	—	—	8	1.875	12	1.922	16	1.938
2 1/4	2.250	4.5	2.031	—	—	—	—	8	2.125	12	2.172	16	2.188
2 1/2	2.500	4	2.250	—	—	—	—	8	2.375	12	2.422	16	2.438
2 3/4	2.750	4	2.500	—	—	—	—	8	2.625	12	2.672	16	2.688
3	3.000	4	2.750	—	—	—	—	8	2.875	12	2.922	16	2.938
3 1/4	3.250	4	3.000	—	—	—	—	8	3.125	12	3.172	16	3.188
3 1/2	3.500	4	3.250	—	—	—	—	8	3.375	12	3.422	16	3.438
3 3/4	3.750	4	3.500	—	—	—	—	8	3.625	12	3.668	16	3.688
4	4.000	4	3.750	—	—	—	—	8	3.875	12	3.922	16	3.938

THREADS PER INCH AND TAP DRILL SIZES

Note: The tap diameter sizes shown are nominal. The class and length of thread will govern the limits on the tapped hole size.

TABLE 8   Inch screw threads.

SERIES WITH GRADED PITCHES — Coarse, Fine
SERIES WITH CONSTANT PITCHES — 4, 3, 2, 1.5, 1.25, 1, 0.75, 0.5, 0.35

Nominal Size Dia (mm) Preferred	Coarse Thread Pitch	Coarse Tap Drill Size	Fine Thread Pitch	Fine Tap Drill Size	4 Thread Pitch	4 Tap Drill Size	3 Thread Pitch	3 Tap Drill Size	2 Thread Pitch	2 Tap Drill Size	1.5 Thread Pitch	1.5 Tap Drill Size	1.25 Thread Pitch	1.25 Tap Drill Size	1 Thread Pitch	1 Tap Drill Size	0.75 Thread Pitch	0.75 Tap Drill Size	0.5 Thread Pitch	0.5 Tap Drill Size	0.35 Thread Pitch	0.35 Tap Drill Size
1.6	0.35	1.25																				
1.8	0.35	1.45																				
2	0.4	1.6																				
2.2	0.45	1.75																				
2.5	0.45	2.05																			0.35	2.15
3	0.5	2.5																			0.35	2.65
3.5	0.6	2.9																			0.35	3.15
4	0.7	3.3																	0.5	3.5		
4.5	0.75	3.7																	0.5	4.0		
5	0.8	4.2																	0.5	4.5		
6	1	5.0															0.75	5.2				
8	1.25	6.7	1	7.0											1	7.0	0.75	7.2				
10	1.5	8.5	1.25	8.8									1.25	8.7	1	9.0	0.75	9.2				
12	1.75	10.2	1.25	10.8							1.5	10.5	1.25	10.7	1	11						
14	2	12	1.5	12.5							1.5	12.5	1.25	12.7	1	13						
16	2	14	1.5	14.5							1.5	14.5			1	15						
18	2.5	15.5	1.5	16.5					2	16	1.5	16.5			1	17						
20	2.5	17.5	1.5	18.5					2	18	1.5	18.5			1	19						
22	2.5	19.5	1.5	20.5					2	20	1.5	20.5			1	21						
24	3	21	2	22					2	22	1.5	22.5			1	23						
27	3	24	2	25					2	25	1.5	25.5			1	26						
30	3.5	26.5	2	28					2	28	1.5	28.5			1	29						
33	3.5	29.5	2	31					2	31	1.5	31.5										
36	4	32	3	33					2	34	1.5	34.5										
39	4	35	3	36					2	37	1.5	37.5										
42	4.5	37.5	3	39	4	38	3	39	2	40	1.5	40.5										
45	4.5	39	3	42	4	41	3	42	2	43	1.5	43.5										
48	5	43	3	45	4	44	3	45	2	46	1.5	46.5										

TABLE 9 Metric screw threads.

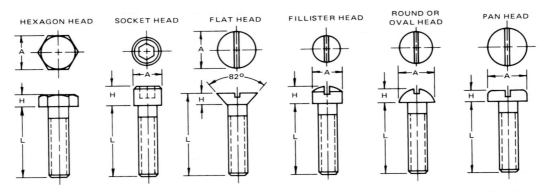

U.S. CUSTOMARY (INCHES)											METRIC (MILLIMETERS)										
Nominal Size	Hexagon Head		Socket Head		Flat Head		Fillister Head		Round or Oval Head		Nominal Size	Hexagon Head		Socket Head		Flat Head		Fillister Head		Pan Head	
	A	H	A	H	A	H	A	H	A	H		A	H	A	H	A	H	A	H	A	H
.250	.44	.17	.38	.25	.50	.14	.38	.24	.44	.19	M3	5.5	2	5.5	3	5.6	1.6	6	2.4	6	1.9
.312	.50	.22	.47	.31	.62	.18	.44	.30	.56	.25	4	7	2.8	7	4	7.5	2.2	8	3.1	8	2.5
.375	.56	.25	.56	.38	.75	.21	.56	.36	.62	.27	5	8.5	3.5	9	5	9.2	2.5	10	3.8	10	3.1
.438	.62	.30	.66	.44	.81	.21	.62	.37	.75	.33	6	10	4	10	6	11	3	12	4.6	12	3.8
.500	.75	.34	.75	.50	.88	.21	.75	.41	.81	.35	8	13	5.5	13	8	14.5	4	16	6	16	5
.625	.94	.42	.94	.62	1.12	.28	.88	.52	1.00	.44	10	17	7	16	10	18	5	20	7.5	20	6.2
.750	1.12	.50	1.12	.75	1.38	.35	1.00	.61	1.25	.55	12	19	8	18	12	23	6.4				
											16	24	10.5	24	16	29	8				
											20	30	13.1			35	9				

TABLE 10   Common machine and cap screws.

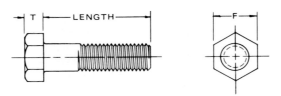

U.S. CUSTOMARY (INCHES)			METRIC (MILLIMETERS)		
Nominal Bolt Size	Width Across Flats F	Thickness T	Nominal Bolt Size and Thread Pitch	Width Across Flats F	Thickness T
.250	.438	.172	M5 x 0.8	8	3.9
.312	.500	.219	M6 x 1	10	4.7
.375	.562	.250	M8 x 1.25	13	5.7
.438	.625	.297			
.500	.750	.344	M10 x 1.5	15	6.8
.625	.938	.422	M12 x 1.75	18	8
.750	1.125	.500	M14 x 2	21	9.3
.875	1.312	.578	M16 x 2	24	10.5
1.000	1.500	.672	M20 x 2.5	30	13.1
1.125	1.688	.750	M24 x 3	36	15.6
1.250	1.875	.844	M30 x 3.5	46	19.5
1.375	2.062	.906	M36 x 4	55	23.4
1.500	2.250	1.000			

TABLE 11   Hexagon-head bolts and cap screws.

TABLE 12 Twelve-spline flange screws.

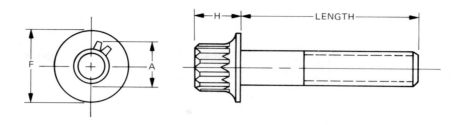

NOMINAL BOLT SIZE AND THREAD PITCH	HEAD SIZES		
	F	A	H
M5 x 0.8	9.4	5.9	5
M6 x 1	11.8	7.4	6.3
M8 x 1.25	15	9.4	8
M10 x 1.5	18.6	11.7	10
M12 x 1.75	22.8	14	12
M14 x 2	26.4	16.3	14
M16 x 2	30.3	18.7	16
M20 x 2.5	37.4	23.4	20

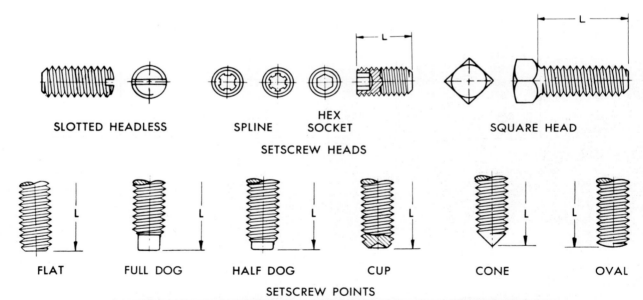

SLOTTED HEADLESS   SPLINE   HEX SOCKET   SQUARE HEAD

SETSCREW HEADS

FLAT   FULL DOG   HALF DOG   CUP   CONE   OVAL

SETSCREW POINTS

U.S. CUSTOMARY (INCHES)		METRIC (MILLIMETERS)	
Nominal Size	Key Size	Nominal Size	Key Size
.125	.06	M1.4	0.7
.138	.06	2	0.9
.164	.08	3	1.5
.190	.09	4	2
.250	.12	5	2
.312	.16	6	3
.375	.19	8	4
.500	.25	10	5
.625	.31	12	6
.750	.38	16	8

TABLE 13 Setscrews.

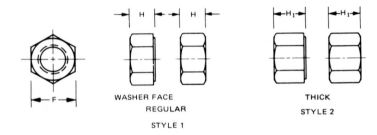

U.S. CUSTOMARY (INCHES)				METRIC (MILLIMETERS)			
	Distance Across Flats F	Thickness Max.			Distance Across Flats F	Thickness Max.	
Nominal Nut Size		Style 1 H	Style 2 H₁	Nominal Nut Size and Thread Pitch		Style 1 H	Style 2 H₁
.250	.438	.218	.281	M4 x 0.7	7	—	3.2
.312	.500	.266	.328	M5 x 0.8	8	4.5	5.3
.375	.562	.328	.406	M6 x 1	10	5.6	6.5
.438	.625	.375	.453	M8 x 1.25	13	6.6	7.8
.500	.750	.438	.562	M10 x 1.5	15	9	10.7
.562	.875	.484	.609	M12 x 1.75	18	10.7	12.8
.625	.938	.547	.719	M14 x 2	21	12.5	14.9
.750	1.125	.641	.812	M16 x 2	24	14.5	17.4
.875	1.312	.750	.906	M20 x 2.5	30	18.4	21.2
1.000	1.500	.859	1.000	M24 x 3	36	22	25.4
1.125	1.688	.969	1.156	M30 x 3.5	46	26.7	31
1.250	1.875	1.062	1.250	M36 x 4	55	32	37.6
1.375	2.062	1.172	1.375				
1.500	2.250	1.281	1.500				

TABLE 14   Hexagon-head nuts.

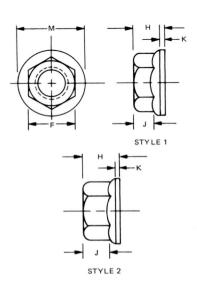

METRIC (MILLIMETERS)							
Nominal Nut Size and Thread Pitch	Width Across Flats F	Style 1				Style 2	
		H	J	K	M	H	J
M6 x 1	10	5.8	3	1	14.2	6.7	3.7
M8 x 1.25	13	6.8	3.7	1.3	17.6	8	4.5
M10 x 1.5	15	9.6	5.5	1.5	21.5	11.2	6.7
M12 x 1.75	18	11.6	6.7	2	25.6	13.5	8.2
M14 x 2	21	13.4	7.8	2.3	29.6	15.7	9.6
M16 x 2	24	15.9	9.5	2.5	34.2	18.4	11.7
M20 x 2.5	30	19.2	11.1	2.8	42.3	22	12.6

TABLE 15   Hex flange nuts.

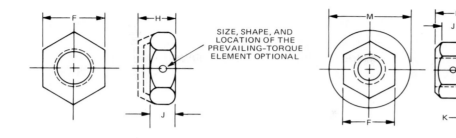

HEX NUTS

HEX FLANGE NUTS

NOMINAL NUT SIZE AND THREAD PITCH	WIDTH ACROSS FLATS F	HEX NUTS				HEX FLANGE NUTS					
		Style 1		Style 2		Style 1				Style 2	
		H max.	J max.	H max.	J max.	H	J	K	M	H	J
M5 × 0.8	8.0	6.1	2.3	7.6	2.9						
M6 × 1	10	7.6	3	8.8	3.7	7.6	3	1	14.2	8.8	3.7
M8 × 1.25	13	9.1	3.7	10.3	4.5	9.1	3.7	1.3	17.6	10.3	4.5
M10 × 1.5	15	12	5.5	14	6.7	12	5.5	1.5	21.5	14	6.7
M12 × 1.75	18	14.2	6.7	16.8	8.2	14.4	6.7	2	25.6	16.8	8.2
M14 × 2	21	16.5	7.8	18.9	9.6	16.6	7.8	2.3	29.6	18.9	9.6
M16 × 2	24	18.5	9.5	21.4	11.7	18.9	9.5	2.5	34.2	21.4	11.7
M20 × 2.5	30	23.4	11.1	26.5	12.6	23.4	11.1	2.8	42.3	26.5	12.6
M24 × 3	36	28	13.3	31.4	15.1						
M30 × 3.5	46	33.7	16.4	38	18.5						
M36 × 4	55	40	20.1	45.6	22.8						

TABLE 16   Prevailing-torque insert-type nuts.

U.S. CUSTOMARY (INCHES)											METRIC (MILLIMETERS)											
NOMINAL SIZE		SLOTTED FLAT COUNTERSUNK HEAD		SLOTTED OVAL COUNTERSUNK HEAD		PAN HEAD		HEX HEAD		HEX WASHER HEAD			NOMINAL SIZE	SLOTTED FLAT COUNTERSUNK HEAD		SLOTTED OVAL COUNTERSUNK HEAD		PAN HEAD			HEX HEAD		HEX WASHER HEAD		
No.	DIA.	A	H	A	H	A	H	A	H	A	B	H		A	H	A	H	A	H	Slot Recess H	A	H	A	B	H
2	.086	.17	.05	.17	.05	.17	.05	.12	.05	.12	.17	.05	2	3.6	1.2	3.6	1.2	3.9	1.4	1.6	3.2	1.3	3.2	4.2	1.3
4	.112	.23	.07	.23	.07	.22	.07	.19	.06	.19	.24	.06	2.5	4.6	1.5	4.6	1.5	4.9	1.7	2	4	1.4	4	5.3	1.4
6	.138	.28	.08	.28	.08	.27	.08	.25	.09	.25	.33	.09	3	5.5	1.8	5.5	1.8	5.8	1.9	1.3	5	1.5	5	6.2	1.5
8	.164	.33	.10	.33	.10	.32	.10	.25	.11	.25	.35	.11	3.5	6.5	2.1	6.5	2.1	6.8	2.3	2.5	5.5	2.4	5.5	7.5	2.4
10	.190	.39	.17	.39	.17	.37	.11	.31	.12	.31	.41	.12	4	7.5	2.3	7.5	2.3	7.8	2.6	2.8	7	2.8	7	9.2	2.8
													5	9.5	2.9	9.5	2.9	9.8	3.1	3.5	8	3	8	10.5	3
													6	11.9	3.6	11.9	3.6	12	3.9	4.3	10	4.8	10	13.2	4.8
													8	15.2	4.4	15.2	4.4	15.6	5	5.6	13	5.8	13	17.2	5.8
													10	1.9	5.4	19	5.4	19.5	6.2	7	15	7.5	15	19.8	7.5
													12	22.9	6.4	22.9	6.4	23.4	7.5	8.3	18	9.5	18	23.8	9.5

TABLE 17   Tapping screws.

KIND OF MATERIAL	THREAD-FORMING								THREAD-CUTTING			SELF-DRILLING	
	Type A	Type B	Type AB	HEX HEAD B	SWAGE FORM*	SWAGE FORM* B	Type U	Type 21	Type F*	Type L	Type B-F*	DRIL-KWICK	TAPITS*
SHEET METAL .0156 to .0469in. thick (0.4 to 1.2mm) (Steel, Brass, Aluminum, Monel, etc.)	✓	✓	✓	✓	✓	✓		✓				✓	✓
SHEET STAINLESS STEEL .0156 to .0469in thick (0.4 to 1.2mm)	✓	✓	✓	✓	✓	✓		✓	✓			✓	✓
SHEET METAL .20 to .50in. thick (1.2 to 5mm) (Steel, Brass, Aluminum, etc.)		✓	✓	✓	✓	✓	✓	✓	✓			✓	
STRUCTURAL STEEL .20 to .50in. thick (1.2 to .5mm)				✓	✓		✓		✓				
CASTINGS (Aluminum, Magnesium, Zinc, Brass, Bronze, etc.)		✓	✓	✓	✓	✓	✓		✓				
CASTINGS (Gray Iron, Malleable Iron, Steel, etc.)					✓		✓		✓				
FORGINGS (Steel, Brass, Bronze, etc.)					✓		✓		✓				
PLYWOOD, Resin Impregnated: Compreg, Pregwood, etc. NATURAL WOODS	✓	✓	✓	✓			✓		✓		✓	✓	✓
ASBESTOS and other compositions: Ebony, Asbestos, Transite, Fiberglas, Insurok, etc.	✓	✓	✓	✓		✓						✓	✓
PHENOL FORMALDEHYDE: Molded: Bakelite, Durez, etc. Cast: Catalin, Marblette, etc. Laminated: Formica, Textolite, etc.		✓	✓	✓		✓	✓		✓		✓		
UREA FORMALDEHYDE: Molded: Plaskon, Beetle, etc. MELAMINE FORMALDEHYDE: Melantite, Melamac						✓	✓				✓		
CELLULOSE ACETATES and NITRATES: Tenite, Lumarith, Plastacele Pyralin, Celanese, etc. ACRYLATE & STYRENE RESINS: Lucite, Plexiglas, Styron, etc.		✓	✓	✓	✓	✓	✓				✓		
NYLON PLASTICS: Nylon, Zytel					✓	✓	✓			✓			

TABLE 18   Selector guide to thread-cutting screws.

FLAT WASHER   LOCKWASHER   SPRING LOCKWASHER

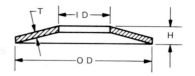

TABLE 19 — U.S. CUSTOMARY (INCHES)

Bolt Size	Flat Washers Type A–N			Lockwashers Regular		
	ID	OD	Thick	ID	OD	Thick
#6	.156	.375	.049	.141	.250	.031
#8	.188	.438	.049	.168	.293	.040
#10	.219	.500	.049	.194	.334	.047
#12	.250	.562	.065	.221	.377	.056
.250	.281	.625	.065	.255	.489	.062
.312	.344	.688	.065	.318	.586	.078
.375	.406	.812	.065	.382	.683	.094
.438	.469	.922	.065	.446	.779	.109
.500	.531	1.062	.095	.509	.873	.125
.562	.594	1.156	.095	.572	.971	.141
.625	.656	1.312	.095	.636	1.079	.156
.750	.812	1.469	.134	.766	1.271	.188
.875	.938	1.750	.134	.890	1.464	.219
1.000	1.062	2.000	.134	1.017	1.661	.250
1.125	1.250	2.250	.134	1.144	1.853	.281
1.250	1.375	2.500	.165	1.271	2.045	.312
1.375	1.500	2.750	.165	1.398	2.239	.344
1.500	1.625	3.000	.165	1.525	2.430	.375

TABLE 19 — METRIC (MILLIMETERS)

Bolt Size	Flat Washers			Lockwashers			Spring Lockwashers		
	ID	OD	Thick	ID	OD	Thick	ID	OD	Thick
2	2.2	5.5	0.5	2.1	3.3	0.5			
3	3.2	7	0.5	3.1	5.7	0.8			
4	4.3	9	0.8	4.1	7.1	0.9	4.2	8	0.3 / 0.4
5	5.3	11	1	5.1	8.7	1.2	5.2	10	0.4 / 0.5 / 0.5
6	6.4	12	1.5	6.1	11.1	1.6	6.2	12.5	0.5 / 0.7
7	7.4	14	1.5	7.1	12.1	1.6	7.2	14	0.5 / 0.8
8	8.4	17	2	8.2	14.2	2	8.2	16	0.6 / 0.9
10	10.5	21	2.5	10.2	17.2	2.2	10.2	20	0.8 / 1.1
12	13	24	2.5	12.3	20.2	2.5	12.2	25	0.8 / 0.9 / 1.5
14	15	28	2.5	14.2	23.2	3	14.2	28	0.9 / 1.5
16	17	30	3	16.2	26.2	3.5	16.3	31.5	1.0 / 1.5 / 1.2
18	19	34	3	18.2	28.2	3.5	18.3	35.5	1.7 / 1.2
20	21	36	3	20.2	32.2	4	20.4	40	2.0 / 1.5 / 2.25 / 1.75
22	23	39	4	22.5	34.5	4	22.4	45	2.5
24	25	44	4	24.5	38.5	5			
27	28	50	4	27.5	41.5	5			
30	31	56	4	30.5	46.5	6			

TABLE 19   Common washer sizes.

TABLE 20 — Belleville washers

U.S. CUSTOMARY (INCHES)

Outside Diameter Max.	Inside Diameter Min.	Stock Thickness T	H Approx.
.187	.093	.010	.015
.250	.125	.009	.017
		.013	.020
.281	.138	.010	.020
		.015	.023
.312	.156	.011	.022
		.017	.025
.343	.164	.013	.024
		.019	.028
.375	.190	.015	.027
		.020	.030
.500	.255	.018	.034
		.025	.038
.625	.317	.022	.042
		.032	.048
.750	.380	.028	.051
		.040	.059
.875	.442	.031	.059
		.045	.067
1.000	.505	.035	.067
		.050	.075
1.125	.567	.038	.073
		.056	.084
1.250	.630	.040	.082
		.062	.092
1.375	.692	.044	.088
		.067	.101

METRIC (MILLIMETERS)

Outside Diameter Max.	Inside Diameter Min.	Stock Thickness T	H Approx.
4.8	2.4	0.16	0.33
		0.25	0.38
6.4	3.2	0.22	0.44
		0.34	0.51
7.9	4	0.27	0.55
		0.42	0.64
9.5	4.8	0.38	0.69
		0.51	0.76
12.7	6.5	0.46	0.86
		0.64	0.97
		0.97	1.20
15.9	8.1	0.56	1.07
		0.81	1.22
19.1	9.7	0.71	1.3
		1.02	1.5
		1.42	1.8
22.2	11.2	0.79	1.5
		1.14	1.7
25.4	12.8	0.89	1.7
		1.27	1.9
		1.85	2.3
28.6	14.4	0.97	1.9
		1.42	2.1
31.8	16	1.02	2.1
		1.58	2.3
34.9	17.6	1.12	2.2
		1.70	2.6
38.1	19.2	1.14	2.4
		1.83	2.7
44.5	22.4	1.45	2.9
		2.16	3.3
50.8	25.4	1.65	3.3
		2.46	3.7
63.5	31.8	2.03	4.1
		3.05	4.6

TABLE 20   Belleville washers. (*Wallace Barnes Co. Ltd.*)

U.S. CUSTOMARY (INCHES)						METRIC (MILLIMETERS)					
Diameter of Shaft		Square Key Nominal Size		Flat Key Nominal Size		Diameter of Shaft		Square Key Nominal Size		Flat Key Nominal Size	
From	To	W	H	W	H	Over	Up To	W	H	W	H
.500	.562	.125	.125	.125	.094	6	8	2	2		
.625	.875	.188	.188	.188	.125	8	10	3	3		
.938	1.250	.250	.250	.250	.188	10	12	4	4		
1.312	1.375	.312	.312	.312	.250	12	17	5	5		
1.438	1.750	.375	.375	.375	.250	17	22	6	6		
1.812	2.250	.500	.500	.500	.375	22	30	7	7	8	7
2.375	2.750	.625	.625			30	38	8	8	10	8
2.875	3.250	.750	.750			38	44	9	9	12	8
3.375	3.750	.875	.875			44	50	10	10	14	9
3.875	4.500	1.000	1.000			50	58	12	12	16	10

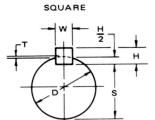
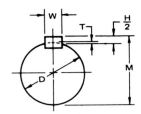

SQUARE   FLAT

C = ALLOWANCE FOR PARALLEL KEYS = .005 in. OR 0.12 mm

$$S = D - \frac{H}{2} - T = \frac{D - H + \sqrt{D^2 - W^2}}{2} \qquad T = \frac{D - \sqrt{D^2 - W^2}}{2}$$

$$M = D - T + \frac{H}{2} + C = \frac{D + H + \sqrt{D^2 - W^2}}{2} + C$$

W = NOMINAL KEY WIDTH (INCHES OR MILLIMETERS)

TABLE 21   Square and flat stock keys.

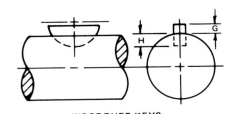

WOODRUFF KEYS

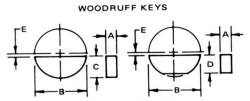

U.S. CUSTOMARY (INCHES)							METRIC (MILLIMETERS)					
Nominal Size	Key			Keyseats		Key No.	Nominal Size	Key			Keyseats	
A × B	E	C	D	H	G		A × B	E	C	D	H	G
.062 × .500	.047	.203	.194	.167	.037	204	1.6 × 12.7	1.5	5.1	4.8	4.24	0.94
.094 × .500	.047	.203	.194	.151	.053	304	2.4 × 12.7	1.3	5.1	4.8	3.84	1.35
.094 × .625	.062	.250	.240	.198	.053	305	2.4 × 15.9	1.5	6.4	6.1	5.03	1.35
.125 × .500	.049	.203	.194	.136	.069	404	3.2 × 12.7	1.3	5.1	4.8	3.45	1.75
.125 × .625	.062	.250	.240	.183	.069	405	3.2 × 15.9	1.5	6.4	6.1	4.65	1.75
.125 × .750	.062	.313	.303	.246	.069	406	3.2 × 19.1	1.5	7.9	7.6	6.25	1.75
.156 × .625	.062	.250	.240	.170	.084	505	4.0 × 15.9	1.5	6.4	6.1	4.32	2.13
.156 × .750	.062	.313	.303	.230	.084	506	4.0 × 19.1	1.5	7.9	7.6	5.84	2.13
.156 × .875	.062	.375	.365	.292	.084	507	4.0 × 22.2	1.5	9.7	9.1	7.42	2.13
.188 × .750	.062	.313	.303	.214	.100	606	4.8 × 19.1	1.5	7.9	7.6	5.44	2.54
.188 × .875	.062	.375	.365	.276	.100	607	4.8 × 22.2	1.5	9.7	9.1	7.01	2.54
.188 × 1.000	.062	.438	.428	.339	.100	608	4.8 × 25.4	1.5	11.2	10.9	8.61	2.54
.188 × 1.125	.078	.484	.475	.385	.100	609	4.8 × 28.6	2.0	12.2	11.9	9.78	2.54
.250 × .875	.062	.375	.365	.245	.131	807	6.4 × 22.2	1.5	9.7	9.1	6.22	3.33
.250 × 1.000	.062	.438	.428	.308	.131	808	6.4 × 25.4	1.5	11.2	10.9	7.82	3.33

NOTE: METRIC KEY SIZES WERE NOT AVAILABLE AT THE TIME OF PUBLICATION. SIZES SHOWN ARE INCH-DESIGNED KEY-SIZES SOFT CONVERTED TO MILLIMETERS. CONVERSION WAS NECESSARY TO ALLOW THE STUDENT TO COMPARE KEYS WITH SLOT SIZES GIVEN IN MILLIMETERS.

TABLE 22   Woodruff keys.

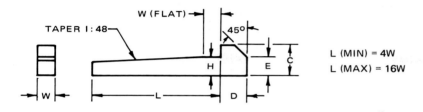

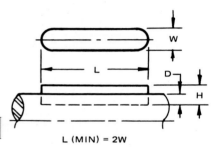

L (MIN) = 2W

U.S. CUSTOMARY (INCHES)										
Shaft Diameter	**Square Type**					**Flat Type**				
	W	**H**	**C**	**D**	**E**	**W**	**H**	**C**	**D**	**E**
.500–.562	.125	.125	.250	.219	.156	.125	.094	.188	.125	.125
.625–.875	.188	.188	.312	.281	.219	.188	.125	.250	.188	.156
.938–1.250	.250	.250	.438	.344	.344	.250	.188	.312	.250	.188
1.312–1.375	.312	.312	.562	.406	.406	.312	.250	.375	.312	.250
1.438–1.750	.375	.375	.688	.469	.469	.375	.250	.438	.375	.312
1.812–2.250	.500	.500	.875	.594	.625	.500	.375	.625	.500	.438
2.312–2.750	.625	.625	1.062	.719	.750	.625	.438	.750	.625	.500
2.875–3.250	.750	.750	1.250	.875	.875	.750	.500	.875	.750	.625

METRIC (MILLIMETERS)										
Shaft Diameter	**Square Type**					**Flat Type**				
	W	**H**	**C**	**D**	**E**	**W**	**H**	**C**	**D**	**E**
12–14	3.2	3.2	6.4	5.4	4	3.2	2.4	5	3.2	3.2
16–22	4.8	4.8	10	7	5.4	4.8	3.2	6.4	5	4
24–32	6.4	6.4	11	8.6	8.6	6.4	5	8	6.4	5
34–35	8	8	14	10	10	8	6.4	10	8	6.4
36–44	10	10	18	12	12	10	6.4	11	10	8
46–58	13	13	22	15	16	13	10	16	13	11
60–70	16	16	27	19	20	16	11	20	16	13
72–82	20	20	32	22	22	20	13	22	20	16

Note: Metric standards governing key sizes were not available at the time of publication. The sizes given in the above chart are "soft conversion" from current standards and are not representative of the precise metric key sizes which may be available in the future. Metric sizes are given only to allow the student to complete the drawing assignment.

TABLE 23   Square and flat gib-head keys.

U.S. CUSTOMARY (INCHES)			
Key No.	**L**	**W**	**H**	**D**
2	.500	.094	.141	.094
4	.625	.094	.141	.094
6	.625	.156	.234	.156
8	.750	.156	.234	.156
10	.875	.156	.234	.156
12	.875	.234	.328	.219
14	1.00	.234	.328	.234
16	1.125	.188	.281	.188
18	1.125	.250	.375	.250
20	1.250	.219	.328	.219
22	1.375	.250	.375	.250
24	1.50	.250	.375	.250
26	2.00	.188	.281	.188
28	2.00	.312	.469	.312
30	3.00	.375	.562	.375
32	3.00	.500	.750	.500
34	3.00	.625	.938	.625

METRIC (MILLIMETERS)			
Key No.	**L**	**W**	**H**	**D**
2	12	2.4	3.6	2.4
4	16	2.4	3.6	2.4
6	16	4	6	4
8	20	4	6	4
10	22	4	6	4
12	22	6	8.4	7
14	25	6	8.4	6
16	28	5	7	5
18	28	6.4	10	6.4
20	32	7	8	5
22	35	6.4	10	6.4
24	38	6.4	10	6.4
26	50	5	7	5
28	50	8	12	8
30	75	10	14	10
32	75	12	20	12
34	75	16	24	16

TABLE 24   Pratt and Whitney keys.

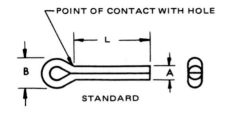

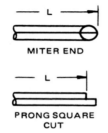

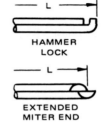

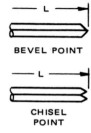

U.S. CUSTOMARY (INCHES)				METRIC (MILLIMETERS)			
Nominal Bolt or Thread Size Range	Nominal Cotter-Pin Size (A)	Cotter-Pin Hole	Min. End Clearance*	Nominal Bolt or Thread-Size Range	Nominal Cotter-Pin Size (A)	Cotter-Pin Hole	Min. End Clearance*
.125	.031	.047	.06	−2.5	0.6	0.8	1.5
.188	.047	.062	.08	2.5−3.5	0.8	1.0	2.0
.250	.062	.078	.11	3.5−4.5	1.0	1.2	2.0
.312	.078	.094	.11	4.5−5.5	1.2	1.4	2.5
.375	.094	.109	.14	5.5−7.0	1.6	1.8	2.5
.438	.109	.125	.14	7.0−9.0	2.0	2.2	3.0
.500	.125	.141	.18	9.0−11	2.5	2.8	3.5
.562	.141	.156	.25	11−14	3.2	3.6	5
.625	.156	.172	.40	14−20	4	4.5	6
1.000−1.125	.188	.203	.40	20−27	5	5.6	7
1.250−1.375	.219	.234	.46	27−39	6.3	6.7	10
1.500−1.625	.250	.266	.46	39−56	8.0	8.5	15
				56−80	10	10.5	20

*End of bolt to center of hole

TABLE 25   Cotter pins.

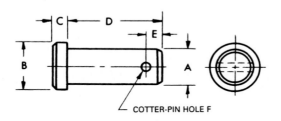

COTTER-PIN HOLE F

U.S. CUSTOMARY (INCHES)						METRIC (MILLIMETERS)					
Pin Dia. A	B	C	Min. D	E	Drill Size F	Pin Dia. A	B	C	Min. D	E	Drill Size F
.188	.31	.06	.59	.11	.078	4	6	1	16	2.2	1
.250	.38	.09	.80	.12	.078	6	10	2	20	3.2	1.6
.312	.44	.09	.97	.16	.109	8	14	3	24	3.5	2
.375	.50	.12	1.09	.16	.109	10	18	4	28	4.5	3.2
.500	.62	.16	1.42	.22	.141	12	20	4	36	5.5	3.2
.625	.81	.20	1.72	.25	.141	16	25	4.5	44	6	4
.750	.94	.25	2.05	.30	.172	20	30	5	52	8	5
1.000	1.19	.34	2.62	.36	.172	24	36	6	66	9	6.3

TABLE 26   Clevis pins.

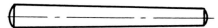

NUMBER	7/0	6/0	5/0	4/0	3/0	2/0	0	1	2	3	4	5	6	7	8	9
U.S. CUSTOMARY (INCHES)																
SIZE (LARGE END)	.062	.078	.094	.109	.125	.141	.152	.172	.193	.219	.250	.289	.314	.409	.492	.591
LENGTH																
.375	x	x														
.500	x	x	x	x	x	x	x									
.625	x	x	x	x	x	x	x									
.750		x	x	x	x	x	x	x	x	x						
.875					x	x	x	x	x	x						
1.000			x	x	x	x	x	x	x	x	x	x				
1.250						x	x	x	x	x	x	x	x			
1.500							x	x	x	x	x	x	x			
1.750								x	x	x	x	x	x			
2.000								x	x	x	x	x	x	x	x	
2.250									x	x	x	x	x	x	x	
2.500									x	x	x	x	x	x	x	
2.750										x	x	x	x	x	x	x
METRIC (MILLIMETERS)																
SIZE (LARGE END)	1.6	2	2.4	2.8	3.2	3.6	4	4.4	4.9	5.6	6.4	7.4	8	10.4	12.5	15
LENGTH																
10	x	x														
12	x	x	x	x	x	x	x									
16	x	x	x	x	x	x	x									
20		x	x	x	x	x	x	x	x	x						
22					x	x	x	x	x	x						
25			x	x	x	x	x	x	x	x	x	x				
30						x	x	x	x	x	x	x	x			
40							x	x	x	x	x	x	x			
45								x	x	x	x	x	x			
50								x	x	x	x	x	x	x	x	
55									x	x	x	x	x	x	x	
65									x	x	x	x	x	x	x	
70										x	x	x	x	x	x	x

TABLE 27   Taper pins.

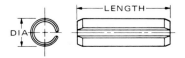

PIN DIAMETER (INCHES)								PIN DIAMETER (MILLIMETERS)										
Length	.062	.094	.125	.156	.188	.250	.312	Length	1.5	2	2.5	3	4	5	6	8	10	12
.250	x	x						5	x	x								
.375	x	x	x					10	x	x	x	x						
.500	x	x	x	x	x			15	x	x	x	x	x	x				
.625	x	x	x	x	x	x		20	x	x	x	x	x	x	x			
.750	x	x	x	x	x	x	x	25	x	x	x	x	x	x	x	x		
.875	x	x	x	x	x	x	x	30		x	x	x	x	x	x	x	x	x
1.00	x	x	x	x	x	x	x	35		x	x	x	x	x	x	x	x	x
1.250		x	x	x	x	x	x	40		x	x	x	x	x	x	x	x	x
1.500		x	x	x	x	x	x	45				x	x	x	x	x	x	x
1.750			x	x	x	x	x	50				x	x	x	x	x	x	x
2.000			x	x	x	x	x	55					x	x	x	x	x	x
2.225				x	x	x	x	60					x	x	x	x	x	x
2.500				x	x	x	x	70							x	x	x	x
3.000						x	x	75							x	x	x	x
3.500						x	x	80								x	x	x

TABLE 28   Spring pins.

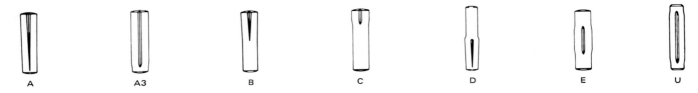

A   A3   B   C   D   E   U

U.S. CUSTOMARY (INCHES) — PIN DIAMETER

Length	.09	.125	.188	.250	.312	.375	.500
.250	x	x					
.375	x	x	x				
.500	x	x	x	x			
.625	x	x	x	x	x		
.750	x	x	x	x	x	x	
.875	x	x	x	x	x	x	
1.000	x	x	x	x	x	x	x
1.250	x	x	x	x	x	x	x
1.500		x	x	x	x	x	x
1.750			x	x	x	x	x
2.000			x	x	x	x	x
2.250			x	x	x	x	x
2.275				x	x	x	x
3.000				x	x	x	x

METRIC (MILLIMETERS) — PIN DIAMETER

Length	2	3	4	5	6	8	10	12
5	x	x	x					
10	x	x	x	x	x			
15	x	x	x	x	x	x		
20	x	x	x	x	x	x	x	
25	x	x	x	x	x	x	x	x
30	x	x	x	x	x	x	x	x
35		x	x	x	x	x	x	x
40			x	x	x	x	x	x
45				x	x	x	x	x
50				x	x	x	x	x
55					x	x	x	x
60					x	x	x	x
65					x	x	x	x
70						x	x	x
75						x	x	x

Note: Metric size pins were not available at the time of publication. Sizes were soft converted to allow students to complete drawing assignments.

TABLE 29   Groove pins.

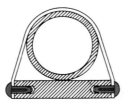

WIDELY USED FOR
FASTENING BRACKETS

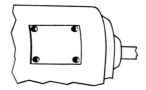

ATTACHING NAMEPLATES,
INSTRUCTION PANELS

U.S. CUSTOMARY (INCHES)

STUD NUMBER	SHANK DIA.	DRILL SIZE	HEAD DIA.	.125	.188	.250	.312	.375	.500
0	.067	51	.130	•	•	•			
2	.086	44	.162	•	•	•			
4	.104	37	.211		•	•	•		
6	.120	31	.260			•	•	•	
7	.136	29	.309				•	•	•
8	.144	27	.309				•	•	
10	.161	20	.359				•	•	
12	.196	9	.408						•
14	.221	2	.457						•
16	.250	¼	.472						•

METRIC (MILLIMETERS)

STUD NUMBER	SHANK DIA.	DRILL SIZE	HEAD DIA.	4	6	8	10	12	14
0	1.7	1.7	3.3	•	•	•			
2	2.2	2.2	4.1	•	•	•			
4	2.6	2.6	5.4		•	•	•		
6	3.0	3.0	6.6			•	•	•	
7	3.4	3.4	7.8				•	•	•
8	3.8	3.8	7.8					•	•
10	4.1	4.1	9.1					•	•
12	5.0	5.0	10.4						•
14	5.6	5.6	11.6						•
16	6.3	6.3	12						•

Note: Metric size studs were not available at the time of publication. Sizes were soft converted to allow students to complete drawing assignments.

TABLE 30   Grooved studs. *(Drive-Lok)*

Use these columns first to locate your correct GRIP LENGTH

Grip = Total thickness of all sheets fastened together

Minimum Grip	Nominal Grip	Maximum Grip
1	2	3
2	3	4
4	5	6
5	6	7
7	8	9
9	10	11
11	12	13
13	14	15
15	16	17
19	20	21
23	24	25
27	28	29

.125in. (3mm) Dia — Part Numbers

Length Under Head L	Universal Head	100° Csk Head	Full Brazier Head
5	✓	✓	✓
6	✓	✓	✓
8	✓	✓	✓
10	✓	✓	✓
12	✓	✓	✓
14	✓	✓	✓
16			
18			
20			
24			
28			
32			

.156in. (4mm) Dia — Part Numbers

Length Under Head L	Universal Head	100° Csk Head
5	✓	
6	✓	✓
8	✓	✓
10	✓	✓
12	✓	✓
14	✓	✓
16	✓	✓
18	✓	✓
20	✓	✓
24		
28		
32		

.188in. (5mm) Dia — Part Numbers

Length Under Head L	Universal Head	100° Csk Head	Full Brazier Head	All Purpose Liner Head
5	✓		✓	
6	✓		✓	
8	✓	✓	✓	✓
10	✓	✓	✓	✓
12	✓	✓	✓	✓
14	✓	✓	✓	✓
16	✓	✓	✓	✓
18	✓		✓	✓
20	✓			✓
24	✓			
28	✓			
32	✓			

.250in. (6mm) Dia — Part Numbers

Length Under Head L	Universal Head	100° Csk Head	Full Brazier Head
5	✓	✓	✓
6	✓	✓	✓
8	✓	✓	✓
10	✓	✓	✓
12	✓	✓	✓
14	✓	✓	✓
16	✓	✓	✓
18	✓	✓	✓
20	✓	✓	✓
24	✓	✓	✓
28	✓	✓	✓
32	✓	✓	✓

IN WOOD

Use "L" Dimension (length under head) instead of grip length. L = M (thickness of metal) + D (hole depth in wood).

IN METAL

Grip Length = Total thickness of sheets to be fastened.

HIT THE PIN

Drive pin flush with rivet head

Expanding prongs clinch sheets tightly, eliminating gaps.

Metal and wood pulled tightly together. Nothing protrudes through wood.

Note: Metric drive rivets were not available at the time of publication. Sizes were soft converted to allow students to complete drawing assignments.

TABLE 31   Aluminum drive rivets. (*Southco Lion Fasteners*)

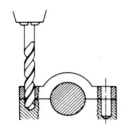
DRILL HOLE SLIGHTLY UNDERSIZE

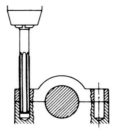
REAM FULL SIZE

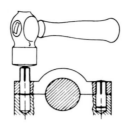
DRIVE OR PRESS LOK DOWELS INTO PLACE

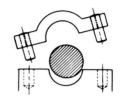
LOK DOWELS LOCK SECURELY AND PARTS SEPARATE EASILY

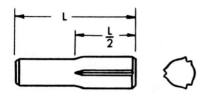

U.S. CUSTOMARY (INCHES)							METRIC (MILLIMETERS)					
	DIAMETER							DIAMETER				
LENGTH	.125	.188	.250	.312	.375	.500	LENGTH	4	6	8	10	12	14
.375	•						10	•					
.500	•	•	•	•			12	•	•	•	•		
.625	•	•	•	•			16	•	•	•	•		
.750	•	•	•	•	•		20	•	•	•	•	•	
.875	•	•	•	•	•		22	•	•	•	•	•	
1.000	•	•	•	•	•	•	26	•	•	•	•	•	•
1.250		•	•	•	•	•	32		•	•	•	•	•
1.500			•	•	•	•	38			•	•	•	•
1.750				•	•	•	44				•	•	•
2.000					•	•	50					•	•

Note: Metric size dowels were not available at the time of publication. Sizes were soft converted to allow students to complete drawing assignments.

TABLE 32 Lok dowels. *(Drive-Lok)*

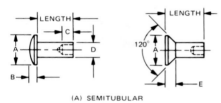

(A) SEMITUBULAR

(B) SPLIT

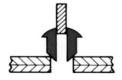

	D	A	B	C	E	MIN. LENGTH
U.S. CUSTOMARY (INCHES)	.062	.125	.031	.062	—	.062
	.094	.156	.031	.062	.031	.078
	.109	.188	.031	.078	—	.078
	.125	.218	.047	.109	.049	.109
	.141	.250	.047	.125	.049	.125
	.188	.312	.062	.141	.062	.156
	.219	.438	.062	.188	.062	.188
	.250	.500	.078	.219	.094	.219
	.312	.562	.109	.250		.250
METRIC (MILLIMETERS)	1.5	2.8	0.4	1.2		1.6
	2.2	3.7	0.6	1.6	0.8	2.0
	2.5	4.7	0.7	2.0		2.0
	3.1	5.5	0.9	2.4	1.0	2.4
	3.6	5.9	1.0	3.2	1.2	3.2
	4.7	7.9	1.5	3.9	1.6	4.0
	5.4	11.1	1.7	4.8	1.8	4.8
	6.3	12.7	2.0	5.6	2.2	5.6
	7.7	14.3	2.4	6.2		6.4

TABLE 33 Semitubular and split rivets.

U.S. CUSTOMARY (INCHES)				METRIC (MILLIMETERS)			
HOLE DIA.	PANEL RANGE	HEAD Dia.	HEAD Height	HOLE DIA.	PANEL RANGE	HEAD Dia.	HEAD Height
.125	.031–.140	.188	.047	3.18	0.8– 3.6	4.8	1.2
	.031–.125	.218	.062		0.8– 3.2	5.5	1.5
.156	.250–.375	.218	.047	4.01	5.9– 9.4	5.5	1.3
.188	.062–.156	.375	.125	4.75	1.6– 4.0	9.5	3.2
	.156–.281	.438	.094		4.0– 7.1	11.1	1.9
.219	.062–.125	.375	.094	5.54	1.6– 3.2	9.5	2.4
	.094–.312		.078		2.4– 8.0		2.0
.250	.094–.219	.625	.125	6.35	2.3– 5.6	16	3.2
	.125–.375	.750	.062		3.2– 9.5	19	1.3
.297	.140–.328	.500	.078	7.14	3.4– 8.1	12.3	1.9
.375	.250–.500	.438	.109	9.53	6.4–12.7	11.1	2.6
.500	.312–.375	.750	.109	12.7	8.1– 9.4	19	2.5

TABLE 34 Plastic rivets.

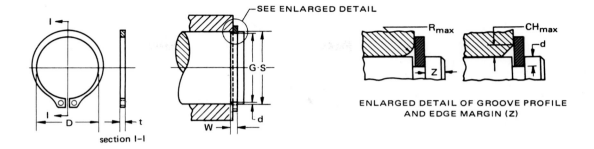

	SHAFT DIA.	EXTERNAL SERIES	RETAINING RING DIMENSIONS		GROOVE DIMENSIONS				MAX. CORNER RADII AND CHAMFER OF RETAINED PARTS		EDGE MARGIN	NOMINAL GROOVE DEPTH (REF)
					DIAMETER		WIDTH					
	S	Size—No.	D	t	G	Tol.	W	Tol.	R Max.	Ch. Max.	z	d
U.S. CUSTOMARY (INCHES)	.188	5100-18	.168	.015	.175	±.0015	.018	+.002	.014	.008	.018	.006
	.250	5100-25	.225	.025	.230	±.0015	.029	+.003	.018	.011	.030	.010
	.312	5100-31	.281	.025	.290	±.002	.029	+.003	.020	.012	.033	.011
	.375	5100-37	.338	.025	.352	±.002	.029	+.003	.026	.015	.036	.012
	.500	5100-50	.461	.035	.468	±.002	.039	+.003	.034	.020	.048	.016
	.625	5100-62	.579	.035	.588	±.003	.039	+.003	.041	.025	.055	.018
	.750	5100-75	.693	.042	.704	±.003	.046	+.003	.046	.027	.069	.023
	.875	5100-87	.810	.042	.821	±.003	.046	+.003	.051	.031	.081	.027
	1.000	5100-100	.925	.042	.940	±.003	.046	+.003	.057	.034	.090	.030
	1.125	5100-112	1.041	.050	1.059	±.004	.056	+.004	.063	.038	.099	.033
	1.250	5100-125	1.156	.050	1.176	±.004	.056	+.004	.068	.041	.111	.037
	1.375	5100-137	1.272	.050	1.291	±.004	.056	+.004	.072	.043	.126	.042
	1.500	5100-150	1.387	.050	1.406	±.004	.056	+.004	.079	.047	.141	.047
METRIC (MILLIMETERS)	4	M5100-4	3.6	0.25	3.80	−0.08	0.32	+0.05	0.35	0.25	0.3	0.10
	6	M5100-6	5.5	0.4	5.70	−0.08	0.5	+0.1	0.35	0.25	0.5	0.15
	8	M5100-8	7.2	0.6	7.50	−0.1	0.7	+0.15	0.5	0.35	0.8	0.25
	10	M5100-10	9.0	0.6	9.40	−0.1	0.7	+0.15	0.7	0.4	0.9	0.30
	12	M5100-12	10.9	0.6	11.35	−0.12	0.7	+0.15	0.8	0.45	1.0	0.33
	14	M5100-14	12.9	0.9	13.25	−0.12	1.0	+0.15	0.9	0.5	1.2	0.38
	16	M5100-16	14.7	0.9	15.10	−0.15	1.0	+0.15	1.1	0.6	1.4	0.45
	18	M5100-18	16.7	1.1	17.00	−0.15	1.2	+0.15	1.2	0.7	1.5	0.50
	20	M5100-20	18.4	1.1	18.85	−0.15	1.2	+0.15	1.2	0.7	1.7	0.58
	22	M5100-22	20.3	1.1	20.70	−0.15	1.2	+0.15	1.3	0.8	1.9	0.65
	24	M5100-24	22.2	1.1	22.60	−0.15	1.2	+0.15	1.4	0.8	2.1	0.70
	25	M5100-25	23.1	1.1	23.50	−0.15	1.2	+0.15	1.4	0.8	2.3	0.75
	30	M5100-30	27.9	1.3	28.35	−0.2	1.4	+0.15	1.6	1.0	2.5	0.83
	35	M5100-35	32.3	1.3	32.9	−0.2	1.4	+0.15	1.8	1.1	3.1	1.05
	40	M5100-40	36.8	1.6	37.7	−0.3	1.75	+0.2	2.1	1.2	3.4	1.15
	45	M5100-45	41.6	1.6	42.4	−0.3	1.75	+0.2	2.3	1.4	3.9	1.3
	50	M5100-50	46.2	1.6	47.2	−0.3	1.75	+0.2	2.4	1.4	4.2	1.4

TABLE 35   Retaining rings—external. (©1965, 1958 Waldes Koh-i-noor, Inc. Reprinted with permission.)

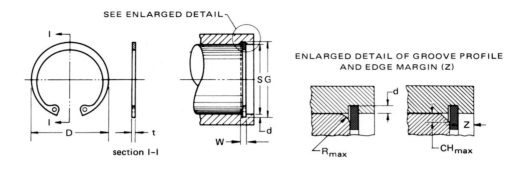

SEE ENLARGED DETAIL

section I–I

ENLARGED DETAIL OF GROOVE PROFILE
AND EDGE MARGIN (Z)

	HOUSING DIA.	INTERNAL SERIES	RETAINING RING DIMENSIONS		GROOVE DIMENSIONS				MAX. CORNER RADII AND CHAMFER OF RETAINED PARTS		EDGE MARGIN	NOMINAL GROOVE DEPTH
					DIAMETER		WIDTH					
	S	Size—No.	D	t	G	Tol.	W	Tol.	R Max.	Ch. Max.	z	d
U.S. CUSTOMARY (INCHES)	.250	N5000-25	.280	.015	.268	±.001	.018	+.002	.011	.008	.027	.009
	.312	N5000-31	.346	.015	.330	±.001	.018	+.002	.016	.013	.027	.009
	.375	N5000-37	.415	.025	.397	±.002	.029	+.003	.023	.018	.033	.011
	.500	N5000-50	.548	.035	.530	±.002	.039	+.003	.027	.021	.045	.015
	.625	N5000-62	.694	.035	.665	±.002	.039	+.003	.027	.021	.060	.020
	.750	N5000-75	.831	.035	.796	±.002	.039	+.003	.032	.025	.069	.023
	.875	N5000-87	.971	.042	.931	±.003	.046	+.003	.035	.028	.084	.028
	1.000	N5000-100	1.111	.042	1.066	±.003	.046	+.003	.042	.034	.099	.033
	1.125	N5000-112	1.249	.050	1.197	±.004	.056	+.004	.047	.036	.108	.036
	1.250	N5000-125	1.388	.050	1.330	±.004	.056	+.004	.048	.038	.120	.040
	1.375	N5000-137	1.526	.050	1.461	±.004	.056	+.004	.048	.038	.129	.043
	1.500	N5000-150	1.660	.050	1.594	±.004	.056	+.004	.048	.038	.141	.047
METRIC (MILLIMETERS)	8	MN5000-8	8.80	0.4	8.40	+0.6	0.5	+0.1	0.4	0.3	0.6	0.2
	10	MN5000-10	11.10	0.6	10.50	+0.1	0.7	+0.15	0.5	0.35	0.8	0.25
	12	MN5000-12	13.30	0.6	12.65	+0.1	0.7	+0.15	0.6	0.4	1.0	0.33
	14	MN5000-14	15.45	0.9	14.80	+0.1	1.0	+0.15	0.7	0.5	1.2	0.40
	16	MN5000-16	17.70	0.9	16.90	+0.1	1.0	+0.15	0.7	0.5	1.4	0.45
	18	MN5000-18	20.05	0.9	19.05	+0.1	1.0	+0.15	0.75	0.6	1.6	0.53
	20	MN5000-20	22.25	0.9	21.15	+0.15	1.0	+0.15	0.9	0.7	1.7	0.57
	22	MN5000-22	24.40	1.1	23.30	+0.15	1.2	+0.15	0.9	0.7	1.9	0.65
	24	MN5000-24	26.55	1.1	25.4	+0.15	1.2	+0.15	1.0	0.8	2.1	0.70
	25	MN5000-25	27.75	1.1	26.6	+0.15	1.2	+0.15	1.0	0.8	2.4	0.80
	30	MN5000-30	33.40	1.3	31.9	+0.2	1.4	+0.15	1.2	1.0	2.9	0.95
	35	MN5000-35	38.75	1.3	37.2	+0.2	1.4	+0.15	1.2	1.0	3.3	1.10
	40	MN5000-40	44.25	1.6	42.4	+0.2	1.75	+0.2	1.7	1.3	3.6	1.20
	45	MN5000-45	49.95	1.6	47.6	+0.2	1.75	+0.2	1.7	1.3	3.9	1.30
	50	MN5000-50	55.35	1.6	53.1	+0.2	1.75	+0.2	1.7	1.3	4.6	1.55

TABLE 36   Retaining rings—internal. (©1965, 1958 Waldes Koh-i-noor, Inc. Reprinted with permission.)

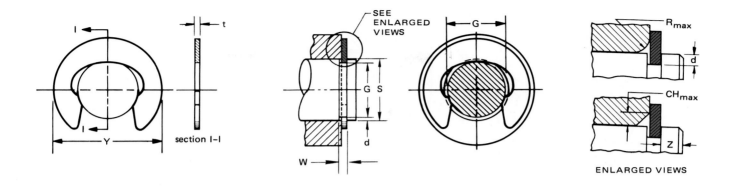

	SHAFT DIA.	EXTERNAL SERIES 11-410	RETAINING RING DIMENSIONS		GROOVE DIMENSIONS				MAXIMUM ALLOWABLE CORNER RADII AND CHAMFER OF RETAINED PARTS		EDGE MARGIN	NOMINAL GROOVE DEPTH (REF)
					Diameter		Width					
	S	Size—No.	Y	t	G	Tol.	W	Tol.	R Max.	Ch. Max.	z	d
U.S. CUSTOMARY (INCHES)	.250	11-410-25	.311	.025	.222	−.004	.029	+.003	.023	.018	.030	.015
	.312	11-410-31	.376	.025	.278	−.004	.029	+.003	.024	.018	.036	.018
	.375	11-410-37	.448	.025	.337	−.004	.029	+.003	.026	.020	.040	.020
	.500	11-410-50	.581	.025	.453	−.006	.039	+.003	.030	.023	.050	.025
	.625	11-410-62	.715	.035	.566	−.006	.039	+.003	.033	.025	.062	.031
	.750	11-410-75	.845	.042	.679	−.006	.046	+.003	.036	.027	.074	.037
	.875	11-410-87	.987	.042	.792	−.006	.046	+.003	.040	.031	.086	.043
	1.000	11-410-100	1.127	.042	.903	−.006	.046	+.003	.046	.035	.100	.050
	1.125	11-410-112	1.267	.050	1.017	−.008	.056	+.004	.052	.040	.112	.056
	1.250	11-410-125	1.410	.050	1.130	−.008	.056	+.004	.057	.044	.124	.062
	1.375	11-410-137	1.550	.050	1.241	−.008	.056	+.004	.062	.048	.138	.069
	1.500	11-410-150	1.691	.050	1.354	−.008	.056	+.004	.069	.053	.150	.075
	1.750	11-410-175	1.975	.062	1.581	−.010	.068	+.004	.081	.062	.174	.087
	2.000	11-410-200	2.257	.062	1.805	−.010	.068	+.004	.091	.070	.200	.100
METRIC (MILLIMETERS)	8	M11-410-080	10	0.6	7	−0.1	0.7	+0.15	0.6	0.45	1.5	0.5
	10	M11-410-100	12.2	0.6	9	−0.1	0.7	+0.15	0.6	0.45	1.5	0.5
	12	M11-410-120	14.4	0.6	10.9	−0.1	0.7	+0.15	0.6	0.45	1.7	0.5
	14	M11-410-140	16.3	1	12.7	−0.1	1.1	+0.15	1	0.8	2	0.65
	16	M11-410-160	18.5	1	14.5	−0.1	1.1	+0.15	1	0.8	2.3	0.75
	18	M11-410-180	20.4	1.2	16.3	−0.1	1.3	+0.15	1.2	0.9	2.6	0.85
	20	M11-410-200	22.6	1.2	18.1	−0.2	1.3	+0.15	1.2	0.9	2.9	0.95
	22	M11-410-220	25	1.2	19.9	−0.2	1.3	+0.15	1.2	0.9	3.2	1.05
	24	M11-410-240	27.1	1.2	21.7	−0.2	1.3	+0.15	1.2	0.9	3.5	1.15
	25	M11-410-250	28.3	1.2	22.6	−0.2	1.3	+0.15	1.2	0.9	3.6	1.2
	30	M11-410-300	33.7	1.5	27	−0.2	1.3	+0.2	1.5	1.15	4.5	1.5
	35	M11-410-350	39.4	1.5	31.5	−0.25	1.6	+0.2	1.5	1.15	5.3	1.75
	40	M11-410-400	45	1.5	36	−0.25	1.6	+0.2	1.5	1.15	6	2
	45	M11-410-450	50.6	1.5	40.5	−0.25	1.6	+0.2	1.5	1.15	6.8	2.25
	50	M11-410-500	56.4	2	45	−0.25	2.2	+0.2	2	1.5	7.5	2.5

TABLE 37   Retaining rings—radial assembly. (©1965, 1958 Waldes Koh-i-noor, Inc. Reprinted with permission.)

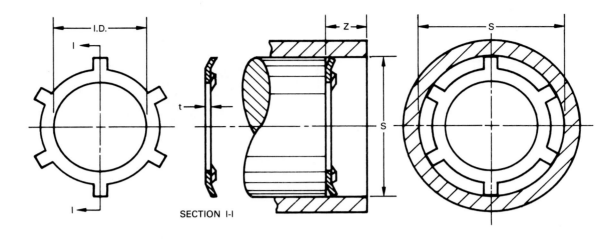

SECTION I-I

SHAFT DIAMETER		INTERVAL SERIES 5005	RING DIMENSIONS			APPLICATION DATA		
							THICKNESS	
Decimal inch			OUTSIDE DIAMETER				.010 ± .001	.015 ± .002	
S						No. of prongs	Allow. thrust load	Allow. thrust load	Z MIN
from	to	SIZE-NO.	OD	tol.				
.311	.313	5005-31	.136		5	80		.040
.374	.376	5005-37	.175	± .005	6	75		.040
.437	.439	5005-43	.237		6	70		.040
.498	.502	5005-50	.258		6	60		.040
.560	.564	5005-56	.312		6	50		.040
.623	.627	5005-62	.390		6	45		.040
.748	.752	5005-75	.500	± .010	8		75	.060
.873	.877	5005-87	.625		8		70	.060
.936	.940	5005-93	.687		10		70	.060
.998	1.002	5005-100	.750		10		75	.060
1.248	1.252	5005-125	.938		10		60	.060
1.371	1.379	Δ5015-137	1.050	± .015	16		150	.125
1.498	1.502	5005-150	1.188		12		60	.060
1.748	1.752	5005-175	1.438	± .010	12		55	.060
1.998	2.002	5005-200	1.600		14		55	.060

TABLE 38   Retaining rings—self-locking, internal. (*©1965, 1958 Waldes Koh-i-noor, Inc. Reprinted with permission.*)

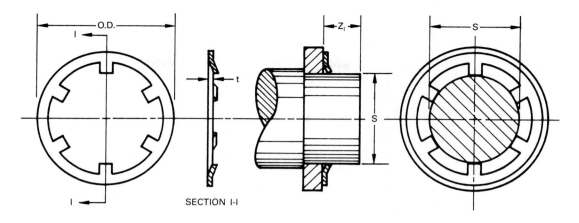

SECTION I-I

SHAFT DIAMETER		EXTERNAL SERIES 5105	RING DIMENSIONS			APPLICATION DATA		
							THICKNESS	
Decimal inch			OUTSIDE DIAMETER				.010	.015
S					No. of prongs	Allow. thrust load	Allow. thrust load	Z MIN
from	to	SIZE-NO.	OD	tol.				
.093	.095	5105-9	.250		3	13		.040
.093	.095	Δ5505D-9	.240		3	10		.040
.124	.126	5105-12	.325		4	20		.040
.155	.157	5105-15	.356		4	25		.040
.187	.189	5105-18	.387		5	35		.040
.218	.220	5105-21	.418	± .005	6	35		.040
.239	.241	5105-24	.460		6		40	.060
.249	.251	5105-25	.450		6	40		.040
.311	.313	5105-31	.512		5	45		.040
.374	.376	5105-37	.575		6	45		.040
.437	.439	5105-43	.638		6		50	.060
.498	.502	5105-50	.750		6		50	.060
.560	.564	5105-56	.812		6		50	.060
.623	.627	5105-62	.875		7		50	.060
.637	.641	Δ5505-63	.875	± .010	8		50	.060
.748	.752	5105-75	1.000		8		55	.060
.873	.877	5105-87	1.125		10		60	.060
.998	1.002	5105-100	1.250		10		65	.060

TABLE 39   Retaining rings—self-locking, external. *(©1965, 1958 Waldes Koh-i-noor, Inc. Reprinted with permission.)*

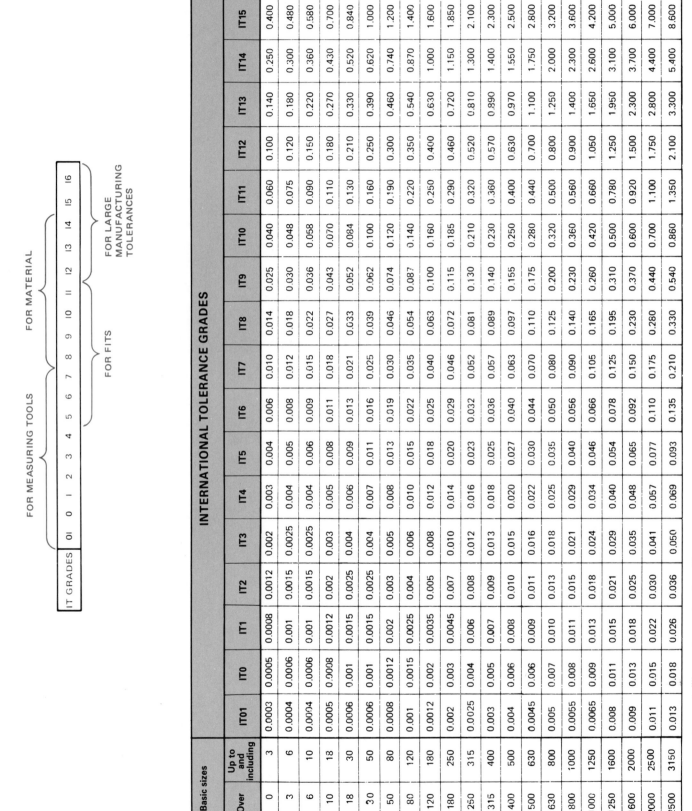

INTERNATIONAL TOLERANCE GRADES

FOR MEASURING TOOLS — IT GRADES 01 0 1 2 3 4 5 6 7 8
FOR FITS — 5 6 7 8 9 10 11 12
FOR MATERIAL
FOR LARGE MANUFACTURING TOLERANCES — 13 14 15 16

Basic sizes Over	Up to and including	IT01	IT0	IT1	IT2	IT3	IT4	IT5	IT6	IT7	IT8	IT9	IT10	IT11	IT12	IT13	IT14	IT15	IT16
0	3	0.0003	0.0005	0.0008	0.0012	0.002	0.003	0.004	0.006	0.010	0.014	0.025	0.040	0.060	0.100	0.140	0.250	0.400	0.600
3	6	0.0004	0.0006	0.001	0.0015	0.0025	0.004	0.005	0.008	0.012	0.018	0.030	0.048	0.075	0.120	0.180	0.300	0.480	0.750
6	10	0.0004	0.0006	0.001	0.0015	0.0025	0.004	0.006	0.009	0.015	0.022	0.036	0.058	0.090	0.150	0.220	0.360	0.580	0.900
10	18	0.0005	0.0008	0.0012	0.002	0.003	0.005	0.008	0.011	0.018	0.027	0.043	0.070	0.110	0.180	0.270	0.430	0.700	1.100
18	30	0.0006	0.001	0.0015	0.0025	0.004	0.006	0.009	0.013	0.021	0.033	0.052	0.084	0.130	0.210	0.330	0.520	0.840	1.300
30	50	0.0006	0.001	0.0015	0.0025	0.004	0.007	0.011	0.016	0.025	0.039	0.062	0.100	0.160	0.250	0.390	0.620	1.000	1.600
50	80	0.0008	0.0012	0.002	0.003	0.005	0.008	0.013	0.019	0.030	0.046	0.074	0.120	0.190	0.300	0.460	0.740	1.200	1.900
80	120	0.001	0.0015	0.0025	0.004	0.006	0.010	0.015	0.022	0.035	0.054	0.087	0.140	0.220	0.350	0.540	0.870	1.400	2.200
120	180	0.0012	0.002	0.0035	0.005	0.008	0.012	0.018	0.025	0.040	0.063	0.100	0.160	0.250	0.400	0.630	1.000	1.600	2.500
180	250	0.002	0.003	0.0045	0.007	0.010	0.014	0.020	0.029	0.046	0.072	0.115	0.185	0.290	0.460	0.720	1.150	1.850	2.900
250	315	0.0025	0.004	0.006	0.008	0.012	0.016	0.023	0.032	0.052	0.081	0.130	0.210	0.320	0.520	0.810	1.300	2.100	3.200
315	400	0.003	0.005	0.007	0.009	0.013	0.018	0.025	0.036	0.057	0.089	0.140	0.230	0.360	0.570	0.890	1.400	2.300	3.600
400	500	0.004	0.006	0.008	0.010	0.015	0.020	0.027	0.040	0.063	0.097	0.155	0.250	0.400	0.630	0.970	1.550	2.500	4.000
500	630	0.0045	0.006	0.009	0.011	0.016	0.022	0.030	0.044	0.070	0.110	0.175	0.280	0.440	0.700	1.100	1.750	2.800	4.400
630	800	0.005	0.007	0.010	0.013	0.018	0.025	0.035	0.050	0.080	0.125	0.200	0.320	0.500	0.800	1.250	2.000	3.200	5.000
800	1000	0.0055	0.008	0.011	0.015	0.021	0.029	0.040	0.056	0.090	0.140	0.230	0.360	0.560	0.900	1.400	2.300	3.600	5.600
1000	1250	0.0065	0.009	0.013	0.018	0.024	0.034	0.046	0.066	0.105	0.165	0.260	0.420	0.660	1.050	1.650	2.600	4.200	6.600
1250	1600	0.008	0.011	0.015	0.021	0.029	0.040	0.054	0.078	0.125	0.195	0.310	0.500	0.780	1.250	1.950	3.100	5.000	7.800
1600	2000	0.009	0.013	0.018	0.025	0.035	0.048	0.065	0.092	0.150	0.230	0.370	0.600	0.920	1.500	2.300	3.700	6.000	9.200
2000	2500	0.011	0.015	0.022	0.030	0.041	0.057	0.077	0.110	0.175	0.280	0.440	0.700	1.100	1.750	2.800	4.400	7.000	11.000
2500	3150	0.013	0.018	0.026	0.036	0.050	0.069	0.093	0.135	0.210	0.330	0.540	0.860	1.350	2.100	3.300	5.400	8.600	13.500

TABLE 40   International tolerance grades. (Values in millimeters.)

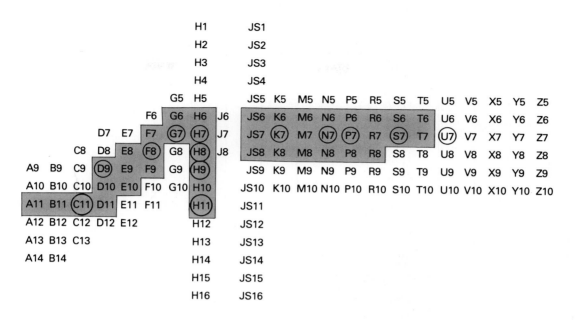

Legend: First choice tolerance zones encircled (ANSI B4.2 preferred)
Second choice tolerance zones framed (ISO 1829 selected)
Third choice tolerance zones open

TOLERANCE ZONES FOR INTERNAL DIMENSIONS (HOLES)

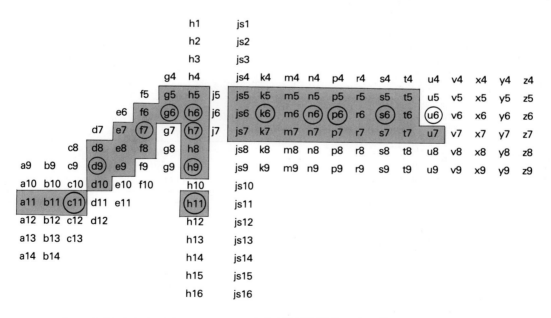

Legend: First choice tolerance zones encircled (ANSI B4.2 preferred)
Second choice tolerance zones framed (ISO 1829 selected)
Third choice tolerance zones open

TOLERANCE ZONES FOR EXTERNAL DIMENSIONS (SHAFTS)

TABLE 40   International tolerance grades. (Values in millimeters.)
(continued)

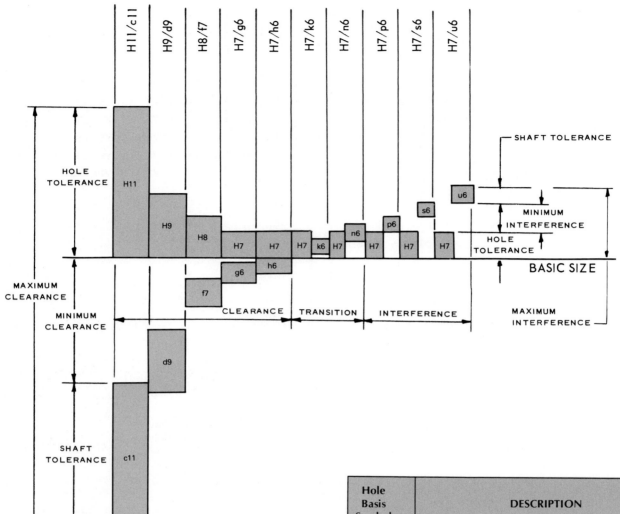

TABLE 41   Preferred hole basis fits description.

Hole Basis Symbol	DESCRIPTION
H11/c11	*Loose-running* fit for wide commercial tolerances or allowances on external members.
H9/d9	*Free-running* fit not for use where accuracy is essential, but good for large temperature variations, high running speeds, or heavy journal pressures.
H8/f7	*Close-running* fit for running on accurate machines and for accurate location at moderate speeds and journal pressures.
H7/g6	*Sliding* fit not intended to run freely, but to move and turn freely and locate accurately.

Hole Basis Symbol	DESCRIPTION
H7/h6	*Locational clearance* fit provides snug fit for locating stationary parts; but can be freely assembled and disassembled.
H7/k6	*Locational transition* fit for accurate location, a compromise between clearance and interference.
H7/n6	*Locational transition* fit for more accurate location where greater interference is permissible.
H7/p6	*Locational interference* fit for parts requiring rigidity and alignment with prime accuracy of location but without special bore pressure requirements.
H7/s6	*Medium drive* fit for ordinary steel parts or shrink fits on light sections, the tightest fit usable with cast iron.
H7/u6	*Force* fit suitable for parts which can be highly stressed or for shrink fits where the heavy pressing forces required are impractical.

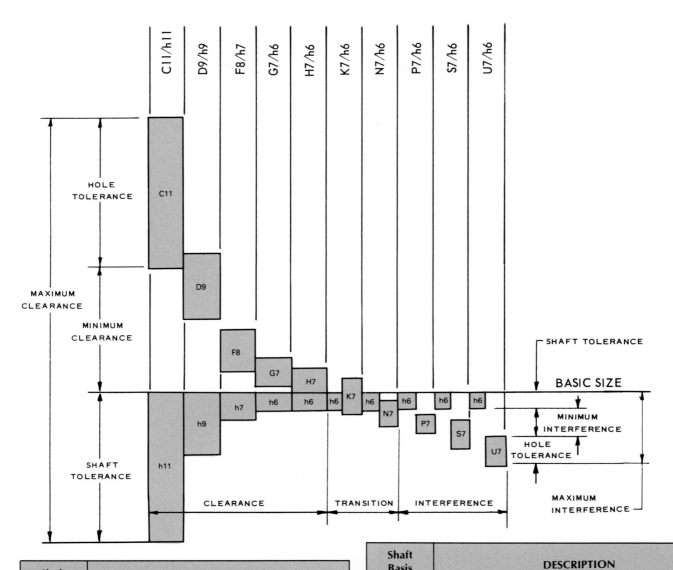

TABLE 42 Preferred shaft basis fits description.

Shaft Basis Symbol	DESCRIPTION
C11/h11	*Loose-running* fit for wide commercial tolerances or allowances on external members.
D9/h9	*Free-running* fit not for use where accuracy is essential, but good for large temperature variations, high running speeds, or heavy journal pressures.
F8/h7	*Close-running* fit for running on accurate machines and for accurate location at moderate speeds and journal pressures.
G7/h6	*Sliding* fit not intended to run freely, but to move and turn freely and locate accurately.
H7/h6	*Locational clearance* fit provides snug fit for locating stationary parts; but can be freely assembled and disassembled.

Shaft Basis Symbol	DESCRIPTION
K7/h6	*Locational transition* fit for accurate location, a compromise between clearance and interference.
N7/h6	*Locational transition* fit for more accurate location where greater interference is permissible.
P7/h6	*Locational interference* fit for parts requiring rigidity and alignment with prime accuracy of location but without special bore pressure requirements.
S7/h6	*Medium drive* fit for ordinary steel parts or shrink fits on light sections, the tightest fit usable with cast iron.
U7/h6	*Force* fit suitable for parts which can be highly stressed or for shrink fits where the heavy pressing forces required are impractical.

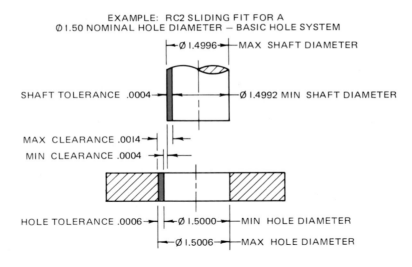

EXAMPLE: RC2 SLIDING FIT FOR A
Ø 1.50 NOMINAL HOLE DIAMETER — BASIC HOLE SYSTEM

	Class RC1 Precision Sliding			Class RC2 Sliding Fit			Class RC3 Precision Running			Class RC4 Close Running		
Nominal Size Range Inches	Hole Tol. GR5	Minimum Clearance	Shaft Tol. GR4	Hole Tol. GR6	Minimum Clearance	Shaft Tol. GR5	Hole Tol. GR7	Minimum Clearance	Shaft Tol. GR6	Hole Tol. GR8	Minimum Clearance	Shaft Tol. GR7
Over　　To	−0		+0	−0		+0	−0		+0	−0		+0
0　　　.12	+0.15	0.1	−0.12	+0.25	0.1	−0.15	+0.4	0.3	−0.25	+0.6	0.3	−0.4
.12　　.24	+0.2	0.15	−0.15	+0.3	0.15	−0.2	+0.5	0.4	−0.3	+0.7	0.4	−0.5
.24　　.40	+0.25	0.2	−0.15	+0.4	0.2	−0.25	+0.6	0.5	−0.4	+0.9	0.5	−0.6
.40　　.71	+0.3	0.25	−0.2	+0.4	0.25	−0.3	+0.7	0.6	−0.4	+1.0	0.6	−0.7
.71　　1.19	+0.4	0.3	−0.25	+0.5	0.3	−0.4	+0.8	0.8	−0.5	+1.2	0.8	−0.8
1.19　　1.97	+0.4	0.4	−0.3	+0.6	0.4	−0.4	+1.0	1.0	−0.6	+1.6	1.0	−1.0
1.97　　3.15	+0.5	0.4	−0.3	+0.7	0.4	−0.5	+1.2	1.2	−0.7	+1.8	1.2	−1.2
3.15　　4.73	+0.6	0.5	−0.4	+0.9	0.5	−0.6	+1.4	1.4	−0.9	+2.2	1.4	−1.4
4.73　　7.09	+0.7	0.6	−0.5	+1.0	0.6	−0.7	+1.6	1.6	−1.0	+2.5	1.6	−1.6
7.09　　9.85	+0.8	0.6	−0.6	+1.2	0.6	−0.8	+1.8	2.0	−1.2	+2.8	2.0	−1.8
9.85　　12.41	+0.9	0.8	−0.6	+1.2	0.8	−0.9	+2.0	2.5	−1.2	+3.0	2.5	−2.0
12.41　　15.75	+1.0	1.0	−0.7	+1.4	1.0	−1.0	+2.2	3.0	−1.4	+3.5	3.0	−2.2

TABLE 43　Running and sliding fits. (Values in thousandths of an inch.)

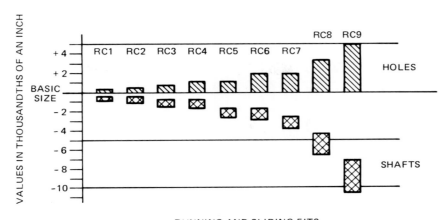

RUNNING AND SLIDING FITS
BASIC HOLE SYSTEM

Class RC5 Medium Running			Class RC6 Medium Running			Class RC7 Free Running			Class RC8 Loose Running			Class RC9 Loose Running		
Hole Tol. GR8	Minimum Clearance	Shaft Tol. GR7	Hole Tol. GR9	Minimum Clearance	Shaft Tol. GR8	Hole Tol. GR9	Minimum Clearance	Shaft Tol. GR8	Hole Tol. GR10	Minimum Clearance	Shaft Tol. GR9	Hole Tol. GR11	Minimum Clearance	Shaft Tol. GR10
−0		+0	−0		+0	−0		+0	−0		+0	−0		+0
+0.6	0.6	−0.4	+1.0	0.6	−0.6	+1.0	1.0	−0.6	+1.6	2.5	−1.0	+2.5	4.0	−1.6
+0.7	0.8	−0.5	+1.2	0.8	−0.7	+1.2	1.2	−0.7	+1.8	2.8	−1.2	+3.0	4.5	−1.8
+0.9	1.0	−0.6	+1.4	1.0	−0.9	+1.4	1.6	−0.9	+2.2	3.0	−1.4	+3.5	5.0	−2.2
+1.0	1.2	−0.7	+1.6	1.2	−1.0	+1.6	2.0	−1.0	+2.8	3.5	−1.6	+4.0	6.0	−2.8
+1.2	1.6	−0.8	+2.0	1.6	−1.2	+2.0	2.5	−1.2	+3.5	4.5	−2.0	+5.0	7.0	−3.5
+1.6	2.0	−1.0	+2.5	2.0	−1.6	+2.5	3.0	−1.6	+4.0	5.0	−2.5	+6.0	8.0	−4.0
+1.8	2.5	−1.2	+3.0	2.5	−1.8	+3.0	4.0	−1.8	+4.5	6.0	−3.0	+7.0	9.0	−4.5
+2.2	3.0	−1.4	+3.5	3.0	−2.2	+3.5	5.0	−2.2	+5.0	7.0	−3.5	+9.0	10.0	−5.0
+2.5	3.5	−1.6	+4.0	3.5	−2.5	+4.0	6.0	−2.5	+6.0	8.0	−4.0	+10.0	12.0	−6.0
+2.8	4.5	−1.8	+4.5	4.0	−2.8	+4.5	7.0	−2.8	+7.0	10.0	−4.5	+12.0	15.0	−7.0
+3.0	5.0	−2.0	+5.0	5.0	−3.0	+5.0	8.0	−3.0	+8.0	12.0	−5.0	+12.0	18.0	−8.0
+3.5	6.0	−2.2	+6.0	6.0	−3.5	+6.0	10.0	−3.5	+9.0	14.0	−6.0	+14.0	22.0	−9.0

TABLE 43   Running and sliding fits. (Values in thousandths of an inch.)
(continued)

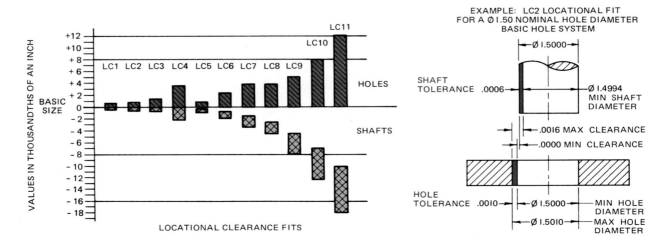

LOCATIONAL CLEARANCE FITS

EXAMPLE: LC2 LOCATIONAL FIT
FOR A Ø 1.50 NOMINAL HOLE DIAMETER
BASIC HOLE SYSTEM

Nominal Size Range Inches		Class LC1			Class LC2			Class LC3			Class LC4			Class LC5		
		Hole Tol. GR6	Minimum Clearance	Shaft Tol. GR5	Hole Tol. GR7	Minimum Clearance	Shaft Tol. GR6	Hole Tol. GR8	Minimum Clearance	Shaft Tol. GR7	Hole Tol. GR10	Minimum Clearance	Shaft Tol. GR9	Hole Tol. GR7	Minimum Clearance	Shaft Tol. GR6
Over	To	−0		+0	−0		+0	−0		+0	−0		+0	−0		+0
0	.12	+0.25	0	−0.15	+0.4	0	−0.25	+0.6	0	−0.4	+1.6	0	−1.0	+0.4	0.1	−0.25
.12	.24	+0.3	0	−0.2	+0.5	0	−0.3	+0.7	0	−0.5	+1.8	0	−1.2	+0.5	0.15	−0.3
.24	.40	+0.4	0	−0.25	+0.6	0	−0.4	+0.9	0	−0.6	+2.2	0	−1.4	+0.6	0.2	−0.4
.40	.71	+0.4	0	−0.3	+0.7	0	−0.4	+1.0	0	−0.7	+2.8	0	−1.6	+0.7	0.25	−0.4
.71	1.19	+0.5	0	−0.4	+0.8	0	−0.5	+1.2	0	−0.8	+3.5	0	−2.0	+0.8	0.3	−0.5
1.19	1.97	+0.6	0	−0.4	+1.0	0	−0.6	+1.6	0	−1.0	+4.0	0	−2.5	+1.0	0.4	−0.6
1.97	3.15	+0.7	0	−0.5	+1.2	0	−0.7	+1.8	0	−1.2	+4.5	0	−3.0	+1.2	0.4	−0.7
3.15	4.73	+0.9	0	−0.6	+1.4	0	−0.9	+2.2	0	−1.4	+5.0	0	−3.5	+1.4	0.5	−0.9
4.73	7.09	+1.0	0	−0.7	+1.6	0	−1.0	+2.5	0	−1.6	+6.0	0	−4.0	+1.6	0.6	−1.0
7.09	9.85	+1.2	0	−0.8	+1.8	0	−1.2	+2.8	0	−1.8	+7.0	0	−4.5	+1.8	0.6	−1.2
9.85	12.41	+1.2	0	−0.9	+2.0	0	−1.2	+3.0	0	−2.0	+8.0	0	−5.0	+2.0	0.7	−1.2
12.41	15.75	+1.4	0	−1.0	+2.2	0	−1.4	+3.5	0	−2.2	+9.0	0	−6.0	+2.2	0.7	−1.4

TABLE 44   Locational clearance fits. (Values in thousandths of an inch.)

Nominal Size Range Inches		Class LT1			Class LT2		
		Hole Tol. GR7	Maximum Interference	Shaft Tol. GR6	Hole Tol. GR8	Maximum Interference	Shaft Tol. GR7
Over	To	−0		+0	−0		+0
0	.12	+0.4	0.1	−0.25	+0.6	0.2	−0.4
.12	.24	+0.5	0.15	−0.3	+0.7	0.25	−0.5
.24	.40	+0.6	0.2	−0.4	+0.9	0.3	−0.6
.40	.71	+0.7	0.2	−0.4	+1.0	0.3	−0.7
.71	1.19	+0.8	0.25	−0.5	+1.2	0.4	−0.8
1.19	1.97	+1.0	0.3	−0.6	+1.6	0.5	−1.0
1.97	3.15	+1.2	0.3	−0.7	+1.8	0.6	−1.2
3.15	4.73	+1.4	0.4	−0.9	+2.2	0.7	−1.4
4.73	7.09	+1.6	0.5	−1.0	+2.5	0.8	−1.6
7.09	9.85	+1.8	0.6	−1.2	+2.8	0.9	−1.8
9.85	12.41	+2.0	0.6	−1.2	+3.0	1.0	−2.0
12.41	15.75	+2.2	0.7	−1.4	+3.5	1.0	−2.2

TABLE 45   Transition fits. (Values in thousandths of an inch.)

TRANSITION FITS

HOLE   = HOLE   = SHAFT

EXAMPLE: LT2 TRANSITION FIT
FOR A Ø1.50 NOMINAL HOLE DIAMETER
BASIC HOLE SYSTEM

	Class LC6			Class LC7			Class LC8			Class LC9			Class LC10			Class LC11		
	Hole Tol. GR9	Minimum Clearance	Shaft Tol. GR8	Hole Tol. GR10	Minimum Clearance	Shaft Tol. GR9	Hole Tol. GR10	Minimum Clearance	Shaft Tol. GR9	Hole Tol. GR11	Minimum Clearance	Shaft Tol. GR10	Hole Tol. GR12	Minimum Clearance	Shaft Tol. GR11	Hole Tol. GR13	Minimum Clearance	Shaft Tol. GR12
	−0		+0	−0		+0	−0		+0	−0		+0	−0		+0	−0		+0
	+1.0	0.3	−0.6	+1.6	0.6	−1.0	+1.6	1.0	−1.0	+2.5	2.5	−1.6	+4.0	4.0	−2.5	+6.0	5.0	−4.0
	+1.2	0.4	−0.7	+1.8	0.8	−1.2	+1.8	1.2	−1.2	+3.0	2.8	−1.8	+5.0	4.5	−3.0	+7.0	6.0	−5.0
	+1.4	0.5	−0.9	+2.2	1.0	−1.4	+2.2	1.6	−1.4	+3.5	3.0	−2.2	+6.0	5.0	−3.5	+9.0	7.0	−6.0
	+1.6	0.6	−1.0	+2.8	1.2	−1.6	+2.8	2.0	−1.6	+4.0	3.5	−2.8	+7.0	6.0	−4.0	+10.0	8.0	−7.0
	+2.0	0.8	−1.2	+3.5	1.6	−2.0	+3.5	2.5	−2.0	+5.0	4.5	−3.5	+8.0	7.0	−5.0	+12.0	10.0	−8.0
	+2.5	1.0	−1.6	+4.0	2.0	−2.5	+4.0	3.6	−2.5	+6.0	5.0	−4.0	+10.0	8.0	−6.0	+16.0	12.0	−10.0
	+3.0	1.2	−1.8	+4.5	2.5	−3.0	+4.5	4.0	−3.0	+7.0	6.0	−4.5	+12.0	10.0	−7.0	+18.0	14.0	−12.0
	+3.5	1.4	−2.2	+5.0	3.0	−3.5	+5.0	5.0	−3.5	+9.0	7.0	−5.0	+14.0	11.0	−9.0	+22.0	16.0	−14.0
	+4.0	1.6	−2.5	+6.0	3.5	−4.0	+6.0	6.0	−4.0	+10.0	8.0	−6.0	+16.0	12.0	−10.0	+25.0	18.0	−16.0
	+4.5	2.0	−2.8	+7.0	4.0	−4.5	+7.0	7.0	−4.5	+12.0	10.0	−7.0	+18.0	16.0	−12.0	+28.0	22.0	−18.0
	+5.0	2.2	−3.0	+8.0	4.5	−5.0	+8.0	7.0	−5.0	+12.0	12.0	−8.0	+20.0	20.0	−12.0	+30.0	28.0	−20.0
	+6.0	2.5	−3.5	+9.0	5.0	−6.0	+9.0	8.0	−6.0	+14.0	14.0	−9.0	+22.0	22.0	−14.0	+35.0	30.0	−22.0

TABLE 44   Locational clearance fits. (Values in thousandths of an inch.)
(continued)

Class LT3			Class LT4			Class LT5			Class LT6		
Hole Tol. GR7	Maximum Interference	Shaft Tol. GR6	Hole Tol. GR8	Maximum Interference	Shaft Tol. GR7	Hole Tol. GR7	Maximum Interference	Shaft Tol. GR6	Hole Tol. GR8	Maximum Interference	Shaft Tol. GR7
−0		+0	−0		+0	−0		+0	−0		+0
+0.4	0.25	−0.25	+0.6	0.4	−0.4	+0.4	0.5	−0.25	+0.6	0.65	−0.4
+0.5	0.4	−0.3	+0.7	0.6	−0.5	+0.5	0.6	−0.3	+0.7	0.8	−0.5
+0.6	0.5	−0.4	+0.9	0.7	−0.6	+0.6	0.8	−0.4	+0.9	1.0	−0.6
+0.7	0.5	−0.4	+1.0	0.8	−0.7	+0.7	0.9	−0.4	+1.0	1.2	−0.7
+0.8	0.6	−0.5	+1.2	0.9	−0.8	+0.8	1.1	−0.5	+1.2	1.4	−0.8
+1.0	0.7	−0.6	+1.6	1.1	−1.0	+1.0	1.3	−0.6	+1.6	1.7	−1.0
+1.2	0.8	−0.7	+1.8	1.3	−1.2	+1.2	1.5	−0.7	+1.8	2.0	−1.2
+1.4	1.0	−0.9	+2.2	1.5	−1.4	+1.4	1.9	−0.9	+2.2	2.4	−1.4
+1.6	1.1	−1.0	+2.5	1.7	−1.6	+1.6	2.2	−1.0	+2.5	2.8	−1.6
+1.8	1.4	−1.2	+2.8	2.0	−1.8	+1.8	2.6	−1.2	+2.8	3.2	−1.8
+2.0	1.4	−1.2	+3.0	2.2	−2.0	+2.0	2.6	−1.2	+3.0	3.4	−2.0
+2.2	1.6	−1.4	+3.5	2.4	−2.2	+2.2	3.0	−1.4	+3.5	3.8	−2.2

TABLE 45   Transition fits. (Values in thousandths of an inch.)
(continued)

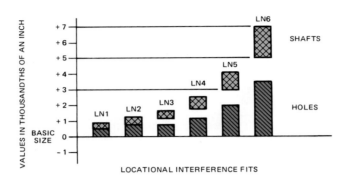

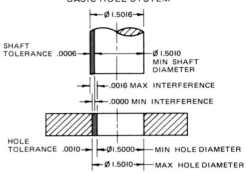

EXAMPLE: LN2 LOCATIONAL INTERFERENCE FIT FOR A Ø 1.50 NOMINAL HOLE DIAMETER BASIC HOLE SYSTEM

Nominal Size Range Inches		Class LN1 Light Press Fit			Class LN2 Medium Press Fit		
		Hole Tol. GR6	Maximum Interference	Shaft Tol. GR5	Hole Tol. GR7	Maximum Interference	Shaft Tol. GR6
Over	To	−0		+0	−0		+0
0	.12	+0.25	0.4	−0.15	+0.4	0.65	−0.25
.12	.24	+0.3	0.5	−0.2	+0.5	0.8	−0.3
.24	.40	+0.4	0.65	−0.25	+0.6	1.0	−0.4
.40	.71	+0.4	0.7	−0.3	+0.7	1.1	−0.4
.71	1.19	+0.5	0.9	−0.4	+0.8	1.3	−0.5
1.19	1.97	+0.6	1.0	−0.4	+1.0	1.6	−0.6
1.97	3.15	+0.7	1.3	−0.5	+1.2	2.1	−0.7
3.15	4.73	+0.9	1.6	−0.6	+1.4	2.5	−0.9
4.73	7.09	+1.0	1.9	−0.7	+1.6	2.8	−1.0
7.09	9.85	+1.2	2.2	−0.8	+1.8	3.2	−1.2
9.85	12.41	+1.2	2.3	−0.9	+2.0	3.4	−1.2
12.41	15.75	+1.4	2.6	−1.0	+2.2	3.9	−1.4

TABLE 46   Locational interference fits. (Values in thousandths of an inch.)

Nominal Size Range Inches		Class FN1 Light Drive Fit			Class FN2 Medium Drive Fit		
		Hole Tol. GR6	Maximum Interference	Shaft Tol. GR5	Hole Tol. GR7	Maximum Interference	Shaft Tol. GR6
Over	To	−0		+0	−0		+0
0	.12	+0.25	0.5	−0.15	+0.4	0.85	−0.25
.12	.24	+0.3	0.6	−0.2	+0.5	1.0	−0.3
.24	.40	+0.4	0.75	−0.25	+0.6	1.4	−0.4
.40	.56	+0.4	0.8	−0.3	+0.7	1.6	−0.4
.56	.71	+0.4	0.9	−0.3	+0.7	1.6	−0.4
.71	.95	+0.5	1.1	−0.4	+0.8	1.9	−0.5
.95	1.19	+0.5	1.2	−0.4	+0.8	1.9	−0.5
1.19	1.58	+0.6	1.3	−0.4	+1.0	2.4	−0.6
1.58	1.97	+0.6	1.4	−0.4	+1.0	2.4	−0.6
1.97	2.56	+0.7	1.8	−0.5	+1.2	2.7	−0.7
2.56	3.15	+0.7	1.9	−0.5	+1.2	2.9	−0.7
3.15	3.94	+0.9	2.4	−0.6	+1.4	3.7	−0.9

TABLE 47   Force and shrink fits. (Values in thousandths of an inch.)

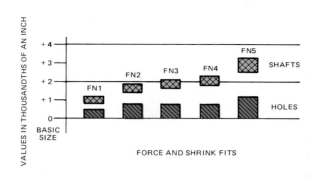

FORCE AND SHRINK FITS

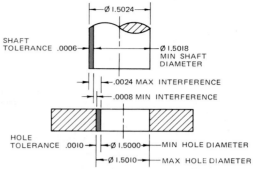

EXAMPLE: FN2 MEDIUM DRIVE
FIT FOR A Ø 1.50 NOMINAL HOLE DIAMETER
BASIC HOLE SYSTEM

	Class LN3 Heavy Press Fit			Class LN4			Class LN5			Class LN6		
	Hole Tol. GR7	Maximum Interference	Shaft Tol. GR6	Hole Tol. GR8	Maximum Interference	Shaft Tol. GR7	Hole Tol. GR9	Maximum Interference	Shaft Tol. GR8	Hole Tol. GR10	Maximum Interference	Shaft Tol. GR9
	−0		+0	−0		+0	−0		+0	−0		+0
	+0.4	0.75	−0.25	+0.6	1.2	−0.4	+1.0	1.8	−0.6	+1.6	3.0	−1.0
	+0.5	0.9	−0.3	+0.7	1.5	−0.5	+1.2	2.3	−0.7	+1.8	3.6	−1.2
	+0.6	1.2	−0.4	+0.9	1.8	−0.6	+1.4	2.8	−0.9	+2.2	4.4	−1.4
	+0.7	1.4	−0.4	+1.0	2.2	−0.7	+1.6	3.4	−1.0	+2.8	5.6	−1.6
	+0.8	1.7	−0.5	+1.2	2.6	−0.8	+2.0	4.2	−1.2	+3.5	7.0	−2.0
	+1.0	2.0	−0.6	+1.6	3.4	−1.0	+2.5	5.3	−1.6	+4.0	8.5	−2.5
	+1.2	2.3	−0.7	+1.8	4.0	−1.2	+3.0	6.3	−1.8	+4.5	10.0	−3.0
	+1.4	2.9	−0.9	+2.2	4.8	−1.4	+4.0	7.7	−2.2	+5.0	11.5	−3.5
	+1.6	3.5	−1.0	+2.5	5.6	−1.6	+4.5	8.7	−2.5	+6.0	13.5	−4.0
	+1.8	4.2	−1.2	+2.8	6.6	−1.8	+5.0	10.3	−2.8	+7.0	16.5	−4.5
	+2.0	4.7	−1.2	+3.0	7.5	−2.0	+6.0	12.0	−3.0	+8.0	19	−5.0
	+2.2	5.9	−1.4	+3.5	8.7	−2.2	+6.0	14.5	−3.5	+9.0	23	−6.0

TABLE 46   Locational interference fits. (Values in thousandths of an inch.)
(continued)

	Class FN3 Heavy Drive Fit			Class FN4 Shrink Fit			FN5 Heavy Shrink Fit		
	Hole Tol. GR7	Maximum Interference	Shaft Tol. GR6	Hole Tol. GR7	Maximum Interference	Shaft Tol. GR6	Hole Tol. GR8	Maximum Interference	Shaft Tol. GR7
	−0		+0	−0		+0	−0		+0
				+0.4	0.95	−0.25	+0.6	1.3	−0.4
				+0.5	1.2	−0.3	+0.7	1.7	−0.5
				+0.6	1.6	−0.4	+0.9	2.0	−0.6
				+0.7	1.8	−0.4	+1.0	2.3	−0.7
				+0.7	1.8	−0.4	+1.0	2.5	−0.7
				+0.8	2.1	−0.5	+1.2	3.0	−0.8
	+0.8	2.1	−0.5	+0.8	2.3	−0.5	+1.2	3.3	−0.8
	+1.0	2.6	−0.6	+1.0	3.1	−0.6	+1.6	4.0	−1.0
	+1.0	2.8	−0.6	+1.0	3.4	−0.6	+1.6	5.0	−1.0
	+1.2	3.2	−0.7	+1.2	4.2	−0.7	+1.8	6.2	−1.2
	+1.2	3.7	−0.7	+1.2	4.7	−0.7	+1.8	7.2	−1.2
	+1.4	4.4	−0.9	+1.4	5.9	−0.9	+2.2	8.4	−1.4

TABLE 47   Force and shrink fits. (Values in thousandths of an inch.)
(continued)

Example ⌀ 50H9/d9

Hole Size ⌀ 50.062 / 50.000

Shaft Size ⌀ 49.920 / 49.858

Clearance Max. 0.204 / Min. 0.080

Example ⌀ 70H7/g6

Hole Size ⌀ 70.030 / 70.000

Shaft Size ⌀ 69.990 / 69.971

Clearance Max. 0.059 / Min. 0.010

		PREFERRED HOLE BASIS CLEARANCE FITS															
UP TO AND INCLUDING		LOOSE RUNNING			FREE RUNNING			CLOSE RUNNING			SLIDING			LOCATIONAL CLEARANCE		
		Hole H11	Shaft c11	Fit	Hole H9	Shaft d9	Fit	Hole H8	Shaft f7	Fit	Hole H7	Shaft g6	Fit	Hole H7	Shaft h6	Fit
1	MAX	1.060	0.940	0.180	1.025	0.980	0.070	1.014	0.994	0.030	1.010	0.998	0.018	1.010	1.000	0.016
	MIN	1.000	0.880	0.060	1.000	0.955	0.020	1.000	0.984	0.006	1.000	0.992	0.002	1.000	0.994	0.000
1.2	MAX	1.260	1.140	0.180	1.225	1.180	0.070	1.214	1.194	0.030	1.210	1.198	0.018	1.210	1.200	0.016
	MIN	1.200	1.080	0.060	1.200	1.155	0.020	1.200	1.184	0.006	1.200	1.192	0.002	1.200	1.194	0.000
1.6	MAX	1.660	1.540	0.180	1.625	1.580	0.070	1.614	1.594	0.030	1.610	1.598	0.018	1.610	1.600	0.016
	MIN	1.600	1.480	0.060	1.600	1.555	0.020	1.600	1.584	0.006	1.600	1.592	0.002	1.600	1.594	0.000
2	MAX	2.060	1.940	0.180	2.025	1.980	0.070	2.014	1.994	0.030	2.010	1.998	0.018	2.010	2.000	0.016
	MIN	2.000	1.880	0.060	2.000	1.955	0.020	2.000	1.984	0.006	2.000	1.992	0.002	2.000	1.994	0.000
2.5	MAX	2.560	2.440	0.180	2.525	2.480	0.070	2.514	2.494	0.030	2.510	2.498	0.018	2.510	2.500	0.016
	MIN	2.500	2.380	0.060	2.500	2.455	0.020	2.500	2.484	0.006	2.500	2.492	0.002	2.500	2.494	0.000
3	MAX	3.060	2.940	0.180	3.025	2.980	0.070	3.014	2.994	0.030	3.010	2.998	0.018	3.010	3.000	0.016
	MIN	3.000	2.880	0.060	3.000	2.955	0.020	3.000	2.984	0.006	3.000	2.992	0.002	3.000	2.994	0.000
4	MAX	4.075	3.930	0.220	4.030	3.970	0.090	4.018	3.990	0.040	4.012	3.996	0.024	4.012	4.000	0.020
	MIN	4.000	3.855	0.070	4.000	3.940	0.030	4.000	3.978	0.010	4.000	3.988	0.004	4.000	3.992	0.000
5	MAX	5.075	4.930	0.220	5.030	4.970	0.090	5.018	4.990	0.040	5.012	4.996	0.024	5.012	5.000	0.020
	MIN	5.000	4.855	0.070	5.000	4.940	0.030	5.000	4.978	0.010	5.000	4.988	0.004	5.000	4.992	0.000
6	MAX	6.075	5.930	0.220	6.030	5.970	0.090	6.018	5.990	0.040	6.012	5.996	0.024	6.012	6.000	0.020
	MIN	6.000	5.855	0.070	6.000	5.940	0.030	6.000	5.978	0.010	6.000	5.988	0.004	6.000	5.992	0.000
8	MAX	8.090	7.920	0.260	8.036	7.960	0.112	8.022	7.987	0.050	8.015	7.995	0.029	8.015	8.000	0.024
	MIN	8.000	7.830	0.080	8.000	7.924	0.040	8.000	7.972	0.013	8.000	7.986	0.006	8.000	7.991	0.000
10	MAX	10.090	9.920	0.260	10.036	9.960	0.112	10.022	9.987	0.050	10.015	9.995	0.029	10.015	10.000	0.024
	MIN	10.000	9.830	0.080	10.000	9.924	0.040	10.000	9.972	0.013	10.000	9.986	0.005	10.000	9.991	0.000
12	MAX	12.110	11.905	0.315	12.043	11.950	0.136	12.027	11.984	0.061	12.018	11.994	0.035	12.018	12.000	0.029
	MIN	12.000	11.795	0.095	12.000	11.907	0.050	12.000	11.966	0.016	12.000	11.983	0.006	12.000	11.989	0.000
16	MAX	16.110	15.905	0.315	16.043	15.950	0.136	16.027	15.984	0.061	16.018	15.994	0.035	16.018	16.000	0.029
	MIN	16.000	15.795	0.095	16.000	15.907	0.050	16.000	15.966	0.016	16.000	15.983	0.006	16.000	15.989	0.000
20	MAX	20.130	19.890	0.370	20.052	19.935	0.169	20.033	19.980	0.074	20.021	19.993	0.041	20.021	20.000	0.034
	MIN	20.000	19.760	0.110	20.000	19.883	0.065	20.000	19.959	0.020	20.000	19.980	0.007	20.000	19.987	0.000
25	MAX	25.130	24.890	0.370	25.052	24.935	0.169	25.033	24.980	0.074	25.021	24.993	0.042	25.021	25.000	0.034
	MIN	25.000	24.760	0.110	25.000	24.883	0.065	25.000	24.959	0.020	25.000	24.980	0.007	25.000	24.987	0.000
30	MAX	30.130	29.890	0.370	30.052	29.935	0.169	30.033	29.980	0.074	30.021	29.993	0.041	30.021	30.000	0.034
	MIN	30.000	29.760	0.110	30.000	29.883	0.065	30.000	29.959	0.020	30.000	29.980	0.007	30.000	29.987	0.000
40	MAX	40.160	39.880	0.440	40.062	39.920	0.204	40.039	39.975	0.089	40.025	39.991	0.050	40.025	40.000	0.041
	MIN	40.000	39.720	0.120	40.000	39.858	0.080	40.000	39.950	0.025	40.000	39.975	0.009	40.000	39.984	0.000
50	MAX	50.160	49.870	0.450	50.062	49.920	0.204	50.039	49.975	0.089	50.025	49.991	0.050	50.025	50.000	0.041
	MIN	50.000	49.710	0.130	50.000	49.858	0.080	50.000	49.950	0.025	50.000	49.975	0.009	50.000	49.984	0.000
60	MAX	60.190	59.860	0.520	60.074	59.900	0.248	60.046	59.970	0.106	60.030	59.990	0.059	60.030	60.000	0.049
	MIN	60.000	59.670	0.140	60.000	59.826	0.100	60.000	59.940	0.030	60.000	59.971	0.010	60.000	59.981	0.000
80	MAX	80.190	79.850	0.530	80.074	79.900	0.248	80.046	79.970	0.106	80.030	79.990	0.059	80.030	80.000	0.049
	MIN	80.000	79.660	0.150	80.000	79.826	0.100	80.000	79.940	0.030	80.000	79.971	0.010	80.000	79.981	0.000
100	MAX	100.220	99.830	0.610	100.087	99.880	0.294	100.054	99.964	0.125	100.035	99.988	0.069	100.035	100.000	0.057
	MIN	100.000	99.610	0.170	100.000	99.793	0.120	100.000	99.929	0.036	100.000	99.966	0.012	100.000	99.978	0.000
120	MAX	120.220	119.820	0.620	120.087	119.880	0.294	120.054	119.964	0.125	120.035	119.988	0.069	120.035	120.000	0.057
	MIN	120.000	119.600	0.180	120.000	119.793	0.120	120.000	119.929	0.036	120.000	119.966	0.012	120.000	119.978	0.000
160	MAX	160.250	159.790	0.710	160.100	159.855	0.345	160.063	159.957	0.146	160.040	159.986	0.079	160.040	160.000	0.065
	MIN	160.000	159.540	0.210	160.000	159.755	0.145	160.000	159.917	0.043	160.000	159.961	0.014	160.000	159.975	0.000

TABLE 48 Preferred hole basis fits. (Dimensions in millimeters.)

Example Ø 10H7/n6

Hole Size Ø $\frac{10.015}{10.000}$

Shaft Size Ø $\frac{10.019}{10.010}$

Max. Clearance 0.005

Max. Interference 0.019

Example Ø 35H7/u6

Hole Size Ø $\frac{35.025}{35.000}$

Shaft Size Ø $\frac{35.076}{35.060}$

Min. Interference 0.035

Max. Interference 0.076

		PREFERRED HOLE BASIS TRANSITION AND INTERFERENCE FITS															
UP TO AND INCLUDING		LOCATIONAL TRANSN.			LOCATIONAL TRANSN.			LOCATIONAL INTERF.			MEDIUM DRIVE			FORCE		
		Hole H7	Shaft k6	Fit	Hole H7	Shaft n6	Fit	Hole H7	Shaft p6	Fit	Hole H7	Shaft s6	Fit	Hole H7	Shaft u6	Fit
1	MAX	1.010	1.006	0.010	1.010	1.010	0.006	1.010	1.012	0.004	1.010	1.020	−0.004	1.010	1.024	−0.008
	MIN	1.000	1.000	−0.006	1.000	1.004	−0.010	1.000	1.006	−0.012	1.000	1.014	−0.020	1.000	1.018	−0.024
1.2	MAX	1.210	1.206	0.010	1.210	1.210	0.006	1.210	1.212	0.004	1.210	1.220	−0.004	1.210	1.224	−0.008
	MIN	1.200	1.200	−0.006	1.200	1.204	−0.010	1.200	1.206	−0.012	1.200	1.214	−0.020	1.200	1.218	−0.024
1.6	MAX	1.610	1.606	0.010	1.610	1.610	0.006	1.610	1.612	0.004	1.610	1.620	−0.004	1.610	1.624	−0.008
	MIN	1.600	1.600	−0.006	1.600	1.604	−0.010	1.600	1.606	−0.012	1.600	1.614	−0.020	1.600	1.618	−0.024
2	MAX	2.010	2.006	0.010	2.010	2.010	0.006	2.010	2.012	0.004	2.010	2.020	−0.004	2.010	2.024	−0.008
	MIN	2.000	2.000	−0.006	2.000	2.004	−0.010	2.000	2.006	−0.012	2.000	2.014	−0.020	2.000	2.018	−0.024
2.5	MAX	2.510	2.506	0.010	2.510	2.510	0.006	2.510	2.512	0.004	2.510	2.520	−0.004	2.510	2.524	−0.008
	MIN	2.500	2.500	−0.006	2.500	2.504	−0.010	2.500	2.506	−0.012	2.500	2.514	−0.020	2.500	2.518	−0.024
3	MAX	3.010	3.006	0.010	3.010	3.010	0.006	3.010	3.012	0.004	3.010	3.020	−0.004	3.010	3.024	−0.008
	MIN	3.000	3.000	−0.006	3.000	3.004	−0.010	3.000	3.006	−0.012	3.000	3.014	−0.020	3.000	3.018	−0.024
4	MAX	4.012	4.009	0.011	4.012	4.016	0.004	4.012	4.020	0.000	4.012	4.027	−0.007	4.012	4.031	−0.011
	MIN	4.000	4.001	−0.009	4.000	4.008	−0.016	4.000	4.012	−0.020	4.000	4.019	−0.027	4.000	4.023	−0.031
5	MAX	5.012	5.009	0.011	5.012	5.016	0.004	5.012	5.020	0.000	5.012	5.027	−0.007	5.012	5.031	−0.011
	MIN	5.000	5.001	−0.009	5.000	5.008	−0.016	5.000	5.012	−0.020	5.000	5.019	−0.027	5.000	5.023	−0.031
6	MAX	6.012	6.009	0.011	6.012	6.016	0.004	6.012	6.020	0.000	6.012	6.027	−0.007	6.012	6.031	−0.011
	MIN	6.000	6.001	−0.009	6.000	6.008	−0.016	6.000	6.012	−0.020	6.000	6.019	−0.027	6.000	6.023	−0.031
8	MAX	8.015	8.010	0.014	8.015	8.019	0.005	8.015	8.024	0.000	8.015	8.032	−0.008	8.015	8.037	−0.013
	MIN	8.000	8.001	−0.010	8.000	8.010	−0.019	8.000	8.015	−0.024	8.000	8.023	−0.032	8.000	8.028	−0.037
10	MAX	10.015	10.010	0.014	10.015	10.019	0.005	10.015	10.024	0.000	10.015	10.032	−0.008	10.015	10.037	−0.013
	MIN	10.000	10.001	−0.010	10.000	10.010	−0.019	10.000	10.015	−0.024	10.000	10.023	−0.032	10.000	10.028	−0.037
12	MAX	12.018	12.012	0.017	12.018	12.023	0.006	12.018	12.029	0.000	12.018	12.039	−0.010	12.018	12.044	−0.015
	MIN	12.000	12.001	−0.012	12.000	12.012	−0.023	12.000	12.018	−0.029	12.000	12.028	−0.039	12.000	12.033	−0.044
16	MAX	16.018	16.012	0.017	16.018	16.023	0.006	16.018	16.029	0.000	16.018	16.039	−0.010	16.018	16.044	−0.015
	MIN	16.000	16.001	−0.012	16.000	16.012	−0.023	16.000	16.018	−0.029	16.000	16.028	−0.039	16.000	16.033	−0.044
20	MAX	20.021	20.015	0.019	20.021	20.028	0.006	20.021	20.035	−0.001	20.021	20.048	−0.014	20.021	20.054	−0.020
	MIN	20.000	20.002	−0.015	20.000	20.015	−0.028	20.000	20.022	−0.035	20.000	20.035	−0.048	20.000	20.041	−0.054
25	MAX	25.021	25.015	0.019	25.021	25.028	0.006	25.021	25.035	−0.001	25.021	25.048	−0.014	25.021	25.061	−0.027
	MIN	25.000	25.002	−0.015	25.000	25.015	−0.028	25.000	25.022	−0.035	25.000	25.035	−0.048	25.000	25.048	−0.061
30	MAX	30.021	30.015	0.019	30.021	30.028	0.006	30.021	30.035	−0.001	30.021	30.048	−0.014	30.021	30.061	−0.027
	MIN	30.000	30.002	−0.015	30.000	30.015	−0.028	30.000	30.022	−0.035	30.000	30.035	−0.048	30.000	30.048	−0.061
40	MAX	40.025	40.018	0.023	40.025	40.033	0.008	40.025	40.042	−0.001	40.025	40.059	−0.018	40.025	40.076	−0.035
	MIN	40.000	40.002	−0.018	40.000	40.017	−0.033	40.000	40.026	−0.042	40.000	40.043	−0.059	40.000	40.060	−0.076
50	MAX	50.025	50.018	0.023	50.025	50.033	0.008	50.025	50.042	−0.001	50.025	50.059	−0.018	50.025	50.086	−0.045
	MIN	50.000	50.002	−0.018	50.000	50.017	−0.033	50.000	50.026	−0.042	50.000	50.043	−0.059	50.000	50.070	−0.086
60	MAX	60.030	60.021	0.028	60.030	60.039	0.010	60.030	60.051	−0.002	60.030	60.072	−0.023	60.030	60.106	−0.057
	MIN	60.000	60.002	−0.021	60.000	60.020	−0.039	60.000	60.032	−0.051	60.000	60.053	−0.072	60.000	60.087	−0.106
80	MAX	80.030	80.021	0.028	80.030	80.039	0.010	80.030	80.051	−0.002	80.030	80.078	−0.029	80.030	80.121	−0.072
	MIN	80.000	80.002	−0.021	80.000	80.020	−0.039	80.000	80.032	−0.051	80.000	80.059	−0.078	80.000	80.102	−0.121
100	MAX	100.035	100.025	0.032	100.035	100.045	0.012	100.035	100.059	−0.002	100.035	100.093	−0.036	100.035	100.146	−0.089
	MIN	100.000	100.003	−0.025	100.000	100.023	−0.045	100.000	100.037	−0.059	100.000	100.071	−0.093	100.000	100.124	−0.146
120	MAX	120.035	120.025	0.032	120.035	120.045	0.012	120.035	120.059	−0.002	120.035	120.101	−0.044	120.035	120.166	−0.109
	MIN	120.000	120.003	−0.025	120.000	120.023	−0.045	120.000	120.037	−0.059	120.000	120.079	−0.101	120.000	120.144	−0.166
160	MAX	160.040	160.028	0.037	160.045	160.052	0.013	160.040	160.068	−0.003	160.040	160.125	−0.060	160.040	160.215	−0.150
	MIN	160.000	160.003	−0.028	160.000	160.027	−0.052	160.000	160.043	−0.068	160.000	160.000	−0.125	160.000	160.190	−0.215

TABLE 48   Preferred hole basis fits. (Dimensions in millimeters.)
(continued)

Example Ø 100C11/h11

Hole Size Ø 100.390 / 100.170

Shaft Size Ø 100.000 / 99.780

Clearance Max. 0.610 / Min. 0.170

Example Ø 11H7/h6

Hole Size Ø 11.018 / 11.000

Shaft Size Ø 11.000 / 10.989

Clearance Max. 0.029 / Min. 0.000

PREFERRED SHAFT BASIS CLEARANCE FITS																	
UP TO AND INCLUDING		LOOSE RUNNING			FREE RUNNING			CLOSE RUNNING			SLIDING			LOCATIONAL CLEARANCE		
		Hole C11	Shaft h11	Fit	Hole D9	Shaft h9	Fit	Hole F8	Shaft h7	Fit	Hole G7	Shaft h6	Fit	Hole H7	Shaft h6	Fit
1	MAX	1.120	1.000	0.180	1.045	1.000	0.070	1.020	1.000	0.030	1.012	1.000	0.018	1.010	1.000	0.016
	MIN	1.060	0.940	0.060	1.020	0.975	0.020	1.006	0.990	0.006	1.002	0.994	0.002	1.000	0.994	0.000
1.2	MAX	1.320	1.200	0.180	1.245	1.200	0.070	1.220	1.200	0.030	1.212	1.200	0.018	1.210	1.200	0.016
	MIN	1.260	1.140	0.060	1.220	1.175	0.020	1.206	1.190	0.006	1.202	1.194	0.002	1.200	1.194	0.000
1.6	MAX	1.720	1.600	0.180	1.645	1.600	0.070	1.620	1.600	0.030	1.612	1.600	0.018	1.610	1.600	0.016
	MIN	1.660	1.540	0.060	1.620	1.575	0.020	1.606	1.590	0.006	1.602	1.594	0.002	1.600	1.594	0.000
2	MAX	2.120	2.000	0.180	2.045	2.000	0.070	2.020	2.000	0.030	2.012	2.000	0.018	2.010	2.000	0.016
	MIN	2.060	1.940	0.060	2.020	1.975	0.020	2.006	1.990	0.006	2.002	1.994	0.002	2.000	1.994	0.000
2.5	MAX	2.620	2.500	0.180	2.545	2.500	0.070	2.520	2.500	0.030	2.512	2.500	0.018	2.510	2.500	0.016
	MIN	2.560	2.440	0.060	2.520	2.475	0.020	2.506	2.490	0.006	2.502	2.494	0.002	2.500	2.494	0.000
3	MAX	3.120	3.000	0.180	3.045	3.000	0.070	3.020	3.000	0.030	3.012	3.000	0.018	3.010	3.000	0.016
	MIN	3.060	2.940	0.060	3.020	2.975	0.020	3.006	2.990	0.006	3.002	2.994	0.002	3.000	2.994	0.000
4	MAX	4.145	4.000	0.220	4.060	4.000	0.090	4.028	4.000	0.040	4.016	4.000	0.024	4.012	4.000	0.020
	MIN	4.070	3.925	0.070	4.030	3.970	0.030	4.010	3.988	0.010	4.004	3.992	0.004	4.000	3.992	0.000
5	MAX	5.145	5.000	0.220	5.060	5.000	0.090	5.028	5.000	0.040	5.016	5.000	0.024	5.012	5.000	0.020
	MIN	5.070	4.925	0.070	5.030	4.970	0.030	5.010	4.988	0.010	5.004	4.992	0.004	5.000	4.992	0.000
6	MAX	6.145	6.000	0.220	6.060	6.000	0.090	6.028	6.000	0.040	6.016	6.000	0.024	6.012	6.000	0.020
	MIN	6.070	5.925	0.070	6.030	5.970	0.030	6.010	5.988	0.010	6.004	5.992	0.004	6.000	5.992	0.000
8	MAX	8.170	8.000	0.260	8.076	8.000	0.112	8.035	8.000	0.050	8.020	8.000	0.029	8.015	8.000	0.024
	MIN	8.080	7.910	0.080	8.040	7.964	0.040	8.013	7.985	0.013	8.005	7.991	0.005	8.000	7.991	0.000
10	MAX	10.170	10.000	0.260	10.076	10.000	0.112	10.035	10.000	0.050	10.020	10.000	0.029	10.015	10.000	0.024
	MIN	10.080	9.910	0.080	10.040	9.964	0.040	10.013	9.985	0.013	10.005	9.991	0.005	10.000	9.991	0.000
12	MAX	12.205	12.000	0.315	12.093	12.000	0.136	12.043	12.000	0.061	12.024	12.000	0.035	12.018	12.000	0.029
	MIN	12.095	11.890	0.095	12.050	11.957	0.050	12.016	11.982	0.016	12.006	11.989	0.006	12.000	11.989	0.000
16	MAX	16.205	16.000	0.315	16.093	16.000	0.136	16.043	16.000	0.061	16.024	16.000	0.035	16.018	16.000	0.029
	MIN	16.095	15.890	0.095	16.050	15.957	0.050	16.016	15.982	0.016	16.006	15.989	0.006	16.000	15.989	0.000
20	MAX	20.240	20.000	0.370	20.117	20.000	0.169	20.053	20.000	0.074	20.028	20.000	0.041	20.021	20.000	0.034
	MIN	20.110	19.870	0.110	20.065	19.948	0.065	20.020	19.979	0.020	20.007	19.987	0.007	20.000	19.987	0.000
25	MAX	25.240	25.000	0.370	25.117	25.000	0.169	25.053	25.000	0.074	25.028	25.000	0.041	25.021	25.000	0.034
	MIN	25.110	24.870	0.110	25.065	24.948	0.065	25.020	24.979	0.020	25.007	24.987	0.007	25.000	24.987	0.000
30	MAX	30.240	30.000	0.370	30.117	30.000	0.169	30.053	30.000	0.074	30.028	30.000	0.041	30.021	30.000	0.034
	MIN	30.110	29.870	0.110	30.065	29.948	0.065	30.020	29.979	0.020	30.007	29.987	0.007	30.000	29.987	0.000
40	MAX	40.280	40.000	0.440	40.142	40.000	0.204	40.064	40.000	0.089	40.034	40.000	0.050	40.025	40.000	0.041
	MIN	40.120	39.840	0.120	40.080	39.938	0.080	40.025	39.975	0.025	40.009	39.984	0.009	40.000	39.984	0.000
50	MAX	50.290	50.000	0.450	50.142	50.000	0.204	50.064	50.000	0.089	50.034	50.000	0.050	50.025	50.000	0.041
	MIN	50.130	49.840	0.130	50.080	49.938	0.080	50.025	49.975	0.025	50.009	49.984	0.009	50.000	49.984	0.000
60	MAX	60.330	60.000	0.520	60.174	60.000	0.248	60.076	60.000	0.106	60.040	60.000	0.059	60.030	60.000	0.049
	MIN	60.140	59.810	0.140	60.100	59.926	0.100	60.030	59.970	0.030	60.010	59.981	0.010	60.000	59.981	0.000
80	MAX	80.340	80.000	0.530	80.174	80.000	0.248	80.076	80.000	0.106	80.040	80.000	0.059	80.030	80.000	0.049
	MIN	80.150	79.810	0.150	80.100	79.926	0.100	80.030	79.970	0.030	80.010	79.981	0.010	80.000	79.981	0.000
100	MAX	100.390	100.000	0.610	100.207	100.000	0.294	100.090	100.000	0.125	100.047	100.000	0.069	100.035	100.000	0.057
	MIN	100.170	99.780	0.170	100.120	99.913	0.120	100.036	99.965	0.036	100.012	99.978	0.012	100.000	99.978	0.000
120	MAX	120.400	120.000	0.620	120.207	120.000	0.294	120.090	120.000	0.125	120.047	120.000	0.069	120.035	120.000	0.057
	MIN	120.180	119.780	0.180	120.120	119.913	0.120	120.036	119.965	0.036	120.012	119.978	0.012	120.000	119.978	0.000
160	MAX	160.460	160.000	0.710	160.245	160.000	0.345	160.106	160.000	0.146	160.054	160.000	0.079	160.040	160.000	0.065
	MIN	160.210	159.750	0.210	160.145	159.900	0.145	160.043	159.960	0.043	160.014	159.975	0.014	160.000	159.975	0.000

TABLE 49 Preferred shaft basis fits. (Dimensions in millimeters.)

Example Ø 16N7/h6

Hole Size Ø $\frac{15.995}{15.977}$

Shaft Size Ø $\frac{16.000}{15.989}$

Max. Clearance 0.006

Max. Interference 0.023

Example Ø 45U7/h6

Hole Size Ø $\frac{44.939}{44.914}$

Shaft Size Ø $\frac{45.000}{44.984}$

Min. Interference 0.045

Max. Interference 0.086

PREFERRED SHAFT BASIS TRANSITION AND INTERFERENCE FITS															
UP TO AND INCLUDING	LOCATIONAL TRANSN.			LOCATIONAL TRANSN.			LOCATIONAL INTERF.			MEDIUM DRIVE			FORCE		
	Hole K7	Shaft h6	Fit	Hole N7	Shaft h6	Fit	Hole P7	Shaft h6	Fit	Hole S7	Shaft h6	Fit	Hole U7	Shaft h6	Fit
1 MAX	1.000	1.000	0.006	0.996	1.000	0.002	0.994	1.000	0.000	0.986	1.000	−0.008	0.982	1.000	−0.012
MIN	0.990	0.994	−0.010	0.986	0.994	−0.014	0.984	0.994	−0.016	0.976	0.994	−0.024	0.972	0.994	−0.028
1.2 MAX	1.200	1.200	0.006	1.196	1.200	0.002	1.194	1.200	0.000	1.186	1.200	−0.008	1.182	1.200	−0.012
MIN	1.190	1.194	−0.010	1.186	1.194	−0.014	1.184	1.194	−0.016	1.176	1.194	−0.024	1.172	1.194	−0.028
1.6 MAX	1.600	1.600	0.006	1.596	1.600	0.002	1.594	1.600	0.000	1.586	1.600	−0.008	1.582	1.600	−0.012
MIN	1.590	1.594	−0.010	1.586	1.594	−0.014	1.584	1.594	−0.016	1.576	1.594	−0.024	1.572	1.594	−0.028
2 MAX	2.000	2.000	0.006	1.996	2.000	0.002	1.994	2.000	0.000	1.986	2.000	−0.008	1.982	2.000	−0.012
MIN	1.990	1.994	−0.010	1.986	1.994	−0.014	1.984	1.994	−0.016	1.976	1.994	−0.024	1.972	1.994	−0.028
2.5 MAX	2.500	2.500	0.006	2.496	2.500	0.002	2.494	2.500	0.000	2.486	2.500	−0.008	2.482	2.500	−0.012
MIN	2.490	2.494	−0.010	2.486	2.494	−0.014	2.484	2.494	−0.016	2.476	2.494	−0.024	2.472	2.494	−0.028
3 MAX	3.000	3.000	0.006	2.996	3.000	0.002	2.994	3.000	0.000	2.986	3.000	−0.008	2.982	3.000	−0.012
MIN	2.990	2.994	−0.010	2.986	2.994	−0.014	2.984	2.994	−0.016	2.976	2.994	−0.024	2.972	2.994	−0.028
4 MAX	4.003	4.000	0.011	3.996	4.000	0.004	3.992	4.000	0.000	3.985	4.000	−0.007	3.981	4.000	−0.011
MIN	3.991	3.992	−0.009	3.984	3.992	−0.016	3.980	3.992	−0.020	3.973	3.992	−0.027	3.969	3.992	−0.031
5 MAX	5.003	5.000	0.011	4.996	5.000	0.004	4.992	5.000	0.000	4.985	5.000	−0.007	4.981	5.000	−0.011
MIN	4.991	4.992	−0.009	4.984	4.992	−0.016	4.980	4.992	−0.020	4.973	4.992	−0.027	4.969	4.992	−0.031
6 MAX	6.003	6.000	0.011	5.996	6.000	0.004	5.992	6.000	0.000	5.985	6.000	−0.007	5.981	6.000	−0.011
MIN	5.991	5.992	−0.009	5.984	5.992	−0.016	5.980	5.992	−0.020	5.973	5.992	−0.027	5.969	5.992	−0.031
8 MAX	8.005	8.000	0.014	7.996	8.000	0.005	7.991	8.000	0.000	7.983	8.000	−0.008	7.978	8.000	−0.013
MIN	7.990	7.991	−0.010	7.981	7.991	−0.019	7.976	7.991	−0.024	7.968	7.991	−0.032	7.963	7.991	−0.037
10 MAX	10.005	10.000	0.014	9.996	10.000	0.005	9.991	10.000	0.000	9.983	10.000	−0.008	9.978	10.000	−0.013
MIN	9.990	9.991	−0.010	9.981	9.991	−0.019	9.976	9.991	−0.024	9.968	9.991	−0.032	9.963	9.991	−0.037
12 MAX	12.006	12.000	0.017	11.995	12.000	0.006	11.989	12.000	0.000	11.979	12.000	−0.010	11.974	12.000	−0.015
MIN	11.988	11.989	−0.012	11.977	11.989	−0.023	11.971	11.989	−0.029	11.961	11.989	−0.039	11.950	11.989	−0.044
16 MAX	16.006	16.000	0.017	15.995	16.000	0.006	15.989	16.000	0.000	15.979	16.000	−0.010	15.974	16.000	−0.015
MIN	15.988	15.989	−0.012	15.977	15.989	−0.023	15.971	15.989	−0.029	15.961	15.989	−0.039	15.956	15.989	−0.044
20 MAX	20.006	20.000	0.019	19.993	20.000	0.006	19.986	20.000	−0.001	19.973	20.000	−0.014	19.967	20.000	−0.020
MIN	19.985	19.987	−0.015	19.972	19.987	−0.028	19.965	19.987	−0.035	19.952	19.987	−0.048	19.946	19.987	−0.054
25 MAX	25.006	25.000	0.019	24.993	25.000	0.006	24.986	25.000	−0.001	24.973	25.000	−0.014	24.960	25.000	−0.027
MIN	24.985	24.987	−0.015	24.972	24.987	−0.028	24.965	24.987	−0.035	24.952	24.987	−0.048	24.939	24.987	−0.061
30 MAX	30.006	30.000	0.019	29.993	30.000	0.006	29.986	30.000	−0.001	29.973	30.000	−0.014	29.960	30.000	−0.027
MIN	29.985	29.987	−0.015	29.972	29.987	−0.028	29.965	29.987	−0.035	29.952	29.987	−0.048	29.939	29.987	−0.061
40 MAX	40.007	40.000	0.023	39.992	40.000	0.008	39.983	40.000	−0.001	39.966	40.000	−0.018	39.949	40.000	−0.035
MIN	39.982	39.984	−0.018	39.967	39.984	−0.033	39.958	39.984	−0.042	39.941	39.984	−0.059	39.924	39.984	−0.076
50 MAX	50.007	50.000	0.023	49.992	50.000	0.008	49.983	50.000	−0.001	49.966	50.000	−0.018	49.939	50.000	−0.045
MIN	49.982	49.984	−0.018	49.967	49.984	−0.033	49.958	49.984	−0.042	49.941	49.984	−0.059	49.914	49.984	−0.086
60 MAX	60.009	60.000	0.028	59.991	60.000	0.010	59.979	60.000	−0.002	59.958	60.000	−0.023	59.924	60.000	−0.057
MIN	59.979	59.981	−0.021	59.961	59.981	−0.039	59.949	59.981	−0.051	59.928	59.981	−0.072	59.894	59.981	−0.106
80 MAX	80.009	80.000	0.028	79.991	80.000	0.010	79.979	80.000	−0.002	79.952	80.000	−0.029	79.909	80.000	−0.072
MIN	79.979	79.981	−0.021	79.961	79.981	−0.039	79.949	79.981	−0.051	79.922	79.981	−0.078	79.879	79.981	−0.121
100 MAX	100.010	100.000	0.032	99.990	100.000	0.012	99.976	100.000	−0.002	99.942	100.000	−0.036	99.889	100.000	−0.089
MIN	99.975	99.978	−0.025	99.955	99.978	−0.045	99.941	99.978	−0.059	99.907	99.978	−0.093	99.854	99.978	−0.146
120 MAX	120.010	120.000	0.032	119.990	120.000	0.012	119.976	120.000	−0.002	119.934	120.000	−0.044	119.869	120.000	−0.109
MIN	119.975	119.978	−0.025	119.955	119.978	−0.045	119.941	119.978	−0.059	119.899	119.978	−0.101	119.834	119.978	−0.166
160 MAX	160.012	160.000	0.037	159.988	160.000	0.013	159.972	160.000	−0.003	159.915	160.000	−0.060	159.825	160.000	−0.150
MIN	159.972	159.975	−0.028	159.948	159.975	−0.052	159.932	159.975	−0.068	159.875	159.975	−0.125	159.785	159.975	−0.215

TABLE 49   Preferred shift basis fits. (Dimensions in millimeters.)
(continued)

MORSE TAPERS		
	TAPER	
No. of Taper	inches per Foot	mm per 100 mm
0	.625	5.21
1	.599	4.99
2	.599	4.99
3	.602	5.02
4	.623	5.19
5	.631	5.26
6	.626	5.22
7	.624	5.20

BROWN AND SHARPE TAPERS		
	TAPER	
No. of Taper	inches per Foot	mm per 100 mm
1	.502	4.18
2	.502	4.18
3	.502	4.18
4	.502	4.18
5	.502	4.18
6	.503	4.19
7	.502	4.18
8	.501	4.18
9	.501	4.18
10	.516	4.3
11	.501	4.18
12	.500	4.17
13	.500	4.17
14	.500	4.17
15	.500	4.17
16	.500	4.17

TABLE 50   Machine tapers.

PRODUCT	OUTSTANDING FEATURES	APPLICATION METHOD	COLOR
1357	A high performance adhesive with long bonding range, excellent initial strength. Meets specification requirements of MMM-A-121 (supersedes MIL-A-1154 C), MIL-A-5092 B, Type II, and MIL-A-21366. Bonds rubber, cloth, wood, foamed glass, paper honeycomb, decorative plastic laminates. Also used with metal-to-metal for bonds of moderate strength.	Spray or Brush	Gray/ Green or Olive
2210	Fast drying, exhibits aggressive tack that allows coated surfaces to knit easily under hand roller pressure. Excellent water and oil resistance. Meets specification requirements of MMM-A-121 (supersedes MIL-A-1154 C), MIL-A-21366, and MMM-A-00130a. Bonds a wide range of materials including rubber, leather, cloth, aluminum, wood, hardboard. Used extensively for bonding decorative plastic laminates.	Brush, Roller, or Trowel	Yellow
2215	Fast drying and has a rapid rate of strength build-up. Its aggressive tack permits adhesive coated surfaces to bond easily with moderate pressure. Bonds decorative plastic laminates to metal, wood, and particle board. Also used for general bonding of rubber, leather, cloth, aluminum, wood, hardboard, etc.	Spray	Light Yellow
2218	Offers rapid strength build-up, high-ultimate strength. Has a high softening point and excellent resistance to plastic flow. Adhesion to steel is especially good. Meets specification requirements of MMM-A-121 (supersedes MIL-A-1154 C). Bonds high density decorative plastic laminates to metal or wood. Widely used to fabricate honeycomb and sandwich-type building panels with various face sheets, including porcelain enamel steel.	Spray	Green
2226	Water dispersed, has high immediate bond strength, long bonding range. Changes color from blue to green while drying. Used to bond foamed plastics, plastic laminate, wood, rubber, plywood, wallboard, wood veneer, plaster, and canvas to themselves and to each other.	Spray or Brush	Wet: Lt. Blue Dry: Green
4420	High performance, fast-drying adhesive designed for application by pressure curtain coating and mechanical roll coating. Bonds decorative plastic laminates to plywood or particle board and is suitable for conveyor line production of laminated panels of various types such as aluminum to wood or hardboard.	Roll Coating	Yellow
4488	Lower viscosity version of Cement 4420 for use specifically with flow-over or Weir-type curtain coaters. Bonds decorative plastic laminates to plywood or particle board and is suitable for conveyor line production of laminated panels of various types, such as aluminum to wood or hardboard.	Curtain and Roll Coating	Yellow
4518	Designed for spray application with automatic or production line equipment. Dries very fast, requires pressure from a niproll (rotary press) or platen press to assure proper bonding. Bonds decorative plastic laminates to plywood or particle board on both flat work and postforming. Also used for conveyor line production of laminated panels of various types such as aluminum to wood or hardboard. Meets requirements of MIL-A-5092 B, Type II.	Spray	Green
4729	Similar to Cement 2218 except that it is formulated with a nonflammable solvent. Requires force drying to prevent blushing.	Spray	Red
5034	Water dispersed, has high immediate bond strength and long bonding range. Bonds foamed plastics, plastic laminate, wood, rubber, plywood, wallboard, wood veneer, plaster, and canvas to themselves and to each other.	Spray or Brush	Neutral

TABLE 51   Physical properties and application data of adhesives. *(3M Co.)*

NORTH AMERICAN GAGES													EUROPEAN GAGES								
Ferrous metals, such as galvanized steel, tin plate						Nonferrous metals, such as copper, brass, aluminum			Steel and iron wire and bare copper piano wire			Galvanized steel, tin plate, copper, strip steel and steel, copper and aluminum tubes						Nonferrous		
U.S. Standard (USS)			U.S. Standard (Revised) Formerly Manufactures Standard			American Standard or Brown and Sharpe (B & S)			United States Steel Wire Gage			Birmingham (BWG)			New Birmingham (BG)			Imperial Wire Gage Imperial Standard (SWG)		
Gage	in.	mm	Gage	in.	mm	Gage	in.	mm	Gage	in.	mm	Gage	in.	mm	Gage	in.	mm	Gage	in.	mm
			3	.240	6.01	3	.229	5.83												
4	.234	5.95	4	.224	5.70	4	.204	5.19	4	.225	5.72	4	.238	6.05	4	.250	6.35	4	.232	5.89
5	.219	5.56	5	.209	5.31	5	.182	4.62	5	.207	5.26	5	.220	5.59	5	.223	5.65	5	.212	5.39
6	.203	5.16	6	.194	4.94	6	.162	4.12	6	.192	4.88	6	.203	5.16	6	.198	5.03	6	.192	4.88
7	.188	4.76	7	.179	4.55	7	.144	3.67	7	.177	4.50	7	.180	4.57	7	.176	4.48	7	.176	4.47
8	.172	4.37	8	.164	4.18	8	.129	3.26	8	.162	4.11	8	.165	4.19	8	.157	3.99	8	.160	4.06
9	.156	3.97	9	.149	3.80	9	.114	2.91	9	.148	3.77	9	.148	3.76	9	.140	3.55	9	.144	3.66
10	.141	3.57	10	.135	3.42	10	.102	2.59	10	.135	3.43	10	.134	3.40	10	.125	3.18	10	.128	3.25
11	.125	3.18	11	.120	3.04	11	.091	2.30	11	.121	3.06	11	.120	3.05	11	.111	2.83	11	.116	2.95
12	.109	2.78	12	.105	2.66	12	.081	2.05	12	.106	2.68	12	.109	2.77	12	.099	2.52	12	.104	2.64
13	.094	2.38	13	.090	2.78	13	.072	1.83	13	.092	2.32	13	.095	2.41	13	.088	2.24	13	.092	2.34
14	.078	1.98	14	.075	1.90	14	.064	1.63	14	.080	2.03	14	.083	2.11	14	.079	1.99	14	.080	2.03
15	.070	1.79	15	.067	1.71	15	.057	1.45	15	.072	1.83	15	.072	1.83	15	.070	1.78	15	.072	1.83
16	.063	1.59	16	.060	1.52	16	.051	1.29	16	.063	1.63	16	.065	1.65	16	.063	1.59	16	.064	1.63
17	.056	1.43	17	.054	1.37	17	.045	1.15	17	.054	1.37	17	.058	1.47	17	.056	1.41	17	.056	1.42
18	.050	1.27	18	.048	1.21	18	.040	1.02	18	.048	1.21	18	.049	1.25	18	.050	2.58	18	.048	1.22
19	.044	1.11	19	.042	1.06	19	.036	0.91	19	.041	1.04	19	.042	1.07	19	.044	1.19	19	.040	1.02
20	.038	0.95	20	.036	0.91	20	.032	0.81	20	.035	0.88	20	.035	0.89	20	.039	1.00	20	.036	0.91
21	.034	0.87	21	.033	0.84	21	.029	0.72	21	.032	0.81	21	.032	0.81	21	.035	0.89	21	.032	0.81
22	.031	0.79	22	.030	0.76	22	.025	0.65	22	.029	0.73	22	.028	0.71	22	.031	0.79	22	.028	0.71
23	.028	0.71	23	.027	0.68	23	.023	0.57	23	.026	0.66	23	.025	0.64	23	.028	0.71	23	.024	0.61
24	.025	0.64	24	.024	0.61	24	.020	0.51	24	.023	0.58	24	.022	0.56	24	.025	0.63	24	.022	0.56
25	.022	0.56	25	.021	0.53	25	.018	0.46	25	.020	0.52	25	.020	0.51	25	.022	0.56	25	.020	0.51
26	.019	0.48	26	.018	0.46	26	.016	0.40	26	.018	0.46	26	.018	0.46	26	.020	0.50	26	.018	0.46
27	.017	0.44	27	.016	0.42	27	.014	0.36	27	.017	0.44	27	.016	0.41	27	.017	0.44	27	.016	0.42
28	.016	0.40	28	.015	0.38	28	.013	0.32	28	.016	0.41	28	.014	0.36	28	.016	0.40	28	.015	0.38
29	.014	0.36	29	.014	0.34	29	.011	0.29	29	.015	0.38	29	.013	0.33	29	.014	0.35	29	.014	0.35
30	.013	0.32	30	.012	0.31	30	.010	0.25	30	.014	0.36	30	.012	0.31	30	.012	0.31	30	.012	0.32
31	.011	0.28	31	.011	0.27	31	.009	0.23	31	.013	0.34	31	.010	0.25	31	.011	0.28			
32	.010	0.26	32	.010	0.25	32	.008	0.20	32	.013	0.33	32	.009	0.23				32	.011	0.27
33	.009	0.24	33	.009	0.23	33	.007	0.18	33	.012	0.30	33	.008	0.20	33	.009	0.22	33	.010	0.25
34	.009	0.22	34	.008	0.21	34	.006	0.16	34	.010	0.26	34	.007	0.18	34	.008	0.20	34	.009	0.23
									35	.010	0.24	35	.005	0.13	35	.007	0.18	35	.008	0.21
36	.007	0.18	36	.007	0.17	36	.005	0.13	36	.009	0.23	36	.004	0.10	36	.006	0.16			
									37	.008	0.22							37	.007	0.17
38	.006	0.16	38	.006	0.15	38	.004	0.10	38	.008	0.20				38	.005	0.12	38	.006	0.15
									39	.008	0.19									
									40	.007	0.18				40	.004	0.10	40	.005	0.12
									41	.007	0.17							42	.004	0.10

NOTE: METRIC STANDARDS GOVERNING GAGE SIZES WERE NOT AVAILABLE AT THE TIME OF PUBLICATION. THE SIZES GIVEN IN THE ABOVE CHART ARE SOFT CONVERSION FROM CURRENT INCH STANDARDS AND ARE NOT MEANT TO BE REPRESENTATIVE OF THE PRECISE METRIC GAGE SIZES WHICH MAY BE AVAILABLE IN THE FUTURE. CONVERSIONS ARE GIVEN ONLY TO ALLOW THE STUDENT TO COMPARE GAGE SIZES READILY WITH THE METRIC DRILL SIZES.

TABLE 52 Wire and sheet-metal gages and thicknesses.

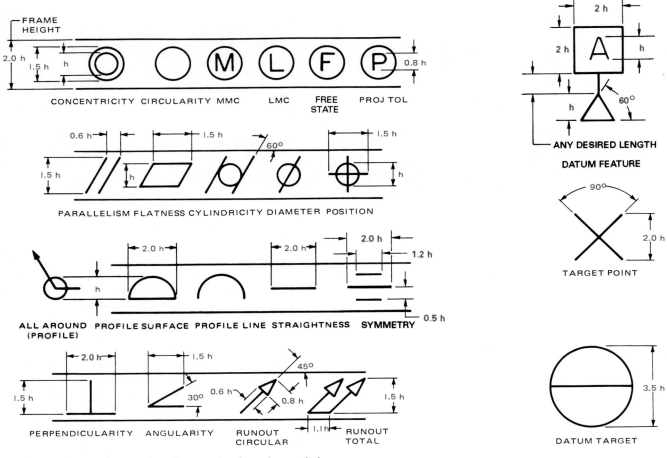

TABLE 53 Form and proportion of geometric tolerancing symbols.

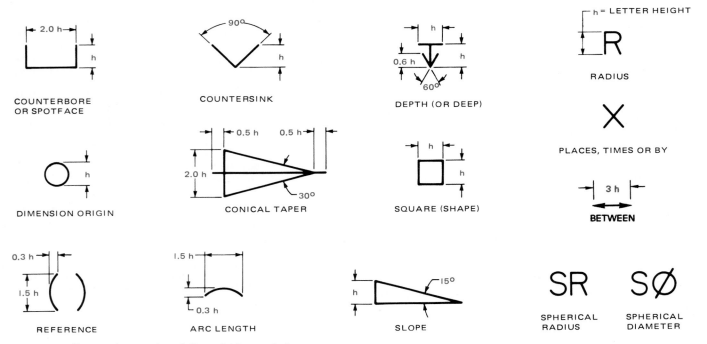

TABLE 54 Form and proportion of dimensioning symbols.

SYMBOL FOR:	ASME Y14.5M–1994	ISO	CAN/CSA–B78.2–M91
STRAIGHTNESS	—	—	—
FLATNESS	▱	▱	▱
CIRCULARITY	○	○	○
CYLINDRICITY	⌭	⌭	⌭
PROFILE OF A LINE	⌒	⌒	⌒
PROFILE OF A SURFACE	⌓	⌓	⌓
ALL AROUND–PROFILE	⟲	⟲	⟲
ANGULARITY	∠	∠	∠
PERPENDICULARITY	⊥	⊥	⊥
PARALLELISM	∥	∥	∥
POSITION	⊕	⊕	⊕
CONCENTRICITY/COAXIALITY	◎	◎	◎
SYMMETRY	⩵	⩵	⩵
CIRCULAR RUNOUT	* ↗	↗	* ↗
TOTAL RUNOUT	* ↗↗	↗↗	* ↗↗
AT MAXIMUM MATERIAL CONDITION	Ⓜ	Ⓜ	Ⓜ
AT LEAST MATERIAL CONDITION	Ⓛ	Ⓛ	Ⓛ (PROPOSED)
REGARDLESS OF FEATURE SIZE	NONE	NONE	NONE
PROJECTED TOLERANCE ZONE	Ⓟ	Ⓟ	Ⓟ
DIAMETER	⌀	⌀	⌀
BASIC DIMENSION	50	50	50
REFERENCE DIMENSION	(50)	(50)	(50)
DATUM FEATURE	*▨ A	*⌷ OR *⌷ A	*▨ A
DATUM TARGET	⌀6/A1	⌀6/A1	⌀6/A1
TARGET POINT	✕	✕	✕
DIMENSION ORIGIN	⊕→	⊕→	⊕→
FEATURE CONTROL FRAME	⊕ ⌀0.5Ⓜ A B C	⊕ ⌀0.5Ⓜ A B C	⊕ ⌀0.5Ⓜ A B C
CONICAL TAPER	▷	▷	▷
SLOPE	◁	◁	◁
COUNTERBORE/SPOTFACE	⌴	⌴ (PROPOSED)	⌴
COUNTERSINK	⌵	⌵ (PROPOSED)	⌵
DEPTH/DEEP	⤓	⤓ (PROPOSED)	⤓
SQUARE (SHAPE)	□	□	□
DIMENSION NOT TO SCALE	15	15	15
NUMBER OF TIMES/PLACES	8X	8X	8X
ARC LENGTH	⏜105	⏜105	⏜105
RADIUS	R	R	R
SPHERICAL RADIUS	SR	SR	SR
SPHERICAL DIAMETER	S⌀	S⌀	S⌀
BETWEEN	↔	NONE	↔ (PROPOSED)

* MAY BE FILLED IN

Table 55 Comparison of ASME (ANSI), ISO, and CSA symbols

A-45

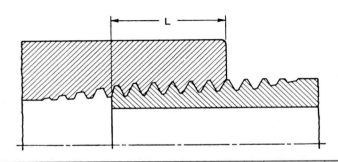

	Nominal Pipe Size Inches	Outside Diameter	Threads Per Inch	Wall Thickness			Approx. Distance Pipe Enters Fitting L	Weight (lbs/ft)		
				Sched. 40 (Standard)	Sched. 80 (Extra Strong)	Sched. 160		Sched. 40 (Standard)	Sched. 80 (Extra Strong)	Sched. 160
U.S. CUSTOMARY (INCHES)	⅛ (.125)	.405	27	.068	.095	—	.188	.24	.31	—
	¼ (.250)	.540	18	.088	.119	—	.281	.42	.54	—
	⅜ (.375)	.675	18	.091	.126	—	.297	.57	.74	—
	½ (.500)	.840	14	.109	.147	.188	.375	.85	1.09	1.31
	¾ (.750)	1.050	14	.113	.154	.219	.406	1.13	1.47	1.94
	1.00	1.315	11.50	.133	.179	.250	.500	1.68	2.17	2.84
	1.25	1.660	11.50	.140	.191	.250	.549	2.27	3.00	3.76
	1.50	1.900	11.50	.145	.200	.281	.562	2.72	3.63	4.86
	2	2.375	11.50	.154	.218	.344	.578	3.65	5.02	7.46
	2.5	2.875	8	.203	.276	.375	.875	5.79	7.66	10.01
	3	3.500	8	.216	.300	.438	.938	7.58	10.25	14.31
	3.5	4.000	8	.226	.318	—	1.000	9.11	12.51	—
	4	4.500	8	.237	.337	.531	1.062	10.79	14.98	22.52
	5	5.563	8	.258	.375	.625	1.156	14.62	20.78	32.96
	6	6.625	8	.280	.432	.719	1.250	18.97	28.57	45.34
	8	8.625	8	.322	.500	.906	1.469	28.55	43.39	74.71

	Nominal Pipe Size Inches	Outside Diameter mm	Threads Per Inch	Wall Thickness			Approx. Distance Pipe Enters Fitting L	Mass (kg/m)		
				Sched. 40 (Standard)	Sched. 80 (Extra Strong)	Sched. 160		Sched. 40 (Standard)	Sched. 80 (Extra Strong)	Sched. 160
METRIC (MILLIMETERS)	⅛ (.125)	10.3	27	1.7	2.4	—	5	0.36	0.46	—
	¼ (.250)	13.7	18	2.2	3.0	—	7	0.63	0.80	—
	⅜ (.375)	17.1	18	2.3	3.2	—	8	0.85	1.10	—
	½ (.500)	21.3	14	2.8	3.7	4.8	10	1.26	1.62	1.95
	¾ (.750)	26.7	14	2.9	3.9	5.6	11	1.68	2.19	2.89
	1.00	33.4	11.50	3.4	4.6	6.4	13	2.50	3.23	4.23
	1.25	42.1	11.50	3.6	4.9	6.4	14	3.38	4.46	5.60
	1.50	48.3	11.50	3.7	5.1	7.1	14	4.05	5.40	7.23
	2.00	60.3	11.50	3.9	5.5	8.7	15	5.43	7.47	11.10
	2.50	73	8	5.2	7.0	9.5	22	8.62	11.40	14.90
	3.00	88.9	8	5.5	7.6	11.1	24	11.28	15.25	21.30
	3.50	101.6	8	5.7	8.1	—	25	13.56	18.62	—
	4.00	114.3	8	6.0	8.6	13.5	27	16.06	22.30	33.51
	5.00	141.3	8	6.6	9.5	15.9	29	21.76	30.92	49.05
	6.00	168.3	8	7.1	11.0	18.3	32	28.23	42.52	67.47
	8.00	219	8	8.2	12.7	23.0	38	42.49	64.57	111.18

TABLE 56   American standard wrought steel pipe.

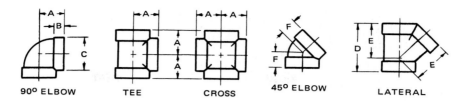

90° ELBOW   TEE   CROSS   45° ELBOW   LATERAL

Nominal Pipe Size in Inches	U.S. CUSTOMARY (INCHES)						METRIC (MILLIMETERS)					
	A	Min B	Min C	D	E	F	A	Min B	Min C	D	E	F
.25	.81	.38	.93	—	—	.73	21	10	24	—	—	19
.375	.95	.44	1.12	—	—	.80	24	11	28	—	—	20
.50	1.12	.50	1.34	2.50	1.87	.88	28	13	34	64	47	22
.75	1.31	.56	1.63	3.00	2.25	.98	33	14	41	76	57	25
1.00	1.50	.62	1.95	3.50	2.75	1.12	38	16	50	89	70	28
1.25	1.75	.69	2.39	4.25	3.25	1.29	44	18	61	108	83	33
1.50	1.94	.75	2.68	4.87	3.81	1.43	49	19	68	124	97	36
2.00	2.25	.84	3.28	5.75	4.25	1.68	57	21	83	146	108	43
2.50	2.70	.94	3.86	6.75	5.18	1.95	69	24	98	171	132	50
3.00	3.08	1.00	4.62	7.87	6.12	2.17	78	25	117	200	155	55
3.50	3.42	1.06	5.20	8.87	6.87	2.39	87	27	132	225	174	61
4.00	3.79	1.12	5.79	9.75	7.62	2.61	96	28	147	248	194	66
5.00	4.50	1.18	7.05	11.62	9.25	3.05	114	30	179	295	235	77
6.00	5.13	1.28	8.28	13.43	10.75	3.46	130	33	210	341	273	88
8.00	6.56	1.47	10.63	16.94	13.63	4.28	167	37	270	430	346	109
10.00	8.08	1.68	13.12	20.69	16.75	5.16	205	43	333	613	425	131

TABLE 57   American standard (125 lb) cast-iron screwed-pipe fittings.

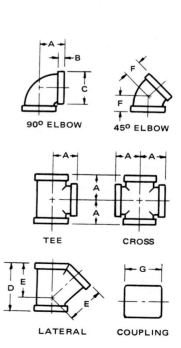

90° ELBOW   45° ELBOW   TEE   CROSS   LATERAL   COUPLING

Nominal Pipe Size in Inches	U.S. CUSTOMARY (INCHES)							METRIC (MILLIMETERS)						
	A	B	C	D	E	F	G	A	B	C	D	E	F	G
.125	.69	.20	.69	—	—	—	.96	18	5.0	18	—	—	—	24
.250	.81	.22	.84	—	—	.73	1.06	21	5	21	—	—	19	27
.375	.95	.23	1.02	1.93	1.43	.80	1.16	24	6	26	49	36	20	29
.500	1.12	.25	1.20	2.32	1.71	.88	1.34	28	6	30	59	43	22	34
.750	1.31	.27	1.46	2.77	2.05	.98	1.52	33	7	37	70	52	25	39
1.00	1.50	.30	1.77	3.28	2.43	1.12	1.67	38	8	45	83	62	28	42
1.25	1.75	.34	2.15	3.94	2.92	1.29	1.93	44	9	55	100	74	33	49
1.50	1.94	.37	2.43	4.38	3.28	1.43	2.15	49	9	62	111	83	36	55
2.00	2.25	.42	2.96	5.17	3.93	1.68	2.53	57	11	75	131	100	43	64
2.50	2.70	.48	3.59	6.25	4.73	1.95	2.88	69	12	91	159	120	50	73
3.00	3.08	.55	4.29	7.26	5.55	2.17	3.18	78	14	109	184	141	55	81
3.50	3.42	.60	4.84	—	—	2.39	3.43	87	15	123	—	—	61	87
4.00	3.79	.66	5.40	8.98		2.61	3.69	96	17	137	228	177	66	94
5.00	4.50	.78	6.58		6.97	3.05	—	114	20	167	—	—	77	—
6.00	5.13	.90	7.77			3.46		130	23	197	—	—	88	—

TABLE 58   American standard (150 lb) malleable-iron screwed-pipe fittings.

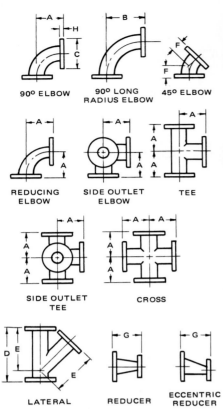

90° ELBOW 90° LONG RADIUS ELBOW 45° ELBOW

REDUCING ELBOW SIDE OUTLET ELBOW TEE

SIDE OUTLET TEE CROSS

LATERAL REDUCER ECCENTRIC REDUCER

NOMINAL PIPE SIZE IN INCHES		A	B	C	D	E	F	G	H
U.S. CUSTOMARY (INCHES)	1.50	4.00	6.00	5.00	9.00	7.00	2.25	—	.56
	2.00	4.50	6.50	6.00	10.50	8.00	2.50	5.00	.62
	2.50	5.00	7.00	7.00	12.00	9.50	3.00	5.50	.69
	3.00	5.50	7.75	7.50	13.00	10.00	3.00	6.00	.75
	3.50	6.00	8.50	8.50	14.50	11.50	3.50	6.50	.81
	4.00	6.50	9.00	9.00	15.00	12.00	4.00	7.00	.94
	5.00	7.50	10.25	10.00	17.00	13.50	4.50	8.00	.94
	6.00	8.00	11.50	11.00	18.00	14.50	5.00	9.00	1.00
	8.00	9.00	14.00	13.50	22.00	17.50	5.50	11.00	1.12
	10.00	11.00	16.50	16.00	25.50	20.50	6.50	12.00	1.19
METRIC (MILLIMETERS)	1.50	102	152	127	229	178	57	—	14
	2.00	114	165	152	267	203	64	127	16
	2.50	127	178	178	305	241	76	140	18
	3.00	140	197	190	330	254	76	152	19
	3.50	153	216	216	368	292	89	165	21
	4.00	165	229	229	381	305	102	178	24
	5.00	190	260	254	432	343	114	203	24
	6.00	203	292	280	457	368	127	229	25
	8.00	229	356	343	559	445	140	279	28
	10.00	279	419	406	648	521	165	305	30

TABLE 59 American standard flanged fittings.

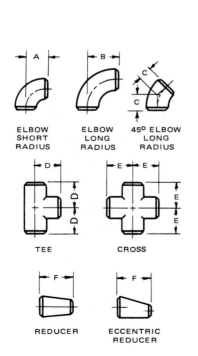

ELBOW SHORT RADIUS ELBOW LONG RADIUS 45° ELBOW LONG RADIUS

TEE CROSS

REDUCER ECCENTRIC REDUCER

NOMINAL PIPE SIZE IN INCHES		A	B	C	D	E	F
U.S. CUSTOMARY (INCHES)	1.50	1.50	2.25	1.12	2.25	2.25	2.50
	2.00	2.00	3.00	1.38	2.50	2.50	3.00
	2.50	2.50	3.75	1.75	3.00	3.00	3.50
	3.00	3.00	4.50	2.00	3.38	3.38	3.50
	3.50	3.50	5.25	2.25	3.75	3.75	4.00
	4.00	4.00	6.00	2.50	4.12	4.12	4.00
	5.00	5.00	7.50	3.12	4.89	4.89	5.00
	6.00	6.00	9.00	3.75	5.62	5.62	5.50
	8.00	8.00	12.00	5.00	7.00	7.00	6.00
	10.00	10.00	15.00	6.25	8.50	8.50	7.00
METRIC (MILLIMETERS)	1.50	38	57	28	57	57	64
	2.00	51	76	35	64	64	76
	2.50	64	95	44	76	76	89
	3.00	76	114	51	86	86	89
	3.50	89	133	57	95	95	102
	4.00	102	152	64	105	105	102
	5.00	127	190	79	124	124	127
	6.00	152	229	95	143	143	140
	8.00	203	305	127	178	178	152
	10.00	254	381	159	216	216	178

TABLE 60 American standard steel butt-welding fittings.

Nominal Pipe Size in Inches	Lift Check A Screwed	Flanged	B	Swing Check C Screwed	Flanged	D
U.S. CUSTOMARY (INCHES)						
2.00	6.50	—	3.50	6.50	8.00	4.25
2.50	7.00	—	4.25	7.00	8.50	4.81
3.00	8.00	9.50	5.00	8.00	9.50	5.06
3.50	—	—	—	9.00	10.50	5.81
4.00	10.00	11.50	6.25	10.00	11.50	6.19
5.00	—	13.00	7.00	11.25	13.00	7.19
6.00	—	14.00	8.25	12.50	14.00	7.50
8.00	—	—	—	—	19.50	10.19
10.00	—	—	—	—	24.50	12.12
METRIC (MILLIMETERS)						
2.00	165	—	89	165	203	108
2.50	178	—	108	178	216	122
3.00	203	241	127	203	241	129
3.50	—	—	—	229	267	148
4.00	254	292	159	254	292	157
5.00	—	330	178	286	330	183
6.00	—	356	210	318	356	190
8.00	—	—	—	—	495	259
10.00	—	—	—	—	622	308

CHECK VALVES

LIFT   SWING

Nominal Pipe Size in Inches	Nonrising Spindle E Screwed	Flanged	F	Rising Spindle G Screwed	Flanged	H
U.S. CUSTOMARY (INCHES)						
2.00	4.75	7.00	10.50	4.75	7.00	13.12
2.50	5.50	7.50	11.19	5.50	7.50	14.50
3.00	6.00	8.00	12.62	6.00	8.00	16.62
3.50	6.62	8.50	13.31	6.62	8.50	18.44
4.00	7.12	9.00	15.25	7.12	9.00	21.06
5.00	8.12	10.00	17.88	8.12	10.00	25
6.00	9.00	10.50	20.19	9.00	10.50	29.25
8.00	10.00	11.50	24.00	10.00	11.50	37.25
10.00	—	13.00	28.19	—	13.00	44.12
METRIC (MILLIMETERS)						
2.00	121	178	267	121	178	333
2.50	140	190	284	140	190	368
3.00	152	203	321	152	203	422
3.50	168	216	338	168	216	468
4.00	181	229	387	181	229	535
5.00	206	254	454	206	254	635
6.00	229	267	513	229	267	743
8.00	254	292	610	254	292	946
10.00	—	330	716	—	330	1121

GATE VALVES

Nominal Pipe Size in Inches	Globe J Screwed	Flanged	K	Angle L Screwed	Flanged	M
U.S. CUSTOMARY (INCHES)						
2.00	4.75	7.00	9.44	3.50	3.88	10.38
2.50	5.50	7.50	11.06	3.88	4.50	12.06
3.00	6.00	8.00	12.38	4.69	4.62	12.69
3.50	6.62	8.50	13.19	5.00	5.38	13.62
4.00	7.12	9.00	15.25	6.00	5.88	14.88
5.00	8.12	10.00	17.25	6.31	6.50	17.69
6.00	9.00	10.50	18.81	8.00	8.00	19.06
8.00	10.00	11.50	22.12	—	9.25	22.75
10.00	—	13.00	24.75	—	10.62	24.94
METRIC (MILLIMETERS)						
2.00	121	178	240	89	98	264
2.50	140	190	281	98	114	281
3.00	152	203	314	119	118	346
3.50	168	216	335	127	136	348
4.00	181	229	387	152	149	378
5.00	206	254	438	160	165	449
6.00	229	267	478	203	203	484
8.00	254	292	562	—	235	578
10.00	—	330	629	—	270	633

GLOBE AND ANGLE VALVES

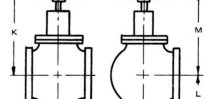

GLOBE   ANGLE

Dimensions taken from manufacturer's catalogs for drawing purposes.

TABLE 61   Common valves.

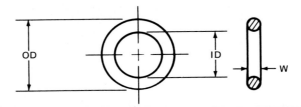

U.S. CUSTOMARY (INCHES)							METRIC (MILLIMETERS)					
	W = .071	W = .087	W = .125	W = .174	W = .209	W = .256		W = 1.5	W = 2	W = 3	W = 4	W = 5	W = 6
ID	OD	OD	OD	OD	OD	OD	ID	OD	OD	OD	OD	OD	OD
.124	x						3	6					
.158	x						4	7					
.132	x						5	8					
.248	x						6	9					
.280	x						7	10					
.295	x	x	x				8	11	12	14			
.354	x	x	x				9	12	13	15			
.394	x	x	x	x	x	x	10	13	14	16	18	20	
.465	x	x	x	x	x	x	12	15	16	18	20	22	24
.551	x	x	x	x	x	x	14	17	18	20	22	24	26
.591	x	x	x	x	x	x	15	18	19	21	23	25	27
.622	x	x	x	x	x	x	16	19	20	22	24	26	28
.630	x	x	x	x	x	x	18	21	22	24	26	28	30
.662	x	x	x	x	x	x	20	23	24	26	28	30	32
.669	x	x	x	x	x	x	22	25	26	28	30	32	34
.787	x	x	x	x	x	x	24	27	28	30	32	34	36
.850	x	x	x	x	x	x	25	28	29	31	33	35	37
1.024	x	x	x	x	x	x	26	29	30	32	34	36	38
1.102	x	x	x	x	x	x	28	31	32	34	36	38	40
1.181	x	x	x	x	x	x	30	33	34	36	38	40	42
1.220	x	x	x	x	x	x	32	35	36	38	40	42	44
1.299	x	x	x	x	x	x	34	37	38	40	42	44	46
1.319	x	x	x	x	x	x	35	38	39	41	43	45	47
1.496	x	x	x	x	x	x	36	39	40	42	44	46	48
1.524	x	x	x	x	x	x	38	41	42	44	46	48	50
1.575	x	x	x	x	x	x	40	43	44	46	48	50	52
1.654	x	x	x	x	x	x	42	45	46	48	50	52	54
1.732	x	x	x	x	x	x	44	47	48	50	52	54	56
1.772	x	x	x	x	x	x	45	48	49	51	53	55	57
1.811	x	x	x	x	x	x	46	49	50	52	54	56	58
1.890	x	x	x	x	x	x	48	51	52	54	56	58	60
2.008	x	x	x	x	x	x	50	53	54	56	58	60	62
2.087	x	x	x	x	x	x	52	55	56	58	60	62	64
2.165	x	x	x	x	x	x	54	57	58	60	62	64	66
2.205	x	x	x	x	x	x	55	58	59	61	63	65	67
2.244	x	x	x	x	x	x	56	59	60	62	64	66	68
2.323	x	x	x	x	x	x	58	61	62	64	66	68	70
2.402	x	x	x	x	x	x	60	63	64	66	68	70	72
2.639	x	x	x	x	x	x	65	68	69	71	73	75	77
2.835	x	x	x	x	x	x	70	73	74	76	78	80	82
3.031	x	x	x	x	x	x	75	78	79	81	83	85	87
3.228	x	x	x	x	x	x	80	83	84	86	88	90	92

TABLE 62   O-rings.

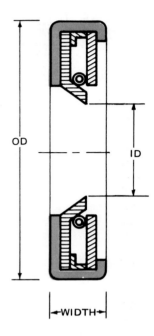

Inside Dia		Outside Dia		Width	
in.	mm	in.	mm	in.	mm
.375	10	.753	19	.25	6
.375	10	.840	21	.31	8
.438	11	1.003	26	.31	8
.438	11	1.128	28	.31	8
.500	12	1.003	26	.31	8
.500	12	1.128	28	.31	8
.500	12	1.254	32	.38	10
.562	14	1.003	26	.31	8
.562	14	1.128	28	.31	8
.625	16	1.250	32	.38	10
.625	16	1.128	28	.31	8
.625	16	1.250	32	.38	10
.688	18	1.379	35	.38	10
.688	18	1.128	28	.31	8
.688	18	1.254	32	.38	10
.750	20	1.379	35	.38	10
.750	20	1.254	32	.38	10
.750	20	1.379	35		10
.750	20	1.503	38	.38	10
.8125	21	1.756	44	.44	12
.8125	21	1.254	32	.38	10
.8125	21	1.379	35	.38	10
.8125	21	1.503	38	.38	10
.875	22	1.756	44	.44	12
.875	22	1.379	35	.38	10
.875	22	1.503	38	.38	10
.875	22	1.628	42	.44	12
.938	24	1.756	44	.44	12
.938	24	1.503	38	.38	10
.938	24	1.628	42	.38	10
1.000	25	1.756	44	.44	12
1.000	25	1.503	38	.38	10
1.000	25	1.756	44	.44	12
1.000	25	1.878	48	.44	12
1.000	25	2.004	50	.44	12

Inside Dia		Outside Dia		Width	
in.	mm	in.	mm	in.	mm
1.062	26	1.503	38	.38	10
	26	1.628	42	.44	12
	26	1.756	44	.44	12
1.125	28	1.628	42	.44	12
	28	1.756	44	.44	12
	28	1.987	50	.50	12
1.188	30	1.832	46	.44	12
	30	1.987	50	.50	12
	30	2.254	58	.50	12
1.25	32	1.756	44	.44	12
	32	1.878	48	.44	12
	32	2.066	52	.50	12
1.312	34	2.060	52	.44	12
	34	2.254	58	.50	12
	34	2.378	60	.50	12
1.375	35	2.066	52	.44	12
	35	2.254	58	.50	12
	35	2.441	62	.50	12
1.438	36	2.254	58	.50	12
	36	2.506	64	.50	12
	36	2.627	66	.50	12
1.500	38	2.254	58	.38	10
	38	2.410	62	.50	12
	38	2.720	70	.50	12
1.562	40	2.441	62	.50	12
	40	2.690	68	.50	12
	40	2.879	74	.50	12
1.625	42	2.441	62	.38	10
	42	2.879	74	.38	10
	42	2.627	66	.50	12
	42	2.879	74	.50	12
	42	3.066	78	.50	12
1.75	44	2.254	58	.50	12
	44	2.441	62	.50	12
	44	2.506	64	.50	12

TABLE 63　Oil seals.

Shaft Dia		Outside Dia		Width		Size of Set Screw	
in.	mm	in.	mm	in.	mm	in.	mm
.375	10	.75	20	.40	10	.250	M6
.500	12	1.00	25	.44	10	.250	M6
.625	16	1.10	28	.50	12	.3125	M8
.750	20	1.20	30	.56	14	.3125	M8
.875	22	1.50	40	.56	14	.3125	M8
1.000	24	1.60	40	.60	16	.3125	M8

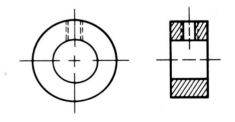

TABLE 64　Setscrew collars.

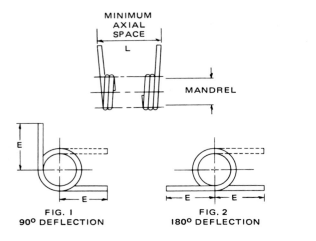

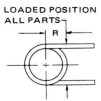

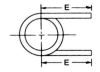

FIGURES SHOW SPRINGS WOUND LEFT-HAND

FIG. I
90° DEFLECTION

FIG. 2
180° DEFLECTION

FIG. 3
270° DEFLECTION

FIG. 4
360° DEFLECTION

U.S. CUSTOMARY (INCHES)								METRIC (MILLIMETERS)							
Wire Dia	Out-side Dia	Pos. of Ends Fig.	Def, Deg.	Torque in · lb	Radius R	E	Min. Axial Space L	Wire Dia	Out-side Dia	Pos. of Ends Fig.	Def, Deg.	Torque N · m	Radius R	E	Min. Axial Space L
.014	.124	1	90	.070	.250	.50	.067	0.35	3.2	1	90	0.008	6.4	13	1.7
	.133	2	180		.250	.50	.105		3.4	2	180		6.4	13	2.7
	.124	3	270		.250	.50	.158		3.2	3	270		6.4	13	4.0
	.194	2	180		.375	.75	.077		4.9	2	180		9.5	19	2.0
	.201	3	270		.375	.75	.102		5.1	3	270		9.5	19	2.6
	.204	4	360		.375	.75	.126		5.2	4	360		9.5	19	3.2
.017	.160	1	90	.117	.250	.50	.081	0.43	4.1	1	90	0.013	6.4	13	2.1
	.172	2	180		.250	.50	.127		4.4	2	180		6.4	13	3.2
	.160	3	270		.250	.50	.192		4.1	3	270		6.4	13	4.9
	.249	2	180		.375	.75	.094		6.3	2	180		9.5	19	2.4
	.259	3	270		.375	.75	.123		6.6	3	270		9.5	19	3.1
	.235	4	360		.375	.75	.170		6.0	4	360		9.5	19	4.3
.020	.191	1	90	.187	.375	.75	.095	0.51	4.9	1	90	0.021	9.5	19	2.4
	.179	2	180		.375	.75	.170		4.6	2	180		9.5	19	4.3
	.175	3	270		.375	.75	.245		4.5	3	270		9.5	19	6.2
	.242	2	180		.500	1.00	.130		6.2	2	180		12.7	25	3.3
	.268	3	270		.500	1.00	.165		6.8	3	270		12.7	25	4.2
	.254	4	360		.500	1.00	.250		6.5	4	360		12.7	25	6.4
.023	.204	1	90	.308	.375	.75	.109	0.59	5.2	1	90	0.035	9.5	19	2.8
	.191	2	180		.375	.75	.196		4.9	2	180		9.5	19	5.0
	.187	3	270		.375	.75	.282		4.8	3	270		9.5	19	7.2
	.259	2	180		.500	1.00	.150		6.6	2	180		12.7	25	3.8
	.251	3	270		.500	1.00	.213		6.4	3	270		12.7	25	5.4
	.271	4	360		.500	1.00	.253		6.9	4	360		12.7	25	6.5
.028	.267	1	90	.515	.500	1.00	.133	0.71	6.8	1	90	0.058	12.7	25	3.4
	.249	2	180		.500	1.00	.238		6.3	2	180		12.7	25	6.0
	.245	3	270		.500	1.00	.344		6.2	3	270		12.7	25	8.8
	.340	2	180		.500	1.00	.182		8.7	2	180		12.7	25	4.6
	.329	3	270		.500	1.00	.259		8.4	3	270		12.7	25	6.6
	.355	4	360		.500	1.00	.308		9.0	4	360		12.7	25	7.9

TABLE 65   Torsion springs. *(Wallace Barnes Co. Ltd.)*

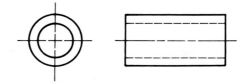

Inside Dia in.	mm	Outside Dia in.	mm	15 .50	20 .75	25 1.00	30 1.25	35 1.50	40 1.75	50 2.00	60 2.50	75 3.00
.375	10	.625	16	•	•	•	•					
		.750	20	•	•	•	•					
.500	12	.625	16	•	•	•	•	•	•			
		.750	20	•	•	•	•	•	•	•		
.625	16	.875	22	•	•	•	•	•	•	•	•	
		1.000	25	•	•	•	•	•	•	•	•	
.750	20	1.000	25	•	•	•	•	•	•	•	•	
		1.125	28	•	•	•	•	•	•	•	•	
.875	22	1.125	28		•	•	•	•	•	•	•	
		1.250	32		•	•	•	•	•	•	•	
1.000	25	1.250	32			•	•	•	•	•	•	•
		1.375	35			•	•	•	•	•	•	•

TABLE 66   Standard plain (journal) bearings.

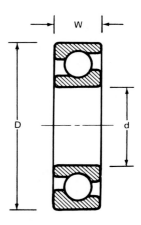

Series	Bearing Number	Bore d in.	mm	Outside Diameter D in.	mm	Width W in.	mm	Basic Load Rating Kips	kN
Light	7205B	0.984	25	2.047	52	0.591	15	2.5	11.4
	7206B	1.181	30	2.441	62	0.630	16	3.4	15.6
	7207B	1.378	35	2.835	72	0.670	17	4.6	20.8
	7208B	1.575	40	3.150	80	0.709	18	5.5	24.5
	7209B	1.772	45	3.347	85	0.748	19	6.2	27.5
	7210B	1.969	50	3.543	90	0.787	20	6.4	28.5
	7211B	2.165	55	3.937	100	0.827	21	8.2	36
	7212B	2.362	60	4.331	110	0.866	22	9.7	43
Medium	7304B	0.787	20	2.047	52	0.591	15	3.1	13.4
	7305B	0.984	25	2.441	62	0.670	17	4.3	19
	7306B	1.181	30	2.835	72	0.748	19	5.4	24
	7307B	1.378	35	3.150	80	0.827	21	6.3	28
	7308B	1.575	40	3.543	90	0.906	23	7.8	34.5
	7309B	1.772	45	3.937	100	0.984	25	10	45
	7310B	1.969	50	4.331	110	1.063	27	11.6	52
	7311B	2.165	55	4.724	120	1.142	29	13.4	61
	7312B	2.362	60	5.118	130	1.221	31	15.6	69.5
Heavy	7405B	0.984	25	3.150	80	0.827	21	6.8	30.5
	7406B	1.181	30	3.543	90	0.906	23	8.3	36.5
	7407B	1.378	35	3.937	100	0.984	25	10.4	46.5
	7408B	1.575	40	4.331	110	1.063	27	12	54
	7409B	1.772	45	4.724	120	1.142	29	14.6	65.5
	7410B	1.969	50	5.118	130	1.221	31	16.6	73.5
	7411B	2.165	55	5.512	140	1.299	33	19	85
	7412B	2.362	60	5.906	150	1.378	35	20.4	91.5

TABLE 67   Angular contact ball bearings. *(SKF Co. Ltd.)*

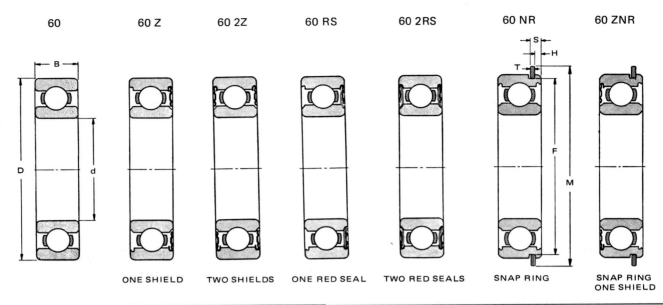

ONE SHIELD   TWO SHIELDS   ONE RED SEAL   TWO RED SEALS   SNAP RING   SNAP RING ONE SHIELD

							Nominal Bearing Dimensions									
							d		D		B		F	M	S	H	T
Bearing Number							in.	mm	in.	mm	in.	mm	Millimeters				
6000	6000 Z	6000 2Z	6000 RS	6000 2RS			.375	10	1.031	26	.312	8					
6001	6001 Z	6001 2Z	6001 RS	6001 2RS			.500	12	1.125	28	.312	8					
6002	6002 Z	6002 2Z	6002 RS	6002 2RS	6002 NR	6002 ZNR	.625	15	1.250	32	.375	9	30	36.5	3	2	1
6003	6003 Z	6003 2Z	6003 RS	6003 2RS			.750	17	1.656	35	.500	10					
6004	6004 Z	6004 2Z	6004 RS	6004 2RS			.875	20	1.750	42	.500	2					
6005	6005 Z	6005 2Z	6005 RS	6005 2RS			1.000	25	1.875	47	.500	12					
6006	6006 Z	6006 2Z	6006 RS	6006 2RS			1.1875	30	2.156	55	.531	13					
6007	6007 Z	6007 2Z	6007 RS	6007 2RS			1.375	35	2.500	62	.562	14					
6008	6008 Z	6008 2Z	6008 RS	6008 2RS			1.500	40	2.688	68	.594	15					
6009							1.750	45	3.000	75	.625	16					
6010	6010 Z	6010 2Z	6010 RS	6010 2RS	6010 NR	6010 ZNR	2.000	50	3.125	80	.625	16	76.8	86	4	2.5	1.6
6011	6011 Z	6011 2Z	6011 RS	6011 2RS	6011 NR	6011 ZNR	2.250	55	3.562	90	.719	18	86.8	96.5	5	3	2.4
6012							2.375	60	3.750	95	.719	18					
6013	6013 Z	6013 2Z			6013 NR	6013 ZNR	2.500	65	4.000	100	.719	18	96.8	106	5	3	2.4
6014							2.750	70	4.375	110	.781	20					
6015	6015 Z	6015 2Z			6015 NR	6015 ZNR	3.000	75	4.500	115	.781	20	112	121	5	3	2.4

Note: When shield is required on same side as snap ring, order as ZNBR.
      Inch size dimensions shown are for inch size bearings.

TABLE 68   Radial ball bearings. *(SKF Co. Ltd.)*

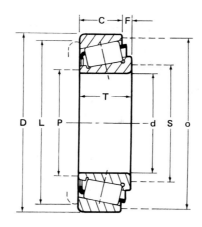

Bearing Number	Nominal External Sizes						U.S. CUSTOMARY (in.)						METRIC (mm)					
	Bore		Outside Dia		Width													
	in.	mm	in.	mm	in.	mm	F	C	S	P	L	O	F	C	S	P	L	O
4059	.590	15	1.375	35	.430	11	.089	.344	.781	.719	1.094	1.250	2	9	20	18	28	32
03062	.625	16	1.625	41	.562	14	.125	.438	.844	.781	1.312	1.469	3	11	21	20	33	37
5062	.625	16	1.850	46	.566	14	.129	.438	.938	.844	1.562	1.688	3	11	24	21	40	43
1749	.688	17	1.570	40	.545	14	.125	.420	.906	.781	1.312	1.438	3	11	23	20	37	36
9195	.695	17.6	1.938	49	.906	20	.219	.562	1.062	.969	1.625	1.750	6	14	27	24	41	44
9070	.695	17.6	1.938	49	.906	23	.219	.688	1.062	.969	1.562	1.750	6	17	27	24	39	44
1774	.748	19	2.240	57	.753	20	.138	.625	1.062	.969	1.875	2.031	4	16	27	24	48	51
1949	.750	20	1.781	45	.610	16	.135	.475	1.000	.922	1.500	1.625	4	12	25	59	38	41
9067	.750	20	1.938	50	.835	21	.148	.688	1.000	.922	1.531	1.750	4	17	25	23	39	44
9078	.750	20	1.938	50	.906	23	.219	.562	1.000	.922	1.625	1.750	6	17	25	23	39	44
7075	.750	20	2.000	51	.591	15	.091	.375	1.000	.938	1.719	1.844	3	12	25	24	44	47
7087	.857	22	1.969	50	.531	14	.156	.375	1.125	1.062	1.719	1.844	4	10	28	27	44	47
1380	.875	22	2.125	54	.762	20	.200	.562	1.125	1.094	1.750	1.906	5	15	28	28	44	48
1280	.875	22	2.250	54	.875	22	.188	.688	1.125	1.094	1.875	2.062	5	17	28	28	48	52
4643	1.000	25	1.980	48	.560	14	.140	.420	1.219	1.156	1.719	1.844	4	10	33	29	44	47
1780	1.000	25	2.240	57	.762	20	.781	.625	1.188	1.172	1.844	2.031	4	16	30	30	48	52

TABLE 69   Tapered roller bearings. *(SKF Co. Ltd.)*

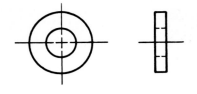

Inside Dia		Thickness		Outside Dia					
in.	mm	in.	mm	in.	mm	in.	mm	in.	mm
.500	12	.12 .18	3 5	.80	20	1.10	25	1.25	30
.625	16	.12 .18	3 5	1.25	30	1.50	40	1.75	46
.750	20	.12 .18	3 5	1.30	34	1.60	40	2.00	50
.875	22	.12 .18	3 5	1.30	34	1.60	40	2.00	50
1.000	25	.12 .18 .25	3 5 6	1.60	40	2.00	50	2.25	60

TABLE 70   Thrust plain bearings.

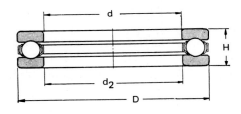

	Nominal Bearing Dimensions									Shoulder Diameter in.	
	d		d₂		D		H		Max. Fillet		
BEARING NUMBER	in.	mm	in.	mm	in.	mm	in.	mm	Radius* in.	Shaft Min.	Housing Max.
51100	.3937	10	.402	10.2	.9449	24	.354	9	.012	.750	.563
51101	.4724	12	.480	12.2	1.0236	26	.354	9	.012	.844	.656
51102	.5906	15	.598	15.2	1.1024	28	.354	9	.012	.938	.750
51103	.6693	17	.677	17.2	1.1811	30	.354	9	.012	1.031	.844
51104	.7874	20	.795	20.2	1.3780	35	.394	10	.012	1.188	.969
51105	.9843	25	.992	25.2	1.6535	42	.433	11	.024	1.438	1.219
51106	1.1811	30	1.189	30.2	1.8504	47	.433	11	.024	1.563	1.469
51107X	1.3780	35	1.386	35.2	2.0472	52	.472	12	.024	1.813	1.656
51108	1.5748	40	1.583	40.2	2.3622	60	.512	13	.024	2.063	1.875
51109	1.7717	45	1.780	45.2	2.5591	65	.551	14	.024	2.281	2.063
51110	1.9685	50	1.976	50.2	2.7559	70	.551	14	.024	2.500	2.250
51111	2.1654	55	2.173	55.2	3.0709	78	.630	16	.024	2.750	2.500
51112	2.3622	60	2.370	60.2	3.3465	85	.669	17	.039	3.000	2.719
51113	2.5591	65	2.567	65.2	3.5433	90	.709	18	.039	3.188	2.875
51114	2.7559	70	2.764	70.2	3.7402	95	.709	18	.039	3.375	3.063
51115	2.9528	75	2.961	75.2	3.9370	100	.748	19	.039	3.625	3.250
51116	3.1496	80	3.157	80.2	4.1339	105	.748	19	.039	3.813	3.438
51117	3.3465	85	3.354	85.2	4.3307	110	.748	19	.039	4.000	3.625

* The maximum fillet on the shaft or in the housing, which will be cleared by the bearing corner.

TABLE 71   Thrust roller bearings. *(SKF Co. Ltd.)*

STEEL AND IRON SPUR GEARS
14.5° PRESSURE ANGLE
Actual Tooth Size (Will not operate with 20° Spurs)

SPUR GEARS STEEL AND IRON
20° PRESSURE ANGLE
(Will not operate with 14.5° Spurs) *Actual Tooth Size*

Style		No. of Teeth	Inch Sizes 8 Pitch 1.25 Face				Metric Sizes (mm) 3.18 Module 30 Face			
			Pitch Dia	Hole	Hub Dia	Hub Proj.	Pitch Dia	Hole	Hub Dia	Hub Proj.
Steel Plain		12	1.500	.750	1.12	.75	38.2	20	28	20
		14	1.750	.750	1.38	.75	44.5	20	35	20
		15	1.875	.875	1.50	.75	47.7	22	40	20
		16	2.000	.875	1.62	.75	50.9	22	40	20
		18	2.250	.875	1.88	.75	57.2	22	48	20
		20	2.500	.875	2.12	.75	63.6	22	54	20
		22	2.750	.875	2.38	.75	70.0	22	60	20
Cast Iron	Web	24	3.000	.875	2.12	1.00	76.3	22	54	25
		28	3.500	.875	2.25	1.00	89.0	22	56	25
		30	3.750	.875	2.25	1.00	95.4	22	56	25
		32	4.000	1.000	2.25	1.00	101.8	25	56	25
		36	4.500	1.000	2.50	1.00	114.5	25	64	25
		40	5.000	1.000	2.50	1.00	127.2	25	64	25
		42	5.250	1.000	2.50	1.00	133.6	25	64	25
		44	5.500	1.000	2.50	1.00	139.9	25	64	25
		48	6.000	1.000	2.50	1.00	152.6	25	64	25
		54	6.750	1.000	2.50	1.00	171.7	25	64	25
	Spoke	56	7.000	1.000	2.50	1.00	178.1	25	64	25
		60	7.500	1.000	2.50	1.00	190.8	25	64	25
		64	8.000	1.000	2.50	1.00	203.5	25	64	25
		72	9.000	1.000	2.50	1.00	229.0	25	64	25
		80	10.000	1.125	3.00	1.12	254.4	28	76	28
		84	10.500	1.125	3.00	1.12	267.1	28	76	28
		88	11.000	1.125	3.00	1.12	279.8	28	76	28
		96	12.000	1.125	3.00	1.12	305.3	28	76	28

Note: Metric size gears were not available at time of publication. Values were soft converted for problem solving only.

TABLE 72   Eight-pitch (3.18 module) spur-gear data. *(Boston Gear Works)*

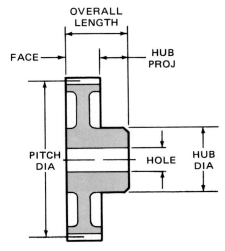

3.18 Module (8 Pitch)

Style		No. of Teeth	Inch Sizes 6 Pitch 1.50 Face				Metric Sizes (mm) 4.23 Module 40 Face			
			Pitch Dia	Hole	Hub Dia	Hub Proj.	Pitch Dia	Hole	Hub Dia	Hub Proj.
Steel Plain		12	2.000	1.00	1.50	.88	50.8	25	38	22
		14	2.333	1.00	1.81	.88	59.2	25	46	22
		15	2.500	1.00	2.00	.88	63.5	25	50	22
		16	2.667	1.00	2.16	.88	67.7	25	55	22
		18	3.000	1.00	2.50	.88	76.1	25	64	22
		20	3.333	1.00	2.84	.88	84.6	25	72	22
		21	3.500	1.00	3.00	.88	88.8	25	76	22
Cast Iron	Web	24	4.000	1.12	2.50	1.00	101.5	28	64	25
		27	4.500	1.12	2.50	1.00	114.2	28	64	25
		30	5.000	1.12	2.50	1.00	126.9	28	64	25
		32	5.333	1.12	2.50	1.00	135.4	28	64	25
		33	5.500	1.12	2.50	1.00	139.6	28	64	25
		36	6.000	1.12	2.50	1.00	152.3	28	64	25
		40	6.667	1.12	2.50	1.00	169.2	28	64	25
		42	7.000	1.12	2.50	1.00	177.7	28	64	25
	Spoke	48	8.000	1.12	2.50	1.00	203.0	28	64	25
		54	9.000	1.12	2.50	1.00	228.4	28	64	25
		60	10.000	1.25	3.00	1.25	253.8	30	76	30
		64	10.667	1.25	3.00	1.25	270.7	30	76	30
		66	11.000	1.25	3.00	1.25	279.2	30	76	30
		72	12.000	1.25	3.00	1.25	304.6	30	76	30
		84	14.000	1.25	3.25	1.25	355.3	30	82	30

Note: Metric size gears were not available at time of publication. Values were soft converted for problem solving only.

TABLE 73   Six-pitch (4.23 module) spur-gear data. *(Boston Gear Works)*

SPUR GEARS STEEL AND IRON
14.5° PRESSURE ANGLE
Actual Tooth Size (Will not operate with 20° Spurs)

SPUR GEARS STEEL AND IRON
20° PRESSURE ANGLE
(Will not operate with 14.5° Spurs) *Actual Tooth Size*

SPUR GEARS STEEL AND IRON
14.5° PRESSURE ANGLE
(Will not operate with 20° Spurs)

Actual Tooth Size

SPUR GEARS STEEL AND IRON
20° PRESSURE ANGLE
(Will not operate with 14.5° Spurs)

Actual Tooth Size

Style	No. of Teeth	Inch Sizes 5 Pitch Pitch Dia	2.00 Face Hole	Hub Dia	Hub Proj.	Metric Sizes (mm) 5.08 Module Pitch Dia	50 Face Hole	Hub Dia	Hub Proj.
Steel Plain	12	2.400	1.06	1.78	.88	61.0	26	45	22
	14	2.800	1.06	2.18	.88	71.1	26	55	22
	15	3.000	1.06	2.38	.88	76.2	26	60	22
	16	3.200	1.06	2.59	.88	81.3	26	65	22
	18	3.600	1.06	3.00	.88	91.4	26	75	22
	20	4.000	1.06	3.38	.88	101.6	26	85	22
Cast Iron Web	24	4.800	1.06	3.00	1.25	121.9	26	75	30
	25	5.000	1.06	3.00	1.25	127.0	26	75	30
	30	6.000	1.06	3.00	1.25	152.4	26	75	30
	35	7.000	1.18	3.00	1.25	177.8	30	75	30
	40	8.000	1.18	3.00	1.25	203.2	30	75	30
	45	9.000	1.18	3.00	1.25	228.6	30	75	30
Cast Iron Spoke	50	10.000	1.18	3.50	1.25	254.0	30	90	30
	55	11.000	1.18	3.50	1.25	279.4	30	90	30
	60	12.000	1.18	3.50	1.25	304.8	30	90	30
	70	14.000	1.18	3.50	1.25	355.6	30	90	30
	80	16.000	1.18	3.50	1.25	406.4	30	90	30
	90	18.000	1.18	3.50	1.25	457.2	30	90	30
	100	20.000	1.31	3.75	1.50	508.0	32	95	38
	110	22.000	1.31	3.75	1.50	558.8	32	95	38
	120	24.000	1.31	4.00	1.50	609.6	32	100	38

Note: Metric size gears were not available at the time of publication. Values were soft converted for problem solving only.

TABLE 74 Five-pitch (5.08 module) spur-gear data. *(Boston Gear Works)*

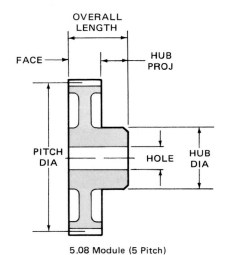

OVERALL LENGTH

FACE — HUB PROJ

PITCH DIA — HOLE — HUB DIA

5.08 Module (5 Pitch)

Style	No. of Teeth	Inch Sizes 4 Pitch Pitch Dia	2.75 Face Hole	Hub Dia	Hub Proj.	Metric Sizes (mm) 6.35 Module Pitch Dia	70 Face Hole	Hub Dia	Hub Proj.
Steel Plain	12	3.000	1.12	2.25	.88	76.2	28	58	22
	14	3.500	1.12	2.75	.88	88.9	28	70	22
	15	3.750	1.12	3.00	.88	95.3	28	76	22
	16	4.000	1.12	3.25	.88	101.6	28	82	22
	18	4.500	1.12	3.75	.88	114.3	28	96	22
	20	5.000	1.12	4.25	.88	127.0	28	108	22
	22	5.500	1.12	4.75	.88	139.7	28	120	22
Cast Iron Web	24	6.000	1.12	3.50	1.50	152.4	28	90	38
	28	7.000	1.25	3.50	1.50	177.8	30	90	38
	30	7.500	1.25	3.50	1.50	190.5	30	90	38
	32	8.000	1.25	3.50	1.50	203.2	30	90	38
	36	9.000	1.25	3.50	1.50	228.6	30	90	38
	40	10.000	1.25	4.00	1.50	254.0	30	100	38
	42	10.500	1.25	4.00	1.50	266.7	30	100	38
	44	11.000	1.25	4.00	1.50	279.4	30	100	38
Cast Iron Spoke	48	12.000	1.25	4.00	1.50	304.8	30	100	38
	54	13.500	1.25	4.00	1.50	342.9	30	100	38
	56	14.000	1.25	4.00	1.50	355.6	30	100	38
	60	15.000	1.25	4.00	1.50	381.0	30	100	38
	64	16.000	1.25	4.00	1.50	406.4	30	100	38
	72	18.000	1.25	4.00	1.50	457.2	30	100	38
	80	20.000	1.38	4.50	1.50	508.0	35	115	38

Note: Metric size gears were not available at the time of publication. Values were soft converted for problem solving only.

TABLE 75 Four-pitch (6.35 module) spur-gear data. *(Boston Gear Works)*

SPUR GEARS STEEL AND IRON
14.5° PRESSURE ANGLE
(Will not operate with 20° Spurs)

Actual Tooth Size

SPUR GEARS STEEL AND IRON
20° PRESSURE ANGLE
(Will not operate with 14.5° Spurs)

Actual Tooth Size

20° PRESSURE ANGLE—STRAIGHT TOOTH

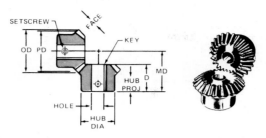

Module (Pitch)	Teeth	U.S. CUSTOMARY (in.)									METRIC (mm)								
		Pitch Dia	Face	OD Approx.	D	MD	Hub Dia	Hub Prov. Approx.	Hole	Keyseat	Pitch Dia	Face	OD Approx.	D	MD	Hub Dia	Hub Prov. Approx.	Hole	Keyseat
2.54 (10)	20	2.000	.44	2.15	1.36	2.00	1.62	.81	.500 .625 .750	.12 × .06 .18 × .09 .18 × .09	50	11.2	54	34	50	42	20	12 16 20	3 × 1.5 5 × 2.5 5 × 2.5
2.54 (10)	25	2.500	.55	2.65	1.62	2.44	2.00	.94	.750 .875 1.000	.18 × .09 .18 × .09 .25 × .12	62	14	68	41	62	50	24	20 22 25	5 × 2.5 5 × 2.5 6 × 3
3.18 (8)	24	3.000	.64	3.18	1.58 1.76 1.76	2.56 2.75 2.75	1.75 2.25 2.50	.81 1.06 1.12	.750 1.000 1.250	.18 × .09 .25 × .12 .25 × .12	76	16	80	40 45 45	65 70 70	45 58 64	20 27 28	20 25 30	5 × 2.5 6 × 3 6 × 3
3.18 (8)	28	3.500	.75	3.68	2.09	3.25	2.50	1.25	1.000 1.188 1.250	.25 × .12	88	16	93	53	82	64	32	25 30 32	6 × 3
4.23 (6)	24 27	4.000 4.500	.86 .96	4.24 4.74	2.31 2.62	3.62 4.12	3.00 3.25	1.31 1.50	1.25 1.50 1.50	.25 × .12 .38 × .18 .38 × .18	100 114	22 24	108 120	58 66	92 104	76 82	33 38	32 38 38	6 × 3 10 × 5 10 × 5
5.08 (5)	25	5.000	1.10	5.29	3.00	4.62	3.50	1.75	1.38 1.50 1.75	.31 × .16 .38 × .18 .38 × .18	128	28	134	76	117	88	44	35 38 44	8 × 4 10 × 5 10 × 5

NOTE: METRIC SIZE GEARS WERE NOT AVAILABLE AT TIME OF PUBLICATION. VALUES WERE SOFT CONVERTED FOR PROBLEM SOLVING ONLY.

TABLE 76   Miter gears. *(Boston Gear Works)*

20° PRESSURE ANGLE

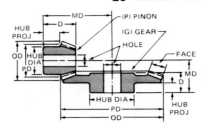

Module (Pitch)	Teeth	U.S. CUSTOMARY (in.)								METRIC (mm)							
		Pitch Dia	Face	Hole	D	MD	Hub Dia	Hub Proj. Approx.	OD Approx.	Pitch Dia	Face	Hole	D	MD	Hub Dia	Hub Proj. Approx.	OD Approx.
2.54 (10)	50	5.000	.70	.750	1.30	2.62	2.00	1.00	5.06	127	18	20	33.0	66.6	50	25	128.5
	25	2.500			1.55	3.38	2.00	.75	2.75	63.5			39.4	85.8	50	20	69.8
	60	6.000	.78	1.00	1.86	2.75	3.00	1.38	6.04	152.4	20	25	47.2	69.8	76	35	153.4
	20	2.000		.875	2.16	4.38	1.75	1.30	2.27	50.8		22	54.9	111.2	45	33	57.6
	60	6.000	.72	.875	1.62	2.25	2.50	1.12	6.03	152.4	18.3	22	41.1	57.2	64	28	153.2
	15	1.500		.625	1.60	3.88	1.44	.84	1.78	38.1		16	40.6	98.6	36	22	45.2
3.18 (8)	40	5.000	.82	1.000	1.84	2.88	3.00	1.25	5.08	127.2	20.8	25	46.7	67.6	76	32	129
	20	2.500			2.28	4.00	2.12	1.40	2.81	63.6			57.9	101.6	54	35	71.4
	48	6.000	.84	.875	1.62	2.38	2.75	1.00	6.05	152.6	21.4	22	41.2	60.4	70	25	153.7
	16	2.000		.750	2.08	4.25	1.75	1.88	2.35	50.9		20	52.8	108	45	48	59.7
	64	8.000	.84	1.000	1.88	2.75	2.75	1.25	8.04	203.5	21.4	25	47.7	69.8	70	32	204.2
	16	2.000		.875	2.09	5.25	1.88	1.22	2.36	50.9		22	53.1	133.4	48	30	60
4.23 (6)	40	5.000	.82	1.000	1.84	2.88	3.00	1.25	5.08	169.2	20.8	25	46.7	67.6	76	32	129
	20	2.500			2.28	4.00	2.12	1.40	2.81	84.6			57.9	101.6	54	35	71.4
	48	6.000	.84	.875	1.62	2.38	2.75	1.00	6.05	203	21.4	22	41.1	60.4	70	25	153.7
	16	2.000		.750	2.08	4.25	1.75	1.18	2.35	67.7		20	52.8	108	45	30	59.7
	64	8.000	.84	1.000	1.88	2.75	2.75	1.25	8.04	270.2	21.4	25	47.7	69.8	70	32	204.2
	16	2.000		.875	2.09	5.25	1.88	1.22	2.36	67.7		22	53.1	133.4	48	30	60
	36	6.000	1.06	1.125	2.25	3.50	3.25	1.50	6.10	152.3	27	28	57.2	88.9	82	38	155
	18	3.000			2.76	4.75	2.50	1.60	3.41	76.1			70.1	120.6	64	40	86.6
	45	7.500	1.06	1.125	2.12	3.00	3.25	1.25	7.57	190.4	27	28	53.8	76.2	82	32	192.3
	15	2.500		.875	2.56	5.25	2.12	1.44	2.95	63.4		22	65.0	133.4	54	36	75

NOTE: METRIC SIZE GEARS WERE NOT AVAILABLE AT TIME OF PUBLICATION. VALUES WERE SOFT CONVERTED FOR PROBLEM SOLVING ONLY.

TABLE 77 Bevel gears. (*Boston Gear Works*)

TYPE P
HEADLESS PRESS-FIT BUSHINGS

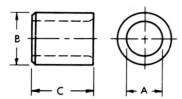

TYPE H
HEAD PRESS-FIT BUSHINGS

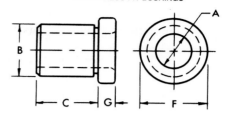

Inside Dia (A)		Outside Dia (B)		Bushing Length C											Technical Data	
From	To	Max.	Min.	.25	.31	.38	.50	.75	1.00	1.38	1.75	2.12	2.50	3.00	F	G
.125	.194	.3141	.3138	•	•	•	•	•	•	•	•				.42	.12
.188	.257	.4078	.4075	•	•	•	•	•	•	•	•	•			.50	.16
.188	.316	.5017	.5014	•	•	•	•	•	•	•	•	•			.60	.22
.312	.438	.6267	.6264	•	•	•	•	•	•	•	•	•	•		.80	.22
.312	.531	.7518	.7515	•	•	•	•	•	•	•	•	•	•	•	.92	.22
.500	.656	.8768	.8765		•	•	•	•	•	•	•	•	•	•	1.10	.25
.500	.766	1.0018	1.0015				•	•	•	•	•	•	•	•	1.24	.32
.625	1.031	1.3772	1.3768				•	•	•	•	•	•	•	•	1.60	.38
1.000	1.390	1.7523	1.7519					•	•	•	•	•	•	•	1.98	.38
				6	8	10	12	16	20	25	30	35	40	45	F	G
3	5	7.978	7.971	•	•	•	•	•	•	•	•	•	•		10	3
5	6.5	10.358	10.351	•	•	•	•	•	•	•	•	•	•		12	4
5	8	12.972	12.736	•	•	•	•	•	•	•	•	•	•		15	5
8	11	15.918	15.911	•	•	•	•	•	•	•	•	•	•		20	6
8	13.5	19.096	19.088	•	•	•	•	•	•	•	•	•	•	•	24	6
12	16	22.271	22.263		•	•	•	•	•	•	•	•	•	•	28	6
12	20	25.446	25.438			•	•	•	•	•	•	•	•	•	32	8
16	26	34.981	34.971			•	•	•	•	•	•	•	•	•	40	10
25	35	44.508	44.498			•	•	•	•	•	•	•	•	•	50	10

NOTE: METRIC SIZE BUSHINGS WERE NOT AVAILABLE AT TIME OF PUBLICATION. VALUES SHOWN WERE SOFT CONVERTED FOR PROBLEM SOLVING ONLY.

TABLE 78 Press-fit drill jig bushings. *(American Drill Bushing Co.)*

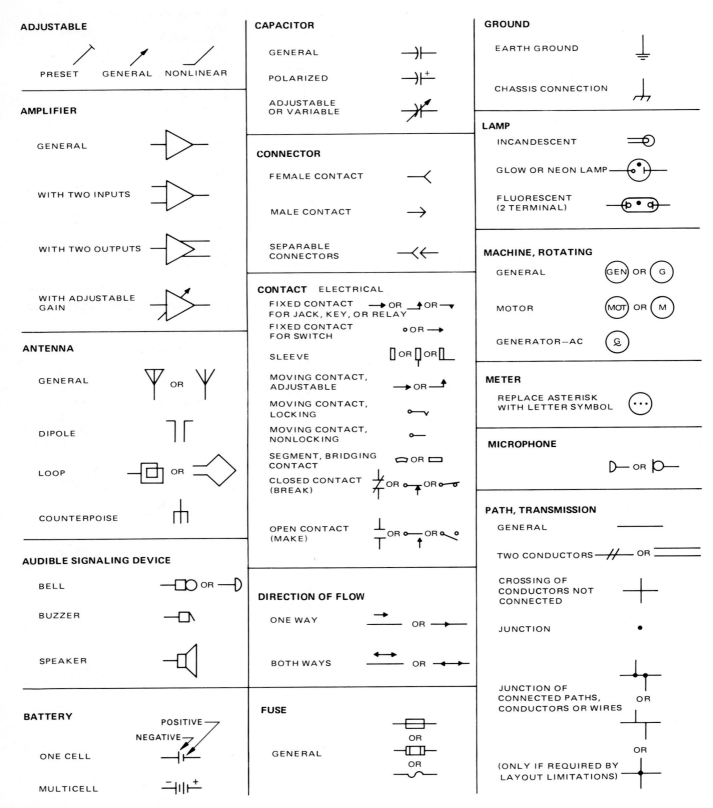

TABLE 79   Graphic symbols for electrical and electronics diagrams.

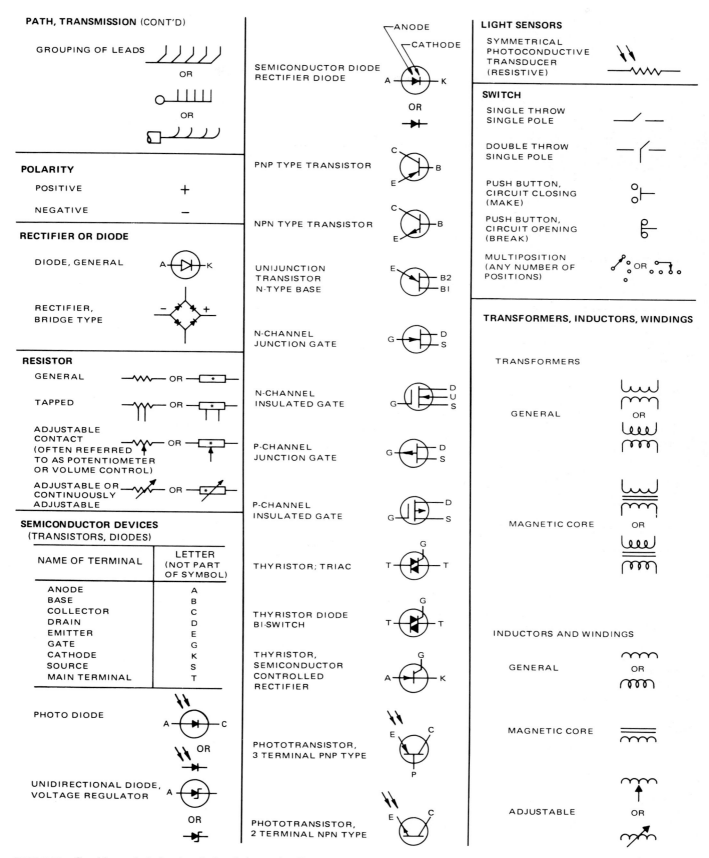

PATH, TRANSMISSION (CONT'D)

GROUPING OF LEADS

OR

OR

POLARITY

POSITIVE +

NEGATIVE −

RECTIFIER OR DIODE

DIODE, GENERAL A—K

RECTIFIER, BRIDGE TYPE − +

RESISTOR

GENERAL

TAPPED

ADJUSTABLE CONTACT (OFTEN REFERRED TO AS POTENTIOMETER OR VOLUME CONTROL)

ADJUSTABLE OR CONTINUOUSLY ADJUSTABLE

SEMICONDUCTOR DEVICES
(TRANSISTORS, DIODES)

NAME OF TERMINAL	LETTER (NOT PART OF SYMBOL)
ANODE	A
BASE	B
COLLECTOR	C
DRAIN	D
EMITTER	E
GATE	G
CATHODE	K
SOURCE	S
MAIN TERMINAL	T

PHOTO DIODE A—C

OR

UNIDIRECTIONAL DIODE, VOLTAGE REGULATOR A

OR

ANODE
CATHODE

SEMICONDUCTOR DIODE RECTIFIER DIODE A K

OR

PNP TYPE TRANSISTOR C B E

NPN TYPE TRANSISTOR C B E

UNIJUNCTION TRANSISTOR N-TYPE BASE E B2 B1

N-CHANNEL JUNCTION GATE G D S

N-CHANNEL INSULATED GATE G D U S

P-CHANNEL JUNCTION GATE G D S

P-CHANNEL INSULATED GATE G D S

THYRISTOR; TRIAC G T T

THYRISTOR DIODE BI-SWITCH G T T

THYRISTOR, SEMICONDUCTOR CONTROLLED RECTIFIER G A K

PHOTOTRANSISTOR, 3 TERMINAL PNP TYPE E C P

PHOTOTRANSISTOR, 2 TERMINAL NPN TYPE E C

LIGHT SENSORS

SYMMETRICAL PHOTOCONDUCTIVE TRANSDUCER (RESISTIVE)

SWITCH

SINGLE THROW SINGLE POLE

DOUBLE THROW SINGLE POLE

PUSH BUTTON, CIRCUIT CLOSING (MAKE)

PUSH BUTTON, CIRCUIT OPENING (BREAK)

MULTIPOSITION (ANY NUMBER OF POSITIONS) OR

TRANSFORMERS, INDUCTORS, WINDINGS

TRANSFORMERS

GENERAL OR

MAGNETIC CORE OR

INDUCTORS AND WINDINGS

GENERAL OR

MAGNETIC CORE

ADJUSTABLE OR

TABLE 79 Graphic symbols for electrical and electronics diagrams. (continued)

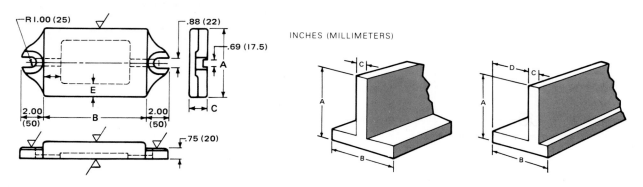

INCHES (MILLIMETERS)

	A	B	C	D	E	Cat. No.	A-B	C	Cat. No.	A-B	C	D
U.S. Customary (in.)	6.00	9.00	1.25	1.88	1.35	ET-300	3.00	.62	UT-300	3.00	.62	1.75
	8.00	12.00	1.38	2.00	1.62	ET-400	4.00	.75	UT-400	4.00	.75	2.62
	10.00	15.00	1.50	2.00	1.75	ET-500	5.00	.75	UT-500	5.00	.75	3.25
	12.00	18.00	1.75	2.25	2.00	ET-600	6.00	1.00	UT-600	6.00	1.00	3.50
						ET-800	8.00	1.25	UT-800	8.00	1.25	4.88
Metric (mm)	150	225	30	45	35	ET-75M	75	15	UT-75M	75	15	45
	200	300	35	50	40	ET-100M	100	20	UT-100M	100	20	65
	250	375	40	50	45	ET-125M	125	20	UT-125M	125	20	85
	300	450	45	55	50	ET-150M	150	25	UT-150M	150	25	90
						ET-200M	200	30	UT-200M	200	30	120

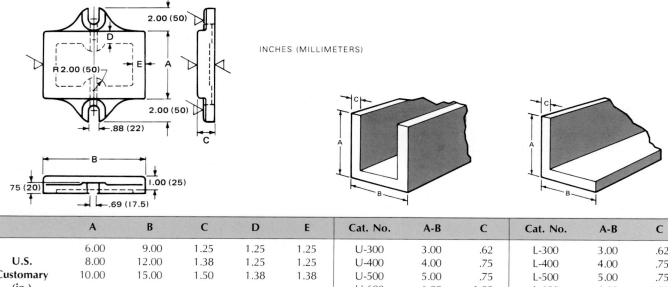

INCHES (MILLIMETERS)

	A	B	C	D	E	Cat. No.	A-B	C	Cat. No.	A-B	C
U.S. Customary (in.)	6.00	9.00	1.25	1.25	1.25	U-300	3.00	.62	L-300	3.00	.62
	8.00	12.00	1.38	1.25	1.25	U-400	4.00	.75	L-400	4.00	.75
	10.00	15.00	1.50	1.38	1.38	U-500	5.00	.75	L-500	5.00	.75
						U-600	6.00	1.00	L-600	6.00	1.00
						U-800	8.00	1.25	L-800	8.00	1.25
Metric (mm)	150	225	30	30	30	U-75M	75	15	L-75M	75	15
	200	300	35	30	30	U-100M	100	20	L-100M	100	20
	250	375	40	35	35	U-125M	125	20	L-125M	125	20
						U-150M	150	25	L-150M	150	25
						U-200M	200	30	L-200M	200	30

TABLE 80   Fixture bases and microsections.

INDEX

PHOTO CREDITS

Table of Contents: pg. iii, (b), Bruno de Hougues/TSW; pg. iii, (t), Doug Martin; pg. iv, (b), courtesy Mayline Company; pg. iv, (t), Theodore Anderson/Stockphotos; pg. v, (b), Autocad/MacIntosh; pg. v, (t), John Terrence Turner/FPG International; pg. vi, (b), Jim Pickerell/Westlight; pg. vi, (t), Glencoe file photo; pg. vii, Glencoe file photo; pg. viii, (b), Glencoe file photo; pg. viii, (t), R. Rathe/FPG International; pg. ix, (b), Andrew Sacks/TSW; pg. ix, (t), E. Alan McGee/FPG International; pg. x, (b), STUDIOHIO; pg. x, (t), Doug Martin; pg. xi, (b), Sylvain Cobbie/TSW; pg. xi, (t), John Garrett/TSW; pg. xii, Doug Martin; pg. xiii, (b), Doug Martin; pg. xiii, (t), Andrew Sacks/TSW; pg. xiv, (b), courtesy International Business Machines Corporation; pg. xiv, (t), Doug Martin; pg. xv, (b), Autocad/MacIntosh; pg. xv, (t), Doug Martin; pg. xvi, Andrew Sacks/TSW.

Part Openers: pg. 1, © 1994 Comstock; pg. 259, ???; pg. 379, ???.

Chapter Openers: pp. 2, 22, 35, 45, 76, 92, 136, 171, 227, 260, 292, 327, 347, 380, 441, Aaron Haupt.